Lecture Notes in Computer Science 1684
Edited by G. Goos, J. Hartmanis and J. van Leeuwen

Springer

Berlin
Heidelberg
New York
Barcelona
Hong Kong
London
Milan
Paris
Singapore
Tokyo

Gabriel Ciobanu Gheorghe Păun (Eds.)

Fundamentals of Computation Theory

12th International Symposium, FCT'99
Iaşi, Romania, August 30 – September 3, 1999
Proceedings

Volume Editors

Gabriel Ciobanu
"A.I.Cuza" University, Faculty of Computer Science
6600 Iaşi, Romania
E-mail: gabriel@info.uaic.ro

Gheorghe Păun
Institute of Mathematics of the Romanian Academy
P.O. Box 1-764, 70700 Bucharest, Romania
E-mail: gpaun@imar.ro

Cataloging-in-Publication data applied for

Die Deutsche Bibliothek - CIP-Einheitsaufnahme

Fundamentals of computation theory : 12th international symposium ; proceedings / FCT '99, Iaşi, Romania, August 30 - September 3, 1999. Gabriel Ciobanu ; Gheorghe Păun (ed.). - Berlin ; Heidelberg ; New York ; Barcelona ; Hong Kong ; London ; Milan ; Paris ; Singapore ; Tokyo : Springer, 1999
(Lecture notes in computer science ; Vol. 1684)
ISBN 3-540-66412-2

CR Subject Classification (1998): F, G.2, I.3.5, E.1

ISSN 0302-9743
ISBN 3-540-66412-2 Springer-Verlag Berlin Heidelberg New York

Typesetting: Camera-ready by author
SPIN: 10704389 06/3142 – 5 4 3 2 1 0 Printed on acid-free paper

Preface

The International Symposium on the Fundamentals of Computation Theory FCT'99 was held in Iaşi, Romania, from August 30 to September 3, 1999. This was the 12th time the conference was held, after Poznan (Poland, 1977), Wendisch-Rietz (Germany, 1979), Szeged (Hungary, 1981), Borgholm (Sweden, 1983), Cottbus (Germany, 1985), Kazan (Soviet Union, 1987), Szeged (Hungary, 1989), Berlin (Germany, 1991), Szeged (Hungary, 1993), Dresden (Germany, 1995), and Kraków (Poland, 1997). This 12th FCT was organized by "Alexandru Ioan Cuza" University of Iaşi, and it was endorsed by the European Association of Theoretical Computer Science and the Romanian Academy. The conference consisted of plenary lectures presented by invited speakers, parallel sessions for technical contributions, and two satellite workshops.

As at previous conferences, the purpose of this FCT conference was to promote high-quality research in all branches of theoretical computer science, and to bring together active specialists in the field. This is especially important now, at the turn of century, when computer science looks for new computing tools and for a better balance between theory and practice. A broad range of topics were considered of interest for submission and most of them are covered by the invited lectures and the papers accepted for presentation. These include: abstract data types, algorithms and data structures, automata and formal languages, categorical and topological approaches, complexity, computational geometry, computer systems theory, concurrency, constructive mathematics, cryptography, distributed computation, fault-tolerant computing, logics in computer science, learning theory, process algebra and calculi, rewriting, semantics, specification, symbolic computation, universal algebra, molecular computing, and quantum computing.

The program committee had a rather difficult task in selecting for presentation only 42 papers out of the 104 submissions. Several papers which have not found themselves a place in the conference program are of high quality and worth accepting. The selection procedure was based on a very efficient and transparent electronic interaction among the members of the program committee by means of web pages and e-mail. We wish to thank all authors who have submitted papers for consideration, all program committee members for their timely and quality work, as well as the referees who have assisted the program committee in the laborious evaluation process.

The present volume contains all 42 of the accepted technical contributions and 4 (of the 5) invited papers.

Two satellite workshops were also organized: Formal Languages and Automata (Chair: A. Mateescu, Bucharest) and Distributed Systems (Chair: G. Ştefănescu, Bucharest).

The organizing committee was chaired by Gabriel Ciobanu, and it includes C. Barbu, A. Bejan, S. Buraga, V.E. Căzănescu, L. Chiran, L. Ibănescu, D. Lucanu, A. Mateescu, C. Mitrofan, R. Negrescu, S. Orzan, G. Păun, and V. Tablan. We would like to thank all these people for their work, and especially Sabin Buraga for his web pages. We express our gratitude to the Dean and to all colleagues from the Department of Computer Science who have supported us in organizing this event. Special thanks are due to Professor G. Popa, Rector of "A.I. Cuza" University, for his decisive help.

We gratefully acknowledge the financial support of our sponsors: the Association for Computing Machinery (many thanks to Claus Unger, its Secretary and Treasurer), the National Agency for Science, Technology and Innovations, Motorola, and other local companies.

Last but not least, we thank Springer-Verlag – in particular Professor Jan van Leuween and Ruth Abraham – for an efficient collaboration in preparing this volume.

June 1999

Gabriel Ciobanu
Gheorghe Păun

FCT'99 Program Committee

S. Bozapalidis (Thessaloniki)
C. Calude (Auckland)
V.E. Căzănescu (Bucharest)
W-N. Chin (Singapore)
G. Ciobanu (Co-Chair, Iaşi)
R. De Nicola (Florence)
M. Dezani (Turin)
J. Diaz (Barcelona)
R. Freivalds (Riga)
Z. Fülöp (Szeged)
J. Gruska (Brno)
J. Karhumäki (Turku)
J. Mitchell (Stanford)
F. Moller (Uppsala)
M. Nielsen (Aarhus)
G. Păun (Co-Chair, Bucharest)
A. Rabinovich (Tel Aviv)
M. Sato (Kyoto)
M. Sudan (MIT)
S. Yu (London, Ontario)

Referees

L. Aceto
A. Alderson
F. Alessi
C. Alvarez
A. Avron
R. Baeza-Yates
A. Baranga
J. Barnard
M. Bartha
G. Belle
J. Bengtsson
Z. Blazsik
R. Bol
R. Bonner
M. Boreale
A. Bouali
S. Bozapalidis
J. Bradfield
M. Bugliesi
C. Calude
V. E. Căzănescu
C. Câmpeanu
M. Cerioli
W.N. Chin
G. Ciobanu
J. Cirulis
H. Cîrstea
M. Coppo
A. Corradini
F. Corradini
J. Csima
E. Csuhaj-Varjú
J. Dassow
U. de'Liguoro
X. Deng
R. de Nicola
F.J. de Vries
M. Dezani
J. Diaz
M.J. Dinneen
T. Erlebach
Z. Esik
M. Falaschi
H. Fernau
R. Freivalds
Z. Fülöp
J. Gabarro
I. Gargantini
G. Georgescu
M. Giesbrecht
R. Gorrieri
B. Grahlmann
S. Gravier
J. Gruska
A. Habel
M. Hanes
T. Harju
M. Hennessy
J. Henriksen
M. Henz
K. Hiroyasu
Y. Hirshfeld
M. Hirvensalo
J. Honkala
S. Huerter
M. Huhn
W. Hui
L. Ilie
H. Ishihara
T. Izumi
P. Jancar
G. Juhas
A. Jung
Y. Kameyama
J. Karhumäki
J. Kärkkäinen
A. Kelemenova
S.C. Khoo
C. Kirchner
B. Kirsig
M. Kochol
U. Kohlenbach
M. Koutny
M. Kretinsky
M. Kudlek
O. Kupferman
T. Kurata
E. Lehtonen
C. Levcopoulos
P. Long
H. Lutfiyya
A. Maggiolo-Schettini
T. Mailund
J. Manuch
A. Masini
A. Mateescu
J. Mazoyer
Y. Metivier
M. Minea
J. Mitchell
V. Mitrana
T. Mizutani
F. Moller
M. Morley
K. Mukai
M. Murakami
G. Navarro
P. Niebert
M. Nielsen
A. Ohori
F. Olariu
N. Olivetti
A. Păun
Gh. Păun
M. Păun
J. Pearson
I. Petre
B. Pierce
M. Pistore
A. Pluhar
K.V.S. Prasad
R. Pugliese
M. Qing

P. Quaglia
D. Remy
S. Riis
J. Sakarovitch
V. Sassone
M. Schwartzbach
I. Stark
M. Sudan
P.S. Thiagarajan
F. Vaandrager
M. Venturini Zilli
R. Voicu
G.J. Woeginger
T. Yokomori
E. Zucca

A. Rabinovich
A. Renvall
L. Roversi
K. Salomaa
M. Sato
M. Serna
M. Steinby
Gh. Ştefănescu
I. Tomescu
F. van Raamsdonk
B. Victor
C. Wei-Ngan
M. Woodward
S. Yu
U. Zwick

G. Rahonis
M. Ribaudo
B. Rozoy
D. Sangiorgi
K. Satoru
J. Shallit
V. Stoltenberg-Hansen
D. Taimina
H. Tsuiki
B. Venneri
W. Vogler
G. Winskel
A. Yamamoto
M. Zakharyaschev

Contents

Invited Lectures

Technical Contributions

Randomized Complexity of Linear Arrangements and Polyhedra *

Marek Karpinski

The Institute for Advanced Study, Princeton, and
Dept. of Computer Science, University of Bonn, 53117 Bonn
Email: marek@cs.uni-bonn.de

Abstract. We survey some of the recent results on the complexity of recognizing n–dimensional linear arrangements and convex polyhedra by randomized algebraic decision trees. We give also a number of concrete applications of these results. In particular, we derive first nontrivial, in fact quadratic, randomized lower bounds on the problems like Knapsack and Bounded Integer Programming. We formulate further several open problems and possible directions for future research.

1 Introduction

Linear search algorithms, algebraic decision trees, and computation trees were introduced early to simulate random access machines (RAM) model. They are also a very useful and simplified abstraction of various other RAM–related computations cf. [AHU74], [DL78], [Y81], [SP82], [M84], [M85a], [KM90], and a useful tool in computational geometry. The same applies for the randomized models of computation. Starting with the papers of Manber and Tompa [MT85], Snir [S85], Meyer auf der Heide [M85a], [M85c] there was an increasing interest, and continuing effort in the last decade to understand the intrinsic power of randomization in performing various computational tasks. We refer to Bürgisser, Karpinski and Lickteig [BKL93], Grigoriev and Karpinski [GK94], Grigoriev, Karpinski, Meyer auf der Heide and Smolensky [GKMS97], and Grigoriev, Karpinski and Smolensky [GKS97] for the recent results (for the corresponding situation in a randomized bit model computation cf., e.g., [KV88], [FK95]). For some new randomized lower bounds on high dimensional geometric problems see also Borodin, Ostrovsky and Rabani [BOR99]. In the retrospective, several algebraic and topological methods introduced for proving lower bounds for deterministic algebraic decision trees turned out to fail utterly for some reason for the randomized model of computation, see the papers on deterministic methods by Yao [Y81], [Y92],

* Research partially supported by the DFG Grant KA 673/4-1, ESPRIT BR Grants 7079, 21726, and EC-US 030, by DIMACS, and by the Max–Planck Research Prize.

[Y94], Steele and Yao [SY82], Ben-Or [B-O83], Björner, Lovász and Yao [BLY92] and Grigoriev, Karpinski and Vorobjov [GKV97]. With the exception of some early results of Bürgisser, Karpinski and Lickteig [BKL93], and Grigoriev and Karpinski [GK93] there were basically no methods available for proving lower bounds on the depth of general randomized algebraic decision trees. In Meyer auf der Heide [M85a], a lower bound has been stated on the depth of randomized linear decision trees (with linear polynomials only) recognizing a linear arrangement. A gap in the proof of the Main Lemma of this paper was closed for the generic case first by Grigoriev and Karpinski in [GK94].

In this paper we survey some of the new methods which yield for the first time nontrivial lower bounds on the depth of randomized algebraic trees. The paper is organized as follows. In Section 2 we give necessary preliminaries for a general reader, and in Section 3 we introduce the underlying models of computation. In Section 4 we formulate the Main Results, and give some concrete applications. Section 5 deals with the phenomenon of a randomized speedup, and an explicit separation of deterministic and randomized depth. Section 6 presents some extensions of the results of Section 4. In Section 7 we formulate some open problems and possible directions for future research.

2 Preliminaries

We refer a general reader to [G67] for basic notions on convex polytopes and linear arrangements, and to [L84] for basic algebraic notions. We refer also to [M64] for basic facts on real varieties and Betti numbers.

For $x, y \in \mathbb{R}^n$ we denote by $<x,y>$ the *scalar product* of x and y,

$$<x,y>= \sum_{i=1}^{n} x_i y_i.$$

A *hyperplane* $H \subseteq \mathbb{R}^n$ is a set defined by $H = \{x \in \mathbb{R}^n | <x,y>= \alpha\}$ for some $y \in \mathbb{R}^n$, $y \neq 0$, and $\alpha \in \mathbb{R}$. A *closed halfspace* $H \subseteq \mathbb{R}^n$ is defined by $H = \{x \in \mathbb{R}^n | <x,y> \geq \alpha\}$ for some $y \in \mathbb{R}^n$, $y \neq 0$, and $\alpha \in \mathbb{R}$.

We call a *finite union* $S = \bigcup_{i=1}^m H_i$ of *hyperplanes* H_i, a ***linear arrangement***, and a *finite intersection* $S^+ = \bigcap_{i=1}^m H_i^+$ of *closed halfspaces* H_i^+, a ***polyhedron.*** A *k-face* L of a linear arrangement S is a ***k-dimensional plane*** defined by intersecting $n - k$ of the hyperplanes H_i. If L is k-dimensional on the boundary of S^+, is is also a *k-face* of S^+. We call a 0–face, a *vertex*.

When $S \subseteq \mathbb{R}^n$ is considered here as a *topological space*, it is with a subspace topology induced by $\mathbb{R}^n$. For any topological space S and an integer $k \geq 0$, $\beta_k(S)$ denotes the *i-th Betti number*, i. e., the ***rank*** of the ***i-th singular homology*** group. The *Euler characteristic* $\chi(S)$ of S is defined by $\chi(S) = \sum_k (-1)^k \beta_k(S)$ provided the Betti numbers of S are finite.

Milnor [M64] and Thom [T65] give fundamental bounds on the sums of Betti numbers $\sum_k \beta_k(S)$ of algebraic sets in $\mathbb{R}^n$ in the function of a degree bound on their defining polynomials: $\sum_k \beta_k(S) \leq d(2d-1)^{n-1}$.

We consider in sequel the following n-dimensional restrictions of NP-complete problems (cf. [DL78], [M84], [M85b]).

A *Bounded Integer Programming Problem* is a problem of *recognizing* a set

$$L_{n,k} = \{x \in \mathbb{R}^n \mid \exists a \in \{0,\ldots,k\}^n [< x, a >= k]\}$$

for a given bound k on the size of *integer solutions.*

The well known *Knapsack Problem* is the problem of recognizing the set $L_{n,1}$.

We consider further the problems of *Element Distinctness*, *Set Disjointness* and the *Resultant (Decision Version)* (cf. [B-O83]).

The *Element Distinctness* problem is the problem of recognizing the complement of the set

$$\{x \in \mathbb{R}^n \mid \exists i, j, i \neq j\, [x_i = x_j]\}\,.$$

The *Set Disjointness* problem is the problem of determining for given two sets $A = \{x_1, \ldots, x_n\}$, $B = \{y_1, \ldots, y_n\} \subseteq \mathbb{R}$ whether or not $A \cap B = \emptyset$, i.e. recognizing the set $\{(x_1,\ldots,x_n,y_1,\ldots,y_n) \in \mathbb{R}^{2n} \mid \forall i,j\,[x_i \neq x_j]\}$

The *Resultant* problem is the problem of computing for given $x, y \in \mathbb{R}^n$ the *resultant* of x and y, $\prod_{i,j}(x_i - y_j)$ (cf. [B-O83]). Any algorithm for the *Resultant* problem can check whether the resultant $\neq 0$, i.e. whether the sets $\{x_i\}$ and $\{y_i\}$ are disjoint, and therefore solve the *Set Disjointness* problem as well.

It is not difficult to prove that the number of *vertices (0-faces)* of the Bounded Integer Programming Problem $L_{n,k}$ is at least $(k+1)^{\frac{n^2}{16}}$, and the number of $\frac{n}{2}-$*faces* (assuming n is even) of the *Element Distinctness* is $(\frac{n}{2})!$ (cf. [GKMS97]).

3 Computational Models

We introduce now our underlying model of *randomized computations, a randomized algebraic decision tree* (RDT).

An *algebraic decision tree* of degree d (d–DT) over $\mathbb{R}^n$ is a *rooted ternary* tree. Its root and inner nodes are labelled by real multivariate polynomials $g \in \mathbb{R}[x_1,\ldots,x_n]$ of degree at most d, its leaves are marked *"accepting"* or *"rejecting"*. A *computation* of a d–DT on an input $x = (x_1,\ldots,x_n) \in \mathbb{R}^n$ consists of a sequence of traverses of a tree from the root to a leaf, always choosing the *left/middle/right* branch from a node labelled by a polynomial g according to the *sign* of $g(x_1,\ldots,x_n)$ (*smaller/equal/greater than* 0). The *inputs* $x \in \mathbb{R}^n$ arriving at accepting leaves *form* the set $S \subseteq \mathbb{R}^n$ *recognized* (or *computed*) by the d–DT.

In this paper we deal with *randomized algebraic decision trees* of *degree d* (*d–RDTs*). A *d–RDT* over $\mathbb{R}^n$ is a finite collection $T = \{T_\alpha\}$ of *d–DTs* T_α with the assigned rational probabilities p_α, $\sum_\alpha p_\alpha = 1$, of *choosing (or randomized compiling)* T_α out of the set $\{T_\alpha\}$.

It is easily seen that the above model is equivalent to the other variant of a randomized algebraic decision tree allowing coin tosses at the special *random* nodes, and not charging for the random bits used. Our model of a *d–RDT* is also easily to be seen equivalent (up to a constant factor in depth) to the "equal probability" model with all trees T_α having equal probabilities $p_\alpha = \frac{1}{|\{T_\alpha\}|}$. The rest of the paper uses this simplified "equal probability" model, and identifies a *d–RDT* with a finite collection $\{T_i\}$ of *d–DTs*. We say that such a *d–RDT* *recognizes (or computes)* a set $S \subseteq \mathbb{R}^n$, if it classifies every $x \in \mathbb{R}^n$ correctly (with respect to S) with probability at least $1 - \varepsilon$ for some $0 < \varepsilon < \frac{1}{2}$. The parameter ε bounds an error probability of *computations* of a *d–RDT*.

It is readily seen that the class of sets $S \subseteq \mathbb{R}^n$ *recognizable* by *d–RDTs* is *closed* under the complement.

The depth of $T = \{T_i\}$ is the *maximum* depth of T_i's in T. It is straightforward to verify that the class of sets $S \subseteq \mathbb{R}^n$ recognizable by *d–RDTs* is depth-invariant under changes of the error probability ε in the interval $(0, \frac{1}{2})$: for any two $\varepsilon_1, \varepsilon_2 \in (0, \frac{1}{2})$, if $S \subseteq \mathbb{R}^n$ is *recognized* by a *d–RDT* with error probability ε_1, and depth t, it is also recognizable by a *d–RDT* with error probability ε_2 and depth $O(t)$ [M85c]. It is also known that a *d–RDT* with a *worst case expected depth t*, a notion used by some authors, can be simulated by a *d–RDT* with depth $O(t)$ ([MT85]).

4 Main Results

We shall deal here with the randomized complexity of linear arrangements, and convex polyhedra. For the first class of sets several topological methods were developed for obtaining lower bounds for deterministic algebraic decision trees, and deterministic computation trees cf. [DL78], [SY82], [B-O83], [BLY92], [Y92] and [Y94]. In Ben-Or [B-O83] a general deterministic lower bound $\Omega(\log C)$ was proven for C being the number of connected components of $S \subseteq \mathbb{R}^n$ or its complement. Yao [Y92] (see also Björner, Lovász and Yao [BLY92]) proved a decade later a deterministic lower bound $\Omega(\log \chi)$ for χ the Euler characteristic of $S \subseteq \mathbb{R}^n$. A stronger lower bound $\Omega(\log B)$ for B the sum of the Betti number of $S \subseteq \mathbb{R}^n$ was proven later in Yao [Y94]. We have obvious inequalities $C, \chi \leq B$. For the second class of sets, convex polyhedra, the above topological methods fail because the invariant $B = 1$. For this class of sets, Grigoriev, Karpinski and Vorobjov [GKV97] introduced a drastically different method of counting the number of faces of $S \subseteq \mathbb{R}^n$ of all dimensions. The new method transforms a set $S \subseteq \mathbb{R}^n$ via "infinitesimal perturbations" into a smooth hypersurface and uses certain new calculus of principal curvatures on it. The resulting lower bound was $\Omega(\log N)$ for N being the number of faces of all dimensions provided N was *large enough.*

All the above mentioned methods did not work, and this for the both classes of sets on the randomized algebraic decision trees, and this for a fundamental reason. In fact, they even did not seem to work for the linear decision trees; the gap in the proof of Meyer auf der Heide [M85a] was firstly closed for the generic case by Grigoriev and Karpinski in [GK94]. The first very special randomized lower bounds were proven in Bürgisser, Karpinski and Lickteig [BKL93], and Grigoriev and Karpinski [GK93].

In this paper we survey some new general methods for proving lower bounds for *d–RDTs* recognizing linear arrangement and convex polyhedra.

Let $H_i \subseteq \mathbb{R}^n$, $1 \leq i \leq m$, $n \leq m$ be the hyperplanes, and $H_i^+ \subseteq \mathbb{R}^n$, $1 \leq i \leq m$, $n \leq m$, the closed halfspaces. Define $S = \bigcup_{i=1}^m H_i$, a linear arrangement, and $S^+ = \bigcap_{i=1}^m H_i^+$, a polyhedron.

In [GKMS97] the following general theorem was proven.

Theorem 1. ([GKMS97]). *Let $\varepsilon, c, \zeta, \delta$ be any constants such that $0 \leq \varepsilon < \frac{1}{2}$, $c > 0$, and $\zeta > \delta \geq 0$. There exists a constant $c^* > 0$ with the following property. If S (S^+) has at least $m^{\zeta(n-k)}$ k-faces for* certain *$0 \leq k < n$, then the* depth *of any d-RDT computing S (S^+) with the error probability ε is greater than $c^*(n-k)\log m$ for any degree $d < cm^\delta$.*

The original idea of this paper uses a nonarchimedean extension of a field, and consequently Tarski's transfer principle [T51], and a *leading term sign* technique combined with a *global labelled flag* construction (attached to all *k-faces* along the path of a decision tree) for counting number of faces of all dimensions of the set S.

We recall now the bounds of Section 2 on the number of *k-faces* of the n-dimensional restrictions of NP-complete problems. The result above yields directly the following concrete applications towards a *Bounded Integer Programming Problem*

$$L_{n,k} = \{x \in \mathbb{R}^n \mid \exists a \in \{0, \ldots, k\}^n [<x, a>= k]\} \text{ (cf. [M85a])},$$

and the *Knapsack Problem* $K_n = L_{n,1}$ (cf. [M85b].

Corollary 1.

(i) $\Omega(n^2 \log(k+1))$ *is a lower bound for the depth of any d-RDT computing the Bounded Integer Programming Problem $L_{n,k}$.*

(ii) $\Omega(n^2)$ is a lower bound for the depth of any d-RDT computing the Knapsack Problem.

Theorem 1 gives in fact much stronger lower bounds for non-constant degree d, *d–RDTs*. In the first case the sufficient condition on degree is $d = \Omega((k+1)^{\delta n})$ for $\delta < \frac{1}{16}$, in the second case $d = \Omega(2^{\delta n})$ for $\delta < \frac{1}{16}$.

It is also not too difficult to derive further randomized lower bounds (cf. [GKMS97]).

Corollary 2. *$\Omega(n \log n)$ is a lower bound for the depth of any d-RDT computing any of the following problems:*

(i) Element Distinctness,
(ii) Set Disjointness,
(iii) Resultant.

Corollary 2 holds also for the non-constant degree *d–RDTs* with $d = \Omega(n^\delta)$ for $\delta < \frac{1}{2}$ (cf. [GKMS97]). This leads us again to the very interesting computational issue of the dependence of the actual computational power of *d–RDTs* on the degree bound d.

It is also interesting to note that the proof method of [GKMS97] gives a new elementary technique for deterministic algebraic decision trees without making use of Milnor-Thom bound on Betti numbers of algebraic varieties.

5 Randomized Speedup

We shall investigate now the computational power of linear degree and sublinear depth *n–RDTs* and compare it with deterministic *n–DTs*. Such models can be easily simulated by ***randomized algebraic computational trees*** (*CTs*) in linear time. Also, it is easy to see that linear time *CTs* and linear time randomized *CTs* correspond to the non-uniform deterministic linear time and randomized linear time classes on the real number machine models (cf. [CKKLW95]).

Let us consider now the following permutational problem $PERM(a) = \{x | x \in \mathbb{R}^n,\ x \text{ is a permutation of } a\}$ for $a = (a_1, ..., a_n) \in \mathbb{R}^n$, $a_i \neq a_j$ for $i \neq j$. The number of connected components of $PERM(a)$ equals $n!$ and of its complement equals 1. By Ben-Or [B-O83] the lower bound of any deterministic CT or any $n - DT$ computing $PERM(a)$ is $\Omega(n \log n)$. However as noticed in [BKL93], there exists an *n–RDT* of constant depth computing $PERM(a)$ as follows. Construct a polynomial $p(\zeta) = \prod_{i=1}^{n}(\zeta - a_i) - \prod_{i=1}^{n}(\zeta - x_i) \in \mathbb{R}[\zeta]$. We have $x = (x_1, ..., x_{n)} \in PERM(a)$ iff $p(\zeta) \equiv 0$. The identity $p(\zeta)$ can be checked probabilistically by randomly chosing ζ from the set $\{1, ..., 4n\}$ and verifying whether $p(\zeta) = 0$. If $p(\zeta) = 0$, we decide that $p(\zeta) \equiv 0$ and $x \in PERM(a)$, otherwise we have a witness that $P(\zeta) \not\equiv 0$ and $x \notin PERM(a)$. The error probability is bounded by $\frac{1}{4}$. Construct now $4n$ many n–DTs T_ζ having a single decision element $p(\zeta)$, $T = \{T_\zeta\}_{\zeta \in \{1,\dots,4n\}}$. T computes $PERM(a)$ with error probability $\frac{1}{4}$.

Lemma 1. ([BKL93]). *There are problems $S \subseteq \mathbb{R}^n$ computable in $O(1)$ depth on n-RDTs which are not computable by any n-DT in depth $o(n \log n)$*

The next separation results will be much more powerful in nature. We extend our underlying decision tree models to allow arbitrary analytic functions as decision elements (cf. [R72]). We denote such decision trees by A–DTs, and A–$RDTs$, respectively.

Let us consider now the Octant Problem $\mathbb{R}^n_+ = \{(x_1, ..., x_n) \in \mathbb{R}^n | x_1 \geq 0, ..., x_n \geq 0\}$, the problem of *testing* the membership to $\mathbb{R}^n_+$.

Rabin [R72] proved the following (see also [GKMS97])

Lemma 2. ([R72]). *Any A-DT computing $\mathbb{R}^n_+$ has depth at least n.*

Grigoriev, Karpinski, Meyer auf der Heide and Smolensky [GKMS97] were able to prove the following degree hierarchy result on randomized decision trees.

Lemma 3. ([GKMS97]). *The depth of any d-RDT computing $\mathbb{R}^n_+$ with error probability $\epsilon \in (0, \frac{1}{2})$ is greater than or equal to $\frac{1}{d}(1 - 2\epsilon)^2 n$.*

The Octant Problem is closely related to the well known MAX Problem: given n real numbers $x_1, \dots, x_n$, $x_i \in \mathbb{R}$, compute the maximum of them. Rabin [R72] proved a sharp bound $n-1$ on depth of any A–DT computing MAX. Ting and Yao [TY94] proved a dramatic improvement on the depth of the *randomized* algebraic decision trees computing MAX for the case of pairwise distinct numbers (the leaves of a decision tree are labelled now by numbers $1, \dots, n$).

Theorem 2. ([TY94]). *There exists an n-RDT computing MAX problem for the case of pairwise distinct numbers in depth $O(log^2 n)$.*

We notice that the problem on whether $x_1 = max\{x_1, ..., x_n\}$ is *equivalent* to the test whether $(x_1 - x_2, ..., x_1 - x_n)$ belongs to the octant $\mathbb{R}^{n-1}_+$.

Grigoriev, Karpinski and Smolensky [GKS97] were able to extend the assertions of Lemma 3 and Theorem 2 to the following.

Theorem 3. ([GKS97]). *There exists an n-RDT computing $\mathbb{R}^n_+$ or deciding whether $x_i = \max\{x_1, \dots, x_n\}$ in depth $O(\log^2 n)$.*

Theorem 4. ([GKS97]). *There exists an n-RDT computing MAX in depth $O(\log^5 n)$.*

One notices a remarkable exponential ***randomized speed-up*** for the above problems having all (!) deterministic linear lower bounds, and this even for the general analytic decision trees ([R72]). An important issue remains whether the randomized speed-up can be carried even further. Interestingly, Wigderson and Yao [WY98] proved the following result connected to the construction of [TY94].

Assume that the decision tree performs only tests of the form "$x < V$", x is *smaller* than all elements in V. We call it a ***subset minimum test***. The test of this form was used in the design of [TY94]. We denote a corresponding randomized decision tree (using the subset minimum test only) by *SM-RDT*.

Theorem 5. ([WY98]). *Every SM-RDT computing MAX problem has depth* $\Omega(\log^2 n / \log\log n)$.

We turn now to the problems of proving lower bounds on the size of algebraic decision trees. Theorem 4 entails the subexponential size of *n-RDTs* computing *MAX*.

In this context Grigoriev, Karpinski and Yao [GKY98] proved the first exponential deterministic size lower bound on (ternary) algebraic decision trees for *MAX*. It should be noted that there was no size lower bound greater than $n-1$ known before.

The method used in this paper depends on the analysis of the so called "touching frequency" of the sets computed along the branches of a decision tree with the special "wall sets" related to the cellular decomposition of the set of $(x_1, \ldots, x_n) \in \mathbb{R}^n$ satisfying $x_1 = max\{x_1, \ldots, x_n\}$.

Theorem 6. ([GKY98]). *Any (ternary) algebraic decision tree of degree d computing MAX problem in dimension n has size* $\Omega(2^{c(d)n})$ *for the constant* $c(d) > 0$ *depending only on d.*

Grigoriev, Karpinski and Yao [GKY98] discovered also a new connection between a cellular decomposition of a set $S \subseteq \mathbb{R}^n$ defined by polynomial constraints of degree d and the maximum number of minimal cutsets $m_{d,n}$ of any rank-d hypergraph on n vertices.

Theorem 7. ([GKY98]). *Any (ternary) algebraic decision tree of degree d computing MAX problem in dimension n has size at least* $2^{n-1}/m_{d,n-1}$.

Interestingly, Theorem 7 gives improvements of the constants $c(d)$ used in Theorem 6. For any *2-DT* computing *MAX* problem, $c(d)$ computed via Theorem 7 is ≈ 0.47, and via Theorem 6 is ≈ 0.18 (cf. [GKY98]).

We are still lacking basic general methods for proving nontrivial lower bounds on the size (number of inner nodes) of both *d-DTs*, and *d-RDTs* with an exception of linear decision trees. In most cases the topological, and face counting methods cannot even deal with the questions about the size lower bounds of the very weak form: "is the size $t+1$ necessary?" for t a known lower bound on the depth of algebraic decision trees.

6 Extensions

We will turn now to the model of a randomized computation tree (*RCT*) modeling straight line computation in which we charge for each arithmetic operation needed to compute its decision elements (cf. [B-O83]).

The papers [GK97], [G98] generalize the results of Section 4 to the case of *RCTs* using some new results on the *border* (generalization of the *border rank* of a tensor) and *multiplicative* complexity of a polynomial.

Theorem 8. ([GK97], [G98]).

(i) $\Omega(n^2 \log(k+1))$ *is a lower bound for the depth of any RCT computing the bounded Integer Programming Problem* $L_{n,k}$.
(ii) $\Omega(n^2)$ *is a lower bound for the depth of any RCT computing the Knapsack.*
(iii) $\Omega(n \log n)$ *is a lower bound for the depth of any RCT computing the Element Distinctness.*

An important issue remains, and this in both cases, deterministic and randomized, about the generalization of algebraic decision trees and computation trees to the "ultimate models" of branching programs obtained by merging together equivalent nodes in a decision tree. An extended research on the boolean model of a branching program was carried throughout the last decade (cf., e.g., Borodin [B93], Razborov [R91] for deterministic programs, and Karpinski [K98a], [K98b], Thathachar [T98] for randomized ones). Much less is known about the model of algebraic branching programs, see also Yao [Y82].

7 Open Problems and Further Research

An important issue of the tradeoffs between the size and the depth of algebraic decision trees, computational trees, and branching programs remains widely open. We are not able at the moment, as mentioned before, to prove any nontrivial lower bound on the size of algebraic decision trees for the n-dimensional restrictions of NP-complete problems like *Knapsack* or *Bounded Integer Programming* (cf. [M84], [M85b], [M93]). Nor can we prove any randomized size upper bounds for these problems better than the best known deterministic ones.

For the recent randomized lower bounds for the *Nearest Neighbor Search* Problem on the related cell probe model see also [BOR99]. It will be very interesting to shed some more light on this model and also other related models capturing hashing and reflecting storage resources required by an actual geometric computation.

Major problems remain open about the randomized decision complexity of concrete geometric problems expressed by ***simultaneous positivity*** of small degree polynomials, like quadratic or cubic ones, or the ***existential*** problems of simultaneous positivity of small degree polynomials, corresponding to an algebraic version of the SAT problem. □

References

[AHU74] A.V. Aho, J.E. Hopcroft and J.D. Ullman, *The Design and Analysis of Computer Algorithms*, Addison-Wesley, 1974.

[B-O83] M. Ben-Or, *Lower Bounds for Algebraic Computation Trees*, Proc. 15th ACM STOC (1983), pp. 80–86.

[BLY92] A. Björner, L. Lovász and A. Yao, *Linear Decision Trees: Volume Estimates and Topological Bounds*, Proc. 24th ACM STOC (1992), pp. 170–177.

[B93] A. Borodin, *Time Space Tradeoffs (Getting Closer to the Barrier?)*, Proc. ISAAC'93, LNCS **762** (1993), Springer, 1993, pp. 209–220.

[BOR99] A. Borodin, R. Ostrovsky and Y. Rabani, *Lower Bounds for High Dimensional Nearest Neighbour Search and Related Problems*, Proc. 31st ACM STOC (1999), pp. 312-321.

[BKL93] P. Bürgisser, M. Karpinski and T. Lickteig, *On Randomized Algebraic Test Complexity*, J. of Complexity **9** (1993), pp. 231–251.

[CKKLW95] F. Cucker, M. Karpinski, P. Koiran, T. Lickteig, K. Werther, *On Real Turing Machines that Toss Coins*, Proc. 27th ACM STOC (1995), pp. 335–342.

[DL78] D.P. Dobkin and R. J. Lipton, *A Lower Bound of $\frac{1}{2}n^2$ on Linear Search Programs for the Knapsack Problem*, J. Compt. Syst. Sci. **16** (1978), pp. 413–417.

[E87] H. Edelsbrunner, *Algorithms in Computational Geometry*, Springer, 1987.

[FK95] R. Freivalds and M. Karpinski, *Lower Time Bounds for Randomized Computation*, Proc. 22nd ICALP'95, LNCS **944**, Springer, 1995, pp. 183–195.

[G67] B. Grünbaum, *Convex Polytopes*, John Wiley, 1967.

[GK93] D. Grigoriev and M. Karpinski, *Lower Bounds on Complexity of Testing Membership to a Polygon for Algebraic and Randomized Computation Trees*, Technical Report TR-93-042, International Computer Science Institute, Berkeley, 1993.

[GK94] D. Grigoriev and M. Karpinski, *Lower Bound for Randomized Linear Decision Tree Recognizing a Union of Hyperplanes in a Generic Position*, Research Report No. 85114-CS, University of Bonn, 1994.

[GK97] D. Grigoriev and M. Karpinski, *Randomized $\Omega(n^2)$ Lower Bound for Knapsack*, Proc. 29th ACM STOC (1997), pp. 76–85.

[GKMS97] D. Grigoriev, M. Karpinski, F. Meyer auf der Heide and R. Smolensky, *A Lower Bound for Randomized Algebraic Decision Trees*, Comput. Complexity **6** (1997), pp. 357–375.

[GKS97] D. Grigoriev, M. Karpinski, and R. Smolensky, *Randomization and the Computational Power of Analytic and Algebraic Decision Trees*, Comput. Complexity **6** (1997), pp. 376–388.

[GKV97] D. Grigoriev, M. Karpinski and N. Vorobjov, *Lower Bound on Testing Membership to a Polyhedron by Algebraic Decision Trees*, Discrete Comput. Geom. **17** (1997), pp. 191–215.

[GKY98] D. Grigoriev, M. Karpinski and A. C. Yao, *An Exponential Lower Bound on the Size of Algebraic Decision Trees for MAX*, Computational Complexity **7** (1998), pp. 193–203.

[G98] D. Grigoriev, *Randomized Complexity Lower Bounds*, Proc. 30th ACM STOC (1998), pp. 219–223.

[K98a] M. Karpinski, *On the Computational Power of Randomized Branching Programs*, Proc. Randomized Algorithms 1998, Brno, 1998, pp. 1–12.

[K98b] M. Karpinski, *Randomized OBDDs and the Model Checking*, Proc. Probabilistic Methods in Verification, PROBMIV'98, Indianapolis, 1998, pp. 35–38.

[KM90] M Karpinski and F. Meyer auf der Heide, *On the Complexity of Genuinely Polynomial Computation*, Proc. MFCS'90, LNCS **452**, Springer, 1990, pp. 362-368.

[KV88] M. Karpinski and R. Verbeek, *Randomness, Provability, and the Separation of Monte Carlo Time and Space*, LNCS **270** (1988), Springer, 1988, pp. 189–207.

[L84] S. Lang, *Algebra*, Addison-Wesley, New York, 1984.

[MT85] U. Manber and M. Tompa, *Probabilistic, Nondeterministic and Alternating Decision Trees*, J. ACM **32** (1985), pp. 720–732.

[M93] S. Meiser, *Point Location in Arrangements of Hyperplanes*, Information and Computation **106** (1993), pp. 286–303.

[M84] F. Meyer auf der Heide, *A Polynomial Linear Search Algorithm for the n–Dimensional Knapsack Problem*, J. ACM **31** (1984), pp. 668–676.

[M85a] F. Meyer auf der Heide, *Nondeterministic versus Probabilistic Linear Search Algorithms*, Proc. IEEE FOCS (1985a), pp. 65–73.

[M85b] F. Meyer auf der Heide, *Lower Bounds for Solving Linear Diophantine Equations on Random Access Machines*, J. ACM **32** (1985), pp. 929–937.

[M85c] F. Meyer auf der Heide, *Simulating Probabilistic by Deterministic Algebraic Computation Trees*, Theoretical Computer Science **41** (1985c), pp. 325–330.

[M64] J. Milnor, *On the Betti Numbers of Real Varieties*, Proc. Amer. Math. Soc. **15** (1964), pp. 275–280.

[R72] M.O. Rabin, *Proving Simultaneous Positivity of Linear Forms*, J. Comput. Syst. Sciences **6** (1972), pp. 639–650.

[R91] A. Razborov, *Lower Bounds for Deterministic and Nondeterministic Branching Programs*, Proc. FCT'91, LNCS **529**, Springer, 1991, pp. 47–60.

[SP82] J. Simon and W.J. Paul, *Decision Trees and Random Access Machines*, L'Enseignement Mathematique. Logic et Algorithmic, Univ. Geneva, 1982, pp. 331–340.

[S85] M. Snir, *Lower Bounds for Probabilistic Linear Decision Trees*, Theor. Comput. Sci. **38** (1985), pp. 69–82.

[SY82] J.M. Steele and A.C. Yao, *Lower Bounds for Algebraic Decision Trees*, J. of Algorithms **3** (1982), pp. 1–8.

[T51] A. Tarski, *A Decision Method for Elementary Algebra and Geometry*, University of California Press, 1951.

[T98] J. S. Thathachar, *On Separating the Read-k-Times Branching Program Hierarchy*, Proc. 30th ACM STOC (1998), pp. 653-662.

[T65] R. Thom, *Sur L'Homologie des Variéetés Algébriques Réelles*, Princeton University Press, Princeton, 1965.

[TY94] H.F. Ting and A.C. Yao, *Randomized Algorithm for finding Maximum with $O((\log n)^2)$ Polynomial Tests*, Information Processing Letters **49** (1994), pp. 39-43.

[WY98] A. Wigderson and A.C. Yao, *A Lower Bound for Finding Minimum on Probabilistic Decision Trees*, to appear.

[Y81] A.C. Yao, *A Lower Bound to Finding Convex Hulls*, J. ACM **28** (1981), pp. 780–787.

[Y82] A.C. Yao, *On the Time-Space Tradeoff for Sorting with Linear Queries*, Theoretical Computer Science **19** (1982), pp. 203–218.

[Y92] A.C. Yao, *Algebraic Decision Trees and Euler Characteristics*, Proc. 33rd IEEE FOCS (1992), pp. 268–277.

[Y94] A.C. Yao, *Decision Tree Complexity and Betti Numbers*, Proc. 26th ACM STOC (1994), pp. 615–624.

Tile Transition Systems as Structured Coalgebras*

Andrea Corradini[1], Reiko Heckel[2], Ugo Montanari[1]

[1] Dipartimento di Informatica, Università degli Studi di Pisa,
Corso Italia, 40, I - 56125 Pisa, Italia, {andrea, ugo}@di.unipi.it
[2] Universität GH Paderborn, FB 17 Mathematik und Informatik,
Warburger Str. 100, D-33098 Paderborn, Germany, reiko@uni-paderborn.de

Abstract. The aim of this paper is to investigate the relation between two models of concurrent systems: tile rewrite systems and coalgebras. Tiles are rewrite rules with side effects which are endowed with operations of parallel and sequential composition and synchronization. Their models can be described as monoidal double categories. Coalgebras can be considered, in a suitable mathematical setting, as dual to algebras. They can be used as models of dynamical systems with hidden states in order to study concepts of observational equivalence and bisimilarity in a more general setting.

In order to capture in the coalgebraic presentation the algebraic structure given by the composition operations on tiles, coalgebras have to be endowed with an algebraic structure as well. This leads to the concept of *structured coalgebras*, i.e., coalgebras for an endofunctor on a category of algebras.

However, structured coalgebras are more restrictive than tile models. Those models which can be presented as structured coalgebras are characterized by the so-called *horizontal decomposition property*, which, intuitively, requires that the behavior is compositional in the sense that all transitions from complex states can be derived by composing transitions out of component states.

1 Introduction

Tile logic [16, 26] relies on certain rewrite rules with side effects, called *basic tiles*, reminiscent of *SOS rules* [30] and *context systems* [21]. Related models are *structured transition systems* [11], as well as *rewriting logic* [22, 23] which extends to concurrent systems with state changes the body of theory developed within the algebraic semantics approach. Tile logic has been conceived with similar aims and similar algebraic structure as rewriting logic, and it extends rewriting logic

* Research partially supported by MURST project *Tecniche Formali per Sistemi Software*, by CNR Integrated Project *Metodi per Sistemi Connessi mediante Reti*, by TMR Network *GETGRATS* and by Esprit WGs *APPLIGRAPH*, *CONFER2* and *COORDINA*.

(in the unconditional case), since it takes into account state changes with side effects and synchronization.

We now briefly introduce tile logic. A tile A is a sequent which has the form:

$$A\colon s \xrightarrow[b]{a} s'$$

and states that the ***initial configuration*** s of the system evolves to the ***final configuration*** s' producing an ***effect*** b. However s is in general ***open*** (not closed) and the rewrite step producing the effect b is actually possible only if the subcomponents of s also evolve producing the ***trigger*** a. Both trigger a and effect b are called ***observations***, and model the interaction, during a computation, of the system being described with its environment. More precisely, both system configurations are equipped with an ***input*** and an ***output interface***, and the trigger just describes the evolution of the input interface from its initial to its final configuration. Similarly for the effect. It is convenient to visualize a tile as a two-dimensional structure (see Figure 1), where the horizontal dimension corresponds to the extension of the system, while the vertical dimension corresponds to the extension of the computation. Actually, we should also imagine a third dimension (the thickness of the tile), which models parallelism: configurations, observations, interfaces and tiles themselves are all supposed to consist of several components in parallel.

The initial configuration of a tile A can also be called north(A), and likewise south(A), west(A) and east(A) stands for final configuration, trigger and effect, respectively.

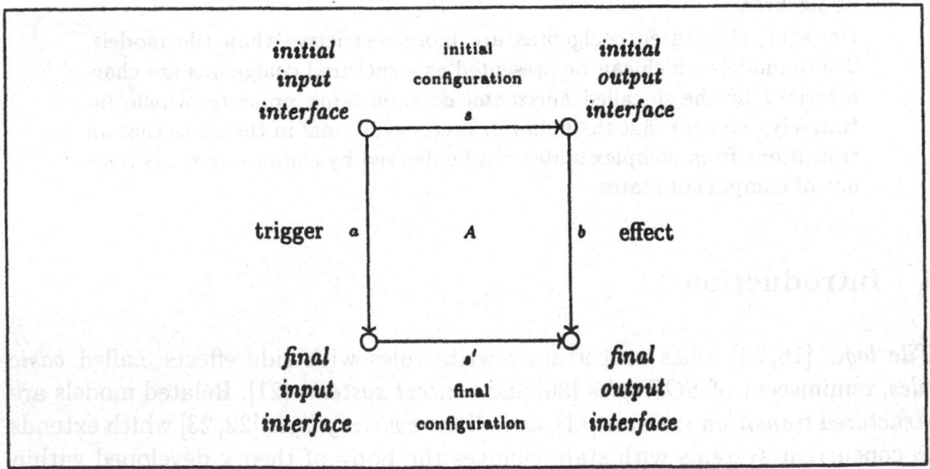

Fig. 1: A tile.

Configurations and observations are sometimes terms over a horizontal and a vertical signature, but for lots of applications to distributed systems it is convenient to employ various kinds of graphs, diagrams and charts [1]. Also suitable structural axioms can be imposed on terms or graphs.

Tiles are equipped with inference rules expressing three operations of composition: parallel ($\_\otimes\_$), horizontal ($\_*\_$), and vertical ($\_\cdot\_$) composition. Similarly, both configurations and observations are assumed to be equipped with operations of parallel and sequential composition, and interfaces with parallel composition only.

The operation of parallel composition is self explanatory. Vertical composition models sequential composition of transitions and computations. Horizontal composition corresponds to synchronization: the effect of the first tile acts as trigger of the second tile, and the resulting tile expresses the synchronized behavior of both.

In tile logic, a *tile rewrite system* provides signatures for configurations and observations and a set of *basic* tiles. Proofs start from basic tiles and apply the composition rules in all possible ways. The structure of tiles entailed in this way is specified by *proof terms*, built from the basic tiles used in the derivation and from the composition operations performed on them, up to certain structural axioms. Following the usual Curry-Howard analogy, proof terms *are* tiles, and their horizontal and vertical sources and targets are their *types*. However, quite often, proof terms are not relevant, and thus are omitted, or equivalently an additional normalizing axiom is introduced stating that two proof terms are the same whenever they have the same sources and targets. This is the case throughout in the paper.

Tile models are monoidal double categories. A *double category* [13] consists of four collections: objects, horizontal arrows, vertical arrows and cells, which correspond respectively to interfaces, configurations, observations and tiles of tile logic. Horizontal arrows with objects form the horizontal 1-category, and cells with vertical arrows form the horizontal 2-category. Contemporary horizontal composition of horizontal arrows and cells corresponds to horizontal composition of tile sequents. Similarly for the vertical dimension. Monoidal double categories have an additional operation which applies to the four collections above and which on tile sequents corresponds to parallel composition. If the tile logic is *flat*, i.e., it forgets about proof terms, it is appropriate to consider only flat models, too. A double category is flat if all the cells with the same sources and targets are identified.

Tile rewrite systems are interpreted as computads, i.e., algebraic structures consisting again of objects, horizontal and vertical arrows, and cells, but where only 1-categories are defined. In the monoidal version of tile logic, configurations and observations are arrows of strict monoidal categories generated by two hypersignatures. A more direct representation of configurations and observations relies on certain hypergraphs analogous to Petri sequential processes [1]. In the initial model, the tiles entailed by a tile rewrite system can now be interpreted as cells. More precisely, the initial model can be constructed as the monoidal double category freely generated by the computad corresponding to the tile rewrite system [26].

Additional operations and axioms can be imposed on proof terms, configurations and observations whenever extra structure is required. Correspondingly, tile models are enriched and the construction of the initial models adapted. Symmetric monoidal, cartesian [3] and cartesian closed versions [4] of tiles have been defined. Moreover, different structures for configurations, observations and proof terms can be introduced to tailor the logic and the models to the specific needs of the applications. An expressive specification language for this purpose is *membership equational logic* [24, 26], which has also been used to map tile logic into rewriting logic for implementation purposes [2]. Of course these enriched tile models are still special cases of the basic models of monoidal double categories. Hence, the results of this paper apply.

As it should be clear from the informal introduction above, the main intended application area of tile logic are distibuted, interactive, open systems. In fact, tiles allow to model directly distribution (via the use of graphs to represent configurations), interaction (via triggers and effects) and openness (via the instantiation of free variables in configurations). In addition, tiles introduce rich forms of compositionality and induction based on their operations. Examples of applications are CCS with localities [2], π-calculus [14] and coordination of distributed systems [29].

The operational semantics of interactive systems is usually given in terms of labeled transition systems, and their abstract semantics in terms of sets of traces, or up to bisimilarity. When compositionality is an issue, semantic equivalence must be shown to be a congruence with respect to composition operations. The labeled transition systems associated to tile models are straightforwardly defined: horizontal arrows are states, tiles are transitions and pairs (trigger, effect) are labels. *Tile bisimilarity* is then defined in the standard way. It is also natural to define *tile congruences* as those equivalences of states, i.e., of horizontal arrows, which are *functorial*, i.e., which preserve their monoidal structure (identities, parallel composition and sequential composition).

The problem of finding sufficient conditions on tile rewrite systems to ensure that bisimilarity is a congruence in the initial model has been considered in [16]. The problem is divided in two parts. First a semantic condition, called *decomposition property* is defined on tile models, and it is shown that decomposition implies that bisimilarity is a congruence. The horizontal decomposition property is defined as follows. Whenever the vertical source h (which is a horizontal arrow) of a cell A can be sequentially decomposed as $h = h_1; h_2$, then also the cell itself must be horizontally decomposable as $A = A_1 * A_2$, where h_1 and h_2 are the vertical sources of A_1 and A_2 respectively. A similar condition is required for parallel composition. Informally, decomposition means that, given any computation A of a system h, and any subsystem h' of h, a computation A' of h' should exist which is a subcomputation of A. The second part consists in providing a syntactic condition, called *basic source property*, on tile rewrite systems: all rewrite rules must have just signature operators (i.e., basic graphs) as initial configurations and effects. It is possible to see that the basic source property ensures the decomposition property in the initial model, provided that there are no structural axioms in the specification.

The aim of this paper is to recast tile models enjoying a slightly stronger decomposition property (which we call *reflective* decomposition) as *structured coalgebras*. The use of coalgebras for the specification of dynamical systems with a hidden state space is receiving more and more attention in the last years, as a valid alternative to algebraic methods based on observational equivalences [31]. Given an endofunctor F on a category $\mathbf{C}$, a coagebra is an arrow $f\colon X \to F(X)$ of $\mathbf{C}$ and a coalgebra morphism from f to f' is an arrow $h : X \to X'$ of $\mathbf{C}$ with $h\,; f' = f\,; F(h)$. Under certain conditions on $\mathbf{C}$ and F, a category of coalgebras admits a final object, which can be considered informally as the minimal realization of the union of all the coalgebras in the category.

Ordinary labeled transition systems (with finite or countable branching) can be represented as coalgebras for a suitable functor on **Set**. Furthermore, the unique morphism to the final coalgebra induces an equivalence which turns out to be exactly bisimilarity. Thus a first (rather straightforward) result of this paper is to show that tile models, seen as transition systems, can be considered as coalgebras and that their bisimilarity can be derived coalgebraically.

However, this representation forgets about the algebraic structure on horizontal arrows, which are seen just as forming a family of sets. As a consequence, the property that bisimilarity is a congruence, which is essential for making abstract semantics compositional, is not reflected in the structure of the model.

The problem of integrating coalgebras and algebras obtaining a model equipped with both structures has been tackled in [32], and an alternative but equivalent approach based on *structured coalgebras* is presented in [8, 10]. Here, the endofunctor determining the coalgebraic structure is lifted from **Set** to the category of Γ-algebras, for some algebraic theory Γ. Morphisms between coalgebras in this category are both Γ-homomorphisms and coalgebra morphisms, and thus the unique morphism to the final coalgebra, which always exists, induces a (coarsest) bisimulation congruence on any coalgebra.

The second result of this paper is to show that by taking as Γ the theory of monoidal categories, a necessary and sufficient condition for the lifting to occur is reflective horizontal decomposition. Thus we obtain in another way that decomposition implies bisimilarity to be a congruence. Reflective decomposition additionally requires that cells with vertical sources which are horizontal identities must be identities for the horizontal composition of cells. It is easy to see that the basic source property implies also this extra condition for the initial model.

The paper is organized as follows. After introducing monoidal double categories in Section 2, Section 3 defines tile rewrite systems and their models, bisimilarity and functoriality, as well as the horizontal decomposition and basic source properties. Section 4 provides the necessary background on structured coalgebras. Section 5 presents tile transition systems as coalgebras over families of sets, while Section 6 provides the lifting of the algebraic structure. Finally in Section 7, as a case study, we show how the *calculus of communicating systems* (CCS) [27] can be recast in the tile framework as well as in terms of structured coalgebras.

2 Monoidal Double Categories

A *double category* is an internal category in the category of categories. Equivalently, double categories can be specified by a theory which is the tensor product[1] of the theory of categories with itself. The theory of *monoidal* double categories can be obtained as the tensor product of the theory of categories (twice) with the theory of monoids. Thus, one can argue that if the desired model of computation must have operations of parallel and horizontal composition (to build more parallel and larger systems) and of vertical composition (to build longer computations), then monoidal double categories are the most natural answer.

Here we give a more direct presentation of double categories, as done by Kelly and Street [19].

Definition 1 (double category). *A* double category $\mathcal{D}$ *consists of a collection* $a, b, c, \ldots$ *of* objects, *a collection* $h, g, f, \ldots$ *of* horizontal arrows, *a collection* $v, u, w, \ldots$ *of* vertical arrows *and a collection* $A, B, C, \ldots$ *of* cells.

Objects and horizontal arrows form the horizontal 1-category $\mathcal{H}$ *(see Figure 2), with identity* id^a *for each object* a*, and composition* $\_ * \_$.

$$a \xrightarrow{h} b * b \xrightarrow{g} c = a \xrightarrow{h*g} c \qquad a \xrightarrow{id^a} a$$

Fig. 2: Composition and identities in the horizontal 1-category.

Objects and vertical arrows form also a category, called the vertical 1-category $\mathcal{V}$ *(see Figure 3), with identity* id_a *for each object* a*, and composition* $\_ \cdot \_$ *(sometimes we will refer to both* id^a *and* id_a *either with the object name* a *or with* id_a*)*

Cells are assigned horizontal source *and* target *(which are vertical arrows) and* vertical source *and* target *(which are horizontal arrows); furthermore sources and targets must be* compatible, *in the sense that, given a cell* A*, with vertical source* h*, vertical target* g*, horizontal source* v*, and horizontal target* u*, then* h *and* v *have the same source,* g *and* u *have the same target, the target of* h *is equal to the source of* u*, and the target of* v *is equal to the source of* g*. These constraints can be represented by the diagram in Figure 4, for which we use the notation* $A : h \xrightarrow[u]{v} g$.

In addition, cells can be composed both horizontally $(\_ * \_)$ *and vertically* $(\_ \cdot \_)$ *as follows: given* $A : h \xrightarrow[u]{v} g$, $B : f \xrightarrow[w]{u} k$*, and* $C : g \xrightarrow[s]{z} h'$*, then* $A * B : (h * f) \xrightarrow[w]{v} (g * k)$*, and* $A \cdot C : h \xrightarrow[u \cdot s]{v \cdot z} h'$ *are cells. Both compositions*

---

[1] Tensor product (see for instance [20]) is a well-known construction for ordinary algebraic (Lawvere) theories. It can be extended to theories with partial operations (e.g. PMEqtl [26]) with essentially the same properties.

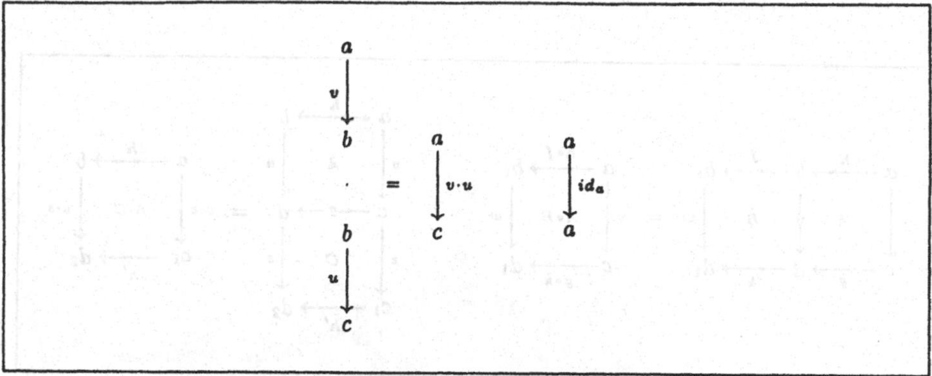

Fig. 3: Composition and identities in the vertical 1-category.

can be pictured by pasting the diagrams in Figure 5. Moreover, given a fourth cell $D : k \xrightarrow[t]{s} f'$, *horizontal and vertical compositions verify the following* exchange law *(see also Figure 6):*

$$(A \cdot C) * (B \cdot D) = (A * B) \cdot (C * D)$$

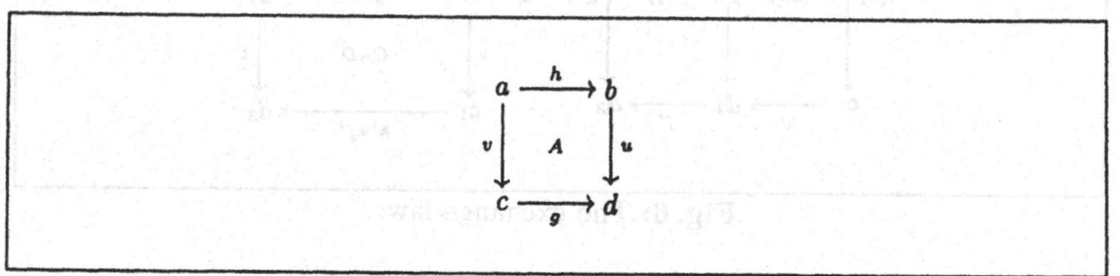

Fig. 4: Graphical representation of a cell.

Under these rules, cells form both a horizontal category $\mathcal{D}^*$ *and a vertical category* $\mathcal{D}^{\cdot}$, *with respective identities* $1_v : a \xrightarrow[v]{v} c$ *and* $1^h : h \xrightarrow[b]{a} h$. *Given* $1^h : h \xrightarrow[b]{a} h$ *and* $1^g : g \xrightarrow[c]{b} g$, *the equation* $1^h * 1^g = 1^{h*g}$ *must hold (and similarly for vertical composition of horizontal identities), as illustrated in Figure 7. Furthermore, horizontal and vertical identities of identities coincide, i.e.,* $1_{id_a} = 1^{id^a}$ *and are denoted by the simpler notation* 1_a *(or just* a*).*

A flat double category satisfies the additional condition that two cells with the same horizontal and vertical sources and targets are the same cell.

Remark 1. As a matter of notation, sometimes we will use $\_;\_$ to denote the composition on both the horizontal and vertical 1-categories.

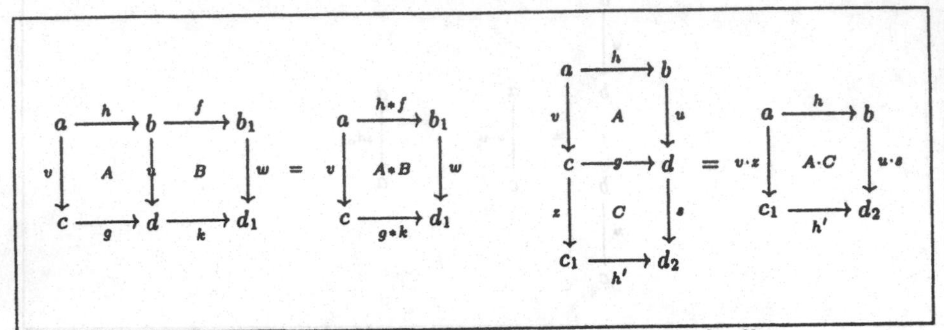

Fig. 5: Horizontal and vertical composition of cells.

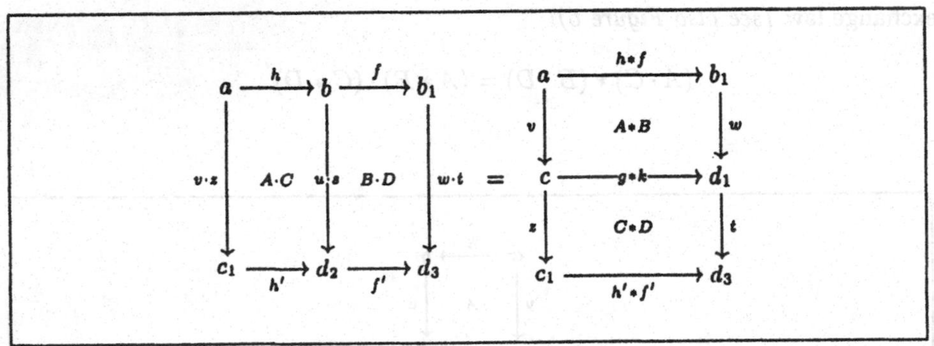

Fig. 6: The exchange law.

Fig. 7: Composition of identities.

Definition 2 (double functor). *Given two double categories $\mathcal{D}$ and $\mathcal{E}$, a* double functor $F : \mathcal{D} \longrightarrow \mathcal{E}$ *is a 4-tuple of functions mapping objects to objects, horizontal and vertical arrows to horizontal and vertical arrows, and cells to cells, preserving identities and compositions of all kinds.*

A *monoidal double category,* for the *strict* case, is defined as follows.

Definition 3 (strict monoidal double category). *A* strict monoidal double category, *sMD in the following, is a triple $(\mathcal{D}, \otimes, e)$, where:*

- *$\mathcal{D}$ is the underlying double category,*
- *$\otimes : \mathcal{D} \times \mathcal{D} \longrightarrow \mathcal{D}$ is a double functor called the* tensor product, *and*
- *e is an object of $\mathcal{D}$ called the* unit object,

such that the following diagrams commute:

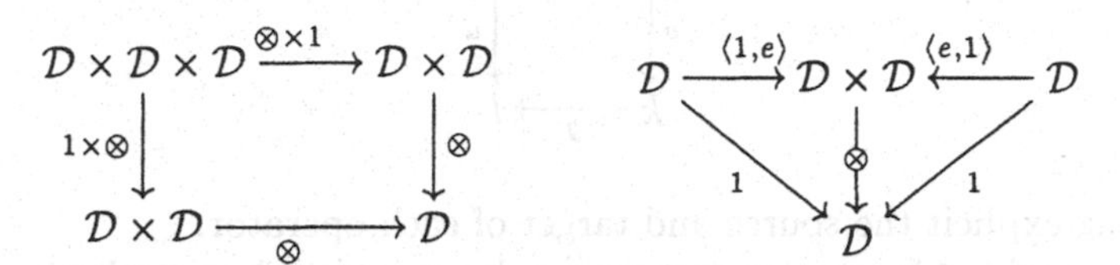

where double functor $1 : \mathcal{D} \longrightarrow \mathcal{D}$ is the identity on $\mathcal{D}$, the double functor $e : \mathcal{D} \longrightarrow \mathcal{D}$ (with some abuse of the notation) is the constant double functor which associates the object e and identities on e respectively to each object and each morphism/cell of $\mathcal{D}$, and $\langle \_, \_ \rangle$ denotes the pairing of double functors induced by the cartesian product of double categories. These equations state that the tensor product $\_ \otimes \_$ is associative on both objects, arrows and cells, and that e is the unit for $\_ \otimes \_$.

A monoidal double functor *is a double functor which preserves tensor product and unit object. We denote by* **fsMDCat** *the category of flat monoidal double categories and monoidal double functors.*

3 Tile Rewrite Systems

We now introduce (the flat monoidal versions of) tile rewrite systems and tile logic. Informally, a tile rewrite system is a set of double cells which, by horizontal, vertical and parallel composition, freely generate a monoidal double category. In the flat, monoidal version, the 1-categories of horizontal and vertical arrows are the strict monoidal categories freely generated by a horizontal and a vertical (hyper-) signature, which share the same set of sorts. The resulting monoidal double category is flat, i.e., two cells with the same horizontal and vertical source and target are identified.

Definition 4 (many-sorted hyper-signature). *Given a set S of sorts, a* (many-sorted, hyper) signature *is an $S^* \times S^*$-indexed family of sets $\Sigma = \{\Sigma_{n,m}\}_{(n,m)\in S^*\times S^*}$, where S^* denotes the free monoid on set S. Each $f \in \Sigma_{n,m}$ is denoted by $f : n \rightarrow m$.*

Definition 5 (monoidal category freely generated by a signature). *Given a signature Σ, $\mathbf{M}(\Sigma)$ is the strict monoidal category freely generated by Σ.*

Definition 6 (tile rewrite systems). *A* monoidal tile rewrite system $\mathcal{T}$ *is a quadruple $\langle S, \Sigma_h, \Sigma_v, R\rangle$, where Σ_h, Σ_v are signatures on the same set S of sorts, and $R \subseteq \mathbf{M}(\Sigma_h) \times \mathbf{M}(\Sigma_v) \times \mathbf{M}(\Sigma_v) \times \mathbf{M}(\Sigma_h)$ is the set of* rewrite rules, *such that for all $\langle h, v, u, g\rangle \in R$, we have $h : n \to m, g : k \to l$ if and only if $v : n \to k, u : m \to l$.*

For $\langle h, v, u, g\rangle \in R$ we use the notation $h \xrightarrow[u]{v} g$, or we depict it as a *tile*

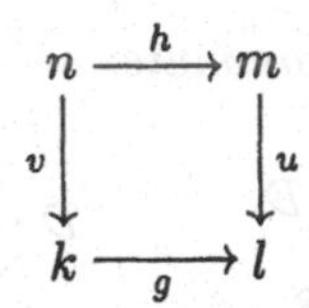

thus making explicit the source and target of each operator.

The rules of a tile rewrite system can be considered as its basic sequents. In the following, we say that h *rewrites to* g, using a *trigger* v and producing an *effect* u, if the (flat) sequent $h \xrightarrow[u]{v} g$ can be obtained by finitely many applications of certain inference rules.

Definition 7 (tile sequents). *Let $\mathcal{T} = \langle S, \Sigma_h, \Sigma_v, R\rangle$ be a monoidal tile rewrite system. We say that $\mathcal{T}$ entails the* tile sequent $h \xrightarrow[u]{v} g$, *written $\mathcal{T} \vdash h \xrightarrow[u]{v} g$, if and only if it can be obtained by a finite number of applications of the inference rules given in Table 1.*

Basic rules provide the generators of the sequents, together with suitable identity arrows, whose intuitive meaning is that an element of $\mathbf{M}(\Sigma_h)$ may stay idle during a rewrite, showing no effect and using no trigger. Similarly, a horizontal identity may be rewritten to itself whenever trigger and effect are equal. Composition rules express the way in which sequents can be combined, either sequentially (*vert*), or executing them in parallel (*par*), or nesting one inside the other (*hor*).

It is easy to see that the tiles entailed by a tile rewrite system are the cells of a double category.

Proposition 1 (from tile rewrite systems to double categories). *Given a monoidal tile rewrite system $\mathcal{T} = \langle S, \Sigma_h, \Sigma_v, R\rangle$, the flat monoidal double category $F_T(\mathcal{T})$ has $\mathbf{M}(\Sigma_h)$ as horizontal 1-category, $\mathbf{M}(\Sigma_v)$ as vertical 1-category, and the flat tile sequents entailed by $\mathcal{T}$ as double cells.*

The models of a tile rewrite systems are defined as follows.

Basic Sequents. *Generators and Identities:*

$$(gen)\ \frac{h \xrightarrow[u]{v} g \in R}{h \xrightarrow[u]{v} g}$$

$$(v\text{-}ref)\ \frac{v : n \to k \in \mathbf{M}(\Sigma_v)}{id_n \xrightarrow[v]{v} id_k} \qquad (h\text{-}ref)\ \frac{h : n \to m \in \mathbf{M}(\Sigma_h)}{h \xrightarrow[id_m]{id_n} h}$$

Composed Sequents. *Parallel, Horizontal and Vertical compositions:*

$$(vert)\ \frac{h \xrightarrow[u_1]{v_1} f,\ f \xrightarrow[u_2]{v_2} g}{h \xrightarrow[u_1;u_2]{v_1;v_2} g}$$

$$(par)\ \frac{h_1 \xrightarrow[u_1]{v_1} g_1,\ h_2 \xrightarrow[u_2]{v_2} g_2}{h_1 \otimes h_2 \xrightarrow[u_1 \otimes u_2]{v_1 \otimes v_2} g_1 \otimes g_2} \qquad (hor)\ \frac{h_1 \xrightarrow[w]{v} g_1,\ h_2 \xrightarrow[u]{w} g_2}{h_1; h_2 \xrightarrow[u]{v} g_1; g_2}$$

Table 1: Inference rules for monoidal tile sequents.

Proposition 2 (models of tile rewrite systems). *Given a tile rewrite system $\mathcal{T}$, its* category of flat models *is the comma category* $(F_T(\mathcal{T}) \downarrow \mathbf{fsMDCat})$.

Notice that the models are themselves double categories and that $F_T(\mathcal{T})$ is initial in the category of models of $\mathcal{T}$. However, in the following we will not be interested in morphisms relating only the models of a single tile rewrite system, since abstraction based on bisimilarity may relate several of them. Thus in the following we will consider *generic* models, i.e., just monoidal double categories.

We now introduce the ordinary notions of transition system and bisimilarity.

Definition 8 (labeled transition systems). *Let L be a fixed set of labels. A* (nondeterministic) labeled transition system (over L), *briefly LTS, is a structure $TS = \langle S, \longrightarrow_{TS} \rangle$, where S is a set of states, and $\longrightarrow_{TS} \subseteq S \times L \times S$ is a labeled transition relation. As usual, we write $s \xrightarrow{l}_{TS} s'$ for $\langle s, l, s' \rangle \in \longrightarrow_{TS}$.*

A transition system morphism *$f : TS \to TS'$ is a function $f : S \to S'$ which "preserves" the transitions, i.e., such that $s \xrightarrow{l}_{TS} t$ implies $f(s) \xrightarrow{l}_{TS'} f(t)$. We will denote by $\mathbf{LTS}_L$ the category of LTS over L and corresponding morphisms.*

Definition 9 (bisimilarity). *Given a LTS $TS = \langle S, \longrightarrow_{TS} \rangle$, an equivalence relation $\equiv$ on S is a* bisimulation *if, whenever $s_1 \equiv s_2$, then for any transition $s_1 \xrightarrow{l}_{TS} s_1'$ there exists a corresponding transition $s_2 \xrightarrow{l}_{TS} s_2'$ with $s_1' \equiv s_2'$. The maximal bisimulation is called* bisimilarity, *and denoted by $\sim_{TS}$.*

Operational bisimilarity of double categories is defined in the straightforward way hinted at in the Introduction.

Definition 10 (bisimilarity for double categories). *Let $\mathcal{D}$ be a double category, and let $TS_{\mathcal{D}}$ be the labeled transition system where labels are pairs (v,u) of vertical arrows, states are horizontal arrows h, and $h \xrightarrow{(v,u)}_{TS} g$ if and only if there is a cell $A\colon h \xrightarrow[u]{v} g$. Two horizontal arrows are* operationally bisimilar, *$h_1 \sim_{op} h_2$, iff $h_1 \sim_{TS_{\mathcal{D}}} h_2$.*

Definition 11 (functorial equivalence relation). *Let $\mathcal{D}$ be a monoidal double category. An equivalence relation $h_1 \cong_{\mathfrak{f}} h_2$ on the horizontal arrows is* functorial *for $\mathcal{D}$ if, whenever $h \cong_{\mathfrak{f}} g, h' \cong_{\mathfrak{f}} g'$ for generic horizontal arrows h, h', g, g', then $h; h' \cong_{\mathfrak{f}} g; g'$ (whenever defined) and $h \otimes h' \cong_{\mathfrak{f}} g \otimes g'$.*

In other words, we are requiring that the *quotient category* of the horizontal 1-category of $\mathcal{D}$ is well-defined, and it is monoidal. In general, it is not true that operational bisimilarity is also functorial. The following results (adapted from [16]) provide a characterization of such a property in terms of *horizontal decomposition*. The results hold for any monoidal double category, in particular for a flat one.

Definition 12 (horizontal tile decomposition). *Let $\mathcal{D}$ be a monoidal double category. We say that it is* horizontally decomposable *(or that it verifies the* horizontal decomposition property*) if*

1. *whenever there is a cell $A\colon\ h_1; h_2 \xrightarrow[u]{v} g$, then there are also cells $A_1\colon$ $h_1 \xrightarrow[w]{v} g_1$ and $A_2\colon h_2 \xrightarrow[u]{w} g_2$ with $A = A_1 * A_2$;*
2. *whenever there is a cell $A\colon\ h_1 \otimes h_2 \xrightarrow[u]{v} g$, then there are also cells $A_1\colon$ $h_1 \xrightarrow[u_1]{v_1} g_1$ and $A_2\colon h_2 \xrightarrow[u_2]{v_2} g_2$ with $A = A_1 \otimes A_2$.*

Category $\mathcal{D}$ verifies the reflective *horizontal decomposition property if, in addition, for each cell $A\colon f \xrightarrow[u]{v} g$, $f = a$ implies $A = 1_v\colon a \xrightarrow[v]{v} a$, and $f = e$ implies $A = e\colon e \xrightarrow[e]{e} e$.*

Proposition 3 (decomposition implies that bisimilarity is functorial). *Let $\mathcal{D}$ be a monoidal double category. If it verifies the horizontal decomposition property, then operational bisimilarity $\sim_{op}$ is functorial.*

The notions of bisimilarity, functoriality and (reflective) horizontal decomposition can be defined also for a tile rewrite system $\mathcal{T}$: it is enough to check if they hold for $F_T(\mathcal{T})$. Notice in particular that for $\mathcal{T}$ the operational bisimilarity $\sim_{op}$ is a relation on configurations.

Given a tile rewrite system $\mathcal{T}$, it may be difficult to check if it is decomposable (and thus if its operational bisimilarity is functorial). We can provide a syntactical property of $\mathcal{T}$ that implies reflective decomposition.

Proposition 4 (basic source property and decomposition). *Let $\mathcal{T} = \langle S, \Sigma_h, \Sigma_v, R\rangle$ be a tile rewrite system such that, for all $h \xrightarrow[u]{v} g \in R$, $h \in \Sigma_h$ and $u \in \Sigma_v$ (hence, both initial configuration and effect are just operators). Then $F_T(\mathcal{T})$ verifies the reflective horizontal decomposition property.*

4 Coalgebras and Structured Coalgebras

As recalled in the introduction, the use of coalgebras for the specification of dynamical systems with a hidden state space is receiving more and more attention in the last years, as a valid alternative to algebraic methods based on observational equivalences [31,32].

In this section we first introduce the standard way to represent labeled transition systems as coalgebras for a suitable powerset functor [31], and then we discuss how this encoding can be lifted to a more structured framework, where the coalgebraic representation keeps the relevant algebraic structure of the states and transition of the encoded system. Let us start introducing the formal definition of coalgebra for a functor.

Definition 13 (coalgebras). *Let $B : \mathcal{C} \to \mathcal{C}$ be an endofunctor on a category $\mathcal{C}$. A* coalgebra for B *or* B-coalgebra *is a pair $\langle A, a\rangle$ where A is an object of $\mathcal{C}$ and $a : A \to B(A)$ is an arrow. A* B-cohomomorphism $f : \langle A, a\rangle \to \langle A', a'\rangle$ *is an arrow $f : A \to A'$ of $\mathcal{C}$ such that*

$$f; a' = a; B(f). \qquad (1)$$

The category of B-coalgebras and B-cohomomorphisms will be denoted B-**Coalg**. *The* underlying functor $U : B\text{-}\mathbf{Coalg} \to \mathcal{C}$ *maps an object $\langle A, a\rangle$ to A and an arrow f to itself.*

Let $P_L : \mathbf{Set} \to \mathbf{Set}$ be the functor defined as $X \mapsto \mathcal{P}(L \times X)$ where L is a fixed set of labels and $\mathcal{P}$ denotes the powerset functor. Then coalgebras for this functor are one-to-one with labeled transition systems over L [31].

Proposition 5 (labeled transition systems as coalgebras). *Category* P_L-**Coalg** *is isomorphic to the sub-category of* $\mathbf{LTS}_L$ *containing all its objects, and all the morphisms $f : TS \to TS'$ which also "reflect" transitions, i.e., such that if $f(s) \xrightarrow{l}_{TS'} t$ then there is a state $s' \in S$ such that $s \xrightarrow{l}_{TS} s'$ and $f(s') = t$.*

It is instructive to spell out the correspondence just stated. For objects, a transition system $\langle S, \longrightarrow\rangle$ is mapped to the coalgebra $\langle S, \sigma\rangle$ where $\sigma(s) = \{\langle l, s'\rangle \mid s \xrightarrow{l} s'\}$, and, vice versa, a coalgebra $\langle S, \sigma : S \to P_L(S)\rangle$ is mapped to the system $\langle S, \longrightarrow\rangle$, with $s \xrightarrow{l} s'$ if $\langle l, s'\rangle \in \sigma(s)$. For arrows, by spelling out condition (1) for functor P_L, we get

$$\forall s \in S . \{\langle l, t\rangle \mid f(s) \xrightarrow{l} t\} = \{\langle l, f(s')\rangle \mid s \xrightarrow{l} s'\},$$

and by splitting this set equality in the conjunction of the two inclusions, one can easily see that inclusion "$\supseteq$" is equivalent to $s \xrightarrow{l} s' \Rightarrow f(s) \xrightarrow{l} f(s')$, showing that f is a transition system morphism, while the left-to-right inclusion is equivalent to $f(s) \xrightarrow{l} t \Rightarrow \exists s' . s \xrightarrow{l} s' \wedge f(s') = t$, meaning that f is a "zig-zag" morphism, i.e., that it reflects transitions.

The property of "reflecting behaviors" enjoyed by cohomomorphisms plays a fundamental rôle, for example, for the characterization of bisimulation relations as spans of cohomomorphisms, for the relevance of final coalgebras, and for various other results of the theory of coalgebras [31]. Given two coalgebras $\langle A, a\rangle$ and $\langle A', a'\rangle$, a *coalgebraic bisimulation* on them is a coalgebra $\langle A \times A', r\rangle$ having as carrier the cartesian product of the carriers, and such that the projections $\pi : A \times A' \to A$ and $\pi' : A \times A' \to A'$ are cohomomorphisms. Interestingly, it is easy to check that two states of a labeled transition system S are bisimilar (in the standard sense, see Definition 9) if and only if there is a coalgebraic bisimulation on S (regarded as a P_L-coalgebra) which relates them.

An even easier definition of categorical bisimilarity can be given if there exists a final coalgebra. In this case, two elements of the carrier of a coalgebra are bisimilar iff they are mapped to the same element of the final coalgebra by the unique cohomomorphism. Unfortunately, due to cardinality reasons, the functor P_L used for the coalgebraic representation of transition systems does not admit a final coalgebra [31]. One satisfactory, alternative solution consists of replacing the powerset functor $\mathcal{P}$ on **Set** by the *countable* powerset functor $\mathcal{P}_c$, which maps a set to the family of its countable subsets. Then defining the functor $P_L^c : \mathbf{Set} \to \mathbf{Set}$ by $X \mapsto \mathcal{P}_c(L \times X)$ one has that coalgebras for this endofunctor are in one-to-one correspondence with transition systems with *countable degree*, i.e., systems where for each state $s \in S$ the set $\{\langle s', l\rangle \mid s \xrightarrow{l} s'\}$ is countable, the correspondence being defined exactly as in Proposition 5. Unlike functor P_L, the functor P_L^c admits cofree and final coalgebras.

Proposition 6 (final and cofree P_L^c-coalgebras). *The obvious underlying functor $U : P_L^c$-**Coalg** $\to$ **Set** has a right adjoint $R :$ **Set** $\to P_L^c$-**Coalg** associating with each set X a cofree coalgebra over X. As a consequence, the category P_L^c-**Coalg** has a final object, which is the cofree coalgebra $R(\mathbf{1})$ over a final set* $\mathbf{1}$.

We shall stick to this functor throughout the rest of the paper, and since there is no room for confusion the superscript c will be understood.

Often transition systems come equipped with some algebraic structure on states, transitions, and/or labels, which plays a relevant role in the corresponding theory. For example, in calculi of the family of process algebras, like CCS [27] and the π-calculus [28], the agents (states) are closed under certain operations that can be interpreted as either structural (like parallel composition) or behavioural (like prefixing and nondeterminstic choice). The same algebraic structure can be extended to the collection of transitions, in a way that is determined by the SOS rules which specify the operational semantics of the calculus. This more structured framework makes it possible to investigate the compositionality properties of relevant equivalences on agents: one typical interesting question is whether bisimilarity is a congruence with respect to the operations defined on states.

Also the ***structured transition systems***, studied for example in [11], are equipped with an (essentially) algebraic structure on both states and transitions. Here the operators are interpreted as structural ones, basic transitions are regarded as local changes on a distributed state, and the algebra on transitions ensures that basic transitions can be fired in any context and also in parallel. It has been shown that programs of many computational formalisms (including, among others, P/T Petri nets in the sense of [25], term rewriting systems, term graph rewriting [7], graph rewriting [15, 9, 18], Horn Clause Logic [6]) can be encoded as ***heterogeneous graphs*** having as collection of nodes algebras with respect to a suitable algebraic specification, and usually a poorer structure on arcs (often they are just a set). ***Structured*** transition systems are defined instead as graphs having a similar algebraic structure both on nodes and on arcs. A free construction associates with each program its induced structured transition system, from which a second free construction is used to generate the free model, i.e., a structured category which lifts the algebraic structure to the transition sequences. This induces an equivalence relation on the computations of a system, which is shown to capture some basic properties of true concurrency. Moreover, since the construction of the free model is a left adjoint functor, it is compositional with respect to operations on programs expressible as colimits.

Last but not least, also the tile transition systems introduced in the previous section have a rich algebraic structure on states, which are the arrows of a monoidal category, and the same structure is also defined on transitions, which are the tiles. Clearly, the general results presented at the end of the previous section, relating basic source and reflective horizontal decomposition properties, make an essential use of this algebraic structure.

For all the systems mentioned above (process algebra, structured transition systems and tile rewrite systems) the coalgebraic representation using functor P_L (for a suitable L) introduced in Proposition 5 is not completely satisfactory, because by definition the carrier is just a set and therefore the algebraic structure on both states and transitions is lost. This calls for the introduction of ***structured coalgebras***, i.e., coalgebras for an endofuctor on a category $\mathcal{A}lg(\Gamma)$ of algebras for a signature (or algebraic specification) Γ which is determined by the structure of states. Since it is natural to require that the structured coalgebraic representation of a system is compatible with the unstructured, set-based one, the following notion will be relevant.

Definition 14 (lifting). *Given endofunctors $B : \mathcal{C} \to \mathcal{C}$, $B' : \mathcal{C}' \to \mathcal{C}'$ and a functor $V : \mathcal{C}' \to \mathcal{C}$, B' is called a* lifting of B along V, *if* $B'; V = V; B$.

In particular, if $V^\Gamma : \mathcal{A}lg(\Gamma) \to \mathbf{Set}$ is the underlying set functor, one will consider typically a functor $B' : \mathcal{A}lg(\Gamma) \to \mathcal{A}lg(\Gamma)$ which is a lifting of P_L along V^Γ.

The structured coalgebraic representation of transition systems has been studied in [32] for the case of CCS and other process algebra whose operational semantics is given by SOS rules in the DeSimone format, and in [8] for structured transition systems. In the first case the lifting of P_L is determined by the SOS rules, while in the second one it is uniquely determined by the specification Γ. In both cases, as well as for the case of tile transition systems addressed in the next sections, the following interesting fact applies [32,8].

Proposition 7 (bisimilarity is a congruence in structured coalgebras). *Let Γ be an algebraic specification, L be a Γ-algebra of labels, and $B_L^\Gamma : Alg(\Gamma) \to Alg(\Gamma)$ be a lifting of $P_L : \mathbf{Set} \to \mathbf{Set}$. If $\langle S, \sigma \rangle$ is a B_L^Γ-coalgebra and $\langle S, \longrightarrow \rangle$ its corresponding structured LTS, then bisimilarity on $\langle S, \longrightarrow \rangle$ is a congruence with respect to the operators in Γ.*

The statement follows by the observation that the right adjoint $R : \mathbf{Set} \to P_L\text{-}\mathbf{Coalg}$ of Proposition 6 lifts to a right adjoint $R^\Gamma : Alg(\Gamma) \to B_L^\Gamma\text{-}\mathbf{Coalg}$ for the forgetful functor U^Γ, with $V^\Gamma ; R = R^\Gamma ; V_B^\Gamma$ (see [32]), as shown in the following diagram.

$$\begin{array}{ccc}
P_L\text{-}\mathbf{Coalg} & \underset{V_B^\Gamma}{\overset{F_B^\Gamma}{\rightleftarrows}} & B_L^\Gamma\text{-}\mathbf{Coalg} \\
R \uparrow \downarrow U & & R^\Gamma \uparrow \downarrow U^\Gamma \\
\mathbf{Set} & \underset{V^\Gamma}{\overset{F^\Gamma}{\rightleftarrows}} & Alg(\Gamma)
\end{array}$$

Now, since R^Γ and V_B^Γ are both right adjoints, B_L^Γ-**Coalg** inherits a final object $R^\Gamma(\mathbf{1})$ from $Alg(\Gamma)$ which is then preserved by V_B^Γ. Hence, bisimilarity induced by the final morphism to $R^\Gamma(\mathbf{1})$ in B^Γ-**Coalg** is determined by the underlying sets and functions, that is, its definition does not use the algebraic structure of states and transitions. Since the final morphisms in B_L^Γ-**Coalg** are Γ-homomorphisms, it follows that bisimilarity is a congruence.

In other words, a structured transition system can be represented as a structured coalgebra only if bisimilarity is a congruence. This property certainly holds, for example, for specifications in GSOS format, which are considered in [32]. Certain structures are used there, called *bialgebras*, which combine aspects of algebras and coalgebras: bialgebras can be regarded as an alternative, equivalent presentation of structured coalgebras [8]. A specification in GSOS format is shown to satisfy a certain diagram called *pentagonal law*, which ensures the existence both of an algebra of transiton systems and of an algebraic structure on their states. The pentagonal law also makes sure that bisimilarity is a congruence, showing that GSOS specifications perfectly fit in the structured coalgebraic framework.

A rather more general specification format is considered in [10], namely, the *algebraic* format [16], where the premise of a rule consists entirely of transitions on variables, and which generalizes rules in deSimone format by allowing complex terms in the source of the conclusion of a rule. In that paper, we first studied under which conditions transition systems can be represented as structured coalgebras on an environment category of algebras. It turned out that the conditions which guarantee a coalgebraic presentation are very similar to the ones which ensure that bisimilarity is a congruence. Essentially they require that the behavior of the system is compositional, in the sense that all transitions from complex states can be derived using the rules from transitions out of component states. Thus one could say that what was considered a methodological convenience, i.e., that in the SOS approach each language construct is defined separately by a few clauses, is in fact mandatory to guarantee a satisfactory algebraic structure.

Next we proposed a general procedure which can be applied also to SOS specifications not satisfying the above property. More precisely, given any SOS specification in algebraic format, we defined its *context closure*, i.e., another specification including also the possible *context transitions*, which are transitions resulting in the addition of some context and labelled by it. We proved that bisimilarity for the context closure corresponds to *dynamic bisimilarity* for the original specification, which is by definition the coarsest bisimulation which is a congruence: as a consequence the context closure of a system is always representable as a structured coalgebra. This result is particularly relevant for *open systems*, for which dynamic bisimilarity seems to be the right notion of equivalence, since it takes into account not only experiments based on communications with the external world, but also experiments consisting of the additions of new components.

A different point of view is taken in [8], where it is argued that the property that bisimilarity is a congruence is too restrictive for structured transition systems, because it implies that basic transitions are defined only on atomic states. As a simple example, let us introduce the structured transition system associated with a simple P/T Petri net N consisting of places $S = \{a, b, c\}$ and of a single transition $T = \{t : a \oplus b \to c\}$ (consuming one token of place a and one of b and producing one in c). According to [25], the relevant algebraic structure is that of commutative monoids: the markings of the net can be regarded as elements of the free commutative monoid $S^{\oplus}$, and the structured transition system associated with N, denoted $TS(N)$, is obtained by adding idle transitions for each place and by extending the parallel composition operation $\oplus$ in an obvious way to transitions. Now, let us assume that idle transitions are not visible (formally, they are labeled with the unit of the monoid of labels). Then it is easily seen that markings a and b are bisimilar (only idle transitions are possible) and clearly b and b are bisimilar as well. However, we have $a \oplus b \xrightarrow{t} c$, while only the idle transition is possible from $b \oplus b$. This shows that these two states are not bisimilar. Therefore one single basic transition having as source a composed state is sufficient to show that bisimilarity is not a congruence. As a consequence, in [8] the notion of *lax coalgebra* is introduced, which weakens the standard definition in order to allow for a full, coalgebra-like representation of structured transition systems.

5 Double and Tile Transition Systems as Coalgebras

In the previous section we have shown that labeled transition systems can be represented as coalgebras for an endofunctor on the category of sets. In this section, a similar representation for tile models (i.e., monoidal double categories) shall be developed.

In analogy to labeled transition systems, *double transition systems* are defined as flat monoidal double categories over a fixed vertical 1-category $\mathcal{V}$ of observations. In Proposition 5 it is shown that coalgebra morphisms correspond to morphisms between labeled transition systems which (preserve and) reflect transitions. Below, an analogous restriction on monoidal double functors is introduced.

Definition 15 (double and tile transition systems). *Given a monoidal category of observations $\mathcal{V}$, a* double transition system *over $\mathcal{V}$ is a flat monoidal double category $\mathcal{D}$ which has $\mathcal{V}$ as vertical 1-category.*

A morphism between double transition systems $\mathcal{D}$ and $\mathcal{D}'$ over $\mathcal{V}$ is a monoidal double functor $F : \mathcal{D} \to \mathcal{D}'$ which acts on $\mathcal{V}$ as identity. It reflects transitions *if*

$$\forall f \in \mathcal{H}.\ F(f) \xrightarrow[b]{a} g' \in \mathcal{D}' \Longrightarrow \exists g \in \mathcal{H}.\ f \xrightarrow[b]{a} g \wedge F(g) = g'$$

We denote by $\mathbf{fsMDCat}_{\mathcal{V}}$ the category of double transition systems over $\mathcal{V}$ and transition reflecting morphisms.

Given a tile rewrite system $\mathcal{T} = \langle S, \Sigma_h, \Sigma_v, R\rangle$, its associated tile transition system *is its free model $F_T(\mathcal{T})$, seen as a double transition system over $\mathbf{M}(\Sigma_v)$.*

The endofunctor whose coalgebras represent double transition systems is only slightly more complex than functor P_L defined in Section 4. Since the states of a double transition system are arrows of a category (the horizontal 1-category $\mathcal{H}$), they are typed by their source and target objects. Consequently, the carrier of the corresponding coalgebra shouldn't be just a set but a family of sets indexed by pairs of objects of $\mathcal{V}$. The endofunctor $P_{\mathcal{V}}$ defined below on the category $\mathbf{Set}^{|\mathcal{V}^2|}$ is therefore a many-sorted version of P_L as defined in Section 4.

Definition 16 (endofunctor $P_{\mathcal{V}}$ for double transition systems). *Given a monoidal category $\mathcal{V} = \langle V, \otimes, e\rangle$, the functor $P_{\mathcal{V}} : \mathbf{Set}^{|\mathcal{V}^2|} \to \mathbf{Set}^{|\mathcal{V}^2|}$ is defined for every $|\mathcal{V}| \times |\mathcal{V}|$-indexed set S by*

$$P_{\mathcal{V}}(S)(n,m) = \mathcal{P}_c\left(\bigcup_{n',m' \in |\mathcal{V}|} \mathcal{V}(n,n') \times \mathcal{V}(m,m') \times S(n',m')\right)$$

On arrows of $\mathbf{Set}^{|\mathcal{V}^2|}$, i.e., $|\mathcal{V}| \times |\mathcal{V}|$-indexed indexed families of functions, the functor is defined analogously.

Notice that, according to the interpretation of tiles as cells, two vertical arrows are provided as observations. Moreover, transitions do not necessarily preserve the type of the state because there may be cells whose vertical source and target (which are horizontal arrows) are arrows of different type.

Proposition 8 (double transition systems as coalgebras). *Given $\mathcal{V}$ and $P_\mathcal{V}$ as above, there exists a functor Clg : $\mathbf{fsMDCat}_\mathcal{V} \to P_\mathcal{V}$-**Coalg** such that for every countably branching double transition system $\mathcal{D} \in \mathbf{fsMDCat}_\mathcal{V}$ with horizontal 1-category $\mathcal{H}$ and all $f, g \in \mathcal{H}$*

$$f \sim_{op} g \text{ iff } \phi(f) = \phi(g)$$

where $\phi : Clg(\mathcal{D}) \to 1_{P_\mathcal{V}}$ is the unique morphism into the final $P_\mathcal{V}$-coalgebra (which exists by similar arguments as in Proposition 6).

Proof. For a double transition system $\mathcal{D}$ as above, the corresponding $P_\mathcal{V}$-coalgebra is $Clg(\mathcal{D}) = \langle (\mathcal{H}(n,m)_{n,m\in|\mathcal{V}|}, \delta(n,m)_{n,m\in|\mathcal{V}|}\rangle$ with

$$\delta(n,m)(f) = \{\langle a,b,g\rangle \mid f \xrightarrow[b]{a} g \in \mathcal{D}\} \tag{2}$$

for all $f \in \mathcal{H}(n,m)$. Notice that $|\mathcal{H}| = |\mathcal{V}|$. □

Thus, the use of many-sorted coalgebras allows us to retain the typing of states of double transition systems by pairs of objects. However, the algebraic structure given by the operations of the horizontal 1-category $\mathcal{H}$ is lost. This problem is solved in the next section by lifting many-sorted to structured coalgebras.

6 Lifting the Algebraic Structure

We follow the outline of Section 4: first, we specify explicitly the algebraic structure on states and transitions. Then, we lift the endofunctor $P_\mathcal{V}$ to the corresponding category of algebras. Finally we show that double transition systems which satisfy the reflective horizontal decomposition property can be actually represented as structured coalgebras for the lifted functor.

In a double transition system $\mathcal{D}$, the algebraic structure of states is given by the monoidal category structure of the horizontal 1-category $\mathcal{H}$. Since we have fixed a monoidal category of observations $\mathcal{V}$ as vertical 1-category, we know in advance the monoid of objects V. This allows to regard a monoidal category as a total algebra for the following signature (given with respect to the fixed monoid $V = \langle O, \otimes, e\rangle$).

signature <u>MonCat(V)</u> =
  sorts
    (n,m) for all $n,m \in O$
  operations
    $;_{n,m,k} : (n,m)(m,k) \to (n,k)$ for all $n,m,k \in O$
    $n : \to (n,n)$ for all $n \in O$
    $e : \to (e,e)$
    $\otimes_{n,m,n'm'} : (n,m)(n',m') \to (n \otimes n', m \otimes m')$ for all $n,m,n'm' \in O$

Algebras for this signature satisfying the usual laws of strict monoidal categories are the objects of the category $\mathbf{MonCat}_V$. Its arrows are given by $\underline{\text{MonCat}(V)}$-homomorphisms, i.e., strict monoidal functors preserving the monoid of objects V. We usually omit the subscripts at operations when this causes no confusion.

As anticipated above, in order to represent a double transition system as a coalgebra in the category $\mathbf{MonCat}_V$, we have to provide a lifting of the endofunctor $P_{\mathcal{V}}$ defined on (families of) sets to the category of algebras. In the definition below, the resulting functor is called $\hat{P}_{\mathcal{V}}$.

Definition 17 (lifting endofunctor $P_{\mathcal{V}}$ to $\mathbf{MonCat}_V$). *The functor $\hat{P}_{\mathcal{V}} : \mathbf{MonCat}_V \to \mathbf{MonCat}_V$ is defined as follows. Given $M = \langle S, ;, id, \otimes, e\rangle \in |\mathbf{MonCat}_V|$, the algebra*

$$\hat{P}_{\mathcal{V}}(M) = PM = \langle P_{\mathcal{V}}(S), ;^{PM}, id^{PM}, \otimes^{PM}, e^{PM}\rangle$$

is given by

$$\begin{aligned}
S ;^{PM} T &= \{\langle a, c, g; h\rangle \mid \langle a, b, g\rangle \in S, \langle b, c, h\rangle \in T\}\\
id^{PM} &= \{\langle a, a, m\rangle \mid a : n \to m \in \mathcal{V}\}\\
S \otimes T &= \{\langle a \otimes a', b \otimes b', g \otimes g'\rangle \mid \langle a, b, g\rangle \in S, \langle a', b', g'\rangle \in T\}\\
e^{PM} &= \{\langle e, e, e\rangle\}
\end{aligned}$$

On arrows of $\mathbf{MonCat}_V$, the functor is defined like $P_{\mathcal{V}}$.

Next we show that a double transition system can be represented as a $\hat{P}_{\mathcal{V}}$-coalgebra iff it satisfies the reflective decomposition property. Denote by $\mathbf{fsMDCat}_{\mathcal{V}}^{\Delta}$ the full subcategory of $\mathbf{fsMDCat}_{\mathcal{V}}$ whose objects are countably branching double transition systems satisfying the reflective decomposition property.

Proposition 9 (double transition systems as structured coalgebras). *Let $\mathcal{D}$ be a countably branching double transition system and $\mathcal{H}$ be its horizontal 1-category. Then, $\hat{Clg}(\mathcal{D}) = \langle \mathcal{H}, \delta\rangle$ with δ defined like in (2) is a $\hat{P}_{\mathcal{V}}$-coalgebra if and only if $\mathcal{D}$ satisfies the reflective horizontal decomposition property.*

Moreover, this translation extends to a functor $\hat{Clg} : \mathbf{fsMDCat}_{\mathcal{V}}^{\Delta} \to \hat{P}_{\mathcal{V}}$-$\mathbf{Coalg}$.

Thus, under the assumption of the reflective horizontal decomposition property, we actually retain the horizontal structure of double transition systems in the coalgebraic presentation. What is lost in any case is the vertical structure which is, however, not relevant for the notion of bisimilarity.

More precisely, we can show that the category $\mathbf{fsMDCat}_{\mathcal{V}}^{\Delta}$ of horizontally decomposable double transition systems is isomorphic to the full subcategory of $\hat{P}_{\mathcal{V}}$-$\mathbf{Coalg}$ whose objects are coalgebras $\langle \mathcal{H}, \delta\rangle$ with transitive and reflexive transition "relations". This means, writing $f \xrightarrow[b]{a} g$ for $\langle a, b, g\rangle \in \delta(f)$, that they have to satisfy the rules *(vert)* and *(v-refl)* of Table 1.

7 An Application to CCS

As a case study, we present in this section a version without recursion of the *calculus of communicating systems* (shortly, CCS) introduced by Robin Milner [27]. The same presentation has appeared in [26]. Essentially the same presentation, but for a cartesian (rather than monoidal) structure of configurations, was reported in [16]. CCS with the replication operator is modeled with tiles in [2]. A concurrent version of CCS with localities is in [14].

Syntax of CCS. Let Δ be the alphabet for basic actions (which is ranged over by α) and $\overline{\Delta}$ the alphabet of complementary actions ($\Delta = \overline{\overline{\Delta}}$ and $\Delta \cap \overline{\Delta} = \emptyset$); the set $\Lambda = \Delta \cup \overline{\Delta}$ will be ranged over by λ. Let $\tau \notin \Lambda$ be a distinguished action, and let $\Lambda \cup \{\tau\}$ (ranged over by μ) be the set of CCS actions.

The syntax of finite CCS agents is defined by the following grammar

$$P ::= \quad nil \quad | \quad \mu.P \quad | \quad P\backslash\alpha \quad | \quad P+P \quad | \quad P \mid P$$

Operational Semantics of CCS. The dynamic behavior of CCS closed agents can be described by a transition system TS_{CCS}, where labels are actions, states are closed CCS agents and the transition relation is freely generated from the following set of inference rules

$$\frac{}{\mu.P \xrightarrow{\mu} P} \qquad \frac{P \xrightarrow{\mu} Q \; \mu \notin \{\alpha, \overline{\alpha}\}}{P\backslash\alpha \xrightarrow{\mu} Q\backslash\alpha}$$

$$\frac{P \xrightarrow{\mu} Q}{P+R \xrightarrow{\mu} Q} \qquad \frac{P \xrightarrow{\mu} Q}{R+P \xrightarrow{\mu} Q}$$

$$\frac{P \xrightarrow{\mu} Q}{P \mid R \xrightarrow{\mu} Q \mid R} \qquad \frac{P \xrightarrow{\lambda} Q, \, P' \xrightarrow{\overline{\lambda}} Q'}{P \mid P' \xrightarrow{\tau} Q \mid Q'} \qquad \frac{P \xrightarrow{\mu} Q}{R \mid P \xrightarrow{\mu} R \mid Q}$$

Given the transition

$$((a.nil + b.nil) \mid \overline{a}.nil)\backslash a \xrightarrow{\tau} (nil \mid nil)\backslash a,$$

its proof is as follows:

$$\cfrac{\cfrac{\cfrac{\cfrac{}{a.nil \xrightarrow{a} nil}}{a.nil + b.nil \xrightarrow{a} nil} \quad \cfrac{}{\overline{a}.nil \xrightarrow{\overline{a}} nil}}{(a.nil + b.nil) \mid \overline{a}.nil \xrightarrow{\tau} nil \mid nil}}{((a.nil + b.nil) \mid \overline{a}.nil)\backslash a \xrightarrow{\tau} (nil \mid nil)\backslash a}$$

Abstract Semantics of CCS. Ordinary bisimilarity between CCS agents, written $P \sim_{ord} Q$, is just the relation $P \sim_{TS_{CCS}} Q$ according to the general Definition 9.

We now present the tile rewrite system for CCS.

Signatures of the CCS tile rewrite system. There is only one sort $\underline{1}$. The free monoid generated by it is represented by underlined natural numbers with $\underline{n} \otimes \underline{m} = \underline{m+n}$ and $e = \underline{0}$. For the horizontal signature the operators are: $nil \in (\Sigma_h^{CCS})_{\underline{0},\underline{1}}$; $\mu.\_$ and $\_\backslash\alpha \in (\Sigma_h^{CCS})_{\underline{1},\underline{1}}$; $\_+\_$ and $\_\mid\_ \in (\Sigma_h^{CCS})_{\underline{2},\underline{1}}$; and $!(\_) \in (\Sigma_h^{CCS})_{\underline{1},\underline{0}}$. The latter constructor, called *eraser*, is needed to discard the rejected alternative after a choice step. Except for the eraser, the operators of the horizontal signature directly correspond to CCS syntactical operators. Thus we will consider CCS agents (not necessarily closed) as suitable arrows of the monoidal category $\mathbf{M}(\Sigma_h^{CCS})$. For the vertical signature the operators are $\underline{\mu}.\_ \in (\Sigma_v^{CCS})_{\underline{1},\underline{1}}$.

Rules of the CCS tile rewrite system.

$$\mathtt{Pref}_\mu : \mu \xrightarrow[\underline{\mu}]{\underline{1}} \underline{1} \qquad \mathtt{Res}^\alpha_\mu : \backslash\alpha \xrightarrow[\underline{\mu}]{\underline{\mu}} \backslash\alpha \quad \text{for} \quad \mu \notin \{\alpha, \overline{\alpha}\}$$

$$\mathtt{Suml}_\mu : \_+\_ \xrightarrow[\underline{\mu}]{\underline{\mu}\otimes\underline{1}} \underline{1}\otimes! \qquad \mathtt{Sumr}_\mu : \_+\_ \xrightarrow[\underline{\mu}]{\underline{1}\otimes\underline{\mu}} !\otimes\underline{1}$$

$$\mathtt{Compl}_\mu : \_\mid\_ \xrightarrow[\underline{\mu}]{\underline{\mu}\otimes\underline{1}} \_\mid\_ \qquad \mathtt{Compr}_\mu : \_\mid\_ \xrightarrow[\underline{\mu}]{\underline{1}\otimes\underline{\mu}} \_\mid\_ \qquad \mathtt{Synch}_\lambda : \_\mid\_ \xrightarrow[\underline{\tau}]{\underline{\lambda}\otimes\underline{\overline{\lambda}}} \_\mid\_.$$

We call $\mathcal{T}_{CCS}$ the CCS tile rewrite system. The rules of $\mathcal{T}_{CCS}$ closely correspond to the SOS rules. For instance rule $\mathtt{Pref}_\mu$ states, as its SOS counterpart, that constructor μ can be deleted, i.e., it can be replaced by the identity $\underline{1}$. Furthermore, the trigger is also the identity, and thus the corresponding SOS rule is an axiom. Finally, the effect is $\underline{\mu}$ and this corresponds to the label of the transition in the SOS case. As another example, rule $\mathtt{Suml}_\mu$ defines left choice. The initial configuration is the constructor $\_+\_ : \underline{2} \to \underline{1}$, while the final configuration is $\underline{1}\otimes! : \underline{2} \to \underline{1}$, which states that the first component is preserved and the second component is discarded. The trigger states that in the first component we must have an action $\underline{\mu}$ while no action (i.e., identity action) is required on the second component. Action $\underline{\mu}$ is then transferred to the effect.

The tile corresponding to the previous example is obtained as follows:

$$A = (((nil * \mathtt{Pref}_a) \otimes (b.nil) * \mathtt{Suml}_a) \otimes (nil * \mathtt{Pref}_{\overline{a}})) * \mathtt{Synch}_a * \mathtt{Res}^a_\tau.$$

The next proposition states the equivalence of the SOS and tile definitions of CCS.

Proposition 10 (SOS/tile equivalence for CCS). *Given two CCS agents P and Q, we have $P \sim_{ord} Q$ if and only if $P \sim_{op} Q$ for the CCS tile rewrite system.*

In order to represent $\mathcal{T}_{CCS}$ as a coalgebra, we have to verify the reflective horizontal decomposition property. Thanks to Proposition 4, it is enough to check by inspection that all rules of $\mathcal{T}_{CCS}$ satisfy the basic source property.

Proposition 11 (coalgebraic presentation of CCS tile rewrite system). *The CCS tile rewrite system $\mathcal{T}_{CCS}$ satisfies the basic source property and hence the reflective horizontal decomposition property. As a consequence, the tile transition system $F_T(\mathcal{T}_{CCS})$ generated by $\mathcal{T}_{CCS}$ can be represented as a $\hat{P}_{\mathbf{M}(\Sigma_v^{CCS})}$-coalgebra*

$$\hat{Clg}(F_T(\mathcal{T}_{CCS})) = \langle \mathbf{M}(\Sigma_h^{CCS}), \sigma \rangle$$

such that $P \sim_{op} Q$ iff $\phi(P) = \phi(Q)$ for each two processes $P, Q \in \mathbf{M}(\Sigma_h^{CCS})$, where $\phi : \hat{Clg}(F_T((\mathcal{T}_{CCS})) \to 1_{\hat{P}_{\mathbf{M}(\Sigma_v^{CCS})}}$ is the unique arrow to the final $\hat{P}_{\mathbf{M}(\Sigma_v^{CCS})}$-coalgebra.

Corollary 1 ($\sim_{op}$ is a congruence). *Tile bisimilarity $\sim_{op}$ for CCS is a congruence w.r.t. the operations of the horizontal 1-category of states.*

Due to Proposition 10, this implies the well-known fact that bisimilarity is a congruence for CCS. Notice however that the above statement is more general, since it does not only apply to closed process terms but also to processes which contain variables.

8 Conclusion

In this paper we have investigated the relation between two models of concurrent systems: tile rewrite systems and coalgebras. Tile rewrite systems consist of rewrite rules with side effects which are reminiscent of SOS rules [30]. For these rules, which can also be seen as elementary transitions, closure operations of parallel and sequential composition and synchronization are defined. The models of tile rewrite systems are monoidal double categories.

Coalgebras can be considered, in a suitable mathematical setting, as dual to algebras. They can be used as models of dynamical systems with hidden states in order to study concepts of observational equivalence and bisimilarity in a more general setting.

We have pointed out that, in order to retain in the coalgebraic presentation of tile models the operations of parallel composition and synchronization, coalgebras have to be endowed with an algebraic structure. This has led us to the concept of ***structured coalgebras***, i.e., coalgebras for an endofunctor on a category of algebras. However, structured coalgebras are a more restricted notion than tile models, since they only allow to represent models where bisimilarity is a congruence. For tile models, this condition corresponds to functoriality of bisimilarity, which is ensured by the horizontal decomposition property.

The insights on the relation between tile rewrite systems and structured coalgebras can be obtained by applying to tile models (seen as double transition systems) the results of [8, 10] on the coalgebraic presentation of transition systems with algebraic structure specified by SOS rules.[2] In that paper, we have characterized those transition systems for which a coalgebraic presentation is possible

[2] In fact, what would be needed is a many-sorted version of the presentation in [10] which, for simplicity, is restricted to the one-sorted case.

and the classes of SOS specifications generating such "well-behaved" systems. It turned out that the conditions which guarantee a coalgebraic presentation are very similar to the ones which ensure that bisimilarity is a congruence. They require that the behavior of the system is compositional in the sense that all transitions from complex states can be derived using the transitions out of component states: this is essentially the statement of the horizontal decomposition property.

In the case without structural axioms, such condition is verified if each basic rule in the tile rewrite system has signature operators both as initial configuration and as effect; indeed this is the common point of many SOS formats (see e.g. [12, 17, 5]). Notice, however, that, with structural axioms, the situation is more complicated, since signature operators can be equivalent to complex terms, and complex states may be decomposed into component states in many different ways.

We think that the coalgebraic presentation of monoidal tile rewrite systems can be a starting point for transferring to the coalgebraic setting concrete applications of tile logic to formalisms like concurrent or located CCS, π-calculus, etc.

9 Acknowledgments

We would like to thank Roberto Bruni for his comments.

References

1. R. Bruni, F. Gadducci and U. Montanari, Normal Forms for Partitions and Relations, Proc. 13th Workshop on Algebraic Development Techniques, Lisbon, April 2-4, 1998, Springer LNCS, 1999, to appear.
2. R. Bruni, J. Meseguer and U. Montanari, Executable Tile Specifications for Process Calculi, in: Jean-Pierre Finance, Ed., FASE'99, Springer LNCS 1577, pp. 60-76.
3. R. Bruni, J. Meseguer and U. Montanari, Symmetric Monoidal and Cartesian Double Categories as a Semantic Framework for Tile Logic, to appear in MSCS.
4. R. Bruni and U. Montanari, Cartesian Closed Double Categories, their Lambda-Notation, and the Pi-Calculus, Proc. LICS'99, to appear.
5. B. Bloom, S. Istrail, and A.R. Meyer. Bisimulation can't be traced. *Journal of the ACM*, 42(1):232–268, 1995.
6. A. Corradini and A. Asperti. A categorical model for logic programs: Indexed monoidal categories. In *Proceedings REX Workshop, Beekbergen, The Netherlands, June 1992*, volume 666 of *LNCS*. Springer Verlag, 1993.
7. A. Corradini and F. Gadducci. A 2-categorical presentation of term graph rewriting. In E. Moggi and G. Rosolini, editors, *Category Theory and Computer Science*, volume 1290 of *LNCS*, pages 87–105. Springer Verlag, 1997.
8. A. Corradini, M. Große-Rhode, and R. Heckel. Structured transition systems as lax coalgebras. In B. Jacobs, L. Moss, H. Reichel, and J. Rutten, editors, *Proc. of First Workshop on Coalgebraic Methods in Computer Science (CMCS'98), Lisbon, Portugal*, volume 11 of *Electronic Notes of TCS*. Elsevier Science, 1998. `http://www.elsevier.nl/locate/entcs`.

9. A. Corradini, M. Große-Rhode, and R. Heckel. An algebra of graph derivations using finite (co–) limit double theories. In J.L. Fiadeiro, editor, *Proc. 13th Workshop on Algebraic Development Techniques (WADT'98)*, volume 1589 of *LNCS*. Springer Verlag, 1999.
10. A. Corradini, R. Heckel, and U. Montanari. From SOS specifications to structured coalgebras: How to make bisimulation a congruence. In B. Jacobs and J. Rutten, editors, *Proc. of Second Workshop on Coalgebraic Methods in Computer Science (CMCS'99), Amsterdam*, volume 19 of *Electronic Notes of TCS*. Elsevier Science, 1999. `http://www.elsevier.nl/locate/entcs`.
11. A. Corradini and U. Montanari. An algebraic semantics for structured transition systems and its application to logic programs. *Theoret. Comput. Sci.*, 103:51–106, 1992.
12. R. De Simone. Higher level synchronizing devices in MEIJE–SCCS. *Theoret. Comput. Sci.*, 37:245–267, 1985.
13. C. Ehresmann, *Catégories Structurées*: I and II, Ann. Éc. Norm. Sup. 80, Paris (1963), 349-426; III, Topo. et Géo. diff. V, Paris (1963).
14. G. Ferrari and U. Montanari, Tile Formats for Located and Mobile Systems, Information and Computation, to appear.
15. F. Gadducci and R. Heckel. A inductive view of graph transformation. In F. Parisi Presicce, editor, *Recent Trends in Algebraic Development Techniques, LNCS 1376*, pages 223 - 237. Springer Verlag, 1998.
16. F. Gadducci and U. Montanari. The tile model. In G. Plotkin, C. Stirling, and M. Tofte, editors, *Proof, Language and Interaction: Essays in Honour of Robin Milner*. MIT Press, 1999. To appear. An early version appeared as Tech. Rep. TR-96/27, Dipartimento di Informatica, University of Pisa, 1996. Paper available from `http://www.di.unipi.it/ gadducci/papers/TR-96-27.ps.gz`.
17. J.F. Groote and F. Vandraager. Structured operational semantics and bisimulation as a congruence. *Information and Computation*, 100:202–260, 1992.
18. R. Heckel. *Open Graph Transformation Systems: A New Approach to the Compositional Modelling of Concurrent and Reactive Systems*. PhD thesis, TU Berlin, 1998.
19. G.M. Kelly and R.H. Street. Review of the elements of 2-categories. In G.M. Kelly, editor, *Sydney Category Seminar*, volume 420 of *Lecture Notes in Mathematics*, pages 75–103. Springer Verlag, 1974.
20. F.W. Lawvere. Some algebraic problems in the context of functorial semantics of algebraic theories. In *Proc. Midwest Category Seminar II*, number 61 in Springer Lecture Notes in Mathematics, pages 41–61, 1968.
21. K.G. Larsen, L. Xinxin, *Compositionality Through an Operational Semantics of Contexts*, in Proc. ICALP'90, LNCS 443, 1990, pp. 526-539.
22. J. Meseguer. Conditional rewriting logic as a unified model of concurrency. *TCS*, 96:73–155, 1992.
23. J. Meseguer, *Rewriting Logic as a Semantic Framework for Concurrency: A Progress Report*, in: U. Montanari and V. Sassone, Eds., *CONCUR'96: Concurrency Theory*, Springer LNCS 1119, 1996, 331-372.
24. J. Meseguer. Membership algebra as logical framework for equational specification. In F. Parisi Presicce, editor, *Recent Trends in Algebraic Development Techniques, LNCS 1376*, pages 18–61. Springer Verlag, 1998.
25. J. Meseguer and U. Montanari. Petri nets are monoids. *Information and Computation*, 88(2):105–155, 1990.

26. J. Meseguer and U. Montanari. Mapping tile logic into rewriting logic. In Francesco Parisi-Presicce, editor, *Recent Trends in Algebraic Development Techniques*, number 1376 in Spinger LNCS, pages 62–91, 1998.
27. R. Milner. *Communication and Concurrency*. Prentice-Hall, 1989.
28. R. Milner, J. Parrow, and D. Walker. A calculus of mobile processes. *Information and Computation*, 100:1–77, 1992.
29. U. Montanari and F. Rossi, Graph Rewriting, Constraint Solving and Tiles for Coordinating Distributed Systems, to appear in Applied Category Theory.
30. G. Plotkin. A structural approach to operational semantics. Technical Report DAIMI FN-19, Aarhus University, Computer Science Deapartment, 1981.
31. J.J.M.M. Rutten. Universal coalgebra: a theory of systems. Technical Report CS-R9652, CWI, 1996. To appear in TCS.
32. D. Turi and G. Plotkin. Towards a mathematical operational semantics. In *Proc. of LICS'97*, pages 280–305, 1997.

Caesar and DNA. Views on Cryptology

Arto Salomaa

Academy of Finland and
Turku Centre for Computer Science
Lemminkäisenkatu 14 A
20520 Turku, Finland
asalomaa@utu.fi

Abstract. The paper discusses some recent aspects of cryptology. Attention is focused on public-key cryptography, in particular, on certain zero-knowledge proofs and the general question of whether and how cryptographic ideas can be realized without computers. Possible impacts of DNA computing on cryptology, as well as recent legislative measures to restrict the marketing of cryptographic products, will also be briefly considered. The paper consists of the following six sections. 1. Digging the bones of Caesar? 2. A big invention. 3. Protocols with or without computers. 4. Truth or consequences? 5. A deck of cards or a computer? 6. Security of security. DNA computing.

1 Digging the bones of Caesar?

The title of our paper might give the impression that we have in mind some kind of a DNA analysis about Caesar, along the lines of the recent analysis concerning Thomas Jefferson, maybe even with the purpose of impeaching somebody. Not at all. The title only reflects the fact that our topics are rather scattered. We discuss several issues on cryptology, mostly ones with very recent interest. I have chosen the issues because of my personal interest and involvement in them. Not all interesting recent issues can be treated in a short paper.

Let us begin with Caesar. A very simple cryptosystem, one of the very oldest, has been named after him. It is based on *substitutions*: each letter of the plaintext is substituted by another letter. The latter is obtained from the former by advancing k steps in the alphabet. At the end of the alphabet one goes cyclically to the beginning.

Thus, for the English alphabet and $k = 4$, the substitutions are as follows.

```
Plain:   A B C D E F G H I J K L M N O P Q R S T U V W X Y Z
Cipher:  E F G H I J K L M N O P Q R S T U V W X Y Z A B C D
```

If we denote by $E_k(w)$ the encryption of the plaintext w, we obtain, for instance,

$$E_4(SANDU) = WERHY,\quad E_{14}(SANDU) = GOBRI,\quad E_{20}(SANDU) = MUHXO.$$

Clearly, there are only 26 such encryption functions E_k, $0 \leq k \leq 25$, the function E_0 being the identity. The encryption function D_k corresponding to E_k satisfies $D_k = E_{26-k}$, for $1 \leq k \leq 25$. Thus we have, for any k satisfying $1 \leq k \leq 25$, $D_k E_k = E_0 = D_0$. Moreover, the functions E_i and D_j have the property of *commutativity*, very relevant for certain aspects of cryptography. For instance,

$$E_3 D_7 E_6 D_{11} = E_6 E_3 D_7 D_{11} = E_9 D_{18} = E_{17} = D_9.$$

The presented basic version of Caesar's system is much too simple for any serious applications. However, it is suitable for illustrating certain basic notions of cryptography. It is a *symmetric* system: the encryption key is either the same as the decryption key, or the former immediately gives away the latter. (Symmetric cryptosystems are often referred to as *classical*.) *Key management* is a major problem in symmetric systems: before any communication can take place, the key has to travel from the sender to the receiver via some secure channel.

Caesar tells in his De Bello Gallico how he sent an encrypted message to Cicero. The encryption was carried out by replacing the Latin letters by Greek ones in a way that is not clear from Caesar's writing. The information that Caesar actually used the cryptosystem described above comes from Suetonius. In fact, according to Suetonius, the shift in the alphabet was always three letters.

Caesar's system as described above is very unsafe simply because the key space is very small. However, the system has numerous variants with large key spaces, [7]. A fundamental idea in varying the system is to change the shift k from letter to letter in the plaintext, according to some specific rule. Very many classical systems, also the most widely used Data Encryption Standard, DES, can be viewed as variants of Caesar's system – especially if your equivalence classes are large enough in this respect.

We want to mention here one variant of Caesar's system that became obsolete with the advent of sufficiently powerful computers. Let us first discuss the idea of an "absolutely secure" cryptosystem. It's reasonable to claim that the system called *one-time pad* is secure. The key is a sequence of bits, say 110101000011, and is communicated to the legal receiver via some secure channel. The key is used for both encryption and decryption. The plaintext, say 010001101011, is encrypted by bitwise addition using bits of the key, resulting in 100100101000. A cryptanalyst knowing the cryptotext but having no information about the key knows really nothing because each bit either comes directly from the plaintext or has been changed by the key. It is essential that the key is used only once, as the name of the system indicates. The big disadvantage is the key management: a key at least as long as the plaintext has at some stage to travel via a secure channel.

This difficulty used to be overcome by specifying the key using a well-known book such as the Bible. The specification indicates the spot in the Bible where the key begins. For instance, Joshua 3, 2, 6 refers to the Book of Joshua, Chapter 3, Verse 2, Letter 6: "came to pass after three days that the officers went through the host and they commanded ..." The key is as long as the plaintext requires, although it can be expressed very briefly. The letters in the alphabetical order are assigned numerical values from 0 to 25. The key is used in a Caesar-like fashion; this variant is often referred to as the Vigenère system. Because the values assigned to the letters of CAME are 2, 0, 13, 4, the shift in the first four letters of the plaintext is 2, 0, 13 and 4, respectively. Thus, the plaintext "Practical perfectly secret systems would cause unemployment among cryptographers" is encrypted as follows.

```
Plain:  P R A C T I C A L P E R F E C T L Y S E C R E T
Key:    C A M E T O P A S S A F T E R T H R E E D A Y S
Crypto: R R M G M W R A D H E W Y I T M S P W I F R C L

Plain:  S Y S T E M S W O U L D C A U S E U N E M P L O
Key:    T H A T T H E O F F I C E R S W E N T T H R O U
Crypto: L F S M X T W K T Z T F G R M O I H G X T G Z I

Plain:  Y M E N T A M O N G C R Y P T O G R A P H E R S
Key:    G H T H E H O S T A N D T H E Y C O M M A N D E
Crypto: E T X U X H A G G G R U R W X M I F M B H R U W
```

While the system was feasible in the past, nowadays an exhaustive search through all the keys (roughly four million in number) presents no problem. Quite another issue - which we will not discuss here - is that redundancies in the plaintext and key languages make the ciphertext far from being random. Indeed, one can question even the security of one-time pad because, even if the key is random, redundancies in the plaintext might still be visible in the ciphertext. They are not visible if the plaintext is short.

We still return to one aspect of Caesar's system already referred to above: commutativity. In general, assume we are dealing with encryption and decryption keys E_i and D_j, i, $j = 1, 2, \ldots$, that commute. Both the sender and the receiver choose their keys from such a commutative pool of keys. It is irrelevant whether we are dealing with a classical or public-key system, that is, whether or not the encryption key E_i gives away the corresponding decryption key D_i.

Denote the sender's (A) and receiver's (B) encryption and decryption keys by E_A, D_A and E_B, D_B, respectively. The following *protocol* can be used in sending a message w.

(i) A sends $E_A(w)$ to B.
(ii) B sends $E_B(E_A(w))$ to A.
(iii) A sends $D_A(E_B(E_A(w))) = D_A(E_A(E_B(w))) = E_B(w)$ to B.
(iv) B decrypts $D_B(E_B(w)) = w$.

The message travels encrypted the whole time but there is no problem in key management because the keys are not distributed at all. Commutativity is essential in point (iii). The protocol can be visualized as sending the message in a box with clasp rings. First A sends the box to B, locked with A's padlock. Then B sends the box back to A, now locked also with B's padlock. Next, A opens the padlock E_A and sends the box back to B. When B receives the box, it is locked only with E_B which B can open. The keys never travel.

2 A big invention

A recent issue of Newsweek reported about a panel of scientists gathered to nominate the biggest invention of the past 2000 years. ***Public-key cryptography*** got at least one vote. Although such voters should perhaps log off for some time, public-key cryptography is undoubtedly a beautiful mathematical idea that has turned out to be also extremely useful. (We refer to [7] for historical references.)

Classical symmetric cryptosystems require the key to be transmitted over a secure channel. This is inconvenient. Moreover, if a secure channel exists, why do we not use it to send the message without encrypting it? The constraint of a secure channel is removed in public-key cryptography. It makes secret communication possible without requiring a secret means of delivering the keys. Public-key (also called asymmetric) systems rely on a pair of keys that are different but complementary. Each key decrypts the message that the other key encrypts. However, the key used to encrypt a message cannot be used to decrypt it. Hence, one of the complementary keys (the public one) can be publicized widely, whereas the other key (the private one) is all the time held only by its owner. If Bob wants to send a secret message to Alice, he uses her public key for encryption, after which she uses her private key for decryption.

Public-key cryptosystems are based on functions that are easy to compute but whose inverses are painfully slow to compute. A very intuitive illustration of such a ***one-way function*** is provided by the telephone directory of a big city. It is easy to find the number of any specific person. On the other hand, it is hard – one might say hopeless! – to find the person who has a certain specified number, in a directory with several thick volumes.

Of course, the telephone directory is only an intuitive example. The most widely used public-key cryptosystem, RSA, is based on the difficulty of factoring. It is straightforward to multiply two large prime numbers together, whereas it is extremely difficult to find the factors from the product. For the encryption, it suffices to know the product, but the factors are needed for the decryption. Or more explicitly: no other method of decryption is known but there is also no proof that this is the only way. Another one-way function widely used in cryptography is modular exponentiation. Let p be a large prime and g a primitive root (mod p). Then g^x (mod p) is computationally easy, whereas the inverse function (discrete logarithm) is intractable.

Another advantage of public-key cryptography is that it can be used for message ***authentication*** and ***digital signatures.*** When Bob sends a message w to Alice, he first applies his private key and, after that, Alice's public key to the result. Then Alice receives $E_A(D_B(w))$. But Alice knows the keys D_A (her own private key) and E_B (Bob's public key). Thus, she obtains first $D_B(w)$ and, finally, w. If the final text is legible, Alice can be confident that Bob actually sent the message: nobody else has the key D_B.

The best known public-key cryptosystem RSA is based on number theory, indeed, on facts known already for centuries. Also other number-theoretic problems have been used as a basis of public-key systems. In general, all these systems are dangerously dependent on number-theoretic problems, such as factoring, whose complexity is not known; there is no proof that they are intractable. In principle, problems from any area of mathematics can be used as a basis for public-key cryptosystems. Some problems in the *theory of formal languages* have been promising in this respect; [2] is a survey of this area. A common drawback in these systems is that the cryptotext tends to be much longer than the plaintext. Such a data expansion is not necessarily present in the different variants of *Tao-Chen* systems, which have also been widely tested in practice, [2].

3 Protocols, with or without computers

A *cryptographic protocol* constitutes an algorithm for communication between different parties, adversaries or not. The algorithm applies cryptographic transformations and is usually based on public-key cryptography. The goal of the protocol is often beyond the simple secrecy of message transmission.

Computing power is normally needed in carrying out a protocol. However, the same goal is often achieved by much simpler means. The cryptographic ideas can be sometimes implemented without computers. This is an area not much investigated.

We now consider a specific task: flipping a coin by telephone, without the assistance of a trusted referee. We first present a protocol, based on number theory. We then show that one has to be careful: there is a rather surprising way of cheating. Finally, we ask whether we could do it simpler, without computers.

Thus, A and B are talking by phone and want to flip a coin, maybe to make a decision. Apparently both have to be involved in this process of generating a random bit. Several methods have been presented. They follow the following general scheme.. Let P be a 50–50 property of integers x such that it is intractable to tell whether or not $P(x)$ holds without having the additional information Q. The protocol runs as follows. B who has the information Q tells A a random number x. A guesses whether or not $P(x)$ holds. B tells A whether the guess was correct (this corresponds to "heads") or wrong ("tails"). Later on, especially if A is suspicious, B gives A the information Q, so A can verify the matter herself.

Let us be explicit. Assume that $n = pq$, where p and q are large primes. Consider numbers a such that $0 < a < n$ and the greatest common divisor of a and n is 1. For exactly half of such numbers a, the Jacobi symbol $(\frac{a}{n})$ satisfies $(\frac{a}{n}) = 1$ and, again, exactly half of the numbers satisfying the latter condition are quadratic residues (mod n). The value of the Jacobi symbol is easily computable. Whether or not a is a quadratic residue (mod n) can be easily computed only if p and q are known. This is the background information; the protocol runs now as follows. B chooses p and q and tells A their product n, as well as a (random) number a such that $(\frac{a}{n}) = 1$. A guesses whether or not a is a quadratic residue (mod n). B tells A whether or not the guess was correct ("heads" or "tails"). Later B discloses p and q, so A can check that the information given previously was correct.

There is the following, rather surprising, way of cheating in this protocol. (This observation is due to Juha Honkala.) It is essential that A also checks that the disclosed p and q are indeed primes. Otherwise, B could cheat as follows. To start with, B chooses *three* large primes p_1, p_2, q_1 and a number a such that

$$(\frac{a}{p_1}) = (\frac{a}{p_2}) = -1 \text{ and } (\frac{a}{q_1}) = +1.$$

If B wants "heads", he proceeds as follows. If A says "residue", B discloses $p = p_1p_2$ and $q = q_1$. If A says "nonresidue", B discloses $p = p_1$ and $q = p_2q_1$.

If B wants "tails", he proceeds as follows. If A says "residue", B discloses $p = p_1$, $q = p_2q_1$. If A says "nonresidue", B discloses $p = p_1p_2$, $q = q_1$.

Clearly, both A and B need a computer to follow the above protocol. For a "harmless" situation, the method suggested gives the impression of killing a fly with heavy artillery – you just are not likely to do things that way. So let us consider an everyday situation: Alice and Bob cannot agree what they are going to do in the evening. Bob would like to go to the opera but Alice likes to see hockey. Alice realizes that it is not good for their relation if they go to different places. Bob would hate to sit in a crowded sports arena, thinking that he missed his favorite opera. So Bob tells Alice over the phone that they should flip a coin but complains that the cryptographic methods for the task are overly complicated. They are not going to compute quadratic residues with respect to a large modulus. But then Alice gets a brilliant idea. Both of them have the same telephone directory. They can flip a coin according to the following protocol.

Step 1. Bob picks a number in the directory (say 2371386) and asks whether the number immediately following it in the directory is even or odd.

Step 2. Alice makes a guess (say "even"). It is indeed a guess because she has to react immediately.

Step 3. Bob tells the result of the guess (here "wrong"). At this stage they can interrupt the protocol and do whatever the result implies (here go to the opera).

Step 4. Bob proves to Alice that he was honest in telling the result. Bob tells that the number 2371386 belongs to Florin Andreescu, after which Alice can immediately check the next number.

Secret balloting systems constitute an area where traditionally, in all important elections, protocols have been carried out without computers (in the election process itself). The basic method of securing ballot secrecy is to make sure that the ballots of individual voters are not counted separately but in aggregates. Typically all the ballots cast in a given election locale are counted simultaneously. The voters are identified before they are allowed to vote; in this way it is also guaranteed that each voter casts only one vote. Initially, during the identification, the officials know the link between a voter and his/her vote but this link is broken in the shuffling and simultaneous counting of the votes. How can the link be broken if there is no election locale but the elections are conducted in a computer network? When elections are conducted in the Internet, as they most likely will be in the future, numerous diverse security and secrecy issues have to be taken into account. However, the basic issues are *soundness* (only legitimate voters should be able to cast votes, and each one only once) and *secrecy* (nobody, including the election officials, should be able to find out the voting strategy of a voter without his/her own consent). Although, at first thought, it might seem an impossible task to combine these two requirements in a network election, several solutions based on public-key cryptography have been presented. A third requirement, that of *verifiability* (each voter should be able to check that his/her vote has been correctly counted) is satisfied in almost all systems proposed and can be viewed, in network elections, as a corollary of the conditions of soundness and secrecy. Observe that the requirement of verifiability is not satisfied in current elections. Although verifiability is likely to increase the motivation of individual voters to vote, it provides also an incentive for selling votes if the buyer can get a proof that the service was actually rendered. In the traditional election protocol, the voting booth does actually more than *permits* a voter to keep his/her vote private; it *forces* the votes to remain secret. This means that a voter can make promises to unreasonable employers, bad dictators, dictator-like family heads or compelling spouses, and yet in the privacy of the voting booth the voter can cast quite a different vote. Thus, booth-type secrecy is desirable also in elections over computer networks; it is essential that the voter does not get a *receipt* of any kind about how he/she actually voted. Although, at a first glance, getting a receipt seems to be a property inherent in every verifiable election scheme, public-key cryptography is often counterintuitive. Election schemes are possible (see [7] for further details) where each of the requirements of soundness, secrecy, verifiability and receipt-freeness are simultaneously satisfied.

4 Truth or consequences

Zero-knowledge proofs constitute a remarkable application of public-key encryption. The *prover* P (also called Peter) possesses some information such as the prime factorization of a large integer, the proof of a long-standing mathematical conjecture, a password or an identification number. In a zero-knowledge proof, the *verifier* V (Vera) becomes convinced, beyond reasonable doubt, about P's knowledge but obtains no knowledge that is new to herself. A good way to visualize this is to think that V could obtain all the information, except her conviction about P's knowledge, entirely without P, that is, V could play the protocol as a solitaire game, without P participating at all.

We do not present here any formal definitions or enter any details of the general theory. This section contains one particular zero-knowledge proof, given in [7]. It is a zero-knowledge protocol for the *satisfiability problem* for propositional calculus. In our estimation, this particular protocol illustrates the basic issues concerning zero-knowledge proofs particularly well.

We need a useful general notion, that of a *lockable box*. The verifier cannot open it because the prover has the key. On the other hand, the prover has to commit himself to the contents of the box, that is, he cannot change the contents when he opens the box. In fact, the verifier may watch when he opens the box. In more formal terms, the hardware of the lockable boxes is replaced by one-way functions. Locking the information x in a box means applying a one-way function f to x. V can now handle $f(x)$ without knowing what x is – only P is in the possession of the inverse f^{-1}. On the other hand, P cannot change $f(x)$, since V knows it. Under the assumption that f is injective, this means that P cannot change x, that is, the contents of the box. When P opens the box by giving x to V, she can verify that $f(x)$ is the number she had before.

We now go to the zero-knowledge protocol for the satisfiability problem for propositional formulas in *3-conjunctive normal form*. The significance of this problem as the basic NP-complete problem is well known, [6]. The setup is that both P and V know a propositional formula α with r variables and s disjunctive clauses in the conjunction. (Thus, each clause is a disjunction of 3 *literals*, a literal being either a variable or its negation.) P wants to convince V that he knows a truth-value assignment g satisfying α, without giving any information about the assignment. The protocol runs as follows.

Step 1. P prepares and gives V three sets of locked boxes, referred to as *variable*, *truth-value* and *assignment boxes*. Variable and truth-value boxes, VAR_i and TV_i, $i = 1, \ldots, 2r$, correspond to pairs (x, y), where x is a variable and y is a truth-value (T or F). The number of such pairs is $2r$. For each pair (x, y), there is an i such that x is locked in VAR_i and y in TV_i, but the ordering of the pairs is random. The indices of the assignment boxes $A_{i,j,k}$ range, independently from each other, from 1 to $2r$ and from ~ 1 to $\sim 2r$. Thus, the number of assignment boxes is $(4r)^3$. (This number can be made considerably smaller by certain natural assumptions concerning α.) Each of the assignment boxes contains the number 0 or 1. Each of the s clauses of α gives rise to exactly one assignment box containing 1, whereas all other boxes contain 0. For the clause $\beta = x \vee y \vee z$, where each of x, y, z is a variable or its negation, the box $A_{i,j,k}$ containing 1 is found as follows. Assume that $x = x_m$ (resp. $x = \sim x_m$) and that n is the index such that x_m is in the box VAR_n and $g(x_m)$ (that is, the truth-value P has assigned for x_m) is in the box TV_n. Then we define $i = n$ (resp. $i = \sim n$). The values j and k are defined in the same way, starting from y and z, respectively.

Step 2. V gives one of the two commands "truth" or "consequences" to P.

Step 3. If V's command was "consequences", P opens all variable and assignment boxes. If her command was "truth", P opens all truth-value boxes and, moreover, all those assignment boxes $A_{i,j,k}$ where each of the three indices is either of the form u with F in TV_u, or of the form $\sim u$ with T in TV_u.

Step 4. V fails P either if in the case of "truth" the number 1 appears in some of the opened assignment boxes, or in the case of "consequences" the assignment boxes containing the number 1 do not yield the original formula α.

Step 5. If P passed in Step 4, V either accepts him (that is, V is convinced) or requests another round of the protocol.

V's decision in Step 5 is based on whether or not she already reached her preset confidence level. If P guesses correctly V's command, then he passes the whole round of the protocol. For the command "truth", P just locks the number 0 in all assignment boxes. For the command "consequences", P just taken care of that the correct α will be found from the assignment boxes. Thus, P passes one round with probability $\frac{1}{2}$, even if he does not know g. But then each round will decrease the probability of P not failing by a factor of $\frac{1}{2}$. Thus, if Vera's (preset) confidence level is an error rate of less than one in a million, she needs twenty rounds of the protocol. On the other hand, Vera can obtain the same information as in the protocol just by herself, without Peter: she just plays the "guessing game", and the opened boxes look exactly the same as in the protocol with P. The only difference is that V's conviction about P knowing the satisfiability assignment g is missing. Why V does not learn anything about g, is illustrated further in the following example. Observe that everything has to be started from scratch in a new round of the protocol. If something remains until the next round, say, from the random order of the pairs, then it might be possible to learn about g.

As an example, consider the following propositional formula α with $r = 5$ and $s = 11$:

$$(\sim x_1 \vee \sim x_2 \vee \sim x_3) \wedge (x_1 \vee x_2 \vee \sim x_4) \wedge (\sim x_1 \vee x_2 \vee x_4)$$
$$\wedge (x_1 \vee x_2 \vee \sim x_5) \wedge (x_1 \vee \sim x_2 \vee \sim x_5) \wedge (x_1 \vee x_3 \vee x_4)$$
$$\wedge (\sim x_1 \vee x_3 \vee \sim x_5) \wedge (x_1 \vee \sim x_4 \vee x_5) \wedge (x_2 \vee \sim x_3 \vee x_4)$$
$$\wedge (x_3 \vee x_4 \vee x_5) \wedge (x_3 \vee \sim x_4 \vee x_5).$$

The formula α is satisfiable, and P knows a truth-value assignment g with

$$g(x_2) = g(x_3) = T, \quad g(x_1) = g(x_4) = g(x_5) = F,$$

for which α assumes the value T. P wants to convince V about his knowledge. For Step 1 of the protocol, he has the following order of the pairs (x, y):

i	1	2	3	4	5	6	7	8	9	10
VAR_i	x_4	x_1	x_5	x_2	x_2	x_3	x_1	x_5	x_3	x_4
TV_i	F	T	F	F	T	T	F	T	F	T

The assignment boxes are no less than 8000 in number. (If we assume that the indices appear in increasing order of magnitude and no index, negated or nonnegated, appears twice, then the smaller total number 960 suffices.) They all contain the number 0, except for the following 11 boxes $A_{i,j,k}$ which contain the number 1. We list only the triples $(i,\ j,\ k)$:

$$(\sim 5, \sim 6, \sim 7), (\sim 1, 5, 7), (1, 5, \sim 7), (\sim 3, 5, 7),$$
$$(\sim 3, \sim 5, 7), (1, 6, 7), (\sim 3, 6, \sim 7), (\sim 1, 3, 7),$$
$$(1, 5, \sim 6), (1, 3, 6), (\sim 1, 3, 6).$$

For readability, the eleven $(s = 11)$ boxes are listed above in the order obtained from the clauses of α. However, P gives the (huge number of) assignment boxes to V in some alphabetic order. (The triples $(i,\ j,\ k)$ are unordered; above we have ordered them increasingly.)

If V commands "consequences", she learns α (in a somewhat permuted form). But she already knew α. Nothing of g is revealed, since TV-boxes are not opened at all. If V commands "truth", she learns in which boxes each T and F is but nothing about their connection with the variables. Moreover, she learns that the assignment g does not assign the value F to any of the clauses of α. This follows because 0 appears in all those 120 assignment boxes, where all three indices come from the set $\{1, 3, 4, 7, 9, \sim 2, \sim 5, \sim 6, \sim 8, \sim 10\}$. This does not reveal anything concerning the association of T and F with the individual variables. Clearly, P fails in one of the two scenarios in Step 4 if he doesn't know a correct truth-value assignment.

5 A deck of cards or a computer

In the early days of cryptography the methods were certainly developed quite independently of computers – there were none around. On the other hand, the whole idea of public-key cryptography and one-way function is difficult to visualize in practice without referring to computing devices. (Methods based on telephone directories are not secure; they can be used only to illustrate ideas.) Proper solutions in cryptography are always tied with complexity. If your method takes unreasonably long, you might as well forget it. The situation in cryptography differs very much from that in classical mathematics. "Impossible tasks", such as factoring a large integer, are not impossible from the point of view of classical mathematics. On the contrary, in principle everything in cryptography is possible, even trivial in classical mathematics. However, in cryptography solutions lose their meaning and significance if they take too long.

But there are different setups and various degrees in the seriousness of the overall situation. Smaller safety measures are adequate if the opponent has little time or resources. We already observed in Section 3 how a protocol, in a particular setup, is naturally carried out entirely without computers. Similarly, the setup in a zero-knowledge proof becomes quite different if the Prover and/or Verifier can somehow observe the computing resources of the other. It is certainly possible to design cryptographic protocols, that is, protocols applying ideas of cryptography, where computing resources are limited or nonexistent. Very little work has so far been done in this direction, although the approach is interesting also from the general mathematical point of view.

There are obvious reasons for investigating cryptographic protocols, where computers are not used. Nonavailability, nonportability, unreliability, mistrust or dislike of computers are certainly among such reasons. Sometimes a protocol without computers is simply better or more natural than one with computers. For instance, assume that a group of people gathered together want to take a secret vote about some important matter. Who would in such a situation consider any sophisticated balloting protocols with computers? It is much easier and more efficient to take ballots (cards or pieces of paper) and a box or boxes. The "cryptographic element" in this protocol will be *shuffling*. When the ballots are shuffled, everybody loses the link between the person and the ballot, although the link could perhaps be observed earlier. It is essential for secrecy that the link is broken before the votes are disclosed. This very simple protocol is in this particular setup better than any other one could think of.

A *deck of cards*, consisting in general of more than 52 cards, is often a very useful tool in cryptographic tasks, even in zero-knowledge proofs. When the cards are face down, information is hidden. In cryptography one wants to shuffle or scramble information but not too much because, otherwise, one might lose everything. If information is hidden in a deck of cards, one cannot shuffle the deck arbitrarily because of this reason. However, one may *cut* the deck without losing everything. As usual, a *cut* of the deck means that some number of topmost cards is moved, without changing their order, to the bottom. An important observation is that the effect of several cuts, made after each other, can always be achieved by one cut. If so many cuts are made in succession that every participant in the protocol has lost the possibility of keeping track of the cutting position, we speak of a *random cut*. Thus, a random cut is a sequence of cuts, viewed as one cut. The unchanged deck constitutes also a cut, being one possibility for a random cut. Our overall cryptographic assumption will be that *it is possible to make a random cut*. The assumption will be made for any number ≥ 1 of participants in the protocol and any number ≥ 2 of cards in the deck.

There are two kinds of cards, white and black. Cards of the same color are indistinguishable. As usual, the back side of each card is identical. White cards are denoted by the bit 0, black cards by the bit 1. A deck of cards can be represented in this way as a word over the alphabet $\{0, 1\}$, using the convention that the *leftmost* letter represents the *topmost* card. Thus, the word 01101 stands for a deck with two white and three black cards, where the topmost and fourth cards are white. We also make a distinction whether the cards are face down or face up.

A *commitment* to the bit 0 (resp. 1) is the deck 10 (resp. 01), cards face down. Thus, a commitment is made using one card of each color, the bottom card telling the bit committed to. It will become apparent below why it is better to use two cards for a commitment, rather than simply a card 0 or 1. The *negations* of bits are understood as $\sim 0 = 1$ and $\sim 1 = 0$. When used as truth-values, the bit 1 (resp. 0) is the truth-value T (resp. F).

Let us consider first the following simple setup. Alice has a secret bit a and Bob has a secret bit b. They want to learn $a \wedge b$ without revealing their secret bits, unless necessary by the definition of conjunction. This means that if $a = 0$ (resp. $b = 0$) then Alice (resp. Bob) learns nothing. If $a = 1$ then Alice actually learns b. The motivation and importance of this setup is discussed in [ ]. The following protocol does the job. Initially, both Alice and Bob have a white card and a black card. An additional black card is put on the table, face down.

Step 1. Alice makes a commitment to her secret bit. Bob makes a commitment to the *negation* of his secret bit.

Step 2. Alice's commitment is put on top of the card on the table, Bob's commitment below it. After that there is a deck of five cards on the table, all cards face down.

Step 3. A random cut is made on the deck.

Step 4. The cards are shown. The conjunction has the value T exactly in case the two white cards are "cyclically next" to each other (that is, either next to each other or one of them is the top and the other the bottom card).

The reader should have no difficulties in verifying the correctness of the above protocol. However, a further step can be taken. We want to compute conjunctions in such a way that the outcome remains in encrypted form. More specifically, we are given two bits x and y in the form of commitments as described above. We want to compute the bit $z = x \wedge y$, also as a commitment. Thus, to start with we have two pairs of cards faces down, one pair being a commitment for x, and the other pair for y. We want to devise a protocol, which now will be a game of *solitaire*, producing two cards faces down, representing a commitment for z. Some additional cards are used in the protocol. But the player of the solitaire does not know or learn later the original bits x and y, and also not the resulting bit z! The reader is referred to [7] or [3] for the details of this, somewhat more complicated protocol. The idea is that z can be used as an input for other protocols.

Note that such a solitaire is obvious for *negation*. Given a commitment for the bit x, we get a commitment for $\sim x$ just by switching the order of cards. This would perhaps not at all be possible if we had defined a commitment to be one card, face down. Another reason for defining a commitment in terms of two cards is that one is then able to *copy a commitment* without learning the bit represented. Such a capability is needed in many protocols; one part of the protocol may cause the loss of the commitment which is still needed later on. The copying protocol is presented below, in the form of a game of solitaire. The only participant is called Verifier, Vera, V. This reflects our final aim of presenting a *non-interactive* (meaning that V does not communicate with P) zero-knowledge proof. The Verifier is not supposed to cheat. In particular, we assume that she makes true random cuts when the protocol so requires and displays cards only if the protocol allows her to. Initially, Vera is given two cards face down, defining a commitment $\sim xx$. In addition, she is given a deck $(01)^{k+1}$ of $2k+2$ cards, for some $k \geq 2$. Also these cards are face down but she may check that they form indeed the deck $(01)^{k+1}$.

Step 1. Vera makes a random cut of the deck $(01)^{k+1}$. She is not any more allowed to look at any card of the resulting deck, but she knows that the deck is of the form $(\sim yy)^{k+1}$, where $y = 0$ or $y = 1$.

Step 2. Vera takes two topmost cards of the deck $(\sim yy)^{k+1}$. She puts these cards under the commitment $\sim xx$, getting the deck $\sim xx \sim yy = Y_4$. She still has also the deck $(\sim yy)^k = Y_{2k}$.

Step 3. Vera makes a random cut of the deck Y_4, after which she looks at the four cards. If they are 0101 or 1010, then she outputs Y_{2k} (face down as it has been the whole time). If they are 0011, 0110, 1100 or 1001, then she outputs $(y \sim y)^k$, obtained from Y_{2k} by moving the topmost card to the bottom (without looking at it).

Step 4. Vera concludes that her output equals $(\sim xx)^k$ and, thus, consists of k copies of the original commitment.

This copying protocol is needed also in the solitaire protocol for conjunction, referred to above. We are now ready for the final step, a simple noninteractive zero-knowledge proof for the satisfiability of propositional formulas.

As before, a propositional formula α with r variables $x_1, \ldots, x_r$ is given. Since every propositional connective can be expressed in terms of conjunction and negation, we assume that these two are the only connectives occurring in α. (Equivalently, we could present a solitaire protocol also for disjunction, after which we could consider α, say, in conjunctive normal form.) The Prover, Peter, knows an assignment for the variables making α true. To convince Vera of his knowledge, Peter gives her his assignment secretly, in the form of r commitments, $2r$ cards. (Peter must indicate also the one-to-one correspondence between the variables and commitments.) In addition, Vera is given a sufficient supply of auxiliary cards, needed in copying the commitments. (Estimates, based on α, for the number of auxiliary cards can be given.) Vera now plays the solitaire, applying the protocols for conjunction and negation. She looks at the final outcome, the commitment for the whole formula α, and accepts iff the commitment is 01. *Only one round is needed in this non-interactive protocol.* Vera's eventual cheating can be revealed if Peter or a person trusted by him stands by, watching Vera's play. One could also imagine a technical device that would have the same effects as card play and would report any wrongdoings of the operator. Finally, the only way Peter can cheat is to give pairs 00 or 11 in place of some commitments. But he would be caught because cards assigned to a variable will be disclosed as an unordered pair whenever the variable takes part in a conjunction.

The theoretical simplicity of the above protocol seems quite fascinating. Further comparisons with protocols such as the one given in Section 4 remain to be done.

6 Security of security. DNA computing

Although secret writing is probably as old as writing itself, interest in cryptology has become widespread only very recently. Earlier there was real need for cryptography only in military and diplomatic circles, as well as among certain criminal organizations. Cryptosystem designers, on one hand, and eavesdroppers and cryptanalysts, on the other hand, constitute two opposing groups of people. (Sometimes it is difficult to tell which group are the "good guys" and which are the "bad guys", for instance, in case of a dictatorial government.) Earlier the fight between these two groups was of little concern to most of the people. Things have become drastically different as a consequence of the information revolution. There is no possibility of any physical data protection in the Internet, that's why one has started to speak about cryptography for the Internet. Strong cryptography is in everyday use to ensure that conversations and transactions remain confidential. Public-key cryptography and the progress in computing technology have made strong encryption methods inexpensive and widespread.

When cryptography was essentially a government matter, "secrecy of secrecy" or "security of security" was natural. The government agencies wanted to restrict and/or prevent the sales and/or export of cryptography. In spite of the total change in the overall picture, some government agencies continue to propagate the regulation of encryption technology, as well as the continuation of all export restrictions. This concerns especially the National Security Agency (NSA) and FBI in the U.S. They are worried that they would be unable to decrypt the messages of potential spies and terrorists. In a recent Wassenaar Agreement, many European countries (including Finland) imposed severe limitations to the export of encryption technology. The effects can be disastrous to certain parts of high technology in small countries where the domestic market is not big enough.

The government agencies are concerned that the "bad guys" will benefit from the new cryptography. This is apparently possible – everything can be used for both good and bad purposes. But a field of technology cannot be discriminated merely for such reasons. R. L. Rivest elaborates this as follows, [5]. "Any U.S. citizen can freely buy a pair of gloves, even though a burglar might use them to ransack a house without leaving fingerprints. Cryptography is a data-protection technology just as gloves are a hand-protection technology. Cryptography protects data from hackers, corporate spies and con artists, whereas gloves protect hands from cuts, scrapes, heat, cold and infection. The former can frustrate FBI wiretapping, and the latter can thwart FBI fingerprint analysis. Cryptography and gloves are both dirt-cheap and widely available. In fact, you can download good cryptographic software from the Internet for less than the price of a good pair of gloves."

It seems that limitations described above will satisfy no one and will only mean a victory for the Big Brother. Also some compromise solutions have been proposed. They would allow strong cryptography to be widely used while still enabling the government agencies to decrypt messages when lawfully authorized to do so. Many models of key-escrow have been developed, where users register their encryption keys with a law-enforcement agency, as well as models of key-recovery, where a law-enforcement agency has a backdoor access to the keys. Typically, together with each encrypted message one sends an encrypted version of the system encryption key. An authorized agency has a "master backdoor key" to decrypt all system encryption keys, after which decryption the agency can decrypt the messages. For various reasons, [5], also such compromises are highly unsatisfactory.

Although it is still too early to predict the significance of new technologies such as *DNA computing*, it is possible that cryptanalysis will be revolutionized, and the complexity issues involved will look quite different. Then agreements such as the Wassenaar Agreement mentioned above, where explicit technologies are referred to, might become quite meaningless. The high hopes for the future of DNA computing, [4], are based on two fundamental features: (i) The massive *parallelism* of DNA strands, (ii) Watson-Crick *complementarity.*

As regards (i), most of the celebrated computationally intractable problems can be solved by an exhaustive search through all possible solutions. However, the insurmountable difficulty lies in the fact that such a search is too vast to be carried out using present technology. On the other hand, the density of information stored in DNA strands and the ease of constructing many copies of them might render such exhaustive searches possible. A very typical example is cryptanalysis: all possible keys can be tried out simultaneously. Indeed, [1] discusses a possible technique of breaking the cryptosystem DES in this fashion – details can be found also in [4]. DNA computing is suitable for cryptanalysis also because deterministic solutions are not required; a low error rate will suffice, see [4] for details.

As regards (ii), Watson-Crick complementarity is a feature provided "for free" by the nature. It gives rise to a powerful tool for computing because, in a well-defined sense, [4], complementarity brings the universal twin-shuffle language to the computing scene. By encoding information in different fashions on the DNA strands subjected to bonding, one is able to make far-reaching conclusions based on the fact that bonding has taken place. Also cryptanalysis can be based on such conclusions, [1], [4].

References

[1] L.M. Adleman, P.W.K. Rothemund, S. Roweiss, E. Winfree, On applying molecular computation to the Data Encryption Standard. In E. Baum et al. (ed.), *DNA based computers.* Proc. of 2nd Ann. Meeting, Princeton (1996) 28–48.

[2] V. Niemi, Cryptology: language-theoretic aspects. In G. Rozenberg, A. Salomaa (ed.), *Handbook of Formal Languages*, Vol. 2, Springer-Verlag (1997) 507–524.

[3] V. Niemi, A. Renvall, Secure multiparty computations without computers. *Theoretical Computer Science* 191 (1998) 173–183.

[4] G. Păun, G. Rozenberg, A. Salomaa, *DNA Computing. New Computing Paradigms.* Springer-Verlag (1998).

[5] R.L. Rivest, The case against regulating encryption technology. *Scientific American*, October 1998, 88–89.

[6] A. Salomaa, *Computation and Automata.* Cambridge University Press (1985).

[7] A. Salomaa, *Public-Key Cryptography, Second Edition.* Springer-Verlag (1996).

Automata and Their Interaction: Definitional Suggestions

Boris A. Trakhtenbrot

Department of Computer Science
School of Mathematical Sciences
Tel Aviv University
Tel Aviv 69978, Israel
trakhte@math.tau.ac.il

Abstract. There is a growing feeling in the community that the current literature on reactive and hybrid systems is plagued by a Babel of models, constructs and formalisms, and by an amazing discord of terminology and notation. Further models and formalisms are engendered, and it is not clear where to stop.
Hence, the urge toward a pithy conceptual/notational setting, supported by a consistent and comprehensive taxonomy for a wide range of formalisms and models.
The paper outlines an automata-based approach to this challenge, which emerged in previous research [PRT, RT] and in teaching experience [T1, T2]. We compare our definitional suggestions with similar background in the current literature, where the subject is sometimes complicated by a premature mixture of semantics, syntax and pragmatics.

1 Introduction

1.1 Mainly Quotations

Hybrid systems, and in particular timed automata, are popular paradigms of "reactive" systems.

This developing area suffers from some methodological and expository embarrassment.

Quotation 1. "The number of formalisms that purportedly facilitate the modeling, specifying and proving of timing properties for reactive systems has exploded over the past few years. The authors, who confess to have added to the confusion by advancing a variety of different syntactic and semantic proposals, feel that it would be beneficial to pause for a second – to pause and look back to sort out what has been accomplished and what needs to be done. This paper attempts such a meditation by surveying logic-based and automata-based real-time formalisms and putting them into perspective" [AH].

Quotation 2. "A new class of systems is viewed by many computer scientists as an opportunity to invent new semantics. A number of years ago, the new class was distributed systems. More recently, it has been real-time systems. The proliferation of new semantics may be fun for semanticists, but developing a practical method for reasoning about systems is a lot of work. It would be unfortunate if every new class of systems required inventing a new semantics, along with proof rules, languages and tools.

Fortunately, these new classes do not require any fundamental change to the old methods for specifying and reasoning about systems" [AL].

So, what is a reactive system?

Quotation 3. "Reactive Systems" is a nice name for a natural class of computational systems. The search is still on for the best precise mathematical concept to explain this [Ab].

It seems, that "Hybrid Systems" share a similar status; moreover, this is the case with other popular terminology in the field.

Quotation 4. "Processes" are not mathematical objects whose nature is universally agreed upon ... not a well understood domain of entities [Mi].

Yet, that does not prevent researchers from developing useful methods of verification for specific structures of "reactive" and "hybrid" systems, such as "reactive programs" and "hybrid automata". Here are some widely accepted informal explanations of those terms:

(i) A reactive program is a program whose role is to maintain an ongoing interaction with its environment.
(ii) A hybrid automaton is a mathematical model for a digital program that interacts with an analog environment.
(iii) Hybrid Systems are interacting networks of digital and continuous systems.

Clearly, precise definitions and taxonomies of the intended entities presume decisions about the choice of appropriate (i) components (programs, automata, processes, etc.) and (ii) interaction architectures.

Note also some expected asymmetry between the components (digital vs. continuous, main object vs. its environment etc.).

Such decisions are suggested in particular by classical automata theory, modestly enriched with standard concurrency and continuous time. This is in full accordance with the appeal to rely on "old-fashioned recipes" (see Quotations 1,2). More about that below, after some comments on the literature on the subject. Let us start with reactive systems.

Quotation 5. "We refer to our object of study as reactive systems... In the literature on the formal development and analysis of reactive systems, we find a rich variety of programming languages...Some of the constructs are based on shared variables, message passing, remote procedure calls, ... semaphores, monitors etc. We introduce a generic (abstract) model for reactive systems, which provides an abstract setting that enables a uniform treatment of all these constructs" [MP2].

Further, in [MP2], according to the generic model, four concrete models are considered:

(i) Transition Diagrams, (ii) Shared Variable Text, (iii) Message Passing Text, (iv) Petri Nets.

Quotation 5 (continued). "Each concrete model is characterized by a programming language consisting of the syntax of the programs in the model, and semantics which maps programs into Basic Transition Systems".

"Petri Nets are a radically different concrete model. The main difference: it is not a programming language, but rather intended to model and specify a broad family of *reactive systems*".

Comments. Clearly, the [MP2] taxonomy of reactive systems is oriented toward programming constructs. That is why it distinguishes between Models 1 and 2, which in essence differ only in syntactic details (graphical syntax for the first vs. textual syntax for the second). On the abstract level of automata theory there are no reasons to distinguish between them, and even less to declare Petri Nets with their suggestive geometry of communication, "a radically different concrete model". As a matter of fact, the automata-based setting supports a broad taxonomy of reactive systems, in which other models (not covered in [MP2]) are identified in a natural way. "The stone that the builders rejected has become the chief cornerstone" (Psalms)!

Now, about existing (and sometimes changing) definitions of Hybrid Automaton. Most of them present some detailed and more precise version of the following descriptions:

Quotation 6. "Informally, a Hybrid Automaton consists of a finite set X of real valued variables, and a labeled multigraph E. The edges E represent discrete jumps and are labeled with guard assignments to variables in X. The vertices V represent continuous flows, and are labeled with differential inequalities over the first derivatives of the variables in X. The state of the automaton changes either instantaneously when a discrete jump occurs ("transition step") or, while time elapses, through a continuous flow ("time step") [HW].

Quotation 7. "Hybrid automata are generalized finite state machines. For each control location ... the continuous activities are governed by a set of differential equations [*this is a labeling of locations, B.T.*] Another label - the invariant condition - must hold while the control resides at the location, and each transition is labeled with a guarded set of assignments" [ACH].

Comments

(i) The definitions do not present explicitly, either the decomposition of the "Hybrid" into separate components, or the relevant interaction architecture.

(ii) They emphasize the role of differential equations, a machinery which is beyond the usual logical based and/or automata-based settings.

(iii) They confine with accepting- mechanisms, and ignore transducers, i.e. input-output mechanisms, which specify/compute functions.

The last remark brings us back to a discussion from the early days of Automata Theory [KT], reconsidered in depth by D. Scott:

Quotation 8. "The author (along with many other people) has come recently to the conclusion that the functions computed by the various machines are more important - or at least more basic - than the sets accepted by these devices. The sets are still interesting and useful, but the functions are needed to understand the nets. In putting the functions first, the relationship between various classes of sets becomes much clearer. This is already done in recursive function theory, and we shall see that the same plan carries over to the general theory" [S].

1.2 Overview

This paper provides an outline of the main features of an automata-based approach to the subject. Note, that it is semantically-oriented, and avoids heavy syntactical ingredients.

The aim of Sections 2-4 is to present enough technical details of the advocated conceptual/notational machinery, and to explain it with appropriate examples and propositions. In view of the many facets of the issues, and because of the lack of generally accepted terminology in the field, an eclectic expository style is adopted: formal definitions are intermixed with less formal explications and comments, and omitted details are left to the imagination of the reader. Hopefully, this will not prevent one from perceiving the full picture. The test will come in Section 5 (Discussion), where the proposed setting is illustrated with respect to (and contrasted with) alternatives in the literature. The discussion focuses on four points:

(i) Networking,
(ii) Taxonomies in fundamental monographs,
(iii) Hybrids in the Theoretical Computer Science community,
(iv) Views of Control Theorists.

Section 2 (Automata) is about the potential components of "reactive systems", and Section 3 (Interaction) - about the ways those components are structured in reactive systems. Note that at this stage, no particular features of Hybridity are on the agenda. Those are postponed to Section 4 (Toward Hybrids).
Now, more details about what is to come.

Section 2: Automata.
The time-domain T of a discrete automaton (with finite or infinite state-space) is the ordered set N of nonnegative integers. For continuous-time automata, T is the real line $R^{\geq 0}$. "Continuity" of automata refers only to their time-domain, and does not concern their data-domains. Hence, no differential equations etc. Of course, algebraic and/or topological properties of the automaton's state-space X may be instrumental for concrete applications, but they are **not** part of the formal automaton concept.

Nondeterministic (unlike probabilistic) automata cannot be implemented directly; they seem to be an useful mathematical abstraction and/or a technical notion. We don't exclude nondeterministic automata, but the most relevant phenomena are best explained in the deterministic setting. In the definition of deterministic automata, we follow [S1].

Input/output. The focus is on the relationship between two entities: the *retrospective* (causal) operator F and the *transducer* which computes/defines it. The operator F outputs at time-instant t a value which depends only on the values inputted not later than t. There are two kinds of transducers: explicit and implicit. Not surprisingly, the transition from explicit to implicit is straightforward, but the opposite direction is less trivial.

Labeled Transition Systems (*LTS*) and other kinds of decorated graphs are not part of the model. Occasionally we may use them for expository convenience, in particular, for comparison with current literature.

Section 3: Interaction.

Four Architectures. We consider two dichotomies, which are routine for discrete time, but some precaution (explained in the main text) is needed in the case of continuous time. Note that hiding issues are not considered.

(i) Synchrony vs. Asynchrony;
Warning: Don't confuse this with the use of "synchronous/asynchronous" in the sense of "tight/buffered communication".
(ii) Nets (communication through ports/wires) vs. Webs (shared memory).

Circuits. The architectures above don't cover specific aspects of input-output interaction. These aspects are well known for discrete-time circuits of logical and delay components, for which Burks and Wright coined the name *Logical Nets* [BW]). The generalization of Logical Nets and their shift to continuous time are formalized as synchronous circuits of transducers; these are relevant, in particular, for the interaction mechanism of hybrid automata. At an appropriate level of abstraction a circuit C with k components, which communicate through their inputs and outputs, is a system Eq of k functional equations, each equation describing one of the components. Eq allows a pictorial representation as a circuit C labeled by the functional symbols which occur in Eq. Hence, the reference to Eq as a circuit of functions. Actually, before the emergence of modern communication/concurrency theory, this was the main (may be even the unique) interaction paradigm in Automata Theory. C is well defined (reliable) iff Eq has an unique solution, φ, which is said to be the input/output behavior of C. This means that the propagation of signals along closed cycles in C is causally motivated (feedback reliable).

The same circuit C also visualizes a circuit of transducers, which adequately implements Eq, i.e. computes φ (in an appropriate sense!).

Note that for continuous time feedback, reliability is more subtle than in the discrete case. Occasionally, in situations related to differential equations, analytical machinery may help. Some automata-based criteria appear in [PRT].

Section 4: Toward Hybrids.

We start with two known properties: Finite Variability (otherwise called 'non-Zeno property') and Duration Independence. As particular cases of FV-automata we define jump, flow, and, finally - jump/flow automata. The last one will appear later as a component named Plant of hybrid automata. A Plant may be described by chance via differential equations; this situation is practically important, but conceptually - irrelevant.

Duration independence means invariance of the behavior under stretching of the time axis. A duration independent (in particular - a finite state) automaton M, usually named Controller, is the second component of an hybrid automaton. Due to duration independence, M can be specified syntactically by a labeled transition system, very close to (but not identical to) those used in classical Automata Theory.

Hybrids. The analysis of Hybridity is best explained on the following three levels:

(i) Plant. This is an explicit transducer with a jump/flow underlying automaton M. No concurrent interaction occurs at this level; "Hybridity" is manifested solely in the coexistence of jumps and flows.

(ii) Hybrid Automaton. This is a reliable synchronous circuit with two components: a Plant (as above) and a Controller, which is a duration independent (in particular a finite-state) transducer.
Note the fundamental asymmetry between the two interacting components. Whereas Controller is a constructive object with almost standard *LTS*-syntax, Plant is a semantical object with a given signature Σ. The interpretation of Σ is supplied by the environment (hence, the term "oracle", used earlier in [RT, T2, P]).

(iii) Hybrid System. We leave this vague term for a structure whose components are hybrid automata which are combined via different interaction combinators.

About Terminology. Standardization of terminology for some of the basic notions in the area has become necessary with the proliferation of models and the often inconsistent and contradictory use of terms in the literature. The confusion is most blatant in the use of the terms "synchronous" and "asynchronous". For example, sometimes it serves to distinguish tight communication from buffered communication. What is a "synchronized" step for one author is "asynchronous" for another; what is an "asynchronous" circuit for one is "nondeterministic" for others (see Section 5). In this context, the author takes the blame for the terminological confusion in the paper ([T3]), where the terms "synchrony" and "simultaneity" were used improperly. Moreover, this confusion was aggravated by the unfortunate notation ϵ for the idling action, which resulted in misleading connotations with different versions of hiding.

Forced to decide now, we have striven to make the most neutral and natural choices. We define "synchronous" interaction to refer to lock-step state transition, and define "asynchronous" composition in interleaving style. We also suggest that the term "net" be used when memory is not shared, and (tentatively) propose the term "web" for the case when ports are private. This leads to a taxonomy for both discrete and real time automata comprising four basic structures: synchronous nets; asynchronous nets (message passing); synchronous webs; and asynchronous webs (shared variable architecture).

2 Automata

2.1 Preliminaries

a) Notational provisos

Z is a space (alphabet), T is the time-domain, i.e. the set N of nonnegative integers, or the set $R^{\geq 0}$ of nonnegative reals. A Z-path, is a function $\tilde{z}$ from T into Z, whose domain is a left-closed (possibly infinite) interval $[t, t+\delta)$ or $[t, t+\delta]$. $\tilde{Z}$ is the set of z-paths. The interval is said to be the *life-time* of $\tilde{z}$; its length δ is said to be the duration $|\tilde{z}|$ of $\tilde{z}$. A path is standard if $t = 0$. The standard shift of $\tilde{z}$ is the standard path whose value at time τ coincides with $\tilde{z}(t+\tau)$. In many respects we do not distinguish between paths with the same standard shift, and we preserve for them the same notation.

Unless stated otherwise, we confine ourselves to paths whose life-times are semi-intervals; their concatenation $\tilde{z}_1.\tilde{z}_2$ is defined in the usual way.

b) A reminder

A discrete-time deterministic automaton M is given by a (state- alphabet) X, an (action-alphabet) U and a map *nextstate* : $X \times U \longrightarrow X$. The associated *terminal transition map* Ψ of type $X \times \tilde{U} \longrightarrow X$ obviously extends *nextstate* from singleton action u to action sequence $\tilde{u} = u_1 ... u_l$. Finally, $\tilde{\Psi} : X \times \tilde{U} \longrightarrow \tilde{X}$ is the associated *full transition map.* When applied to state x and an action sequence $\tilde{u}$ it returns the state-sequence of length $l+1$, that starts with x and leads to the terminal state $x' = \Psi(x, \tilde{u})$.

Comments: In general, there are no restrictions on the cardinalities of U, X. For example, let X be the Euclidean space R^n, and let U consists of (names of) appropriate matrices. Then, *nextstate* may perform the corresponding linear transforms of R^n. But algebraic assumptions about X, U are not part of the model. The terminal and full transition maps are uniquely determined as soon as *nextstate* is given. But for continuous time domain $R^{\geq 0}$, *nextstate* does not make sense; hence, the need for a direct definition of terminal/full transitions.

2.2 Deterministic Automata

a) Basics

Even though the forthcoming formulations are general (hence, applicable for discrete time as well), they are intended mainly for the less routine continuous time-domain. (In [S1] these are also called Dynamic Systems or Machines).

An automaton M is given by a state- space X, an action-space U and a partial
terminal transition map $\Psi : X \times \tilde{U} \rightarrow X$.

The intended semantics is: if state x occurs at time t, then $\tilde{u}$ with life-time $[t, t+\delta)$ produces state x' at time $t+\delta$.

A finite path $\tilde{u}$ is admissible for x iff $\Psi(x, \tilde{u})$ is defined.

An infinite path $\tilde{u}$ is admissible for x if all its finite prefixes are admissible for x.

Below, $\tilde{u}_1, \tilde{u}_2 \in \tilde{U}$ are finite paths, $\tilde{u}_1.\tilde{u}_2$ designates their concatenation, and ϵ is the empty action path.

Axioms

(i) Non-triviality. For each state x there is a non-empty action path $\tilde{u}$, which is admissible for x.

(ii) Semi-group.

$$\Psi(x, \epsilon) = x$$

$$[\Psi(x, \tilde{u}_1) = x' \;\&\; \Psi(x', \tilde{u}_2) = x''] \rightarrow \Psi(x, \tilde{u}_1.\tilde{u}_2) = x''$$

(iii) Restriction (Density). Assume that $\Psi(x, \tilde{u}) = x''$. If $\tilde{u} = \tilde{u}_1.\tilde{u}_2$ then there exists x' such that $\Psi(x, \tilde{u}_1) = x' \;\&\; \Psi(x', \tilde{u}_2) = x''$.

Semi-group closure. An automaton M may be defined as follows:

(i) Consider Ψ for some set Ω of admissible paths, and check the non-triviality and restriction axioms;

(ii) Extend Ψ (and preserve this notation!) via concatenations of paths from Ω.

Full transition map of M. This is a function $\tilde{\Psi} : X \times \tilde{U} \longrightarrow \tilde{X}$, which returns a path $\tilde{x}$ with the same life-time $[0, \delta)$ as that of $\tilde{u}$. Namely,

$$\tilde{\Psi}(x, \tilde{u}) = \tilde{x} \;\; iff \;\; \forall t \in [0, \delta).\Psi(x, \tilde{u}|t) = \tilde{x}(t)$$

It follows from the axioms, that for (deterministic!) M, the definition is correct, i.e. $\tilde{\Psi}$ is indeed a function.

The pair $(\tilde{u}, \tilde{x})$ is a finite *trajectory* of M, and $\tilde{x}$ is a finite *state-path*

Note, that unlike Ψ, the extension of $\tilde{\Psi}$ (hence also of state-paths and of trajectories) to infinite duration is straightforward.

b) Flows

Definition 1. A flow on the state-space X is a function $f : X \times T \longrightarrow X$, that meets the following conditions:
(i) $f(x, 0) = x$; (ii) if $f(x, t)$ is defined, so is $f(x, t')$ for each $0 < t' < t$ and (iii) additivity:

$$f(x, t_1) = x' \;\& f(x', t_2) = x'' \;\; \longrightarrow f(x, t_1 + t_2) = x''$$

Notation. *nil* is the (polymorphic) trivial flow: $\forall t[nil(x, t) = x]$

Example 1. $f(x, t) \stackrel{\text{def}}{=} x + t.$

Clearly, a flow is nothing but the transition map Ψ of an automaton M, whose action-space is a singleton. *Flow* is a pure semantical notion, without any commitment to specific syntax. Note, however, that in Control Theory the favorite way to describe flows is via differential equations.

Example 2. Consider a finite-dimensional differential equation

$$\dot{x} = F(x), \tag{1}$$

where x is an n-dimensional vector. Assume the existence of unique solutions of this equation for arbitrary initial conditions and for arbitrary time intervals. (This can be guaranteed under appropriate technical assumptions, which are not relevant here.) The associated flow f is defined as follows:

Given a time interval $[t_0, t]$ and an initial state $x(t_0) = x_0$, consider the corresponding solution x of equation (1) on that interval. Then, $f(x_0, t) \stackrel{\text{def}}{=} x(t)$.

c) Structured spaces (alphabets)

Assume $W = V \times Z$, so that each path $\tilde{w}$ is uniquely defined by the pair $\{\tilde{v}, \tilde{z}\}$ of its projections into V and Z; we do not distinguish between path $\tilde{w}$ and that pair, for which we use notation $\tilde{v}\tilde{z}$. Specific delimiters and/or ordering restrictions may be used for additional information. For example $\tilde{v}/\tilde{z}$ or $\tilde{v} \longrightarrow \tilde{z}$ may characterize $\tilde{v}$ as an argument value, and $\tilde{z}$ as the corresponding function value.

In the sequel, the state-space X will be structured only as the Cartesian product of a set of *components* $X_1, X_2, \ldots, X_m$. On the other hand, the action space U may be structured via a set of *components* $U_1, U_2, \ldots, U_k$ in two different formats:

a) Multiplicative (Cartesian) format: $U = U_1 \times U_2 \ldots \times U_k$
b) Additive (disjoint sum) format: $U = U_1 + U_2 \ldots + U_k$

Respectively, two formats of admissible action paths will be considered.

(i) In the multiplicative format such a path belongs to $\tilde{U}$.
(ii) In the additive format an additional assumption is needed:

Interleaving structure. Each admissible action-path is the concatenation of "straight" paths $\tilde{u}_i \in \tilde{U}_i$. Hence (by semigroup closure), it suffices to define transitions only for *straight* paths.

Clearly, for discrete (but not for continuous) time the interleaving structure is guaranteed for all paths that belong to $\tilde{U}$.

2.3 Behavior of Initialized Automata

a) Retrospection

Notation. $\alpha|\tau$ is the prefix of path α restricted to life-time $[0, \tau)$.

Definition 2. f is a retrospective operator (shorthand: retrooperator) of type $\tilde{U} \rightarrow \tilde{Y}$ if it is defined on a prefix-closed subset of $\tilde{U}$ and satisfies the condition:

$$\text{If } \tilde{y} = f(\tilde{u}), \text{ then } \tilde{y}|\tau = f(\tilde{u}|\tau)\,.$$

In other words, for each t the value $\tilde{y}(t)$ depends only on the values of $\tilde{u}$ on the right closed interval $[0,t]$. When $\tilde{y}(t)$ does not depend on $\tilde{u}(t)$, the operator f is said to be ***strongly retrospective***.

Clearly, for each initial state $x_0 \in X$, the full transition map $\tilde{\Psi}$ induces a strong retrospective operator X, which is called the input/state behavior (i/o behavior) of the initialized automaton $< M, x_0 >$.

b) Accepted sets

The behavior of $< M, x_0 >$ can also be characterized by a set (language) which consists of all (or of an appropriate part of) the action paths (of the trajectories or the state-paths, respectively) admissible at x_0. This is the action (respectively, trajectory or state) set, accepted by $< M, x_0 >$. Most important is infinite behavior that obeys some reasonable fairness conditions. Fairness is beyond our subject, so, when referring to infinite paths, all admissible infinite paths are assumed.

If the action-space is structured as $U \times V \times ...$ then the accepted set is presented naturally by a characteristic relation $L(\tilde{u}, \tilde{v}, ...)$, called ***relational behavior*** of the automaton.

It may happen that for some partition of the arguments in L, the relational behavior is the graph of a function, so one could refer to the corresponding ***functional behavior***. Note that for a given L there may happen to be different "functional" partitions of this kind.

c) Transducers

Implicit transducers. Consider, for example, a retrooperator F of type $\tau = \tilde{U} \times \tilde{A} \longrightarrow \tilde{B}$. Let $< M, x_{init} >$ be an initialized automaton with state-space X and action alphabet $U \times A \times B$.

Definition 3. $< M, x_{init} >$ is a implicit transducer of type τ with input/output behavior (i/o behavior) F iff it accepts the graph of F (hence $\Psi_M(x_{init}, \tilde{u}\tilde{a}\tilde{b})$ is defined iff $\tilde{b} = F(\tilde{u}\tilde{a})$).

Remark. In the case above it might be convenient to use the mnemonic notation $\Psi_M(x_{init}, \tilde{u}\tilde{a} \;/\; \tilde{b})$ which points out the type of the intended behavior. Note, that an automaton M may happen to be typable in different ways as an implicit transducer.

Explicit transducers: strong and weak readouts. Let M be an automaton with spaces X, U. Consider in addition: (i) a space Y of output (or measurement) values, and (ii) a map $h : X \longrightarrow Y$. The pair $< M, h >$ is said to be an explicit transducer with underlying automaton M and ***strong readout*** map h. Let $G : \tilde{U} \longrightarrow \tilde{X}$ be the ***i/s*** behavior of the underlying automaton M. Then, the ***i/o***-behavior of the transducer $< M, h >$ is the retrooperator $F : \tilde{U} \longrightarrow \tilde{Y}$, defined as follows:

Assume $\tilde{x} = G(\tilde{u})$; then $F(\tilde{u})$ returns $\tilde{y}$ such that

$$\forall t \leq |\tilde{u}|.\ \tilde{y}(t) = h(\tilde{x}(t)) \quad (2)$$

Clearly, like G, the retrooperator F is also strongly retrospective.

A *weak readout* map h (not considered in [S1]!) is of type $X \times U \longrightarrow Y$. The definition of i/o-behavior in (2) should be modified, namely, $\tilde{y}(t) = h(\tilde{x}(t))$ is replaced by $\tilde{y}(t) = h(\tilde{x}(t), \tilde{u}(t))$. The i/o behavior is still retrospective, but strong retrospection is no longer guaranteed.

d) Comparing transducers and operators

Implicit transforms. Consider an explicit transducer $< M, h >$ with i/o behavior $F\colon\ \tilde{U} \times \tilde{A} \longrightarrow \tilde{B}$.

Assume that $< M, h >$ is specified by $\Psi : Q \times \tilde{U} \times \tilde{A} \longrightarrow Q$ and (for simplicity) by a strong readout $h : Q \longrightarrow B$.

Definition 4. Let M' be the automaton with transition map $\Psi' : Q \times \tilde{U} \times \tilde{A} \times B \longrightarrow Q$, defined below ($\delta$ is the common length of the paths):

$$\Psi'(q, \tilde{u}\tilde{a}/\tilde{b}, \delta) = q' \text{ iff } \tilde{\Psi}(q, \tilde{u}\tilde{a}, \delta) = \tilde{q} \ \ \& \ \ q' = \tilde{q}(\delta) \ \ \& \ \ \forall\, t < \delta.\ \tilde{b}(t) = h(\tilde{q}(t)).$$

Say that M' is the implicit transform of $< M, h >$, and $< M, h >$ is the explicit transform of M'.

It is easy to see that M' is an implicit transducer with i/o behavior F (the same as that of $< M, h >$).

Clearly, an implicit transducer is not necessarily the implicit transform of an explicit transducer.

Consider three properties of an operator F:

(i) F is a retrospective operator,
(ii) F is the i/o-behavior of an implicit transducer,
(iii) F is the i/o-behavior of an explicit transducer.

Proposition 5. *These properties are equivalent.*

That (ii) and (iii) imply (i) is trivial. The implication (iii) $\longrightarrow$ (ii) is due to implicit transforms. The other directions are not trivial.

2.4 Nondeterminism

Nondeterministic automata. If nondeterminism is not excluded a priori, then, instead of the terminal transition map Ψ one might expect a terminal transition relation $M(x, \tilde{u}, x')$. However, we confine ourselves below to the full-transition format $\tilde{M}(q, \tilde{u}, \tilde{q})$.

A particular case: $\exists$-automata. Assume that M is a deterministic automaton with input alphabet $U = U_1 \times U_2$ (hence, with input paths $\tilde{u}_1 \times \tilde{u}_2$) and with full transition map $\tilde{\Psi}$. The non-deterministic automaton (call it M') such that

$$\tilde{M}'(q, \tilde{u}_1, \tilde{q}) \text{ iff } \exists \tilde{u}_2[\tilde{\Psi}(q, \tilde{u}_1, \tilde{u}_2) = \tilde{q}]$$

is said to be a $\exists$-automaton.

Remark. Note that two $\exists$-automata may have the same terminal transitions but different full transitions.

Nondeterministic retrooperator f^*

Definition 6. A multi-valued function f^* is an $\exists$-*retrooperator* iff there exists an appropriate retrooperator f such that $\tilde{y} \in f^*(\tilde{u})$ iff $\exists\tilde{b}[\tilde{y} = f(\tilde{b}, \tilde{u})]$.

3 Interaction

Interacting agents are (possibly infinite) automata with structured spaces (alphabets).

For discrete-time automata the binary interaction combinators are defined in a routine way, via derivation of the *nextstate*-map (*nextstate*-relation) for the composition M from the *nextstate*-maps (*nextstate*-relations) of the components M_i. However, one should be careful with continuous time, when one needs to deal directly with terminal, or even with full transition maps. For these reasons we start with *nextstate* for discrete time, and then provide motivated definitions for terminal and full transition maps, which cover both discrete and continuous time.

Note that in all cases below, if the components are deterministic, so is their composition. Finally, the combinators are commutative and associative; hence, compositions may be considered for arbitrary sets $\{M_1, M_2, M_3, \ldots\}$ of components.

3.1 First Dichotomy of Interaction: Synchrony (multiplicative version) vs. Asynchrony (additive version)

a) Synchrony

For simplicity consider M_i with state-space $X_i \times X_0$ and action-space $U_i \times U_0$. Call the X_i – ports, and the U_i – registers. Note, that the components are allowed to share registers (here - X_0) and ports (here - U_0).

Definition 7. (Synchronous composition: $M = M_1 \times M_2$). The state-space of M is $X_1 \times X_0 \times X_2$, the action-space is $U_1 \times U_0 \times U_2$, and the transitions are as follows:

$$M(x_1x_0x_2, u_1u_0u_2, x_1'x_0'x_2') \stackrel{\text{def}}{=} M_1(x_1x_0, u_1u_0, x_1'x_0') \;\&\; M_2(x_2x_0, u_2u_0, x_2'x_0') \tag{3}$$

In particular, for deterministic M_i

$$\begin{aligned} &\mathit{nextstate}(x_1x_0x_2, u_1u_0u_2) = x_1'x_0'x_2' \text{ iff} \\ &\mathit{nextstate}_1(x_1x_0, u_1u_0) = x_1'x_0' \;\&\; \mathit{nextstate}_2(x_2x_0, u_2u_0) = x_2'x_0' \end{aligned} \tag{4}$$

Warning. Note, that (4) cannot be extended to terminal transition maps in the form

$$\Psi(x_1x_0x_2, \tilde{u}_1\tilde{u}_0\tilde{u}_2) = x_1'x_0'x_2' \text{ iff } \Psi_1(x_1x_0, \tilde{u}_1\tilde{u}_0) = x_1'x_0' \;\&\; \Psi_2(x_2x_0, \tilde{u}_2\tilde{u}_0) = x_2'x_0' \tag{5}$$

The reason is that in (5), for given $\tilde{u}_1\tilde{u}_0\tilde{u}_2$, the M_i are required only to reach the same terminal (!) value x_0', which would guarantee that $\Psi(x_1x_0x_2, \tilde{u}_1\tilde{u}_0\tilde{u}_2)$ is defined. However, the Restriction Axiom for Automata requires more, namely, definedness for all prefixes of $\tilde{u}_1\tilde{u}_0\tilde{u}_2$.

One possible remedy might be to use full transitions instead of terminal ones.

Definition 8. (For both discrete or continuous time).

$$\tilde{\Psi}(x_1x_0x_2, \tilde{u}_1\tilde{u}_0\tilde{u}_2) = \tilde{x}_1\tilde{x}_0\tilde{x}_2 \text{ iff } \tilde{\Psi}_1(x_1x_0, \tilde{u}_1\tilde{u}_0) = \tilde{x}_1\tilde{x}_0 \;\&\; \tilde{\Psi}_2(x_2x_0, \tilde{u}_2\tilde{u}_0) = \tilde{x}_2\tilde{x}_0 \tag{6}$$

The way to handle terminal transitions is as follows:

Definition 9. (For both discrete and continuous time). If the components M_i don't share registers, then

$$\Psi(x_1x_2, \tilde{u}_1\tilde{u}_0\tilde{u}_2) = x_1'x_2' \text{ iff } \Psi_1(x_1, \tilde{u}_1\tilde{u}_0) = x_1' \;\&\; \Psi_2(x_2, \tilde{u}_2\tilde{u}_0) = x_2' \tag{7}$$

Let $act(M)$ denote the action set accepted by M.

Proposition 10. (Restorability of $act(M)$). *Assume that the M_i don't share registers.*

Let $L_i(\tilde{u}_i, \tilde{u}_0)$ be the characteristic predicate of $act(M_i)$. Then

$$L(\tilde{u}_1, \tilde{u}_0, \tilde{u}_2) = L_1(\tilde{u}_1, \tilde{u}_0) \;\&\; L_2(\tilde{u}_2, \tilde{u}_0)$$

is the characteristic predicate of $act(M)$

b) Asynchrony

This interaction combinator is considered for automata with interleaving structure (see Sec. 2.2c).

Consider first discrete time.

Definition 11. (Asynchronous composition: $M = M_1 \| M_2$). Assume M_1 with spaces $X_1 \times X_0$, $U_1 + U_0$, and M_2 with spaces $X_2 \times X_0$, $U_2 + U_0$. Then M has spaces $X_1 \times X_0 \times X_2$, $U_1 + U_0 + U_2$ and $nextstate(x_1x_0x_2, u_i) = x_1'x_0'x_2'$ holds in one of the cases:

$$nextstate_1(x_1x_0, u_i) = x_1'x_0' \;\&\; x_2' = x_2 \quad if \;\; i = 1 \tag{a}$$

$$nextstate_2(x_2x_0, u_i) = x_2'x_0' \;\&\; x_1' = x_1 \quad if \;\; i = 2 \tag{b}$$

$$nextstate_1(x_1x_0, u_i) = x_1'x_0' \;\&\; nextstate_2(x_2x_0, u_i) = x_2'x_0' \quad if \;\; i = 0 \tag{c}$$

Again, one should be careful about the extension from *nextstate* to terminal and full transition maps Ψ and $\tilde{\Psi}$.

The way to deal with terminal transitions, is as follows:

Definition 12. (No shared ports). Assume M_1 with spaces $X_1 \times X_0$, U_1, and M_2 with spaces $X_2 \times X_0$, U_2. Then M has spaces $X_1 \times X_0 \times X_2$, U_1+U_2, and (up to semi-group closure) the following terminal transition map: $\Psi(x_1x_0x_2, \tilde{u}_i) = x'_1x'_0x'_2$ holds in one of the cases:

$$\Psi_1(x_1x_0, \tilde{u}_i) = x'_1x'_0 \;\&\; x'_2 = x_2 \quad if \quad i = 1 \tag{a}$$

$$\Psi_2(x_2x_0, \tilde{u}_i) = x'_2x'_0 \;\&\; x'_1 = x_1 \quad if \quad i = 2 \tag{b}$$

Definition 13. (No shared registers). Assume M_1 with spaces X_1, U_1+U_0 and M_2 with spaces X_2, U_2+U_0. Then $\Psi(x_1x_2, u_i) = x'_1x'_2$ holds in one of the cases:

$$\Psi_1(x_1, \tilde{u}_i) = x'_1 \;\&\; x'_2 = x_2 \quad if \quad i = 1 \tag{a}$$

$$\Psi_2(x_2, \tilde{u}_i) = x'_2 \;\&\; x'_1 = x_1 \quad if \quad i = 2 \tag{b}$$

$$\Psi_1(x_1, \tilde{u}_i) = x'_1 \;\&\; \Psi_2(x_2, \tilde{u}_i) = x'_2 \quad if \quad i = 0 \tag{c}$$

3.2 Second Dichotomy of Interaction: Nets vs. Webs

a) Basic architectures

This dichotomy reflects available communication mechanisms:

Nets: communication, if any, is via shared ports; all registers of a component are private.

Webs: communication, if any, is via shared registers; all ports of a component are private.

Note that for nets, because of the privacy of registers, it can be assumed (without loss of generality) that components have unique registers. Similarly for ports in a web. The two dichotomies induce four "architectures":

	×	∥
private registers	synchronous net	asynchronous net
private ports	synchronous web	asynchronous web

Remark. Terminal transition maps do not fit directly with architecture 3 (synchronous webs). In this case the formulation would require first the consideration of full transitions (see Warning in Sect. 3.1a).

b) Petri graphs

This is a bipartite graph G (possibly directed) with k circle-nodes and m box-nodes. G is an atomic net if it consists of a single circle with its neighboring boxes; it is an atomic web if it consists of an unique box with its neighboring circles. Hence, two dual decompositions of G: into k atomic subnets or into m atomic subwebs. G is said to be a net or a web if it is equipped with the corresponding decomposition.

Figure 1 shows an atomic net with circle labeled Q and an atomic web with box labeled A.

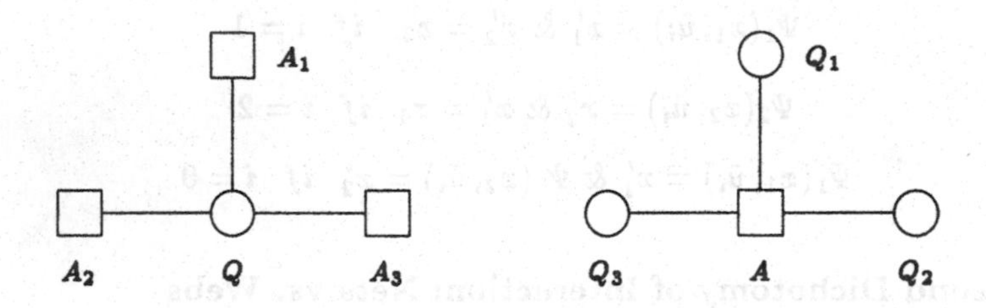

Fig. 1. Nets versus Webs.

Example 3. Consider the Petri graph shown in Figure 2. It has two dual decompositions. Figure 3 shows the decomposition into three atomic nets which "communicate" via shared boxes. Figure 4 shows the decomposition into two atomic webs which "communicate" via shared circles.

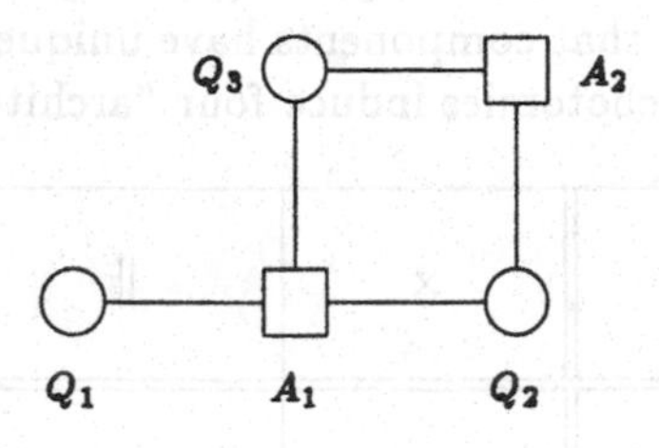

Fig. 2. A Petri graph.

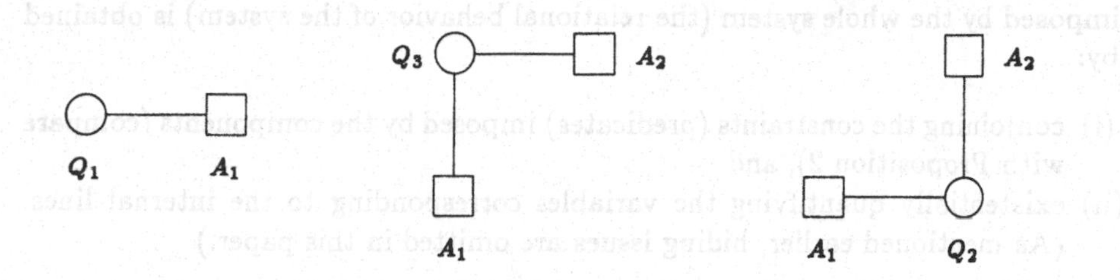

Fig. 3. Decomposition into three atomic nets.

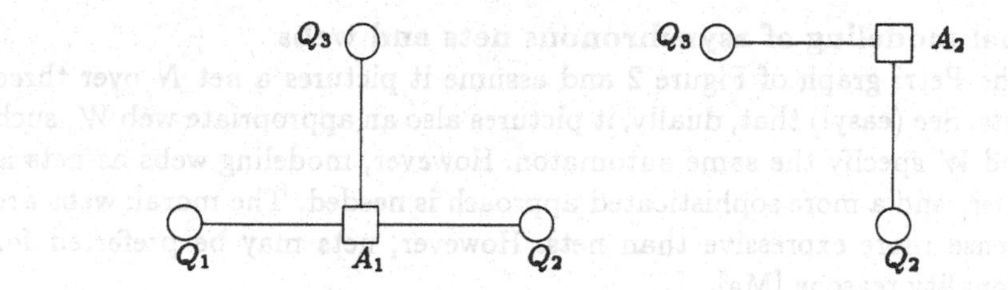

Fig. 4. Decomposition into two atomic webs.

c) Nets and webs of automata

Appropriately labeled nets or webs offer a suggestive graphical syntax for paradigms which involve sharing and/or communication.

A net of automata $N(M_1, \ldots, M_k)$ is an entity given by:

(i) A net N with numbered circles: $c_i : i = 1, 2, \ldots, k$.

(ii) A map *env* which *correctly* assigns to each c_i an automaton M_i; this means that the ports of M_i are in $1-1$ correspondence with the neighboring boxes of N. It is assumed that each M_i has a single register, which is its private register.

Semantics: if $M_1 \times \ldots \times M_k = M$, then say that M is decomposed as (specified by) the synchronous net $N(M_1, \ldots, M_k)$. Similarly, for asynchrony.

Webs of automata $W(M_1, \ldots, M_m)$ are handled in the dual way. In particular:

(i) $b_1, \ldots, b_m$ is an enumeration of the boxes in W.

(ii) Each of the automata has an unique (private!) port, and *env* correctly assigns automata to boxes.

Remark. Synchronous nets are commonly used in hardware specification, where ports correspond to 'pins' of the physical component devices ([G]). The *relational behavior* of a device may be specified by defining a predicate $\text{Dev}(a_1, \ldots)$, which holds iff $a_1, \ldots$. are allowable values on the corresponding lines (ports). Note that the values on the lines can be modelled with infinite paths. The constraint imposed by the whole system (the relational behavior of the system) is obtained by:

(i) conjoining the constraints (predicates) imposed by the components (compare with Proposition 2), and
(ii) existentially quantifying the variables corresponding to the internal lines. (As mentioned earlier, hiding issues are omitted in this paper.)

Recall that in the Petri Net community circles are called places, and boxes are called transitions. However, having in mind the *env* maps above, we will refer occasionally to the circles (boxes) of a Petri graph as registers (ports).

d) Mutual modeling of asynchronous nets and webs
Look at the Petri graph of Figure 2 and assume it pictures a net N over three components. See (easy!) that, dually, it pictures also an appropriate web W, such that N and W specify the same automaton. However, modeling webs as nets is much harder, and a more sophisticated approach is needed. The moral: webs are in some sense more expressive than nets. However, nets may be preferred for compositionality reasons [Ma].

e) Directed Petri-graphs
Assume that the edges in G are oriented (directed). A box b is an output-box of circle c, if there is a directed edge (channel) from c to b. Similarly, for input boxes of c. Hence, for given c we have the partition $< in(c), out(c) >$ into its input and output boxes. If b is an output-box of some $c \in G$, then it is said to be an output box of G; otherwise it is an input-box of G. Hence, we also have a partition of all boxes of G into $in(G)$ and $out(G)$. We call this partition the port-type of G, and use for it the notation $in(G) \longrightarrow out(G)$. Similarly, is defined the register type of G. In the sequel we focus on directed nets of automata, whose ports are partitioned into input and output ports, and require that *env* respects the status of these ports (i.e. input ports correspond to neighboring input boxes etc.).

Orientation of the edges may be used for additional information about the components of the net or of the web. But note that it does not affect the semantics of synchrony and asynchrony.

For example, can consider a directed synchronous net of implicit transducers (see Section 2.3c). But note that (meanwhile) it would specify an automaton without commitments to specific i/o behavior. The term "Circuit" will be used below for directed nets, whenever the type of the net is semantical relevant.

3.3 Circuits

Circuits offer a suggestive pictorial representation for specific systems of equations and for related nets of functions, or of transducers.

a) Systems of equations vs. directed nets of functions
Consider a system Eq of equations Eq_i with the format $x_i = f_i(y_1, ..., y_{n_i}); i = 1, ..., k$. The variables occurring only on the right hand side of an equation (say

$v_1, ..., v_m$) are declared as input variables of Eq, the others (say $w_1, ..., w_n$) - as its output variables.

It is appealing to represent Eq as a directed net of functions $N_{Eq}(f_1, ..., f_k)$, whose registers $r_1, ..., r_k$ are labeled by the function symbols $f_1, \ldots, f_k$, and whose ports are labeled by the corresponding argument symbols.

Clearly, the input (output) ports of N_{Eq} will be labeled by the input (output) variables of Eq. We refer to $N_{Eq}(f_1, ..., f_k)$ also as a circuit of functions, with port-type $v_1, ..., v_m \longrightarrow w_1, ..., w_n$.

Under appropriate interpretation I of the functional symbols, the conjunction $\bigwedge Eq_i$ defines an $m+n-$dimensional relation, which may happen to be the graph of a function ϕ with type $v_1, ..., v_m \longrightarrow w_1, ..., w_n$. In this case say that Eq and also the net of functions N_{Eq} are *reliable* (under the intended interpretation I) and define ϕ. Say also that they define each of the corresponding m-place functions ϕ_i with type $v_1, ..., v_m \longrightarrow w_i$; $1 \leq i \leq n$

b) Circuits $C(f_1, ..., f_k)$ of retrospective operators

In the seminal paper [BW], Burks & Wright initiated the investigation of the fundamental case, when the functions f_i in Eq (and in N_{Eq}) are retrospective operators (see Def. 2); moreover, the defined function ϕ is also required to be retrospective.

Say that p is a confluence port (register) in a directed Petri Graph G, if it the output port of different registers (ports). It was observed in [BW] that confluent ports have the semantical effect of "backward passage of causal influence", which harms reliability. Hence, the following constraint is assumed from now on:

(i) For circuits: no confluent ports.

(ii) For Eq: no two equations in Eq may share left-hand side variable.

For discrete time retrooperators the following is well known and easy;

Proposition 14. *If every oriented cycle in the circuit $C(f_1, ..., f_k)$ (confluence constraint assumed) passes through a strong retrospective operator f_i, then C is reliable.*

Remark. For continuous time this claim is wrong, and the search for sufficient conditions of reliability is a non trivial task. In [PRT] it is considered for specific retrospective operators (see right-continuous operators in the next section).

c) Circuits of $\exists$-retrooperators (For $\exists$-retrooperators, see Sec. 2.4).

These circuits, their input and output variables, as well as their semantics, differ from those for single-valued operators in the obvious way: instead of equations with format " variable = term " one considers inclusions with format " variable $\in$ term ". Say that the circuit $C(f_1^*, ..., f_k^*)$ is reliable iff the relation it defines is the graph of a $\exists$-*retrooperator* f.

d) Circuits of transducers

Circuit $C(M_1', ..., M_k')$ of implicit transducers. This is a synchronous directed net of transducers, which obeys the confluence constraint.

Assume that the M_i' define respectively the retrooperators f_i, and let $C(f_1, .., f_k)$ be the corresponding circuit of retrooperators.

Proposition 15. (i) *The circuit* $C(M'_1, ..., M'_k)$ *specifies a deterministic automaton* M'; (ii) M' *is an implicit transducer (of the respective type) iff the circuit of operators* $C(F_1, .., F_k)$ *is reliable.*

Circuit $C(< M_1, h_1 >, ..., < M_k, h_k >)$ **of explicit transducers.** We omit the formal definition, but consider the circuit of the corresponding implicit transforms (see Section 2.3) $C(< M'_1, ..., M'_k >)$ as well as the automaton M' it specifies.

Proposition 16. *Assume that* M' *is an implicit transducer, which defines a retrooperator* ϕ. *Then* M' *is the implicit transform of an explicit transducer* $< M, h >$ *which defines* F.

Comments

(i) Proposition 5 (see, sect. 2.3) guarantees only that there exists an explicit transducer $< M, h >$ which has the same *i/o*-behavior as M', but $< M, h >$ is not claimed to be the explicit transform of M'.

(ii) In the general case (reliability not guaranteed), the pair $< M, h >$ above with underlying $\exists$-automaton M might be considered as some kind of non-deterministic explicit transducer, whose i/o-behavior is $\exists$-retrospective.

(iii) Clearly, for circuits with no feedback cycles, reliability problems don't occur at all. This is essentially the case with the "deterministic timed automata" from [AFH].

4 Toward Hybrids

4.1 Finite Variability

Say that $\tilde{z}$ is an elementary path with duration δ iff for some $a, b \in Z$ there holds:

$$\tilde{z}(0) = a;\ \tilde{z}(\tau) = b \text{ for } 0 < \tau < \delta$$

The corresponding notation is $\tilde{z} =< a \bullet b, \delta >$. A path $\tilde{z}$:

(i) is continuous at time-instant t if there is an open time -interval, containing t, in which $\tilde{z}$ is constant; otherwise it changes (is discontinuous) at t.

(ii) has finite variability (is a FV-path) if on each finite subinterval (of its lifetime) it changes at most at a finite set of time instances.

(iii) has variability $< k$ if it changes less than k times in each subinterval with length ≤ 1.

(iv) has latency $\geq \alpha$ if whenever it changes at instants t_1, t_2 then $|t_1 - t_2| \geq \alpha$.

(v) is J-sparse (for some $J \subseteq Z$) if values from J occur in each finite subinterval at most at a finite set of time-instances.

A retrooperator f has finite variability if it maps FV-paths into FV-paths. Let out_f be the set of all paths outputted by f. Say that f has bounded variability (bounded latency) iff for some constant c all paths in out_f have variability $< c$ (latency $> c$).

Remark. See that bounded latency implies bounded variability, but the inverse fails.

Clearly, a path $\tilde{z}$ with finite (respectively - infinite) duration has ***finite variability*** if it can be represented (we say also - *scanned*) as the concatenation of a finite (respectively - infinite) sequence of elementary path-components. In the case of infinite duration it is usually called the *non-Zeno* property; it means that there exists an increasing sequence $t_i \longrightarrow \infty$ such that $\tilde{z}$ is constant on every interval (t_i, t_{i+1}).

In particular, $\tilde{z}$ is a ***burst-path*** if everywhere beyond the t_i it has a fixed value, usually designated as *nil*. Note, that burst-paths encode ***timed sequences*** ([AD]). $\tilde{z}$ is a ***right-continuous*** path if all the elementary paths in the scanning are constant, i.e. have the format $a \bullet a$.

M is a FV- automaton (RC-automaton) if its action paths have finite variability (are right-continuous). In this case, up to semigroup closure, it suffices to define the terminal transition function Ψ only for elementary action-paths. Use for them the self-explanatory notations $\Psi(x, u \bullet u', \delta)$ and $\Psi(x, u, \delta)$.

We consider now some kinds of FV-automata with finite action-space (alphabet).

4.2 Flows and Jumps

a) Flow automata

Definition 17. M is a flow-automaton if to each $u_i \in U$ there corresponds a flow (see, Def. 1) denoted $||u_i||$, such that whenever $\Psi(x, u_j \bullet u_i, \delta) = x'$ there holds (for arbitrary u_j !) $x' = ||u_i||(x, \delta)$.

It is easy to see that each $RC-$automaton M is a flow-automaton.

Proposition 18. (trivial). *Flow-automata have the following property: If $\Psi(x, \tilde{u}_i) = x'_i$, and the $\tilde{u}_i$ differ in at most a finite set of time instants, then the x'_i coincide.*

Example 4. Consider the differential equation

$$\dot{x} = F(x, u), \tag{8}$$

where x, u are respectively n and m dimensional vectors. Under appropriate conditions, equation (8) has unique solutions for a rich class of $U-$paths $\tilde{u}$, including all FV-paths. Compare with *Flows* in 1.1, and see that the automaton associated with the equation above is indeed a flow automaton.

b) Jump/flow automaton M

Definition 19. (the sequential version). Assumptions about the action paths:

(i) U is the disjoint union $JUMP \bigcup FLOW$ of so called jump-alphabet $JUMP = \{j_1, ..., j_k\}$ and flow-alphabet $FLOW = \{f_1, ..., f_m\}$.

(ii) Admissible action-paths are $JUMP$-sparse (see, sect. 4.1). Hence, the elementary action-paths have the format (up to duration δ): $f_i \bullet f_s$ *(pure flow)* or $j_l \bullet f_s$ *(flow after jump)*

(iii) Semantics: with each j_l is associated a jump, i.e. a function $\|j_l\| : X \longrightarrow X$, and with each f_s a flow. The elementary transitions are as follows (note that the value of f_i does not matter):

$$\Psi(x, f_i \bullet f_s, \delta) = f_s(x, \delta); \quad \Psi(x, j_l \bullet f_s, \delta) = f_s(j_l(x), \delta)$$

Definition 20. (Jump/flow automaton M': the parallel version). The intuitive idea is that M' behaves essentially as M, up to the following modification:

M' inputs in parallel a path for flows and a path for jumps. Here is the formal definition:

Let $JUMP, FLOW$ be as in the sequential version, and let $JUMP' = JUMP \bigcup \{nil\}$. The input-space of M' is $U' \stackrel{\text{def}}{=} JUMP' \times FLOW$. The admissible elementary action-paths are pairs (α, β), where

(i) β is an elementary path over $FLOW$;

(ii) α is $< nil \bullet nil, \delta >$ (trivial) or $< j_l \bullet nil, \delta >$ (jump).

The transition map Ψ' is defined as follows:

$$\Psi'(x, nil \bullet nil, f_i \bullet f_s, \delta) = f_s(x, \delta); \quad \Psi'(x, j_l \bullet nil, f_i \bullet f_s, \delta) = f_s(j_l(x), \delta) \quad (9)$$

Remark. Consider the following transform $\varphi : JUMP' \times FLOW \longrightarrow JUMP \bigcup FLOW$:

$$\varphi(nil, f_i) = f_i; \quad \varphi(j_l, f_i) = j_l$$

See that the definition of M' is chosen in such a way that the transform φ induces an isomorphism between M, M'. Namely, the transition map Ψ' is reduced to the map Ψ of M as follows:

$$\Psi'(x, nil \bullet nil, f_i \bullet f_s, \delta) = \Psi(x, f_i \bullet f_s, \delta); \quad \Psi'(x, j_l \bullet nil, f_i \bullet f_s, \delta) = \Psi(x, j_l \bullet f_s, \delta); \quad (10)$$

c) Examples of jump/flow automata

Example 5. (Two clocks with resetting).

The AD-clock (see [AD]): State-space $X = R^1$; initial state 0.
Unique flow f with $[\|\|f\|\|(x)](t) = x + t$ and unique jump *reset* with $\|reset\|(x) = 0$.

The P-clock (periodic clock, see [P]): $X = [0, 1]$; initial state 0.
Unique flow f with $f(x, t) = x + t$ *if* $x + t \leq 1$; $f(x, t) = x + t(mod\ 1)$ *if* $x + t > 1$. Unique jump j with $j(x) = 0$.

Example 6. (A dynamic system with control u and disturbance j).

This is specified as a jump/flow automaton (parallel version) by equation

$$\dot{x} = f(x, u, j) \quad (11)$$

where u provides the paths for flows, and j for jumps.

d) Jump-automata

Clearly, flow-automata present the extreme case of jump-flow automata when $JUMP$ is empty, or, said otherwise, it is the singleton $\{j_0\}$ with $||j_0||(x) = x$. Another extreme case is presented by *Jump-automata*, where $F = \{nil\}$. Hence, elementary action-paths are $JUMP$-sparse paths of two kinds:

(i) $(nil \bullet nil, \delta)$. This is the trivial flow, i.e $\Psi(x, nil \bullet nil, \delta) = x$ for arbitrary δ.
(ii) $(j_i \bullet nil, \delta)$ with. $\Psi(x, j_l \bullet nil, \delta) = ||j_l||(x)$.

Mutual modeling of a discrete-time automaton M and a jump-automaton M^j. Assume M with state-space X, action-alphabet $J = \{j_1, ..., j_k\}$ and nextstate-function Ψ. Then M^j has state space X, action-alphabet $J' \stackrel{\text{def}}{=} J \bigcup \{nil\}$, and the following elementary transitions for arbitrary δ

$$\Psi^j(x, j \bullet nil, \delta) = x' \text{ iff } \Psi(x, j) = x'; \quad \Psi^j(x, nil \bullet nil, \delta) = x.$$

Clearly, the modeling above is reversible; it points to the natural correspondence between discrete-time automata and continuous-time jump-automata. Moreover, it is faithful w.r.t. asynchrony interaction in the following sense.

Proposition 21. *Let M_i be discrete-time automata, and M_i^j the corresponding jump-automata. Then, for both nets and webs there holds:*

$$(M_1||M_2)^j \;=\; M_1^j||M_2^j$$

4.3 Duration Independence

a) Basics

Definition 22. An automaton M with finite variability is *duration independent* if whenever $M(q_1, a \bullet b, \delta, q_2)$ holds for some duration δ, it holds for arbitrary δ.

Proposition 23. *If M is duration independent, then it respects continuity, i.e. if $\tilde{u}$ is continuous at some $t > 0$, then the state-path $\tilde{\Psi}(q_0, \tilde{u})$ is also continuous at t.*

Proposition 24. (Consequence.) *The i/s behavior of M is a retrooperator with finite variability; moreover, it improves variability in the sense: on each time subinterval the state does not change more times than the action.*

Proposition 25. ([R]) *Every automaton M with finite variability and with finite state-space is duration independent.*

Proposition 26. (trivial). *If M is a jump-automaton then it is duration independent; but the inverse implication is not true.*

b) Graphical representation

The graphical representation of a transition with elementary action-path is as in Fig. 5(i). If the action-alphabet is structured, the standard representation is as in Fig. 5(iv), but Fig. 5(vi-vii) are more suggestive when $a_1 \bullet b_1$, $a_2 \bullet b_2$ are intended respectively as the argument and the value of a function.

Warning. Some authors prefer *bilabeling* graphical representations, in which both edges and nodes are labeled. For example, in [ACHP, H] instead of Fig. 5(i) would appear Fig. 5(ii), and instead of each of the Figures 5(iv), 5(vi) or 5(vii) would appear 5(v). The direct translation of our standard labeling to bilabeling is not always possible, as hinted by Fig. 5(iii). In this sense bilabeling does not seem to be enough universal. Nevertheless, a more indirect translation is still possible.

Labeling Transition Systems (LTS) for duration independent) automata.

Example 7. Fig. 6(i) presents such an LTS for a duration independent automaton with states l_0, l_1, and with five values in the action-space $H = \{h_1, h_2, h_3, h_4, h_5\}$. Clearly, there is no need to include duration-labels. Here, $H' = \{h_1, h_2, h_3\}$; $H'' = \{h_3, h_4, h_5\}$, and self-explanatory shorthands are used to save explosion of labeled edges. For example

$$H' \bullet H'' \stackrel{\text{def}}{=} \{h_i \bullet h_j\} \ for \ \ h_i \in H', h_j \in H'' \tag{12}$$

It is easy to see that M is deterministic. However it is not complete. See that among 25 possible elementary inputs, only 12 are enabled at state l_1; namely, 3 in $h_4 \bullet H''$ and 9 in $H' \bullet H'$.

4.4 Controllers and Plants

These are two kinds of transducers, intended to serve as components of hybrid automata. The simplest form of hybridity is displayed in the way jumps and flows coexist (apart) in Plants and Controllers, even before their including into an interacting system.

a) FV-transducers

Let T be a transducer, which arises when a FV-automaton M is equipped with appropriate readout h.

For a finite measurement space $H = \{h_1, ..., h_k\}$, the input map $h : X \longrightarrow H$ induces a partition of the state space X into a family of k disjoint subsets, called regions induced by h. By abuse of notation we designate by h_i also the region $\{x | h(x) = h_i\}$ and its characteristic predicate; hence, we don't distinguish between: $h_i(x)$ and $x \in h_i$.

In general, the output paths of transducer T don't necessarily obey the FV-requirement. If indeed they do, say that T is a FV-transducer; in this case its i/o behavior is a FV-retrooperator (see, sect. 4.1).

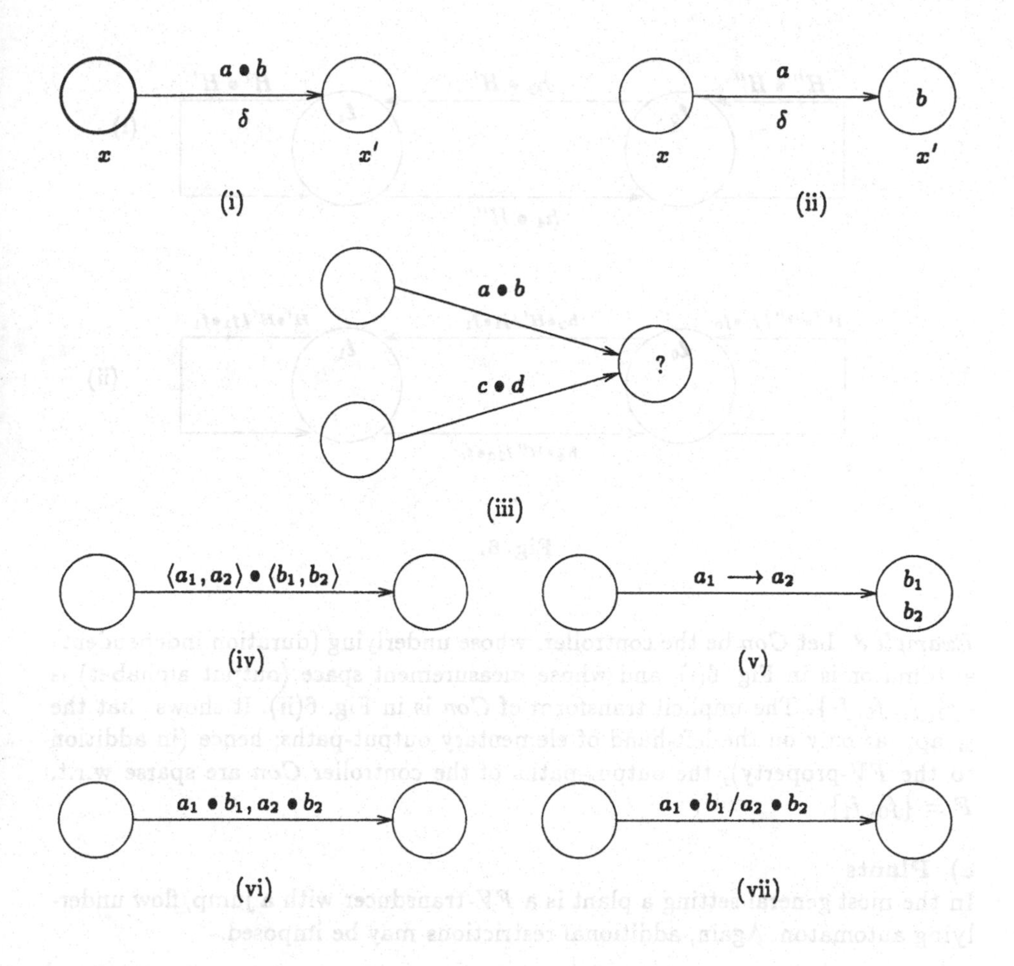

Fig. 5.

Remark. Consider the implicit transform M' of T (see, sect. 2.3d), and assume that the action apace of the underlying automaton M is $\{u_1, u_2, \ldots\}$. It is easy to see that the elementary transitions of M' have the format:
$M'(x_0, u_m \bullet u_n / h_s \bullet h_r, \delta, x_1)$.

Below are examples of FV-transducers. Occasionally, the outputs may meet properties that are even stronger than FV, like bounded variability/latency, spareness w.r.t. some sets (of values) (see, sect. 4.1).

b) Controllers

It follows from Proposition 24 that duration independent transducers are FV-transducers, and, moreover, they improve variability.

In the most general setting, a controller is a duration independent transducer. Occasionally, additional restrictions may be imposed, say finite state-space may be required.

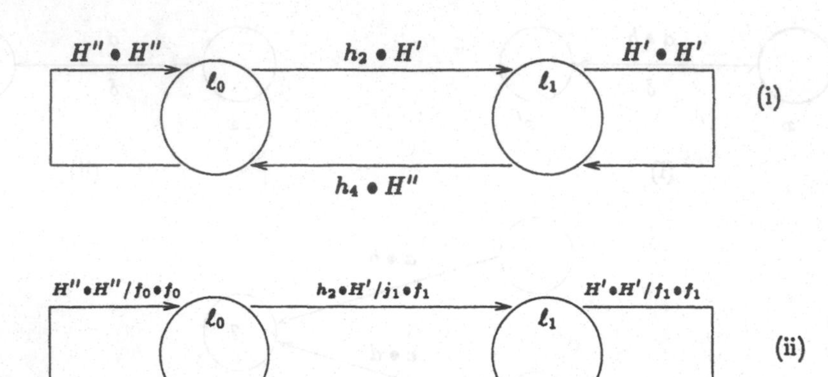

Fig. 6.

Example 8. Let *Con* be the controller, whose underlying (duration independent) automaton is in Fig. 6(i), and whose measurement space (output alphabet) is $\{j_0, j_1; f_0, f_1\}$. The implicit transform of *Con* is in Fig. 6(ii). It shows that the j_l appear only on the left-hand of elementary output-paths; hence (in addition to the *FV*-property), the output-paths of the controller *Con* are sparse w.r.t. $F = \{f_0, f_1\}$.

c) Plants

In the most general setting a plant is a *FV*-transducer with a jump/flow underlying automaton. Again, additional restrictions may be imposed.

Example 9. Timers.

Timers are transducers which appear when clocks with resetting are equipped with appropriate readouts. In particular:

(i) AD-timer ([AD]): this is the AD-clock with the following weak readout: $h(x, reset) = 0$; $h(x, f) = 1$ *for* $x > 1$, *and* $= 0$, *otherwise*.
Easy to see that AD-timer has output with bounded variability, but not with bounded latency. Its i/o behavior is retrospective, but not strongly retrospective.

(ii) P-timer ([P]): this is the P-clock with the following strong readout: $h(1) = 1$, $h(x) = 0$ *for* $x \neq 1$

Note that P-Timer is strongly retrospective (see, Def. 12), and has output with bounded latency.

Example 10. The transducer P.

This is a transducer $< M, h >$, whose underlying automaton M is a jump/flow automaton, equipped with a strong readout h.

Signature of $\langle M,h\rangle$: $\langle X; j_0, j_1; f_0, f_1; h_1, h_2, h_3, h_4, h_5\rangle$

Interpretation.

(i) State-space $X = R$(Euclidean)
(ii) Jumps. $j_0(x) = x + 1$; $j_1(x) = x + 2$.
(iii) Flows. $f_0(x,t) = x.e^{-\kappa t}$ (specified by the differential equation $\dot{x} = -\kappa x$), and f_1 (specified by the equation $\dot{x} = \kappa(h - x)$).
(iv) Strong readout h with following regions: $h_1(x)$ iff $x < 5$; $h_2(x)$ iff $x = 5$; $h_3(x)$ iff $5 < x < 10$; $h_4(x)$ iff $x = 10$; $h_5(x)$ iff $x > 10$.

Remark. **The i/o behavior of P is strongly retrospective, and the output-paths have bounded variability.**

The elementary transitions of the implicit transform of P (see Sect. 2.3d) have the format: $P(x_0, j_m \bullet f_n/h_s \bullet h_r, \delta, x_1)$ or $P(x_0, f_m \bullet f_n/h_s \bullet h_r, \delta, x_1)$.

4.5 Hybrid Automaton

This is (essentially!) a circuit $C(Plant, Controller)$ of specific transducers. We preserve notation C for the deterministic automaton specified by the circuit, i.e.

$$C = Plant \times Controller \qquad (*)$$

Some comments are in order.

Note that synchronous composition preserves finite variability (FV).

Hence, C is a FV-automaton (see, Sect. 4.1) and it accepts (see Sect. 2.3b) a set (language) L of FV-paths. Here is the point where reliability considerations enter the play: is L the graph of a FV-retrooperator which fits the type of the circuit C?

Look, for example, at the case when Controller and Plant are the transducers P and Con above (see 4.4b and 4.4c). Note, that they have a common alphabet $A \stackrel{\text{def}}{=} H \bigcup JUMP \bigcup FLOW$, and, of course, different state-spaces. In this particular case, C is outputless; hence, reliability (if any) would guarantee the existence and uniqueness of a FV-trajectory in the space $A \times X$. Otherwise, we would know only, that C is deterministic, i.e. for each A-path there cannot be more than one X-path.

Reliability holds indeed, and that can be argued with the help of the following facts:

(i) P is strong retrospective, and displays bounded variability of the output.
(ii) The outputs of Con are sparse (see, sect. 4.1) w.r.t. $J = \{j_0, j_1\}$.

Arguments of this kind underly some (sufficient) reliability conditions considered in [PRT].

Even though we focus on two components, the extension to many Plants (Interfaces, Timers, etc.) and/or many controllers is straightforward. Note also, that duration independence of controllers is a more general feature than discreteness or finite memory. Indeed, in the continuous time setting, discrete automata

are modeled by jump-automata (see Sect. 4.2a), and each finite-state automaton is duration independent (see Sect. 4.3a).

Timed Automata are a particular case of Hybrid Automata, whose Plants are different kinds of Timers. Actually, the seminal paper [AD], promoted versions of $AD-$timer, which is not strongly retrospective. The consideration of P-timer is motivated by its properties (strong retrospection and bounded output latency), which support reliability. Note that the "classical" timed-automata were developed as accepting devices, and, hence were not committed to reliability requirements.

Note also that if the controller is nondeterministic it is handled as an appropriate $\exists$-automaton.

Below we list some popular examples of Hybrid Automata ([ACHP]) and point out their decomposition into components.

Example A. Temperature Controller.
The Plant: state space R, no jumps, decreasing flow f_1 and increasing flow f_0.

Controller: deterministic, with two states.

Comment. Actually, our main example with Plant P and Controller Con is an adaptation of Example A, up to the following deliberate deviations from the [ACHP] text:

(i) Jumps j_0, j_1 added;

(ii) The controller is represented by a detailed labeled transition system (Fig.6(ii)), which displays all the (relevant) elementary transitions.

Example B. Water Level Monitor.
Plant: state space R^2, $jump(x, y) = (0, y)$, two flows.

Controller: deterministic, with four states.

Example C. Leaking Gas Burner.
Plant: state space R^3, $jump(t, x, y) = (t, y)$, two flows. $flow_1(t', x, y)(t) = (t' + t, x + t, y + t)$; $flow_2(t', x, y) = (t', x + t, y + t)$

Controller: nondeterministic, with two states.

4.6 Hybrid System

This is a structure whose components are Hybrid Automata $H_1, ..., H_k$, which are combined via different interaction combinators.

Examples of Hybrid Systems ([ACHP])

Example D. *MUTEX* - the Mutual Exclusion Protocol.
Allegation: Considered *asynchronous shared – memory system* consisting of two processes P_1, P_2, which share variable k. The system modeled by the *product* of two hybrid systems presented in Fig. 4 of [ACHP]

We handle *MUTEX* as the asynchronous web of two hybrid automata HA_1 and HA_2, each of them being structured as $Timer \times Controller \times Counter$. Note that in this case the Plant is structured as the composition of two sub-plants, *Timer* and *Counter*.

Example E. Railroad Gate Control.
This system consists of three components. The Controller Automaton and the Gate Automaton are hybrid automata. The Train Automaton is not; instead of manipulating explicit flows it refers only to "flow conditions" imposed on flows.

5 Discussion

We comment briefly on some papers and focus on the comparison with the conceptual/terminological/notational background presented in this paper.

5.1 Versions of Synchrony and Asynchrony

Here are two natural generalizations of Synchrony and Asynchrony, skipped in the main text.

Synchrony w.r.t. a given ***consistency check*** $\bigotimes$ **[H].** Assume M_i with disjoint registers Q_i, and with action-spaces V, W. The result of the operation is expected to be an automaton M with state-space $Q_1 \times Q_2$ and with some appropriate action-space Z. The consistency check is an associative partial function $\bigotimes$: $\tilde{V} \times \tilde{W} \longrightarrow \tilde{Z}$, which maps action-paths of the components into the action-space Z of the result.

Whereas the original treatment of $\times$ relies on the specific consistency check $\bigotimes(\tilde{u}_1\tilde{u}_0, \tilde{u}_2\tilde{u}_0) = \tilde{u}_1\tilde{u}_0\tilde{u}_2$, in the general case the definition is:

$$M(q_1q_2, \bigotimes(\tilde{v}, \tilde{w}), q_1'q_2') \text{ iff } M_1(q_1, \tilde{v}, q_1') \;\&\; M_2(q_2, \tilde{w}, q_2')$$

Multifiring asynchrony. The definition of $\|$ assumes strong interleaving of the U_i (single firings). The generalization to "multifirings" is straightforward. It allows the simultaneous firing for arbitrary subsets of $\{U_1, ...,\}$, whereas the empty subset is handled as *idling* ([AD]).

Reduction of asynchrony to synchrony primitives is a known issue in concurrency (see for example [Mi]). Here is one more version,

Proposition 27. *Let* M_1, M_2 *be automata with disjoint registers. Then, for multifiring* $\|$ *and appropriate consistency check there holds*

$$M_1 \| M_2 = M_1 \times M_2$$

5.2 Networking

Table 1 compares different concepts of networks and circuits according to the automata-oriented taxonomy. It is worthwhile to add some comments about the cases when the same term is qualified in different ways.
About two versions of "Asynchronous Circuit". These are indeed two different (even contradictory) entities. The model in [MP1] (unlike that from [DC]) does not fit the asynchrony criteria of our taxonomy, even though it reflects

Table 1

Author	Term	Architecture	Components	Input/output behavior
1) Burks & Wright [BW]	Logical Nets	×-net (circuit)	transducers	Discrete time retrospective operator
2) Maler & Pnueli [MP1]	Asynchronous Circuit	×-net (circuit)	∃-transducers	Continuous time ∃-retrooperators
3) Dill &Clarke [DC]	Asynchronous Circuit	||-web	deterministic automata	
4) Kahn [K]	Dataflow Networks	||-net (circuit)	read/wait automata which compute "sequential" stream functionals	Discrete time stream operators
5) Manna & Pnueli [MP2]	Safe Petri Nets	||-web	deterministic automata	
6) Mazurkiewicz [Ma]	Safe Petri Nets	||-nets	deterministic automata	
7) Control Theorists [A,S2]	Hybrid Automaton	×-net (circuit)	transducers: Plant, Controller	Discrete/continuous time retrooperator

Remarks:
2)-3) – the same term for different models
5)-6) – different architectures for the same entity

the intuition that synchrony means "precisely predictable amount of time", while asynchrony means just the opposite. The "asynchrony" in [MP1] is supported by nondeterministic (asynchronous?) delay operators with lower bound l and upper bound u for the delay size (see the operator $\Delta_{l,u}$ in their Def. 4). Moreover, a "circuit" (Def. 5) is presented as a system of inclusions (called there "equations"), whose components are Δ_{l_i,u_i} delays and pointwise extensions of booleans.

The following facts (not mentioned in [MP1]) may be easily checked:

Fact 1. Each $\Delta_{l,u}$ is a ∃-retrooperator, and the system of inclusions is actually a synchronous circuit C of ∃-retrooperators.

Fact 2. This circuit C is feedback reliable, and defines a ∃-retrooperator.

Hence, the [MP1]-asynchrony raises questions which are best explained and solved in the setting of synchrony.

About two versions of "Safe Petri Nets". Both versions handle the same Petri graph, but they analyze it faithfully in two dual ways.

(i) Decomposing Petri Nets as Webs.

The corresponding components are illustrated in Figure 7. Each register has two possible values, empty (0) or full (1), and each port has two possible values, active (1) or idling (0). The transition

$$q_1q_2p_1p_2 \xrightarrow{active} q_1'q_2'p_1'p_2'$$

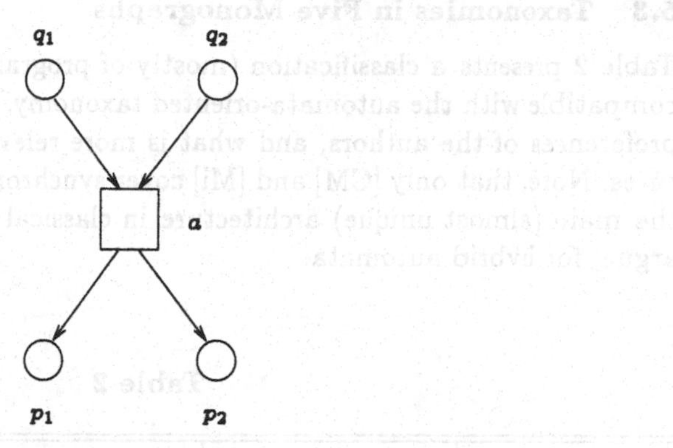

Fig. 7. A component of a Petri Net which is decomposed as a Web.

is enabled iff $q_1 = q_2 = 1$; $p_1 = p_2 = 0$; $q'_1 = q'_2 = 0$; and $p'_1 = p'_2 = 1$.
(ii) Decomposing Petri Nets as Nets.

Now the component is shown in Figure 8. The corresponding automaton has

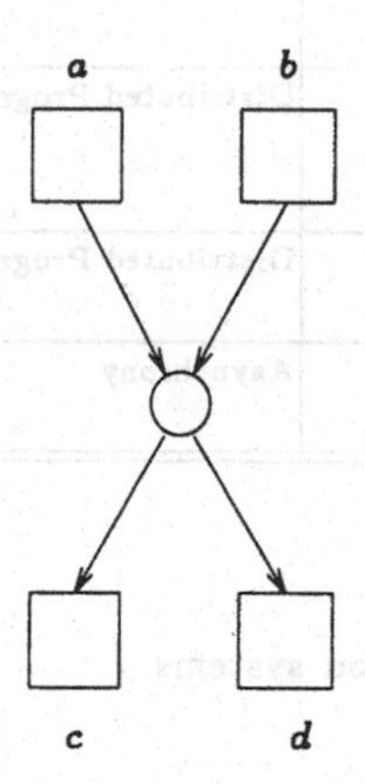

Fig. 8. A component of a Petri Net which is decomposed as a Net.

the transitions $q \xrightarrow{abcd} q'$ iff

- $q = 1 \land q' = 0 \land [a = b = 0] \land [c = 1 \Leftrightarrow d = 0]$; or
- $q = 0 \land q' = 1 \land [c = d = 0] \land [a = 1 \Leftrightarrow b = 0]$.

Paradoxically, the "Petri Web" version reflects the original token-game semantics, and, hence has historical priority. The later version was proposed by Mazurkiewicz for the sake of modularity.

5.3 Taxonomies in Five Monographs

Table 2 presents a classification (mostly of programming paradigms!), which is compatible with the automata-oriented taxonomy. It exposes the terminological preferences of the authors, and what is more relevant, their preferred architectures. Note, that only [CM] and [Mi] cover synchronous nets. And this is exactly the main (almost unique) architecture in classical automata theory and, as we argue, for hybrid automata.

Table 2

Author	×-nets	‖-nets	×-webs	‖-webs
1) Chandy & Misra [CM]	Synchronous Parallel 1) circuits 2) systolic arrays Synchronous Processes	Asynchronous Distributed 1) message passing 2) stream processing Asynchronous Processes	Synchronous Shared variables (under write/read consistency)	Asynchronous Shared variables
2) Manna & Pnueli [MP2]		Message Passing Text		1) Transition Diagrams 2) Shared Variable Text
3) Francez [F]		Distributed Programs		Shared Variable Programs
4) Apt & Olderog [AO]		Distributed Programs		Parallel Programs
5) Milner [Mi]	Synchrony	Asynchrony		

Comments:

1) "union" for pure ‖
2) a) Generic model: basic transition systems
b) Petri Nets out of classification
c) × out of classification
3) a) "parallel" used as "generic"
b) Sometimes: concurrent ⇔ shared variables
c) Interleaving "called" asynchronous computation
4) "concurrent" used as "generic"
5) Primitives distinct from shared variables

Some observations about the basic models in [MP2] are in order. The generic model is, up to syntactical details, an automaton with structured alphabets (enriched later with fairness conditions). Four concrete models of reactive systems (reactive programs) are considered: (i) Transition Diagrams, (ii) Shared Variable Text, (iii) Message-Passing Text, (iv) Petri Nets.

1. Up to syntax, models (i) and (ii) are essentially the same, namely - asynchronous webs. Model (iii) is an asynchronous net.
2. Only asynchrony (||) composition is considered: shared variables for (i-ii) and disjoint variables for (iii). Synchrony not is covered.
3. For Petri Nets the firing semantics is explained; hence, implicitly, the Petri Web architecture is assumed.

5.4 Hybrids in [H]

The conceptual/notational approach in [H] (which may be consulted for further references) differs from ours in the following points:

(i) No consideration of operators/transducers, feedback reliability.
The alleged Hybrids specify sets of trajectories (i.e. languages, as opposed to input/output behavior specified by transducers), the challenge being to verify relevant properties of these languages.

(ii) No explicit representation of a Hybrid Automaton as a pair of interacting components.[1]

(iii) Use of instantaneous transitions. Unlike discrete time, when an execution of a transition is considered as an instantaneous event, for continuous time the duration aspects become more relevant. In ([AHH]), graphs in which both edges and nodes are labeled are used. The intended semantics is that edges represent instantaneous activities (jumps), whereas nodes represent lasting activities (flows). Remember, however, that , according to the axioms of automata, the only instantaneous transition is the identity.

(iv) Inclusion of asynchrony in the basic model of Hybrid Automata.
Some examples (Railroad Crossing, Mutex) handle what we called ***Hybrid Systems***, i.e. synchronous or asynchronous composition of Hybrid Automata, but most of them (Temperature Controller, Water Level Monitor, Leaking Gas Burner etc.) can easily be analyzed in the synchronous model of Hybrid Automata.

5.5 Graphical Representation of Hybrid Automata

Actually, a representation of this kind underlies the [H]-model of hybrid automata (see quotations 6-7 in the introduction), in which the behaviors of the plant and of the controller are indivisibly coupled. From the perspective of the

[1] I learned recently that in [OD] Olderog and Dierks also advocate the explicit decomposition of timed automata (particular case of hybrid automata):
Quotation 9. ...real-time system can be decomposed into an untimed system communicating with suitable timers. Both synchronous and asynchronous communications are considered... At first sight it seems that the decomposition...is already solved by the model of timed automata... However, the main difference... is that in the timed automata model the clock operations are indivisibly coupled with the transitions whereas here we present a clear separation of untimed system and timers with explicit communication between them.

circuit paradigm, the essence of this representation is that in the Labeled Transition System of *Controller* the edges are decorated with relevant information about the semantics of *Plant*. Clearly, since *Plant* is a semantical object, this cannot be done in a systematic way, so we confine with the illustration w.r.t. our particular example. Namely, consider $C = P \times Con$, and ask: what is the detailed structure of an (elementary!) transition

$$C(l_0 x_0, \tilde{u}, l_1 x_1) \tag{13}$$

in C?

Note first that the components have a common action-alphabet $H \bigcup JUMP \bigcup FLOW$ and different state spaces: X and $\{l_0, l_1\}$. Below, referring to P and Con we have, actually, in mind their implicit transforms.

Clearly, (13) is the result of the simultaneous performing of an elementary transition of Con, say

$$Con(l_0, h_2 \bullet h_3 / j_1 \bullet f_1, l_1) \tag{14}$$

and a companion elementary transition of P (see Sect. 4.4c), which must have the format

$$P(x_0, j_1 \bullet f_1 / h_2 \bullet h_3, \delta, x_1) \tag{15}$$

Since the Controller is duration independent, no information about the duration δ was included in (14). However, the parameter δ is crucial for the companion transition (15). Note also, that in (14) the symbols h_2, h_3, j_1, f_1 are handled only as values in the corresponding space (alphabet) $H \bigcup JUMP \bigcup FLOW$, whereas in (15) they are (according to the concrete interpretation I of the P) names (codes) of, respectively, predicates, jumps and flows. Hence, below the corresponding constraints on the intended state-path $\tilde{x}$, which evolves in P. They reflect the essential difference between Con, which is equipped with a clear *LTS*-syntax and P, whose only syntactical commitments are reflected in its signature. Here is the "constraining package," which interprets Con in the semantical environment of P:

(i) Values on the ends of the time interval: $\tilde{x}(0) = x_0$ & $\tilde{x}(\delta) = x_1$
(ii) Region constraint: $\tilde{x}(0) \in h_2$ & $\tilde{x}(t) \in h_3$ *for* $t \in (0, \delta)$
Hence: $\tilde{x}(0) = 5$ & $5 < \tilde{x}(t) < 10$ *for* $t \in (0, \delta)$
(iii) Flow-after-jump constraint: $\tilde{x}(t) = f_1(j_1(x(0), t)$ *for* $t \in (0, \delta)$
Hence, $\tilde{x}(t) = f_1(x(0) + 2, t)$ *for* $t \in (0, \delta)$.

According to [H] one might use (up to minor details) the following simplified notations:

(i) The region constraint is replaced by $x = 5 \bullet 5 < x < 10$
(ii) The flow-after-jump constraint is replaced by $x := x + 2 \bullet f_1 = ...$

Hence, the "Constraining Package" above may be disguised as a "transition" of C with the format

$$Con(l_0, x = 5 \bullet 5 < x < 10 / x := x + 2 \bullet f_1 = ..., l_1) \tag{16}$$

which is represented in Fig. 9(i). Actually, according to [H], instead of Fig. 9 (i) would appear Fig. 9 (ii), which is the bilabeled version of Fig. 9 (i) (Consult Warning in Sect. 4.3b).

Finally, remind that the [H]-hybrid automaton is an acceptor of state-paths $\tilde{x}$; hence no commitments to reliability issues. The operational semantics of Fig. 9(ii) is as follows. Assume that

(i) at time t the component Con is in state l_0;
(ii) $x = 5$ at time t, and up to (but not necessarily including) $t + \delta$ there holds $5 < x < 10$.

Then C leaves location l_0 at time t and, in the (right-closed!) life-time interval $(t, t + \delta]$, it can reside in l_1; moreover, in that interval, the evolving of $\tilde{x}$ occurs according to flow f_1 after $x = 5 + 2$.

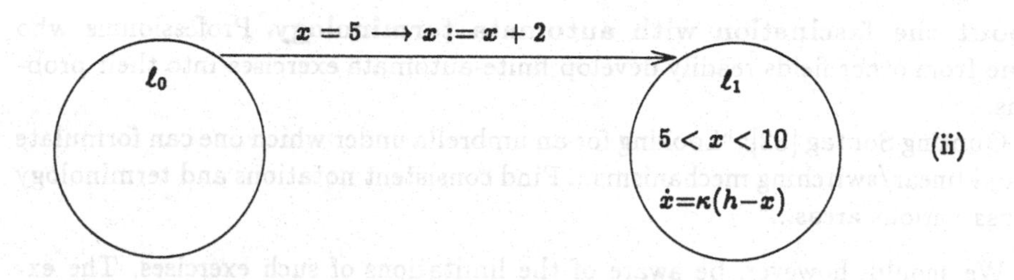

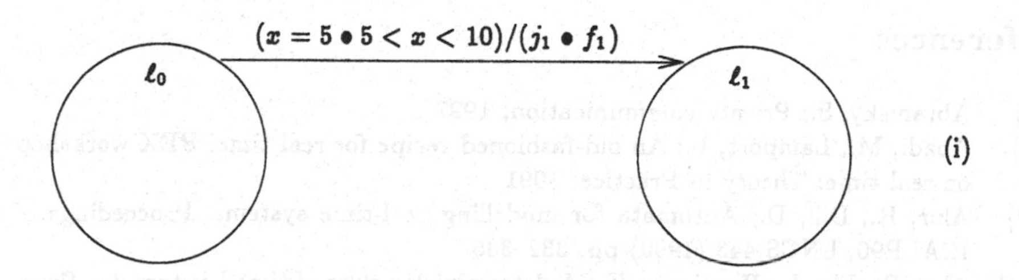

Fig. 9.

5.6 A View of Control Theorists

It has been encouraging for the author to learn that in ([A,S2]) hybrids are indeed treated as synchronous circuits of appropriate transducers: plants, interfaces, controllers.

In order to interact with finite-memory controllers, the plants have to be equipped with discrete output. This seems to be a rather sophisticated part of the whole enterprise. For example, stabilization requires that the state-trajectories imposed by the controller converge to 0. How should this be achieved via finite input/output alphabets? In [A] this is actually done through the inclusion in the network of a very non-trivial interface. Inventing the interface, proving the reliability of the circuit (not to mention the very treatment of the differential equations) is in the full competence of control theorists, and beyond concepts and techniques of automata-theory and logic.

In [A,S2] the controller is implemented as a timed automaton. According to our view, this means that, beyond the main Plant and the interface, other auxiliary plants are also used (Timers, maybe something else). From this perspective the controller is again a finite (and hence – a duration-independent) automaton.

Note also the difference: [A] uses (implicitly) finite-state controllers, whereas [S2] allows controllers with infinite state-space. This kind of infinity is compensated by the fact, that the transitions are defined via a *piecewise-linear* relation.

About the fascination with automata terminology. Professionals who came from other fields readily develop finite-automata exercises into their problems.

Quoting Sontag [S2]: "Looking for an umbrella under which one can formulate mixed linear/switching mechanisms... Find consistent notations and terminology across various areas..."

We should, however, be aware of the limitations of such exercises. The expectations from the core results in Automata Theory seem to be exaggerated.

References

[Ab] Abramsky, S.: Private communication, 1997

[AL] Abadi, M., Lamport, L.: An old-fashioned recipe for real time. REX workshop on real time: Theory in Practice, 1991

[AD] Alur, R., Dill, D.: Automata for modelling real-time systems. Proceedings of ICALP90, LNCS 443 (1990) pp. 332–335

[AFH] Alur, R., Fix, L., Henzinger, T.: A deterministic class of timed automata. Proc. CAV'94, LNCS 818, pp. 1-13

[AH] Alur, R., Henzinger, T.: Logics and models for real time: theory in practice. LNCS 600 (1992) pp. 74-106

[ACHP] Alur, A., Courcoubetis, C., Henzinger, T., Pei-Sin, Ho.: Hybrid automata: approach to the specification and verification of hybrid systems. LNCS 736 (1993) pp. 209-229

[AO] Apt, K. Olderog, E.R.: Verification of Sequential and Concurrent Programs, Springer Verlag, 1991

[A] Artstein, Z.: Examples of stabilization with hybrid feedback. LNCS 1066 (1996) pp. 173-185

[Br] Broy, M.: Semantics of finite and infinite networks of concurrent communicating agents. Distributed Computing 2 (1987) 13-31

[BW] Burks, A.W., Wright, J.B.: Theory of logical nets. Proceedings of the I.R.E., 1953

[CM] Chandy, K.M., Misra, J.: Parallel Program Design: A Foundation, Addison-Wessley, 1988

[DC] Dill, D., Clarke, E.: Automatic verification of asynchronous circuits using temporal logic. IEEE Proc. 133 (1986) 276-

[F] Francez, N.: Programm Verification, Addison-Wesley P.C., 1992

[G] Gordon, M.: Why higher-order logic is a good formalism for specifying and verifying hardware. Formal Aspects of VLSI design, Elsevier Science Publ., pp.153-177, 1986

[H] Henzinger, T.: The theory of hybrid automata. Proceedings of the IEEE (1996) 278–292

[Ka] Kahn, G.: The semantics of a simple language for parallel programming, IFIP 74

[KT] Kobrinsky, N., Trakhtenbrot, B.A.: Introduction to the Theory of Finite Automata (Russian edition, 1962), Transl. North Holland, 1965

[MP1] Maler, O., Pnueli, A.: Timing analysis of asynchronous circuits using timed automata. Proceedings CHARME'95, LNCS 987, pp. 189-205

[MP2] Manna, Z., Pnueli, A.: The Temporal Logic of Reactive and Concurrent Systems, Springer Verlag, 1992

[MP3] Manna, Z., Pnueli, A.: Models for reactivity. Acta Informatica 30 (1993) 609-678.

[Ma] Mazurkiewicz, A.: Semantics of concurrent systems: A modular fixed point trace approach. In "Advances in Petri Nets", LNCS 188, 1984

[Mi] Milner, R.: Communication and Concurrency, Prentice Hall, 1989

[OD] Olderog, E.R., Dierks, H.: Decomposing real-time specifications. Intern. Symp. COMPOS'97. LNCS 1536, pp. 465-489

[P] Pardo, D.: Timed Automata: Transducers and Circuits, M.Sc. Thesis, Tel-Aviv Univ., 1997

[PRT] Pardo, D., Rabinovich, A., Trakhtenbrot, B.A.: On synchronous circuits over continuous time, Technical Report, Tel Aviv University, 1997

[R] Rabinovich, A.: Finite automata over continuous time, Technical Report, Tel Aviv University, 1996

[RT] Rabinovich, A., Trakhtenbrot, B.: From finite automata toward hybrid systems. Proceedings FCT'97, LNCS

[S] Scott, D.: Some definitional suggestions for automata theory. J. of Computer and System Science (1967) 187-212

[S1] Sontag, E.: Mathematical Control Theory: Deterministic Finite Dimensional Systems, Springer, N.Y., 1990

[S2] Sontag, E.: Interconnected Automata and Linear Systems. LNCS, No. 1066 (1996) pp. 436–448

[T1] Trakhtenbrot, B.A.: Lecture notes on a course on verification of software and hardware systems, Tel Aviv University, Fall 1994

[T2] Trakhtenbrot, B.A.: Automata and hybrid systems. Lecture Notes on a course at Uppsala University, Fall 1997. See also updating Appendix 1, 1998, Tel Aviv University

[T3] Trakhtenbrot, B.A.: Compositional proofs for networks of processes. Fundamenta Informaticae 20(No. 1,2,3) (1994) 231-275

[T4] Trakhtenbrot, B.A.: Origins and metamorphose of the trinity: Logics, nets, automata, Proceedings, LICS'95

Axiomatising Asynchronous Process Calculi

Matthew Hennessy

University of Sussex

Abstract. Most semantic theories for process calculi presuppose that communication is **synchronous**; the sending and receiving processes must rendezvous for the communication to occur. However there is now considerable interest in process languages based on **asynchronous** communication, where the sender is not blocked but may transmit a message even in the absence of a waiting receiver. On the one hand this communication paradigm is much easier to implement and consequently has been adopted by numerous recently developed process languages, [4, 7]. On the other hand it has been argued in papers such as [2, 5] that, at least for pi-calculus based theories, asynchrony is a more basic concept in terms of which theories of synchronous communication can be established.
Despite this interest in asynchrony there has been little research into axiomatising process calculi based on this form of communication. In this talk I will survey existing results, such as those in [1, 3], and discuss equational theories for synchronous versions of both value-passing CCS [6] and the pi-calculus [2].

References

1. R. Amadio, I. Castellani, and D. Sangiorgi. On bisimulations for the asynchronous π-calculus. *Theoretical Computer Science*, 195:291–324, 1998.
2. G. Boudol. Asynchrony and the π-calculus. Research Report 1702, INRIA, Sophia-Antipolis, 1992.
3. I. Castellani and M. Hennessy. Testing theories for asynchronous languages. In *Proc. of FST-TCS 98,* LNCS 1530, 1998.
4. C. Fournet, G. Gonthier, J. J. Levy, L. Marganget and D. Remy. A Calculus of Mobile Agents. In *Proc. CONCUR'96,* LLNCS 1119, 1996.
5. K. Honda and M. Tokoro. On asynchronous communication semantics. In *Proc. Object-Based Concurrent Computing,* LNCS 612, 1992.
6. R. Milner. *Communication and Concurrency.* Prentice Hall, 1989.
7. Benjamin C. Pierce and David N. Turner. Pict: A programming language based on the pi-calculus. Technical Report CSCI 476, Computer Science Department, Indiana University, 1997. To appear in *Proof, Language and Interaction: Essays in Honour of Robin Milner*, Gordon Plotkin, Colin Stirling, and Mads Tofte, editors, MIT Press.

A Polynomial Time Approximation Scheme for Dense MIN 2SAT

Cristina Bazgan[1] W. Fernandez de la Vega[2]

[1] Université Paris-Sud, LRI, bât.490, F–91405 Orsay, France, bazgan@lri.fr
[2] CNRS, UMR 8623, Université Paris-Sud, LRI, F–91405 Orsay, France, lalo@lri.fr

Abstract. It is proved that everywhere-dense MIN 2SAT and everywhere-dense MIN EQ both have polynomial time approximation schemes.

1 Introduction

The approximability theory of dense instances of maximization problems such as MAX CUT, MAX 2SAT has had many recent successes, starting with [1] and [2]. (See [5] for a recent review.) In [1] it is proved in particular that the dense instances of any problem in MAX SNP have a *polynomial-time approximation scheme*. In [4], it is proved that many of these problems can be approximated in constant time with an additive error ϵn^2 where n is the size of the input in a certain probe model (implying that the dense versions have constant-time approximation schemes).

The case of dense instances of minimization problems (or edge-deletion problems) seems to be harder. The case of BISECTION was settled in [1]. In this paper, we bring a further contribution to this case by proving that everywhere-dense MIN 2SAT and everywhere-dense MIN EQ both have polynomial-time approximation schemes. Our main tool is a constrained version of BISECTION, which we call PAIRED BISECTION: a pairing Π of the vertices is given and we look only at the bisections which split each pair of vertices in Π. The key step in the proof is an L-reduction from MIN EQ to PAIRED BISECTION. Then we adapt the algorithm of [1] for BISECTION to PAIRED BISECTION. This yields a polynomial-time approximation scheme for everywhere-dense MIN EQ. A density preserving L-reduction from MIN 2SAT to MIN EQ concludes the proof.

2 Preliminaries

We begin with some basic definitions.

Approximability. Let us recall a few definitions about approximability. Given an instance x of an optimization problem A and a feasible solution y of x, we denote by $m(x, y)$ the value of the solution y, and by $opt_A(x)$ the value of an optimum solution of x. In this paper we consider mainly minimization problems.

The *performance ratio* of the solution y for an instance x of a minimization problem A is

$$R(x,y) = \frac{m(x,y)}{opt_A(x)}.$$

For a constant $c > 1$, an algorithm is a *c-approximation* if for any instance x of the problem it returns a solution y such that $R(x,y) \leq c$. We say that an optimization problem is *constant approximable* if, for some $c > 1$, there exists a polynomial-time c-approximation for it. APX is the class of optimization problems that are constant approximable. An optimization problem has a *polynomial-time approximation scheme* (a ptas, for short) if, for every constant $\varepsilon > 0$, there exists a polynomial-time $(1+\varepsilon)$-approximation for it.

L-reduction. The notion of L-reduction was introduced by Papadimitriou and Yannakakis in [6]. Let A and B be two optimization problems. Then A is said to be *L-reducible* to B if there are two constants $\alpha, \beta > 0$ such that

1. there exists a function, computable in polynomial time, which transforms each instance x of A into an instance x' of B such that $opt_B(x') \leq \alpha \cdot opt_A(x)$,
2. there exists a function, computable in polynomial time, which transforms each solution y' of x' into a solution y of x such that $|m(x,y) - opt_A(x)| \leq \beta \cdot |m(x',y') - opt_B(x')|$.

For us the important property of this reduction is that it preserves ptas's; that is, if A is L-reducible to B and B has a ptas then A has a ptas as well.

Equivalence. Given n variables, an *equivalence* is an expression of the form $l_i \equiv l_j$ where l_i, l_j are literals. The equivalence $l_i \equiv l_j$ is true under an assignment A iff A gives the same truth value (*true* or *false*) to l_i and l_j.

Graphs. As usual, we write $G = (V(G), E(G))$ for the undirected graph with vertex set $V(G)$ and edge set $E(G)$. The vertices are indexed by the integers $1, ..., n = |V(G)|$. For two vertices u and v, uv denotes the edge linking u to v. We denote by $\Gamma(u)$ the set of neighbors of u. If S and T are two disjoint subsets of $V(G)$, we denote by $e(S,T)$ the number of edges linking S to T.

Bisection. Let $G = (V(G), E(G))$ be an undirected graph with an even number of vertices. A *bisection* of G is a partition of vertex set $V(G)$ in two equal size sets R and L. The value of the bisection is the number of edges between R and L.

Paired Bisection. Let $G = (V(G), E(G))$ be a graph with $|V| = 2n$ and let a pairing Π of $V(G)$ be fixed, $\Pi = \{\{u_1, v_1\}, ..., \{u_n, v_n\}\}$, say, (with $\cup_{1 \leq i \leq n}\{u_i, v_i\}$ $= V(G)$). We say that a bisection $\{R, L\}$ of G is *admissible* with respect to Π, (admissible for short), iff it splits each pair $\{u_i, v_i\}$, (i.e., for $i = 1, \ldots, n$, either $u_i \in R$ and $v_i \in L$ or $v_i \in R$ and $u_i \in L$). We call Paired Bisection the problem of minimizing the value of an admissible bisection where of course Π is part of the data. (See the formal definition below.)

Dense Instances. A graph with n vertices is *δ-dense* if it has at least $\delta n^2/2$ edges. It is *everywhere-δ-dense* if the minimum degree is at least δn. Similarly, a

2CNF formula or a set of equivalences on n variables is *everywhere-δ-dense* if for each variable the total number of occurrences of the variable and its negation is at least δn. A 2CNF formula (a set of equivalences) on n variables is *δ-dense* if the number of clauses (equivalences) is at least δn^2. A set of instances is *dense* if there is a constant $\delta > 0$ such that it is δ-dense and a set of instances is everywhere-dense if there is a constant $\delta > 0$ such that it is everywhere-δ-dense. So, everywhere-dense implies dense but the converse is not true.

We now define the problems in question formally.

MIN 2SAT

Input: A 2CNF formula F.

Solution: A truth assignment for the variables.

Value: The number of clauses satisfied by the assignment.

MIN EQ

Input: A set of equivalences.

Solution: A truth assignment for the variables.

Value: The number of equivalences satisfied by the assignment.

BISECTION

Input: A graph $G = (V(G), E(G))$.

Solution: A bisection of G.

Value: The number of edges in the bisection.

PAIRED BISECTION

Input: A graph $(V(G), E(G))$ with $|V| = 2n$ and a pairing Π of $V(G)$, $\Pi = \{\{u_1, v_1\}, ..., \{u_n, v_n\}\}$.

Solution: A bisection of G which splits each pair $\{u_i, v_i\}$.

Value: The number of edges in the bisection.

All these problems are minimization problems, i.e., a solution with value as small as possible is sought in each case.

3 The Results

Our main result is

Theorem 1. *Everywhere-dense* MIN 2SAT *and everywhere-dense* MIN EQ *both have ptas.*

In the course of proving Theorem 1, we also obtain the next result, which has some interest in view of the fact that the approximability status of BISECTION is wide open.

Theorem 2. PAIRED BISECTION *is APX-hard.*

Remark. It is easy to see that (simply) dense instances of MIN 2SAT or MIN EQ *do not* have a ptas if $P \neq NP$. As far as we know, these are the only problems which are known to have a ptas in the everywhere-dense case but not in the dense case.

The proof of Theorem 1 occupies the rest of the paper. First, we give a density preserving L-reduction from MIN 2SAT to MIN EQ (Lemma 1). As already mentioned, the key step in the proof of Theorem 1 is an L-reduction from MIN EQ to PAIRED BISECTION (Lemma 2). The proof is then easily completed by adapting the ptas for everywhere-dense BISECTION of [1] to obtain a ptas for everywhere-dense PAIRED BISECTION.

4 The Proofs

Lemma 1. *There is an L-reduction from* MIN 2SAT *to* MIN EQ.

Proof. Let F be a set of clauses with at most two literals on n variables $x_1, \ldots, x_n$. We construct a set of equivalences E as follows. We add a new variable y and we replace each clause $l_i \vee l_j$ in F by the following set of equivalences: $l_i \equiv \neg l_j, l_i \equiv \neg y, l_j \equiv \neg y$. By inspection, one sees that if $l_i \vee l_j$ is satisfied by some assignment, at most 2 of these 3 equivalences are true, so that the inequality $opt(E) \leq 2opt(F)$ holds, showing that the first condition of the definition of the L-reduction is satisfied. Now, suppose that we have a solution of E (an assignment A for the variables that appear in E). We can suppose that $y = false$ in A since the complementary assignment satisfies the same number of equivalences. We consider the same assignment for the variables in F. Let B denote this second assignment. Now one sees that if $l_i \vee l_j$ is satisfied by B then exactly 2 of the equivalences in E corresponding to $l_i \vee l_j$ are satisfied, so that the values satisfy $m(F, B) = m(E, A)/2$, showing that the second condition of the definition of the L-reduction is also satisfied.

Lemma 2. MIN EQ *and* PAIRED BISECTION *are mutually L-reducible one to the other.*

Proof. Firstly we construct a L-reduction from PAIRED BISECTION to MIN EQ. Let $G = (V(G), E(G))$ be a graph and $\Pi = \{\{u_1, v_1\}, ..., \{u_n, v_n\}\}$ a pairing of $V(G)$. For convenience, we consider that each pair in Π is ordered. We can then represent a bisection (L, R) of G by a vector of n logical variables $\{x_1, ..., x_n\}$ with the understanding that, if $x_i = true$ then we put u_i in L and v_i in R and if $x_i = false$ then we put u_i in R and v_i in L.

Now, for each edge $u_i v_j \in E(G)$ we introduce the equivalence $x_i \equiv x_j$. For each edge $u_i u_j \in E(G)$ or $v_i v_j \in E(G)$ we introduce the equivalence $x_i \equiv \neg x_j$. Call E the set of all these equivalences. By inspection, one can see that an edge of G contributes to the bisection (L, R) exactly when the the corresponding equivalence holds. This implies clearly $opt(E) = opt(G)$ and the L-reduction in one direction.

For the reduction in the other direction, we replace each equivalence $x_i \equiv \neg x_j$ by the edges $u_i u_j$ and $v_i v_j$ and each equivalence $x_i \equiv x_j$ by the edges $u_i v_j$ and $v_i u_j$.

It is straightforward to check that the reductions of Lemma 3 and Lemma 4 map an everywhere-dense set of instances into another everywhere-dense set of instances.

Corollary 1. PAIRED BISECTION *is APX-hard.*

Proof. In [3] it is proved that the following problem is APX-hard: given a set of equivalences find an assignment that minimize the number of equivalences that we have to remove such that the new set of equivalences is satisfiable. There is a simple L-reduction between the above problem and MIN EQ (we replace each equivalence $\ell_i \equiv \ell_j$ by $\ell_i \equiv \neg \ell_j$) that implies that MIN EQ is APX-hard. The Lemma follows immediately from Lemma 1.

Theorem 3. *Everywhere-dense* PAIRED BISECTION *has a ptas.*

As already mentioned, our ptas is a rather straightforward modification of the ptas of [1] for everywhere-dense BISECTION. The main difference is the fact that in our case we don't have to care of the "equal sides" condition which is implicit in the pairing Π, and our algorithm is in fact simpler than that of [1]. Let the input be (G, Π) with $\Pi = \{\{u_1, v_1\}, ..., \{u_n, v_n\}\}$ and $G = (V(G), E(G))$. Let ϵ be the allowed error and $\alpha = \frac{4\delta^2 \epsilon}{25}$. As in [1] we run two distinct algorithms and select the solution with the smallest value. The first algorithm gives a good solution for the instances whose minimum value is at least αn^2 and the second for the instances whose minimum value is less than αn^2.

1. First algorithm (Algorithm for the case of "large" bisection)

Let y_i indicate the side (0 for Left, 1 for Right) of the vertex u_i in the bisection (L, R). [1] use smooth polynomial integer programming for the instances with large optimum value. We just have to check that we can express the value of PAIRED BISECTION as a degree 2 polynomial in the y_i's:

$$\sum a_{ij} y_i y_j + \sum b_i y_i + d$$

where each $|a_{ij}| \leq c, |b_i| \leq cn, |d| \leq cn^2$ for some fixed constant c. We can use

$$\begin{aligned}\text{Paired Bisection} = \min & \sum_{u_i v_j \in E(G)} (y_i(1 - y_j) + y_j(1 - y_i)) \\ & + \sum_{u_i u_j \in E(G)} [1 - (y_i y_j + (1 - y_i)(1 - y_j))] \\ & + \sum_{v_i v_j \in E(G)} [1 - (y_i y_j + (1 - y_i)(1 - y_j))]\end{aligned}$$

This program can be solved approximately in polynomial time by an algorithm of [1].

2. Second algorithm (Algorithm for the case of a "small" bisection)

This second algorithm is again similar to that of [1]. However, it will be seen that important differences appear and we felt the need for a new (albeit sketched) correctness proof although this proofs relies heavily on [1]. Actually, the proof of correctness of the algorithm of [1] for the case of small bisection relies on the property that one can assume that the vertices in one side of the bisection have no negative bias. (Lemma 5.1 in [1]. We define the bias of a vertex u as the difference between the number of edges it sends to his side in the bisection and the number of edges it sends to the other side.) There is apparently no analogue to this property in our case. [1] use exhaustive sampling for the case of a "small" bisection. Here we sample the set of pairs Π rather than the set of vertices. Actually, we will work with pairs all along the way. Let S be the set theoretical union of $m = O((\log n)/\delta)$ pairs picked randomly from Π. We can assume by renaming that $S = \cup_{i=1}^{m}\{u_i, v_i\}$.

Let (L_o, R_o) be an optimal admissible bisection of G and let $S_L = S \cap L_o, S_R = S \cap R_o$. Again by renaming, we can assume that $S_L = \{u_1, ..., u_m\}$ and $S_L = \{v_1, ..., v_m\}$. (Actually, the algorithm which does not know the partition (S_L, S_R), will be run for each of the 2^{m-1} admissible partitions of S.)

As in [1], the placement is done in two stages. In the first stage, pairs of vertices are placed on the basis of their links with S. (An important difference with the algorithm of [1] occurs here: in the algorithm of [1], only "right" vertices are placed at this step.) In the second step, the remaining pairs are placed on the basis of their links with the vertices placed during the first step and with S. In the description below, we let L and R denote the current states of the left-hand side (resp. right-hand side) of the bisection constructed by the algorithm. Thus, we start with $L = S_L,\ R = S_R$.

1. Let

   $$T_1 = \{i > m : |\Gamma(u_i) \cap S_R| + |\Gamma(v_i) \cap S_L| \le (|\Gamma(u_i) \cap S_L| + |\Gamma(v_i) \cap S_R|)/2\}$$

   $$T_2 = \{i > m : |\Gamma(u_i) \cap S_L| + |\Gamma(v_i) \cap S_R| \le (|\Gamma(u_i) \cap S_R| + |\Gamma(v_i) \cap S_L|)/2\}$$

   For each $i \in T_1$ we put u_i in L and v_i in R. For each $i \in T_2$ we put u_i in R and v_i in L.
2. Let $L_1 = S_L \cup (\cup_{i \in T_1}\{u_i\}) \cup (\cup_{i \in T_2}\{v_i\})$ denote the set of vertices placed on the left side after the completion of stage 1, and similarly, let $R_1 = S_R \cup (\cup_{i \in T_1}\{v_i\}) \cup (\cup_{i \in T_2}\{u_i\})$ denote the "right" vertices.
   Let $J = \{m+1, ..., n\} \backslash (T_1 \cup T_2)$. For each $i \in J$
   (a) if $|\Gamma(u_i) \cap R_1| + |\Gamma(v_i) \cap L_1| \le |\Gamma(u_i) \cap L_1| + |\Gamma(v_i) \cap R_1|$ then we add u_i in L and v_i in R;
   (b) otherwise we add v_i in L and u_i in R.

Let us sketch now a proof of correctness of the second algorithm. (algorithm for "small" bisection). We denote by $opt(G) = opt(G, \Pi)$ the value of an optimum admissible bisection of G.

Lemma 3. *With high probability,*

1. *T_1 contains each index i with the property that* $|\Gamma(u_i) \cap R_o| + |\Gamma(v_i) \cap L_o| \leq (|\Gamma(u_i) \cap L_o| + |\Gamma(v_i) \cap R_o|)/4$
2. *T_2 contains each index i with the property that* $|\Gamma(u_i) \cap L_o| + |\Gamma(v_i) \cap R_o| \leq (|\Gamma(u_i) \cap R_o| + |\Gamma(v_i) \cap L_o|)/4$

Also with high probability, each pair in the set $\{(u_i, v_i) : i \in T_1 \cup T_2\}$ is placed as in the optimum solution.

Proof. The proof is completely similar to that of Lemma 5.2 in [1] and is omitted. We remark in passing that sample size $O(\sqrt{\log n}/\delta)$ suffices (instead of $m = O((\log n)/\delta)$ used in [1]).

Lemma 4. $n - (m + |T_1| + |T_2|) \leq \frac{5\text{opt}(G)}{2\delta n}$.

Proof. The proof of this Lemma is again very similar to the proof of Lemma 5.3 in [1] and is omitted.

Lemma 5. *If $opt(G) < \alpha n^2$ then with high probability the value of the bisection given by algorithm 2 is at least $(1+\varepsilon)opt(G)$ where $\varepsilon = \frac{25\alpha}{4\delta^2}$.*

Proof. We need first some notations. Let $U = \cup_{i \in J}\{u_i, v_i\}$ and let $u = |J|$. (U is the set of vertices which are placed during step 2.) Let $U_L = U \cap L$, $U_R = U \cap R$, $U_L^{opt} = U \cap L_o$ and $U_R^{opt} = U \cap R_o$. Let $m(G, sol)$ denote the value of the bisection given by the algorithm, Let $d(U) = e(U_L, U_R)$ and $d_{opt}(U) = e(U_L^{opt}, U_R^{opt})$. For each $i \in J$, we define

$$val(i) = |\Gamma(v_i) \cap L_1| - |\Gamma(v_i) \cap R_1|$$

if the case (a) of stage 2 of the algorithm occurs for the index i, and

$$val(i) = |\Gamma(u_i) \cap L_1| - |\Gamma(u_i) \cap R_1|$$

otherwise. We denote by d_1 the number of edges of G with exactly one extremity in R_1. Let us check that we have

$$m(G, sol) = d_1 + \sum_{i \in J} val(i) + d(U). \tag{1}$$

Indeed, assume that case (a) occurs for the index i. (The treatment of case (b) is similar.) This means that $u_i \in L$, $v_i \in R$. Then, apart from edges linking U_L to U_R (which are separately counted), $|\Gamma(v_i) \cap L_1|$ new edges contribute to the bisection, and $|\Gamma(v_i) \cap R_1|$ are to be subtracted, since they are counted in d_1 and do not contribute to the bisection.

We see, using Lemma 3, that with high probability, the optimum value of an admissible bisection is

$$\text{opt}(G) = d_1 + \sum_{i \in J} val_{\text{opt}}(i) + d_{\text{opt}}(U)$$

where

$$val_{\text{opt}}(i) = |\Gamma(v_i) \cap L_1| - |\Gamma(v_i) \cap R_1|$$

if $u_i \in U_L^{opt}$ and

$$val_{\text{opt}}(i) = |\Gamma(u_i) \cap L_1| - |\Gamma(u_i) \cap R_1|$$

if $u_i \in U_R^{opt}$. The bisection of U constructed in stage 2 minimizes $\sum_{i \in J} val(i)$. We have thus

$$\sum_{i \in J} val(i) \leq \sum_{i \in J} val_{\text{opt}}(i).$$

This implies, with (1)

$$\begin{aligned} m(G, sol) &\leq d_1 + \sum_{i \in J} val_{\text{opt}}(i) + d_{opt}(U) - d_{opt}(U) + d(U) \\ &= opt(G) - d_{opt}(U) + d(U) \leq opt(G) + d(U) \\ &\leq opt(G) + u^2 \leq opt(G) + \frac{25 opt(G)^2}{4\delta^2 n^2} \\ &\leq opt(G)\left(1 + \frac{25\alpha}{4\delta^2}\right) \end{aligned}$$

using Lemma 8.

The correctness follows now from our choice of α.

5 Open Problems

The major open problem is of course the approximability or inapproximability of BISECTION. Can our approximation hardness theorem for PAIRED BISECTION help?

The true complexity of approximate MIN 2SAT in the dense case is another interesting question. It is known that the case of "large" bisection can be done in constant time (see [4]). Can overall constant time be achieved?

References

1. S. Arora, D. Karger, M. Karpinski, *Polynomial time approximation schemes for dense instances of NP-hard problems*, Proc. of 27th STOC, 1995, 284–293. The full paper will appear in Journal of Computer and System Sciences.
2. W. Fernandez de la Vega, *Max-Cut has a Randomized Approximations Scheme in Dense Graphs*, Random Structures and Algorithms, 8(3) (1996), 187–198.

3. N. Garg, V. V. Vazirani and M. Yannakakis, *Approximate max-flow min-(multi)cut theorems and their applications*, SIAM Journal on Computing 25 (1996), 235–251.
4. O. Goldreich, S. Goldwasser and D. Ron, *Property Testing and its Connection to Learning and Approximation*, Proc. of 37th IEEE FOCS, 1996, 339–348. The full paper has appeared in Journal of the ACM, 45 (4) (1998), 653–750.
5. M. Karpinski, *Polynomial Time Approximation Schemes for Some Dense Instances of NP-Hard Optimization Problems*, , Randomization and Approximation Techniques in Computer Science, LNCS 1269, 1–14.
6. C. Papadimitriou and M. Yannakakis, *Optimization, Approximation and Complexity Classes*, Journal of Computer and System Science 43 (1991), 425–440.

Decidable Classes of the Verification Problem in a Timed Predicate Logic

Danièle Beauquier and Anatol Slissenko

Dept. of Informatics, University Paris-12
61 Av. du Gén. de Gaulle
94010, Créteil, France
E-mail: { beauquier, slissenko}@univ-paris12.fr

Abstract. We consider a first order timed logic that is an extension of the theory of real addition and scalar multiplications (by rational numbers) by unary functions and predicates of time. The time is treated as non negative reals. This logic seems to be well adapted to a direct, full-scale specification of real-time systems. It also suffices to describe runs of timed algorithms that have as inputs functions of time. Thus it permits to embed the verification of timed systems in one easily understandable framework. But this logic is incomplete, and hence undecidable. To develop an algorithmic support for the verification problem one theoretical direction of research is to look for reasonable decidable classes of the verification problem. In this paper we describe such classes modeling typical properties of practical systems such as dependence of behavior only on a small piece of history and periodicity.

1 Introduction

Requirements specifications of real-time systems involve various timed properties and timed constraints often containing arithmetical operations. Most formalisms used for the verification of real-time systems (e. g., temporal or duration logics [Eme90,Han94,Hen98,Rab98]) are rather limited (e. g. temporal logics do not have arithmetic operations) and can express the initial requirements specification neither directly nor completely. Usually they *model* some properties of the specification. However, these restricted formalisms are used as for them there are developed algorithms of verification (usually, of model checking) which computer implementation sometimes works. But limited modeling has two shortcomings: first, it does not cover the specification entirely, and second, it cannot justify that the modeling itself is correct. Another type of approaches to the verification is based on powerful logics like that of PVS [PVS]. Within this framework it is very hard to find reasonable decidable classes, not to speak about feasible ones, because of too abstract formalisms used.

In the present paper we continue the study of First Order Timed Logic (FOTL) with explicit continuous time [BS97b,BS97a]. A concrete FOTL is an extension of a decidable theory of reals (here we consider the theory of real addition and scalar multiplications by rational numbers) by timed predicates and functions. For a large class of problems we can state the following properties of such a logic. First, it is sufficiently expressible from the user's point of view to rewrite *directly* and *entirely* the requirements specification of the problem under consideration usually given in a natural language. For example, the property "two x-events are never separated by exactly 1 time unit" can be directly rewritten as the FOTL formula $\neg\exists tt' \, (x(t) \wedge x(t') \wedge |t - t'| = 1)$. Another example, the property "the average value of the clocks $x_1, ..., x_n$ does not exceed d" can be directly rewritten as the formula $\forall t \, (x_1(t) + ... + x_n(t) \leq n \cdot d)$ with t being a time variable. The presence of arithmetics permits to easily specify also such problems as clock synchronization that is impossible for commonly used temporal logics. For other examples see [BS97a,BS98]. Second, FOTL permits to represent rather easily the set of runs of timed programs (e. g. the runs of Gurevich Abstract State Machines [BS97a] or timed automata [BS98]). Third, we can describe decidable classes of the verification problem based on the fact that the underlying theory of reals is decidable. And we have in fact only one logic to consider as compared with numerous temporal logics. (The unifying framework [HR98,HR99] for temporal logics or the version of Büchi's second order monadic logic for continuous time [Tra98,Rab97b,Rab97a] neither give sufficient power of expressibility, though preserve the decidability.)

We look for decidable classes of the verification problem modeling some "finiteness" properties of practical systems of control. The verification problem can be treated as establishing the validity of some proposition $(\Phi \rightarrow \Psi)$, where Ψ describes the requirements on functioning such as safety, dependability (liveness) and Φ describes the environment (constraints on inputs, relations between functions external to the controller etc...) and the runs of the controller to verify. The "finiteness" properties are formulated in terms of formula interpretations and are called ***finite satisfiability*** and ***finite refutability***. Finite satisfiability of a formula Φ says that every "finite piece" of any model of Φ is extendable to a "finitely definable" model of Φ. Finite refutability says that if a formula is refutable, i. e. possesses a counter-model, the contradiction given by this counter-model is concentrated on a finite piece of a fixed size; and it must be so for every counter-model. The class of implications of our logic where the premise is finitely satisfiable and the conclusion is finitely refutable with a fixed complexity is decidable if the existence of the respective finitely definable counter-model is decidable. Thus, we have a decidable class of verification problems of the mentioned form $(\Phi \rightarrow \Psi)$, where Φ is finitely satisfiable and Ψ is finitely refutable (with a fixed complexity). In the examples we looked at, the finite refutability of requirements of functioning, such as safety or dependability are easy to check. But on the whole to find interesting decidable sufficient conditions for this property remains an open problem. Concerning finite satisfiability, the situation is studied better, for example in [BS98] we show that formulas describing runs of reducible timed automata are finitely satisfiable, and that their reducibility is ***decidable***. We consider a notion of finite satisfiability which concerns properties with the periodicity flavor. This notion is ***ultimate repetitiveness***. Concatenations of ultimate repetitive models are also studied.

The structure of the paper is as follows. In section 2 we describe the FOTL we consider here. Section 3 contains the definitions of finiteness properties. In section 4 we prove that the existence of an ultimately repetitive model of a given complexity is decidable. This gives a decidable class of verification problems. Though the formalism under consideration is destined to analyze general type algorithms, e. g. Gurevich Abstract State Machines [Gur95], we use as examples timed automata to simplify the presentation and to compare our decidable classes with existing ones treated in the setting of temporal logics.

2 First Order Timed Logic (FOTL)

Syntax of FOTL.
The vocabulary W of a FOTL consists of a finite set of *sorts*, a finite set of *function symbols* and a finite set of *predicate symbols*. To each sort there is attributed a set of variables. Some sorts are predefined, i. e. have fixed interpretations. Here the predefined sorts are the real numbers $\mathbb{R}$ and time $\mathcal{T} =_{df} \mathbb{R}_{\geq 0}$ treated as a subsort of $\mathbb{R}$. The other sorts are finite. One can use also boolean combinations of these sorts but here we will not do it.

Some functions and predicates are also predefined. As predefined constants we take *Bool* for boolean values and $\mathbb{Q}$ for rational numbers. Addition $+$, subtraction $-$ and scalar multiplications of reals by rational numbers are predefined functions of the vocabulary. The predicates $=, \leq, <$ over reals are predefined predicates of W. The vocabulary contains $=$ for all types of objects, and the identity function *id* of the type $\mathcal{T} \to \mathcal{T}$ to represent the current time. The part of the vocabulary concerning reals, rational and boolean constants will be sometimes called *standard*.

Any *abstract function* (i. e. without any a priori fixed interpretation) is of the type $\mathcal{T} \times \mathcal{X} \to R$, and any *abstract predicate* is of the type $\mathcal{T} \times \mathcal{X} \to Bool$, where $\mathcal{X}$ is a direct product of finite sorts and R is an arbitrary sort. The sets of abstract functions and predicates are denoted respectively by V_{Funct} and V_{Pred}; we set $V =_{df} V_{Funct} \cup V_{Pred} \cup \{id\}$.

A vocabulary W being fixed, the notion of *term* and that of *formula* over W are defined in a usual way.

In this paper we consider a subclass of FOTL, that we will call FOTL_0, where *all the sorts are predefined*, and all abstract functions and predicates are respectively of the types $\mathcal{T} \to R$ and $\mathcal{T} \to Bool$. Notice that for a fixed interpretation of abstract sorts of a given FOTL (if not to care about complexity) any FOTL-formula can be replaced by an equivalent FOTL_0-formula over another vocabulary (of larger cardinality in the general case).

Example 1 (Vocabulary for Timed Automaton.) Before defining the semantics of our logic, we give a vocabulary W_A to describe total runs of a timed automaton $\mathcal{A} = (S, s_{init}, C, \mu, E)$. In this notation the set S is a set of locations, and $s_{init} \in S$ is the initial location. The set C is the set of clocks of $\mathcal{A}$, μ is a function which ascribes to each location s a guard $\mu(s)$, the latter being a formula constructed from atoms *true*, *false*, $c\omega n$, where $c \in C, n \in \mathbb{N}$ and $\omega \in \{>,<,=\}$, with the help of $\wedge$ and $\vee$. And the set E is the set of edges of $\mathcal{A}$ which describe possible transitions. Each edge $e \in E$ is of the form (s, s', X, δ), where s is the location of departure, s' is the location of arrival, $X \subseteq C$ is the set of clocks to reset to 0, and δ is a guard that must be satisfied in order to fire this transition. The clocks that are not reset to 0, continue to augment. For details see [AD94]. An example of timed automaton is shown on Figure 1 and will be commented later.

The vocabulary $W_{\mathcal{A}}$ contains, in addition to the standard part and *id*, the sort S consisting of elements of S, and the set C of abstract symbols constituted of symbols c for clocks, each one of the type $\mathcal{T} \to \mathcal{T}$, and of symbol *loc* of the type $\mathcal{T} \to S$. The value $c(t)$ represents the value of the clock c at the moment t and $loc(t)$ represents the location of the automaton at the moment t.

Semantics of FOTL$_0$.

The ***admissible interpretations*** introduced below can be motivated by arguments related to hybrid systems (cf. [ACHH93,Tra98]). As admissible interpretations of the closed formulas of our logic we consider functions and predicates that are piecewise finitely defined. For predicates this means that they are piecewise constant. Before giving a precise definition of admissible interpretations take as an example interpretations of function symbols describing runs of a timed automaton $\mathcal{A}$. Each unary function symbol c will be interpreted as a piecewise linear function of time with coefficient one, i. e. $c(\tau) = \tau - a$ for $\tau \in [t, t')$, where $t \geq a$ and $[t, t')$ is an interval of the partition defining these linear pieces, and the function symbol *loc* will be interpreted as piecewise constant function.

We assume that for every abstract function f of type $\mathcal{T} \to R$ there is fixed a term U_f with values of type R constructed only from constants, variables and predefined functions. The vocabulary of FOTL$_0$ does not give many possibilities to construct U_f. We will limit even these possibilities, in fact not essentially, and will consider the following types of terms: first, those of the form z with z being a variable for an abstract sort (representing abstract constants of the type R) if R is an abstract sort, and second, the terms of the form $\xi_0 t + \xi_1 a + z$, where (ξ_0, ξ_1) can be chosen from a finite set $\Xi_f \subset \mathbb{Q}^2$ and t, a and z are real variables which role is fixed as follows: t is the time variable standing for the time argument, a is the left end of the interval on which we consider our function, and z is a real parameter. (We cannot make ξ_0 and ξ_1 variables as adding the sort $\mathbb{Q}$ to our vocabulary destroys the decidabilities we wish to prove.) For a particular vocabulary we fix the sets Ξ_f for abstract f of the type $\mathcal{T} \to \mathbb{R}$.

For a time interval ζ we denote by ζ^- and ζ^+ respectively its *left and right ends.*

We will write U_f also as $U_f(t, a, \xi_0, \xi_1, z)$ to make explicit the parameters, the dummy or fixed parameters will be omitted. If a concrete value z_0 of z is given and some $(\xi_0, \xi_1) \in \Xi_f$ is chosen then for $t \in \zeta$ the value $f(t)$ will be defined as $f(t) = U_f(t, \zeta^-, \xi_0, \xi_1, z_0)$. For example, the interpretations of *loc* are defined by the term $U_{loc}(z) = z$, where z is a variable of type S. To say that an interpretation loc^* of *loc* has the value s_i on ζ, we say that on ζ the function *loc* is defined by its term $U_{loc}(z)$ for the value of the parameter $z = s_i$, and write $loc^*(t) = U_{loc}(s_i) = s_i$ for $t \in \zeta$. An interpretation c^* of a clock c which says that c has been reset to 0 at the moment t_0 is defined to the right of t_0 for example by the term $U_c(t, \zeta^-, 1, 0, -t_0) = t - t_0$ until the next reset or until ∞.

Define also U_{id} as t, $U_P(t)$ for $P \in V_{Pred}$ as B, where B is a Boolean variable, and thus, U_f is attributed to every $f \in V$.

A *partition* of $\mathcal{T}$ is a sequence $\pi = (\zeta_i)_{i \in \overline{N}}$ of non empty disjoint intervals where: (1) $\overline{N}$ is a prefix of $\mathbb{N}$, (2) $\bigcup_{i \in \overline{N}} \zeta_i = \mathcal{T}$, (3) $\zeta_i^+ = \zeta_{i+1}^-$ for $0 \leq i \leq |\overline{N}| - 1$, (4) $\zeta_0 = 0$, $\zeta_k^+ = \infty$ if $\overline{N}$ is finite and k is its last element.

Consider an abstract $f \in V$ (its type is $\mathcal{T} \to R$). An interpretation f^* of f is *admissible* if there exists a partition $\pi = (\zeta_i)_{i \in \overline{N}}$ and a function $\delta : \overline{N} \to \Xi_f \times R$ (if R is an abstract sort then Ξ_f is useless and can be dropped off) such that $f^*(t) = U_f^*(t, \zeta_i^-, \delta(i))$ for $t \in \zeta_i$ and $i \in \overline{N}$. Such a partition π is an *admissible partition* for the interpretation f^*.

Example 2 The pair of sequences
$\pi = (([0,1), [1,3.7), [3.7,\infty)),\ \delta = (s_0, s_1, s_2))$ defines the piecewise interpretation of *loc* such that $loc^*(t) = U_{loc}^*(\delta(0)) = \delta(0) = s_0$ for $t \in [0,1)$, $loc^*(t) = U_{loc}^*(\delta(1)) = s_1$ for $t \in [1, 3.7)$ and $loc^*(t) = U_{loc}^*(\delta(2)) = s_2$ for $t \in [3.7, \infty)$.
Now take $\Xi_c = \{(1,-1),(1,0)\}$ for a clock c. The same π and $\delta_c = ((1,-1,0),(1,-1,0),(1,0,-3.7))$ give the following interpretation c^* of c: $U_c(t,0,1,-1,0) = t$ for $t \in [0,1)$, $U_c(t,1,1,-1,0) = t-1$ for $t \in [1,3.7)$ and $U_c(t,3.7,1,0,-3.7) = t - 3.7$ for $t \in [3.7,\infty)$. This interpretation says that the clock is reset to 0 at the beginning of each interval of the partition.

An *(admissible) interpretation* of V is a set of admissible interpretations, one for each abstract element of V. Together with the interpretation of sorts this gives an interpretation of the entire vocabulary W. As V is finite we may assume that any its interpretation is defined by a common partition which is admissible for all these interpretations of functions and predicates. Such an interpretation can be described as $(\pi, (\delta_f)_{f \in V})$ or $(\zeta_i, \Sigma_i)_{i \in \overline{N}}$, where $\pi = (\zeta_i)_{i \in \overline{N}}$ is a partition and $\Sigma_i = ((\delta_f(i))_{f \in V})$ is a list $\delta_f(i)$ of parameters defining f on ζ_i, $i \in \overline{N}$. We will call such a sequence $(\zeta_i, \Sigma_i)_{i \in \overline{N}}$ *an interpretation over the partition* $\pi = (\zeta_i)_{i \in \overline{N}}$.

Let V^* be an interpretation of V. For any $t \in \mathcal{T}$ we denote by $V^*(t)$ the vector composed by the values of functions and predicates of V at t given by V^*. Each such vector will be called a *state* over V or W. For $t \in \zeta_i$ we have $V^*(t) = (f^*(t))_{f \in V} = (U_f(t, \zeta_i^-, \delta(i))_{f \in V})$.

The closed formulas of our logic are evaluated only over just defined admissible interpretations. The notations $\mathcal{M} \models F$, $\mathcal{M} \not\models F$ and $\models F$ will respectively mean that the admissible interpretation $\mathcal{M}$ is a model of a formula F, is a counter-model of F and that F is valid, i. e. every *admissible* interpretation is a model of F.

In this logic we can describe the runs of Gurevich Abstract State Machines [BS97a] or runs of timed automata [BS98].

3 Finiteness Properties

Finite Partial Interpretations (FPI).
A *(partial) interpretation element* with *support* ζ is pair of the form (ζ, Σ), where ζ is an arbitrary interval, and Σ is a list of values of parameters, each value defining an interpretation of an abstract symbol of V over ζ.

A *partial interpretation* of V is a set of disjoint interpretation elements $(\zeta_i, \Sigma_i)_{i \in I}$, $I \subseteq \mathbb{N}$. Its *support* is the set $\bigcup_i \zeta_i$. A *finite partial interpretation (FPI)* of V is a partial interpretation with finite number of elements. A (total) interpretation with finite number of elements will be called *finite interpretation.* The *complexity* of a FPI is the number of its intervals, i. e. $|I|$. "FPI of complexity k" or "k-FPI" will mean that the complexity of the FPI is k.

A k-FPI which support is $\mathcal{T}$ and that together with the interpretation of sorts and predefined symbols constitutes a model of a formula G, will be called a *k-model* of G.

A partial interpretation $\mathcal{M}'$ is an *extension* of a partial interpretation $\mathcal{M}$ if every interval of $\mathcal{M}$ is contained in an interval of $\mathcal{M}'$, and the restriction of functions of $\mathcal{M}'$ on intervals of $\mathcal{M}$ gives functions of $\mathcal{M}$.

We will define the *finiteness* properties in terms of FPI contained in models or counter-models $\mathcal{M}$ of the formulas under consideration.

Finite Refutability and Finite Satisfiability.
A formula G is *finitely refutable with complexity k* if for every counter-model $\mathcal{M}$ of G there is a k-FPI $\mathcal{M}_1$ that is a restriction of $\mathcal{M}$ such that every extension of $\mathcal{M}_1$ remains a counter-model of G.

Let $\mathcal{C}$ be a class of interpretations. Let α be a total computable function from $\mathbb{N}$ to complexity of interpretations that can be represented either by one or two natural numbers depending on the class under consideration. A formula G is *$\mathcal{C}$-satisfiable with augmentation α* if for every k-FPI $\mathcal{M}$ extendable to a model of G there is a model from $\mathcal{C}$ with complexity $\alpha(k)$ which is an extension of $\mathcal{M}$.

An interpretation $\mathcal{M}$ is *ultimately repetitive of complexity K and of period h* if it can be represented as $(\zeta_i, \Sigma_i)_{i \in \overline{N}}$, where either $|\overline{N}| = K$, or

$$\overline{N} = \mathbb{N} \text{ and } \forall n > 0 \left(\sum_{i=0}^{K-1} |\zeta_{(nK+i)}| = h \right.$$

$$\left. \wedge \bigwedge_{i=0}^{K-1} \left(\Sigma_{(nK+i)} = \Sigma_{((n+1)K+i)} \wedge |\zeta_{nK+i}| = |\zeta_{(n+1)K+i}| \right) \right).$$

If $|\overline{N}| = K$ then $\mathcal{M}$ is a *finite interpretation of complexity k* (this case was considered in [BS98]).

Notice that if each U_f is of the form z for an abstract z or of the form $U_f(t, a, \xi_0, \xi_1, z) = \xi(t - a) + z$ then ultimately repetitive means ultimately periodic.

We denote by $\mathcal{UR}$ and $\mathcal{F}$ respectively the classes of ultimately repetitive and finite interpretations. Finite satisfiability means $\mathcal{C}$-satisfiability for $\mathcal{C} = \mathcal{UR}, \mathcal{F}$.

Example 3 Consider n clocks c_i, $1 \leq i \leq n$ and 2 formulas Ψ and Φ, the first one describing the behavior of these clocks and the second one saying that infinitely often the average value of the clocks is 1:

$\Psi =_{df} \bigwedge_i \forall t \exists t_1 \left(0 \leq t_1 \leq t \wedge \forall \tau \in [t_1, t] \, c_i(\tau) = t - t_1 \right),$

$\Phi =_{df} : \forall t \exists t' > t \sum_i c_i(t') = n.$

The conjunction $F =_{df} (\Phi \wedge \Psi)$ is $\mathcal{UR}$-satisfiable with augmentation $\alpha(k) = k(n+2)$. Indeed, take any model $\mathcal{M}$ of F, and let $\mathcal{M}_1$ be some restriction of complexity k. Let t_0 be a time moment that lies to the right of the support of $\mathcal{M}_1$. We can extend $\mathcal{M}_1$ on $[0, t_0)$ to a $k(n+2)$-FPI by suppressing for each clock all its resets to 0 between 2 consecutive connected components of $\mathcal{M}_1$ except the last one and all its resets to 0 between the last connected component of $\mathcal{M}_1$ and t_0. Then we reset each clock to 0 at time moments $t_0 + 2k$ for $k \in \mathbb{N}$, and thus, we get a model of F in the class $\mathcal{UR}$ of complexity $\alpha(k)$ if $(1, -1) \in \Xi_{c_i}$ for all i. It is clear that F is not $\mathcal{F}$-satisfiable whatever be the choice of the Ξ_{c_i}.

Remark 1 In [BS98] we used more general notions of finite refutability and finite satisfiability where the involved models were considered up to some equivalence over states. The notions of this paper can be also extended in this way. In [BS98] we proved that the formula representing the runs of a reducible timed automata is $\mathcal{F}$-satisfiable, and the reducibility is decidable. A timed automaton is reducible with a threshold L if any its run having more than L changes of states can be replaced by an equivalent run having not more than L changes.

Remark 2 Finite refutability of properties of functioning of real time systems often takes place. E. g. safety and dependability (liveness) properties of the Generalized Railroad Crossing Problem (see [BS97a]) are clearly finitely refutable. Concerning, say, the critical section problem the safety is always finitely refutable. As for liveness it is not the case unless we bound the waiting time of each process and suppose that the density of changes of functions and predicates is bounded. The latter hypotheses are justified from practical point of view.

Remark 3 Intuitively finite satisfiability of an algorithm means that every its run is reducible in the following sense: every interval of the run can be replaced by a piece of bounded complexity with respect to the class C under consideration. Many control algorithms possess this property which is, in a way, a finite memory property (which holds for the controllers of the problems mentioned in the previous remark).

4 Decidable Classes of the Verification Problem

Remind that the satisfiability and validity of FOTL$_0$-formulas are defined for *fixed* U_f. Denote by $VERIF(C, k, \alpha)$ the class of FOTL$_0$-formulas of the form $(\Phi \to \Psi)$, where Ψ is finitely refutable with complexity k and Φ is C-satisfiable with augmentation α.

The initial observation to describe the decidable classes is the following one:

Proposition 1 *Let C be a class of interpretations. Suppose that the existence of a counter-model of a fixed complexity from C is decidable for closed $FOTL_0$-formulas. Given k, the validity of formulas from $VERIF(C,k,\alpha)$ is decidable: such a formula has a counter-model iff it has a counter-model of complexity $\alpha(k)$ in C.*

Proof. A counter-model of an implication is a model of the premise and a counter-model of the conclusion. The conclusion Ψ is finitely refutable. Suppose there is a counter-model $\mathcal{M}$ of $F = (\Phi \to \Psi)$. Take its k-FPI restriction $\mathcal{M}_1$ which extensions are counter models of Ψ. For this FPI there exists an extension in C which is a model of the premise Φ with complexity $\alpha(k)$. It gives the desired counter-model of $(\Phi \to \Psi)$. ■

V. Weispfenning's Quantifier Elimination Theorem.
In [Wei99] V. Weispfenning gives a quantifier elimination for theory L'' with *mixed* variables, namely variables over reals and variables over integers. The vocabulary of L'' consists of two just mentioned sorts: reals $\mathbb{R}$ and integers $\mathbb{Z} \subset \mathbb{R}$, rational numbers $\mathbb{Q}$ as constants, (binary) addition, scalar (unary) multiplication by rational numbers, integer part $\lfloor\,\rfloor$, congruences $\equiv_n$ modulo concrete natural numbers n. We consider the vocabulary without congruences as the latter can be eliminated, see [Wei99]. The first part of the Corollary 3.4 of [Wei99] says that there is an algorithm assigning to a given L''-formula Φ a quantifier-free L''-formula that is equivalent to Φ.

Existence of Ultimately Repetitive Models.
Remark that each abstract function of time with abstract values can be modeled by a finite number of predicates, and, hence, eliminated. Thus we assume that *all abstract functions are of the type $\mathcal{T} \to \mathbb{R}$.*

Let G be a closed FOTL$_0$-formula and $K > 0$ be a number limiting the complexity of ultimately repetitive models. We consider the difficult case whether there exists a model with non trivial repetitive part (the existence of a finite model was considered in [BS98]).

The atomic formulas of G are either of the form $P(\theta(T))$ or of the form $\theta(T)\omega\theta'(T')$ where T and T' are lists of real or time variables, $\theta(T)$ and $\theta'(T')$ are terms and ω is an arithmetic relation $(=, <, \leq, \ldots)$. Time variables can be trivially eliminated; thus we assume that there are only real variables. To simplify the atomic formulas containing abstract symbols we first apply the equivalences:

$$P(\theta(T)) \leftrightarrow \forall\tau\,(\tau = \theta(T) \to P(\tau)),$$
$$\theta(T)\omega\theta'(T') \leftrightarrow \forall\tau\tau'\,((\tau = \theta(T) \wedge \tau' = \theta'(T')) \to \tau\omega\tau').$$

Every term $\theta(T)$ is either of the form $\varphi(\eta(T))$, where φ is a unary function and $\eta(T)$ is a term, or of the form $\eta_1(T_1)+\ldots+\eta_m(T_m)$, where $\eta_i(T_i)$ are terms, $1 \leq i \leq m$. Now using the equivalences

$$\tau = \varphi(\eta(T)) \leftrightarrow \forall\tau'(\tau' = \eta(T) \to \tau = \varphi(\tau')),$$
$$\tau = \eta_1(T_1)+\ldots+\eta_m(T_m) \leftrightarrow \forall\tau_1'\ldots\forall\tau_m'\,(\textstyle\bigwedge_{i=1}^{m} \tau_i' = \eta_i(T_i) \to \tau = \tau_1'+\ldots+\tau_m'),$$

all atomic formulas containing abstract symbols can be reduced to ones either of the form $P(t)$, where P is a unary predicate and t is a variable, or of the form $t = \varphi(\tau)$, where φ is a unary function and t, τ are variables.

The fact that the set Ξ_f of parameters which appear in U_f is fixed and finite for every $f \in V$ permits to express the existence of an ultimately repetitive model of complexity K of G as disjunction $\bigvee_{f \in V} \bigvee_{(\xi_0,\xi_1) \in \Xi_f}$ of the existence of reals $0 = \alpha'_0 < \alpha'_1 < \ldots < \alpha'_K$ and $0 = \alpha_0 < \alpha_1 < \ldots < \alpha_K = h$ that define a partition of time, and lists of values defining the functions and predicates on the intervals of this partition. It suffices to consider only the latter existence of reals for fixed parameters (ξ_0, ξ_1) for each f. The lists of values are as follows. For each predicate P of G there exist lists $\gamma'_{P,1}, \ldots, \gamma'_{P,K}$ and $\gamma_{P,1}, \ldots, \gamma_{P,K}$ that define the values of $P(t)$ respectively as $\gamma'_{P,j}$ on $[\alpha'_{j-1}, \alpha'_j)$ and as $\gamma_{P,i}$ on $[\alpha_{k,(i-1)}, \alpha_{k,i}) =_{df} [\alpha'_K + k \cdot h + \alpha_{i-1}, \alpha'_K + k \cdot h + \alpha_i)$ for $k \in \mathbb{N}$ and $1 \leq i \leq K$. For each function f of G there exist $\lambda'_{f,1}, \ldots, \lambda'_{f,K}$ and $\lambda_{f,1}, \ldots, \lambda_{f,K}$ such that $f(t)$ is defined as $U_f(t, \alpha'_{j-1}, \xi_0, \xi_1, \lambda'_{f,j})$ on $[\alpha'_{j-1}, \alpha'_j)$ and as $U_f(t, \alpha_{k,(j-1)}, \xi_0, \xi_1, \lambda_{f,j})$ on $[\alpha_{k,(i-1)}, \alpha_{k,(i)})$.

Eliminate from G all the abstract symbols in the following way. Assume that all bounded variables of G are pairwise distinct (one can consider that G is in a prenex form).

Replace every occurrence of each atomic formula of the form $P(t)$ by

$$\bigwedge_i \left(t \in [\alpha'_{i-1}, \alpha'_i) \to \gamma'_{P,i} = 1 \right)$$

$$\wedge \forall k \bigwedge_i \left(t \in [\alpha_{k,(i-1)}, \alpha_{k,i}) \to \gamma_{P,i} = 1 \right) \qquad (1)$$

and every occurrence of each atomic formula of the form $x = f(t)$ by

$$\bigwedge_i \left(t \in [\alpha'_{i-1}, \alpha'_i) \to x = U_f(t, \alpha'_{i-1}, \xi_0, \xi_1, \lambda'_{f,i}) \right)$$

$$\wedge \forall k \bigwedge_i \left(t \in [\alpha_{k,(i-1)}, \alpha_{k,i}) \to x = U_f(t, \alpha_{k,(i-1)}, \xi_0, \xi_1, \lambda_{f,i}) \right) \qquad (2)$$

Denote by $\widetilde{G}$ the formula obtained from G after these transformations. Denote by R the formula which is a conjunction of inequalities $0 = \alpha'_0 < \alpha'_1 < \ldots < \alpha'_K$ and $0 = \alpha_0 < \alpha_1 < \ldots < \alpha_K = h$, and of the disjunctions $(\gamma'_{P,i} = 0 \vee \gamma'_{P,i} = 1)$ and $(\gamma_{P,i} = 0 \vee \gamma_{P,i} = 1)$ for all i and P. Denote by Π the list of all variables α'_i, α_i, $\gamma_{P,i}$, $\gamma'_{P,i}$, $\lambda_{f,i}$, $\lambda'_{f,i}$ except h.

Denote by G_0 the formula $\exists h \exists \Pi \, \widetilde{G}_0$, where $\widetilde{G}_0 =_{df} (R \wedge \widetilde{G})$.

Proposition 2 *A formula G has an ultimately repetitive model of complexity $K \Leftrightarrow \models G_0$.*

The formula G_0 is not a L''-formula as it contains mixed binary multiplications in atoms of the form $t\omega(k \cdot h + z_1 + z_2)$, where $\omega \in \{<, \leq, >, \geq\}$, t and z are variables for reals, h is a variable mentioned above and k is a variable for integers. All the other atoms are of the form $a_1 \cdot z_1 + \ldots + a_n \cdot z_n \omega c$ with $a_i, c \in \mathbb{Q}$ and z_i being real variables. Divide the both types of inequalities by h. Underline that $h > 0$ and is common for the whole formula. The bijection $z \leftrightarrow \frac{z}{h}$

preserves the order relations and commutes with the operations over reals. Replacing expressions $\frac{z}{h}$, where z is a variable, by new variables we get a formula $G_1 =_{df} \exists h \exists \Pi \widetilde{G}_1$, that is valid iff G_0 is valid and such that all atoms of $\widetilde{G}_1$ are of the form $t\omega(k + z_1 + z_2)$ and $a_1 \cdot z_1 + \ldots + a_n \cdot z_n \omega \frac{c}{h}$, where c is a concrete rational number. Now separate $\frac{c}{h}$ from this formula introducing new variables x_c, one for each constant term. Formula G_1 is equivalent to the formula

$$G_2 =_{df} \exists h \exists \overline{x}_c \, (\bigwedge_c x_c = \frac{c}{h} \wedge \widetilde{G}_2),$$

where $\overline{x}_c$ is a list of variables x_c for all constant terms c, and $\widetilde{G}_2$ is an L''-formula not containing h (remark that a more natural way is to use universal quantifiers over $\overline{x}_c$ but here one can use the existential ones). Now eliminate quantifiers in $\widetilde{G}_2$ by Weispfenning's elimination. We get a formula

$$G_3 =_{df} \exists h \exists \overline{x}_c \, (\bigwedge_c x_c = \frac{c}{h} \wedge \widetilde{G}_3),$$

where $\widetilde{G}_3$ is quantifier free. Replacing back x_c by $\frac{c}{h}$ we get an equivalent formula

$$G_4 =_{df} \exists h \, \widetilde{G}_4,$$

with $\widetilde{G}_4$ having as atoms inequalities of the form

$$\sum_i d_i \left\lfloor \frac{a_i}{h} + b_i h + c_i \right\rfloor + \frac{A}{h} + Bh + C \;\omega\; 0, \tag{3}$$

where all the letters different from h stand for concrete rational numbers. Thus the initial decidability problem is reduced to the existence of a solution of a system of inequalities of the form (3) with one real unknown h. One can prove that the latter problem is decidable. Hence,

Theorem 1 *The existence of an ultimately repetitive model (or counter-model) of a given complexity is decidable for closed* $FOTL_0$*-formulas.*

Chains of Repetitive Interpretations.
We say that an interpretation is a *chain of ultimately repetitive interpretations of complexity* (K, M) if it is a concatenation of at most $(M - 1)$ prefixes of repetitive interpretations and of one (infinite) ultimately repetitive interpretation, each of complexity K. Sure, such a concatenation demands an appropriate time translation. We denote the class of chains of ultimately (quasi)-repetitive interpretations by $\mathcal{UR}^*$.

Two other notations: $\mathcal{UR}^*(K, M, \Pi)$ is the set of interpretations of $\mathcal{UR}^*$ with complexity (K, M) and with all period lengths in a set $\Pi \subset \mathbb{Q}$; $\mathcal{UR}^*(\Pi)$ is the set $\bigcup_{K,M} \mathcal{U}(\mathcal{Q})\mathcal{R}^*(K, M, \Pi)$.

Proposition 3 *Given a finite set of period lengths* Π *and natural numbers* K, M *the existence of a (counter-)model from* $\mathcal{UR}^*(K, M, \Pi)$ *is decidable for* $FOTL_0$*-formulas.*

Together with Proposition 1 it gives

Theorem 2 *Given a k, α and a finite set $\Pi \subset \mathbb{Q}$, the validity of formulas from $VERIF(C, k, \alpha)$ is decidable for $C = \mathcal{UR}, \mathcal{UR}^*(\Pi)$.*

The *complexity* of the decision procedure is determined by the complexity of Weispfenning's Quantifier Elimination and by the complexity of our reductions. The complexity of Weispfenning's Quantifier Elimination is $2^{l^{n^{O(a)}}}$, where l is the length of the formula (presumed to be in a prenex form), n is the number of variables and a is the number of blocks of alternating quantifiers. Our reductions add $O(\alpha(k)|V|)$ variables and (together with transforming the formula into a prenex form) augment the size of the initial formula exponentially in the general case.

Example 4 ($\mathcal{UR}^*$-satisfiable but not $\mathcal{UR}$-satisfiable formula.)
Consider the set of runs of the automaton on Figure 1. It is described by some

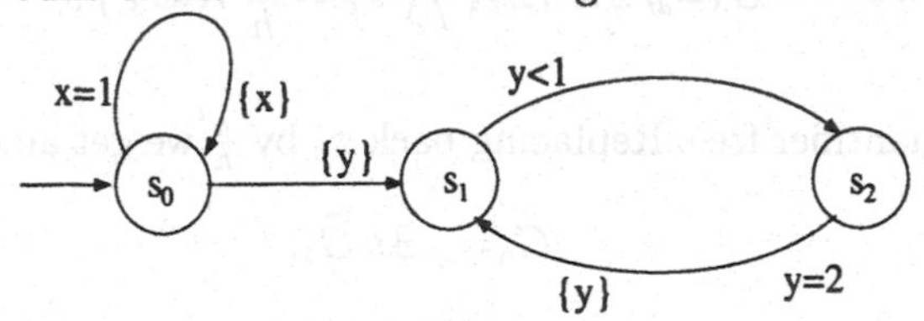

Fig. 1. An automaton with runs being chains of repetitive interpretations.

formula, but we will speak about the set. This set of runs is not $\mathcal{UR}$-satisfiable. Suppose that it is $\mathcal{UR}$-satisfiable with augmentation α, and $\alpha(1) = N_0$. Take a model $\mathcal{M}$ such that at moment $t_0 = N_0 + \frac{1}{2}$ we have $loc(t_0) = s_1$, $x(t_0) = \frac{1}{2}$ and $y(t_0) = 0$. The 1-FPI with support t_0 can be extended to a model, but any such extension has complexity at least $N_0 + 1$ whatever be the choice of Ξ_x and Ξ_y.

In a similar way one can prove that the set of runs is $\mathcal{UR}^*(\{1, 2\})$-satisfiable.

References

[ACHH93] R. Alur, C. Courcoubetis, T. Henzinger, and P.-H. Ho. Hybrid automata: an algorithmic approach to the specification and verification of hybrid systems. In R.L. Grossman, A. Nerode, A.P. Ravn, and H. Rischel, editors, *Workshop on Theory of Hybrid Systems, 1992*, pages 209–229. Springer Verlag, 1993. Lect. Notes in Comput. Sci, vol. 736.

[AD94] R. Alur and D. Dill. A theory of timed automata. *Theoretical Computer Science*, 126:183–235, 1994.

[BS97a] D. Beauquier and A. Slissenko. On semantics of algorithms with continuous time. Technical Report 97–15, Revised version., University Paris 12, Department of Informatics, 1997. Available at http://www.eecs.umich.edu/gasm/ and at http://www.univ-paris12.fr/lacl/.

[BS97b] D. Beauquier and A. Slissenko. The railroad crossing problem: Towards semantics of timed algorithms and their model-checking in high-level languages. In M. Bidoit and M. Dauchet, editors, *TAPSOFT'97: Theory and Practice of Software Development*, pages 201–212. Springer Verlag, 1997. Lect. Notes in Comput. Sci., vol. 1214.

[BS98] D. Beauquier and A. Slissenko. Decidable verification for reducible timed automata specified in a first order logic with time. Technical Report 98–16, University Paris 12, Department of Informatics, 1998. Available at http://www.univ-paris12.fr/lacl/.

[Eme90] A. Emerson. Temporal and modal logic. In J. van Leeuwen, editor, *Handbook of Theoretical Computer Science. Vol. B: Formal Models and Sematics*, pages 995–1072. Elsevier Science Publishers B.V., 1990.

[Gur95] Y. Gurevich. Evolving algebra 1993: Lipari guide. In E. Börger, editor, *Specification and Validation Methods*, pages 9–93. Oxford University Press, 1995.

[Han94] H. A. Hansson. *Time and Probability in Formal Design of Distributed Systems.* Elsevier, 1994. Series: "Real Time Safety Critical System", vol. 1. Series Editor: H. Zedan.

[Hen98] T. Henzinger. It's about time: real-time logics reviewed. In *Proc. 10th CONCUR*, pages 439–454. Springer-Verlag, 1998. Lect. Notes in Comput. Sci., vol. 1466.

[HR98] Y. Hirshfeld and A. Rabinovich. Quantitative temporal logic. Manuscript, 11 p., 1998.

[HR99] Y. Hirshfeld and A. Rabinovich. A framework for decidable metrical logics. Manuscript, 13 p., 1999.

[PVS] PVS. WWW site of PVS papers. http://www.csl.sri.com/sri-csl-fm.html.

[Rab97a] A. Rabinovich. Decidability in monadic logic of order over finitely variable signals. Manuscript, 15 p., 1997.

[Rab97b] A. Rabinovich. On the decidability of continuous time specification formalisms. Manuscript, 15 p., 1997. To appear in *J. of Logic and Comput.*

[Rab98] A. Rabinovich. Expressive completeness of duration calculus. Manuscript, 33 p., 1998.

[Tra98] B. Trakhtenbrot. Automata and hybrid systems. Lecture Notes 153, Uppsala University, Computing Science Department, 1998. Edited by F. Moller and B. Trakhtenbrot.

[Wei99] V. Weispfenning. Mixed real-integer linear quantifier elimination. In *Proc. of the 1999 Int. Symp. on Symbolic and Algebraic Computations (ISSAC'99)*, ACM Press, 1999. To appear.

Interpretations of Extensible Objects and Types

Viviana Bono[1] and Michele Bugliesi[2]

[1] School of Computer Science, The University of Birmingham
Edgbaston, Birmingham, B152 TT, UK
V.Bono@cs.bham.ac.uk

[2] Dipartimento di Informatica, Università "Ca' Foscari" di Venezia
Via Torino 155, I-30173 Mestre, Italy
michele@dsi.unive.it

Abstract. We present a type-theoretic encoding of extensible objects and types. The ambient theory is a higher-order λ-calculus with polymorphic types, recursive types and operators, and subtyping. Using this theory, we give a type preserving and computationally adequate translation of a full-fledged object calculus that includes object extension and override. The translation specializes to calculi of nonextensible objects and validates the expected subtyping relationships.

1 Introduction

The attempt to reduce object-oriented programming to procedural or functional programming is motivated by the desire to give sound and formal foundations to object-oriented languages and their specific constructs and techniques. The research in this area initiated with Cook's work [Coo87,Coo89] on the *generator model*, and Kamin's *self-application semantics* [Kam88]. Refined formulations of the generator model were later proposed by Bruce [Bru94] to give interpretations of *class-based* object calculi. A number of encodings for *object-based* calculi have then been formulated by Pierce and Turner [PT94], Abadi, Cardelli and Viswanathan [AC96,ACV96,Vis98], Bruce, Pierce and Cardelli [BCP97], and by Crary [Cra98]. These interpretations apply to a rich variety of object calculi with primitives of object formation, message send and (functional) method override: they succeed in validating the operational semantics of these calculi as well as the expected subtyping relations.

None of these proposals, however, scales to calculi of extensible objects, where primitives are provided for modifying the size of an object with the addition of new methods. Method addition poses two major problems: the first is the need for MyType polymorphic typing of methods, to allow method types to be specialized when methods are inherited; the second arises from the combination of subtyping and object extension [FM95].

The interpretation we present in this paper addresses both these problems. Our source calculus features extensible objects in the spirit of the *Lambda Calculus of Objects* [FHM94] and subsequent calculi [FM95,BL95,BB98]. MyType polymorphism is rendered via *match*-bounded polymorphism, as in the system we developed in [BB98]. Subtyping, is accounted for by distinguishing extensible from nonextensible objects as suggested by Fisher and Mitchell in [FM95].

As in other papers on encodings, our interpretation is a translation of the source object calculus into a polymorphic λ-calculus with recursive types and (higher-order) subtyping. In the encoding, extensible objects are represented as recursive records that include "selectable" methods, "method updaters" invoked upon override, as well as "method generators" that reinstall selectable methods upon extension. The contributions of our approach can be summarized as follows.

Firstly, it constitutes the first[1] interpretation of extensible objects into a fully formal functional calculus. The interpretation is faithful to the source calculus, as it is computationally adequate and validates the typing of terms.

Secondly, the translation specializes to the case of nonextensible objects, validating the expected subtypings: although we focus on one particular calculus – specifically, on one approach to combining object extension with subtyping – the translation is general enough to capture other notions of subtyping over object types (notably, the notions of covariant and invariant subtyping of [AC96]).

The rest of the paper is organized as follows. In Sections 2 and 3 we review the object and functional calculi used in the translation. In Sections 4 and 5 we describe the translation of extensible objects. In Section 6 we discuss the interpretation of nonextensible objects and various forms of subtyping relationships. In Section 7 we discuss related work and some final remarks.

2 Ob$^+$: Extensible Objects and Types

The source calculus of our translation, called Ob$^+$, is essentially a typed version of the *Lambda Calculus of Objects* of [FHM94]. There are two differences from the original proposal of [FHM94]: (*i*) the syntax of Ob$^+$ is typed, and (*ii*) methods are ς-abstractions instead of the λ-abstractions of [FHM94]. The typed syntax is useful in the translation, as it ensures that well-typed objects have unique types. The choice of ς-binders makes the syntax of Ob$^+$ a proper extension of the the typed ς-calculus of [AC96], and thus it facilitates comparisons with previous translations in the literature.

Types and Terms. An object type has the form $\mathtt{pro}(\mathtt{X})\langle m_i{:}B_i\{\mathtt{X}\}^{i\in[1..n]}\rangle$: it denotes the collection of objects with methods $m_1,\ldots,m_n$ that, when invoked, return values of types $B_1,\ldots,B_n$, respectively, with every free occurrence of X substituted by the pro-type itself. Types include type variables, denoted by X, U, ... The syntax of terms is defined by the following productions:

$a, b ::= x,$	variable
$\varsigma(\mathtt{U}, A)\langle m_i = \varsigma(x:\mathtt{U})b_i\rangle^{i\in[1..n]}$	object (m_i distinct)
$a \leftarrow\!+\, m{=}\varsigma(\mathtt{U}, A)\varsigma(x:\mathtt{U})b$	object extension
$a \leftarrow m{=}\varsigma(\mathtt{U}, A)\varsigma(x:\mathtt{U})b$	method override
$a \Leftarrow m$	method invocation

[1] But see [BDZ99] in these proceedings, for a similar approach.

An object is a collection of labelled methods: each method has a bound variable that represents self, and a body. In the above productions, the type A is the type of the object, and the type variable U is MyType, the type of self. This format of terms is inspired by [Rem97] and [Liq97]. Unlike those proposals, however, we use two operators for overriding and extension: this choice is well motivated, as the two operations are distinguished by our interpretation. The construct for extension allows the addition of a single method: a simple generalization of the syntax (and of the typing rules) would allow multiple simultaneous additions. The relation of top-level reduction (cf. App. A) extends the reduction relation of [AC96], with a clause for method additions (this clause simplifies the corresponding clause used in [Rem97]). The reflexive and transitive congruence generated by reduction is denoted by $\xrightarrow{obj}$; results are terms in object form (cf. App. A). We say that a closed term a converges – written $a \Downarrow_{obj}$ – if there exists a result v such that $a \xrightarrow{obj} v$.

Type System. The type system of Ob^+ relies on the same form of (implicit) match-bounded polymorphism we studied in [BB98] for the Lambda Calculus of Objects [FHM94]. The typing rules (cf. App A) generalize the corresponding typing rules of [AC96] for nonextensible objects. (Val Extend) requires the object a being extended to be a pro-type: method addition is thus typed with *exact* knowledge of the type of a. (Val Send) and (Val Override), instead, are both *structural*, in the sense of [ACV96]. In both rules, the type A may either unknown (i.e. a type variable), or a pro-type. When A is a pro-type, the operation (invocation or override) is *external*; when it is a type variable, the operation is *self-inflicted*: in both cases, A, (hence the object a), is required to have a method m with type B. In (Val Override), the typing of the method ensures that the new body has the same type as the original method: the bound for the type variable U, denoted by $\Gamma\langle A\rangle$, is either A, if A is a pro-type, or the current bound for A declared in the context Γ.

3 The Functional Calculus $F_{\omega<:\mu}$

The target calculus of the translation is $F_{\omega<:\mu}$, a variant of the omega-order polymorphic λ-calculus $F^{\omega}_{<:}$ with (higher-order) subtyping, extended with recursive types and operators, recursive functions and records, and local definitions. Types and type operators are collectively called *constructors*. A type operator is a function from types to types. The notation $\mathbb{A} :: \mathcal{K}$ indicates that the constructor $\mathbb{A}$ has kind $\mathcal{K}$, where $\mathcal{K}$ is either $\mathcal{T}$, the kind of types, or $\mathcal{K} \Rightarrow \mathcal{K}$, the kind of type operators. The typing rules are standard (see [AC96], Chap. 20). The following notation is used throughout: Op stands for the kind $\mathcal{T} \Rightarrow \mathcal{T}$; $\mathbb{A} \leq \mathbb{B}$ denotes subtyping over type operators; if $\mathbb{A}$ is a constructor of kind Op, $\mathbb{A}^*$ denotes the fixed point $\mu(\mathtt{X})\mathbb{A}(\mathtt{X})$ of $\mathbb{A}$; dually, for $\mathbb{A} :: \mathcal{T} \equiv \mu(\mathtt{X})\mathbb{B}(\mathtt{X})$, $\mathbb{A}^{\mathrm{OP}}$ is the type operator $\lambda(\mathtt{X})\mathbb{B}(\mathtt{X}) :: Op$ corresponding to $\mathbb{A}$. The syntax of types and terms, and the reduction rules for $F_{\omega<:\mu}$ are standard (cf. App. B). Evaluation, denoted by $\xrightarrow{fun}$, is the transitive and reflexive congruence generated by reduction; results include λ-abstractions and records. We say that a closed term a converges – written $a \Downarrow_{fun} v$ – if there exists a result v such that $a \xrightarrow{fun} v$.

4 Overview of the Translation

Looking at the typing rules of $\mathtt{Ob}^+$, we may identify two distinguished views of methods: the internal view, in which methods are concrete values, and the external view where methods may be seen as "abstract services" that can be accessed via message sends. The polymorphic typing of methods reflects the internal view, while the external view is provided by the types of methods in the object types. Based on this observation, our translation splits methods into two parts, in ways similar to, but different from, the translation of [ACV96]. Each method m_i is represented by two components: m_i^{poly}, associated with the actual method body, and m_i^{sel} which is selected by a message send.

Given $A \equiv \mathtt{pro}(\mathtt{X})\langle m_i : B_i\{\mathtt{X}\}\rangle^{i\in[1..n]}$, the m_i^{sel} components are collected in the *abstract interface* associated with A, which is represented by the type operator $\mathbb{A}^{\mathrm{IN}} \equiv \lambda(\mathtt{X})[m_i^{sel} : \mathbb{B}_i\{\mathtt{X}\}]^{i\in[1..n]}$ (here, and below, $\mathbb{B}_i$ is the translation of B_i). The type A, instead, is represented as the recursive record type $\mathbb{A} = \mu(\mathtt{X})[m_i^{poly} : \forall(\mathtt{U} \leq \mathbb{A}^{\mathrm{IN}})\mathtt{U}^*\rightarrow\mathbb{B}_i\{\mathtt{U}^*\},\ m_i^{sel} : \mathbb{B}_i\{\mathtt{X}\}]^{i\in[1..n]}$. Note that the polymorphic components are exposed in the type, as they will be needed in the interpretation of object extension. The translation of objects parallels this interpretation of object types. Letting $\mathbb{A}^{\mathrm{OP}} \equiv \lambda(\mathtt{X})[m_i^{poly} : \forall(\mathtt{U} \leq \mathbb{A}^{\mathrm{IN}})\mathtt{U}^*\rightarrow\mathbb{B}_i\{\mathtt{U}^*\},\ m_i^{sel} : \mathbb{B}_i\{\mathtt{X}\}]^{i\in[1..n]}$, the translation of an object $\varsigma(\mathtt{X}, A)\langle m_i = \varsigma(x : \mathtt{X})b_i\rangle^{i\in[1..n]}$ is the recursive record satisfying the equation $a = [m_i^{poly} = \Lambda(\mathtt{U} \leq \mathbb{A}^{\mathrm{IN}})\lambda(x : \mathtt{U}^*)\,[\![\, b_i \,]\!]\,, m_i^{sel} = a.m_i^{poly}(\mathbb{A}^{\mathrm{OP}})(a)]^{i\in[1..n]}$, where $[\![\, b_i \,]\!]$ is the translation of the body b_i. Method bodies, labelled by the m_i^{poly}'s, are represented as polymorphic functions of the self parameter, whose type is $\mathtt{U}^*$, the fixed point of the type operator $\mathtt{U}$. The constraint $\mathtt{U} \leq \mathbb{A}^{\mathrm{IN}}$ ensures that $\mathtt{U}^*$ contains all the m_i^{sel}'s, thus allowing each method to invoke its sibling methods via self. The m_i^{sel} components, in turn, are formed by self-application: method invocation for each m_i may then safely be interpreted as record selection on m_i^{sel}.

Method Override. Method override is accounted for by extending the interpretation of objects with a collection of *updaters*, as in [ACV96]. In the new translation, each method m_i is split in three parts, introducing the updater m_i^{upd}. The function of the updater is to take the method body supplied in the override and return a new object with the new body installed in place of the original: overriding m_i is thus translated by a simple call to m_i^{upd}. The typing of updaters requires a different, and more complex definition of the abstract interface. The problem arises from self-inflicted overrides: if a self-inflicted override is to be translated as a call to the updater, the updater itself must be exposed in the interface $\mathbb{A}^{\mathrm{IN}}$ used in the type of the polymorphic components. But then, since the polymorphic components and the updaters must be typed consistently, the updaters must be exposed in the interface $\mathbb{A}^{\text{IN}}$ used in the type of the updaters themselves. This leads to a definition of the interface as the type operator that satisfies the equation $\mathbb{A}^{\text{IN}} = \lambda(\mathtt{X})[m_i^{upd} : (\forall(\mathtt{U} \le \mathbb{A}^{\text{IN}})\mathtt{U}^* \to \mathbb{B}_i\{\mathtt{U}^*\}) \to \mathtt{X},\ m_i^{sel} : \mathbb{B}_i\{\mathtt{X}\}]$.

5 The Translation, Formally

The translation is given parametrically on contexts. Parameterization on contexts is required to ensure a well-defined translation of type variables.

Table 1: Translation of Types

$$A \equiv \mathtt{pro}(\mathtt{X})\langle m_i : B_i\{\mathtt{X}\}\rangle^{i\in[1..n]}$$

$$\begin{aligned}
\llbracket \Gamma', \mathtt{X} \lessdot\!\# A, \Gamma'' \rhd \mathtt{X} \rrbracket^{\text{IN}} &\triangleq \mathtt{X}\\
\llbracket \Gamma \rhd A \rrbracket^{\text{IN}} &\triangleq \mu(\mathtt{Y})\lambda(\mathtt{X})[\, m_i^{upd} : (\forall(\mathtt{U} \le \mathtt{Y})\mathtt{U}^* \to \llbracket \Gamma, \mathtt{X} \rhd B_i\{\mathtt{X}\} \rrbracket^{\text{TY}}\{\mathtt{X}{:=}\mathtt{U}^*\}) \to \mathtt{X},\\
&\qquad\qquad\qquad\; m_i^{sel} : \llbracket \Gamma, \mathtt{X} \rhd B_i\{\mathtt{X}\} \rrbracket^{\text{TY}}]^{i\in[1..n]})\\
\llbracket \Gamma', \mathtt{X} \lessdot\!\# A, \Gamma'' \rhd \mathtt{X} \rrbracket^{\text{OP}} &\triangleq \mathtt{X}\\
\llbracket \Gamma \rhd A \rrbracket^{\text{OP}} &\triangleq \lambda(\mathtt{X})[\, ext : \forall(\mathtt{U} \le \llbracket \Gamma \rhd A \rrbracket^{\text{IN}})\mathtt{U}^* \to \mathtt{U}^*\\
&\qquad\qquad m_i^{poly} : \forall(\mathtt{U} \le \llbracket \Gamma \rhd A \rrbracket^{\text{IN}})\mathtt{U}^* \to \llbracket \Gamma, \mathtt{X} \rhd B_i\{\mathtt{X}\} \rrbracket^{\text{TY}}\{\mathtt{X}{:=}\mathtt{U}^*\},\\
&\qquad\qquad m_i^{upd} : (\forall(\mathtt{U} \le \llbracket \Gamma \rhd A \rrbracket^{\text{IN}})\mathtt{U}^* \to \llbracket \Gamma, \mathtt{X} \rhd B_i\{\mathtt{X}\} \rrbracket^{\text{TY}}\{\mathtt{X}{:=}\mathtt{U}^*\}) \to \mathtt{X},\\
&\qquad\qquad m_i^{sel} : \llbracket \Gamma, \mathtt{X} \rhd B_i\{\mathtt{X}\} \rrbracket^{\text{TY}}]^{i\in[1..n]}\\
\llbracket \Gamma', \mathtt{X}, \Gamma'' \rhd \mathtt{X} \rrbracket^{\text{TY}} &\triangleq \mathtt{X}\\
\llbracket \Gamma', \mathtt{X} \lessdot\!\# A, \Gamma'' \rhd \mathtt{X} \rrbracket^{\text{TY}} &\triangleq \mathtt{X}^*\\
\llbracket \Gamma \rhd A \rrbracket^{\text{TY}} &\triangleq \mu(\mathtt{X})[\, ext : \forall(\mathtt{U} \le \llbracket \Gamma \rhd A \rrbracket^{\text{IN}})\mathtt{U}^* \to \mathtt{U}^*\\
&\qquad\qquad m_i^{poly} : \forall(\mathtt{U} \le \llbracket \Gamma \rhd A \rrbracket^{\text{IN}})\mathtt{U}^* \to \llbracket \Gamma, \mathtt{X} \rhd B_i\{\mathtt{X}\} \rrbracket^{\text{TY}}\{\mathtt{X}{:=}\mathtt{U}^*\},\\
&\qquad\qquad m_i^{upd} : (\forall(\mathtt{U} \le \llbracket \Gamma \rhd A \rrbracket^{\text{IN}})\mathtt{U}^* \to \llbracket \Gamma, \mathtt{X} \rhd B_i\{\mathtt{X}\} \rrbracket^{\text{TY}}\{\mathtt{X}{:=}\mathtt{U}^*\}) \to \mathtt{X},\\
&\qquad\qquad m_i^{sel} : \llbracket \Gamma, \mathtt{X} \rhd B_i\{\mathtt{X}\} \rrbracket^{\text{TY}}]^{i\in[1..n]}
\end{aligned}$$

The translation of types is by structural induction. As in [AC95], the treatment of object types depends on the context where they are used: in certain contexts they are interpreted as type operators, while in other contexts they are interpreted as types. From the translation of contexts and judgments (cf. Table 3), we see that $\llbracket\cdot\rrbracket^{\text{IN}}$ and $\llbracket\cdot\rrbracket^{\text{TY}}$ are used, respectively, in typing statements of the form $a : A$, and matching statements of the form $A \lessdot\!\# B$. The translation $\llbracket\cdot\rrbracket^{\text{OP}}$ is used in the translation of terms in Table 2 below, which also explains the presence of the *ext* field in $\llbracket\cdot\rrbracket^{\text{TY}}$ and $\llbracket\cdot\rrbracket^{\text{OP}}$.

For the translation of terms, we first introduce a recursive function that forms the (recursive fold of) the record with the m_i^{poly}, m_i^{sel} and m_i^{upd} components, together with the *ext* field needed to encode object extension. There is one such function for each type object type A.

$$\begin{array}{l}
\texttt{letrec}\ mkobj_A(f_i : \forall(\mathtt{U} \leq [\![\Gamma \rhd A]\!]^{\mathtt{IN}})\mathtt{U}^* \to \mathbb{B}_i\{\mathtt{U}^*\}^{i \in [1..n]}) : [\![\Gamma \rhd A]\!]^{\mathtt{TY}} = \\
\quad \texttt{let SELF} : [\![\Gamma \rhd A]\!]^{\mathtt{TY}} = mkobj_A(f_1)\dots(f_n)\ \texttt{in} \\
\quad \texttt{fold}([\![\Gamma \rhd A]\!]^{\mathtt{TY}}, \\
\qquad [ext = \Lambda(\mathtt{U} \leq [\![\Gamma \rhd A]\!]^{\mathtt{IN}})\lambda(x : \mathtt{U}^*)x \\
\qquad m_i^{poly} = f_i, \\
\qquad m_i^{upd} = \lambda(g : \forall(\mathtt{U} \leq [\![\Gamma \rhd A]\!]^{\mathtt{IN}})\mathtt{U}^* \to \mathbb{B}_i\{\mathtt{U}^*\})mkobj_A(f_1)\dots(g)\dots(f_n), \\
\qquad m_i^{sel} = \texttt{unfold}(\texttt{SELF}).m_i^{poly}([\![\Gamma \rhd A]\!]^{\mathtt{OP}})(\texttt{SELF})]^{i \in [1..n]})
\end{array}$$

where $A \equiv \mathtt{pro}(\mathtt{X})\langle m_i : B_i\{\mathtt{X}\}\rangle^{i \in [1..n]}$, and $\mathbb{B}_i\{\mathtt{U}^*\} \equiv [\![\Gamma, \mathtt{X} \rhd B_i\{\mathtt{X}\}]\!]^{\mathtt{TY}}\{\mathtt{X}{:=}\mathtt{U}^*\}$.

Table 2: Translation of Terms

$$[\![\Gamma \rhd \varsigma(\mathtt{U}, A)\langle m_i = \varsigma(x : \mathtt{U})b_i\rangle^{i \in [1..n]}]\!] \triangleq$$
$$mkobj_A(\Lambda(\mathtt{U} \leq [\![\Gamma \rhd A]\!]^{\mathtt{IN}})\lambda(s : \mathtt{U}^*)[\![\Gamma, \mathtt{U} \lessdot\!\!\# A, s : \mathtt{U} \rhd b_i]\!]^{i \in [1..n]})$$
where $A \equiv \mathtt{pro}(\mathtt{X})\langle m_i : B_i\{\mathtt{X}\}\rangle^{i \in [1..n]}$

$$[\![\Gamma \rhd a \leftarrow\!\!+\ m_{n+1} = \varsigma(\mathtt{U}, A^+)\varsigma(x : \mathtt{U})b]\!] \triangleq$$
$$\overline{a}.ext\ ([\![\Gamma \rhd A]\!]^{\mathtt{OP}})\ (mkobj_{A^+}(\overline{a}.m_1{}^{poly})\cdots(\overline{a}.m_n{}^{poly})\ (\Lambda(\mathtt{U} \leq [\![\Gamma \rhd A^+]\!]^{\mathtt{IN}})\overline{b}))$$
where $A \equiv \mathtt{pro}(\mathtt{X})\langle m_i : B_i\{\mathtt{X}\}\rangle^{i \in [1..n]}$, $A^+ \equiv \mathtt{pro}(\mathtt{X})\langle m_i : B_i\{\mathtt{X}\}\rangle^{i \in [1..n+1]}$,
$\overline{a} \equiv [\![\Gamma \rhd a]\!]$, and $\overline{b} \equiv \lambda(x{:}\mathtt{U}^*)[\![\Gamma, \mathtt{U} \lessdot\!\!\# A^+, x{:}\mathtt{U} \rhd b]\!]$

$$[\![\Gamma \rhd a \leftarrow m = \varsigma(\mathtt{U}, A)\varsigma(x : \mathtt{U})b]\!] \triangleq$$
$$\texttt{unfold}([\![\Gamma \rhd a]\!]).m^{upd}(\Lambda(\mathtt{U} \leq [\![\Gamma \rhd \Gamma\langle A\rangle]\!]^{\mathtt{IN}})\lambda(x : \mathtt{U}^*)\ [\![\Gamma, \mathtt{U} \lessdot\!\!\# \Gamma\langle A\rangle, x{:}\mathtt{U} \rhd b]\!])$$

$$[\![\Gamma \rhd a \Leftarrow m]\!] \triangleq \texttt{unfold}([\![\Gamma \rhd a]\!]).m^{sel}$$

In the clause for object formation, the typing of the m_i^{sel} components requires the relation $[\![\Gamma \rhd A]\!]^{\mathtt{OP}} <: [\![\Gamma \rhd A]\!]^{\mathtt{IN}}$, which is derived by first unrolling the fixed-point, and then applying the rules for constructor subtyping.

A method addition forms a new object by applying $mkobj_{A^+}$ (A^+ is the type of the extended object) to the (translation of) the method bodies of the original object a, and to the newly added method. Selecting the ext field from $\overline{a}$, – the object being extended – guarantees that $\overline{a}$ is evaluated prior to the extension: this is required for computational adequacy as the reduction rules of $\mathtt{Ob}^+$ do require a to be in object form prior to reducing a method addition. The call to $mkobj_{A^+}$ is well typed, as every $m_i^{poly} : \forall(\mathtt{U} \leq \mathbb{A}^{\mathtt{IN}})\mathtt{U}^* \to \mathbb{B}_i\{\mathtt{U}^*\}$ may be given, by subsumption, the type $\forall(\mathtt{U} \leq (\mathbb{A}^+)^{\mathtt{IN}})\mathtt{U}^* \to \mathbb{B}_i\{\mathtt{U}^*\}$, using $[\![\Gamma \rhd \mathbb{A}^+]\!]^{\mathtt{IN}} \leq [\![\Gamma \rhd \mathbb{A}]\!]^{\mathtt{IN}}$, which holds as $[\![\Gamma \rhd \mathbb{A}]\!]^{\mathtt{IN}}$ is covariant in the bound variable $\mathtt{Y}$.

The translation of method invocation and override on a method m are translated by a call to the corresponding components, m^{sel} or m^{upd}. In both cases, a recursive unfold is required prior to accessing the desired component.

The translation of contexts and judgments is obtained directly from the translation of types and terms.

Table 3: Translation of Contexts and Judgments

$[\![\Gamma \vdash *]\!] \triangleq [\![\Gamma]\!] \vdash \diamond$	$[\![\Gamma \vdash A \lessdot\!\!\# B]\!] \triangleq [\![\Gamma]\!] \vdash [\![\Gamma \rhd A]\!]^{\mathtt{IN}} \leq [\![\Gamma \rhd B]\!]^{\mathtt{IN}}$
$[\![\Gamma \vdash A]\!] \triangleq [\![\Gamma]\!] \vdash [\![\Gamma \rhd A]\!]^{\mathtt{TY}}$	$[\![\Gamma \vdash a : A]\!] \triangleq [\![\Gamma]\!] \vdash [\![\Gamma \rhd a]\!] : [\![\Gamma \rhd A]\!]^{\mathtt{TY}}$

We note that the translation of a judgment does not depend on its derivation in $\mathtt{Ob}^+$: as in [ACV96], we can thus avoid coherence issues in our proofs.

Theorem 1 (Validation of Typing). *If $\Gamma \vdash a$ is derivable in $\mathtt{Ob}^+$, then:*

1. $[\![\Gamma]\!] \vdash [\![\Gamma \triangleright a]\!] : [\![\Gamma \triangleright A]\!]^{\mathrm{TY}}$ *is derivable in* $\mathrm{F}_{\omega<:\mu}$.
2. *if* $a \xrightarrow{obj} b$, *then* $[\![\Gamma \triangleright a]\!] \xrightarrow{fun} [\![\Gamma \triangleright b]\!]$.

Theorem 2 (Computational Adequacy). *Let a be an $\mathtt{Ob}^+$ term such that $\emptyset \vdash a : A$ is derivable in $\mathtt{Ob}^+$. Then $a \Downarrow_{obj}$ if and only if $[\![\emptyset \triangleright a]\!] \Downarrow_{fun}$.*

6 Subtyping and Nonextensible Objects

The combination of object extension with subtyping has been studied from two orthogonal points of view in the literature: either limit subtyping in the presence of object extension, or distinguish extensible from nonextensible objects and disallow subtyping on the former while allowing it on the latter. Below, we focus on the second approach, deferring a discussion on the first to the full paper.

The idea of distinguishing between extensible and nonextensible objects was first proposed by Fisher and Mitchell in [FM95], to which the reader is referred to for details. Below, instead, we show that this idea allows different subtype relations to be formalized uniformly within the same framework.

Nonextensible objects are accounted for in $\mathtt{Ob}^+$ by introducing new types, contexts, and judgments as in the system $\mathtt{Ob}^+_{<:}$ (cf. Appendix A).

Table 4: Translation for $\mathtt{Ob}^+_{<:}$.

TYPES AND CONTEXTS	JUDGMENTS
$[\![\Gamma', \mathtt{X} <: A, \Gamma'' \triangleright \mathtt{X}]\!]^{\mathrm{TY}} \triangleq \mathtt{X}$	$[\![\Gamma \vdash A <: B]\!] \triangleq [\![\Gamma]\!] \vdash [\![\Gamma \triangleright A]\!]^{\mathrm{TY}} <: [\![\Gamma \triangleright B]\!]^{\mathrm{TY}}$
$[\![\Gamma, \mathtt{X} <: A]\!] \triangleq [\![\Gamma]\!], \mathtt{X} <: [\![\Gamma \triangleright A]\!]^{\mathrm{TY}}$	

A further clause handles the translation of nonextensible object types: the format of this clause depends on how these types and the corresponding subtyping relation are defined. Below, we illustrate two cases.

Covariant Subtyping à la Fisher & Mitchell'95. The new types have the form $\mathtt{obj}(\mathtt{X})\langle\!\langle m_i{:}B_i\{\mathtt{X}\}\rangle\!\rangle^{i\in[1..n]}$, and their reading is similar to that of the pro-types of Section 2: unlike pro-typed objects, however, obj-typed objects may *not* be modified or extended from the outside. pro and obj types are ordered by subtyping, as established by the rule (Sub probj FM95) (in Appendix). Informally, pro-types may only be promoted to obj-types, not to other pro-types: hence only reflexive subtyping is available for pro-type, as required for the soundness of method addition and override. This subtyping rule allows subtyping both in *width* and *depth*: since elements of obj-types may not be overridden or extended, this powerful form of subtyping is sound. We note that the covariance condition $\Gamma, \mathtt{Y}, \mathtt{X} <: \mathtt{Y} \vdash B_i\{\mathtt{X}\} <: B'_i\{\mathtt{Y}\}$ is required also for the subtyping

$\mathtt{pro(X)}\langle m_i{:}B_i\{\mathtt{X}\}\rangle^{i\in[1..n]} <: \mathtt{obj(Y)}\langle m_i{:}B_i\{\mathtt{Y}\}\rangle^{i\in[1..n]}$: as discussed in [FM95] covariance is crucial for subject-reduction: our translation, given below, explains why it is generally required for soundness.

Translation for obj types.

$$[\![\Gamma \triangleright \mathtt{obj(X)}\langle m_i : B_i\{\mathtt{X}\}\rangle^{i\in[1..n]}]\!]^{\mathrm{TY}} \triangleq \mu(\mathtt{X})[m_i^{sel} : [\![\Gamma, \mathtt{X} \triangleright B_i\{\mathtt{X}\}]\!]^{\mathrm{TY}}]^{i\in[1..n]}$$

The translation (which coincides with the standard recursive-record encoding) explains why obj-typed objects may not be extended or overridden: this is easily seen once we note that their type hides the polymorphic methods and the updaters. Self-inflicted updates, instead, are still allowed, as in [FM95]. This also explains why subtyping between pro and obj types is only allowed to covariant occurrences of the recursion variable. To exemplify, consider a term $e_1 : [\![\mathtt{pro}(x)\langle m : \mathtt{X}{\to}B\rangle]\!]^{\mathrm{TY}}$, and assume that we allow e_1 to be viewed as an element of $[\![\mathtt{obj}(x)\langle m : \mathtt{X}{\to}B\rangle]\!]^{\mathrm{TY}}$. Now, given $e_2 : [\![\mathtt{obj}(x)\langle m : \mathtt{X}{\to}B\rangle]\!]^{\mathrm{TY}}$, the interpretation of $e_1 \Leftarrow m(e_2)$ is not sound, as the code of m in e_1 could use a *self-inflicted* update that is not available in the code for m in e_2 (consider that e_2 may not have the polymorphic methods available in e_1).

Theorem 3 (Validation of *Fisher-Mitchell* Subtyping). *If $\Gamma \vdash A <: B$ is derivable in $\mathtt{Ob}^{+}_{<:}$, (using (Sub* probj FM95*) for object subtyping) then the judgment $[\![\Gamma]\!] \vdash [\![\Gamma \triangleright A]\!]^{\mathrm{TY}} <: [\![\Gamma \triangleright B]\!]^{\mathrm{TY}}$ is derivable in $\mathrm{F}_{\omega<:\mu}$.*

Invariant Subtyping for Covariant Self Types à la Abadi & Cardelli'96. Covariant Self Types, denoted here by the type expression $\mathtt{obj_{AC}(X)}\langle m_i : B_i\{\mathtt{X}\}\rangle^{i\in[1..n]}$ are described in [AC96] (cf. Chaps. 15, 16). They share several features with the obj-types of [FM95], notably the fact that both describe collections of nonextensible objects. However, they have important specificities: (*i*) method override is a legal operation on elements of $\mathtt{obj_{AC}}$ types, and (*ii*) subtyping over $\mathtt{obj_{AC}}$ types is only allowed in *width*, and defined by the rule (Sub probj AC96) (cf. Appendix). A translation that validates that rule is given below:

Translation of $\mathtt{obj_{AC}}$ Types

Let $A \equiv \mathtt{obj_{AC}(X)}\langle m_i : B_i\{\mathtt{X}\}\rangle^{i\in[1..n]}$, and let $[\![\Gamma \triangleright A]\!]^{\mathrm{IN}}$ be defined as in Table 1.

$$[\![\Gamma \triangleright A]\!]^{\mathrm{TY}} \triangleq \mu(\mathtt{X})[\ m_i^{upd} : (\forall(\mathtt{U} \le [\![\Gamma \triangleright A]\!]^{\mathrm{IN}})\mathtt{U}^* \to [\![\Gamma, \mathtt{X} \triangleright B_i\{\mathtt{X}\}]\!]^{\mathrm{TY}}\{\mathtt{X}{:=}\mathtt{U}^*\}) \to \mathtt{X},\ m_i^{sel} : [\![\Gamma, \mathtt{X} \triangleright B_i\{\mathtt{X}\}]\!]^{\mathrm{TY}}]^{i\in[1..n]}$$

Note how the updaters are exposed by the translation, thus making the translation of overrides well typed. Each of the component B_i is invariant in the translated type, as a result of a contravariant occurrence in the updater's type, and of a covariant occurrence in the selector's type.

Theorem 4 (Validation of *Abadi-Cardelli* Subtyping). *If $\Gamma \vdash A <: B$ is derivable in $\mathtt{Ob}^{+}_{<:}$, (using (Sub* probj AC96*) for object subtyping) then the judgment $[\![\Gamma]\!] \vdash [\![\Gamma \triangleright A]\!]^{\mathrm{TY}} <: [\![\Gamma \triangleright B]\!]^{\mathrm{TY}}$ is derivable in $\mathrm{F}_{\omega<:\mu}$.*

Invariant Subtyping. In [ACV96], an encoding is presented that validates invariant subtyping for object types, without requiring the covariance restriction for the component types. However, as discussed in [AC96], covariance is critical for sound method invocations: briefly, the problem arises with binary methods, since the use of bounded abstraction in the coding of the binder $\mathtt{obj}_{\mathtt{AC}}$ makes the type of self *unique*, hence different from any other type. The same problem affects the coding of [ACV96]: only covariant methods may be effectively invoked.

An interpretation with the same properties may be obtained from our translation. Given the type $\mathtt{obj}_{\mathtt{AC}}(\mathtt{X})\langle\!\langle m_i{:}B_i\{\mathtt{X}\}\rangle\!\rangle^{i\in[1..n]}$, invariant subtyping may be rendered by exposing the updaters of all the m_i's methods, while hiding the selectors of all the m_i's whose type B_i is not covariant in the bound variable. This translation would be the exact equivalent of that proposed in [ACV96]: it would validate invariant subtyping, and allow invocation only for covariant methods.

7 Related Work

The idea to split methods into different components is inspired by the object encoding of [ACV96]. That translation applies only to nonextensible objects, which are encoded by a combined use of recursive and bounded existential types, subsequently named ORBE encoding [BCP97]. Our translation, instead, uses a combination of recursion and universal quantification to render MyType polymorphism. We are then able to obtain a corresponding translation for nonextensible objects with essentially equivalent results as [ACV96].

A variant of the ORBE encoding that does not use existential types is proposed in [AC96] (Chap. 18): our translation can be viewed as an extension of that encoding to handle primitives of method addition.

Other, more recent papers have studied object encodings. In [Cra98], Crary proposed a simpler alternative to the ORBE encoding for nonextensible objects based on a combination of existential and intersection types. In [Vis98] Visvanathan gives a full-abstract translation for first-order objects with recursive types (but no Self Types). Again, the translation does not handle extensible objects. In [BDZ99], Boudol and Dal-Zilio study an encoding for extensible objects that relies on essentially the same idea used in our interpretation, namely the representation of extensible objects as a pair of a generator and a non extensible object. The difference is that [BDZ99] uses extensible records in the target calculus to model object generators in ways similar to [Coo89].

References

[AC95] M. Abadi and L. Cardelli. On Subtyping and Matching. In *Proceedings of ECOOP'95: European Conference on Object-Oriented Programming*, volume 952 of *LNCS*, pages 145–167. Springer–Verlag, August 1995.

[AC96] M. Abadi and L. Cardelli. *A Theory of Objects.* Monographs in Computer Science. Springer, 1996.

[ACV96] M. Abadi, L. Cardelli, and R. Viswanathan. An Iterpretation of Objects and Object Types. In *Proc. of POPL'96*, pages 396–409, 1996.

[BB98] V. Bono and M. Bugliesi. Matching for the Lambda Calculus of Objects. *Theoretical Computer Science*, 1998. To appear.

[BCP97] K. Bruce, L. Cardelli, and B. Pierce. Comparing Object Encodings. In *Proc. of TACS'97*, volume 1281 of *Lecture Notes in Computer Science*, pages 415–438. Springer-Verlag, 1997.

[BDZ99] G. Boudol and S. Dal-Zilio. An interpretation of extensible objects. In *Proceedings of FCT'99*, 1999.

[BL95] V. Bono and L. Liquori. A Subtyping for the Fisher-Honsell-Mitchell Lambda Calculus of Objects. In *Proc. of CSL*, volume 933 of *Lecture Notes in Computer Science*, pages 16–30. Springer-Verlag, 1995.

[Bru94] K.B. Bruce. A Paradigmatic Object-Oriented Programming Language: Design, Static Typing and Semantcs. *Journal of Functional Programming*, 1(4):127–206, 1994.

[Coo87] W. Cook. A Self-ish Model of Inheritance. Manuscript, 1987.

[Coo89] W.R. Cook. *A Denotational Semantics of Inheritance.* PhD thesis, Brown University, 1989.

[Cra98] K. Crary. Simple, efficient object encoding using intersection types. Technical report, Cornell University, April 1998.

[FHM94] K. Fisher, F. Honsell, and J. C. Mitchell. A Lambda Calculus of Objects and Method Specialization. *Nordic Journal of Computing*, 1(1):3–37, 1994.

[FM95] K. Fisher and J. C. Mitchell. A Delegation-based Object Calculus with Subtyping. In *Proc. of FCT*, volume 965 of *Lecture Notes in Computer Science*, pages 42–61. Springer-Verlag, 1995.

[Kam88] S. Kamin. Inheritance in Smalltalk-80: a denotational definition. In *Proc. of POPL'88*, pages 80–87. ACM Press, 1988.

[Liq97] L. Liquori. An Extended Theory of Primitive Objects: First Oder System. In *Proc. of ECOOP*, volume 1241 of *Lecture Notes in Computer Science*, pages 146–169. Springer-Verlag, 1997.

[PT94] B. Pierce and D. Turner. Simple type-theoretic foundations for object-oriented programming. *Journal of Functional Programming*, 4(2):207–248, 1994.

[Rem97] D. Remy. From classes to objects, via subsumption. Technical report, INRIA, 1997. Also in Proceeding of ESOP'98.

[Vis98] R. Viswanathan. Full abstraction for first-order objects with recursive types and subtyping. In *Proc. of LICS'98*, pages 380–391, 1998.

A The Source Calculus

Reduction

$$a \equiv \varsigma(\mathtt{U}, A)\langle m_i = \varsigma(x : \mathtt{U})b_i\rangle^{i\in[1..n]}$$

(Call) $(j \in [1..n])$

$$a \Leftarrow m_j \;\succ\; [a/x]b_j$$

(Extend) $m_{n+1} \notin \{m_1, .., m_n\}$

$$a \hookleftarrow m_{n+1} = \varsigma(\mathtt{U}, A')\varsigma(x : \mathtt{U})b \;\succ\; \varsigma(\mathtt{U}, A')\langle m_i = \varsigma(x : \mathtt{U})b_i\rangle^{i\in[1..n+1]}$$

(Override) $(j \in [1..n])$

$$a \leftarrow m_j = \varsigma(\mathtt{U}, A)\varsigma(x : \mathtt{U})b \;\succ\; \varsigma(\mathtt{U}, A)\langle m_i = \varsigma(x : \mathtt{U})b_i, m_j = \varsigma(x : \mathtt{U})b\rangle^{i\in[1..n]-\{j\}}$$

Results $\quad v ::= \varsigma(\mathtt{U}, A)\langle m_i = \varsigma(x : \mathtt{U})b_i\rangle^{i\in[1..n]}$

Context Formation – $\mathtt{Ob}^+$

(Ctx $\emptyset$)
$$\frac{}{\emptyset \vdash *}$$

(Ctx x)
$$\frac{\Gamma \vdash A \quad x \notin Dom(\Gamma)}{\Gamma, x : A \vdash *}$$

(Ctx Match)
$$\frac{\Gamma \vdash A \quad \mathtt{U} \notin Dom(\Gamma)}{\Gamma, \mathtt{U} <\!\# A \vdash *}$$

(Ctx $\mathtt{X}$)
$$\frac{\Gamma \vdash * \quad \mathtt{X} \notin Dom(\Gamma)}{\Gamma, \mathtt{X} \vdash *}$$

Type formation – $\mathtt{Ob}^+$

(Type Match $\mathtt{U}$)
$$\frac{\Gamma', \mathtt{U} <\!\# A, \Gamma'' \vdash *}{\Gamma', \mathtt{U} <\!\# A, \Gamma'' \vdash \mathtt{U}}$$

(Type $\mathtt{X}$)
$$\frac{\Gamma', \mathtt{X}, \Gamma'' \vdash *}{\Gamma', \mathtt{X}, \Gamma'' \vdash \mathtt{X}}$$

(Type pro)
$$\frac{\Gamma, \mathtt{X} \vdash A_i}{\Gamma \vdash \mathtt{pro}(\mathtt{X})\langle m_i : A_i \rangle^{i \in I}}$$

Term Formation – $\mathtt{Ob}^+$ **The notation $\Gamma\langle A \rangle$ in (Val Override) is defined as follows:**

$$\begin{array}{lll} \Gamma\langle A \rangle & \equiv A & \text{if } A \text{ is a pro-type;} \\ \Gamma\langle A \rangle & \equiv A' & \text{if } A \equiv \mathtt{U}' \text{ and } \mathtt{U}' <\!\# A' \in \Gamma. \end{array}$$

(Val x)
$$\frac{\Gamma', x : A, \Gamma'' \vdash *}{\Gamma', x : A, \Gamma'' \vdash x : A}$$

(Val Override)
$$\frac{\Gamma \vdash a : A \quad \Gamma \vdash A <\!\# \mathtt{pro}(\mathtt{X})\langle m : B\{\mathtt{X}\}\rangle \qquad \Gamma, \mathtt{U} <\!\# \Gamma\langle A \rangle, x : \mathtt{U} \vdash b : B\{\mathtt{U}\}}{\Gamma \vdash a \leftarrow m = \varsigma(\mathtt{U}, A)\varsigma(x : \mathtt{U})b : A}$$

(Val Send)
$$\frac{\Gamma \vdash e : A \qquad \Gamma \vdash A <\!\# \mathtt{pro}(\mathtt{X})\langle m : B\{\mathtt{X}\}\rangle}{\Gamma \vdash e \Leftarrow m : B\{A\}}$$

(Val Extend)
$$\frac{\begin{array}{l}(A \equiv \mathtt{pro}(\mathtt{X})\langle m_i : B_i\{\mathtt{X}\}\rangle^{i \in [1..n]} \\ A^+ \equiv \mathtt{pro}(\mathtt{X})\langle m_i : B_i\{\mathtt{X}\}\rangle^{i \in [1..n+1]}) \\ \Gamma \vdash a : A \quad \Gamma, \mathtt{U} <\!\# A^{ext}, x : \mathtt{U} \vdash b : B_{n+1}\{\mathtt{U}\}\end{array}}{\Gamma \vdash a \leftarrow\!+ \varsigma(\mathtt{U}, A^+)m_{n+1} = \varsigma(x : \mathtt{U})b : A^+}$$

(Val Object)
$$\frac{\begin{array}{l}(A \equiv \mathtt{pro}(\mathtt{X})\langle m_i : B_i\{\mathtt{X}\}\rangle^{i \in [1..n]}) \\ \Gamma, \mathtt{U} <\!\# A, x : \mathtt{U} \vdash b_i : B_i\{\mathtt{U}\} \quad \forall i \in [1..n]\end{array}}{\Gamma \vdash \varsigma(\mathtt{U}, A)\langle m_i = \varsigma(x : \mathtt{U})b_i \rangle^{i \in [1..n]} : A}$$

Matching – $\mathtt{Ob}^+$

(Match $\mathtt{U}$)
$$\frac{\Gamma', \mathtt{U} <\!\# A, \Gamma'' \vdash *}{\Gamma', \mathtt{U} <\!\# A, \Gamma'' \vdash \mathtt{U} <\!\# A}$$

(Match Refl)
$$\frac{\Gamma', \mathtt{U} <\!\# A, \Gamma'' \vdash \mathtt{U}}{\Gamma', \mathtt{U} <\!\# A, \Gamma'' \vdash \mathtt{U} <\!\# \mathtt{U}}$$

(Match Trans)
$$\frac{\Gamma \vdash u <\!\# B \quad \Gamma \vdash B <\!\# A}{\Gamma \vdash \mathtt{U} <\!\# A}$$

(Match pro)
$$\frac{\Gamma \vdash \mathtt{pro}(\mathtt{X})\langle m_i : B_i\{\mathtt{X}\}\rangle^{i \in [1..n+k]}}{\Gamma \vdash \mathtt{pro}(\mathtt{X})\langle m_i : B_i\{\mathtt{X}\}\rangle^{i \in [1..n+k]} <\!\# \mathtt{pro}(\mathtt{X})\langle m_i : B_i\{\mathtt{X}\}\rangle^{i \in [1..n]}}$$

Context and Type Formation – $\mathtt{Ob}^+_{<:}$

(Ctx Sub)
$$\frac{\Gamma \vdash A \quad \mathtt{U} \notin Dom(\Gamma)}{\Gamma, \mathtt{U} <: A \vdash *}$$

(Type obj)
$$\frac{\Gamma, \mathtt{X} \vdash B_i \quad \forall i \in [1..n]}{\Gamma \vdash \mathtt{obj}(\mathtt{X})\langle m_i : B_i \rangle^{i \in [1..n]}}$$

Term Formation – $\mathtt{Ob}^+_{<:}$

(Val Send Obj)
$$\frac{\Gamma \vdash e : A \quad (A \equiv \mathtt{obj}(\mathtt{X})\langle m_i : B_i\{\mathtt{X}\}\rangle^{i \in [1..n]}, \;\; j \in [1..n])}{\Gamma \vdash e \Leftarrow m_j : B_j\{A\}}$$

(Val Subsumption)
$$\frac{\Gamma \vdash e : A \quad \Gamma \vdash A <: B}{\Gamma \vdash e : B}$$

Subtyping – $\mathtt{Ob}^{+}_{<:}$

(Sub U)

$$\frac{\Gamma', \mathtt{U} <: A, \Gamma'' \vdash *}{\Gamma', \mathtt{U} <: A, \Gamma'' \vdash \mathtt{U} <: A}$$

(Sub Refl)

$$\frac{\Gamma \vdash A}{\Gamma \vdash A <: A}$$

(Sub Trans)

$$\frac{\Gamma \vdash A_1 <: A_2 \quad \Gamma \vdash A_2 <: A_3}{\Gamma \vdash A_1 <: A_3}$$

(Sub **probj FM95**)

$$\frac{\Gamma, \mathtt{Y}, \mathtt{X} <: \mathtt{Y} \vdash B'_i\{\mathtt{X}\} <: B_i\{\mathtt{Y}\} \qquad \forall i \in [1..n]}{\Gamma \vdash \mathtt{probj}(\mathtt{X})\langle m_i{:}B'_i\{\mathtt{X}\}\rangle^{i \in [1..n+k]} <: \mathtt{obj}(\mathtt{Y})\langle B_i\{\mathtt{Y}\}\rangle^{i \in [1..n]}}$$

(Sub **probj AC96**)

$$\frac{\Gamma, \mathtt{X} \vdash B_i\{\mathtt{X}\} \quad B_i \ \textit{covariant in}\ \mathtt{X} \quad \forall i \in [1..n+k]}{\Gamma \vdash \mathtt{probj}_{\mathtt{AC}}(\mathtt{X})\langle m_i : B_i\{\mathtt{X}\}\rangle^{i \in [1..n+k]} <: \mathtt{obj}_{\mathtt{AC}}(\mathtt{X})\langle m_i : B_i\{\mathtt{X}\}\rangle^{i \in [1..n]}}$$

B The Target Calculus $F_{\omega<:\mu}$

Syntax

$\mathcal{K} ::=$	*Kinds*	
	$\mathcal{T}$	types
	$\mathcal{K} \Rightarrow \mathcal{K}$	type operators
$\mathbb{A}, \mathbb{B} ::=$	*Constructors*	
	$\mathtt{X}$	constructor variable
	$\top$	greatest constructor of kind $\mathcal{T}$
	$\mathbb{A} \to \mathbb{B}$	function type
	$[m_1 : \mathbb{B}, \ldots, m_k : \mathbb{B}]$	record type
	$\forall(\mathtt{X} <: \mathbb{A} :: \mathcal{K})\mathbb{A}$	bounded universal type
	$\mu(\mathtt{X})\mathbb{A}$	recursive type
	$\lambda(\mathtt{X} :: \mathcal{K})\mathbb{B}$	operator
	$\mathbb{B}(\mathbb{A})$	operator application
$e ::=$	*Expressions*	
	x	variables
	$\lambda(x : \mathbb{A})\ e$	abstraction
	$e_1\, e_2$	application
	$\Lambda(\mathtt{X} <: \mathbb{A} :: \mathcal{K})\ e$	type-abstraction
	$e\ \mathbb{A}$	type-application
	$[m_1 = e_1, \ldots, m_k = e_k]$	record
	$e.m$	record selection
	$\mathtt{fold}(\mathbb{A}, e)$	recursive fold
	$\mathtt{unfold}(e)$	recursive unfold
	$\mathbf{let}\ x = e_1\ \mathbf{in}\ e_2$	local definition
	$\mathbf{letrec}\ f(x : \mathbb{A}) : \mathbb{B} = e_1\ \mathbf{in}\ e_2$	recursive local definition

Reduction

$(\beta_1)\quad (\lambda(x : \mathbb{A})e_1)\, e_2 \leadsto [e_2/x]\, e_1 \qquad (\textit{select}) \quad [m_i = e_i]^{i \in [1..n]}.m_j \leadsto e_j \quad (j \in [1..n])$

$(\beta_2)\quad (\Lambda(\mathtt{X} \leq \mathbb{A})e_1)\, \mathbb{B} \leadsto [\mathbb{B}/\mathtt{X}]\, e_1 \qquad (\textit{unfold}) \quad \mathtt{unfold}(\mathtt{fold}(\mathbb{A}, e)) \leadsto e$

Results $\quad v ::= \lambda(x : \mathbb{A})\ e \mid [m_1 = e_1, \ldots, m_k = e_k] \mid \mathtt{fold}(\mathbb{A}, e) \mid \Lambda(\mathtt{X} <: \mathbb{A} :: \mathcal{K})\ e$

Restrictive Acceptance Suffices for Equivalence Problems

Bernd Borchert[1], Lane A. Hemaspaandra[2], and Jörg Rothe[3*]

[1] Mathematisches Institut, Universität Heidelberg, 69120 Heidelberg, Germany
[2] Dept. of Computer Science, University of Rochester, Rochester, NY 14627, USA
[3] Institut für Informatik, Friedrich-Schiller-Universität Jena, 07740 Jena, Germany

Abstract. One way of suggesting that an NP problem may not be NP-complete is to show that it is in the class UP. We suggest an analogous new approach—weaker in strength of evidence but more broadly applicable—to suggesting that concrete NP problems are not NP-complete. In particular we introduce the class EP, the subclass of NP consisting of those languages accepted by NP machines that when they accept always have a number of accepting paths that is a power of two. Since if any NP-complete set is in EP then all NP sets are in EP, it follows—with whatever degree of strength one believes that EP differs from NP—that membership in EP can be viewed as evidence that a problem is not NP-complete.
We show that the negation equivalence problem for OBDDs (ordered binary decision diagrams [17, 12]) and the interchange equivalence problem for 2-dags are in EP. We also show that for boolean negation [20] the equivalence problem is in $\mathrm{EP}^{\mathrm{NP}}$, thus tightening the existing $\mathrm{NP}^{\mathrm{NP}}$ upper bound. We show that FewP [2], bounded ambiguity polynomial time, is contained in EP, a result that is not known to follow from the previous SPP upper bound. For the three problems and classes just mentioned with regard to EP, no proof of membership/containment in UP is known, and for the problem just mentioned with regard to $\mathrm{EP}^{\mathrm{NP}}$, no proof of membership in $\mathrm{UP}^{\mathrm{NP}}$ is known. Thus, EP is indeed a tool that gives evidence against NP-completeness in natural cases where UP cannot currently be applied.

1 Introduction

NP languages can be defined via machines that reject by having zero accepting paths, and that accept by having their number of accepting paths belong to the set $\{1, 2, 3, \ldots\}$. A number of researchers have sought to refine the class NP by shrinking the path-cardinality set signifying acceptance, while retaining the requirement that rejection be associated with having zero accepting paths. We will call any such class a *restricted counting class*. The most common restricted counting classes in the literature

* Corresponding author. Email: `rothe@informatik.uni-jena.de`.

are random polynomial time (usually denoted R or RP) and ambiguity-bounded classes such as UP and FewP. Ambiguity-bounded classes will be of central interest to us in the present paper.

Valiant's class UP (unambiguous polynomial time) [32], which is known to differ from P exactly if one-way functions exist [19], has the acceptance set $\{1\}$, and so is a restricted counting class. Acceptance sets of the forms $\{1, 2, 3, \ldots, n^{\mathcal{O}(1)}\}$ and $\{1, 2\}$, $\{1, 2, 3\}$, ... define, respectively, the class FewP [2] and the classes $\mathrm{UP}_{\leq 2}$, $\mathrm{UP}_{\leq 3}$, ... [3], and thus these too are restricted counting classes. (Note: $\mathrm{UP} \subseteq \mathrm{UP}_{\leq 2} \subseteq \mathrm{UP}_{\leq 3} \subseteq \cdots \subseteq \mathrm{UP}_{\mathcal{O}(1)} \subseteq \mathrm{FewP} \subseteq \mathrm{NP}$, where $\mathrm{UP}_{\mathcal{O}(1)} = \bigcup_{k \geq 1} \mathrm{UP}_{\leq k}$.) These classes are also connected to the existence of one-way functions and have been extensively studied in a wide variety of contexts, such as class containments [16, 25], complete sets [23], reducibilities [22], boolean hierarchy equivalences [24], complexity-theoretic analogs of Rice's Theorem [11], and upward separations [28].

Of course, the litmus test of NP refinements such as UP, $\mathrm{UP}_{\leq k}$, and FewP is *the extent to which they allow us to refine the upper bounds on the complexity of natural NP problems.* Of these classes, UP has been most successful in this regard. UP is known to provide an upper bound on the complexity of (a language version of) the discrete logarithm problem [19], and UP (indeed $\mathrm{UP} \cap \mathrm{coUP}$) is known to provide an upper bound on the complexity of primality testing [15].

However, there are certain NP problems whose richness of structure has to date defied attempts to put them in UP or even FewP, yet that nonetheless intuitively seem to use less than the full generality of NP's acceptance mechanism. To try to categorize these problems, we introduce the class EP, which is intermediate between FewP and NP: $\mathrm{FewP} \subseteq \mathrm{EP} \subseteq \mathrm{NP}$. In particular, EP is the NP subclass whose acceptance set is $\{2^i \mid i \in \mathbb{N}\}$, $\mathbb{N} = \{0, 1, 2, 3, \ldots\}$.

In Section 2, we provide improved upper bounds on the complexity of the problems OBDD (Ordered Binary Decision Diagram) Negation Equivalence, 2-Dag Interchange Equivalence, and Boolean Negation Equivalence. These three problems are trivially in, respectively, NP, NP, and $\mathrm{NP}^{\mathrm{NP}}$. We provide, respectively, EP, EP, and $\mathrm{EP}^{\mathrm{NP}}$ upper bounds. The problems are not known to belong to (and do not seem to obviously belong to), respectively, FewP, FewP, and $\mathrm{FewP}^{\mathrm{NP}}$.

In Section 3, we prove a general result regarding containment of FewP in certain restricted counting classes. In particular, we establish a sufficient condition for when restricted counting classes contain FewP. From our result it follows that EP contains FewP and, moreover, our result subsumes as special cases some previously known results from the literature.

In Section 4, we list some open questions related to our work.

2 Concrete Problems and EP

In this section, we provide concrete problems known to be in NP (or NP^{NP}), and we prove they are in fact in EP (or EP^{NP}). We now define the class EP (mnemonic: the number of accepting computation paths is restricted to being either 0 or some power (some *exponentiation*) of 2). For any nondeterministic polynomial-time Turing machine N and any string x, let $\#acc_N(x)$ denote the number of accepting computation paths of N on input x. Our alphabet Σ will be $\{0,1\}$. For any string $x \in \Sigma^*$, let $|x|$ denote the length of x.

Definition 1. EP *denotes the class of all languages L for which there is a nondeterministic polynomial-time Turing machine N such that, for each input* $x \in \Sigma^*$,

$$x \notin L \Longrightarrow \#acc_N(x) = 0, \text{ and}$$
$$x \in L \Longrightarrow \#acc_N(x) \in \{2^i \mid i \in \mathbb{N}\}.$$

Consider the following well-known problem.

Problem: Boolean Negation Equivalence (BNE) (see the survey by Harrison [20] and the bibliography provided after the references in the paper by Borchert, Ranjan, and Stephan [10])
Input: Two boolean functions (input as boolean formulas using variable names and the symbols $\{\wedge, \vee, \neg, (,)\}$), $f(x_1, \ldots, x_n)$ and $g(x_1, \ldots, x_n)$, over the same n boolean variables.
Question: Are f and g negation equivalent? That is, can one negate some of the inputs of g such that f and the modified function g' are equivalent?[1]

For concreteness as a language problem, BNE $= \{(f, g) \mid f$ and g are negation equivalent$\}$.

For example, the two boolean functions described by the formulas $x_1 \vee x_2 \vee x_3$ and $x_1 \vee \neg x_2 \vee \neg x_3$ are negation equivalent by negating x_2 and x_3.

[1] The notion of boolean function equivalence underlying the definition of negation equivalence is the standard one. Two boolean functions (over the same variables) are equivalent if they have the same truth value for every assignment to their variables. Testing equivalence of pairs of boolean formulas is in coNP.

Regarding lower bounds, Borchert, Ranjan, and Stephan [10] have shown that BNE is US-hard [8], and thus in particular is coNP-hard. Regarding upper bounds, BNE $\in \mathrm{NP}^{\mathrm{NP}}$ [10] and BNE $\in \mathrm{coAM}^{\mathrm{NP}}$ (combining [10] and [1]). It follows from the latter that BNE is not $\mathrm{NP}^{\mathrm{NP}}$-complete unless the polynomial hierarchy collapses ([1], in light of [10, 29]). Interestingly, neither of these two upper bounds—$\mathrm{NP}^{\mathrm{NP}}$ and $\mathrm{coAM}^{\mathrm{NP}}$—is known to imply the other.

We now prove BNE $\in \mathrm{EP}^{\mathrm{NP}}$, which is neither known to imply nor known to be implied by the $\mathrm{coAM}^{\mathrm{NP}}$ upper bound, but which clearly improves the $\mathrm{NP}^{\mathrm{NP}}$ upper bound as $\mathrm{EP}^{\mathrm{NP}} \subseteq \mathrm{NP}^{\mathrm{NP}}$. The proof of Theorem 1 can be found in the full version of this paper.

Theorem 1. BNE $\in \mathrm{EP}^{\mathrm{NP}}$.

There are ways of describing boolean functions such that the equivalence problem is in P. The most prominent such way is by ordered binary decision diagrams (OBDDs).[2] So, essentially by the same type of discussion found in the proof of Theorem 1, the following computational problem, OBDD Negation Equivalence, is in (nonrelativized) EP: Given a pair (e, f) of OBDDs, are the boolean functions described by e and f negation equivalent?

If we consider the special case that for the two OBDDs (e, f) above the order of the variables is required to be the same, we see that the following graph-theoretic problem is in (nonrelativized) EP. A *2-dag* is a directed acyclic graph (without labels) with a unique root and either 0 or 2 ordered successors for each node. For a 2-dag each node is assigned a depth, namely the distance to the root. Now consider the following computational problem (2-Dag Interchange Equivalence): Given two 2-dags F and G, is there a sequence of natural numbers $(i_1, \ldots, i_m)$ such that, if in G for each node of depth $i_1, \ldots, i_m$ its two successors (if they exist) are interchanged, then the modified 2-dag G' equals F? This problem can be shown to be in EP (similarly to the argument above). Moreover, the problem can easily be reduced to Graph Isomorphism. The authors know of no P algorithm for the general case of 2-Dag Interchange Equivalence, though the special case of this problem with binary trees instead of general 2-dags has an easy deterministic polynomial-time algorithm.

[2] Fortune, Hopcroft, and Schmidt [17] were the ones who proved that equivalence for OBDDs is in P. OBDDs have recently become a structure of interest to theoretical computer scientists in a variety of settings, see, e.g., [30, 14]. For general background on OBDDs see, for example, the survey by Bryant [12].

3 Location of EP

3.1 Result

We state a general result that our technique gives, regarding the containment of FewP in restricted counting classes. We need some additional definitions.

Definition 2. *Let S be any set of positive integers. Define the restricted counting class* RC_S *as follows.* $L \in \mathrm{RC}_S$ *if and only if there exists a nondeterministic polynomial-time Turing machine N such that, for every* $x \in \Sigma^*$,

1. *if* $x \in L$ *then* $\#acc_N(x) \in S$, *and*
2. *if* $x \notin L$ *then* $\#acc_N(x) = 0$.

For example, Valiant's extensively studied class UP equals $\mathrm{RC}_{\{1\}}$, and, for each $k \geq 2$, the class $\mathrm{ModZ}_k\mathrm{P}$ of Beigel, Gill, and Hertrampf [6] equals $\mathrm{RC}_{\mathbb{N}-\{a \mid (\exists b \in \mathbb{N})\, [a = b \cdot k]\}}$.

A set is non-gappy if it has only small holes.

Definition 3. *Let S be any set of positive integers. We say S is* non-gappy *if* $S \neq \emptyset$ *and* $(\exists k > 0)(\forall n \in S)(\exists m \in S)[m > n \wedge m/n \leq k]$.

Definition 4. [21] *Let L be any subset of Σ^*. We say L is* P-printable *if there is a deterministic Turing machine M that runs in polynomial-time such that, for every nonnegative integer n, $M(0^n)$ prints out the set* $\{x \mid x \in L \wedge |x| \leq n\}$.

Theorem 2. *Let T be any set of positive integers such that T has a non-gappy,* P*-printable subset. Then* $\mathrm{FewP} \subseteq \mathrm{RC}_T$.

Our proof technique builds (e.g., by adding a rate-of-growth argument) on that used by Cai and Hemachandra [13] to prove $\mathrm{FewP} \subseteq \oplus\mathrm{P}$, where $\oplus\mathrm{P}$ [18, 27] is the class of languages L such that for some nondeterministic polynomial-time Turing machine N, on each x it holds that $x \in L \iff \#acc_N(x) \equiv 1 \pmod 2$. We note that Köbler, Schöning, Toda, and Torán [25] interestingly built on that technique in their proof that $\mathrm{FewP} \subseteq \mathrm{C_{=}P}$, where $\mathrm{C_{=}P}$ [34] is the class of languages L such that there is a polynomial-time function f and a nondeterministic polynomial-time Turing machine N such that for each x, $x \in L$ if and only if $\#acc_N(x) = f(x)$.

Proof of Theorem 2. Let S be a non-gappy, P-printable subset of T. Let $k > 0$ be, for S, some constant satisfying Definition 3.

Let L be any language in FewP. Let $\hat{N}$ be a machine witnessing that $L \in \mathrm{FewP}$, and let p be a polynomial bounding the nondeterministic ambiguity of $\hat{N}$, i.e., for each input x, $\#\mathrm{acc}_{\hat{N}}(x) \leq p(|x|)$. To show that $L \in \mathrm{RC}_T$, we describe a nondeterministic polynomial-time Turing machine N that accepts L via the RC_T acceptance mechanism.

On input x, N chooses $p(|x|)$ natural numbers $c_1, c_2, \ldots, c_{p(|x|)}$ as follows. Initially, we assume that c_1, which is defined to be the least element of S, is hard-coded into the program of N. Successively, for $i = 2, \ldots, p(|x|)$, machine N on input x does the following:

- Let $c_1, \ldots, c_{i-1}$ be the constants that have already been chosen. Define

$$b_i = \binom{i}{1} c_1 + \binom{i}{2} c_2 + \cdots + \binom{i}{i-1} c_{i-1}.$$

- Let a_i be the least element of S such that $b_i \leq a_i$.
- Set $c_i = a_i - b_i$.

After having chosen these constants, N (still on input x) will do the following: Nondeterministically guess an integer $i \in \{1, 2, \ldots, p(|x|)\}$ and, for each i guessed, nondeterministically guess each (unordered) i-tuple of distinct paths of $\hat{N}(x)$. On each path α resulting from such a guess series, $N(x)$ sees whether the i paths of $\hat{N}(x)$ that were guessed on α are all accepting paths. If all are accepting paths, then path α, via trivial nondeterministic guesses, splits itself into c_i accepting paths. On the other hand, if at least one of the i guessed paths is a rejecting path, then path α simply rejects. This completes the description of N.

The intuition behind the construction of N is that for each input x the following holds. $N(x)$ has c_1 accepting paths for each accepting path of $\hat{N}(x)$; $N(x)$ has c_2 additional accepting paths for each pair of distinct accepting paths of $\hat{N}(x)$; and so on. So, if $x \in L$, $N(x)$ has $c_{\#\mathrm{acc}_{\hat{N}}(x)}$ additional accepting paths for the (one) $\#\mathrm{acc}_{\hat{N}}(x)$-tuple of distinct accepting paths of $\hat{N}(x)$. However, if for some z with $\#\mathrm{acc}_{\hat{N}}(x) < z \leq p(|x|)$ a z-tuple of distinct paths of $\hat{N}(x)$ was guessed on a path α of $N(x)$, then α must contain a rejecting path of $\hat{N}(x)$, and thus $N(x)$ will have no accepting paths related to c_z. This intuition is expressed formally by:

$$\#\mathrm{acc}_N(x) = \binom{\#\mathrm{acc}_{\hat{N}}(x)}{1} c_1 + \binom{\#\mathrm{acc}_{\hat{N}}(x)}{2} c_2 + \cdots + \binom{\#\mathrm{acc}_{\hat{N}}(x)}{\#\mathrm{acc}_{\hat{N}}(x)} c_{\#\mathrm{acc}_{\hat{N}}(x)}.$$

Assume $x \in L$. Thus, $0 < \#\mathrm{acc}_{\hat{N}}(x) \leq p(|x|)$. Since $c_{\#\mathrm{acc}_{\hat{N}}(x)}$ was chosen such that

$$\#\mathrm{acc}_{\hat{N}}(x) = 1 \Longrightarrow \#\mathrm{acc}_N(x) = c_1, \text{ and}$$
$$\#\mathrm{acc}_{\hat{N}}(x) \geq 2 \Longrightarrow \#\mathrm{acc}_N(x) = b_{\#\mathrm{acc}_{\hat{N}}(x)} + c_{\#\mathrm{acc}_{\hat{N}}(x)} = a_{\#\mathrm{acc}_{\hat{N}}(x)},$$

and since both c_1 and $a_{\#\mathrm{acc}_{\hat{N}}(x)}$ are elements of S, it follows that $\#\mathrm{acc}_N(x) \in T$. On the other hand, if $x \notin L$ then $\#\mathrm{acc}_{\hat{N}}(x) = 0$, and so $\#\mathrm{acc}_N(x) = 0$.

So now, to prove that $L \in \mathrm{RC}_T$, it suffices to establish an exponential (in $|x|$) upper bound on the value of $\max_{i \leq p(|x|)} c_i$.

We will consider, for $j \geq 2$, what bounds hold on the value of c_j. By construction of N and since S is non-gappy, we have $c_j \leq a_j \leq kb_j$. Regarding the latter inequality, note that b_j is not necessarily an element of S. However, for each j, $c_1 \leq b_j$; so for each j, there exists a $\hat{b}_j \in S$ such that $\hat{b}_j \leq b_j$ and $\hat{b}_j$ is the greatest such integer in S. Since a_j is defined to be the least element of S such that $b_j \leq a_j$, we have $a_j \leq k\hat{b}_j \leq kb_j$.

From the above and the definition of b_j, we have:

$$c_j \leq k\left(\binom{j}{1}c_1 + \binom{j}{2}c_2 + \cdots + \binom{j}{j-1}c_{j-1}\right)$$

$$\text{(1)} \qquad \leq k(j-1)\binom{j}{\lceil \frac{j}{2} \rceil} \max_{1 \leq i \leq j-1} c_i.$$

The factor $j-1$ in inequality (1) is the number of terms in b_j, and the coefficient $\binom{j}{\lceil \frac{j}{2} \rceil}$ is the biggest binomial coefficient of any term in b_j.

Recall that once we were given $S \subseteq T$ we fixed k. For all sufficiently large j the following holds:

$$\text{(2)} \qquad k(j-1)\binom{j}{\lceil \frac{j}{2} \rceil} \leq \binom{j}{\lceil \frac{j}{2} \rceil}^2 \leq \left(2^j\right)^2.$$

In particular, let $j_{bad} = j_{bad}(k)$ be the largest j for which the above inequality fails to hold (if it always holds, set $j_{bad} = 1$). Let $I_{bad} = \max_{1 \leq i \leq j_{bad}} c_i$. From inequalities (1) and (2), we clearly have that, for $j > j_{bad}$:

$$c_j \leq I_{bad} \cdot \prod_{j_{bad} < i \leq j} 2^{2i},$$

and, for $j \leq j_{bad}$, $c_j \leq I_{bad}$. This implies that $c_j = 2^{\mathcal{O}(j^2)}$.

Thus, for the fixed k associated with $S \subseteq T$, the value of $\max_{i \leq p(|x|)} c_i$ indeed is bounded by an exponential function in $|x|$. Hence, $L \in \mathrm{RC}_T$, and thus $\mathrm{FewP} \subseteq \mathrm{RC}_T$. ∎

It is immediate from its definition that EP $\subseteq$ NP. It is also clear that the quantum-computation-related class $\mathrm{C_{==}P[half]}$ of Berthiaume and Brassard [7] is contained in EP.[3] From Theorem 2 it immediately follows that FewP $\subseteq$ EP, since $\mathrm{EP} = \mathrm{RC}_{\{2^i \mid i \in \mathbb{N}\}}$ and $\{2^i \mid i \in \mathbb{N}\}$ is clearly a P-printable, non-gappy set.

Corollary 1. FewP $\subseteq$ EP.

The comments attached to our on-line technical report version [9] give some of the history of the proof of our results and of some valuable comments made by Richard Beigel, in particular that FewP is also contained in the EP analog based on any integer n (note that the acceptance sets for such classes are P-printable and non-gappy).

Cai and Hemachandra's result FewP $\subseteq \oplus$P [13] has been generalized to FewP $\subseteq \mathrm{ModZ}_k\mathrm{P}$, for each $k \geq 2$ [6]. This generalization also follows as a special case of Theorem 2, since $\mathrm{ModZ}_k\mathrm{P} = \mathrm{RC}_{\mathbb{N} - \{a \mid (\exists b \in \mathbb{N})\, [a = b \cdot k]\}}$ as mentioned above.

Corollary 2. [6] *For each* $k \geq 2$, FewP $\subseteq \mathrm{ModZ}_k\mathrm{P}$.

3.2 Discussion

An immediate question is how Corollary 1 relates to known results about FewP. Clearly, Corollary 1 represents an improvement on the trivial inclusion FewP $\subseteq$ NP. However, how does it compare with the nontrivial result of Köbler et al. [25] and Fenner, Fortnow, and Kurtz [16] that FewP $\subseteq$ Few $\subseteq$ SPP $\subseteq \oplus\mathrm{P} \cap \mathrm{C_{=}P}$? Informally stated, Few [13] is what a P machine can do given one call to a #P function that obeys the promise that its value is always at most polynomial. SPP [16, 26] is the class of sets L such that for some nondeterministic polynomial-time Turing machine N it holds that if $x \notin L$ then $N(x)$ has one fewer accepting path than it has rejecting paths, and if $x \in L$ then the numbers of accepting and rejecting paths of $N(x)$ are equal. Curiously, note that the nontrivial

[3] $\mathrm{C_{==}P[half]}$, introduced by Berthiaume and Brassard [7] in their study of quantum complexity, is a variant of the class WPP of Fenner, Fortnow, and Kurtz [16]. Namely, $\mathrm{C_{==}P[half]}$ is the class of languages L such that there is some nondeterministic Turing machine such that if the input is in L exactly half of the paths are accepting paths and if the input is not in L none of the paths are accepting paths.

result that FewP $\subseteq$ SPP itself neither is known to imply nor is known to be implied by the trivial result FewP $\subseteq$ NP.

There are a number of related aspects to the question raised above. First, is SPP $\subseteq$ EP? This inclusion—which would make Corollary 1 a trivial consequence of the known result FewP $\subseteq$ SPP—seems unlikely, as if SPP $\subseteq$ EP, then SPP $\subseteq$ NP, and SPP $\subseteq$ NP is considered unlikely (see [16, 31]). Second, is EP $\subseteq$ SPP? (This inclusion would make Corollary 1 a strengthening of the known result that FewP $\subseteq$ SPP.) We do not know. Third, notice that we proved FewP $\subseteq$ EP but that the Köbler et al. [25] and Fenner, Fortnow, and Kurtz [16] work shows that Few $\subseteq$ SPP. Can our result be extended to show Few $\subseteq$ EP? The reason we mention this is that often it is the case that when one can prove something about FewP, then one can also prove it about the slightly bigger class Few. For example, Cai and Hemachandra, after showing that FewP is in $\oplus$P, then easily applied their technique to show that even Few is in $\oplus$P [13]. Similarly, it is immediately clear that FewP has Turing-complete sets if and only if Few has Turing-complete sets, and so the proof that there is a relativized world in which FewP lacks Turing-complete sets [23] implicitly proves that there is a world in which Few lacks Turing-complete sets (see also [33]). However, in the case of Corollary 1, it is unlikely that by modifying the technique in a way similar to that done by Cai and Hemachandra one could hope to establish the slightly stronger result that EP even contains Few. Why? Clearly coUP $\subseteq$ Few and EP $\subseteq$ NP, so the assumption Few $\subseteq$ EP would imply (along with other even more unlikely things) coUP $\subseteq$ NP.

Fourth, one might wonder directly, since FewP $\subseteq \oplus$P is known, about the relationship between EP and $\oplus$P. That is, how is EP (powers-of-two acceptance) related to $\oplus$P (multiples-of-two acceptance)?[4] We note the following. By a diagonalization so routine as to not be worth including here, one can show $(\exists A)\,[\mathrm{coUP}^A \not\subseteq \mathrm{EP}^A]$. It follows immediately, since (for each B) $\mathrm{coUP}^B \subseteq \mathrm{Few}^B$, that $(\exists A)\,[\mathrm{Few}^A \not\subseteq \mathrm{EP}^A]$ and $(\exists A)\,[\oplus\mathrm{P}^A \not\subseteq \mathrm{EP}^A]$. Similarly, if one looks at the test language inside the proof of Proposition 12 of Beigel's 1991 "mod classes" paper [4], one can see that for his case "$k = 2$" the test language is in (relativized) $\mathrm{C_{=}P[half]}$, and thus as a corollary to his proof one can claim $(\exists A)\,[\mathrm{C_{=}P[half]}^A \not\subseteq \oplus\mathrm{P}^A]$. It follows immediately that $(\exists A)\,[\mathrm{EP}^A \not\subseteq \oplus\mathrm{P}^A]$. Since these are standard diagonalizations that can easily be interleaved, it is easy to see that there is a relativized world in which EP and $\oplus$P are incomparable (i.e., neither

[4] However, one should keep in mind the contrasting rejection sets of these two classes.

is contained in the other). Concerning $\mathrm{Mod}_q\mathrm{P}$ classes [13, 6] for values $q > 2$, see the discussion in our on-line technical report version [9].

Fifth and finally, to complete this discussion, what is the relation between EP and $\mathrm{C_{=}P}$? Proposition 1 below shows that EP is contained in $\mathrm{C_{=}P}$;[5] the proof of this result can be found in the full version of this paper. Thus, Corollary 1 improves upon Köbler et al.'s result that FewP $\subseteq$ $\mathrm{C_{=}P}$ [25]—an improvement that seems to neither imply nor be implied by other improvements of their result such as Few $\subseteq$ SPP ([25], see also [16]).

Proposition 1. EP $\subseteq$ $\mathrm{C_{=}P}$.

4 Open Questions

Does EP equal NP? It would be nice to give evidence that such an equality would, for example, collapse the polynomial hierarchy. However, UP $\subseteq$ EP $\subseteq$ NP, and at the present time, it is open whether even the stronger assumption UP = NP implies any startling collapses. Also, does EP, in contrast to most promise classes, have complete sets? We conjecture that EP lacks complete sets (of course, if EP equals NP then EP has complete sets). It is clear that EP is closed under conjunctive reductions and under disjoint union, and (thus) under intersection. Is EP closed under disjunctive reductions or union?

Acknowledgments

We thank Richard Beigel, Frank Stephan, and Gerd Wechsung for interesting comments or discussions, and Dieter Kratsch, Haiko Müller, and Johannes Waldmann for very kindly letting us use their office's computers to type in part of this paper.

The second author was supported in part by grant NSF-CCR-9322513. The second author and the third author were supported in part by grants NSF-INT-9513368/DAAD-315-PRO-fo-ab and NSF-INT-9815095/DAAD-315-PPP-gü-ab. The third author was supported in part by a NATO Postdoctoral Science Fellowship from the Deutscher Akademischer Austauschdienst ("Gemeinsames Hochschulsonderprogramm III von Bund und Ländern"). This work was done in part while the second

[5] After seeing an earlier draft of this paper, Richard Beigel has communicated (February, 1998) to the authors that he observed that EP is even contained in the class LWPP [16]. Since it is known from the work of Fenner, Fortnow, and Kurtz [16] that SPP $\subseteq$ LWPP $\subseteq$ $\mathrm{C_{=}P}$, this improves upon our result and in particular shows that EP is PP-low (i.e., $\mathrm{PP} = \mathrm{PP}^{\mathrm{EP}}$), where PP denotes probabilistic polynomial time.

author was visiting Friedrich-Schiller-Universität Jena and while the third author was visiting the University of Rochester and Le Moyne College, and we thank the host institutions for their hospitality.

References

1. M. Agrawal and T. Thierauf. The boolean isomorphism problem. In *Proceedings of the 37th IEEE Symposium on Foundations of Computer Science*, pages 422–430. IEEE Computer Society Press, October 1996.
2. E. Allender and R. Rubinstein. P-printable sets. *SIAM Journal on Computing*, 17(6):1193–1202, 1988.
3. R. Beigel. On the relativized power of additional accepting paths. In *Proceedings of the 4th Structure in Complexity Theory Conference*, pages 216–224. IEEE Computer Society Press, June 1989.
4. R. Beigel. Relativized counting classes: Relations among thresholds, parity, and mods. *Journal of Computer and System Sciences*, 42(1):76–96, 1991.
5. R. Beigel, R. Chang, and M. Ogiwara. A relationship between difference hierarchies and relativized polynomial hierarchies. *Mathematical Systems Theory*, 26(3):293–310, 1993.
6. R. Beigel, J. Gill, and U. Hertrampf. Counting classes: Thresholds, parity, mods, and fewness. In *Proceedings of the 7th Annual Symposium on Theoretical Aspects of Computer Science*, pages 49–57. Springer-Verlag *Lecture Notes in Computer Science #415*, February 1990.
7. A. Berthiaume and G. Brassard. The quantum challenge to structural complexity theory. In *Proceedings of the 7th Structure in Complexity Theory Conference*, pages 132–137. IEEE Computer Society Press, June 1992.
8. A. Blass and Y. Gurevich. On the unique satisfiability problem. *Information and Control*, 55:80–88, 1982.
9. B. Borchert, L. Hemaspaandra, and J. Rothe. Powers-of-two acceptance suffices for equivalence and bounded ambiguity problems. Technical Report TR96-045, Electronic Colloquium on Computational Complexity, `http://www.eccc.uni-trier.de/eccc/`, August 1996.
10. B. Borchert, D. Ranjan, and F. Stephan. On the computational complexity of some classical equivalence relations on boolean functions. *Theory of Computing Systems*, 31:679–693, 1998.
11. B. Borchert and F. Stephan. Looking for an analogue of Rice's Theorem in circuit complexity theory. In *Proceedings of the 1997 Kurt Gödel Colloquium*, pages 114–127. Springer-Verlag *Lecture Notes in Computer Science #1289*, 1997.
12. R. Bryant. Symbolic boolean manipulation with ordered binary decision diagrams. *ACM Computing Surveys*, 24(3):293–318, 1992.
13. J. Cai and L. Hemachandra. On the power of parity polynomial time. *Mathematical Systems Theory*, 23(2):95–106, 1990.
14. J. Feigenbaum, S. Kannan, M. Vardi, and M. Viswanathan. Complexity of problems on graphs represented as OBDDs. In *Proceedings of the 15th Annual Symposium on Theoretical Aspects of Computer Science*, pages 216–226. Springer-Verlag *Lecture Notes in Computer Science #1373*, February 1998.
15. M. Fellows and N. Koblitz. Self-witnessing polynomial-time complexity and prime factorization. In *Proceedings of the 7th Structure in Complexity Theory Conference*, pages 107–110. IEEE Computer Society Press, June 1992.

16. S. Fenner, L. Fortnow, and S. Kurtz. Gap-definable counting classes. *Journal of Computer and System Sciences*, 48(1):116–148, 1994.
17. S. Fortune, J. Hopcroft, and E. Schmidt. The complexity of equivalence and containment for free single program schemes. In *Proceedings of the 5th International Colloquium on Automata, Languages, and Programming*, pages 227–240. Springer-Verlag *Lecture Notes in Computer Science #62*, 1978.
18. L. Goldschlager and I. Parberry. On the construction of parallel computers from various bases of boolean functions. *Theoretical Computer Science*, 43:43–58, 1986.
19. J. Grollmann and A. Selman. Complexity measures for public-key cryptosystems. *SIAM Journal on Computing*, 17(2):309–335, 1988.
20. M. Harrison. Counting theorems and their applications to classification of switching functions. In A. Mukhopadyay, editor, *Recent Developments in Switching Theory*, pages 4–22. Academic Press, 1971.
21. J. Hartmanis and Y. Yesha. Computation times of NP sets of different densities. *Theoretical Computer Science*, 34:17–32, 1984.
22. E. Hemaspaandra and L. Hemaspaandra. Quasi-injective reductions. *Theoretical Computer Science*, 123(2):407–413, 1994.
23. L. Hemaspaandra, S. Jain, and N. Vereshchagin. Banishing robust Turing completeness. *International Journal of Foundations of Computer Science*, 4(3):245–265, 1993.
24. L. Hemaspaandra and J. Rothe. Unambiguous computation: Boolean hierarchies and sparse Turing-complete sets. *SIAM Journal on Computing*, 26(3):634–653, 1997.
25. J. Köbler, U. Schöning, S. Toda, and J. Torán. Turing machines with few accepting computations and low sets for PP. *Journal of Computer and System Sciences*, 44(2):272–286, 1992.
26. M. Ogiwara and L. Hemachandra. A complexity theory for closure properties. *Journal of Computer and System Sciences*, 46(3):295–325, 1993.
27. C. Papadimitriou and S. Zachos. Two remarks on the power of counting. In *Proceedings 6th GI Conference on Theoretical Computer Science*, pages 269–276. Springer-Verlag *Lecture Notes in Computer Science #145*, 1983.
28. R. Rao, J. Rothe, and O. Watanabe. Upward separation for FewP and related classes. *Information Processing Letters*, 52(4):175–180, 1994.
29. U. Schöning. Probabilistic complexity classes and lowness. *Journal of Computer and System Sciences*, 39(1):84–100, 1989.
30. Y. Takenaga, M. Nouzoe, and S. Yajima. Size and variable ordering of OBDDs representing threshold functions. In *Proceedings of the 3rd Annual International Computing and Combinatorics Conference*, pages 91–100. Springer-Verlag *Lecture Notes in Computer Science #1276*, August 1997.
31. S. Toda and M. Ogiwara. Counting classes are at least as hard as the polynomial-time hierarchy. *SIAM Journal on Computing*, 21(2):316–328, 1992.
32. L. Valiant. The relative complexity of checking and evaluating. *Information Processing Letters*, 5(1):20–23, 1976.
33. N. Vereshchagin. Relativizable and nonrelativizable theorems in the polynomial theory of algorithms. *Russian Academy of Sciences-Izvestiya-Mathematics*, 42(2):261–298, 1994.
34. K. Wagner. The complexity of combinatorial problems with succinct input representations. *Acta Informatica*, 23:325–356, 1986.

Grammar Systems as Language Analyzers and Recursively Enumerable Languages

Henning Bordihn[1], Júrgen Dassow[1], and Gyorgy Vaszil[2]

[1] Otto-von-Guericke-Universität Magdeburg, Fakultät für Informatik
PSF 4120, D–39016 Magdeburg, Germany
email: {bordihn,dassow}@iws.cs.uni-magdeburg.de

[2] Computer and Automation Research Institute, Hungarian Academy of Sciences
Kende u. 13 – 17, H–1111 Budapest, Hungary
email: vaszil@sztaki.hu

Abstract. We consider parallel communicating grammar systems which consist of several grammars and perform derivation steps, where each of the grammars works in a parallel and synchronized manner on its own sentential form, and communication steps, where a transfer of sentential forms is done. We discuss accepting and analyzing versions of such grammar systems with context-free productions and present characterizations of the family of recursively enumerable languages by them.

In accepting parallel communicating grammar systems rules of the form $\alpha \to A$ with a word α and a nonterminal A are applied as in the generating case, and the language consists of all terminal words which can derive the axiom. We prove that all types of these accepting grammar systems describe the family of recursively enumerable languages, even if λ-rules are forbidden.

Moreover, we study analyzing parallel communicating grammar systems, the derivations of which perform the generating counterparts backwards. This requires a modification of the generating derivation concept to strong-returning parallel communicating grammar systems which also generate the family of recursively enumerable languages.

1 Introduction

Parallel communicating grammar systems are introduced in [15] and are widely investigated nowadays (see [4, 6, 7, 10, 18, 19]). They consist of several grammars (called components or processors of the system) which work in a parallel and synchronized manner, each on its own sentential form. In a derivation step of the system each component transforms its sentential form according to its production rules. Moreover, the components cooperate by communication. A communication step is done by request through special nonterminals, called query symbols. Whenever a component has introduced a query symbol, the rewriting process is interrupted, and the components to which the query symbols point immediately send their current sentential forms to those components which have requested the communication step. There, each query symbol is rewritten by the corresponding received sentential form. Finally, the system generates a common language, namely the set of all terminal strings which can be obtained in one certain component.

Parallel communicating grammar systems are of interest because they form a syntactic model of a problem solving system called classroom model in artificial intelligence and a model of parallel and distributed computation as it appears, e.g., in computer networks. This approach has several advantages, e.g., it allows to compare the power of distinct communication structures (parallel architectures) in a way in which other (computing) models have not been able to establish results (for a discussion of the latter item, see [10]).

On the other hand, in many applications one is interested in analyzing (recognizing) formal languages instead of generating them. One concept for this purpose is the notion of analyzing grammars as it has been considered for the well-known type-n grammars of the Chomsky hierarchy, $n \in \{0, 1, 2, 3\}$, by Arto Salomaa in [16, Chapter 1]. Here, the general idea is the following: a given (terminal) word w is *accepted* iff one can derive the axiom (which therefore is seen as a goal symbol instead of a start symbol) by iterated applications of productions to w, where an analyzing production, is defined as in the generating case but the left-hand sides are interchanged with the right-hand sides. It is proved that the families of languages described by accepting type-n grammars, $n \in \{0, 1, 2, 3\}$, trivially coincide with the families of languages generated by grammars of the corresponding type.

This concept of analyzing grammars can be seen in two different ways: at first, one might gain the intuition that derivations of analyzing grammars just mimic the derivations of their generating counterparts step by step, simply performing any possible derivation of the generating mechanism backwards. In this case, one has to look for an appropriate definition of the corresponding analyzing device such that the language families generated and analyzed in this way automatically coincide. This is not always as straightforward as in case of Chomsky grammars, e.g., for some grammars with controlled derivations or for grammar systems (for an example of such approach see [13]).

A second interpretation of the idea of analyzing grammars is in line with the research on accepting grammars and systems as it has been done, e.g., in [1–3, 8, 9]. Here, the yield relation is defined by textually transferring the definition of a derivation step from the generating case to the accepting one. That is, we take the well-known mechanism, now, in order to recognize words and not to generate them, and we investigate which language class can be described by these means. It has been shown that the trivial equivalence result known from Chomsky grammars does not hold any longer for several types of grammars and systems.

In this paper, we follow both interpretations of language analyzing grammar tools for the case of parallel communicating grammar systems. In the next section we shall treat the idea of textual transfer of the definition of the yield relation (both for rewriting steps and for communication steps). Here, generating and accepting derivations cannot trivially simulate each other since the communication has a different effect. We show that all types of accepting parallel communicating grammar systems describe the family of recursively enumerable languages. This result is surprising because it also holds for accepting grammar systems where λ-rules are forbidden.

In Section 3 we follow the other approach. In a derivation step of a generating parallel communicating grammar system the terminal strings are not changed. This leads to a problem if we want to reverse the generation mechanism by an analyzing grammar since terminal strings have to be changed in some cases and have not to be changed in some other steps although the rules of the component can be applied. In order to overcome this unnatural behavior, in Section 3 we modify the working of a generating parallel communicating grammar system as follows: if a terminal word is obtained by a component (i.e., the component has finished its work), it starts a new derivation from its axiom. First, we note that this modification has a meaningful interpretation in the classroom model: if a component has obtained a solution (i.e., one possible solution) for its subproblem, then – independently of whether or not there is a request for the obtained solution by some other component – the component starts a new derivation process in order to get a possibly different solution for the subproblem. Furthermore, this modification allows a definition of an analyzing grammar with derivations that can mimic generating ones and vice versa. Moreover, this modification does not affect the generative power of generating parallel communicating grammar systems.

In what follows, we assume that the reader is familiar with basic notions and basic knowledge of formal language and automata theory. Concerning our notation, we mostly follow [7]: $\subseteq$ denotes inclusion, $\subset$ denotes strict inclusion, $|M|$ is the number of elements in the set M. The empty word is denoted by λ. We consider two languages L_1 and L_2 to be equal iff $L_1 \setminus \{\lambda\} = L_2 \setminus \{\lambda\}$, and we simply write $L_1 = L_2$ in this case. We term two devices describing languages equivalent if the two described languages are equal. The length of a word x is denoted by $|x|$. If $x \in V^*$, where V is some alphabet, and if $W \subseteq V$, then $|x|_W$ denotes the number of occurrences of letters from W in x. The families of context-free, λ-free context-free and type-0 Chomsky grammars are denoted by CF, CF-λ and RE, respectively. If X is one of these families, $\mathcal{L}(X)$ denotes the family of languages generated (accepted) by some device from family X.

2 Accepting parallel communicating grammar systems

We now give a definition for parallel communicating grammar systems which covers the language generating and accepting case. We restrict ourselves to systems with context-free productions.

Definition 1. *A generating [accepting] parallel communicating grammar system (PC grammar system, for short) with n context-free components, where* $n \geq 1$, *is an* $(n+3)$*-tuple* $\Gamma = (N, K, T, G_1, G_2, \ldots, G_n)$, *where* N, K, *and* T *are pairwise disjoint alphabets of nonterminal symbols, query symbols, and* *terminal symbols, respectively. For* $1 \leq i \leq n$, $G_i = (N \cup K, T, P_i, S_i)$ *is a generating [accepting] context-free Chomsky grammar with nonterminal alphabet* $N \cup K$, *terminal alphabet* T, *a set of rewriting rules* $P_i \subseteq (N \cup K) \times (N \cup T \cup K)^*$ $[P_i \subseteq (N \cup T \cup K)^* \times (N \cup K)]$ *and an axiom* S_i.

The grammars $G_1, G_2, \ldots G_n$ are called components of Γ, G_1 is said to be the master component (or master grammar) of Γ. The total alphabet $N \cup K \cup T$ of Γ is denoted by V_Γ.

Definition 2. *Let* $\Gamma = (N, K, T, G_1, G_2, \ldots, G_n)$, $n \geq 1$, *be a PC grammar system as above. An n-tuple* $(x_1, x_2, \ldots, x_n)$, *where* $x_i \in V_\Gamma^*$, $1 \leq i \leq n$, *is called a configuration of* Γ. *If* Γ *is generating,* $(S_1, S_2, \ldots, S_n)$ *is said to be the initial configuration, in accepting case it is called goal configuration, whereas an initial configuration in accepting mode is given by a tuple in* $T^* \times (V_\Gamma^*)^{n-1}$.

PC grammar systems change their configurations by performing direct derivation steps.

Definition 3. *Let* $\Gamma = (N, K, T, G_1, G_2, \ldots, G_n)$, $n \geq 1$, *be a parallel communicating grammar system, and let* $(x_1, x_2, \ldots, x_n)$ *and* $(y_1, y_2, \ldots, y_n)$ *be two configurations of* Γ. *We say that* $(x_1, x_2, \ldots, x_n)$ *directly derives* $(y_1, y_2, \ldots, y_n)$, *denoted by* $(x_1, x_2, \ldots, x_n) \Longrightarrow (y_1, y_2, \ldots, y_n)$, *if one of the next three cases holds:*

1a. Γ *is a generating PC grammar system, and there is no* x_i *which contains any query symbol. Then, for* $1 \leq i \leq n$, *either* $x_i \in (N \cup T)^* \setminus T^*$ *and* $x_i \underset{G_i}{\Longrightarrow} y_i$ *or* $x_i \in T^*$ *and* $y_i = x_i$.

1b. Γ *is an accepting PC grammar system, and there is no* x_i *which contains any query symbol. Then, for* $1 \leq i \leq n$, *either* $x_i \in (N \cup T)^* \setminus \{S_i\}$ *and* $x_i \underset{G_i}{\Longrightarrow} y_i$ *or* $x_i = y_i = S_i$.

2. Γ *is a generating or accepting PC grammar system, and there is some* x_i, $1 \leq i \leq n$, *which contains at least one occurrence of query symbols. Let* x_i *be of the form* $x_i = z_1 Q_{i_1} z_2 Q_{i_2} \ldots z_t Q_{i_t} z_{t+1}$, *where* $z_j \in (N \cup T)^*$, $1 \leq j \leq t+1$, *and* $Q_{i_l} \in K$, $1 \leq l \leq t$. *In this case* $y_i = z_1 x_{i_1} z_2 x_{i_2} \ldots z_t x_{i_t} z_{t+1}$, *if* x_{i_l}, $1 \leq l \leq t$, *does not contain any query symbol. In so-called returning systems,* $y_{i_l} = \alpha_{i_l}$, *for* $1 \leq l \leq t$, *where* α_{i_l} *is the* (i_l)*-th sentential form of the initial configuration* $(\alpha_1, \alpha_2, \ldots, \alpha_n)$ *of the current derivation. In non-returning systems* $y_{i_l} = x_{i_l}$, $1 \leq l \leq t$. *If some* x_{i_l} *contains at least one occurrence of query symbols, then* $y_i = x_i$. *For all* i, $1 \leq i \leq n$, *for which* y_i *is not specified above,* $y_i = x_i$ *holds.*

The first case is the description of a rewriting step in generating mode (1a) and in accepting mode (1b): if no query symbols are present in any of the sentential forms, then each component grammar uses one of its rewriting rules except those which have already produced a terminal string (the latter point will be changed in the systems considered in Section 3) or the axiom, respectively. The derivation is blocked if a sentential form is not a terminal string or the axiom, respectively, but no rule can be applied to it.

The second case describes a communication step which has priority over effective rewriting: if some query symbol, say Q_j, appears in a sentential form, then the rewriting process stops and a communication step must be performed. The symbol Q_j has to be replaced by the current sentential form of component G_j, say x_j, supposing that x_j does not contain any query symbol. If this sentential form also contains query symbols, then first these symbols must be replaced with the requested sentential forms and so on. If this condition cannot be fulfilled (a circular query appeared), then the derivation is blocked.

If the sentential form of a component was communicated to another one, this component can continue its own work in two ways: in so-called *returning* systems, the component must return to its sentential form of the initial configuration and begin to derive a new string. In *non-returning* systems the components do not return to their initial sentential forms but continue to process their current string.

In the following, by $\stackrel{*}{\Longrightarrow}$ the reflexive and transitive closure of the yield relation $\Longrightarrow$ is denoted.

Definition 4. *Let $\Gamma = (N, K, T, G_1, G_2, \ldots, G_n)$ be a PC grammar system as above. If Γ is a generating PC grammar system, its language defined by Γ is*

$$L_{\mathrm{gen}}(\Gamma) = \{x_1 \in T^* \mid (S_1, S_2, \ldots, S_n) \stackrel{*}{\Longrightarrow} (x_1, x_2, \ldots, x_n),\ x_i \in V_\Gamma^*,\ 2 \leq i \leq n\}.$$

If Γ is an accepting PC grammar system, its language is

$$L_{\mathrm{acc}}(\Gamma) = \{x_1 \in T^* \mid (x_1, x_2, \ldots, x_n) \stackrel{*}{\Longrightarrow} (S_1, S_2, \ldots, S_n),\ x_i \in V_\Gamma^*,\ 2 \leq i \leq n\}.$$

Thus, the *generated* language consists of the terminal strings appearing as sentential forms of the master grammar G_1 in a derivation which started off with the initial configuration $(S_1, S_2, \ldots, S_n)$, whereas the *accepted* language consists of all terminal words appearing as sentential forms of the master grammar G_1 in the initial configuration of an arbitrary derivation which yields the goal configuration $(S_1, S_2, \ldots, S_n)$.

Finally, we define a special variant of PC grammar systems where the ability to ask for communication is restricted to the master component. A PC grammar system $\Gamma = (N, K, T, G_1, G_2, \ldots, G_n)$ (in both generating and accepting case) is referred to as *centralized* if, for $2 \leq i \leq n$, $P_i \subseteq (N \cup T)^* \times (N \cup T)^*$. Otherwise (in the unrestricted case) it is called *non-centralized.*

We shall denote the families of languages generated [accepted] by returning and non-returning PC grammar systems with context-free components by $\mathcal{L}_{\mathrm{gen}}(\mathrm{PC}_*\mathrm{CF})$ and $\mathcal{L}_{\mathrm{gen}}(\mathrm{NPC}_*\mathrm{CF})$ [$\mathcal{L}_{\mathrm{acc}}(\mathrm{PC}_*\mathrm{CF})$ and $\mathcal{L}_{\mathrm{acc}}(\mathrm{NPC}_*\mathrm{CF})$], respectively. When only centralized PC grammar systems are used, we add the letter C coming, e.g., to families $\mathcal{L}_{\mathrm{gen}}(\mathrm{CPC}_*\mathrm{CF})$, $\mathcal{L}_{\mathrm{gen}}(\mathrm{NCPC}_*\mathrm{CF})$ and so on. Furthermore, we replace CF by CF-λ in that notation if λ-rules are forbidden in any component of the systems.

We now prove that in all accepting cases – also those where λ-rules are forbidden – we describe the family of recursively enumerable languages. Together with the relations given above we also know the relations between generating and accepting variants.

Theorem 1. *For* $X \in \{\text{PC}, \text{CPC}, \text{NPC}, \text{NCPC}\}$, $Y \in \{\text{CF}-\lambda, \text{CF}\}$,

$$\mathcal{L}_{\text{acc}}(X_*Y) = \mathcal{L}(\text{RE}).$$

Proof. The inclusions $\mathcal{L}_{\text{acc}}(X_*Y) \subseteq \mathcal{L}(\text{RE})$ trivially hold by Turing machine constructions. The converse inclusions are proved by the following simulations of type-0 grammars which are assumed to be given in an appropriate normal form.

Let $L \in \mathcal{L}(\text{RE})$. Then there is a type-0 grammar $G = (V_N, V_T, P, S)$ generating L, where the set P of productions can be assumed to contain only rules of the forms $A \to BC$, $A \to a$, $AB \to CD$, and $Z \to \lambda$, where $A, B, C, D \in V_N$, $a \in V_T$, and Z is a special nonterminal symbol. This can be seen by combining the idea of the proof of Theorem 9.9 in [16] with the usual construction of Kuroda normal form (cf., e.g., [12]).

Let the total alphabet $V_N \cup V_T$ of G contain r symbols, say $V_N \cup V_T = \{x_1, x_2, \ldots, x_r\}$, and let the number of (pairwise different) productions in P of the form $AB \to CD$ be n. Moreover, let us assume a unique label r_i, $1 \leq i \leq n$, being attached to each production of this form.

We now consider the accepting parallel communicating grammar system

$$\Gamma = (N, K, T, G_1, G_2, \ldots, G_{n+2r+1})$$

with $n + 2r + 1$ components, where

$$N = V_N \cup \{S_2, S_3, \ldots, S_{n+2r+1}\},\ K = \{Q_2, Q_3, \ldots, Q_{n+2r+1}\},\ T = V_T$$

with additional symbols $S_2, S_3, \ldots S_{n+2r+1}, Q_2, Q_3, \ldots, Q_{n+2r+1}$, and the components are constructed as follows:

$$\begin{aligned} G_1 &= (N \cup K, T, P_1, S) \text{ with} \\ P_1 &= \{BC \to A \mid A \to BC \in P,\ A, B, C \in V_N\} \\ &\quad \cup \{a \to A \mid A \to a \in P,\ A \in V_N,\ a \in V_T\} \\ &\quad \cup \{CD \to Q_{i+1} \mid r_i : AB \to CD \in P,\ 1 \leq i \leq n\} \\ &\quad \cup \{x_j \to Q_{n+1+j} \mid 1 \leq j \leq r\} \cup \{x_j \to Q_{n+1+r+j} \mid 1 \leq j \leq r\} \end{aligned}$$

and, for $2 \leq i \leq n+1$, if $r_i : AB \to CD$,

$$\begin{aligned} G_i &= (N \cup K, T, P_i, S_i) \text{ with} \\ P_i &= \{A \to A, AB \to S_i\}. \end{aligned}$$

Furthermore, for any symbol $x_j \in V_N \cup V_T$, $1 \leq j \leq r$, two additional components G_{n+1+j} and $G_{n+1+r+j}$ are introduced providing the strings Zx_j and x_jZ, respectively. More precisely, for $1 \leq j \leq r$, we have

$$\begin{aligned} G_{n+1+j} &= (N \cup K, T, P_{n+1+j}, S_{n+1+j}) \text{ with} \\ P_{n+1+j} &= \{Z \to Z, Zx_j \to S_{n+1+j}\}, \\ G_{n+1+r+j} &= (N \cup K, T, P_{n+1+r+j}, S_{n+1+r+j}) \text{ with} \\ P_{n+1+r+j} &= \{Z \to Z, x_jZ \to S_{n+1+r+j}\}. \end{aligned}$$

Obviously, a derivation with initial configuration $(w, \alpha_2, \alpha_3, \ldots, \alpha_{n+2r+1})$ yields the goal configuration if

- $S \stackrel{*}{\Longrightarrow} w$ in G,
- for $2 \leq i \leq n+1$, $\alpha_i = AB$ if AB is the left-hand side of production r_i in P, and
- for $1 \leq j \leq r$, $\alpha_{n+1+j} = Zx_j$ and $\alpha_{n+1+r+j} = x_jZ$.

Then the master component can directly simulate the (reverse) application of context-free productions from P whereas the "real" monotone productions as well as the λ-productions are simulated by communication steps. Clearly, in those accepting derivations, the components G_i, $i \geq 2$, must behave such that the α_i's are simultaneously rewritten to S_i exactly in the moment (tact) when the master grammar derives its axiom S. Otherwise, the derivation might be blocked, since there is no component which can rewrite any S_i, $i \geq 2$, but an application of a rule $\beta \to S_i$ at a "wrong moment" does not allow the master to derive words which are not in $L_{\text{gen}}(G)$.

Hence, $L_{\text{acc}}(\Gamma) = L_{\text{gen}}(G) = L$ for the (centralized) PC grammar system Γ both in returning and in non-returning mode. Note that all productions occurring in a component of Γ are accepting context-free and that no λ-rules are needed. □

Unfortunately, by this construction, both the number of nonterminals and the number of components in the simulating PC grammar system depend on the size of the type-0 grammar to be simulated. We do not know whether or not any given type-0 grammar can be simulated by a PC grammar system with a bounded number of nonterminals and/or components.

In conclusion, we list the following relationships between generating and accepting PC grammar systems which are can be seen by the results given in the present paper and in [4], [6], [7], and [11], where the relations of the families of generated languages and $\mathcal{L}(\text{RE})$ are stated.

Corollary 1. *Let* $X \in \{\text{CF}, \text{CF}-\lambda\}$ *and* $Y \in \{\text{N}, \lambda\}$. *The following relations hold:*

(i) $\mathcal{L}_{\text{gen}}(\text{PC}_*X) = \mathcal{L}_{\text{acc}}(\text{PC}_*X)$.
(ii) $\mathcal{L}_{\text{gen}}(\text{NPC}_*, X) = \mathcal{L}_{\text{acc}}(\text{NPC}_*, X)$.
(iii) $\mathcal{L}_{\text{gen}}(Y\text{CPC}_*, \text{CF}-\lambda) \subset \mathcal{L}_{\text{acc}}(Y\text{CPC}_*, \text{CF}-\lambda)$
(iv) $\mathcal{L}_{\text{gen}}(Y\text{CPC}_*, \text{CF}) \subseteq \mathcal{L}_{\text{acc}}(Y\text{CPC}_*, \text{CF})$ □

Note that in case of both generating and accepting PC grammar systems, query symbols can be introduced in some sentential form only by rewriting steps and they can be replaced only by means of communication. Hence, we can assume without loss of generality that query symbols never appear on left-hand sides of productions of the components. Thus the generative power of generating parallel communicating grammar systems is not altered if we restrict to rules from $N \times (N \cup K \cup T)^*$. However, if one would already require this restriction in the definition of a generating parallel communicating grammar system, then the associated accepting parallel communicating grammar systems would have productions from $(N \cup K \cup T)^* \times N$. It is easy to prove that this restricted form of accepting PC grammar systems only accept context-free languages, and trivially, any context-free language can be accepted by such an restricted accepting PC grammar system.

3 Analyzing parallel communicating grammar systems

In this section we follow the other interpretation of analyzing grammars. We look for a definition of analyzing PC grammar systems in such a way, that analyzing derivations mimic their generative counterparts performing the same derivation steps backwards. Our goal is to use exactly the same system in both ways, to generate and to analyze a language. The generating and accepting versions of a parallel communicating grammar system as considered in the preceding section do not satisfy this requirement as one can see from the results and proofs in that section. In order to distinguish the grammars considered in this section from those considered in the preceding section we call them analyzing grammars.

We mention a problem arising from the treatment of terminal strings in generating PC grammar systems. If a component generates a terminal string, this string remains unchanged through the rest of the derivation. Thus, in the analyzing derivation a terminal string can remain unchanged simulating a generating derivation step on a terminal string or it can be changed simulating a generating derivation step backwards. This is artificial since by the productions of the component in any moment a change is possible.

In order to eliminate this feature from analyzing derivations we have to eliminate the feature from the generating derivations. Thus we make a slight modification in defining derivation steps in the generative mode, a modification which will enable us to find analyzing counterparts to each generative derivation and vice versa. Therefore the equivalence of the generated and accepted language classes will be obvious. After this, we show that the modification of the generating derivation step does not effect the power of returning PC grammar systems in the generative case, so analyzing grammar systems defined this way accept the same class of languages that is generated in the conventional returning generating mode.

Let us start with defining the modified derivation step for the generative mode.

Definition 5. *Let $\Gamma = (N, K, T, G_1, G_2, \ldots, G_n)$, $n \geq 1$, be a generating PC grammar system as above, with initial configuration $(S_1, S_2, \ldots, S_n)$. The configuration $(x_1, x_2, \ldots, x_n)$ directly derives the configuration $(y_1, y_2, \ldots, y_n)$ in strong-returning mode, denoted by $(x_1, x_2, \ldots, x_n) \underset{sr}{\Longrightarrow} (y_1, y_2, \ldots, y_n)$, if one of the following three cases holds.*

1. There is no x_i which contains any query symbol, and there is no x_i which is a terminal word, that is, $x_i \in (N \cup T)^* \setminus T^*$ for $1 \leq i \leq n$. Then $x_i \underset{G_i}{\Longrightarrow} y_i$.

2. There is no x_i which contains any query symbol, that is, $x_i \in (N \cup T)^*$, $1 \leq i \leq n$. Then $y_j = S_j$ if $x_j \in T^*$, and $y_j = x_j$ if $x_j \in (N \cup T)^* \setminus T^*$.

3. There is some x_i, $1 \leq i \leq n$, which contains at least one occurrence of query symbols, that is $x_i = z_1 Q_{i_1} z_2 Q_{i_2} \dots z_t Q_{i_t} z_{t+1}$ where $z_j \in (N \cup T)^*$, $1 \leq j \leq t+1$ and $Q_{i_l} \in K$, $1 \leq l \leq t$. Then $y_i = z_1 x_{i_1} z_2 x_{i_2} \dots z_t x_{i_t} z_{t+1}$, where x_{i_l}, $1 \leq l \leq t$ does not contain any query symbol, and $y_{i_l} = S_{i_l}$, $1 \leq l \leq t$. If some x_{i_l} contains at least one occurrence of query symbols, then $y_i = x_i$. For all i, $1 \leq i \leq n$, for which y_i is not specified above, $y_i = x_i$ holds.

The first point is the description of a rewriting step, where no terminal strings are present among the sentential forms. This is a usual rewriting step known from returning PC grammar systems.

The second point is the description of a derivation step, after at least one terminal string appeared among the sentential forms. In this case, the terminal strings are changed to the start symbol, the other ones remain the same.

The third point again is describing a usual returning communication step.

The language generated by systems in strong-returning mode is defined as before, now of course using strong-returning steps during the derivations.

Again, by $\overset{*}{\underset{sr}{\Longrightarrow}}$ we denote the reflexive and transitive closure of $\underset{sr}{\Longrightarrow}$.

Definition 6. *Let $\Gamma = (N, K, T, G_1, G_2 \dots, G_n)$ be a (generating) PC grammar system with master grammar G_1, and let $(S_1, S_2, \dots, S_n)$ denote the initial configuration of Γ. The language generated by the PC grammar system Γ in strong-returning mode is*

$$L_{sr}(\Gamma) = \{x_1 \in T^* \mid (S_1, S_2, \dots, S_n) \overset{*}{\underset{sr}{\Longrightarrow}} (x_1, x_2, \dots, x_n), x_i \in V_\Gamma^*, 2 \leq i \leq n\}.$$

Let the class of languages generated by PC grammar systems in the strong-returning mode with context-free components be denoted by $\mathcal{L}_{\text{gen}}(\text{PC}_*\text{CF}, sr)$.

As a first result we mention that strong-returning PC grammar systems generate the family of recursively enumerable languages (which is generated by returning PC grammar systems, too).

Lemma 1. $\mathcal{L}(\text{RE}) = \mathcal{L}_{\text{gen}}(\text{PC}_*\text{CF}, sr)$. □

By reasons of space the proof (which uses earlier results from [5] and the concept of rule-synchronization, see [14] and [17], and a technically demanding simulation) is omitted here.

Now we define analyzing derivations by "turning around" strong-returning derivation steps.

Definition 7. *Let $\Gamma = (N, K, T, G_1, G_2, \dots, G_n)$ be a (generating) parallel communicating grammar system as above with axioms S_i, $1 \leq i \leq n$, and let $(x_1, x_2, \dots, x_n)$ and $(y_1, y_2, \dots, y_n)$ be two configurations of Γ. We say that $(x_1, x_2, \dots, x_n)$ directly derives $(y_1, y_2, \dots, y_n)$ in analyzing mode, denoted by $(x_1, x_2, \dots, x_n) \underset{ana}{\Longrightarrow} (y_1, y_2, \dots, y_n)$, if one of the following three cases holds.*

1. For $1 \leq i \leq n$, $x_i = z_1 \alpha z_2$ for some $z_1, z_2 \in (N \cup T)^*$, $\alpha \in (N \cup T \cup K)^*$, $y_i = z_1 X z_2$ and $X \to \alpha \in P_i$.

2. If $x_i \in (N \cup T)^*$ for $1 \leq i \leq n$, then either $y_i = x_i$ or $y_i \in T^*$ and $x_i = S_i$.

3. Let there be at least one j, $1 \leq j \leq n$, with $x_j = S_j$. For $1 \leq i \leq n$, if $|x_i|_K > 0$, then $y_i = x_i$, and if $|x_i|_K = 0$, then either $y_i = x_i$ or $y_i = z_1 Q_{i_1} z_2 Q_{i_2} \ldots z_t Q_{i_t} z_{t+1}$ for some $z_l \in (N \cup T)^*$, $1 \leq l \leq t+1$, and some $Q_{i_k} \in K$, $1 \leq k \leq t$, such that the following condition holds: if $y_i = z_1 Q_{i_1} z_2 Q_{i_2} \ldots z_t Q_{i_t} z_{t+1}$ then $x_i = z_1 y_{i_1} z_2 y_{i_2} \ldots z_t y_{i_t} z_{t+1}$ and $y_{i_k} \in (N \cup T)^*$ and $x_{i_k} = S_{i_k}$ for $1 \leq k \leq t$.

The first point is the description of an analyzing rewriting step, each grammar uses one of its rules "backwards" (therefore analyzing grammars work as accepting grammars).

The second point describes the analyzing counterpart of the strong-returning feature: if an axiom is present at some component, it can be replaced with an arbitrary terminal string while the other sentential forms remain unchanged.

The third point describes a communication step, which is possible to perform if the sentential form of at least one of the components is its axiom. In this case, the other components send subwords of their sentential forms to these components (the ones which have the axiom as their current string), and replace the subword they have sent, with the appropriate query symbol (Q_j for example, if the subword was sent to component G_j, for some j, $1 \leq j \leq n$). According to the classroom model this can be interpreted as a distribution of subtasks to agents who have finished their assignments and protocolling the distribution by the corresponding query symbol.

By $\stackrel{*}{\underset{ana}{\Longrightarrow}}$ we denote the reflexive and transitive closure of $\underset{ana}{\Longrightarrow}$.

Definition 8. *Let $\Gamma = (N, K, T, G_1, G_2, \ldots, G_n)$ be a PC grammar system. The language analyzed by the PC grammar system Γ is*

$$L_{ana}(\Gamma) = \{x_1 \in T^* \mid (x_1, x_2, \ldots, x_n) \stackrel{*}{\underset{ana}{\Longrightarrow}} (S_1, S_2, \ldots, S_n), x_i \in V_\Gamma^*, 2 \leq i \leq n\}.$$

Let the class of languages analyzed by PC grammar systems with context-free components be denoted by $\mathcal{L}(\mathrm{PC}_*\mathrm{CF}, ana)$.

Note the following difference between the generating process (with and without strong return) or accepting process on one side and the analyzing process on the other side. In a generating and accepting derivation the current sentential form determines uniquely whether or not a usual derivation step, a derivation step with strong return or a communication step has to be done. In an analyzing derivation we have to make a choice what type of step we want to make backwards. By our motivation we cannot avoid to choose a derivation step or a communication step. On the other hand, if we restrict to parallel communicating grammar systems where, for any production, the axiom does not occur in the word on the right-hand side, then the generative power of generating systems (with and without strong return) is not changed (as one can easily see) and in analyzing grammars there is no choice between doing backwards usual and strong returning derivation steps.

Now, from the definitions it can easily be seen that all strong-returning derivations have an analyzing counterpart, and that, similarly, all analyzed strings can be generated in the strong-returning mode. Thus $L_{sr}(\Gamma) = L_{ana}(\Gamma)$ holds for any PC grammar system Γ and we obtain the following statement.

Lemma 2. $\mathcal{L}_{\text{gen}}(\text{PC}_*\text{CF}, sr) = \mathcal{L}(\text{PC}_*\text{CF}, ana)$. □

Summarizing, we obtain the following theorem.

Theorem 2. $\mathcal{L}(\text{RE}) = \mathcal{L}_{\text{gen}}(\text{PC}_*\text{CF}) = \mathcal{L}_{\text{gen}}(\text{PC}_*\text{CF}, sr) = \mathcal{L}(\text{PC}_*\text{CF}, ana)$. □

Note that with respect to analysis we have only considered non-centralized returning PC grammar systems. In a certain sense this is natural by the definition of strong return. However, we have not taken into consideration analyzing PC grammar systems in the centralized and/or non-returning case.

References

1. H. Bordihn and H. Fernau, Accepting grammars and systems. Technical Report 9/94, Universität Karlsruhe, Fakultät für Informatik, 1994.
2. H. Bordihn and H. Fernau, Accepting grammars with regulation. *Intern. J. Comp. Math.* **53** (1994) 1–18.
3. H. Bordihn and H. Fernau, Accepting grammars and systems via context condition grammars. *Journal of Automata, Languages and Combinatorics* **1** (1996) 97–112.
4. E. Csuhaj-Varjú, J. Dassow, J. Kelemen and Gh. Păun, *Grammar Systems: A Grammatical Approach to Distribution and Cooperation.* Volume 5 of *Topics in Computer Mathematics.* Gordon and Breach, 1994.
5. E. Csuhaj-Varjú and Gy. Vaszil, On context-free parallel communicating grammar systems: synchronization, communication, and normal forms. Accepted for publication in *Theoretical Computer Science.*
6. E. Csuhaj-Varjú and Gy. Vaszil, On the computational completeness of context-free parallel communicating grammar systems. *Theor. Comp. Sci.* **215** 1-2 (1999), 349-358.
7. J. Dassow, Gh. Păun and G. Rozenberg, Grammar systems. In: A. Salomaa and G. Rozenberg (eds.), *Handbook of Formal Languages,* Vol. 2, Chapter 4, Springer-Verlag, Berlin-Heidelberg, 1996, 155–213.
8. H. Fernau and H. Bordihn, Remarks on accepting parallel systems. *Intern. J. Comp. Math.* **56** (1995) 51–67.
9. H. Fernau, M. Holzer and H. Bordihn, Accepting multi-agent systems: the case of cooperating distributed grammar systems. *Computers and Artificial Intelligence* **15** (1996) 123–139.
10. J. Hromkovič, On the communication complexity of distributive language generation. In: J. Dassow, G. Rozenberg and A. Salomaa (eds.), *Developments in Language Theory II,* World Scientific Publ. Co. Pte. Ltd., 1995, 237–246.
11. N. Mandache, On the computational power of context-free PCGSs. Submitted.
12. A. Mateescu and A. Salomaa, Aspects of classical language theory. In: A. Salomaa and G. Rozenberg (eds.), *Handbook of Formal Languages,* Vol. 1, Chapter 4, Springer-Verlag, Berlin-Heidelberg, 1996, 175–251.
13. V. Mihalache, Accepting cooperating distributed grammar systems with terminal derivation. *EATCS Bulletin* **61** (1997) 80–84.
14. Gh. Păun, On the synchronization in parallel communicating grammar systems. *Acta Informatica* **30** (1993) 351–367.

15. Gh. Păun and L. Santean, Parallel communicating grammar systems: the regular case. *Ann. Univ. Buc. Ser. Mat.-Inform.* **37** (1989) 55–63.
16. A. Salomaa, *Formal Languages*. Academic Press, 1973.
17. F. L. Ţiplea, C. Ene, C. M. Ionescu and O. Procopiuc. Some decision problems for parallel communicating grammar systems. *Theor. Comp. Sci.* **134** (1994) 365–385.
18. Gy. Vaszil, On simulating non-returning PC grammar systems with returning systems. *Theor. Comp. Sci.* **209** (1998), 319–329.
19. Gy. Vaszil, On parallel communicating Lindenmayer systems, In: Gh. Păun and A. Salomaa (eds.), *Grammatical Models of Multi-Agent Systems*, Gordon and Breach, 1999, 99–112.

An Interpretation of Extensible Objects

Gérard Boudol and Silvano Dal-Zilio

INRIA Sophia Antipolis
BP 93 – 06902 Sophia Antipolis Cedex

Abstract. We provide a translation of Fisher-Honsell-Mitchell's delegation-based object calculus with subtyping into a λ-calculus with extensible records. The target type system is an extension of the system $\mathcal{F}^\omega$ of types depending on types with recursion, extensible records and a form of bounded universal quantification. We show that our translation is computationally adequate, that the typing rules of Fisher-Honsell-Mitchell's calculus can be derived in a rather simple and natural way, and that our system enjoys the standard subject reduction property.

1 Introduction

The theoretical foundations of object-oriented programming have been intensively explored in the recent past, the main purpose being to design type systems that would be safe, allowing in particular to prevent some run-time errors like *message not understood*, while being at the same time flexible enough to support object-oriented programming idioms. This has proven to be actually quite difficult. Some early works, following Cardelli's pioneering paper [9], developed *encodings* of object-oriented constructions into calculi with records. For instance, Cook & al. in [12] proposed an interpretation of class-based programming. In their model, an object is the fixed-point of a function (of self) returning a record. An object can only be invoked, while inheritance acts on the object's "generator", by adding or updating fields of the resulting record, and extending the scope of the self parameter. The types, based on F-bounded quantification (or else higher-order quantification and recursive types, see [19]), allow in particular to address the property identified as *method specialization* by Mitchell [20], by which is meant the fact that the type of a method is updated whenever the hosting class is inherited.

Mitchell's work deals with delegation-based object-oriented programming, where a single entity, called a *prototype*, may be both invoked, by messages calling for some methods to be executed, and inherited, to build a new prototype by adding a new method or modifying an existing one. Although this looks simpler than the class-based approach, the property of method specialization is not easy to obtain in this setting. For this reason, Mitchell & al. proposed in [13] a primitive object calculus, with a specific typing construct for prototypes, that is not derived from an encoding. Indeed, the authors conclude that "[*method specialization*] *seems very difficult to achieve directly with any calculus of records*". Several other primitive object calculi were developed at about the same time by Abadi and Cardelli (see for instance [1, 2]) to formalize abstractly what can be expected from the typing of objects, while avoiding the difficulties of discovering adequate encodings in λ-calculi with records. Later on however, these authors, together with R. Viswanathan [3], solved the encoding problem for a delegation-based calculus without object extension (hence with no non-trivial method specialization), and Viswanathan improved their results in [21], where the target of the translation is a first-order calculus.

Our purpose in this paper is to introduce an encoding of Mitchell & al. calculus, extended with subtyping [15], into a λ-calculus with extensible records equipped with a rich but fairly standard type system, that has been recognized as a suitable framework for studying the typing of objects, see [19, 7]. We show in particular how the typing rules for prototypes can be derived in a natural way. The idea of the encoding is very simple. To explain it, let us first come back to the difficulty of typing method specialization for prototypes: there are two seemingly opposite requirements for types in this setting. One is that, when the prototype is invoked, its type should not tell too much, because the prototype can be revealed upon invocation, for instance by an identity method. In particular, the type should only exhibit the method names that are actually available, since otherwise runtime errors could occur. On the other hand, when the prototype is inherited, its type – or more precisely the type of the self parameter – should be "open" to potential extensions (see [15] where the authors distinguish a "client interface" from the "inheritance interface" for an object). This tension is solved in Cook's model, simply by fulfilling separately these two requirements.

Then the idea of our encoding is to *separate the two usages* of a prototype by means of record field selection: in our interpretation, a prototype is a recursive record with two fields. The first one, that we call inht, contains the current value of the prototype generator, that is the function of self that returns the record of methods. The other field, called invk, contains the application of the generator to the prototype itself. The first field is selected to inherit the prototype, while the second is selected to invoke it. We show how to type such a prototype in a system involving extensible record types, types depending on types and a limited form of bounded universal quantification, thus deriving naturally the typing rules of [13] – or more precisely rules given by Fisher in her thesis [16]. We also show that the subtyping of prototypes proposed in [15] arises in a very simple way: while a "pro" type of a prototype is a record with two fields, the "obj" type only contains the field invk, thus allowing a "sealed prototype" to be only invoked. The standard subtyping rule for recursive types allows to derive the subtyping relations of [15].

We think that our interpretation validates, and even justifies the (non-trivial) rules designed for primitive type constructors in Fisher and Mitchell's systems, and allows to reuse results that can be established in the target calculus. Bruce in [6] has shown that, from a semantical point of view, Cook and Mitchell's models are equivalent. Our interpretation confirms this view from a syntactical perspective, and also confirms that Bruce's matching – that is width subtyping of the record of methods [8] – is the form of subtyping we need to type extensible objects, as far as the goal is to retrieve Fisher and Mitchell's systems. Our interpretation also shows that it is possible to do a little better, allowing the type of an updated method to be specialized to a subtype, by using the standard subtyping relation, together with width subtyping which is useful to type self-inflicted method update.

2 The Calculus of Prototypes

The calculus of prototypes introduced by Fisher, Honsell and Mitchell in [13] is a λ-calculus extended with constructions for building prototypes from the empty one, namely prototype extension, denoted $\langle P \leftarrow\!\!+\, \ell = Q\rangle$, and method update $\langle P \leftarrow \ell = Q\rangle$. There is also an operation of method invocation, written $P \Leftarrow \ell$. In [13] the notation $\langle P \leftarrow\!\circ\, \ell = Q\rangle$ is introduced, to mean either $\langle P \leftarrow\!\!+\, \ell = Q\rangle$ or $\langle P \leftarrow \ell = Q\rangle$. Then the evaluation of prototypes mainly consists in the following transition:

$$\langle P \leftarrow\!\circ\, \ell = Q\rangle \Leftarrow \ell \;\rightarrow\; Q\langle P \leftarrow\!\circ\, \ell = Q\rangle$$

There are actually several possible ways to define the evaluation mechanism for prototypes. In [17] an auxiliary operation is used, which is a combination of method extraction and self-application. Moreover, in that paper no distinction is made between the two ways of building prototypes, and $\langle P \leftarrow \ell = Q\rangle$ stands for prototype extension as well as method update – or override. Since from the operational point of view our translation mimics the prototypes exactly as they are described in [17], we adopt the syntax of this paper. We assume given a denumerable set $\mathcal{X}$ of *variables*, and a denumerable set $\mathcal{K}$, disjoint from $\mathcal{X}$, of *keys*, or labels, used as method names. We use x, y, $z\ldots$ and ℓ, k to range over $\mathcal{X}$ and $\mathcal{K}$ respectively. The grammar for terms is as follows:

$$\begin{array}{rcll}
P, Q\ldots & ::= & x \mid W \mid (PQ) & \textit{λ-calculus} \\
& \mid & (P \Leftarrow \ell) & \textit{method invocation} \\
& \mid & \mathsf{S}(P, \ell, Q) & \textit{subsidiary operation} \\
W & ::= & \lambda x P \mid O & \textit{values} \\
O & ::= & \langle\rangle & \textit{empty prototype} \\
& \mid & \langle P \leftarrow \ell = Q\rangle & \textit{prototype extension/update}
\end{array}$$

As usual $\langle \ell_1 = P_1, \ldots, \ell_n = P_n\rangle$ denotes $\langle\langle\langle\rangle \leftarrow \ell_1 = P_1\rangle \cdots \leftarrow \ell_n = P_n\rangle$. The meaning of the auxiliary operation $\mathsf{S}(P, \ell, Q)$ is that it extracts from the prototype P the method of name ℓ and applies it to Q. Due to space limitations, we cannot give the rules for evaluation. We denote by $\mathsf{eval}(P)$ the (unique) value of P, if any.

There are several versions of the type system for the calculus of prototypes. The system presented in [15] is essentially the one of [13], enriched with a form of subtyping. Here we use a simplified version of the system given in Fisher's thesis [16]. The simplifications are as follows: we adopt the system of Chapter 3 of [16], which does not involve variance annotation nor existential quantification, with a restriction on the assumptions regarding row variables given below. Moreover, we do not take subsumption into account in the simplified system (this will be dealt with in a next section). Then the typing system we consider here is defined as follows. The grammar of kinds, types and rows is

$$\begin{array}{rcll} kind & ::= & \mathrm{T} \mid \kappa & \\ \kappa & ::= & \mathrm{M} \mid \mathrm{T} \to \mathrm{M} & \text{where } \mathrm{M} = \{\ell_1, \ldots, \ell_n\} \\ \tau & ::= & t \mid (\tau_1 \to \tau_2) \mid \mathsf{pro}\, t.\rho & \\ \rho & ::= & r \mid [] \mid [\rho, \ell : \tau] \mid (\lambda t.\rho) \mid (\rho\tau) & \end{array}$$

The kind T is that of well-formed types (the only constraint is on $\mathsf{pro}\, t.\rho$). The kind $\{\ell_1, \ldots, \ell_n\}$ is that of a row which does not contain the keys $\ell_1, \ldots, \ell_n$. The typing contexts are

$$\Gamma ::= \emptyset \mid \Gamma, x : \tau \mid \Gamma, t :: \mathrm{T} \mid \Gamma, r <:_w \rho$$

Notice that an assumption about a row variable r only involves the "width subtyping" relation $<:_w$. Moreover, in Fisher's thesis, there is a kind annotation, that is the assumption is $r <:_w \rho :: \kappa$; here we omit this annotation since in our simplified system this κ is always $\mathrm{T} \to \emptyset$. The type judgements are

$\Gamma \vdash *$	*well-formed context*	$\Gamma \vdash \rho <:_w \rho'$	*subrow*
$\Gamma \vdash \tau :: \mathrm{T}$	*type is well-formed*	$\Gamma \vdash P : \tau$	*term has type*
$\Gamma \vdash \rho :: \kappa$	*row has kind*		

For lack of space, we omit most of the rules of the system, which are quite standard. For instance the only interesting rule for well-formedness of types is

$$\frac{\Gamma, t :: \mathrm{T} \vdash \rho :: \mathrm{M}}{\Gamma \vdash \mathsf{pro}\, t.\rho :: \mathrm{T}}$$

The interesting rules for typing terms are:

$$\frac{\Gamma \vdash *}{\Gamma \vdash \langle\rangle : \mathsf{pro}\, t.[]} \ (\textit{empty pro}) \qquad \frac{\Gamma \vdash P : \mathsf{pro}\, t.\rho \quad , \quad \Gamma, t :: \mathrm{T} \vdash \rho <:_w [\ell : \tau]}{\Gamma \vdash P \Leftarrow \ell : [\mathsf{pro}\, t.\rho/t]\tau} \ (\textit{pro} \Leftarrow)$$

where $[\ell : \tau]$ denotes $[[], \ell : \tau]$,

$$\frac{\begin{array}{l} \Gamma \vdash P : \mathsf{pro}\, t.\rho \\ \Gamma, t :: \mathrm{T} \vdash \rho :: \{\ell\} \\ \Gamma, r <:_w \lambda t.[\rho, \ell : \tau] \vdash Q : [\mathsf{pro}\, t.(rt)/t](t \to \tau) \end{array}}{\Gamma \vdash \langle P \leftarrow \ell = Q \rangle : \mathsf{pro}\, t.[\rho, \ell : \tau]} \ r \notin \mathsf{fv}(\tau) \ (\textit{pro ext})$$

and

$$\frac{\begin{array}{l} \Gamma \vdash P : \mathsf{pro}\, t.\rho \\ \Gamma, t :: \mathrm{T} \vdash \rho <:_w [\ell : \tau] \\ \Gamma, r <:_w \lambda t.\rho \vdash Q : [\mathsf{pro}\, t.(rt)/t](t \to \tau) \end{array}}{\Gamma \vdash \langle P \leftarrow \ell = Q \rangle : \mathsf{pro}\, t.\rho} \ r \notin \mathsf{fv}(\tau) \ (\textit{pro over})$$

(there is also a rule for the subsidiary operation). As we said, there are two distinct rules for typing the construction $\langle P \leftarrow \ell = Q \rangle$. In the first one, the premise $\Gamma, t :: \mathrm{T} \vdash \rho :: \{\ell\}$ indicates that in the prototype P there is no method named ℓ, and then this rule allows to type a true extension. On the other hand, the premise $\Gamma, t :: \mathrm{T} \vdash \rho <:_w [\ell : \tau]$ says that a method with name ℓ is already provided by P, with result type τ. Then one may override this method, with another one having the same result type. We refer to [13, 14] for examples of typing.

3 Encoding Prototypes with Extensible Records

As a target for our encoding of prototypes, we use a λ-calculus enriched with extensible records. We also find it convenient to use an explicit fixpoint construct. The syntax of the calculus is as follows:

$$
\begin{array}{rcll}
M, N \ldots & ::= & x \mid V \mid (MN) & \textit{λ-calculus} \\
& \mid & \mathsf{fix}\, x.M & \textit{fixpoint} \\
& \mid & (M.\ell) & \textit{field selection} \\
V & ::= & \lambda x M \mid R & \textit{values} \\
R & ::= & [] & \textit{empty record} \\
& \mid & [M, \ell = N] & \textit{record extension}
\end{array}
$$

where $x \in \mathcal{X}$ and $\ell \in \mathcal{K}$. For ease of readability, one quite often writes $(\lambda x M)$ for $\lambda x M$, and MN for (MN), and similarly $M.\ell$ for $(M.\ell)$. We denote the record $[\cdots[[], \ell_1 = M_1] \cdots \ell_n = M_n]$ by $[\ell_1 = M_1, \ldots, \ell_n = M_n]$. Again, due to space limitations, we do not give the rules for evaluation. Let us just say that for extensible records we have

$$
\begin{array}{ll}
[M, \ell = N].\ell \to N & \\
[M, k = N].\ell \to M.\ell & \quad k \neq \ell
\end{array}
$$

We denote by $\mathsf{eval}(M)$ the unique value of M, if any. Now we turn to the interpretation of the calculus of prototypes. To represent a prototype, say for instance $\langle \ell_1 = P_1, \ldots, \ell_n = P_n \rangle$, where the P_i's are functions of the prototype itself (usually in the form of a self parameter), we shall use a function returning a record

$$G = \lambda \mathsf{self}\, [\ell_1 = M_1 \mathsf{self}, \ldots, \ell_n = M_n \mathsf{self}]$$

This is an "object definition" in Cook's model (see [12, 14]), also called "object generator" (see [2, 3]). As we said in the introduction, in our interpretation of a prototype, we separate the two usages we have of it, namely invocation and inheritance. This idea of separating the usages of an object is embodied in the "split-method" of [3, 21], but we cannot use this specific approach here because the set of methods of a prototype is not fixed once for all. To give our translation, let us introduce a notation: we define

$$\mathsf{proto} =_{\mathsf{def}} \mathsf{fix}\, p.\lambda z.[\,\mathsf{inht} = z\,,\ \mathsf{invk} = z(pz)\,]$$

Then the translation $[\![.]\!]$ from the λ-calculus of prototypes to the λ-calculus of extensible records is given by – omitting the part of the translation that regards the λ-calculus, which is trivial:

$$
\begin{aligned}
[\![\langle\rangle]\!] &= \mathsf{proto}(\lambda\,\mathsf{self}\,[])\\
[\![\langle P \leftarrow \ell = Q\rangle]\!] &= \mathsf{proto}(\lambda\,\mathsf{self}\,[[\![P]\!].\mathsf{inht}\,\mathsf{self}, \ell = [\![Q]\!]\,\mathsf{self}])\\
[\![P \Leftarrow \ell]\!] &= [\![P]\!].\mathsf{invk}.\ell\\
[\![\mathsf{S}(P,\ell,Q)]\!] &= ([\![P]\!].\mathsf{inht}\,[\![Q]\!]).\ell
\end{aligned}
$$

In the translation we obviously assume that the name self is not free in the source terms. To shorten the notation, we shall replace self by the ordinary λ-variable x in the sequel. As one can see, our interpretation is very close to Cook's model, separating objects from object generators, with inheritance acting on the generators; in particular, extending and updating a prototype are operationally the same. There is a difference with Cook's model, however, since prototypes feature runtime extension. A first result we can prove about our translation is its operational soundness:

PROPOSITION (COMPUTATIONAL ADEQUACY) 0.1. *A prototype P has a value if and only if its translation has a value. More precisely,* $\mathsf{eval}([\![P]\!]) = \mathsf{eval}([\![\mathsf{eval}(P)]\!])$.

4 Deriving the Typing Rules

In this section we show how to derive the typing rules for prototypes. We first introduce a type system $\mathcal{T}$ for our λ-calculus with extensible records. This system features arrow types and extensible record types – also called row expressions, following [22] –, as well as recursive types, types depending on types [18] and a limited form of bounded universal quantification [10]. Moreover, we will only allow types which are well-formed with respect to a system of kinds. There are two basic kinds, □, the kind of record types, and ◇, the kind of types. Since we also have type operators, the syntax of kinds and types is the following:

$$
\begin{aligned}
\kappa &::= \square \mid \diamond \mid (\kappa \to \kappa)\\
\tau, \sigma \ldots &::= t \mid (\tau \to \sigma) \mid [] \mid [\sigma, \ell\colon \tau] \mid (\mu t^{\kappa}.\tau) \mid (\Lambda t^{\kappa}.\tau) \mid (\tau\sigma) \mid (\forall t \lhd \tau.\sigma)
\end{aligned}
$$

For instance $\diamond \to \square$ is the kind of interfaces, that is functions from types to records. In $(\forall t \lhd \tau.\sigma)$, the variable t is bound in σ, but not in τ (hence we do not have F-bounded polymorphism). Dependent types are used as prototype's *interface* (see [14, 19]), which typically are of the form $\Lambda t^{\diamond}.[\ell_1 : \tau_1, \ldots, \ell_n : \tau_n]$. Bounded quantification is used to model the fact that such an interface may be extended.

The preorder in $(\forall t \lhd \tau.\sigma)$ is *width subtyping* (of rows, not records), as this is precisely the notion of subtyping one needs in order to derive the typing of method specialization (see the comments at the end of this section). We use the notation $\lhd$ since our axiomatization of this preorder makes it slightly more generous than $<:_w$ of [16]. For instance we have $[\ell : \tau', \ell : \tau] \lhd [\ell : \tau]$, but the first of these record types is not even well-formed in Fisher's system. In our setting, we allow a row $[[\sigma, \ell : \tau'], \ell : \tau]$ to be well-formed, and this will simplify the type system, but the reader should notice that this is a (width) subtype of $[\sigma, \ell : \tau']$ only if $\tau' = \tau$. The typing contexts of $\mathcal{T}$ are as follows:

$$\Gamma ::= \emptyset \mid \Gamma, x : \tau \mid \Gamma, t \lessdot \tau \mid \Gamma, t :: \kappa$$

and the type judgements are $\Gamma \vdash *$, $\Gamma \vdash \tau :: \kappa$, $\Gamma \vdash \tau \lessdot \sigma$, $\Gamma \vdash M : \tau$, with the same meaning as above, plus $\Gamma \vdash \tau <: \sigma$. In the type system we use a notion of type equality, which is the *$\beta\mu$-conversion*, denoted $=_{\beta\mu}$. This is the congruence generated by the laws:

$$((\Lambda t^{\kappa}.\sigma)\tau) =_{\beta\mu} [\tau/t]\sigma$$
$$(\mu t^{\kappa}.\tau) =_{\beta\mu} [\mu t^{\kappa}.\tau/t]\tau$$

All the rules of the system, for well-formedness of contexts, kinding of types, subtyping judgements $\Gamma \vdash \tau \lessdot \sigma$ and for typing terms, collected in the appendices, are fairly standard (except perhaps for what regards width subtyping of rows). More precisely, system $\mathcal{T}$ consists in the rules collected in the first three appendices. Let us comment on some of the rules: regarding the subtyping of rows, $\Gamma \vdash \sigma \lessdot [\ell : \tau] \;\Rightarrow\; \Gamma \vdash [\sigma, \ell : \tau] \lessdot \sigma$ expresses the fact that if we know that a row σ contains an ℓ field with type τ, then extending it with $\ell : \tau$ actually does not modify it. This rule will be used for typing method update. One should remark that, although in $\mathcal{T}$ we have the usual *subsumption rule* this is of limited use, since the only way to infer $\Gamma \vdash \tau <: \sigma$ in $\mathcal{T}$ is by means of the rule $\Gamma \vdash \tau \lessdot \sigma \;\Rightarrow\; \Gamma \vdash \tau <: \sigma$. That is, in $\mathcal{T}$ we only have "width subsumption". One should also observe that we have no subtyping rule for bounded quantification. However, using the rules of instanciation, subsumption and generalization, one can achieve the same effect. That is, the following inference is valid:

$$\frac{\Gamma \vdash M : (\forall t \lessdot \tau.\sigma) \quad , \quad \Gamma \vdash \tau' \lessdot \tau \quad , \quad \Gamma \vdash \sigma \lessdot \sigma'}{\Gamma \vdash M : (\forall t \lessdot \tau'.\sigma')} \qquad (\star)$$

To see how the typing rules for prototypes are mimicked in our system, let us first give some valid inferences. To this end, we introduce some notations. Let us define:

$$\delta =_{\mathsf{def}} \diamond \rightarrow \Box$$
$$\gamma =_{\mathsf{def}} (\diamond \rightarrow \Box) \rightarrow \Box$$
$$\pi =_{\mathsf{def}} \mu p^{\gamma}.\Lambda s^{\delta}.\mu o^{\Box}.[\,\mathsf{inht} : (\forall s' \lessdot s.ps' \rightarrow s(ps'))\,,\, \mathsf{invk} : so]$$
$$\varsigma =_{\mathsf{def}} \Lambda t^{\delta}.(\forall s' \lessdot t.\pi s' \rightarrow t(\pi s'))$$

It is easy to see that, for any σ:

$$(\pi\sigma) =_{\beta\mu} [\,\mathsf{inht} : (\varsigma\sigma)\,,\, \mathsf{invk} : \sigma(\pi\sigma)\,] \qquad (1)$$

and therefore the following rule is admissible:

$$\frac{\Gamma \vdash G : (\varsigma\sigma)}{\Gamma \vdash (\mathsf{proto}\, G) : (\pi\sigma)} \qquad (2)$$

This holds in particular whenever $G = \lambda\mathsf{self}\,[\ell_1 = M_1\mathsf{self}, \ldots, \ell_n = M_n\mathsf{self}]$ is an object generator, which has type $\varsigma(\Lambda t^{\diamond}.\rho)$ where $\rho = [\ell_1 : \tau_1, \ldots, \ell_n : \tau_n]$, using the assumption that self has type $(\pi s')$ with constraint $s' \lhd (\Lambda t^\diamond.\rho)$, and that the M_i's have type $t \to \tau_i$. In this case the type of $(\text{proto}\, G)$ is

$$\pi(\Lambda t^\diamond.\rho) =_{\beta\mu} [\mathsf{inht}\colon (\forall s' \lhd \sigma.\pi s' \to [\pi s'/t]\rho)\,,\, \mathsf{invk}\colon [\pi\sigma/t]\rho] \qquad \text{with} \quad \sigma = \Lambda t^\diamond.\rho$$

One can see that invk is a record of type $[\ell_1 : \tau_1', \ldots, \ell_n : \tau_n']$, thus exhibiting only the method names that are actually available in the prototype, while inht is a function that takes arguments of type $\pi\sigma'$ with $\sigma' \lhd \sigma$, that is any possible extension to the original prototype, and returns the record of methods "specialized" to this extension. Another admissible rule, derived using the law (1), is:

$$\frac{\Gamma \vdash M\colon (\pi\sigma)\ ,\ \Gamma \vdash \sigma \lhd \Lambda t^\diamond.[\ell\colon \tau]}{\Gamma \vdash M.\mathsf{invk}.\ell\colon [\pi\sigma/t]\tau} \tag{3}$$

Again using (1), one can check that for $G = \lambda x[M.\mathsf{inht}\, x, \ell = Nx]$ the following is a valid inference:

$$\frac{\Gamma \vdash M\colon (\pi\sigma)\quad ,\quad \Gamma \vdash \sigma' \lhd \sigma\quad ,\quad \Gamma \vdash N\colon (\forall s \lhd \sigma'.\pi s \to [\pi s/t]\tau)}{\Gamma \vdash G\colon (\forall s \lhd \sigma'.\pi s \to [\sigma(\pi s), \ell\colon [\pi s/t]\tau])}$$

Now if we try to exploit the rule (2), to build a prototype out of G, we have to find a way to replace $[\sigma(\pi s), \ell\colon [\pi s/t]\tau]$ by $\sigma'(\pi s)$. An obvious solution is $\sigma' = \Lambda t^\diamond.[\sigma t, \ell\colon \tau]$, since then $[\sigma(\pi s), \ell\colon [\pi s/t]\tau] =_{\beta\mu} \sigma'(\pi s)$, and this gives us another admissible rule:

$$\frac{\Gamma \vdash M\colon (\pi\sigma), \Gamma \vdash \Lambda t^\diamond.[\sigma t, \ell\colon \tau] \lhd \sigma, \Gamma \vdash N\colon (\forall s \lhd \Lambda t^\diamond.[\sigma t, \ell\colon \tau].\pi s \to [\pi s/t]\tau)}{\Gamma \vdash (\text{proto}\, \lambda x[M.\mathsf{inht}\, x, \ell = Nx])\colon \pi(\Lambda t^\diamond.[\sigma t, \ell\colon \tau])} \tag{4}$$

There is another possibility, however, which is $\sigma' = \sigma$ and $\sigma \lhd \Lambda t^\diamond.[\ell\colon \tau]$, since in this case the following inference is valid:

$$\frac{\dfrac{\Gamma \vdash \sigma \lhd \Lambda t^\diamond.[\ell\colon \tau]}{\Gamma \vdash [\sigma(\pi s), \ell\colon [\pi s/t]\tau] \lhd \sigma(\pi s)}}{\Gamma \vdash \pi s \to [\sigma(\pi s), \ell\colon [\pi s/t]\tau] \lhd \pi s \to \sigma(\pi s)}$$

Using the inference $(\star)$ above, this gives us the following admissible rule:

$$\frac{\Gamma \vdash M\colon (\pi\sigma)\quad ,\quad \Gamma \vdash \sigma \lhd \Lambda t^\diamond.[\ell\colon \tau]\quad ,\quad \Gamma \vdash N\colon (\forall s \lhd \sigma.\pi s \to [\pi s/t]\tau)}{\Gamma \vdash (\text{proto}\, \lambda x[M.\mathsf{inht}\, x, \ell = Nx])\colon (\pi\sigma)} \tag{5}$$

Now the translation from the simplified version of Fisher's system sketched in Sect. 2 to our type system should be clear. As far as types are concerned, it is given by

$$\boxed{[\![\text{pro}\, t.\rho]\!] = \pi(\Lambda t^\diamond.[\![\rho]\!])}$$

(the rest is trivial, e.g. $[\![\lambda t.\rho]\!] = \Lambda t^\diamond.[\![\rho]\!]$). Regarding the kinds, we let $[\![\mathrm{M}]\!] = \Box$ and $[\![\mathrm{T}]\!] = \diamond$. Translating the contexts in the obvious way (where $r <:_w \rho$ is translated into $r \lhd [\![\rho]\!]$), we have:

THEOREM. *If $\Gamma \vdash M : \tau$ can be inferred in the type system for prototypes, then $[\![\Gamma]\!] \vdash [\![M]\!] : [\![\tau]\!]$ can be proved in the type system $\mathcal{T}$ for the λ-calculus with extensible records.*

The derived inferences (2) to (5) allow us to retrieve the rules (*empty pro*), (*pro* $\Leftarrow$), (*pro ext*) and (*pro over*) respectively, with $\sigma = \Lambda t^{\diamond}.[\![\rho]\!]$ in the last three cases. One should notice that the only use of subsumption we made is in deriving the rule (*pro over*), via the inference ($\star$). One should also observe that our inference (4) is actually slightly more general than the corresponding rule (*pro ext*), since for the premise $\Gamma \vdash \sigma \lessdot \Lambda t^{\diamond}.[\sigma t, \ell : \tau]$ to be valid with $\sigma = \Lambda t^{\diamond}.[\![\rho]\!]$ there may actually be two cases: either ρ does not contain the field ℓ, in which case we are typing a true extension, as in the rule (*pro ext*), or ρ does contain the field ℓ, but then it must be, thanks to the axiomatization of $\lessdot$, of type τ, and this is actually a particular case of inference (5). Remark that in the latter the premise regarding N is stronger (a bounded quantification is contravariant in the bound) than in (4), and this is needed to type self-inflicted update, as in the standard "movable point" object – see [13].

Our result is a *completeness* result similar to the one obtained by Abadi & al. in [3]. Its converse does not hold, for some interesting reasons: there is a slight difference in the modelling of prototypes between Fisher-Honsell-Mitchell's calculus and our encoding – or Cardelli's [9] and Cook's [12] models, for that matter –, which is that the underlying record representing the prototype is, in the former, a record of pre-methods, which are functions of self, while it is a record of methods, where the self parameter is free, in the latter. In other word, the self parameter is "late bound" in the recursive record or generator model, while it is bound earlier – that is, before extension – in Fisher-Honsell-Mitchell's prototypes. Then for instance the prototype

$$\langle\langle\langle\rangle \leftarrow k = \lambda \mathsf{self}\,(\mathsf{self}.\ell)\rangle \leftarrow \ell = \lambda \mathsf{self}\,(\mathsf{self}.\ell)\rangle$$

is not typable in Fisher-Honsell-Mitchell's type system (see [13] for a similar example), while its translation is typable in our system. In fact, this prototype has "semantically" (modulo permutation of fields of different names in a record) the same translation as

$$\langle\langle\langle\rangle \leftarrow \ell = \lambda \mathsf{self}\,(\mathsf{self}.\ell)\rangle \leftarrow k = \lambda \mathsf{self}\,(\mathsf{self}.\ell)\rangle$$

because both are represented by the "same" generator, and the latter is typable in Fisher-Honsell-Mitchell's calculus.

Finally, one may notice that, when we specialize the inferences (3), (4) and (5) with $\sigma = \Lambda t^{\diamond}.[\![\rho]\!]$, where ρ is supposed to be of record kind, we could reformulate these inferences using a preorder introduced by Bruce in [6] and now called *matching*, denoted $<\#$, which may be defined as follows:

$$\frac{\Gamma, t :: \diamond \vdash \rho :: \Box \quad , \quad \Gamma \vdash \rho \lessdot \rho'}{\Gamma \vdash \mathsf{pro}\, t.\rho <\# \mathsf{pro}\, t.\rho'}$$

Then the reformulated admissible rules would be nothing else than the rules recently proposed for Fisher-Honsell-Mitchell's calculus of prototypes by Bono and Bugliesi in [5]. As shown by Bruce in [6], matching is precisely the kind of subtyping we need in typing the operations of extension and update on prototypes (though the paper [6] actually deals with classes). This explains why we need, at the lower level of records, to use width subtyping.

5 Objects and Subtyping

As we said, our type system $\mathcal{T}$ for the λ-calculus of extensible records only makes use of a particular form of subtyping. It has been remarked by Fisher and Mitchell in [14] that adding the subsumption rule in the typing of prototypes, one actually does not gain anything, since "***in pure delegation based languages, no subtyping is possible***". This is easy to explain looking at our translation: one easily sees that ς is invariant in its type parameter (the bound τ in $\forall t \lessdot \tau.\sigma$ is in a contravariant position), and therefore π is invariant too. For this reason we cannot deal with the preorder proposed by Bono and Liquori [4] for the calculus of prototypes, nor with the width subtyping of prototypes of fixed size as it is done by Abadi and Cardelli in [2], at least if we stick to the translation given here – it might be possible to combine it with the translation of [3, 21], allowing their "assembled" objects to be part of the syntax, but this looks a bit ad hoc.

In a subsequent paper [15], Fisher and Mitchell enriched their calculus by distinguishing, at the level of types, prototypes from *objects*, which are "sealed prototypes". The distinction is that objects feature only method invocation, but no extension or update. Then a new type construct $\mathsf{obj}\, t.\rho$ is introduced together with three new rules, which are essentially the following, where the typing contexts are enriched with new constraints $t <: \tau$:

$$\frac{\Gamma, t <: t' \vdash \rho <: \rho'}{\Gamma \vdash \mathsf{pro}\, t.\rho <: \mathsf{obj}\, t'.\rho'} \qquad \frac{\Gamma, t <: t' \vdash \rho <: \rho'}{\Gamma \vdash \mathsf{obj}\, t.\rho <: \mathsf{obj}\, t'.\rho'}$$

$$\frac{\Gamma \vdash P : \mathsf{obj}\, t.\rho \quad , \quad \Gamma, t : \mathrm{T} \vdash \rho <:_w [\ell : \tau]}{\Gamma \vdash P \Leftarrow \ell : [\mathsf{obj}\, t.\rho/t]\tau}$$

Now consider our typing system $\mathcal{T}$ enriched with subtyping. That is, we introduce an extension, called $\mathcal{T}_{<:}$, of $\mathcal{T}$ in which we have a new kind of contexts, namely $\Gamma, t <: \tau$, and new rules for inferring the judgements $\Gamma \vdash \tau <: \sigma$ – the rules are given in the fourth appendix. If we let

$$[\![\mathsf{obj}\, t.\rho]\!] = \omega(\Lambda t^{\diamond}.\rho) \qquad \text{where} \quad \omega =_{\mathsf{def}} \Lambda s^{\delta}.\mu o^{\square}.[\mathsf{invk} : so]$$

then it is easy to see, using the standard rule for subtyping recursive types, that the three rules above can be derived in our system with subtyping. Notice that $[\![\mathsf{obj}\, t.\rho]\!] =_{\beta\mu} \mu t^{\diamond}.[\mathsf{invk} : \rho]$ and therefore this is essentially what Bruce & al. [7] call the "classical recursive record encoding" (see also [14] § 6.3), which is the natural typing of Cardelli's recursive records [9].

Regarding the full type system with subtyping $\mathcal{T}_{<:}$ for our λ-calculus with extensible records, our main result is *type safety*. Indeed, we can prove the subject reduction property:

THEOREM (SUBJECT REDUCTION). *If $\Gamma \vdash M : \tau$ is provable in $\mathcal{T}_{<:}$ and $M \rightarrow M'$ then $\Gamma \vdash M : \tau'$ is provable in $\mathcal{T}_{<:}$.*

Our proof, which, not surprisingly, is quite long and technical, differs from Compagnoni's one [11]; in particular, since we have recursive types, we cannot rely on a strong normalization result. However, since we have no subtyping rule for bounded quantification, we can make a direct proof, analysing the typing of a compound term in typings of its components.

It is possible to exploit subsumption to gain another typing rule for method update, where the type of the method is specialized to a subtype (this is valid in Cook's system, as observed in [14]). To this end one would use a more standard bounded quantification ($\forall t <: \tau.\sigma$), and then the derived inference (4) could be generalized to one involving the premiss $\Gamma \vdash \Lambda t^{\diamond}.[\sigma t, \ell : \tau] <: \sigma$ (a similar remark is made by Bruce in [6]). To type "self-inflicted" method update however, we still need width subtyping $\lessdot$. One may regard this improved system as the right one to adopt for typing prototypes.

References

1. M. ABADI, *Baby Modula-3 and a theory of objects*, J. of Functional Programming Vol 4, No 2 (1994) 249-283.
2. M. ABADI, L. CARDELLI, *A theory of primitive objects: untyped and first-order systems*, TACS'94, Lecture Notes in Comput. Sci. 789 (1994) 296-320.
3. M. ABADI, L. CARDELLI, R. VISWANATHAN, *An interpretation of objects and object types*, POPL'96 (1996) 396-409.
4. V. BONO, L. LIQUORI, *A subtyping for the Fisher-Honsell-Mitchell lambda calculus of objects*, CSL'94, Lecture Notes in Comput. Sci. 933 (1994) 16-30.
5. V. BONO, M. BUGLIESI, *Matching constraints for the lambda calculus of objects*, TLCA'97, Lecture Notes in Comput. Sci. 1210 (1997) 46-63.
6. K. BRUCE, *The equivalence of two semantic definitions for inheritance in object-oriented languages*, MFPS'92, Lecture Notes in Comput. Sci. 598 (1992) 102-124.
7. K. BRUCE, L. CARDELLI, B. PIERCE, *Comparing object encodings*, TACS'97, Lecture Notes in Comput. Sci. 1281 (1997) 415-438.
8. K. BRUCE, L. PETERSEN, A. FIECH, *Subtyping is not a good "match" for object-oriented languages*, ECOOP'97, Lecture Notes in Comput. Sci. 1241 (1997) 104-127.
9. L. CARDELLI, *A semantics of multiple inheritance*, Semantics of Data Types, Lecture Notes in Comput. Sci. 173 (1984) 51-67. Also published in Information and Computation, Vol 76 (1988).
10. L. CARDELLI, P. WEGNER, *On understanding types, data abstraction, and polymorphism*, Computing Surveys 17 (1985) 471-522.
11. A. COMPAGNONI, *Subject reduction and minimal types for higher order subtyping*, Tech. Rep. ECS-LFCS-97-363, University of Edinburgh (1997).
12. W. COOK, W. HILL, P. CANNING, *Inheritance is not subtyping*, POPL'90 (1990) 125-135.
13. K. FISHER, F. HONSELL, J. MITCHELL, *A lambda calculus of objects and method specialization*, LICS'93 (1993) 26-38.
14. K. FISHER, J. MITCHELL, *Notes on typed object-oriented programming*, TACS'94, Lecture Notes in Comput. Sci. 789 (1994) 844-885.

15. K. FISHER, J. MITCHELL, *A delegation-based object calculus with subtyping*, FCT'95, Lecture Notes in Comput. Sci. 965 (1995) 42-61.
16. K. FISHER, *Types Systems for Object-Oriented Programming Languages*, PhD Thesis, Stanford University (1996).
17. P. DI GIANANTONIO, F. HONSELL, L. LIQUORI, *A lambda-calculus of objects with self-inflicted extension*, OOPSLA'98, ACM SIGPLAN Notices Vol 33, No 10 (1998) 166-178.
18. J.-Y. GIRARD, *Interprétation fonctionnelle et élimination des coupures dans l'arithmétique d'ordre supérieur*, Thèse d'État, Université Paris 7 (1972).
19. M. HOFMANN, B. PIERCE, *A unifying type-theoretic framework for objects*, J. of Functional Programming Vol. 5, No 4 (1995) 593-635.
20. J. MITCHELL, *Toward a typed foundation for method specialization and inheritance*, POPL'90 (1990) 109-124.
21. R. VISWANATHAN, *Full abstraction for first-order objects with recursive types and subtyping*, LICS'98 (1998).
22. M. WAND, *Complete type inference for simple objects*, LICS'87 (1987) 37-44.

Appendix 1: well-formedness of contexts and kinding

$$\emptyset \vdash * \qquad \frac{\Gamma \vdash \tau :: \diamond}{\Gamma, x:\tau \vdash *}\ (\dagger) \qquad \frac{\Gamma \vdash *}{\Gamma, t::\kappa \vdash *}\ (\ddagger) \qquad \frac{\Gamma \vdash \tau :: \kappa}{\Gamma, t \lessdot \tau \vdash *}\ (\ddagger)$$

$$\frac{\Gamma \vdash \sigma :: \square}{\Gamma \vdash \sigma :: \diamond} \qquad \frac{\Gamma \vdash *}{\Gamma \vdash t :: \kappa}\ t::\kappa \in \Gamma \qquad \frac{\Gamma \vdash \tau :: \kappa}{\Gamma \vdash t :: \kappa}\ t \lessdot \tau \in \Gamma$$

$$\frac{\Gamma \vdash \tau :: \diamond\ ,\ \Gamma \vdash \sigma :: \diamond}{\Gamma \vdash (\tau \to \sigma) :: \diamond} \qquad \frac{\Gamma \vdash *}{\Gamma \vdash [] :: \square} \qquad \frac{\Gamma \vdash \sigma :: \square\ ,\ \Gamma \vdash \tau :: \diamond}{\Gamma \vdash [\sigma, \ell : \tau] :: \square}$$

$$\frac{\Gamma, t::\kappa \vdash \tau :: \kappa}{\Gamma \vdash (\mu t^{\kappa}.\tau) :: \kappa}\ (\ddagger) \qquad \frac{\Gamma, t::\kappa \vdash \tau :: \chi}{\Gamma \vdash (\Lambda t^{\kappa}.\tau) :: \kappa \to \chi}\ (\ddagger)$$

$$\frac{\Gamma \vdash \tau :: \chi \to \kappa\ ,\ \Gamma \vdash \sigma :: \chi}{\Gamma \vdash (\tau\sigma) :: \kappa} \qquad \frac{\Gamma, t \lessdot \tau \vdash \sigma :: \diamond}{\Gamma \vdash (\forall t \lessdot \tau.\sigma) :: \diamond}\ (\ddagger)$$

($\dagger$) $x \notin \mathrm{dom}(\Gamma)$, ($\ddagger$) $t \notin \mathrm{fv}(\Gamma)$

Appendix 2: width subtyping

$$\frac{\Gamma \vdash \tau, \sigma :: \kappa}{\Gamma \vdash \tau \lessdot \sigma}\ \tau =_{\beta\mu} \sigma \qquad \frac{\Gamma \vdash \tau \lessdot \theta\ ,\ \Gamma \vdash \theta \lessdot \sigma}{\Gamma \vdash \tau \lessdot \sigma} \qquad \frac{\Gamma \vdash *}{\Gamma \vdash t \lessdot \tau}\ t \lessdot \tau \in \Gamma$$

$$\frac{\Gamma \vdash \tau \lessdot \sigma}{\Gamma \vdash \tau <: \sigma} \qquad \frac{\Gamma \vdash \tau, \sigma :: \diamond\ ,\ \Gamma \vdash \tau' \lessdot \tau\ ,\ \Gamma \vdash \sigma \lessdot \sigma'}{\Gamma \vdash \tau \to \sigma \lessdot \tau' \to \sigma'} \qquad \frac{\Gamma \vdash \sigma :: \square}{\Gamma \vdash \sigma \lessdot []}$$

$$\frac{\Gamma \vdash \sigma \lessdot \sigma'\quad ,\quad \Gamma \vdash \sigma :: \square, \tau :: \diamond}{\Gamma \vdash [\sigma, \ell : \tau] \lessdot [\sigma', \ell : \tau']}\ \tau' =_{\beta\mu} \tau \qquad \frac{\Gamma \vdash \sigma \lessdot [\ell : \tau]}{\Gamma \vdash [\sigma, \ell : \tau] \lessdot \sigma}$$

$$\frac{\Gamma \vdash \sigma :: \Box \quad , \quad \Gamma \vdash \tau, \tau' :: \Diamond}{\Gamma \vdash [[\sigma, k : \tau'], \ell : \tau] \lessdot [[\sigma, \ell : \tau], k : \tau']}\ k \neq \ell$$

$$\frac{\Gamma, t :: \kappa \vdash \tau \lessdot \sigma}{\Gamma \vdash (\Lambda t^{\kappa}.\tau) \lessdot (\Lambda t^{\kappa}.\sigma)}\ t \notin \mathsf{fv}(\Gamma) \qquad \frac{\Gamma \vdash (\tau\sigma) :: \kappa \, , \, \Gamma \vdash \tau \lessdot \tau'}{\Gamma \vdash (\tau\sigma) \lessdot (\tau'\sigma)}$$

Appendix 3: typing

$$\frac{\Gamma \vdash *}{\Gamma \vdash x : \tau}\ x : \tau \in \Gamma \qquad \frac{\Gamma, x : \tau \vdash M : \sigma}{\Gamma \vdash \lambda x M : \tau \to \sigma} \qquad \frac{\Gamma \vdash M : \tau \to \sigma \, , \, \Gamma \vdash N : \tau}{\Gamma \vdash (MN) : \sigma}$$

$$\frac{\Gamma, x : \tau \vdash M : \tau}{\Gamma \vdash \mathsf{fix}\, x.M : \tau} \qquad \frac{\Gamma \vdash M : [\ell : \tau]}{\Gamma \vdash M.\ell : \tau} \qquad \frac{\Gamma \vdash R : \sigma \, , \, \Gamma \vdash M : \tau}{\Gamma \vdash [R, \ell = M] : [\sigma, \ell : \tau]}$$

$$\frac{\Gamma \vdash *}{\Gamma \vdash [] : []} \qquad \frac{\Gamma \vdash \tau' \lessdot \tau \quad , \quad \Gamma \vdash M : (\forall t \lessdot \tau.\sigma)}{\Gamma \vdash M : [\tau'/t]\sigma} \qquad \frac{\Gamma, t \lessdot \tau \vdash M : \sigma}{\Gamma \vdash M : (\forall t \lessdot \tau.\sigma)}$$

$$\frac{\Gamma \vdash M : \tau \, , \, \Gamma \vdash \tau <: \sigma}{\Gamma \vdash M : \sigma} \qquad \frac{\Gamma \vdash M : \tau \, , \, \Gamma, \Gamma' \vdash *}{\Gamma, \Gamma' \vdash M : \tau}$$

Appendix 4: subtyping

$$\frac{\Gamma \vdash \tau :: \kappa}{\Gamma, t <: \tau \vdash *}\ t \notin \mathsf{fv}(\Gamma) \qquad \frac{\Gamma \vdash \tau :: \kappa}{\Gamma \vdash t :: \kappa}\ t <: \tau \in \Gamma \qquad \frac{\Gamma \vdash *}{\Gamma \vdash t <: \tau}\ t <: \tau \in \Gamma$$

$$\frac{\Gamma \vdash \tau <: \theta \, , \, \Gamma \vdash \theta <: \sigma}{\Gamma \vdash \tau <: \sigma} \qquad \frac{\Gamma \vdash \tau, \sigma :: \Diamond \, , \, \Gamma \vdash \tau' <: \tau \, , \, \Gamma \vdash \sigma <: \sigma'}{\Gamma \vdash \tau \to \sigma <: \tau' \to \sigma'}$$

$$\frac{\Gamma \vdash \sigma :: \Box, \tau :: \Diamond \, , \, \Gamma \vdash \sigma <: \sigma' \, , \, \Gamma \vdash \tau <: \tau'}{\Gamma \vdash [\sigma, \ell : \tau] <: [\sigma', \ell : \tau']}$$

$$\frac{\Gamma, t <: s \vdash \tau <: \sigma \quad , \quad \Gamma, t :: \kappa \vdash \tau :: \kappa \quad , \quad \Gamma, s :: \kappa \vdash \sigma :: \kappa}{\Gamma \vdash \mu t^{\kappa}.\tau <: \mu s^{\kappa}.\sigma}\ t \neq s$$

$$\frac{\Gamma, t :: \kappa \vdash \tau <: \sigma}{\Gamma \vdash (\Lambda t^{\kappa}.\tau) <: (\Lambda t^{\kappa}.\sigma)} \qquad \frac{\Gamma \vdash (\tau\sigma) :: \kappa \, , \, \Gamma \vdash \tau <: \tau'}{\Gamma \vdash (\tau\sigma) <: (\tau'\sigma)}$$

Modeling Operating Systems Schedulers with Multi-Stack-Queue Grammars

Luca Breveglieri[1], Stefano Crespi Reghizzi[1] and Alessandra Cherubini[2]

[1] Politecnico di Milano, Dipartimento di Elettronica e Informazione, e-mail: Stefano.Crespi-Reghizzi@Elet.PoliMi.IT, Luca.Breveglieri@Elet.PoliMi.IT
[2] Politecnico di Milano, Dipartimento di Matematica, e-mail: Alessandra.Cherubini@Mate.PoliMi.IT

Abstract. This original method for specifying and checking the sequences of events taking place in process scheduling brings the classical syntax-directed approach of compilation to this new area. The formal language of scheduling events cannot be specified by BNF grammars, but we use instead the Augmented BNF grammars, which combine breadth-first and depth-first derivations. Their recognizers feature one or more FIFO or LIFO tapes. The basic scheduling policies are covered: FCFS, time-slicing, mutex. Combined policies, such as readers/writers and background/foreground, are obtainable by composition. Constraints on the minimum number of data structures (i.e. queues) for priority scheduling policies may be proved by using a pumping lemma. The construction of schedule checkers is presented in the form of augmented LL(1) parsers. For scheduling algorithms, such as shortest job first, which depend on parameters and in particular on time, a syntax-directed approach is proposed, which adds semantic attributes and functions to the underlying augmented BNF grammar.

1 Introduction

This work presents a new syntax-directed approach for specifying and implementing schedulers, based on grammars. Scheduling, a ubiquitous operation of computers, consists in ordering incoming service requests according to some policy (such as First Come First Served, Round-Robin, or Shortest Job First), producing a totally ordered sequence called a schedule. By considering requests and services (to be named events) as symbols of an alphabet, a schedule is a string of events. The sets of schedules for a given policy can then be viewed as a formal language. One can thus consider the scheduling language associated to a given policy. We propose to specify scheduling languages by multi- stack-queue grammars, a family of grammars and related automata using both stacks and queues as their storage [1] [2] [3] [4] [5], much as programming languages are defined by BNF grammars.

Motivation for this investigation: syntax-directed methods and accompanying parsing techniques have been extremely successful for specifying languages and designing their processors. It would be desirable to apply similar techniques to the specification of scheduling algorithms used in operating systems, communication/peripheral drivers, protocols, graphic user interfaces, etc. Among the potential benefits we foresee: more rigorous description of scheduling algorithms than currently found in the literature; a reduced effort for implementing experimental scheduling algorithms as parsers, starting from their grammars; the possibility to add various functionalities, e.g. tracing, to scheduling algorithms in a syntax-directed manner. Returning to the technical problem, it is well-known that BNF grammars fall short of the capacity needed to generate the simplest scheduling languages. Intuitively this inadequacy is caused by the fact that BNF based recognizers, the push-down automata, make use of LIFO storage, whereas scheduling algorithms typically require one or more FIFO queues. The use of queues in abstract computing devices has been investigated by theoreticians: it suffices to mention the queue automaton (called Post machine in [12]), the work by Brandenburg on multi-queue machines [1], and by Citrini et alii on real–time queue machines [6]. Recently our language theory group introduced a new family of grammars [2] [3] [5], involving queues and stacks, called Augmented BNF (ABNF), which preserve the important features of BNF grammars, that are essential for practical usability. Since such grammars are schemata involving queues and stacks, the same data structures engineers use when they describe the scheduling policies by other notations, we performed the present investigation in order to assess their suitability for modeling schedules.

We considered the elementary scheduling algorithms found in operating systems (e.g. [11] and [13]), and succeeded in specifying by formal grammars all the policies that do not involve the computation of process parameters. For the latter policies (e.g. Shortest Job First) we propose a syntax-directed approach that views process parameters as semantic attributes. Grammars for less elementary cases, such as Readers–Writers, can be constructed by a combinational approach, using the basic grammars as building blocks and applying language transformations. We believe the relationships thus discovered shed a new light on the interdependencies between scheduling policies.

Section 2 introduces ABNF grammars and their recognizers (in [2] [3] [5] the formal definitions and essential properties of ABNF grammars and languages can be found), and presents the grammars for elementary algorithms. Section 3 covers more complex scheduling algorithms and highlights the syntactical relations between scheduling policies. Section 4 shows a formal result on the minimum number of queues necessary for a given scheduling language. Section 5 shows a syntax directed schedule checker built as a deterministic parser, and discusses the parsing problem. The Conclusion 6 discusses limits and applicability of the proposal, and raises some theoretical questions.

2 ABNF grammars of basic schedulers

We recall some elementary properties of context-free grammars as the base for the analogy that defines the new grammars to be used. Consider a familiar BNF grammar $G = (V, \Sigma, P, S)$ as defined e.g. in [4]. Without loss of generality the productions of P can be assumed to be in one of the following forms:

$$A \rightarrow bA_1A_2 \ldots A_k \quad \text{and} \quad k \geq 0$$

where b is a (possibly null) symbol from the terminal alphabet Σ and A_i is in the nonterminal alphabet V. If b is missing and $k = 0$ the production becomes $A \rightarrow \epsilon$, with ϵ the empty string.

The language generated by G, L(G), is the set of terminal strings derived from the axiom S by repetitive application of the productions. Considering now the recognizer of L(G), a production can be interpreted as the following instruction of a pushdown automaton:

if the next input char is 'b' and the top symbol of the LIFO tape is 'A' *then*
advance the input head;
erase 'A';
write '$A_1A_2 \ldots A_k$' onto the LIFO tape.

Note that if b is missing from the production, the instruction does not check or affect the input tape. By applying such instructions to a given string x, starting with the axiom S as the tape contents, the string is accepted if the tape is empty when the whole string has been scanned. Notice that this recognizer is in general nondeterministic, accepts by empty tape and does not have a control unit, meaning that it does not make use of internal states.

The new Augmented BNF grammars are a family including the BNF type since they use more general data-structures. Each type of grammar inside the family is characterized by the organization (called *disposition*) of its memory tape. As we know the BNF disposition is a LIFO tape, i.e. a push-down Stack (shortened S). But the disposition can be a FIFO tape, i.e. a Queue (shortened Q), or in general any finite sequence of Q's and S's.

Consider now an ABNF grammar with disposition equal to Q, in short a Q-grammar. Its productions are written as

$$A \rightarrow b(A_1A_2 \ldots A_k)_Q \quad \text{and} \quad k \geq 0$$

to make explicit that the tape is a queue. Similarly, a usual BNF production is now written

$$A \rightarrow b(A_1A_2 \ldots A_k)_S$$

The recognizer associated to a Q-grammar is a queue automaton with the following instruction corresponding to the previous production (as before b can be missing from the production):

if the next input char is 'b' and the queue head symbol is 'A' *then*
advance the input head;
erase 'A' from the queue head;
write '$A_1A_2 \ldots A_k$' into the queue (FIFO) tail.

Figure 1 shows the recognizers for a S-grammar and a Q-grammar. The instruction above, combined with the acceptance condition by empty tape, gives a precise operational semantics to Q-productions, but we need to consider how such productions can generate a language. This will be gradually explained in the next section, as we move into scheduling languages and their grammars. Formal definitions are in [4].

3 Grammars of elementary scheduling languages

We show that the languages of the simplest scheduling disciplines are formally generated by ABNF grammars, with one queue. Then for convenience we introduce also a stack and explain how the stack and queue interact in the ABNF model. For such simple cases, semantic attributes and functions are not needed. The detailed examples are also intended to familiarize the reader with the unusual features of breadth- and depth–first derivations of ABNF grammars.

3.1 First Come First Served

The simplest kind of scheduler (e.g. in [13]) enforces the FCFS discipline. First we introduce the classes of events to be scheduled.
Arrival: We denote by a symbol in $\{a, b, \ldots\}$ the arrival of a client requesting a service: distinct letters denote different types of requests.
Service: We denote by a symbol in $\{a'', b'', \ldots\}$ the execution of the requested service. Execution is assumed to be atomic. In later developments other classes of events will be introduced, for modeling non-atomic services.
Therefore the terminal alphabet is $\Sigma = \{a, b, \ldots, a'', b'', \ldots\}$. In FCFS scheduling clients are serviced in order of arrival, as illustrated in Figure 5.

In all our examples modeling is based on a non-blocking approach for the services given to the processes; for instance the string $aba''ab''$ is not considered a valid FCFS schedule because the second request of service a is not satisfied. This is not a restriction (the language with a blocking policy is obviously a prefix of the language with a non-blocking one). The language of FCFS schedules is generated by the following Q-grammar, with nonterminal alphabet $\{S, A'', B'', \ldots\}$ and axiom S, called $\mathrm{G_Q}$:

1. $\mathrm{S} \rightarrow (SS)_\mathrm{Q}$
2. $\mathrm{S} \rightarrow a(\mathrm{SA}'')_\mathrm{Q}$
3. $\mathrm{S} \rightarrow b(\mathrm{SB}'')_\mathrm{Q}$
4. $\mathrm{S} \rightarrow \epsilon$ —short for $\mathrm{S} \rightarrow (\epsilon)_\mathrm{Q}$
5. $\mathrm{A}'' \rightarrow a''(\mathrm{S})_\mathrm{Q}$
6. $\mathrm{B}'' \rightarrow b''(\mathrm{S})_\mathrm{Q}$

Next we introduce the concept of string derivation. It is important to say that only leftmost derivations are considered (i.e. the leftmost nonterminal has to be rewritten at each step). Figure 2 depicts the derivation tree corresponding to the derivation:

$$\mathrm{S} \Rightarrow (\mathrm{SS})_\mathrm{Q} \Rightarrow a(\mathrm{SSA}'')_\mathrm{Q} \Rightarrow ab(\mathrm{SA''SB}'')_\mathrm{Q} \Rightarrow ab(\mathrm{A''SB}'')_\mathrm{Q} \Rightarrow aba''(\mathrm{SB''S})_\mathrm{Q} \Rightarrow$$
$$\Rightarrow aba''a(\mathrm{B''SSA}'')_\mathrm{Q} \Rightarrow aba''ab''(\mathrm{SSA''S})_\mathrm{Q} \Rightarrow aba''ab''(\mathrm{SA''S})_\mathrm{Q} \Rightarrow$$
$$\Rightarrow aba''ab''(\mathrm{A''S})_\mathrm{Q} \Rightarrow aba''ab''a''(\mathrm{SS})_\mathrm{Q} \Rightarrow aba''ab''a''(\mathrm{S})_\mathrm{Q} \Rightarrow aba''ab''a''$$

summarized by: $S \Rightarrow^{+} aba''ab''a''$.
It is easy to check that the language generated $L_{FCFS} = L(G_Q)$ is the set of event sequences associated to the FCFS policy. Notice that production (2) models an arrival of request a; the corresponding move of the automaton enqueues the nonterminal A", as a prediction that a service a'' will come. By way of the breadth–first order of derivation, the enqueued predictions will be serviced in FCFS order, as required.

The language LFCFS is also known as AntiDyck [5], owing its name to the fact that it is in some sense the antinomy of the Dyck language (the set of well–parenthesized strings), since the pairs (a, a'') and (b, b'') are never well–nested inside each other.

3.2 Separate generation of arrivals

In the previous grammar generation of arrivals is performed by production (1), that inserts one or more S's into the queue, and by (2), that generates the terminal a (or b). This simple solution results in non–deterministic behaviour, that we wish to avoid (more of that in sect. 4). Therefore we prefer another solution that separates generation of arrivals from their processing, by using a superior class of ABNF grammars.

As we mentioned such grammars are classified by their disposition, a string in $(Q, S)^{+}$, that specifies the ordering of queues and stacks used in the memory tape. The disposition SQ is shown in Figure 3 along with the effect of a production.

The stack will be used as a finite-state device to generate arrivals, an approach that will be applied consistently to all schedules. In more advanced schedulers the stack will also be used for storing the state of the server (busy, etc.). G_{FCFS} :

1. $S \rightarrow a(SS)_S(A'')_Q$
2. $S \rightarrow b(S)_S(B'')_Q$
3. $S \rightarrow \epsilon$ –short form for $S \rightarrow (\epsilon)_S(\epsilon)_Q$
4. $A'' \rightarrow a''(S)_S$ –short form for $A'' \rightarrow (S)_S(\epsilon)_Q$
5. $B'' \rightarrow b''(S)_S$ –short form for $A'' \rightarrow (S)_S(\epsilon)_Q$

Generation proceeds as follows. The nonterminal S generates a string of arrivals and stores the corresponding service requests as a string of A" or B" in the queue. Then S disappears, using (2), and the first nonterminal, say B", is taken from the queue, which by production (4) generates the event b'' and recreates S into the stack. From there the same process is iterated. The derivation of $aba''b''$ is shown in Figure 4. Notice that the symbols in the queue cannot be rewritten until the stack is empty.

3.3 Round-Robin or time-slicing

In the RR policy each process of type a consumes a series of quanta (in particular CPU bursts) of service, denoted by a_q. The arrival of a client, i.e. a job request, is denoted by $\{a, b, \ldots\}$; therefore the process associated to client a is specified by the regular expression $a(a_q)^+$. Service is provided to clients in order of arrival, with quanta assigned in turn to all waiting processes, as exemplified in Figure 5. Here process a consumes three quanta and process b two quanta, but other occurrences of say a could take different numbers of quanta. Notice that here too several arrivals can occur in a row.

The grammar can be obtained from G_{FCFS} by small changes: the nonterminals A and B in the queue denote pending requests that will generate one or more quanta, when they advance to the stack, changing into a nonterminal S (productions 4 and 3). G_{RR} :

1. $S \to a(S)_S(A)_Q$
2. $S \to b(S)_S(B)_Q$
3. $S \to \epsilon$
4. $A \to a_q(S)_S(A)_Q$
5. $B \to b_q(S)_S(B)_Q$
6. $A \to a_q(S)_S$
7. $B \to b_q(S)_S$

A derivation is:

$$S \Rightarrow a(S)_S(A)_Q \Rightarrow a(A)_Q \Rightarrow aa_q(S)_S(A)_Q \Rightarrow aa_qb(S)_S(AB)_Q \Rightarrow$$
$$\Rightarrow aa_qb(AB)_Q \Rightarrow aa_qba_q(BA)_Q \Rightarrow aa_qba_qb_q(S)_S(A)_Q \Rightarrow aa_qba_qb_q(A)_Q \Rightarrow$$
$$\Rightarrow aa_qba_qb_qa_q(S)_S \Rightarrow aa_qba_qb_qa_q$$

Actually one could also write a one queue grammar, but the SQ grammar is to be preferred for the reasons stated above.

3.4 Other scheduling disciplines

Several other scheduling languages can be generated by means of multi-stack-queue grammars. Here we give only a short account thereof.
Service duration and mutual exclusion: Widespread mutual exclusion policies prevent two or more clients from simultaneously accessing the same resource. This constraint is only meaningful if services lasts a time interval, between the start and end events. Here we study the basic mutex case, and in later sections we study the readers/writers discipline. The events occurring in a mutex problem are:

a_r, b_r : client $a, b, \ldots$ requests service
a_s, b_s : client $a, b, \ldots$ starts service
a_e, b_e : client $a, b, \ldots$ ends service

Mutex schedules are exemplified in Figure 5. Of course any number of requests may arrive at any time. A grammar with disposition SQ exists that generates this language [4].
Cascaded services of decreasing priority: In this priority discipline a client a needs two cascaded services, denoted by a' and a'', with the stipulation that the second is less urgent than the first service of any other client. Apart from that,

services are handled in FCFS order. Such schedules as the one shown in Figure 5 are generated by the following grammar with disposition QQ [4].

1. $S \to a(SS)_Q(\epsilon)_Q$	5. $A' \to a'(S)_Q(\epsilon)_Q$
2. $S \to a(SA')_Q(A'')_Q$	6. $A'' \to a''(S)_Q(\epsilon)_Q$
3. $S \to b(SB')_Q(B'')_Q$	7. $B' \to b'(S)_Q(\epsilon)_Q$
4. $S \to \epsilon$	8. $B'' \to b''(S)_Q(\epsilon)_Q$

Static priorities: The classical priority discipline for $n \geq 2$ priority levels involves n classes of clients, ordered by decreasing priority. A client a requests a service denoted by a''. Services are ordered by priority classes and granted in FCFS order within the same class. The n-priority schedule language is denoted by PRI_n. For simplicity we consider two levels: urgent jobs are denoted by $\{a, b, \ldots\}$ and background jobs are denoted by $\{p, q, \ldots\}$. A typical event string in PRI_2 is shown in Figure 5.

$S \to (SS)_Q$ – short for $S \to (S)_Q(\epsilon)_Q$ $\quad S \to \epsilon$

$S \to a(SA'')_Q$ $\quad A'' \to a''(S)_Q$

...

$S \to p(S)_Q(P'')_Q$ $\quad P'' \to p''(S)_Q$

It is immediate to derive the grammar of PRI_2 (see above) from that of Cascaded Priorities, by exploiting a formal closure property of the family of ABNF languages: for any disposition d (in particular for d = QQ) the family of d-languages is closed w.r.t. any alphabetical mapping. This ability to reuse existing specifications (and corresponding implementations) is an attractive feature of the proposed approach.

Combined policies: This family of schedulers (called multi-level queues in [13]) apply more articulated policies than the ones so far considered. Typical examples are: the combination of priorities with preemption, time-slicing, or mutex. We refer to such situations as combined scheduling policies, because they put together two or more basic disciplines. We only give a list: Preemption; FCFS and Round-Robin; Readers/Writers; and Multi- level feedback queues (or dynamic priority). The details of the various constructions are in [3]: they rely upon formal closure properties of the multi–stack–queue grammars.

4 A formal result on the number of queues

We formally prove that priority schedules cannot be generated by a one–queue ABNF grammar. In the theory of languages, the pumping lemma is the main tool for proving that a language is not context-free. Similar lemmas have been proved for subfamilies of ABNF languages. We use the version for Q-grammars from [2]. A *list* over an alphabet Σ is an ordered set of (possibly empty) strings separated by semicolons (a special character not in the alphabet): $\underline{x} = x_1; x_2; \ldots; x_r$, where $r \geq 1$ is the number of components of the list. The flattening of the list $\underline{x}$ is the string: $x = x_1 x_2 \ldots x_r$. On a pair of lists: $\underline{x} = x_1; x_2; \ldots; x_r$ and $\underline{y} =$

$y_1; y_2; \ldots; y_s$, we define the following two operations. The *merge* of x and y is the list: $merge(x, y) = x_1y_1; x_2y_2; \ldots; x_ry_r; y_{r+1}; \ldots; y_s$ if $r \leq s$ (and similarly if $r \geq s$). Since merge is associative, it can be written with any number of arguments. The catenation of $\underline{x}$ and $\underline{y}$ is the list defined by: $\underline{x} \cdot \underline{y} = x_1; x_2; \ldots; x_r; y_1; y_2; \ldots; y_s$.
Statement 1 (Pumping lemma for Q-languages) Let L be a ABNF language with disposition Q. There exist two constants p and q depending only on L, such that for every word t of L, with $|t| > p$, the following properties hold. There exist strings u, x, y, w, z, $v \in \Sigma^*$ and corresponding finite lists $\underline{u}, \ldots, \underline{v}$ such that:

$$\underline{t} = merge(\underline{u}, \underline{x} \bullet merge(\underline{y}, \underline{w}, \underline{z}), \underline{v}) \quad \text{where } |xywz| < q \text{ and } xyz \neq \epsilon$$
$$merge(\underline{u}, \underline{x} \bullet merge(\underline{y}, \underline{x} \bullet merge(\underline{y}, \underline{w}, \underline{z}), \underline{z}), \underline{v}) \in L$$

By this statement we prove the following statement.
Statement 2: The language PRI_2 of two priorities services is not a Q–language.
Proof: PRI_2 contains an infinity of words of the form $p^n a^m (a'')^m (p'')^n$. Then, renaming for clarity the terminal symbols, we can extract the language: $L = \{a^m b^n c^n d^m | m \text{ and } n \text{ non} - \text{negative integers}\} \subset PRI_2$. It suffices to prove that L is not a Q-language. Suppose by contradiction that L is a Q-language, then a long enough string can be decomposed as: $merge(\underline{u}, \underline{x} \bullet merge(\underline{y}, \underline{w}, \underline{z}), \underline{v})$, with xyz non–empty string. Let r be the number of components of the list $merge(\underline{u}, \underline{x} \bullet merge(\underline{y}, \underline{w}, \underline{z}), \underline{v})$. Consider the first letter of $\underline{x} \bullet merge(\underline{y}, \underline{w}, \underline{z})$, if it is not a d. Applying r times the pumping lemma we surely construct a list $\underline{t}'$ where a letter different from d follows a d, hence the string t' is not in the language. Then let d be the first letter of $\underline{x} \bullet merge(\underline{y}, \underline{w}, \underline{z})$. After the first application of the pumping lemma we increase the number of d's without increasing the numbers of a, b, c's, and again we obtain a string not in the language. QED

Since we know that PRI_2 is a Q^2-language [3], it follows that 2-priority schedules can be handled by two - but not by one - queue. We are not aware of other formal proofs of this intuitively obvious result.

5 Schedulers as parsers

ABNF grammars can be used to implement schedule checkers in the (parsers). At present the theory of parsing for ABNF grammars has not reached the same maturity as the one for BNF grammars, so that the account we give is based by necessity on examples.

In a computer system concurrent activities are serialized for execution by a scheduler. External events causing interrupt signals, as well as internal events (e.g. the termination of a program) are the sources of requests for computing services. The scheduler, an essential part of any operating system, provides data structures for storing the requests, the states of the tasks, as well as the state of the computer, and implements the scheduling decisions in accordance with the selected policies. Simple schedulers are included in the real-time kernels, small system packages widely used for embedded systems. More complex schedulers have been developed for the larger operating systems. Concurrent programming languages, e.g. Ada or Java, include a scheduler in their run-time support. The ubiquitous presence of schedulers warrants a more systematic approach for their design. Here we indicate that a scheduler is essentially the parser of the corresponding scheduling language.

We model a scheduler as an automaton equipped with queues and stacks, taking as input the incoming events, and making state transitions in accordance with the prescribed discipline, which is specified by a grammar. Performance reasons require the automaton to be deterministic. To be more precise, two descriptions of a scheduler are possible, as a recognizer or as a transducer. A recognizer is the acceptor of a scheduling language, i.e. a *schedule checker*; this stresses the temporal order of events and actions. A transducer defines a mapping from events to actions. A transducer would be a more realistic model than a recognizer, since it describes the algorithm for dispatching, suspending and resuming requests; but for simplicity we decided to focus on the recognizer model, in agreement with the established approach to study decision problems.

Next we present the design method of schedule checkers, by adapting a well-known parser generation method in use for compilers. We restrict our attention to top-down deterministic LL(1) parsing algorithms, because they are more natural to consider for ABNF grammars. It is straightforward to extend the notion of LL(1) grammar to our case. Recall that a BNF grammar is LL(1) iff there do not exist two distinct leftmost derivations: $S \rightarrow \ldots uAv \rightarrow u\alpha v \rightarrow \ldots uzv$ and $S \rightarrow \ldots \rightarrow uAw \rightarrow u\alpha v' \rightarrow \ldots \rightarrow uz'v'$, where $A \rightarrow \alpha | \alpha'$ are alternative productions, such that the first character of zv and of $z'v'$ is the same. The same definition applies to ABNF grammars, provided derivations are taken according to the ABNF manner.

For an ABNF grammar with a tape disposition d in $(Q, S)^+$, the parser can be constructed in the same manner, provided the LIFO store is replaced by the FIFO and/or LIFO stores required by the disposition. For instance a SQ-grammar requires a stack and a queue. We illustrate this approach by implementing a LL(1) parser for the language LMUTEX generated by the SQ-grammar GMUTEX, which is reproduced together with the Look-ahead sets:

	Productions	Look-ahead sets
1	$S \rightarrow a_r(S)_S(A_S A_e)_Q$	$\{a_r\}$
2	$S \rightarrow \epsilon$	$\{a_s, \perp\}$
3	$A_S \rightarrow a_s(S)_S$	$\{a_s\}$
4	$A_S \rightarrow a_e(S)_S$	$\{a_e\}$

Note that the set of production (2) contains the *follow set* of $S \rightarrow \epsilon$, i.e. the first character that may follow (2) in some derivation tree. This can be: a_s, the first character generated by A_S; and $\perp$, the end-marker of the event string. The grammar is LL(1) since the look-ahead sets of alternatives 1|2 are disjoint. For this grammar the computation of the look-ahead sets is straightforward. The computation for generic ABNF grammars is beyond the scope of this paper. To implement the parser we use recursive descent for the LIFO store and an external queue for the FIFO store. Other implementations would of course be possible, including the use of the system queues provided by operating systems for remote procedure calls or by the Ada run-time systems for entry calls. The queue is provided with three classic methods: head, enqueue, and isQempty. The three syntactic procedures and then the main program are listed in Figure 6. The parser is deterministic and clearly operates in linear time. The case when the original grammar is not deterministic is briefly discussed in [4].

6 Conclusion

We have demonstrated that formal grammars can be used to specify the basic scheduling algorithms found in operating systems. The Augmented BNF model preserves the spirit of BNF grammars and is only moderately more complex: with some practice ABNF grammars can be written and understood almost as easily as context-free ones. The proposed method is compositional: compound scheduling policies are obtained by composing basic grammars. Descriptional complexity is directly related to the complexity of the scheduling policy.

Grammars are an operational specification that permits to mechanically construct language recognizers, in our case schedulers or schedule checkers. In [4] we give an example of how to construct a parser for scheduling languages, based on multi stack-queue grammars extended with semantic attributes. A systematic study of ABNF parsing techniques has still to be done, but some results are available for specific tape dispositions [7].

We have seen that essentially all CPU scheduling policies which are based on events can be modeled by purely syntactic specifications. On the other hand, time-driven schedulers require the computation of parameters, that can be conveniently handled by semantic attributes. The classical model of attribute grammars conveniently fits ABNF grammars. We believe this approach can handle any reasonable on-line scheduling policy [4].

Related work: we know of very little research on specification of schedulers (apart from general methods such as algebraic ones). Hemmendinger [10] proposed to use the intersection of context-free languages for readers/writers. In an early work [2] we specified schedulers by automata theory, using quasi-real-time queue automata. Disregarding the presence of stacks, ABNF grammars are a restricted kind of queue automata, that are finite-state machines equipped with one or more independent FIFO tapes. The ancestor of the queue machines is usually considered to be the Post machine (Manna [12]), which is computationally equivalent to the Turing machine. More tractable models, subject to the real-time constraint, have been later studied by Brandenburg [1] and by Citrini et alii [6]. Floyd and Beigel [8] offer a systematic approach to the study of abstract machines equipped with various data structures. ABNF are simpler than general queue automata because the tape segments have a fixed order and because internal states are not allowed (however the first stack can be used for storing some state information). The effect of the simplification is that such languages can be generated by grammars that in many respects are as convenient as the context-free ones.

On the application side, this research sets the ground for experimenting with syntax-directed schedulers. Potentially interesting directions are: algorithms evaluation and experimentation especially for new policies; generation of test suites and workloads as strings of scheduling languages; computation of performance indicators such as average waiting time; customization of operating system schedulers. On the theoretical side we hope the formal relationships between scheduling languages can improve our understanding of scheduling theory, and allow to prove properties. It is hoped that, as the classical formal language theory lead, after some years, to algorithms for solving practical problems such as compilation, useful results will be attained using the presented model.

References

1. Brandenburg F. J.: On the Intersection of Stacks and Queues, in Theoretical Computer Science, Elsevier, vol. 58, pp. 69-80, 1988
2. Breveglieri, L., Cherubini, A., Crespi Reghizzi, S: Real-Time Scheduling by Queue Automata, in LNCS Formal Techniques in Real-Time and Fault-Tolerant Systems '92, n. 571, (J. Vytopil, ed.), pp. 131-148, Springer Verlag, 1992
3. Breveglieri, L., Cherubini, A. Citrini, C., Crespi Reghizzi, S.: Multi-Pushdown Stack Languages and Grammars, International Journal of Foundation of Computer Science, World Scientific, n., pp., 1996
4. Breveglieri, L., Cherubini, A., Crespi Reghizzi, S: Syntax-Directed Scheduling, Int. Rep. 98–048, Dip. di Elettronica e Informazione, Politecnico di Milano, 1998
5. Cherubini, A., Citrini, C., Crespi Reghizzi, S., Mandrioli, D.: Breadth and Depth Grammars and Dequeue Automata International Journal of Foundations of Computer Science, World Scientific, vol. 1, n. 3, pp. 219–232, 1990
6. Cherubini, A., Citrini, C., Crespi Reghizzi, S., Mandrioli, D.: Quasi-Real-Time FIFO Automata, Breadth-first Grammars and their Relations, in Theoretical Computer Science, Elsevier, n. 85, pp.171–203, 1991
7. Cherubini, A., San Pietro, P.: A Polynomial-Time Parsing Algorithm for k-Depth Languages, Journal of Computer Systems Science, vol. 52, n. 1, pp. 61–79, 1996
8. Floyd, R., Beigel, R.: The Language of Machines, Freeman, 1994
9. Ginsburgh, S.: The mathematical Theory of Context-free Languages, McGraw-Hill, New York, 1968
10. Hemmendinger, D.: Specifying Ada Server Tasks with executable formal Grammars, in IEEE Transactions on Software Engineering, n. 16, pp. 741–754, 1990
11. Kang, S., Lee, H.: Analysis and Solution of non-Preemptive Policies for Scheduling Readers and Writers, in Operating Systems Review, ACM Press, vol. 32, n. 3, pp. 30–50, July, 1998
12. Manna, Z.: Mathematical Theory of Computation, McGraw Hill, New York, 1974
13. Galvin, P., Silberschatz, A.: Operating Systems Concepts, Addison Wesley, Reading, 1994
14. Vauquelin, B., Franchi–Zannettacci, P.: Automates à File, in Theoretical Computer Science, Elsevier, n. 11, pp. 221–225, 1980

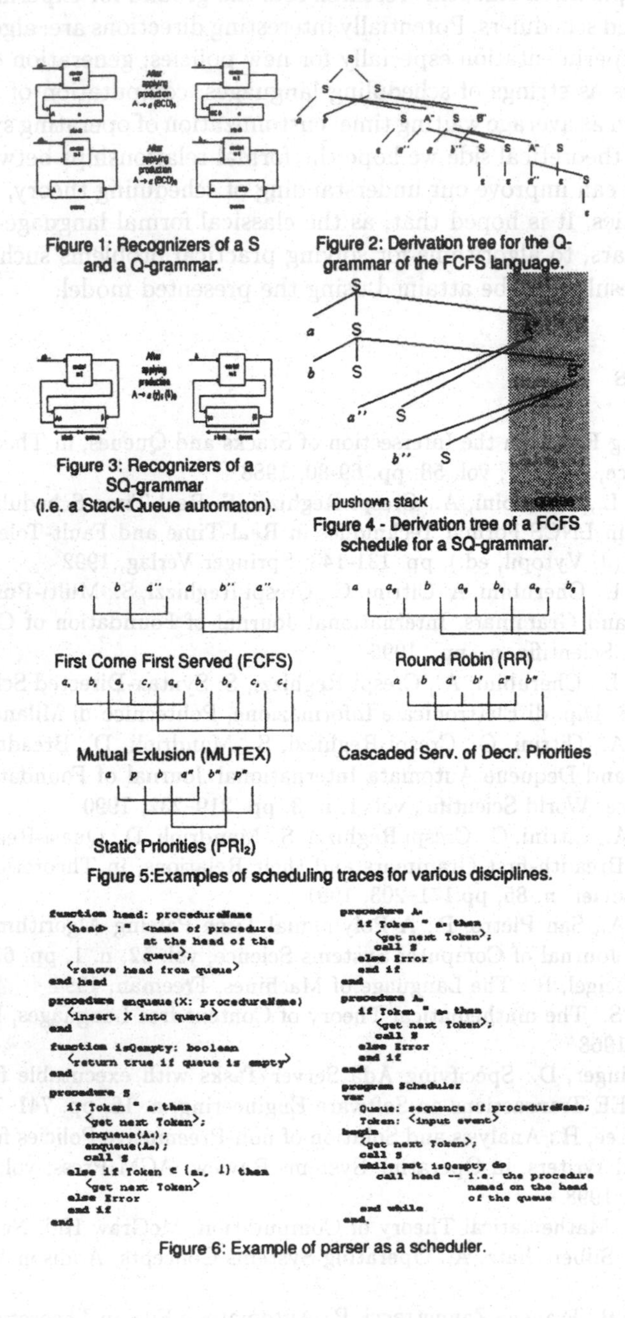

Figure 1: Recognizers of a S and a Q-grammar.

Figure 2: Derivation tree for the Q-grammar of the FCFS language.

Figure 3: Recognizers of a SQ-grammar (i.e. a Stack-Queue automaton).

Figure 4 - Derivation tree of a FCFS schedule for a SQ-grammar.

Figure 5: Examples of scheduling traces for various disciplines.

Figure 6: Example of parser as a scheduler.

Iterative Arrays with a Wee Bit Alternation

Thomas Buchholz, Andreas Klein, and Martin Kutrib
Institute of Informatics, University of Giessen
Arndtstr. 2, D-35392 Giessen, Germany
{buchholz,kutrib}@informatik.uni-giessen.de

Abstract. An iterative array is a line of interconnected interacting finite automata. One distinguished automaton, the communication cell, is connected to the outside world and fetches the input serially symbol by symbol. We are investigating iterative arrays with an alternating communication cell. All the other automata are deterministic. The number of alternating state transitions is regarded as a limited resource which depends on the length of the input.
We center our attention to real-time computations and compare alternating IAs with nondeterministic IAs. By proving that the language families of the latter are not closed under complement for sublogarithmic limits it is shown that alternation is strictly more powerful than nondeterminism. Moreover, for these limits there exist infinite hierarchies of properly included alternating language families. It is shown that these families are closed under boolean operations.

1 Introduction

Devices of interconnected parallel acting automata have extensively been investigated from a language theoretic point of view. The specification of such a system includes the type and specification of the single automata, the interconnection scheme (which sometimes implies a dimension to the system), a local and/or global transition function and the input and output modes. One-dimensional devices with nearest neighbor connections whose cells are deterministic finite automata are commonly called iterative arrays (IA) if the input mode is sequential to a distinguished communication cell.

Especially for practical reasons and for the design of systolic algorithms a sequential input mode is more natural than the parallel input mode of so-called cellular automata. Various other types of acceptors have been investigated under this aspect (e.g., the iterative tree acceptors in [7]).

In connection with formal language recognition IAs have been introduced in [6] where it was shown that the language families accepted by real-time IAs form a Boolean algebra not closed under concatenation and reversal. Moreover, there exists a context-free language that cannot be accepted by any d-dimensional IA in real-time. On the other hand, in [5] it is shown that for every context-free grammar a 2-dimensional linear-time IA parser exists. Compared with e.g., Turing machines there are essential differences in the recognition power. For example, the language of palindromes needs a lower bound of n^2 time steps for Turing machines but is acceptable in real-time by IAs.

In [8] a real-time acceptor for prime numbers has been constructed. Pattern manipulation is the main aspect in [1]. A characterization of various types of IAs by restricted Turing machines and several results, especially speed-up theorems, are given in [9, 10, 11].

Various generalizations of IAs have been considered. In [12] IAs are studied in which all the finite automata are additionally connected to the communication cell. Several more results concerning formal languages can be found e.g., in [13, 14, 15].

Sometimes completely nondeterministic arrays have been studied. In [4] arrays with restricted nondeterminism have been introduced. There it has been shown that the number of nondeterministic transitions can be reduced by a constant factor and that there exists an infinite hierarchy of properly included language families for necessarily sublogarithmic limits. Some closure properties for such families are given.

Here we continue the work initiated in [4] by making a further generalization step. We introduce arrays with restricted alternation. Our interest focuses on the question how much alternation is required, if at all, to enhance the power of a particular (nondeterministic) class. Thereby we are trying to identify the power and limitations of commonly known iterative arrays. In order to define alternations as limited resource we restrict the ability to perform alternating transitions to the communication cell, all the other automata are deterministic ones. Moreover, we limit the number of allowed alternating transitions which additionally have to appear at the beginning of the computation. Our attention is centered on real-time computations.

The basic notions and the model in question are defined in the next section. Section 3 is devoted to technical results mainly. By generalizing a method in [6] an equivalence relation is used to define a necessary condition for real-time languages. Another result states that for a given alternating iterative array one can always find another one that accepts the same language and that uses existential and universal states by turns at every time step. In Section 4 the closure under Boolean operations is investigated. Comparing alternating iterative arrays to nondeterministic ones for sublogarithmic limits in Section 5 it is shown that the former are strictly more powerful. The properness of the inclusion is proved at the hand of different closure properties. In particular the nondeterministic families are not closed under complement, a question left open in [4]. Finally we obtain infinite hierarchies of properly included language families varying the amount of allowed alternation.

2 Model and Notions

We denote the positive rational numbers by $\mathbb{Q}_+$, the integers by $\mathbb{Z}$, the positive integers $\{1, 2, \ldots\}$ by $\mathbb{N}$, the set $\mathbb{N} \cup \{0\}$ by $\mathbb{N}_0$ and the powerset of a set S by 2^S. The empty word is denoted by ϵ and the reversal of a word w by w^R.

An iterative array with alternating communication cell is an infinite linear array of finite automata, sometimes called cells, each of them is connected to its both nearest neighbors to the left and to the right. For our convenience we identify the cells by integers. Initially they are in the so-called quiescent state. The input is supplied sequentially to the distinguished communication cell at the origin. For this reason we have two local transition functions. The state transition of all cells but the communication cell depends on the actual state of the cell itself and the actual states of its both neighbors. The state transition of the communication cell additionally depends on the actual input symbol (or if the whole input has been consumed on a special end-of-input symbol). The finite automata work synchronously at discrete time steps. Their states are partitioned into existential and universal ones. What makes a, so far, nondeterministic computation to an alternating computation is the mode of acceptance, which will be defined with respect to the partitioning. More formally:

Definition 1. *An* iterative array with alternating communication cell *(*A-IA*) is a system* $(S, \delta, \delta_{nd}, s_0, \#, A, F)$*, where*

1. S *is the finite, nonempty set of* states *which is partitioned into* existential S_e *and* universal S_u *states:* $S = S_e \cup S_u$*,*
2. A *is the finite, nonempty set of* input symbols,
3. $F \subseteq S$ *is the set of* accepting states,
4. $s_0 \in S$ *is the* quiescent state,
5. $\# \notin A$ *is the* end-of-input symbol,
6. $\delta : S^3 \to S$ *is the deterministic* local transition function for non-communication cells *satisfying* $\delta(s_0, s_0, s_0) = s_0$*,*
7. $\delta_{nd} : S^3 \times (A \cup \{\#\}) \to 2^S$ *is the* local transition function for the communication cell *satisfying* $\forall s_1, s_2, s_3 \in S, a \in A \cup \{\#\} : \delta_{nd}(s_1, s_2, s_3, a) \neq \emptyset$*.*

Let $\mathcal{M}$ be an A-IA. A configuration of $\mathcal{M}$ at some time $t \geq 0$ is a description of its global state, which is actually a pair (w, c_t), where $w \in A^*$ is the remaining input sequence and $c_t : \mathbb{Z} \to S$ is a mapping that gives the actual states of the single cells. The configuration (w, c_0) at time 0 is defined by the input word w and the mapping $c_0(i) := s_0$, $i \in \mathbb{Z}$, while subsequent configurations are chosen according to the global transition Δ_{nd}: Let (w, c) be a configuration then the possible successor configurations (w', c') are as follows:

$$(w', c') \in \Delta_{nd}\big((w, c)\big) \iff \begin{array}{l} c'(i) = \delta\big(c(i-1), c(i), c(i+1)\big), i \in \mathbb{Z} \setminus \{0\}, \\ c'(0) \in \delta_{nd}\big(c(-1), c(0), c(1), a\big) \end{array}$$

where $a = \#$ and $w' = \epsilon$ if $w = \epsilon$, and $a = w_1$ and $w' = w_2 \cdots w_n$ if $w = w_1 \cdots w_n$. Thus, the global transformation Δ_{nd} is induced by δ and δ_{nd}. The i-fold composition of Δ_{nd} is defined as follows:

$$\Delta_{nd}^0\big((w, c)\big) := \big\{(w, c)\big\}, \qquad \Delta_{nd}^{i+1}\big((w, c)\big) := \bigcup_{(w', c') \in \Delta_{nd}^i((w, c))} \Delta_{nd}\big((w', c')\big)$$

The evolution of $\mathcal{M}$ is represented by its computation tree.

The *computation tree* $T_{\mathcal{M},w}$ of $\mathcal{M}$ under input $w \in A^+$ is a tree whose nodes are labeled by configurations. The root of $T_{\mathcal{M},w}$ is labeled by (w, c_0). The children of a node labeled by a configuration (w, c) are the nodes labeled by the possible successor configurations of (w, c). Thus, the node (w, c) has exactly $|\Delta_{nd}((w, c))|$ children.

A configuration (w, c) is *accepting* iff $c(0) \in F$, it is *universal* iff $c(0) \in S_u$ and it is said to be *existential* iff $c(0) \in S_e$.

In order to define *accepting computations* on input words we need the notion of accepting subtrees.

Definition 2. *Let $\mathcal{M} = (S, \delta, \delta_{nd}, s_0, \#, A, F)$ be an A-IA and $T_{\mathcal{M},w}$ be its computation tree for an input word $w \in A^n$, $n \in \mathbb{N}$. A finite subtree T' of $T_{\mathcal{M},w}$ is said to be an* accepting subtree *iff it fulfills the following conditions:*

1. *The root of T' is the root of $T_{\mathcal{M},w}$.*
2. *If a non-leaf node of T' is labeled by an universal configuration then all its successors belong to T'.*
3. *If a non-leaf node of T' is labeled by an existential configuration then exactly one of its successors belongs to T'.*
4. *The leafs of T' are labeled by accepting configurations.*

From the computational point of view an accepting subtree is built by considering one possible successor (a guessed successor) if the communication cell is in an existential state and by considering all successors if the communication cell is in an universal state.

Now we are prepared to define the language accepted by an A-IA.

Definition 3. *Let $\mathcal{M} = (S, \delta, \delta_{nd}, s_0, \#, A, F)$ be an A-IA.*

1. *A word $w \in A^+$ is* accepted *by $\mathcal{M}$ iff there exists an accepting subtree of $T_{\mathcal{M},w}$.*
2. *$L(\mathcal{M}) = \{w \in A^+ \mid w$ is accepted by $\mathcal{M}\}$ is the* language accepted *by $\mathcal{M}$.*
3. *Let $t : \mathbb{N} \to \mathbb{N}$, $t(n) > n$, be a mapping. Iff for all $w \in L(\mathcal{M})$ there exists an accepting subtree of $T_{\mathcal{M},w}$ the height of which is less than $t(|w|)$, then L is said to be of* time complexity *t.*

An A-IA $\mathcal{M}$ has a *nondeterministic* communication cell if the state set consists of existential states only. An accepting subtree is now a list of configurations which corresponds to a possible computation path of $\mathcal{M}$. Iterative arrays with nondeterministic communication cell are denoted by G-IA.

A G-IA is deterministic if $\delta_{nd}(s_1, s_2, s_3, a)$ is a singleton for all states $s_1, s_2, s_3 \in S$ and all input symbols $a \in A \cup \{\#\}$. In these cases the course of computation is unique for a given input word w and, thus, the whole computation tree is a list of configurations. Deterministic iterative arrays are denoted by IA.

If the state set is a Cartesian product of some smaller sets $S = S_0 \times S_1 \times \cdots \times S_r$, we will use the notion *register* for the single parts of a state. The concatenation of a specific register of all cells forms a *track*.

The family of all languages which can be accepted by an A-IA with time complexity t is denoted by $\mathscr{L}_t$(A-IA). In the sequel we will use a corresponding notion for other types of acceptors. If $t(n)$ equals $n+1$ acceptance is said to be in real-time and we write $\mathscr{L}_{rt}$(A-IA). The *linear-time* languages $\mathscr{L}_{lt}$(A-IA) are defined according to $\mathscr{L}_{lt}(\text{A-IA}) := \bigcup_{k\in\mathbb{Q}_+, k>1} \mathscr{L}_{k\cdot n}(\text{A-IA})$.

There is a natural way to restrict the alternation of the arrays. One can limit the number of allowed alternating state transitions of the communication cell. Note, here we do not limit the number of alternations (i.e., transitions from an universal to an existential state or vice versa) but the number of time steps at which alternating transitions may occur. For this reason a deterministic local transition function $\delta_d : S^3 \times (A \cup \{\#\}) \to S$ for the communication cell is provided and the global transition induced by δ and δ_d is denoted by Δ_d. Let $f : \mathbb{N} \to \mathbb{N}_0$ be a mapping that gives the number of allowed alternating transitions dependent on the length of the input.

The resulting system $(S, \delta, \delta_{nd}, \delta_d, s_0, \#, A, F)$ is a fA-IA (f alternating IA) if starting with the initial configuration (w, c_0) the possible configurations at some time t are given by the global transition as follows:

$\{(w, c_0)\}$ if $t = 0$, $\Delta_{nd}^t((w, c_0))$ if $t \leq f(|w|)$ and

$$\bigcup_{(w',c')\in\Delta_{nd}^{f(|w|)}((w,c_0))} \Delta_d^{t-f(|w|)}((w',c')) \quad \text{otherwise}$$

Observe that all alternating transitions have to be applied before the deterministic ones. Up to now we have f not required to be computable at all. Of course for almost all applications we will have to do so but some of our general results can be developed without such a requirement.

3 Equivalence Classes and Normalization

Definition 4. *Let $L \subseteq A^*$ be a language over an alphabet A and $l \in \mathbb{N}$ be a constant. Two words w and w' are* l-equivalent *with respect to L iff $ww_l \in L \iff w'w_l \in L$ for all $w_l \in A^l$. The* number of l-equivalence classes of words of length n with respect to L *are denoted by $N(n, l, L)$ (i.e. $|ww_l| = n$).*

The following lemma gives a necessary condition for a language to be real-time acceptable by an fA-IA.

Lemma 5. *Let $f : \mathbb{N} \to \mathbb{N}_0$, $f(n) \leq n$, be a mapping. If $L \in \mathscr{L}_{rt}(f\text{A-IA})$ then there exist constants $p, q \in \mathbb{N}$ such that $N(n, l, L) \leq p^{l \cdot q^{f(n)}}$.*

Proof. Let $\mathcal{M} = (S, \delta, \delta_{nd}, \delta_d, s_0, \#, A, F)$ be a real-time fA-IA which accepts L. We define $q := \max\{|\delta_{nd}(s_1, s_2, s_3, a)| \mid s_1, s_2, s_3 \in S, a \in A\}$.

In order to determine an upper bound to the number of l-equivalence classes at first we consider the possible configurations of $\mathcal{M}$ after reading all but l input symbols. The remaining computation depends on the last l input symbols and the states of the cells $-l-1, \ldots, 0, \ldots, l+1$. For the $2l+3$ states there are

$|S|^{2l+3}$ different possibilities. Let $p_1 := |S|^5$ then due to $|S|^{2l+3} = |S|^{2l} \cdot |S|^3 = (|S|^2)^l \cdot |S|^3 \leq (|S|^2)^l \cdot (|S|^3)^l = (|S|^2 \cdot |S|^3)^l = p_1^l$ we have at most p_1^l different possibilities.

Now we consider the computation trees of $\mathcal{M}$. Since the number of alternating steps is bounded by $f(n)$ in each computation tree there are at most $q^{f(n)}$ internal nodes that are labeled by existential or universal configurations (all the others are part of the deterministic computation) we have to distinguish $2^{q^{f(n)}}$ different labelings. Each computation tree of finite height has at most $q^{f(n)}$ leafs. Each leaf at level $n-l$ can be labeled with one of the p_1^l different configurations. Since the number of equivalence classes is not affected by the last l input symbols altogether one can distinguish $(p_1^l)^{q^{f(n)}} \cdot 2^{q^{f(n)}}$ different computation trees of height $n-l$. Correspondingly, there are at most $p_1^{l \cdot q^{f(n)}} \cdot 2^{q^{f(n)}}$ classes. For a suitable $p \in \mathbb{N}$ this is less than $p^{l \cdot q^{f(n)}}$. □

If $\mathcal{M}$ is a fG-IA for the number of equivalence classes we need not to take the labelings into account. Thus, we obtain less than $p_1^{l \cdot q^{f(n)}}$ classes.

Now we are going to extend the previous lemma. The question is how the number of l-equivalence classes is affected if we concatenate each word of L by another arbitrary l symbols from A.

Lemma 6. *Let $f : \mathbb{N} \to \mathbb{N}_0$, $f(n) \leq n$, be an increasing mapping that satisfies $f(2n) \leq 2f(n)$. If the number of l-equivalence classes with respect to a language $L \subseteq A^*$ is not bounded according to Lemma 5 then $L \bullet A^l \notin \mathscr{L}_{rt}(f\text{G-IA})$.*

Proof. At first we prove $N(n+l+1, 2 \cdot l+1, L \bullet A^l) = N(n, l, L)$.

From $ww_l \in L$ for an arbitrary $w_l \in A^l$ it follows $ww_l \bullet w_l' \in L \bullet A^l$ for all $w_l' \in A^l$ and $w'w_l \in L$. From $w'w_l \in L$ it follows $w'w_l \bullet w_l' \in L \bullet A^l$ for all $w_l' \in A^l$.

Conversely, let w and w' be $(2l+1)$-equivalent with respect to $L \bullet A^l$. From $ww_l \bullet w_l' \in L \bullet A^l$ for an arbitrary $w \in A^l$ and all $w_l' \in A^l$ it follows $ww_l \in L$ and $w'w_l \bullet w_l' \in L \bullet A^l$. From the latter we obtain $w'w_l \in L$.

Secondly, there exist n and l such that we have $N(n, l, L) > p^{l \cdot q^{f(n)}}$ for every $p, q \in \mathbb{N}$, since the number of l-equivalence classes with respect to L is not bounded according to Lemma 5 (i.e., $L \notin \mathscr{L}_{rt}(f\text{G-IA})$).

On the other hand, a real-time fG-IA can distinguish at most $p^{(2 \cdot l+1) \cdot q^{f(n+l+1)}}$ equivalence classes with respect to $L \bullet A^l$. Since $l < n$ it follows $p^{(2 \cdot l+1) q^{f(n+l+1)}} \leq p^{(2 \cdot l+1) q^{2f(n)}} \leq p'^{l \cdot q'^{f(n)}} < N(n, l, L) = N(n+l+1, 2 \cdot l+1, L \bullet A^l)$.

Thus, $L \bullet A^l \notin \mathscr{L}_{rt}(f\text{G-IA})$ by Lemma 5. □

In order to reduce the technical effort for proofs it is often useful to be able to start with devices that meet a certain normal-form. For our purposes it is convenient to consider iterative arrays which are *alternation normalized* as follows: $s_0 \in S_e$ and $\forall\, s_1, s_2, s_3 \in S, a \in A \cup \{\#\} : \delta_{nd}(s_1, s_2, s_3, a) \subseteq S_e$ if $s_2 \in S_u$ and $\delta_{nd}(s_1, s_2, s_3, a) \subseteq S_u$ if $s_2 \in S_e$.

Thus the communication cell changes continually from an existential state into an universal state and vice versa.

Lemma 7. *Let $f : \mathbb{N} \to \mathbb{N}_0$, $f(n) \leq n$, and $t : \mathbb{N} \to \mathbb{N}$, $t(n) > n$, be two mappings. If $L \in \mathscr{L}_t(f\text{A-IA})$ then exists an alternation normalized fA-IA which accepts L with time complexity t.*

Proof. The proof can be found in [3].

4 Closure under Boolean Operations

Lemma 8. *Let $f : \mathbb{N} \to \mathbb{N}_0$, $f(n) \leq n$, and $t : \mathbb{N} \to \mathbb{N}$, $t(n) > n$, be two mappings. $\mathscr{L}_t(f\text{A-IA})$ is closed under union and intersection.*

Proof. Let $\mathcal{M}_1$ and $\mathcal{M}_2$ be two t-time fA-IAs. By Lemma 7 we may assume that $\mathcal{M}_1$ and $\mathcal{M}_2$ are alternation normalized. Due to the normalized behavior we can construct a t-time fA-IA $\mathcal{M}'$ that simulates $\mathcal{M}_1$ and $\mathcal{M}_2$ on different tracks in parallel. It is easy to see that the computation tree of $\mathcal{M}'$ contains an accepting subtree if $\mathcal{M}_1$ or $\mathcal{M}_2$ accept simply by considering the corresponding track only. The closure under union follows.

In order to find an accepting subtree for the intersection we have to use the successor that contains guesses of $\mathcal{M}_1$ and $\mathcal{M}_2$ which lead to acceptance in existential steps, respectively. Clearly in universal steps all successor configurations of $\mathcal{M}'$ contain all successor configurations of $\mathcal{M}_1$ and $\mathcal{M}_2$ and vice versa. The closure under intersection follows. □

The comparison between nondeterministic and alternating IAs in the next section is done at the hand of closure under complement. It is easy to prove the closure of A-IAs but hard to disprove it for G-IAs. Here is the easy part:

Lemma 9. *Let $f : \mathbb{N} \to \mathbb{N}_0$, $f(n) \leq n$, be a mapping then $\mathscr{L}_{rt}(f\text{A-IA})$ is closed under complement.*

Proof. The meaning of an existential transition is that there must exist one successor configuration which leads to acceptance. In order to accept the complement this can be replaced by the meaning that all successor configurations do not lead to acceptance. On the other hand, the meaning of an universal step that all successors must lead to acceptance can be replaced by the meaning that one successor does not lead to acceptance. The negation in the new meaning is simply realized as follows: if the communication cell has consumed the whole input it now accepts if it would have rejected before and vice versa. Thus, final and non-final configurations are exchanged. □

5 Alternating Hierarchy

5.1 Comparison with Nondeterministic Iterative Arrays

In the following we incorporate some results of a previous work [4] concerning IAs with nondeterministic communication cell.

In order to define an important language let $f : \mathbb{N} \to \mathbb{N}_0$ be an increasing mapping such that $f \in o(\log)$. We define another mapping $h : \mathbb{N} \to \mathbb{N}$ by $h(n) := 2^{f(n)}$. It is increasing since f is. Moreover, since $f \in o(\log)$ for all $k \in \mathbb{Q}_+$ it holds $\lim_{n\to\infty} \frac{h(n)}{n^k} = \lim_{n\to\infty} \frac{2^{f(n)}}{2^{\log(n)\cdot k}} = 0$ and therefore $h \in o(n^k)$. Especially for $k = \frac{1}{2}$ it follows that the mapping $m(n) := \max\{n' \in \mathbb{N}_0 \mid (h(n)+1)\cdot(n'+1) \leq n\}$ is unbounded, and for large n we obtain $m(n) > h(n)$. The following language depends on f only.

$$\begin{aligned} L_f := \{\$^r w_1\$w_2\$\cdots\$w_j¢y¢ \mid \exists\, n \in \mathbb{N} : j = h(n) \wedge w_i \in \{0,1\}^{m(n)}, 1 \leq i \leq j, \\ \wedge\, r = n - (h(n)+1)\cdot(m(n)+1) \\ \wedge\, \exists\, 1 \leq i' \leq j : w_{i'} = y^R\} \end{aligned}$$

The words of length n of L_f consist of $2^{f(n)}$ subwords w_i and one subword y which is the reversal of one of the w_i. The number of subwords is fixed for a given n. The lengths of the subwords is as large as possible.

The next theorem follows immediately from a theorem shown in [4] in order to prove a nondeterministic hierarchy.

Theorem 10. *Let $f : \mathbb{N} \to \mathbb{N}_0$ and $g : \mathbb{N} \to \mathbb{N}_0$ be two increasing mappings such that $f \in o(\log)$ and $g \in o(f)$ then $L_f \in \mathscr{L}_{rt}(f\text{G-IA})$ and $L_f \notin \mathscr{L}_{rt}(g\text{G-IA})$.*

Since for $g \in o(f)$ the language L_f is not a real-time gG-IA language but, on the other hand, it can be accepted in real-time by a fG-IA, and the number of guesses can be reduced by a constant factor [4] one obtains the following corollary. Moreover, it holds for A-IAs too, since our approximation of the numbers of equivalence classes are identical regardless of whether or not nondeterministic or alternating IAs are in question:

Corollary 11. *Let $f : \mathbb{N} \to \mathbb{N}_0$ and $g : \mathbb{N} \to \mathbb{N}_0$ be two increasing mappings such that $f \in o(\log)$ then $L_f \in \mathscr{L}_{rt}(g\text{G-IA}) \Longrightarrow g \in \Omega(f)$ and $L_f \in \mathscr{L}_{rt}(g\text{A-IA}) \Longrightarrow g \in \Omega(f)$.*

The next theorem is the main result of the present section. It states that under some preconditions the real-time alternating IAs are strictly more powerful than the real-time nondeterministic IAs. For the proof we need a closure property concerning marked iteration.

Definition 12. *Let L be a language over an alphabet A and $\bullet \notin A$ be a distinguished marking symbol. The language $(L\bullet)^+$ is the* marked iteration *of L.*

Here we have to require f to be in some sense computable. This can be done in terms of deterministic real-time IA languages. It should be mentioned that the family $\mathscr{L}_{rt}(\text{IA})$ is very rich.

Theorem 13. *Let $f : \mathbb{N} \to \mathbb{N}_0$ be an increasing, unbounded mapping such that $f \in o(\log)$ and $\{a^{f(m)}b^{m-f(m)} \mid m \in \mathbb{N}\} \in \mathscr{L}_{rt}(\text{IA})$ then $\mathscr{L}_{rt}(f\text{G-IA}) \subset \mathscr{L}_{rt}(f\text{A-IA})$.*

Proof. Since a fG-IA is just a fA-IA with only existential states we have the inclusion $\mathscr{L}_{rt}(f\text{G-IA}) \subseteq \mathscr{L}_{rt}(f\text{A-IA})$.

It remains to show $\mathscr{L}_{rt}(f\text{G-IA}) \neq \mathscr{L}_{rt}(f\text{A-IA})$. The idea is to prove the inequality at the hand of different closure properties.

By Lemma 9 the family $\mathscr{L}_{rt}(f\text{A-IA})$ is closed under complement. We are going to show that $\mathscr{L}_{rt}(f\text{G-IA})$ is not closed under complement.

In order to do so suppose $\mathscr{L}_{rt}(f\text{G-IA})$ is not closed under marked iteration which will be shown by Lemma 14.

Let $L \in \mathscr{L}_{rt}(f\text{G-IA})$ be a language over an alphabet A. If $\overline{L}$ does not belong to $\mathscr{L}_{rt}(f\text{G-IA})$ we are done.

Assume now $\mathscr{L}_{rt}(f\text{G-IA})$ is closed under complement and let $\mathcal{M}$ be a fG-IA that accepts $\overline{L}$ in real-time. Now we construct a real-time fG-IA $\mathcal{M}'$ that accepts $\overline{(L\bullet)^+}$.

In [2, 4] the real-time simulation of stacks by deterministic IAs has been shown. Thereby the communication cell contains the symbol at the top of the stack. We will use the ability of IAs to simulate such data structures at the construction.

One deterministic regular task of $\mathcal{M}'$ is to check whether the input is of the form $x_1\bullet x_2\bullet\cdots\bullet x_k\bullet$ where $x_i \in A^+$, $1 \le i \le k$. All words that do not fit this form are accepted.

A word $x_1\bullet x_2\bullet\cdots\bullet x_k\bullet$ belongs to $\overline{(L\bullet)^+}$ iff at least one x_i, $1 \le i \le k$, belongs to $\overline{L}$. In order to accept such words $\mathcal{M}'$ simulates $\mathcal{M}$ on x_1 directly and additionally during its nondeterministic steps the $f(|x_i|)$ nondeterministic steps of $\mathcal{M}$ on input x_i for $i > 1$. Since f is increasing $\mathcal{M}'$ has at least as many nondeterministic steps as $\mathcal{M}$. The guessing is done by choosing nondeterministically one of the (finite) local transition functions at each time step and pushing it onto a stack.

When the direct simulation of $\mathcal{M}$ on x_1 succeeds the job of $\mathcal{M}'$ is done. Otherwise it starts the following task every time a $\bullet$ appears in the input.

A signal is sent through the stack which copies the content of the stack to a second stack cell by cell. Additionally, $\mathcal{M}'$ simulates $\mathcal{M}$ on the next subword x_i. In order to simulate a nondeterministic step one mapping is popped from the second stack (leaving the first stack unchanged) and is applied to the local configuration. So the communication cell can simulate a nondeterministic step of $\mathcal{M}$ deterministically by applying a previously nondeterministically determined deterministic local transition. Again, if one of the simulations succeeds $\mathcal{M}'$ accepts otherwise it rejects.

Up to now we kept quiet about a crucial point. The number $f(|x_i|)$ of simulated nondeterministic transitions may be incorrect. Therefore, the decision of $\mathcal{M}'$ depends on corresponding verifications additionally: In order to perform this task an acceptor for the language $L' = \{a^{f(m)}b^{m-f(m)} \mid m \in \mathbb{N}\}$ is simulated in parallel whenever a $\bullet$ appears in the input. Thereby an input symbol a is assumed for each nondeterministic step (up to the guessed time $f(|x_i|)$) and an input symbol b for each deterministic step (up to the end of input x_i). So the number x resp. y of simulated nondeterministic resp. deterministic transitions corresponds to a word $a^x b^y$ belonging to L' iff there exists an $m \in \mathbb{N}$ such that $x = f(m)$ and $y = m - f(m)$. Thus, iff $|x_i| = x + y = f(m) + m - f(m) = m$.

Altogether $\mathcal{M}'$ accepts $\overline{(L\bullet)^+}$ in real-time.

Since we have assumed that $\mathscr{L}_{rt}(f\text{G-IA})$ is closed under complement it follows $(L\bullet)^+ \in \mathscr{L}_{rt}(f\text{G-IA})$. But we have supposed $\mathscr{L}_{rt}(f\text{G-IA})$ is not closed under marked iteration. From the contradiction it follows that $\mathscr{L}_{rt}(f\text{G-IA})$ is not closed under complement if $\mathscr{L}_{rt}(f\text{G-IA})$ is really not closed under marked iteration. This will be proved in the next theorem. □

The next lemma has already been used to prove a previous one.

Lemma 14. *Let $f : \mathbb{N} \to \mathbb{N}_0$ be an increasing, unbounded mapping such that $f \in o(\log)$ and $\{a^{f(m)}b^{m-f(m)} \mid m \in \mathbb{N}\} \in \mathscr{L}_{rt}(\text{IA})$ then $\mathscr{L}_{rt}(f\text{G-IA})$ is not closed under marked iteration.*

Proof. By Theorem 10 L_f belongs to $\mathscr{L}_{rt}(f\text{G-IA})$. Now we are going to show that the marked iteration $(L_f\bullet)^+$ of L_f does not belong to $\mathscr{L}_{rt}(f\text{G-IA})$ from which the lemma follows.

Assume in contrast there exists a fG-IA $\mathcal{M} = (S, \ldots)$ which accepts $(L_f\bullet)^+$ in real-time. We consider words $x_1\bullet x_2\bullet \cdots \bullet x_k\bullet \in (L_f\bullet)^+$ for a $k \in \mathbb{N}$. Let x_k be an arbitrary word in L_f and n_k be its length: $n_k := |x_k|$. Since m is an unbounded mapping we can find a smallest $n_i \in \mathbb{N}$ such that $m(n_i) \geq |x_i\bullet x_{i+1}\bullet \cdots \bullet x_k\bullet|$ respectively, for $1 \leq i \leq k-1$. Obviously, there exist words of length n_i in L_f. Let x_i be one of them respectively. For the lengths l_i of the subwords $x_i\bullet \cdots x_k\bullet$ we obtain $l_k = n_k + 1$ and for $1 \leq i \leq k-1$: $l_i = n_i + 1 + l_{i+1}$.

In what follows let k_j be appropriated constants. Since $h(n_i) \leq m(n_i)$ and $r_i \leq m(n_i)$ it holds $n_i = \big(m(n_i)+1\big)\big(h(n_i)+1\big) + r_i \leq \big(m(n_i)+1\big)^2 + m(n_i) \leq k_8 \cdot m(n_i)^2$. For l_i we obtain:

$$
\begin{aligned}
l_i &= r_i + \big(h(n_i)+1\big)\big(m(n_i)+1\big) + 1 + l_{i+1} \\
&\leq k_5 \cdot l_{i+1} + \big(h(n_i)+1\big)\big(k_5 \cdot l_{i+1} + 1\big) + 1 + l_{i+1} \text{ since } r_i \leq m(n_i) \leq k_5 \cdot l_{i+1} \\
&\leq k_6 \cdot h(n_i) \cdot l_{i+1} \\
&\leq k_6 \cdot h(k_7 \cdot l_{i+1}^2) \cdot l_{i+1} \\
&\leq k_i' \cdot l_{i+1}^{\epsilon} l_{i+1} \text{ since } h(n) \in o(n^{\epsilon/2}) \text{ for all } \epsilon \in \mathbb{Q}_+
\end{aligned}
$$

It follows $l_1 \leq k_1' \cdot l_2^{1+\epsilon} \leq \cdots \leq k_1' \cdot \ldots \cdot k_{k-1}' \cdot l_k^{(1+\epsilon)^{k-1}}$.

If we choose $\epsilon \in \mathbb{Q}_+$ such that $(1+\epsilon)^{k-1} < 2$ then for large n we obtain that $l_1 \leq \frac{1}{2} \cdot l_k^2 = \frac{1}{2} \cdot (n_k+1)^2 \leq n_k^2$.

Thus for processing $x_1\bullet \cdots x_k\bullet$ $\mathcal{M}$ performs at most $f(n_k^2)$ nondeterministic transitions. Since $f \in o(\log)$ there exists $k_1 \in \mathbb{N}$ such that $k_1 \cdot f(n_k) \geq f(n_k^2)$ for large n_k. Therefore, for large n at most $k_1 \cdot f(n_k)$ nondeterministic steps are performed by $\mathcal{M}$. (note that k_1 does not depend on k).

Now we consider the equivalence classes that appear if we cut $x_1\bullet \cdots x_k\bullet$ after the first symbol ¢ in x_i respectively. Since $x_2\bullet \cdots x_k\bullet$ is at most as long as the y_1 in x_1 we have $N(|x_1\bullet \cdots x_k\bullet|, 2|y_1|+1, L_f\bullet A^{|y_1|})$ different equivalence classes for the cut in x_1. By Lemma 6 this number equals $N(|x_1|, |y_1|, L_f)$. By Corollary 11 there exists a constant such that at least $k_2 \cdot f(n_1)$ guesses are necessary in order to accept languages with such a number of equivalence classes. Define $q_m := \max\{|\delta_{nd}(s_1, s_2, s_3, a)| \mid s_1, s_2, s_3 \in S, a \in A\}$. Thus, the computation of $\mathcal{M}$ on input $x_1\bullet$ contains at least $q_m^{k_2 \cdot f(n_1)}$ different paths.

Now we consider all computation paths of $\mathcal{M}$. For all $x_1 \in L_f$ there exists a class of paths that are accepting for words of the form $x_1\bullet\cdots$ Since for computations on $x_1\bullet$ there are at least $q_m^{k_2 \cdot f(n_1)}$ different paths we have now at least $q_m^{k_2 \cdot f(n_1)}$ disjoint classes.

If we cut $x_1\bullet\cdots x_k\bullet$ after the first symbol ¢ in x_2, again, it results in $N(|x_1\bullet x_2|, |y_2|, L_f)$ equivalence classes for which $k_2 \cdot f(n_2)$ different computations paths are necessary. These paths are all in the same class for x_1. Therefore, every class contains at least $q_m^{k_2 \cdot f(n_2)}$ paths. Since at least $q_m^{k_2 \cdot f(n_1)}$ classes are disjoint there are at least $q_m^{k_2 \cdot f(n_1)} \cdot q_m^{k_2 \cdot f(n_2)}$ different paths.

Proceeding inductively we conclude that there are at least $q_m^{k_2 \cdot f(n_1)} \cdot \ldots \cdot q_m^{k_2 \cdot f(n_k)} \geq \left(q_m^{k_2 \cdot f(n_k)}\right)^k$ different paths. To realize the paths $\mathcal{M}$ at least needs to perform $k \cdot k_2 \cdot f(n_k)$ nondeterministic steps (here we need $q_m > 1$ what follows since f is unbounded). For a k such that $k \cdot k_2 > k_1$ we get a contradiction because $\mathcal{M}$ performs at most $k_1 \cdot f(n_k)$ nondeterministic transitions. □

Corollary 15. *Let $f : \mathbb{N} \to \mathbb{N}_0$ be an increasing mapping such that $f \in o(\log)$ then $\mathscr{L}_{rt}(f\text{G-IA})$ is not closed under complement.*

5.2 The Hierarchy

In [4] the following nondeterministic hierarchies have been shown: Let $f : \mathbb{N} \to \mathbb{N}_0$, $f \in o(\log)$, and $g : \mathbb{N} \to \mathbb{N}_0$, $g \in o(f)$, be two increasing mappings such that $\forall\, m, n, \in \mathbb{N} : f(m) = f(n) \Longrightarrow g(m) = g(n)$. If $L = \{a^{g(m)} b^{f(m)-g(m)} \mid m \in \mathbb{N}\}$ belongs to the family $\mathscr{L}_{lt}(\text{IA})$ then $\mathscr{L}_{rt}(g\text{G-IA}) \subset \mathscr{L}_{rt}(f\text{G-IA})$.

By the results of the previous subsection we obtain an alternating hierarchy, too.

Theorem 16. *Let $f : \mathbb{N} \to \mathbb{N}_0$, $f \in o(\log)$, and $g : \mathbb{N} \to \mathbb{N}_0$, $g \in o(f)$ be two increasing mappings such that $\forall\, m, n, \in \mathbb{N} : f(m) = f(n) \Longrightarrow g(m) = g(n)$. If $\{a^{g(m)} b^{f(m)-g(m)} \mid m \in \mathbb{N}\} \in \mathscr{L}_{lt}(\text{IA})$ and $\{a^{f(m)} b^{m-f(m)} \mid m \in \mathbb{N}\} \in \mathscr{L}_{rt}(\text{IA})$ then $\mathscr{L}_{rt}(g\text{A-IA}) \subset \mathscr{L}_{rt}(f\text{A-IA})$.*

Proof. Due to the assumption $L := \{a^{g(m)} b^{f(m)-g(m)} \mid m \in \mathbb{N}\} \in \mathscr{L}_{lt}(\text{IA})$ a real-time fG-IA can limit its nondeterministic transitions up to the guessed time step $g(n)$ and verify its guess. For a deterministic real-time IA language this technique has been used in the proof of Lemma 14. It is known that deterministic linear-time IAs can be sped up to $2 \cdot n$ time [10]. Since $f \in o(\log)$ we can assume $f \leq \frac{n}{2}$ and, hence, during n time steps a $(2 \cdot n)$-time IA for L can be simulated.

By this constructibility property and for structural reasons we obtain $\mathscr{L}_{rt}(g\text{A-IA}) \subseteq \mathscr{L}_{rt}(f\text{A-IA})$.

Since g is of order $o(f)$ but by Corollary 11 it has to be of order $\Omega(f)$ in order to accept L_f in real-time we conclude $L_f \notin \mathscr{L}_{rt}(g\text{A-IA})$.

On the other hand by Theorem 10 L_f belongs to $\mathscr{L}_{rt}(f\text{G-IA})$. We obtain $\mathscr{L}_{rt}(f\text{G-IA}) \not\subseteq \mathscr{L}_{rt}(g\text{A-IA})$.

By Theorem 13 it holds $\mathscr{L}_{rt}(f\text{G-IA}) \subset \mathscr{L}_{rt}(f\text{A-IA})$.

It follows $\mathscr{L}_{rt}(g\text{A-IA}) \subset \mathscr{L}_{rt}(f\text{A-IA})$. □

On a first glance the preconditions of the hierarchy seem to be rather complicated but the following natural functions meet them. Let $i > 1$ be a constant then $f(n) := \log^i(n)$ and $g(n) := \log^{i+1}(n)$ ($\log^i$ denotes the i-fold composition of log).

References

[1] Beyer, W. T. *Recognition of topological invariants by iterative arrays*. Technical Report TR-66, MIT, Cambridge, Proj. MAC, 1969.

[2] Buchholz, Th. and Kutrib, M. *Some relations between massively parallel arrays*. Parallel Comput. 23 (1997), 1643–1662.

[3] Buchholz, Th., Klein, A., and Kutrib, M. *Iterative arrays with a wee bit alternation*. IFIG Report 9905, Institute of Informatics, University of Giessen, 1999.

[4] Buchholz, Th., Klein, A., and Kutrib, M. *Iterative arrays with limited nondeterministic communication cell*. IFIG Report 9901, Institute of Informatics, University of Giessen, 1999.

[5] Chang, J. H., Ibarra, O. H., and Palis, M. A. *Parallel parsing on a one-way array of finite-state machines*. IEEE Trans. Comput. C-36 (1987), 64–75.

[6] Cole, S. N. *Real-time computation by n-dimensional iterative arrays of finite-state machines*. IEEE Trans. Comput. C-18 (1969), 349–365.

[7] Čulik II, K. and Yu, S. *Iterative tree automata*. Theoret. Comput. Sci. 32 (1984), 227–247.

[8] Fischer, P. C. *Generation of primes by a one-dimensional real-time iterative array*. J. Assoc. Comput. Mach. 12 (1965), 388–394.

[9] Ibarra, O. H. and Jiang, T. *On one-way cellular arrays*. SIAM J. Comput. 16 (1987), 1135–1154.

[10] Ibarra, O. H. and Palis, M. A. *Some results concerning linear iterative (systolic) arrays*. J. Parallel and Distributed Comput. 2 (1985), 182–218.

[11] Ibarra, O. H. and Palis, M. A. *Two-dimensional iterative arrays: Characterizations and applications*. Theoret. Comput. Sci. 57 (1988), 47–86.

[12] Seiferas, J. I. *Iterative arrays with direct central control*. Acta Inf. 8 (1977), 177–192.

[13] Seiferas, J. I. *Linear-time computation by nondeterministic multidimensional iterative arrays*. SIAM J. Comput. 6 (1977), 487–504.

[14] Smith III, A. R. *Real-time language recognition by one-dimensional cellular automata*. J. Comput. System Sci. 6 (1972), 233–253.

[15] Terrier, V. *On real time one-way cellular array*. Theoret. Comput. Sci. 141 (1995), 331–335.

Secret Sharing Schemes with Detection of Cheaters for a General Access Structure *

Sergio Cabello, Carles Padró and Germán Sáez

Dep. Matemàtica Aplicada i Telemàtica
Universitat Politècnica de Catalunya
Mòdul C3, Campus Nord, c/Jordi Girona 1–3, 08034 Barcelona
matcpl@mat.upc.es,german@mat.upc.es

Abstract. In a secret sharing scheme, some participants can lie about the value of their shares when reconstructing the secret in order to obtain some illicit benefits. We present in this paper two methods to modify any linear secret sharing scheme in order to obtain schemes that are unconditionally secure against that kind of attack. The schemes obtained by the first method are robust, that is, cheaters are detected with high probability even if they know the value of the secret. The second method provides secure schemes, in which cheaters that do not know the secret are detected with high probability. When applied to ideal linear secret sharing schemes, our methods provide robust and secure schemes whose relation between the probability of cheating and the information rate is almost optimal. Besides, those methods make it possible to construct robust and secure schemes for any access structure.

Keywords: Cryptography, Secret sharing schemes, Detection of cheaters, Robust and secure schemes.

1 Introduction

In a *secret sharing scheme*, a secret value is distributed into shares among the participants in a set P in such a way that only qualified subsets of P can reconstruct the secret from their shares. Such a scheme is said to be *perfect* if the subsets that are not qualified to reconstruct the secret have absolutely no information on it. All the secret sharing schemes in this paper are considered to be perfect. See [16, 18, 14] for a comprehensive introduction to secret sharing schemes.

The family of qualified subsets, $\Gamma \subset 2^P$, of a secret sharing scheme is called the *access structure* of the scheme. We consider only *monotone* access structures, that is, any subset that contains a qualified subset must be qualified. Then, the access structure Γ is determined by the family of minimal qualified subsets, Γ_0, which is called the *basis* of Γ. For example, the access structure of a (t, N)-*threshold scheme* consists of all the subsets with at least t of N participants. Threshold schemes, that were independently introduced by Blakley [1] and Shamir [13] in 1979, were the first considered secret sharing schemes.

* This work was partially supported by spanish CICYT under project TIC97-0963.

The size of the shares given to the participants is one of the main parameters to be taken into account in the design of secret sharing schemes. The ***information rate*** ρ of a secret sharing scheme is the ratio between the length (in bits) of the secret and the maximum length of the shares given to the participants. That is, if $\mathcal{K}$ is the set of secrets and, for any $p \in P$, the share of the participant p is taken from the set S_p, then

$$\rho = \frac{\log |\mathcal{K}|}{\max \log |S_p|}.$$

A secret sharing scheme is said to be ***ideal*** if its information rate is equal to one, which is the maximum possible value of this parameter. Several authors have given upper and lower bounds on the information rate of secret sharing schemes realizing different access structures [4, 17, 3, 10].

There exists the possibility that some participants in a secret sharing scheme provide false shares during the reconstruction of the secret in order to obtain some illicit benefit. Therefore, the security against this kind of attack is a key point in the implementation of secret sharing schemes. We are concerned only in ***unconditional security*** against cheaters, that is, the probability of cheating successfully must not depend on the computational resources available to the participants.

We consider in this paper the following scenario: the participants in a coalition in the form $A - \{p_t\}$, where $A = \{p_1, \ldots, p_{t-1}, p_t\} \in \Gamma_0$ is a minimal qualified subset, forge their shares in order to deceive the honest participant p_t. That is, the participants in $A - \{p_t\}$ try to find a set of false shares $\{s_1^*, \ldots, s_{t-1}^*\}$ such that a false value $k^* \neq k$ of the secret is reconstructed from the shares $\{s_1^*, \ldots, s_{t-1}^*, s_t\}$. In this case, we say that the ***participant*** p_t ***is cheated by the false shares*** $\{s_1^*, \ldots, s_{t-1}^*\}$. Two different cases are considered. In the first one, we suppose that the cheaters somehow know the value of the secret $k \in \mathcal{K}$. In the second one, we assume that the cheaters have no information about the value of the secret.

The unconditional security of a scheme against this kind of attack is measured by the ***probability of cheating***, that was first formally defined in [7] for threshold schemes. Let us consider $A \in \Gamma_0$ a minimal qualified subset and a participant $p \in A$. If we suppose that the participants in $A - \{p\}$ know the secret, we define the ***probability that the participants in*** $A - \{p\}$ ***deceive the participant*** p, denoted by $\mathrm{PC}_1(A - \{p\})$, as

$$E_{b,k}\left(\max_{b'} \Pr\,(p \text{ is cheated by } b' | A - \{p\} \text{ have } b, \text{ the secret is } k)\right)$$

where b denotes the shares received by the participants in $A - \{p\}$ and b' denotes the forged shares used by the participants in $A - \{p\}$ in order to deceive the participant p. If we suppose that the cheaters in the coalition $A - \{p\}$ have no information about the value of the secret, the ***probability that the participants in*** $A - \{p\}$***, that do not know the secret, deceive the participant*** p, $\mathrm{PC}_2(A - \{p\})$, is defined analogously [8].

A $(\Gamma, \epsilon)-$*robust scheme* is a secret sharing scheme with access structure Γ such that $\mathrm{PC}_1(A - \{p\}) \leq \epsilon$ for any minimal qualified subset $A \in \Gamma_0$ and for any participant $p \in A$. A $(\Gamma, \delta)-$*secure scheme* is a secret sharing scheme with access structure Γ such that, for any minimal qualified subset $A \in \Gamma_0$ and for any participant $p \in A$, $\mathrm{PC}_2(A - \{p\}) \leq \delta$.

All ideal schemes are insecure against cheating. If $A \in \Gamma_0$ is a minimal qualified subset in an ideal scheme, it is not difficult to prove that, for any $p \in A$, the coalition of all participants in $A - \{p\}$ can forge their shares in order to obtain the correct secret k and deceive p with a false secret k^*.

The vector space construction [5] is a method to construct ideal secret sharing schemes for a family of access structures, the *vector space access structures*, that includes threshold structures as a particular case. A generalization of this method makes it possible to find secret sharing schemes, that are not ideal in general, for any access structure [15]. The schemes constructed in this way are called *linear secret sharing schemes*. Since in such a scheme the secret value is a linear function of the shares of the participants in any qualified subset, linear schemes are very vulnerable to the action of cheaters. More information about this kind of schemes is given in Section 2.

Unconditional security is obtained in general by adding redundant information to the shares. Several lower bounds on the length of the shares in robust and secure schemes have been found [7,8,11,2]. One can see from these bounds that the information rate decreases with the probability of cheating. That leads to a problem that has received considerable attention: to find robust and secure schemes with information rate as high as possible.

Most of the secure schemes [12,8] and robust schemes [19,11] that have been proposed until now are threshold schemes. The design of schemes with detection of cheaters for a different kind of access structures was first considered in [11]. A method to find secure schemes that realize any vector space access structure is presented in that work. In fact, although Ogata and Kurosawa [8] have considered only threshold structures, their secure scheme can be implemented for vector space access structures too. The first robust scheme that can be applied on any vector space access structure has been proposed in [9].

The aim of this work is to present two methods to modify a linear secret sharing scheme with access structure Γ and set of secrets $\mathcal{K}$, a vector space over a finite field $GF(q)$, in order to obtain a (Γ, ϵ)-robust scheme and a (Γ, δ)-secure scheme with the same set of secrets and $\epsilon = \delta = 1/|\mathcal{K}|$. In the case of secure schemes, the characteristic of the field cannot be equal to 2. Since there exists a linear secret sharing scheme for any access structure, our methods provide robust and secure schemes for a general access structure. The information rate of the robust and secure schemes we present here is equal to, respectively, one third and one half of the information rate of the original linear scheme. When applied to vector space access structures, our methods provide robust and secure schemes whose information rates appear to be almost optimal when compared to the bounds given in [8]. The robust schemes we present here improve the robust schemes proposed in [19,11] for threshold access structures and the one given in [9] for vector space access structures. Our secure schemes are based on the secure schemes proposed in [11] for vector space access structures.

An introduction to linear secret sharing schemes is given in Section 2. Section 3 deals with the construction of robust schemes. The method to construct secure schemes is given in Section 4.

2 Linear secret sharing schemes

All vector spaces we consider here have finite dimension. Let $P = \{p_1, p_2, \ldots, p_N\}$ be a set of participants and consider a special participant $D = p_0 \notin P$ called *dealer*. Let E be a vector space over the finite field $GF(q)$. For every $p_i \in P \cup \{D\}$, let us consider E_i, a vector space over $GF(q)$, and a surjective linear mapping $\pi_i : E \to E_i$. For any $i = 0, 1, \ldots, N$, let $F_i = \ker \pi_i$ be the kernel of the linear mapping π_i. Let us suppose that these linear mappings verify that, for any $A \subset P$,

$$\bigcap_{p_i \in A} F_i \subset F_0 \text{ or } \bigcap_{p_i \in A} F_i + F_0 = E.$$

In this situation, we can define a secret sharing scheme in the set of participants P as follows: for a secret value $k \in E_0$, a vector $v \in E$ such that $\pi_0(v) = k$ is taken at random and every participant $p_i \in P$ receives as its share the vector $s_i = \pi_i(v) \in E_i$. It is not difficult to prove that this is a perfect secret sharing scheme with access structure

$$\Gamma = \left\{ A \subset P : \bigcap_{p_i \in A} F_i \subset F_0 \right\}$$

The information rate of this scheme is $\rho = \dim E_0 / (\max_{1 \le i \le N} \dim E_i)$. Secret sharing schemes constructed in this way are called *linear secret sharing schemes.*

Linear secret sharing schemes were first introduced by Brickell [5], who considered only ideal linear schemes with $\dim E_i = 1$ for any $p_i \in P \cup \{D\}$. In that case, we can consider that the surjective linear mappings π_i are non-zero elements of the dual space E^*. In such an ideal linear scheme, a subset $A \subset P$ is qualified if and only if the vector $\pi_0 \in E^*$ can be expressed as a linear combination of the vectors $\{\pi_i \mid p_i \in A\}$. The access structures that can be defined in this way are called *vector space access structures*. Threshold structures are a particular case of vector space access structures. Effectively, if Γ is a (t, N)-threshold structure, we can take $q > N$ a prime power and $x_i \in GF(q)$, for any $p_i \in P$, non zero and pairwise distinct elements and consider $E = GF(q)^t$, $\pi_0 = (1, 0, \ldots, 0) \in E^*$ and $\pi_i = (1, x_i, x_i^2, \ldots, x_i^{t-1}) \in E^*$ for any $i = 1, \ldots, N$. The ideal linear scheme we obtain in this way is in fact equivalent to the Shamir's threshold scheme [13].

Simmons, Jackson and Martin [15] proved that any access structure Γ can be realized with a linear secret sharing scheme that, in general, is not ideal. Besides, an algorithm that provides a linear secret sharing scheme for any access structure is given in [15]. The main handicap of the schemes constructed in such a way is that its information rate is very small. Nevertheless, for many access structures, it is possible to find a linear scheme with much better information rate by using decomposition techniques such as the Stinson's *λ-decomposition construction* [17], which is one of the most powerful tools to construct secret sharing schemes with good information rate. It can be seen that a linear secret sharing scheme is obtained when linear secret sharing schemes are combined in a λ-decomposition construction.

Finally, we remark that vector space secret sharing schemes are very vulnerable to the action of cheaters. Effectively, for any minimal qualified subset $A = \{p_{i_1}, \dots, p_{i_t}\} \in \Gamma_0$, the secret can be computed from the shares of the participants in A by using a linear map $\chi_A : E_{i_1} \times \dots \times E_{i_t} \to E_0$. We observe that the map χ_A is publicly known, that is, it can be determined from the mappings π_i. It is not difficult to see that, for any $\epsilon \in E_0$, the participant p_{i_1} can compute a false share $s^*_{i_1}$ such that $\chi_A(s^*_{p_1} - s_{p_1}, 0, \dots, 0) = \epsilon$. Then, if $\epsilon \neq 0$, a wrong secret $k^* = k + \epsilon = \chi_A(s^*_{p_1}, s_{p_2}, \dots, s_{p_t})$ is recovered. In this way, the participant p_1 can deceive the remaining participants and, besides, he or she can obtain the correct value of the secret, $k = k^* - \epsilon$.

3 Robust schemes

We present in this section a method to modify any linear secret sharing scheme with access structure Γ and set of secrets E_0 in order to obtain a (Γ, ϵ)-robust scheme with the same set of secrets and $\epsilon = 1/|E_0|$.

Let Σ be a linear secret sharing scheme with access structure Γ on the set of participants $P = \{p_1, \dots, p_N\}$. Let $\pi_i : E \to E_i$, where $i = 0, 1, \dots, N$ and E and E_i are vector spaces over a finite field $GF(q)$, be the surjective linear mappings that define Σ. If $n = \dim E_0$, then E_0 and the finite field $GF(q^n)$ are isomorphic as $GF(q)$-vector spaces. Therefore, we can suppose that the set of secrets of Σ is the finite field $GF(q^n)$. In this way, we can consider the product xy of two elements $x, y \in E_0$.

We define next the scheme Σ_1 that will be proved to be a $(\Gamma, 1/q^n)$-robust scheme. Given a secret value $k \in E_0 = GF(q^n)$, the dealer takes at random a vector $r \in E_0$ and three vectors $v_1, v_2, v_3 \in E$ such that $\pi_0(v_1) = k$, $\pi_0(v_2) = r$ and $\pi_0(v_3) = kr$, where the product is computed in the field $GF(q^n)$. For any $i = 1, \dots, N$, the share of the participant p_i is equal to $\mathbf{s}_i = (s_{i1}, s_{i2}, s_{i3})$, where $s_{ij} = \pi_i(v_j)$. That is, in the scheme Σ_1, three secret values, namely, the secret k, a random element r and the product kr, are distributed into shares by using the linear secret sharing scheme Σ. When the participants in a qualified subset try to reconstruct the secret value, they obtain $(x_1, x_2, x_3) \in E_0^3$ from their shares. If $x_3 = x_1 x_2$, they take $k = x_1$ as the correct value of the secret. They are warned about the existence of cheaters if $x_3 \neq x_1 x_2$.

Proposition 1 *The scheme Σ_1 is a (Γ, ϵ)-robust scheme with set of secrets E_0 and information rate $\rho/3$, where $\epsilon = 1/|E_0|$ and ρ is the information rate of scheme Σ.*

Proof. First of all, we have to prove that Σ_1 is a secret sharing scheme with access structure Γ. It is obvious that the participants in any qualified subset can reconstruct the secret from their shares. On the other hand, the shares of the participants in a non-qualified subset do not provide any information on the value of the secret. In effect, let $B \notin \Gamma$ be a non-qualified subset. We can suppose that $B = \{p_1, \ldots, p_\ell\}$. If the participants in B want to obtain some information about the value of the secret, they have to find it out from the system of equations

$$\begin{aligned} s_{i1} &= \pi_i(v_1) \\ s_{i2} &= \pi_i(v_2) \\ s_{i3} &= \pi_i(v_3) \\ \pi_0(v_3) &= \pi_0(v_1)\pi_0(v_2) \end{aligned}$$

where $i = 1, \ldots, \ell$ and the unknowns are the vectors $v_1, v_2, v_3 \in E$. Since

$$\bigcap_{p_i \in B} F_i + F_0 = E$$

where $F_i = \ker \pi_i$, it is not difficult to see that, from this system of equations, the secret $k = \pi_i(v_1)$ can take any value in $E_0 = GF(q^n)$ with the same probability.

We prove next that the probability of cheating $PC_1(A - \{p\})$ is equal to $\epsilon = 1/q^n$ for any minimal qualified subset $A \in \Gamma_0$ and for any participant $p \in A$. Let $A = \{p_1, \ldots, p_{t-1}, p_t\}$ be a minimal qualified subset and let us suppose that the participants in $A - \{p_t\}$ somehow know the value of the secret $k \in K$ and try to deceive the honest participant p_t. The cheaters know that the shares correspond to a vector $(x_1, x_2, x_3) \in E_0^3$ such that $x_1 = k$ and $x_3 = x_1 x_2$, but they do not know the values of $x_2, x_3 \in E_0$. For any $(\epsilon_1, \epsilon_2, \epsilon_3) \in E_0^3$, the participants in $A - \{p_t\}$ can compute forged shares $\mathbf{s}_1^*, \ldots, \mathbf{s}_{t-1}^*$ such that, in the reconstruction process, the vector (x_1^*, x_2^*, x_3^*) is computed, where $x_i^* = x_i + \epsilon_i$. Of course, if they want to deceive the participant p_t, they have to take $\epsilon_1 \neq 0$. The cheaters are not detected if and only if $x_3^* = x_1^* x_2^*$, that is, if and only if

$$\epsilon_1 x_2 + \epsilon_2 x_1 + \epsilon_1 \epsilon_2 - \epsilon_3 = 0 \tag{1}$$

Then, for every choice of $(\epsilon_1, \epsilon_2, \epsilon_3)$, with $\epsilon_1 \neq 0$, there exist a unique $x_2 \in E_0$ that satisfy equation (1). Therefore, for any forged shares b' used by the participants in $A - \{p_t\}$ in order to deceive the participant p_t,

$$\Pr(p_t \text{ is cheated by } b' | A - \{p_t\} \text{ have } b, \text{ the secret is } k) = \frac{1}{q^n}$$

Then, $\mathrm{PC}_1(A - \{p_t\}) = 1/q^n$. □

Ogata and Kurosawa [8] found a lower bound on the size of the set S_p of possible shares of any participant $p \in P$ in a (Γ, ϵ)-robust scheme, where Γ is a threshold structure,

$$|S_p| \geq M_1(|\mathcal{K}|, \epsilon) = \frac{|\mathcal{K}| - 1}{\epsilon^2} + 1 \tag{2}$$

where $\mathcal{K}$ is the set of possible values of the secret. It is not difficult to prove that this bound is valid for any access structure Γ.

If Σ is an ideal linear scheme with access structure Γ and set of secrets $GF(q)$, then Σ_1 is a $(\Gamma, 1/q)$-robust scheme such that $|\mathcal{K}| = q$ and $|\mathcal{S}_p| = q^3$ for any $p \in P$. Observe that the cardinality of the set of shares is very close to the lower bound (2), that in this case is $M_1(q, 1/q) = q^3 - q^2 + 1$.

A (Γ, ϵ)-robust scheme was proposed in [9] for any vector space access structure. Actually, this robust scheme can be used to modify any linear secret sharing scheme with access structure Γ, set of secrets $GF(q)^n$, where q is a power of a prime different from 2, and information rate ρ, into a $(\Gamma, 2/q^n)$-robust scheme with the same set of secrets and information rate $\rho/3$. Observe that the robust scheme Σ_1 has the same information rate and a lower probability of cheating. Besides, the set of secrets of Σ_1 can be a field with characteristic 2.

Finally, it is proved in [9] that the robust scheme proposed in that paper presents a much better relation between the information rate and the probability of cheating than the robust threshold schemes presented in [19] and [11]. Then, our robust scheme improves those schemes too.

The robust scheme Σ_1 can be generalized in order to obtain (Γ, ϵ)-robust schemes whose probability of cheating is less than $1/|\mathcal{K}|$. Given a secret $k \in E_0 = GF(q^n)$ each entry of the vector $(k, r_1, \ldots, r_\ell, kr_1, \ldots, kr_\ell) \in E_0^{2\ell+1}$, where $r_1, \ldots, r_\ell$ are random elements of E_0, is distributed by using a linear secret sharing scheme. When the participants in a minimal qualified subset try to recover the secret from their shares, a vector $(x_1, \ldots, x_{2\ell+1})$ is computed. They take $k = x_1$ as the correct value of the secret if $x_{i+\ell} = x_1 x_i$ for $i = 2, \ldots, \ell + 1$, else, they are warned about the existence of cheaters. It is not difficult to check that a (Γ, ϵ)-robust scheme with $\epsilon = 1/q^\ell$ is obtained in this way. When applied to ideal linear secret sharing schemes with $E_0 = GF(q)$, we observe that the size of the shares in this robust scheme, $|\mathcal{S}_p| = q^{2\ell+1}$, is very close to the lower bound $M_1(q, 1/q^\ell) = q^{2\ell+1} - q^{2\ell} + 1$. Of course, for a given $\epsilon > 0$ and a linear secret sharing scheme Σ for Γ with information rate ρ, we can take $\ell = \lceil -\log \epsilon / \log q \rceil$ in order to construct a (Γ, ϵ)-robust scheme with information rate $\rho/(2\ell + 1)$.

4 Secure schemes

In this section we will show how any linear secret sharing scheme with access structure Γ and set of secrets E_0, a vector space over a finite field $GF(q)$ with characteristic different from 2, can be modified in order to obtain a (Γ, δ)-secure scheme with the same set of secrets and $\delta = 1/|E_0|$. This construction is based on the secure scheme introduced in [11] for vector space access structures.

Let us consider E and E_i, where $0 \leq i \leq N$, vector spaces over a finite field $GF(q)$, where q is odd, and surjective linear maps $\pi_i : E \to E_i$ that define a linear secret sharing scheme Σ with access structure Γ and set of secrets E_0 on the set of participants $P = \{p_1, \ldots, p_N\}$. As before, we can suppose that E_0 is equal to the finite field $GF(q^n)$, where $n = \dim E_0$.

We define next a $(\Gamma, 1/q^n)$-secure scheme Σ_2 from the linear secret sharing scheme Σ. For a given secret $k \in E_0$, the dealer takes at random two vectors $v_1, v_2 \in E$ such that $\pi_0(v_1) = k$ and $\pi_0(v_2) = k^2$. Every participant $p_i \in P$ receives the share $\mathbf{s}_i = (s_{i1}, s_{i2}) \in E_i \times E_i$, where $s_{ij} = \pi_i(v_j)$. That is, the linear secret sharing scheme Σ is used to distribute two secret values, namely k and its square k^2. In the recovering process, the participants in a minimal qualified subset $A \in \Gamma_0$, compute $(x_1, x_2) \in E_0 \times E_0$ from their shares. If $x_2 = x_1^2$, the participants in A take $k = x_1$ as the correct value of the secret and they are warned about the existence of cheaters if $x_2 \neq x_1^2$.

Proposition 2 *The scheme Σ_2 is a (Γ, δ)-secure scheme with set of secrets E_0 and information rate $\rho/2$, where $\delta = 1/|E_0|$ and ρ is the information rate of scheme Σ.*

Proof. First we will prove that Σ_2 is a perfect secret sharing scheme with access structure Γ. Obviously, the participants in any qualified subset can reconstruct the secret using their shares. We should prove now that the shares of the participants in a non qualified subset do not reveal any information about the value of the secret. Let $B \subset P$ be a non qualified subset. Suppose that $B = \{p_1, \ldots, p_s\}$. The information that participants of B have about the value of the secret is reflected on the equation system

$$\begin{cases} \pi_i(v_1) = s_{i1} & i = 1, \ldots, s \\ \pi_i(v_2) = s_{i2} & i = 1, \ldots, s \\ (\pi_0(v_1))^2 = \pi_0(v_2) \end{cases}$$

where the unknowns are vectors $v_1, v_2 \in E$. Since $B \notin \Gamma$, we have that $\bigcap_{p_i \in B} F_i + F_0 = E$, where $F_i = \ker \pi_i$. Then, it is easy to see that this equation system does not reveal any information about the value of the secret, that is, all the secret values $k = \pi_0(v_1)$ are equiprobable.

Finally, we have to prove that the probability of cheating $\mathrm{PC}_2(A - \{p\})$ is at most $1/q^n$. Let $A = \{p_1, \ldots, p_t\} \in \Gamma_0$ be a minimal qualified subset and let us suppose that participants in $A - \{p_t\}$, that do not have any information on the value of the secret, try to deceive the participant p_t about the value of the secret. For any $(\epsilon_1, \epsilon_2) \in E_0^2$, the cheaters can compute a set of false shares $b' = \{(s_{11}^*, s_{12}^*), \ldots, (s_{t-1\,1}^*, s_{t-1\,2}^*)\}$ such that $\chi_A(s_{11}^*, \ldots, s_{t-1\,1}^*, s_{t1}) = k + \epsilon_1$ and $\chi_A(s_{12}^*, \ldots, s_{t-1\,2}^*, s_{t2}) = k^2 + \epsilon_2$. The participant p_t is deceived by b' if and only if the cheaters have chosen $(\epsilon_1, \epsilon_2) \in E_0^2$ such that $\epsilon_1 \neq 0$ and $(k + \epsilon_1)^2 = k^2 + \epsilon_2$, that is, such that $2k\epsilon_1 + \epsilon_1^2 = \epsilon_2$. Observe that, for every $(\epsilon_1, \epsilon_2) \in E_0^2$ with $\epsilon_1 \neq 0$, there is exactly one value of $k \in E_0$ such that $2k\epsilon_1 + \epsilon_1^2 = \epsilon_2$. Therefore,

$$\Pr(p_t \text{ is cheated by } b' | A - \{p_t\} \text{ have } b) = \frac{1}{q^n}$$

Then, $\mathrm{PC}_2(A - \{p_t\}) = 1/q^n$. □

If Σ is an ideal secret sharing scheme with vector space access structure Γ and set of secrets $GF(q)$, where q is odd, Σ_2 is the $(\Gamma, 1/q)$-secure scheme that was proposed in [11] for this kind of access structures. The set of secrets of Σ_2 is $GF(q)$ and $|\mathcal{S}_p| = q^2$ for any $p \in P$. We can compare that with the lower bound given in [8, 11] for the size of the set of shares in a (Γ, δ)-secure scheme,

$$|\mathcal{S}_p| \geq M_2(|\mathcal{K}|, \delta) = \frac{|\mathcal{K}| - 1}{\delta} + 1 \quad (3)$$

In our case, $M_2(q, 1/q) = q^2 - q + 1$.

Ogata and Kurosawa [8] propose a threshold $(\Gamma, 1/q)$-secure scheme achieving this bound, that is, with optimal information rate. In fact, this scheme can be defined for vector space access structures too. Then, for this kind of structures, the Ogata and Kurosawa's scheme has slightly better information rate than the secure scheme Σ_2 but, on the other hand, it is computationally less efficient. See [11] for a complete comparison between these two schemes.

Finally, in the case of general access structures, the Ogata and Kurosawa's secure scheme can be extended only to linear secret sharing schemes with $E_0 = \mathbf{Z}_q$, where $q = q_1^2 + q_1 + 1$ is a prime number and q_1 is a prime power. That is, it can be applied only to linear schemes with $\dim(E_0) = 1$. In many access structures, linear schemes with $\dim(E_0) > 1$ have better information rate than those with $\dim(E_0) = 1$. In this case, our method will provide a secure scheme with better information rate.

References

1. G.R. Blakley. Safeguarding cryptographic keys. *AFIPS Conference Proceedings* 48 (1979) 313–317.
2. C. Blundo and A. De Santis. Lower Bounds for Robust Secret Sharing Schemes. *Information Processing Letters* 63 (1997) 317–321.
3. C. Blundo, A. De Santis, L. Gargano and U. Vaccaro. Tight Bounds on the Information Rate of Secret Sharing Schemes. *Designs, Codes and Cryptography* 11 (1997) 107–122.
4. C. Blundo, A. De Santis, D.R. Stinson and U. Vaccaro. Graph Decompositions and Secret Sharing Schemes. *J. Cryptology* 8 (1995) 39–64.
5. E.F. Brickell. Some ideal secret sharing schemes. *J. Combin. Math. and Combin. Comput.* 9 (1989) 105–113.
6. E. F. Brickell and D. M. Davenport. On the Classification of Ideal Secret Sharing Schemes. *J. Cryptology*, 4 (1991) 123–134.
7. M. Carpentieri, A. De Santis and U. Vaccaro. Size of shares and probability of cheating in threshold schemes. *Advances in Cryptology, EUROCRYPT 93*, Lectures Notes in Computer Science 765, Springer-Verlag (1994) 118–125.
8. W. Ogata and K. Kurosawa. Optimum Secret Sharing Scheme Secure against Cheating. *Advances in Cryptology, EUROCRYPT 96, Lecture Notes in Computer Science* **1070** (1996) 200–211.
9. C. Padró. Robust vector space secret sharing schemes. *Information Processing Letters* **68**(1998) 107–111.
10. C. Padró and G. Sáez. Secret sharing schemes with bipartite access structure. *Advances in Cryptology, EUROCRYPT'98, Lecture Notes in Computer Science* **1403** (1998) 500–511.

11. C. Padró, G. Sáez and J.L. Villar. Detection of cheaters in vector space secret sharing schemes. *Designs, Codes and Cryptography* **16** (1999) 75–85.
12. J. Rifà-Coma. How to avoid cheaters succeeding in the key sharing scheme. *Designs, Codes and Cryptography* **3** (1993) 221–228.
13. A. Shamir. How to share a secret. *Commun. of the ACM* **22** (1979) 612–613.
14. G.J. Simmons. An Introduction to Shared Secret and/or Shared Control Schemes and Their Application. *Contemporary Cryptology. The Science of Information Integrity*. IEEE Press (1991) 441-497.
15. G.J. Simmons, W. Jackson and K. Martin. The geometry of secret sharing schemes. *Bulletin of the ICA* 1 (1991) 71–88.
16. D.R. Stinson. An explication of secret sharing schemes. *Designs, Codes and Cryptography* 2 (1992) 357–390.
17. D.R. Stinson. Decomposition Constructions for Secret-Sharing Schemes. *IEEE Trans. on Information Theory* 40 (1994) 118–125.
18. D.R. Stinson. *Cryptography: Theory and Practice*. CRC Press Inc., Boca Raton (1995).
19. M. Tompa and H. Woll. How to share a secret with cheaters. *J. Cryptology* 1 (1988) 133–139.

Constructive Notes on Uniform and Locally Convex Spaces

Luminiţa Dediu[1] and Douglas Bridges[2]

[1] Department of Mathematics and Statistics, University of Canterbury, Christchurch, New Zealand
E-mail: lde15@student.canterbury.ac.nz; ldediu@math.ugal.ro
[2] Department of Mathematics and Statistics, University of Canterbury, Christchurch, New Zealand

Abstract. Some elementary notions in the constructive theory of uniform and locally convex spaces are introduced, and a number of basic results established. In particular, it is shown that if the unit ball of a locally convex space X is totally bounded, then so is the intersection of that ball with the kernel of any nonzero continuous linear functional on X.

1 Uniform Spaces

Although Errett Bishop considered that "in most cases of interest it seems to be unnecessary to make use of any deep facts from the general theory of locally convex spaces", recent developments in constructive analysis (in particular, operator algebra theory) increasingly depend on such a theory. In turn, that theory draws on the general theory of uniform spaces, the beginnings of which were outlined in Problems 17–21 on pages 110–111 of [2]. (Some basic definitions in the theory of locally convex spaces also appear in Chapter 8 of [9].)

By **constructive mathematics** we mean mathematics carried out with intuitionistic logic, without any restriction on the type of objects considered (see [12]). By using this logic, we obtain results that not only hold classically but also can be reinterpreted in any reasonable model for computable analysis, such as recursive mathematics [11] or Weihrauch's TTE ([16], see also [14]). Moreover, intuitionistic logic facilitates the recognition of certain distinctions of meaning that are obscured by classical logic. For background material on constructive mathematics, see [1], [2], [3], [6], [15].

We now introduce the basic terminology and establish some fundamental facts about uniform spaces; in general, we do not define notions, or prove facts, that carry over unchanged from the classical to the constructive setting.

Definition 1. *A* ***uniform space*** *is a set X together with a family $(\rho_i)_{i\in I}$ of pseudometrics on X. The* ***equality*** *and* ***inequality*** *on X are defined, respectively, as follows:*

$$x = y \text{ if and only if } \forall i \in I\ (\rho_i(x,y) = 0),$$
$$x \neq y \text{ if and only if } \exists i \in I\ (\rho_i(x,y) > 0).$$

The corresponding **uniform topology** *on X is the topology in which, for each $x_0 \in X$, the sets*

$$V(x_0, F, \varepsilon) = \left\{ x \in X : \sum_{i \in F} \rho_i(x, x_0) < \varepsilon \right\},$$

with $\varepsilon > 0$ and F a finitely enumerable[1] subset of I, form a basis of neighbourhoods of x_0; the pseudometrics ρ_i are called the **defining pseudometrics** *for this topology.*

Metric and normed spaces are viewed as uniform spaces in the obvious way. For our purposes, a more important type of uniform space is a **locally convex space**, which consists of a linear space X over $\mathbb{F}$ ($\mathbb{R}$ or $\mathbb{C}$), together with a family $(p_i)_{i \in I}$ of seminorms for which the corresponding family $\mathcal{F}$ of pseudometrics $(x, y) \mapsto p_i(x - y)$ defines the topology (and, incidentally, the inequality) on X. In this case we refer to the seminorms p_i as **defining seminorms** for the **locally convex topology**—that is, the uniform topology defined by $\mathcal{F}$; and we call the set

$$X_1 = \{x \in X : \forall i \in I \ (p_i(x) \leq 1)\}$$

the **unit ball** of the locally convex space X.

In the rest of this section, unless we specify otherwise, $\left(X, (\rho_i)_{i \in I}\right)$ and $\left(Y, (\sigma_j)_{j \in J}\right)$ are uniform spaces.

Definition 2. *A mapping $f : X \to Y$ is* **uniformly continuous** *on X if for each $\varepsilon > 0$ and each finitely enumerable subset G of J there exist $\delta > 0$ and a finitely enumerable subset F of I, such that if $x, y \in X$ and $\rho_i(x, y) < \delta$ for all $i \in F$, then $\sigma_j(f(x), f(y)) < \varepsilon$ for all $j \in G$.*

Each defining pseudometric ρ_i is uniformly continuous.

Definition 3. *A subset S of X is bfseries totally bounded with respect to the subset F of I if for each $\varepsilon > 0$ there exists a finitely enumerable subset S_ε of S such that for each $x \in S$ there exists $s \in S_\varepsilon$ with $\sum_{i \in F} \rho_i(x, s) < \varepsilon$. The set S_ε is called a* **finitely enumerable** *ε***-approximation** *to S relative to F. If S is totally bounded with respect to each finitely enumerable subset of I, then we say that S is* **totally bounded.**

We omit the proofs of the next proposition and its corollary, since they are very similar to those of the corresponding results on page 94 of [3].

Proposition 1. *If X is totally bounded and $f : X \to Y$ is uniformly continuous, then $f(X)$ is totally bounded.*

---

[1] A set A is said to be **finitely enumerable** (respectively, **finite**) if there exist a nonnegative integer n and a mapping (respectively, one–one mapping) of $\{1, \ldots, n\}$ onto A.

Many of the most important results in classical analysis depend on the least–upper–bound principle, which, in its full classical form, is not constructive. Fortunately, we have the following **constructive least–upper–bound principle**:

> Let S be a nonempty subset of $\mathbb{R}$ that is bounded above; then $\sup S$ exists if and only if for all real numbers α, β with $\alpha < \beta$, either β is an upper bound for S or else there exists $x \in S$ such that $x > \alpha$ ([3], page 37, Proposition (4.3)).

Using this as on page 94 of [3], we obtain one important application of total boundedness.

Corollary 1. *A uniformly continuous mapping of a uniform space X into $\mathbb{R}$ has a supremum and an infimum.*

The notion of a located subset—one from which one can measure the distance to any point of the ambient space—plays a very significant role in the constructive theory of metric spaces. We now generalise that notion to the present context.

Definition 4. *A subset S of X is **located** if*

$$\inf\left\{\sum_{i\in F}\rho_i(x,y) : y \in S\right\}$$

exists for each $x \in X$ and each finitely enumerable subset F of I.

It follows from the constructive least–upper–bound principle that S is located if and only if for each $x \in X$, each finitely enumerable subset F of I, and all real numbers α, β with $0 \le \alpha < \beta$,

either $\sum_{i\in F}\rho_i(x,y) > \alpha$ for all $y \in S$
or else there exists $y \in S$ such that $\sum_{i\in F}\rho_i(x,y) < \beta$.

Proposition 2. *A totally bounded subset of a uniform space is located.*

Proof. Consider a totally bounded subset S of X. Let $x \in X$, let F be a finitely enumerable subset of I, and let $0 \le \alpha < \beta$. Writing $\varepsilon = \frac{1}{2}(\beta - \alpha)$, construct a finitely enumerable ε-approximation $\{s_1, \ldots, s_n\}$ to S relative to F. Let

$$d = \inf\left\{\sum_{i\in F}\rho_i(x,s_k) : 1 \le k \le n\right\},$$

which exists as the infimum of a finitely enumerable subset of $\mathbb{R}$. Either $d > \alpha+\varepsilon$ or $d < \beta$. In the first case, given $y \in S$ and choosing k $(1 \le k \le n)$ such that $\sum_{i\in F}\rho_i(y,s_k) < \varepsilon$, we have

$$\sum_{i\in F}\rho_i(x,y) \ge \sum_{i\in F}\rho_i(x,s_k) - \sum_{i\in F}\rho_i(y,s_k) > d - \varepsilon > \alpha \; .$$

In the second case, there exists k $(1 \le k \le n)$ such that $\sum_{i\in F}\rho_i(x,s_k) < \beta$.

□

Proposition 3. *A located subset of a totally bounded uniform space is totally bounded.*

Proof. Assuming that X is totally bounded, let S be a located subset of X. Given $\varepsilon > 0$ and a finitely enumerable subset F of I, choose a finitely enumerable $\frac{\varepsilon}{3}$–approximation $\{x_1, \ldots, x_n\}$ to X. Since S is located, we can write $\{1, \ldots, n\}$ as a union of subsets P, Q such that

if $k \in P$, then $\sum_{i \in F} \rho_i(s, x_k) > \varepsilon/3$ for all $s \in S$, and
if $k \in Q$, then there exists $s \in S$ such that $\sum_{i \in F} \rho_i(s, x_k) < 2\varepsilon/3$.

For each $k \in Q$ choose $s_k \in S$ such that $\sum_{i \in F} \rho_i(s_k, x_k) < 2\varepsilon/3$. Given $s \in S$, choose k $(1 \leq k \leq n)$ such that $\sum_{i \in F} \rho_i(s, x_k) < \varepsilon/3$. Then $k \in Q$ and so

$$\sum_{i \in F} \rho_i(s, s_k) \leq \sum_{i \in F} \rho_i(s, x_k) + \sum_{i \in F} \rho_i(x_k, s_k) < \varepsilon .$$

Thus $\{s_k : k \in Q\}$ is a finitely enumerable ε–approximation to S. □

We omit the proofs of the next three results, since they are simple adaptations of (4.7) on page 30 of [6], (4.8) on page 31 of [6], and (4.9) on page 98 of [3], respectively.

Theorem 1. *Let (E, ρ) be a totally bounded pseudometric space, x_0 a point of E, and r a positive number. Then there exists a closed, totally bounded subset K of E such that $B(x_0, r) \subset K \subset B(x_0, 8r)$.*

Corollary 2. *If E is a totally bounded pseudometric space, then for each $\varepsilon > 0$ there exist totally bounded sets $K_1, \ldots, K_n$, each of diameter less than ε, such that $E = \bigcup_{i=1}^{n} K_i$.*

Proposition 4. *Let f be a uniformly continuous mapping on a totally bounded subset S of a pseudometric space E. Then for all but countably many real numbers $t > m = \inf\{f(x) : x \in S\}$ the set*

$$S_t = \{x \in E : |f(x)| \leq t\}$$

is totally bounded; in other words, there exists a sequence $(t_n)_{n=1}^{\infty}$ in the interval (m, ∞) such that S_t is totally bounded whenever $t > m$ and $t \neq t_n$ for each n.

2 Continuous Linear Functionals on Locally Convex Spaces

In the rest of this paper, X denotes a locally convex space with its topology defined by the family $(p_i)_{i \in I}$ of seminorms. Our main result, Theorem 2, identifies certain useful totally bounded subsets of the kernel of a continuous linear functional on X. This requires some preliminaries; for the first of these, we recall that a mapping f between linear spaces is **homogeneous** if $f(\lambda x) = \lambda f(x)$ for all scalars λ and vectors x.

Proposition 5. *Let (E,p) be a seminormed space, and let S be a balanced, totally bounded subset of E. If $f : E \to \mathbb{F}$ is a homogeneous mapping, uniformly continuous on S, then for each $t > 0$ the set*

$$S_t = \{x \in S : |f(x)| \leq t\}$$

is totally bounded.

Proof. Since S is balanced, it contains 0, and therefore $\inf\{|f(x)| : x \in S\} = 0$. Being totally bounded, S is bounded: there exists $M > 0$ such that $p(x) \leq M$ for all $x \in S$. Let $t > 0$ and $0 < \varepsilon < 1$. By Proposition 4, there exists $t' < t$ such that

$$\frac{t}{t'} < \min\left\{2, 1 + \frac{\varepsilon}{2M}\right\}$$

and $S_{t'}$ is totally bounded. Let $\{x_1, \ldots, x_n\}$ be an $\frac{\varepsilon}{4}$–approximation to $S_{t'}$. If $x \in S_t$, then $\frac{t'}{t}x \in S_{t'}$, so there exists j $(1 \leq j \leq n)$ such that

$$p\left(\frac{t'}{t}x - x_j\right) < \frac{\varepsilon}{4}.$$

Then

$$p\left(x - \frac{t}{t'}x_j\right) < \frac{t}{t'}\frac{\varepsilon}{4} < \frac{\varepsilon}{2}$$

and so

$$\begin{aligned} p(x - x_j) &\leq p\left(x - \frac{t}{t'}x_j\right) + p\left(\left(\frac{t}{t'} - 1\right)x_j\right) \\ &< \frac{\varepsilon}{2} + \left(\frac{t}{t'} - 1\right)p(x_j) \\ &< \frac{\varepsilon}{2} + \left(\frac{t}{t'} - 1\right)M < \frac{\varepsilon}{2} + \frac{\varepsilon}{2} = \varepsilon. \end{aligned}$$

Thus the set $\{x_1, \ldots, x_n\}$ is a finitely enumerable ε–approximation to S_t. □

The following criterion for the continuity of linear functionals in terms of families of defining seminorms enables us to show that in a locally convex space the sets S_t of Proposition 5 are totally bounded with respect to any finitely enumerable family of defining seminorms. (Here, *continuity* means continuity at each point of X, relative to the topology associated with the defining family of seminorms on X.)

Proposition 6. *A linear functional f on the locally convex space X is continuous if and only if there exist a positive real number C and a finitely enumerable subset F of I such that*

$$|f(x)| \leq C \sup_{i \in F} p_i(x) \tag{1}$$

for each $x \in X$.

Proof. We include the (slightly adapted) well-known argument for the sake of completeness. Since f is continuous and $f(0)=0$, the set $\{x \in X : |f(x)| < 1\}$ is open in X; so there exist $\delta > 0$ and a finitely enumerable subset F of I such that if $\sum_{i\in F} p_i(x) < \delta$, then $|f(x)| < 1$. It follows that for each $x \in X$ and each $\varepsilon > 0$,

$$\left| f\left(\frac{\delta x}{\sum_{i\in F} p_i(x) + \varepsilon}\right)\right| < 1$$

and therefore

$$|f(x)| < \delta^{-1}\left(\sum_{i\in F} p_i(x) + \varepsilon\right) .$$

Since $\varepsilon > 0$ is arbitrary, we see that (1) holds with $C = \delta^{-1}$. □

In the presence of linearity we can improve Proposition 4 substantially.

Proposition 7. *Let f be a continuous linear functional on the locally convex space X, and S a totally bounded subset of X. Then for all $t > 0$ the sets*

$$S_t = \{x \in S : |f(x)| \leq t\}$$

are totally bounded.

Proof. Choose a finitely enumerable subset F of I such that (1) holds for some $C > 0$ and all $x \in X$, and let G be an arbitrary finitely enumerable subset of I. Since

$$|f(x)| \leq C \sum_{i\in F\cup G} p_i(x) \quad (x \in X),$$

f is uniformly continuous with respect to the seminorm $\sum_{i\in F\cup G} p_i$ on X. It follows from Proposition 5 that for each $t > 0$ the set S_t is totally bounded with respect to $F \cup G$. Given $\varepsilon > 0$, choose a finitely enumerable ε-approximation $\{x_1, \ldots, x_n\}$ to S_t relative to $F \cup G$. Then for each $x \in S_t$ we have

$$\sum_{i\in G} p_i(x - x_j) \leq \sum_{i\in F\cup G} p_i(x - x_j) < \varepsilon$$

for some j $(1 \leq j \leq N)$. Since $\varepsilon > 0$ is arbitrary, we conclude that S_t is totally bounded relative to G. □

Theorem 2. *Let $\left(X, (p_i)_{i\in I}\right)$ be a locally convex space, S a balanced, convex, totally bounded subset of X, and f a nonzero linear functional on X that is uniformly continuous on S. Then $S \cap \ker f$ of f is totally bounded.*

Proof. By Corollary 1, the real number

$$C = \sup\{|f(x)| : x \in S\}$$

exists. Since f is nonzero, we can choose $x \in S$ with $|f(x)| > \frac{C}{2}$. Then

$$x_0 = \frac{C}{2|f(x)|}x$$

belongs to S, and $|f(x_0)| = \frac{C}{2}$. Let ε be a positive number, F a finitely enumerable subset of I, and t a positive number such that

$$0 < t < \frac{\varepsilon}{1 + 4C^{-1}|F|},$$

where $|F|$ is the cardinality of F. Since (by Proposition 7)

$$S_t = \{x \in S : |f(x)| \leq t\}$$

is totally bounded, it has a finitely enumerable t-approximation $\{s_1, \ldots, s_n\}$ relative to F. Setting

$$x_k = \left(1 + 2C^{-1}t\right)^{-1}\left(s_k - 2C^{-1}f(s_k)x_0\right) \quad (1 \leq k \leq n),$$

we have $x_k \in \ker f$. Also, for each $i \in I$,

$$\begin{aligned} p_i(x_k) &\leq \left(1 + 2C^{-1}t\right)^{-1}\left(p_i(s_k) + 2C^{-1}|f(s_k)|\, p_i(x_0)\right) \\ &\leq \left(1 + 2C^{-1}t\right)^{-1}\left(1 + 2C^{-1}t\right) \\ &= 1, \end{aligned}$$

so x_k belongs to $S \cap \ker f$. Now consider any element x of $S \cap \ker f$. Since $x \in S_t$, there exists k $(1 \leq k \leq n)$ such that $\sum_{i \in F} p_i(x - s_k) < t$ and therefore

$$\begin{aligned} \sum_{i \in F} p_i(x - x_k) &\leq \sum_{i \in F} p_i(x - s_k) + \sum_{i \in F} p_i(s_k - x_k) \\ &\leq t + 2\left(C + 2t\right)^{-1} \sum_{i \in F} p_i(ts_k + f(s_k)x_0) \\ &\leq t + 2C^{-1}t \sum_{i \in F} \left(p_i(s_k) + p_i(x_0)\right) \\ &\leq t(1 + 4C^{-1}|F|) \\ &< \varepsilon. \end{aligned}$$

Thus, relative to F, the set $\{x_1, \ldots, x_n\}$ is a finitely enumerable ε-approximation to $S \cap \ker f$. □

Under the hypotheses of Theorem 2, if $f = 0$, then $S \cap \ker f$ equals X_1 and is totally bounded. The following Brouwerian example shows that we cannot expect to prove that $S \cap \ker f$ is totally bounded unless we know that $f = 0$ or $f \neq 0$.

Let $a \in \mathbb{R}$, and define a linear functional $f : \mathbb{R} \to \mathbb{R}$ by $f(x) = ax$. Then f is bounded—it has norm equal to a—and the unit ball $[-1, 1]$ of $\mathbb{R}$ is balanced, convex, and totally bounded. Suppose that $K = [-1, 1] \cap \ker f$ is totally bounded, so that $s = \sup K$ exists. Either $s > 0$ or $s < 1$. In the first case there exists $x \neq 0$ in $\mathbb{R}$ with $f(x) = 0$, so $a = 0$; in the second we have $\neg (f(1) = 0)$, so $\neg (a = 0)$. Thus Theorem 2 without the hypothesis $f = 0 \vee f \neq 0$ implies that

$$\forall x \in \mathbb{R}\ (x = 0 \vee \neg (x = 0)),$$

a statement that is known to be essentially nonconstructive.

Theorem 2 has an important application in the context of a Hilbert space H. In this application, $X = \mathcal{B}(H)$ is the space of all bounded operators H, and the defining seminorms are the mappings of the form $T \mapsto \langle Tx, y \rangle$ with $x, y \in H$; the corresponding locally convex topology is called the **weak–operator topology** τ_w on $\mathcal{B}(H)$. The unit ball of $\mathcal{B}(H)$ is the set

$$\mathcal{B}_1(H) = \{T \in \mathcal{B}(H) : \forall x \in H \ (\|Tx\| \leq \|x\|)\},$$

which is τ_w–totally bounded.

We use Theorem 2 to avoid invoking an unnecessary separability hypothesis in the following result.

Theorem 3. *Let H be a Hilbert space, $\mathcal{R}$ a linear subset of $\mathcal{B}(H)$ whose unit ball$\mathcal{R}_1 = \mathcal{R} \cap \mathcal{B}_1(H)$ is τ_w–totally bounded, and f a linear functional on $\mathcal{R}$ that is τ_w-uniformly continuous on $\mathcal{R}_1$. Then f extends to a linear functional on $\mathcal{B}(H)$ that is τ_w-uniformly continuous on $\mathcal{B}_1(H)$ and has the form*

$$f(T) = \sum_{n=1}^{\infty} \langle Tx_n, y_n \rangle$$

with (x_n), (y_n) elements of the direct sum $\bigoplus_{n=1}^{\infty} H$ of a sequence of copies of H.

The proof of this theorem is found in [5].

References

1. Beeson, M.: *Foundations of Constructive Mathematics* . Springer-Verlag Heidelberg (1985)
2. Bishop, E.: *Foundations of Constructive Analysis.* McGraw-Hill New York (1967)
3. Bishop, E., Bridges, D.: *Constructive Analysis.* Grundlehren der math. Wissenschaften **279** Springer-Verlag (1985)
4. Bridges, D.: *Constructive Mathematics—its Set Theory and Practice.* D.Phil. Thesis, Oxford University (1975)
5. Bridges, D., Dediu, L.: Constructing Extensions of Ultraweakly Continuous Linear Functionals. University of Canterbury New Zealand (preprint)
6. Bridges, D., Richman, F.: *Varieties of Constructive Mathematics.* Cambridge University Press (1987)
7. Dieudonné, J.: *Foundations of Modern Analysis.* Academic Press New York (1960)
8. Holmes, R. B.: *Geometric Functional Analysis and its Applications.* Springer-Verlag New York (1975)
9. Hajime Ishihara, *Normability and Compactness of Constructive Linear Mappings.* Ph.D. Thesis, Tokyo Institute of Technology (1990)
10. Kadison, R.V., Ringrose, J. R.: *Fundamentals of the Theory of Operator Algebras.* **Vol.1** Academic Press New York (1983)
11. Kushner, B. A.: *Lectures on Constructive Mathematical Analysis.* Amer. Math. Soc. Providence RI (1985)
12. Richman, F.: Intuitionism as a Generalization. Philosophia Math. **5** (1990) 124–128

13. Rudin, W.: *Functional Analysis*. Second Edition McGraw-Hill New York (1991)
14. Troelstra, A. S.: Comparing the theory of representations and constructive mathematics. In *Computer Science Logic: Proceedings of the 5th Workshop, CSL'91, Berne Switzerland* (E. Börger, G. Jäger, H. Kleine Büning, and M.M. Richter, eds), Lecture Notes in Computer Science **626** Springer–Verlag Berlin (1992) 382–395
15. Troelstra, A. S., van Dalen, D.: *Constructivity in Mathematics: An Introduction* (two volumes). North Holland Amsterdam (1988)
16. Weihrauch, K.: *Computability*. Springer-Verlag Heidelberg (1987)

Graph Automorphisms with Maximal Projection Distances

H.N. de Ridder and H.L. Bodlaender

Department of Computer Science, Utrecht University, P.O. Box 80.089, 3508 TB Utrecht, the Netherlands, email: hnridder@cs.uu.nl, and hansb@cs.uu.nl

Abstract. This paper introduces diametrical graph automorphisms. A graph automorphism is called diametrical if it has the property that the distance between each vertex and its image is equal to the diameter of the graph. The structure of diametrically automorphic graphs is examined. The complexity of recognizing these graphs is shown to be NP-complete in general, while efficient algorithms for cographs and circular arc graphs are developed. The notion of distance lower bounded automorphism is introduced in order to apply the results on diametrical automorphisms to a wider range of automorphisms.

1 Introduction

The problem to decide whether a given graph possesses a non-trivial automorphism is well studied (see [3] for a survey). In this paper a specific kind of automorphisms, called diametrical automorphisms, are studied. These are automorphisms satisfying an additional requirement, which is defined below.

The following notation and terminology is used: A graph G with set of vertices $V(G)$ and set of edges $E(G) \subset V \times V$ is written $G = (V(G), E(G))$. Only connected simple graphs (undirected, no multi-edges, no loops) are considered. The distance between two vertices v and w is written $d(v, w)$, possibly using a subscript to specify the graph in which the distance is measured, eg. $d_G(v, w)$. The diameter of a graph G is denoted $d(G)$. The *(closed) neighbourhood* of v is $\mathrm{N}[v] = \{v\} \cup \{w \mid (v, w) \in E\}$. Disjoint union is written $\uplus$. Two vertices v and w are said to be in the same *cell* if there is an automorphism such that $\pi.v = w$. An automorphism is said to be *stable* on a set of vertices if it projects the set onto itself.

Definition 1. *A graph G is said to be* diametrically automorphic *or D.A. if there exists an automorphism π on G with the property that $d(v, \pi.v) = d(G)$ for all vertices v. π is then called a diametrical automorphism.*

For some classes of graphs known from the literature all members are D.A. Examples are: the cycles C_n; the circulants $C_n(a_1, a_2, \ldots, a_r)$; the complete graphs K_n; the complete multipartite graphs $K_{n_0,\ldots,n_r}$, with $n_i \geq 2$ for all $0 \leq i \leq r$; the octahedra $O_n = \overline{nK_2}$.

Classes for which no member is D.A. include trees with more than 2 vertices and chordal graphs (except K_n). The latter fact is established in Cor. 5.

The related notions of fixed-point-free and (co)affine automorphisms have been described before ([12] and [14, 13], respectively) and are defined as:

Definition 2. *Let π be an automorphism on a graph $G = (V, E)$. Then*
(i) π is said to be fixed-point-free *when $\forall v \in V : v \neq \pi.v$.*
(ii) π is said to be affine *when $\forall v \in V : (v, \pi.v) \in E$.*
(iii) π is said to be coaffine *when π is affine on $\overline{G}$. Then $\forall v \in V : d(v, \pi.v) > 1$.*

Just as for D.A. graphs, a graph is called fixed-point-free etc. if there is an automorphism on the graph that is fixed-point-free etc. These definitions can be unified by introducing the notion of distance lower bounded automorphisms.

Definition 3. *An automorphism π on a graph G is said to be* k distance lower bounded *or dl_k, if $d(v, \pi.v) \geq k$ for all vertices v.*

Then D.A. is $dl_{d(G)}$, coaffine is dl_2 and fixed-point-free is dl_1. Furthermore every automorphism is dl_0 and no automorphism is $dl_{d(G)+1}$. Obviously, if π is dl_{k+1}, then π is dl_k. Similarly, the class of dl_{k+1} graphs is a subset of the class of dl_k graphs.

Note however, that D.A.-ness does not truly fit into this hierarchy: A lower bound on $d(v, \pi.v)$ is fixed over all graphs in a dl_k automorphism, whereas it depends on the graph for a diametrical automorphism.

2 The Structure of D.A. Graphs

An obvious constraint on the structure of a D.A. graph is that the graph should be equal to its center – it is self-centered or equi-eccentric – and hence 2-connected. Next, define the *eccentric graph* $A(G)$ of G as follows: $A(G)$ has the same set of vertices as G, and $(u, v) \in E(A(G))$ iff $d_G(u, v) = d(G)$. If π is a D.A. on G, then $\pi.v, \pi^2.v, \ldots, \pi^{|v|}.v$ is a cycle in $A(G)$. Moreover, every vertex of G is part of precisely one such cycle, so a diametrical automorphism on G induces a cycle cover on $A(G)$; the converse does not hold. The following theorem gives more information on the cycles of G itself.

Theorem 4. *Let G be a graph with diameter d, let π be a dl_k automorphism on G with $k > 0$, and let v be a vertex of G. Then:*
(i) A chordless cycle of length at least $2k$ exists containing both v and $\pi.v$.
(ii) If $k = d$, then a cycle of length $2d$ or $2d + 1$ exists containing both v and $\pi.v$.

Proof. (i) Let p be a shortest path from v to $\pi.v$. Then $p, \pi.p, \pi^2.p, \ldots$ is on a simple cycle c, with c having length at least $2|p| \geq 2k$. If c does not contain a chord, then the theorem becomes true.

If c does contain a chord (r, s), then the distance between r and s measured along c is at least $2k - 1$: If it were less, then $d(r, \pi.r) < k$. Hence, there is a l such that both $\pi^l.v$ and $\pi^{l+1}.v$ are on the cycle $r, \ldots, s, r$, with $r, \ldots, s$ on c.

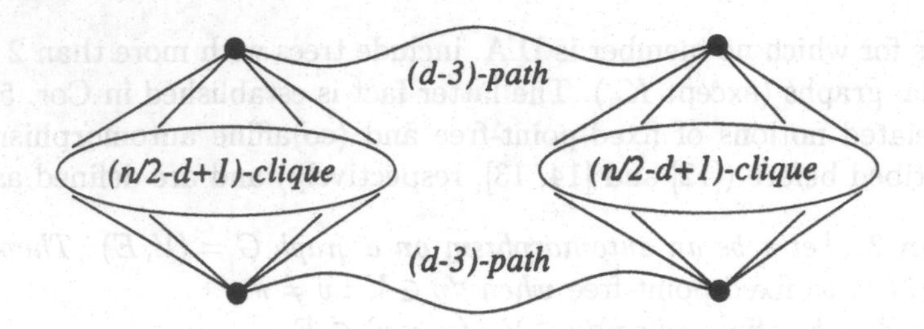

Fig. 1. A conjectured edge-maximal D.A. graph.

Let (x, y) be a chord of c, with the distance between x and y measured along c minimal over all chords. By this minimality, the cycle $x, \ldots, y, x$ is chordless. By the previous paragraph, it has length at least $2k$ and contains both $\pi^l.v$ and $\pi^{l+1}.v$, for some l. The theorem follows. □

Corollary 5. *(i) The only chordal graphs that are diametrically automorphic are the complete graphs.*
(ii) For every automorphism π on a chordal graph, there is a vertex v such that either $v = \pi.v$ or $(v, \pi.v) \in E$.

Proposition 6 ([4]). *Let e be the number of edges of a D.A. graph with n vertices and diameter $d > 1$, then: $e \geq (nd - 2d - 1)/(d - 1)$*

Theorem 7. *Let e be the number of edges of a D.A. graph with n vertices and diameter $d > 1$, then:*

$$e \leq \begin{cases} \frac{n}{2}(n-2) & \text{if } d = 2 \\ \frac{n}{2}\left(\frac{n}{2} - d + 2\right) & \text{if } d \geq 3 \end{cases}$$

Proof. The number of edges is at most $n/2$ times the maximal degree of a vertex. An upper bound on the degree of a vertex v can be established by picking a shortest cycle (of length $2d$ or $2d + 1$) through v and $\pi.v$ and distributing all vertices that are not on this cycle evenly among v and $\pi.v$. □

The lower bound of Prop. 6 is the minimal number of edges in a self-centered graph. The upper bound of Th. 7 is straightforward, but for $d \lll n$ still better than the general $(n^2 - 4nd + 5n - 4d^2 - 6d)/2$ upper bound for self-centered graphs [4]. Although the upper bound of Th. 7 is tight for $d \leq 3$, it is not expected to be tight for $d > 3$, because it may not be possible for all vertices to be of maximal degree. The following conjecture gives an upper bound that is expected to be tight.

Conjecture 8. Let e be the number of edges of a D.A. graph with n vertices and diameter $d \geq 3$, then:

$$e \leq \left(\frac{n}{2} - d + 1\right)\left(\frac{n}{2} - d\right) + 2(n - d)$$

The motivation for this bound is the following:

When partitioning n vertices into cliques such that the total number of edges is maximal, it is beneficial to create as few cliques as possible and make these cliques as big as possible. Furthermore, every clique is vertex-disjoint to its image under a diametrical automorphism. Hence, for every k, a D.A. graph contains either at least two cliques of size k, or none. Combining these two observations, construct a graph that consists of two cliques of maximal size, connected by two paths, see Fig. 1. This graph contains $2(n/2-d+1)+4$ vertices of the maximal possible degree $n/2-d+2$ and $2(d-3)$ vertices of the minimal possible degree 2. This gives the conjectured bound.

3 Recognizing D.A. Graphs

Define the D.A. problem as the problem of deciding whether a given graph possesses a diametrical automorphism. In this section it is proved that the D.A. problem is NP-complete. Next, two algorithms are given that can be used to solve the problem in polynomial time for a specific class of graphs.

First, it is proved that the D.A. problem is NP-complete. This is done by reducing the fixed-point-free automorphism problem [12] to the D.A. problem: A graph G' is constructed from G in such a way that G' has a diametrical automorphism iff G has a fixed-point-free automorphism.

Definition 9. *Let $G = (V, E)$ be a graph. Then the graph $G' = (V', E')$ is defined by*

$$V' = V \cup E \cup \{t_V, u_V\} \cup \{t_G, u_G\} \tag{1}$$
$$E' = \{(v_i, e_j) \mid e_j \in E \text{ is incident to } v_i \in V \text{ in } G\} \tag{2}$$
$$\cup \{(t_G, v), (u_G, v) \mid v \in V \cup E\} \tag{3}$$
$$\cup \{(t_V, v), (u_V, v) \mid v \in V\} \tag{4}$$

The construction is the following: Each edge in G is split in two, adding a vertex in the middle (2). Every vertex in this subdivision graph of G is connected to new vertices t_G and u_G (3). Furthermore every vertex that was present in G is connected to new vertices t_V and u_V (4). The resulting graph is depicted in Fig. 2. This construction ensures (a) that the distance in G' between any two vertices in V is equal to $d(G') = 2$, and (b) that π is an automorphism of G' iff the restriction of π to V is an automorphism of G.

Theorem 10. *Recognizing D.A. graphs is NP-complete.*

Proof. Note that D.A. is in NP. We claim that G' constructed as in Def. 9 is D.A. iff G is fixed-point-free. This proves NP-completeness of D.A, because recognizing fixed-point-free graphs is NP-complete [12], and G' can be constructed in polynomial time.

When examining the automorphism partition of G' it becomes apparent that $\{t_G, u_G\}$ form a cell because they have the same set of neighbours $V \cup E$ and

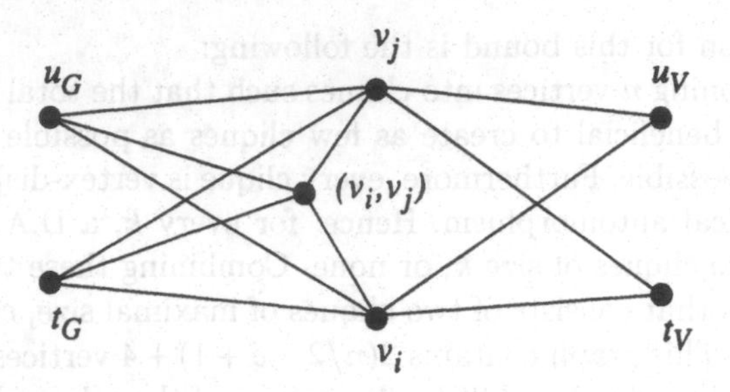

Fig. 2. The graph G'.

no other vertex in G' has degree $|V| + |E|$. Furthermore, $\{t_V, u_V\}$ form a cell because these vertices have the same set of neighbours V and they are the only vertices in $V(G') \setminus \{t_G, u_G\}$ that are not adjacent to t_G, u_G. Because of this cell every automorphism of G' must be stable on both V and E. Hence, if π' is defined by

$$\begin{array}{rlr} & \pi'.v = \pi.v & v \in V \\ & \pi'.(u,v) = (\pi.u, \pi.v) & (u,v) \in E \\ \pi'.t_V = u_V & \pi'.u_V = t_V & \\ \pi'.t_G = u_G & \pi'.u_G = t_G & \end{array}$$

then π' is an automorphism on G' iff π is an automorphism on G. Examination of the distances between vertices in G' shows that π' is D.A. iff $\forall v \in V : \pi.v \neq v$, in other words iff π is fixed-point-free. □

Because $d(G') = 2$, this also proves that recognizing coaffine graphs is NP-complete. The general dl_k problem can be proved NP-complete through a simple reduction:

Theorem 11. *Recognizing dl_k graphs is NP-complete.*

Proof. Let G be a graph with n vertices and let $3 \leq k \leq d(G) - 1$. The cases $k = 1$, $k = 2$ and $k = d(G)$ are already proved. Create a graph G' by connecting an instance of the gadget of Fig. 3 to each pair of vertices u, v with the property that $d_G(u,v) = d(G)$, then: $d_G(u,v) = d(G) \Longrightarrow d_{G'}(u,v) = k$.

Because G cannot contain an n-clique, any automorphism of G' permutes these gadgets among themselves. It follows that G' is dl_k iff G is D.A. □

3.1 Diametrically Automorphic Cographs

The *Zykov-join* $G * H$ of G and H is formed taking the disjoint union of G and H and adding all possible edges between G and H. Cographs are defined as:

Definition 12. *The class of cographs is defined recursively as: (i) K_1 is a cograph. (ii) If $G_i, i \in I$ are cographs, then $\ast_{i \in I} G_i$ is a cograph. (iii) If $G_i, i \in I$ are cographs, then $\biguplus_{i \in I} G_i$ is a cograph.*

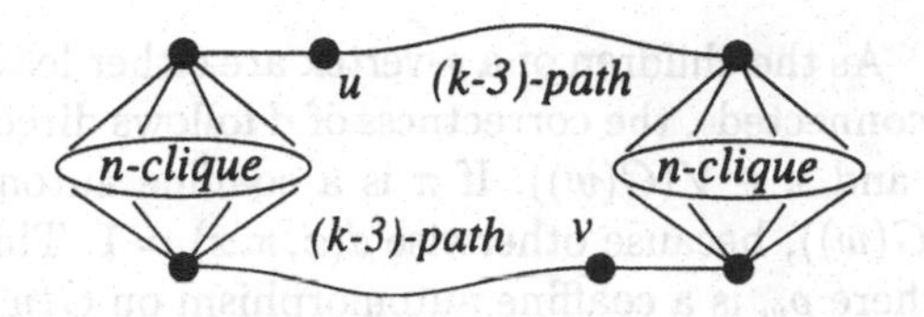

Fig. 3. A gadget to reduce dl_k automorphism to D.A.

Proposition 13. *Let* $G = \ast_{i \in I} G_i$ *with* $|I| \geq 2$, *then* $d(G) = 1$ *if* $\forall i \in I$: $d(G_i) \leq 1$, *and* $d(G) = 2$ *otherwise.*

For a background on cographs, see [6]. Associated with every cograph G is a unique rooted tree, the *cotree* T_G of G. If v is a vertex in T_G, the tree rooted at v is written $T(v)$. The cograph represented by $T(v)$ is $G(v)$. The set of children of v is $C(v)$. Each leaf of T_G represents a vertex of G, while an internal vertex v of T_G is either a $\ast$-vertex or a $\uplus$-vertex. The former represents $\ast_{w \in C(v)} G(w)$, the latter $\biguplus_{w \in C(v)} G(w)$. $\ast$-vertices and $\uplus$-vertices alternate on every path from the root to a leaf.

Like many other algorithms on cographs (cf. [7]), the algorithm to determine whether a cograph G is D.A. proceeds from the leaves of T_G to the root. To every vertex v of T_G three attributes are assigned, whose values are computed from the attributes of v's children.

Lemma 14. *Let* v *be a vertex in a cotree, then* v *is assigned three attributes* l, d *and* a:

$l(v)$, *a label such that* $l(v) = l(v') \Leftrightarrow G(v) \simeq G(v')$,
$d(v)$, *with* $d(v) \equiv d(G(v))$,
$a(v)$, *a boolean which indicates whether a coaffine automorphism on* $G(v)$ *exists.*

The following rules correctly maintain the meaning of the attributes d *and* a:

v is a leaf:
$$d(v) = 0$$
$$a(v) = \mathit{False}$$

v is a $\ast$-vertex:
$$d(v) = \begin{cases} 1 & \text{if } \forall w \in C(v) : d(w) = 0, \\ 2 & \text{otherwise.} \end{cases}$$
$$a(v) = \bigwedge_{x \in C(v)} a(x) \tag{5}$$

v is a $\uplus$-vertex:
$$d(v) = \infty$$
$$a(v) = \bigwedge_{x \in C(v)} (a(x) \vee \exists y \in C(v) : l(x) = l(y)) \tag{6}$$

Proof. The case for a leaf is trivial. Assume the attributes have been determined for all children of an internal vertex v.

1. v is a $*$-vertex. As the children of a $*$-vertex are either leaves (diameter 0) or $\uplus$-vertices (disconnected), the correctness of d follows directly from Prop. 13. Let $w \in C(v)$ and $x \in V(G(w))$. If π is a coaffine automorphism on $G(v)$ then $\pi.x \in V(G(w))$, because otherwise $d(x, \pi.x) = 1$. This means that $\pi = \bigcup_{w \in C(v)} \rho_w$, where ρ_w is a coaffine automorphism on $G(w)$. The correctness of a follows.
2. v is a $\uplus$-vertex. Since $G(v)$ is disconnected, the value of $d(v)$ is correct. A coaffine automorphism can project a connected component $G(x)$ of $G(v)$ either onto itself, when $a(x) =$ *True*, or onto another connected component $G(y)$, when $l(x) = l(y)$.

□

Theorem 15. *Determining whether a cograph G is D.A. can be done in $O(|V| + |E|)$ time.*

Proof. First the cotree T_G, with root r, is computed using the algorithm from [8] in $O(|V| + |E|)$ time. Using Lemma 14 the attributes $a(v)$ and $d(v)$ can be computed in $O(|C(v)|)$ time, hence in $O(|V(G)|)$ time in total. The common tree isomorphism algorithm (see eg. [1]) calculates the labels l in the same bound. The theorem follows because G is D.A. iff $d(r) \leq 1 \vee a(r)$. □

Because of the limited diameter of cographs, this covers the general dl_k, $k > 1$ problems. The case $k = 1$ can be solved by replacing formula (5) by (6).

3.2 Diametrically Automorphic Circular Arc Graphs

Circular arc graphs are graphs that possess an intersection model consisting of arcs of a circle (see eg. [10]). Interval graphs are graphs that possess an intersection model consisting of intervals of a line. The interval graphs form a proper subset of the circular arc graphs. A circular arc graph that possesses an interval model is called *degenerated*. As every interval graph is chordal, it follows from Cor. 5 that an interval graph is D.A. iff it is complete. In this section an algorithm is developed that decides whether a circular arc graph is D.A.

If v is a vertex of a circular arc graph, the arc representing v is written $A(v) = (\alpha(v), \omega(v))$, with $\alpha(v)$ denoting the counterclockwise endpoint of the arc, and $\omega(v)$ the clockwise endpoint. The same notation is used for an arc x directly: $x = (\alpha(x), \omega(x))$. All endpoints are assumed to be distinct. The relative positioning of two different arcs is named as follows:

Definition 16. *Let x, y be different vertices of a circular arc graph G and consider a circular arc model of G. Then $A(x)$ and $A(y)$ are positioned in one of the following ways:*

1. *The endpoints appear in the cyclic order $\alpha(x)\ \omega(x)\ \alpha(y)\ \omega(y)$.* *$A(x)$ and $A(y)$ are said to be* independent.
2. *The endpoints appear in the cyclic order $\alpha(y)\ \alpha(x)\ \omega(x)\ \omega(y)$.* *$A(x)$ is said to be* contained *in $A(y)$, which is also written as $x \subset y$.*

3. The endpoints appear in the cyclic order $\alpha(x)\ \omega(y)\ \alpha(y)\ \omega(x)$.
$A(x)$ and $A(y)$ are said to cover the circle.
4. The endpoints appear in the cyclic order $\alpha(x)\ \alpha(y)\ \omega(x)\ \omega(y)$.
$A(x)$ and $A(y)$ are said to strictly overlap. *If this order is clockwise the notation $x \prec y$ is used.*

Let $N[v]$ be the closed neighbourhood of v, then the intersection models used in this section are assumed to be normalized in the following sense [11]:

Definition 17. *An intersection model of a circular arc graph G is said to be* normalized *if for all vertices x, y of G the following conditions hold:*

1. *$A(x)$ and $A(y)$ are independent iff $(x,y) \notin E(G)$.*
2. *$A(x)$ is contained in $A(y)$ iff $N[x] \subset N[y]$.*
3. *$A(x)$ and $A(y)$ cover the circle iff*
   – *$(x,y) \in E(G)$ and both $N[x] - N[y]$ and $N[y] - N[x]$ are non-empty*
   – *and $\forall w \notin N[x] : N[w] \subset N[y]$ and $\forall w \notin N[y] : N[w] \subset N[x]$.*
4. *$A(x)$ and $A(y)$ strictly overlap iff*
   – *$(x,y) \in E(G)$ and both $N[x] - N[y]$ and $N[y] - N[x]$ are non-empty*
   – *and not ($\forall w \notin N[x] : N[w] \subset N[y]$ and $\forall w \notin N[y] : N[w] \subset N[x]$).*

There are two limitations on the existence of a normalized model for a non-degenerated circular arc graph G. The first is that a vertex v with $N[v] = V(G)$ should not exist: all other arcs must then be contained in $A(v)$. However, the only D.A. circular arc graphs that contain such a vertex are the complete graphs, which are degenerated. The second limitation is that there should not be two vertices x and y such that $N[x] = N[y]$. These vertices can be fit into the model by representing them by the same arc.

Note that normalized models need not be unique. In this section all circular arc graphs are assumed to be normalized and not degenerated.

Diametrical automorphisms on a circular arc graph have the nice property that they preserve vertex ordering; a diametrical automorphism can be seen as a rotation of the model of the graph. In the next theorem this property is proved.

Proposition 18. *A D.A. circular arc graph does not contain a pair of vertices whose corresponding arcs cover the circle in a normalized model.*

Theorem 19. *Let v, w be vertices of a circular arc graph G, and π a D.A. on G. Then the order of the endpoints of $\pi.v, \pi.w$ is precisely the same as the order of the endpoints of v, w.*

Proof. Consider vertices v and w such that w is the first vertex encountered clockwise from v, ie. $\neg\exists x : \alpha(x) \in (\alpha(v), \alpha(w))$. The case where $v \supset w$ is trivially true, so only the cases where $v \prec w$ and where $(v,w) \notin E(G)$ need to be considered. The latter case can only exist if v and w are contained in a common vertex (otherwise G would be degenerated), so the theorem can be proved by examining the following three cases:

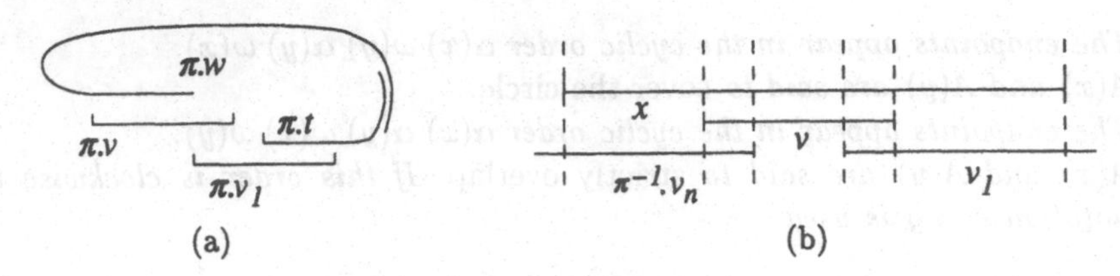

Fig. 4. Illustrating Theorem 19.

1. $v \prec w$ and w is on a shortest path from v to $\pi.v$.
   As $d(w, \pi.v) = d(G) - 1$ and $(\pi.v, \pi.w) \in E(G)$ it follows that $w, \ldots, \pi.v, \pi.w$ is a shortest path. But if $\pi.w \prec \pi.v$, then $\pi.v$ would not be on this shortest path. So $\pi.v \prec \pi.w$.
2. $v \prec w$ and w is not on shortest path from v to $\pi.v$.
   Let $v, v_1, v_2, \ldots, \pi.v$ be a shortest path, then $v \prec v_1$. Because w is not on a shortest v–$\pi.v$-path, it follows that $(w, v_2) \notin E(G)$, and $w \prec v_1$. Therefore, a vertex t exists such that $t \notin N[v]$, $t \in N[w]$ and $t \in N[v_1]$. Then $\pi.t \notin N[\pi.v]$, $\pi.t \in N[\pi.w]$, $\pi.t \in N[\pi.v_1]$ and, by the previous case, $\pi.v \prec \pi.v_1$. In order to establish a contradiction, suppose $\pi.w \prec \pi.v$ (see Fig. 4(a)). Then the overlap between $\pi.w$ and $\pi.t$ is outside the segment $(\omega(\pi.v), \alpha(\pi.t))$, implying that either $\pi.w \in N[w]$ or $\pi.t \in N[t]$ which cannot hold if G is not complete.
3. $(v, w) \notin E(G)$, $v \subset x$ and $w \subset x$.
   Let $v, v_1, \ldots, v_n, \pi.v$ be a shortest path, then $x, v_1, \ldots, v_n, \pi.x$ is also a shortest path. Because $(v_n, \pi.v) \in E(G)$ and $(\pi.v, \pi.v_1) \in E(G)$, the arc x can be partitioned into five consecutive segments (see Fig. 4(b)):

$$\alpha(x) \cdots \alpha(v) \cdots \omega(\pi^{-1}.v_n) \cdots \alpha(v_1) \cdots \omega(v) \cdots \omega(x)$$

   The position of w in these segments determines the position of $\pi.w$ uniquely, thereby preserving the order of v and w.

□

Corollary 20. *Let the vertices of a circular arc graph G be numbered from $0 \ldots |V(G)| - 1$ in the clockwise order of their counterclockwise endpoints (ie. as they are discovered by a clockwise scan of the model, starting at v_0).*

Then a diametrical automorphism π on G is of the form $\pi : v_i \to v_{i \oplus k}$, where $\oplus$ denotes addition mod $|V(G)|$.

For brevity, a vertex will be identified with its number, writing $v < w$ instead of $v = v_i, w = v_j, i < j$.

The shortest path from v to w is contained in either $(\alpha(v), \omega(w))$ or in $(\alpha(w), \omega(v))$. The shortest path contained in such a segment $(\alpha(x), \omega(y))$ is called the *clockwise shortest path* between x and y and its length is the *clockwise distance* $d^c(x, y)$, with the equality $d(x, y) = \min(d^c(x, y), d^c(y, x))$ holding.

Consider a circular arc graph G with its vertices numbered clockwise, as in Cor. 20. Let $v, v_1, \ldots, v_n, w$ be a clockwise shortest path, with $v < w$. Then $v_i < v_{i+1}$, except possibly $v_1 < v$. This exception occurs when $v \subset v_1$. This implies that for $k > 1$, the vertices t such that $d^c(v,t) = k$ are sequential, so by defining for all $v \in V(G)$

$$\mathrm{dl}(v) \equiv \min\{w \mid d^c(v,w) = d(G)\} \quad \mathrm{dh}(v) \equiv \max\{w \mid d^c(v,w) = d(G)\} \quad (7)$$

$$\mathrm{dl}(G) \equiv \max_v\{\mathrm{dl}(v) - v\} \quad \mathrm{dh}(G) \equiv \min_v\{\mathrm{dh}(v) - v\} \quad (8)$$

we get for any $1 < k \leq d(G)$

$$(\forall v : d(v, v \oplus k) = d(G)) \Leftrightarrow \mathrm{dl}(G) \leq k \leq \mathrm{dh}(G) .$$

Following this, determining whether a circular arc graph is D.A. can be done by first determining the clockwise distance between all pairs of vertices, next calculating $d(G)$, $\mathrm{dl}(G)$ and $\mathrm{dh}(G)$ and finally checking whether a $k \in [\mathrm{dl}(G), \mathrm{dh}(G)]$ exists, such that $\pi_k : v \to v \oplus k$ is an automorphism.

The final step requires an efficient way to compare rotations of a circular arc model. What makes up a circular arc model is the order and length of the arcs. The length of an arc can be measured in the number of counterclockwise endpoints it contains: $\mathrm{length}(v) = |\{w \mid \alpha(w) \in (\alpha(v), \omega(v))\}|$

By labelling every arc with its length and concatenating these labels in the clockwise order of the arcs, a string representation of the model is created. The precise order of the clockwise endpoints is lost in this representation[1], but the intersection relationship between arcs and hence the model is not affected by this loss. This string representation allows for an efficient D.A. algorithm:

Theorem 21. *Determining whether a circular arc graph is D.A. can be done in* $O(|V|^2)$ *time.*

Proof. First, a circular arc model of G is created in $O(|V|^2)$ time using [9] and it is normalized in the same time using the algorithm in [11, p.415]. Next, the clockwise distances between all pairs of vertices are calculated in the same bound by [2]. Then $\mathrm{dl}(G)$, $\mathrm{dh}(G)$ and $d(G)$ are calculated by a nested loop implementing (8). The time consuming phase in the creation of a string representation r of the model is the calculation of the lengths of the arcs. This takes $O(|V|)$ for each arc and $O(|V|^2)$ in total. Finally, the string matching algorithm of Knuth, Morris and Pratt (see [5, Ch.34.4]) is used to determine in $O(|r|) = O(|V|)$ time whether r occurs in rr with shift $k, \mathrm{dl}(G) \leq k \leq \mathrm{dh}(G)$. □

The general dl_k problem can be solved by the same algorithm by making the appropriate changes to (7).

4 Discussion

Although the bound on the distance between a vertex and its image is fixed in dl_k automorphisms, and depends on the graph in diametrical automorphisms,

[1] The order of the clockwise endpoints can easily be stored by inserting a special token followed by a length into the string.

results could readily be adapted from one to the other. The structure of D.A. graphs and the precedent set by coaffine automorphisms [13, 14] suggest that D.A. or dl_k automorphisms can be used for investigating the properties of cycles or graph operators. The operators most vulnerable to this attack are probably those defined in terms of distances between vertices, or in terms of cycles.

The lower bound on the number of edges in a D.A. graph with given diameter was shown to be $\Omega(|V|)$, while the upper bound is $O(|V|^2)$. This would suggest many D.A. graphs exist, but the enumeration problem was not tackled — how many D.A. graphs on a given number of vertices (and, optionally, edges) do exist? And is the conjectured tight upper bound on the number of edges correct?

References

[1] Alfred V. Aho, John E. Hopcroft, and Jeffrey D. Ullman. *The Design and Analysis of Computer Algorithms.* Addison-Wesley, 1974.

[2] M.J. Atallah, D.Z. Chen, and D.T. Lee. An optimal algorithm for shortest paths on weighted interval and circular-arc graphs, with applications. *Algorithmica*, 14:429–441, 1995.

[3] László Babai. Automorphism groups, isomorphism, reconstruction. In R. Graham, M. Grötschel, and L. Lovász, editors, *Handbook of Combinatorics*, chapter 27, pages 1447-1540. Elsevier Science, 1995.

[4] Fred Buckley. Self-centered graphs. In *Graph Theory and its Applications: East and West : Proceedings of the First China-USA International Graph Theory Conference*, volume 576 of *Annals of the New York Academy of Sciences*, pages 71–78, 1989.

[5] Thomas H. Cormen, Charles E. Leiserson, and Ronald L. Rivest. *Introduction to Algorithms.* The MIT Press, 1990.

[6] D.G. Corneil, H. Lerchs, and L. Stewart Burlingham. Complement reducible graphs. *Discrete Appl. Math.*, 3(1):163–174, 1981.

[7] D.G. Corneil, Y. Perl, and L. Stewart. Cographs: Recognition, applications and algorithms. *Congressus Numerantium*, 43:249–258, 1984.

[8] D.G. Corneil, Y. Perl, and L. K. Stewart. A linear recognition algorithm for cographs. *SIAM J. Comput.*, 14(4):926–934, November 1985.

[9] E.M. Eschen and J.P. Spinrad. An $O(n^2)$ algorithm for circular-arc graph recognition. In *Proc. of the 4th Annual ACM-SIAM Symposium on Discrete Algorithms*, pages 128–137, 1993.

[10] Martin Charles Golumbic. *Algorithmic Graph Theory and Perfect Graphs.* Academic Press, 1980.

[11] Wen-Lian Hsu. $O(M \cdot N)$ algorithms for the recognition and isomorphism problems on circular-arc graphs. *SIAM J. Comput.*, 24(3):411–439, June 1995.

[12] Anna Lubiw. Some NP-complete problems similar to graph isomorphism. *SIAM J. Comput.*, 10(1):11–21, February 1981.

[13] Victor Neumann-Lara. On clique-divergent graphs. In *Problèmes Combinatoires et Théorie des Graphes*, number 260 in Colloques Internationaux C.N.R.S., pages 313–315, 1978.

[14] Erich Prisner. *Graph Dynamics.* Number 338 in Pitman Research Notes in Mathematics. Longman, 1995.

Kleene Theorems for Event-Clock Automata

Cătălin Dima *

Bucharest University, Department of Foundations of Computer Science

Abstract. We define some class of regular expressions equivalent to event-clock automata. It is shown that regular expressions cannot be given a compositional semantics in terms of timed state sequences. We introduce a modified version of timed state sequences supporting a *partial* operation of concatenation on which we may build the semantics of regular expressions. A *forgetting* map then induces a semantics in terms of the classic version of timed state sequences. We also define several types of languages of automata in terms of classic or modified timed state sequences. Two Kleene theorems, one for each type of timed state sequences, relating expressions and event-clock automata are proved.

1 Introduction

Automata models of real-time systems have proved instrumental in the verification of real-time systems [AD94,AFH94]. They have been studied with the hope that the classical decidability results and the relationship to temporal logics can be adapted in the real-time framework, hence giving powerful decidable real-time logics. However the program has not reached its scope, since there is no fully satisfactory class of automata fitting the above requirements. Timed automata of [AD94] are too powerful, having a decidable emptiness problem but an undecidable complementation problem. Real-time automata [Di2], in spite of their decidability of the complementation problem, have too little expressive power, being unable to capture behaviors of distributed real-time systems [Di3]. Another proposal are event-clock automata [AFH94] (ECA for short), which, in their introductory paper, were shown to have a decidable complementation problem as they are determinizable. These automata have also sufficient expressive power and have served as a guide for defining some real-time logic [RS97,HRS98].

Another subject for research was the relationship of automata to some regular expressions. As yet just timed automata [ACM97] and real-time automata [Di2] have been the subject of this question, but only real-time automata have proved satisfactory since as (languages accepted by) timed automata have a weak Kleene theorem in the sense that it involves renaming, and it is still an open question if regular expressions in [ACM97] are equally expressive to timed automata.

Here we investigate the class of languages accepted by ECA for their relationship to some regular expressions. The expressions in study have the same form as the labels of the transitions on ECA, i.e. are composed of a letter or the empty word ϵ and a constraint on the history and prophecy clocks. We call

* This research was done during the author's visit to TIFR, Bombay, supported by an extension of a UNU/IIST fellowship. Mailing address: str. Academiei 14, Bucharest, Romania, R0-70109, e-mail: `cdima@funinf.math.unibuc.ro`

concatenation of these expressions *chop*. It turns out that the constraints on the left-hand side of chop may refer to points after the chopping point. We are lead to consider a modified version of timed state sequences of [AD94], in which we add information about some interval to which attention is restricted (we call them TSS *with limited observation*). This allows us to define the chop of two TSS with limited observation only iff they have the same underlying TSS, hence chop being a partial operation. This is then lifted to a total operation on sets of TSS with limited observation. We also show that, if we assume the idea that constraints on one part of chop may refer to points on the other part of chop, we cannot define semantics of the regular expressions in terms of a concatenation operation on TSS.

We also study here ECA with ϵ-transitions, not yet defined in the litterature. Why ϵ-transitions, if ECA are not closed under renaming and hiding [AFH94]? The reason lies in their relationship to the logic proposed in [RS97,HRS98]. Take e.g. the formula $\Diamond\rhd_{=1}p$ (which is to be read as "there is some time point t such that the first time point $t' > t$ at which p starts to hold must satisfy $t' - t = 1$"). The formula might be associated to the state-clock automaton in figure 1 (see [RS97,HRS98] for details).

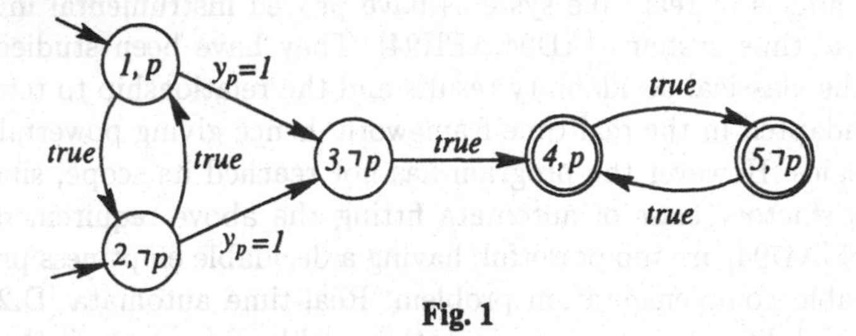

Fig. 1

Intuitively the automaton may be in states 1 or 4 at some time point τ only if the value of variable p is true at τ, while during the stay in states 2, 3 or 5 the value of p must be false. Some transition may be taken at time point τ iff the values of the variable p before and after τ fit the respective labels of the states and the constraint on the transition is true at τ. The constraint $y_p = 1$ is true at some time point τ iff the next time point τ' at which p becomes *true* satisfies $\tau' - \tau = 1$ (this implies that p is false at τ). The automaton is built such that it stays in states 1 or 2 before the time point t; at moment t it goes to 3 and at time point t' it goes to 4; after this it stays either in 4 or in 5. Hence it is equivalent to the above formula. Note that transitions 1 to 2, 2 to 1 etc. involve also a change in the value of p; but there is also the transition 2 to 3 which is identified only by the moment at which the constraint becomes true. We argue that, when transforming this automaton into an ECA, the transition from 2 to 3 has to be an ϵ-transition by the intuition that *an action (i.e. letter) occurs only when some variable changes its value.*

Note then that in the presence of ϵ-transitions the algorithm in [AFH94] does not give a deterministic ECA as the ϵ-transitions have to be removed first - the same holds in the classical setting [HU]. In fact the notion of deterministic automaton of [AFH94] has to be redefined. The problem of the elimination of ϵ-transitions is solved in [Di4].

The paper runs as follows: we introduce in the next section extended TSS (ETSS for short) which allow us to define the language of an ECA with ϵ-transitions and ETSS with limited observation (LETSS for short) with the partial concatenation ("chop") on them which allows defining the semantics of regular expressions. In the third section we define ECA with ϵ-transitions and ϵ-regular expressions. We associate two types of semantics to regular expressions, one in terms of LETSS and one in terms of TSS. Only the first one is compositional, the other one is not and comes via some "forgetting" map. Also we associate to each ECA four languages: the first language, in terms of ETSS, is the natural one, as it extends the definition of [AFH94] to ECA with ϵ-transitions; the second language, in terms of TSS comes via the forgetting map. But, since we cannot prove the Kleene theorems unless we define chop on automata too we also need some language in terms of LETSS, which is the third type. The fourth type language is again in terms of TSS and comes from the third via some forgetting map too. An important property showed here is that the second and the fourth types of languages, though being not the same for a given ECA, have equal expressive power. The fourth section contains the proof of the Kleene theorem relating the first type semantics of some ϵ-regular expression and the third type language of some ECA. As a corollary we obtain the (desired) Kleene theorem saying that ϵ-regular expression and ECA have the same expressive power w.r.t. their TSS semantics. We end with some comments and topics for further study.

2 Preliminaries

We fix a finite set of letters Σ. For a word $w \in \Sigma^*$ we denote by $|w|$ its length. $[n]$ is the set of natural numbers $\{1, \ldots, n\}$. Notations $\mathbb{R}_{\geq 0}$ and $\mathbb{R}_{>0}$ are for the sets of nonnegative, resp. positive reals. Similarily $\mathbb{Q}_{\geq 0}$ stands for the set of nonnegative rationals.

Definition 1. *An extended timed state sequence* (ETSS *for short) is a tuple* (w, τ, I) *consisting of a word* $w \in \Sigma^*$*, a strictly increasing function* $\tau : [m] \longrightarrow \mathbb{R}_{\geq 0}$ *with* $m \geq |w|$*, called the* ***sequence of time points of the actions*** *and a subset* I *of* $[m]$ *having card*$(I) = |w|$.

We denote $dom(\tau)$ the domain $[m]$ of the sequence of time points τ and by $range(\tau)$ the range of τ.

An ETSS intuitively models the behavior of a real-time system in the following sense: the letters in Σ are the *observable actions*, the empty word ϵ represents an *inobservable action* and τ holds the time points at which actions occur. Moreover observable actions occur only at $\tau(i)$ with $i \in I$. Hence, if we consider that I consists of $j_1 < j_2 < \ldots < j_{|w|}$ we can define, for each $i \in dom(\tau)$, the i-th symbol of the ETSS as being ϵ (the empty word) iff $i \notin I$ or the k-th letter in w iff $i = j_k \in I$ and denote this by $w[i]$. Then $w = w[1]w[2] \ldots w[card(I)]$.

A natural equivalence arises on ETSS: two ETSS (w_i, τ_i, I_i), $i \in [2]$ are equivalent, denoted $(w_1, \tau_1, I_1) \simeq (w_2, \tau_2, I_2)$, iff $w_1 = w_2$ and the time points of observable actions are the same, i.e. if we list each I_i in increasing order as $I_i = \{j_1^i, \ldots, j_{|w^1|}^i\}$ then $\tau_1(j_k^1) = \tau_2(j_k^2)$ for all $k \in [|w_1|]$.

A representative in each equivalence class can be defined as follows: for an ETSS $\alpha = (w, \tau, I)$, where I is listed in increasing order as $I = \{j_1, \ldots, j_{|w|}\}$), we define a function $\theta : [|w|] \longrightarrow \mathbb{R}_{>0}$ with $\theta(k) = \tau(j_k)$ for all $k \in [|w|]$. If we pair the word w with θ we actually get the following definition:

Definition 2. *A timed state sequence [AD94] (TSS for short) is a pair (w, θ) consisting of a word $w \in \Sigma^*$ and a strictly increasing function $\theta : [|w|] \longrightarrow \mathbb{R}_{\geq 0}$.*

Hence in a TSS we know only the time points of the observable actions. Clearly they are special kind of ETSS, in which $I = dom(\tau)$. Simple verification shows that the map $[\cdot]$ from the set of ETSS to the set of TSS defined as $[(w, \tau, I)] = (w, \theta)$ where θ is defined above maps two equivalent ETSS to the same TSS. Thus TSS can be thought of as representatives for each equivalence class of ETSS.

On ETSS we can also define concatenation as an extension of the concatenation on words: given two ETSS (w_1, τ_1, I_1) and (w_2, τ_2, I_2) with $dom(\tau_i) = [m_i]$ $(i \in [2])$ define $(w_1, \tau_1, I_1) \cdot (w_2, \tau_2, I_2) = (w_1 w_2, \tau, I)$ where $dom(\tau) = [m_1 + m_2]$ and

$$\tau(i) = \begin{cases} \tau_1(i) & \text{iff } i \in [m_1] \\ \tau_1(m_1) + \tau_2(i - m_1) & \text{iff } m_1 + 1 \leq i \leq m_1 + m_2 \end{cases}$$

Concatenation can then be extended to sets of ETSS as usual. We denote by $\cdot$ this extension. Concatenation can be defined on TSS too, as TSS are special kind of ETSS.

A third basic notion is the following:

Definition 3. *An ETSS with limited observation (LETSS for short) is a tuple (w, τ, I, t_1, t_2) where (w, τ, I) is an ETSS and $0 \leq t_1 \leq t_2 < \infty$*

The intuition is that observation of the system is restricted to the interval $[t_1, t_2]$. LETSS are essential in giving semantics to regular expressions.

The *underlying* ETSS of a LETSS is defined by $sub((w, \tau, I, t_1, t_2)) = (w, \tau, I)$. The equivalence relation on ETSS can be straightforwardly extended to LETSS by putting two LETSS equivalent when their underlying ETSS are. Then we can define a map $[[\cdot]]$ as the composition of *sub* with $[\cdot]$. This map also makes TSS as representatives of equivalent LETSS.

We can also define concatenation of LETSS, but in a rather different manner than on ETSS: it is a *partial* operation, defined when both operands have the same underlying ETSS and when the endpoint of the interval of observation of the first equals the initial point of the interval of observation of the second:

$$(w, \tau, I, t_1, t_2); (w', \tau', I', u_1, u_2) = \begin{cases} (w, \tau, I, t_1, u_2) & \text{iff } w = w', \tau = \tau', I = I', t_2 = u_1 \\ \text{undefined} & \text{otherwise} \end{cases}$$

This operation is then extended to a *total* operation on sets of LETSS:

$$T_1;T_2 = \{(w,\tau,I,t_1,t_3) \mid \exists t_2 \in (t_1,t_3) \text{ such that } (w,\tau,I,t_1,t_2) \in T_1 \text{ and } (w,\tau,I,t_2,t_3) \in T_2\}$$

Note that $\{(w,\tau,I,t_1,t_2)\};\{(w,\tau,I,t_2,t_3)\} = \{(w,\tau,I,t_1,t_3)\}$ while $\{(w,\tau,I)\} \cdot \{(w,\tau,I)\} = \{(ww,\tau',I')\}$ for some τ' and I'. Hence

Remark 1. Given two sets of LETSS T_1 and T_2, $sub(T_1;T_2)$ does not equal $sub(T_1) \cdot sub(T_2)$ in general.

In the sequel we remind the use of clocks in association with an ETSS:

We associate with each letter a of Σ two symbols, a **history clock** x_a and a **prophecy clock** y_a. Denote then $H(\Sigma) = \{x_a \mid a \in \Sigma\}$ and $P(\Sigma) = \{y_a \mid a \in \Sigma\}$. the respective copies of Σ that arise this way[1]. The use of the clocks is connected to each ETSS as follows: at each time point $t \in \mathbb{R}_{\geq 0}$ we can define the value of the clock x_a as the time elapsed since the last occurence of letter a in the ETSS; if no such occurences, put $x_a = \bot$. Similarily, y_a holds the time remaining till the next occurence of a and is $\bot$ if there is no such occurence. Hence some ETSS (w,τ,I) defines a **trajectory of clocks** which is a function $Ck = Ck_{((w,\tau,I),t_1)} : (H(\Sigma) \cup P(\Sigma)) \times \mathbb{R}_{\geq 0} \longrightarrow \mathbb{R}_{\geq 0} \cup \{\bot\}$ defined as:

$$Ck(x_a,t) = \begin{cases} t-\tau(i) & \text{iff, denoting } A = \{k \in I \mid w[k]=a \text{ and } \tau(k)<t\} \text{ we have} \\ & A \neq \emptyset \text{ and } i = \max A \\ \bot & \text{otherwise} \end{cases}$$

$$Ck(y_a,t) = \begin{cases} \tau(i)-t & \text{iff, denoting } B = \{k \in I \mid w[k]=a \text{ and } \tau(k)>t\} \text{ we have} \\ & B \neq \emptyset \text{ and } i = \min B \\ \bot & \text{otherwise} \end{cases}$$

A **clock constraint** is a boolean combination of atomic formulas of the type $x_a < c$, $x_a \leq c$, $x_a = \bot$, and $y_a < c$, $y_a \leq c$ or $y_a = \bot$ where c is some positive rational $c \in \mathbb{Q}_{\geq 0}$. The set of clock constraints over Σ is denoted $Constr(\Sigma)$. For each initialized ETSS $((w,\tau,I),t_1)$ the trajectory of clocks Ck defines at each time point $t \in \mathbb{R}_{\geq 0}$ an interpretation of the clock constraints. We use the notation $(w,\tau,I,t_1,t) \models C$ to encode that the interpretation $Ck(\cdot,t)$, maps the constraint C to true.

For each clock $K \in H(\Sigma) \cup P(\Sigma)$ define the **elementary clock constraints** as all the formulas $K = \bot$, $c < K$, $c \leq K$, $c_1 < K \wedge K < c_2$, $c_1 \leq K \wedge K < c_2$, $c_1 < K \wedge K \leq c_2$ or $c_1 \leq K \wedge K \leq c_2$, for some $c, c_1, c_2 \in \mathbb{Q}_{\geq 0}$. A **simple constraint** is a formula $\bigwedge_{a\in\Sigma}(C(x_a) \wedge C(y_a))$ where $C(x_a)$ and $C(y_a)$ are elementary constraints. We denote $SC(\Sigma)$ the set of simple constraints.

It is clear that any clock constraint can be brought to a disjunction of simple clock constraints (a normal form without negation). E.g. $\neg(x_a < 2) \equiv (x_a \geq 2) \vee (x_a = \bot)$. This property will be used in simplifying event-clock automata.

[1] Note that we do not use clocks x_ϵ or y_ϵ. These clocks may be regarded as event-clock versions for the automata clocks in [HRS98].

3 Event-clock automata and regular expressions

In this his section we introduce event-clock automata with ϵ-transitions and regular expressions which match these automata. The introduction of ϵ-transitions in event-clock automata justifies the introduction of extended time state sequences.

Definition 4. *An* ***event-clock automaton*** *(ECA for short) is a tuple* $\mathcal{A} = (Q, \Sigma, \delta, q_0, F)$ *where* Q *is the set of* states, $Q_0, F \subseteq Q$ *are the sets of* initial *resp.* final states *and* δ *is the* transition relation $\delta \subseteq Q \times (\Sigma \cup \{\epsilon\}) \times Constr(\Sigma) \times Q$ *with* $|\delta| < \infty$.

A **run** of length n is a sequence of transitions $((q_{i-1}, x_i, C_i, q_i))_{i \in [n]}$ ($x_i \in \Sigma \cup \{\epsilon\}$) such that $q_0 \in Q_0$ and $q_n \in F$. A zero-length run consists of some initial and final state. A **partial run** is a sequence of transitions like above but which may start or end in any location.

We will give four types of languages to ECA. The first one is in terms of ETSS: an ETSS (w, τ, I) with $dom(\tau) = [n]$ is accepted by $\mathcal{A}$ iff there is a run (of the same length) $((q_{i-1}, x_i, C_i, q_i))_{i \in [n]}$ such that the word defined by the run is w and at the $i-th$ action in the ETSS the respective constraint is true. Formally, for all $i \in [n]$ we have that $w[i] = x_i$ and $(w, \tau, I, \tau(i)) \models C_i$. We call some run ***admissible*** iff it is associated to some ETSS in the above sense. We denote $L_e(\mathcal{A})$ the set of ETSS accepted by $\mathcal{A}$ and call it the **extended language** of $\mathcal{A}$.

The second semantics comes via the map $[\cdot]$: the **abstract language** of $\mathcal{A}$, is the set of TSS which are the representatives of some ETSS accepted by $\mathcal{A}$: $L_a(\mathcal{A}) = [L_e(\mathcal{A})] = \{[(w, \tau, I)] \mid (w, \tau, I) \in L_e(\mathcal{A})\}$

The third semantics is in terms of LETSS: a LETSS (w, τ, I, t_1, t_2) with $dom(\tau) = [n]$ is accepted by $\mathcal{A}$ iff there is a run $((q_{i-1}, x_i, C_i, q_i))_{i \in [m]}$ such that m is the number of actions that occur within the interval $[t_1, t_2]$ and at the i-th action occuring in $[t_1, t_2]$ constraint C_i is satisfied. Formally, $m = card(\{i \in [n] \mid \tau(i) \in [t_1, t_2]\}$ and, denoting $j = \min\{i \in [m] \mid \tau(i) \in [t_1, t_2]\}$ we have that $(w, \tau, I, \tau(j + i - 1)) \models C_i$ for all $i \in [n]$. We denote $L_l(\mathcal{A})$ the set of LETSS accepted by $\mathcal{A}$ and call it the **limited observation language**.

The last semantics, the **language of abstract limited observation** is $L_{al}(\mathcal{A}) = [[L_l(\mathcal{A})]] = \{[[(w, \tau, I, t_1, t_2)]] \mid (w, \tau, I, t_1, t_2) \in L_l(\mathcal{A})\}$.

Note that $L_a(\mathcal{A}) \neq L_{al}(\mathcal{A})$ in general, a counterexample being the ECA $\mathcal{A} = (\{q, r\}, \{a\}, \{(q, a, true, r)\}, \{q\}, \{r\})$ for which L_{al} is the set of all TSS having at least one letter, while $L_a(\mathcal{A})$ is a singleton. However

Proposition 1. *The class of sets of TSS which are the abstract language of some ECA equals the class of sets of TSS which are the language of abstract limited observation of some ECA.*

The proof of this is based on the following normal form of automata:

Proposition 2. *For each ECA* $\mathcal{A}$ *there exist an ECA* $\mathcal{B}$ *such that with* $L_e(\mathcal{A}) = L_e(\mathcal{B})$ *and such that all transitions in* $\mathcal{B}$ *have simple constraints (call this automaton* ***simple ECA****).*

Proof. We rely on the decomposition of each constraint into a disjunction of simple constraints mentioned at the end of the previous section. We decompose first each constraint C on some transition into n simple constraints and then split that transition into n transitions with the same source and destination, each labeled with one of the simple constraints C is decomposed into. □

The proof of Proposition 1 can be done as follows: First observe that in a simple ECA an initial transition (i.e. starting in an initial state) which is labeled with a simple constraint C is in an admissible run iff C constrains all the history clocks to $\perp$ and similarily for final transitions (i.e. leading to some final state) and prophecy clocks. Note that this does not hold for runs associated to LETSS: e.g. the ECA $(\{q,r\},\{a\},\{(q,\epsilon,x_a=2,r)\},\{q\},\{r\})$ has no admissible run but has some run associated to some LETSS (w,τ,I,u_1,u_2) in which $range(\tau)\cap[t_1,t_2] = \{\tau(i)\}$, $w[i]=\epsilon$ and $(w,\tau,I,\tau(i)) \models x_a = 2$.

Hence, for the left-to-right passage transform the given ECA $\mathcal{A}$ (assumed simple) such that initial states have no incoming transitions and final states have no outgoing transitions (by the state-splitting technique). Then remove all initial transitions whose (simple) constraints do not check some history clock x_a to $\perp$; similarily remove all final transitions on which some prophecy clock y_a is not constrained to $\perp$. Note that the resulting ECA has the property that it accepts a LETSS (w,τ,I,t_1,t_2) iff the interval $[t_1,t_2]$ contains the domain of τ. Finally, for all the remaining initial transitions remove from their constraints all conjuncts that refer to history clocks (which may be only $K=\perp$), keeping the conjuncts that refer to prophecy clocks; similarily, for all the remaining final transitions remove from their constraints the conjuncts that refer to prophecy clocks. We get an ECA $\mathcal{B}$ with $L_{al}(\mathcal{B}) = L_a(\mathcal{A})$.

What for the reverse, starting with some ECA $\mathcal{A}$ (not necessarily simple) transform it first such that initial states do not have incoming transitions and final states do not have outgoing transitions. Then, at all initial and final states, add loops labeled with all letters in Σ and constraint *true*. The abstract language of the resulting automaton will be equal to the language of abstract limited observation of $\mathcal{A}$. □

Definition 5. *An ECA is **ϵ-free** iff there is no tuple in the transition relation labeled with the empty word. An ϵ-free ECA is **deterministic** iff it has a single initial state and for each two distinct transitions having the same source and labeled with the same symbol the conjunction of their constraints is a nonsatisfiable formula.*

An ϵ-free ECA can be brought to a deterministic one by the algorithm in [AFH94], which is an adaptation of the subset construction. As we have noted in the introduction this construction does not work for ECA that have ϵ-transitions: this is similarily to the situation in finite automata theory. Bringing an ECA to an ϵ-free one is the subject of [Di4].

Definition 6. *The set $\epsilon RE(\Sigma)$ of **ϵ-regular expressions** over Σ is defined by the following grammar where $a \in \Sigma$ and $C \in Constr(\Sigma)$:*

$$E ::= \emptyset \mid () \mid (\epsilon, C) \mid (a, C) \mid E + E \mid E;E \mid E^*$$

The rules without ϵ give the set of regular expressions.

We call *atomic* expressions of the type (x, C) with $x \in \Sigma \cup \{\epsilon\}$. Also () is called the *empty expression*.

We define two types of semantics of ϵ-regular expressions, namely one in terms of LETSS (the *concrete* semantics) and another one in terms of TSS (the *abstract* semantics). The *concrete semantics* of an ϵ-regular expression E is denoted $|E|$ and for atoms is:

- $|()|$ containts all the LETSS (w, τ, I, t_1, t_2) such that the interval $[t_1, t_2]$ does not contain any action (i.e. $range(\tau) \cap [t_1, t_2] = \emptyset$).
- $|(\epsilon, C)|$ contains some LETSS (w, τ, I, t_1, t_2) iff there is a unique time point $\tau(i) \in [t_1, t_2]$ at which the inobservable action occurs, C holds there and no other action occurs in $[t_1, t_2]$ (i.e. $range(\tau) \cap [t_1, t_2] = \{\tau(i)\}$, $(w, \tau, I, \tau(i)) \models C$ and $w[i] = \epsilon$).
- $|(a, C)|$ contains some LETSS (w, τ, I, t_1, t_2) iff there is a unique time point $\tau(i) \in [t_1, t_2]$ at which the observable action a occurs, C holds there and no other action occurs in the interval $[t_1, t_2]$ (i.e. $range(\tau) \cap [t_1, t_2] = \{\tau(i)\}$, $(w, \tau, I, \tau(i)) \models C$ and $w[i] = a$).

The semantics is then easily extended to ϵ-regular expressions by structural induction:

$$|E_1; E_2| = |E_1|; |E_2| \qquad |E_1 + E_2| = |E_1| \cup |E_2| \qquad |\emptyset| = \emptyset$$
$$|E^*| = \bigcup_{n \in \mathbb{N}} |E^n| \text{ where } E^0 = () \text{ and } E^{n+1} = E^n; E$$

The *abstract* semantics, denoted $\|E\|$ comes via the map $[[\cdot]]$ defined in the previous section: we put $\|E\| = [[|E|]]$.

Hence there are two notions of equivalence of ϵ-regular expressions: a concrete one and an abstract one. Clearly two ϵ-regular expressions which are concrete equivalent are abstract equivalent too. The expressions $(\epsilon, (x_a = 2) \wedge (y_a = 4))$ and $(\epsilon, (x_a = 4) \wedge (y_a = 2))$ show the reverse does not hold.

Proposition 3. *Each ϵ-regular expression is concrete equivalent to an ϵ-regular expression in which all constraints are simple.*

We conclude this section by pointing out that the semantics of chop on ϵ-regular expressions cannot rely on concatenation of ETSS as defined above, hence ETSS cannot serve for defining some "natural" semantics of ϵ-regular expression. The exact meaning of *natural* has to be studied further, but we rely on some properties concatenation should have based on our intuition that constraints on the lhs of ";" may refer to time points after the chopping point. One such property is that the semantics of $(a, y_b = \bot); (b, C)$ should be the empty set regardless of the constraint C, but the semantics of (b, C) should be the empty set iff C would be some unsatisfiable constraint. Hence the operation that gives the semantics of chop should be able to capture the situation when applying it to nonempty sets gives the empty set, situation which is not captured by any operation which is defined at the level of elements (like "·" on ETSS).

4 The two Kleene theorems

Theorem 1 (Concrete Kleene theorem). *The class of sets of LETSS which are the limited observation semantics of some ECA equals the class of LETSS which are the concrete semantics of some ϵ-regular expression.*

Proof. For the right-to-left inclusion we proceed by structural induction. Note first that, for each atomic ϵ-regular expression (x, C) $(x \in \Sigma \cup \{\epsilon\})$ the ECA $\mathcal{A}_{x,C} = (\{q, r\}, \Sigma, \{(q, x, C, r)\}, \{q\}, \{r\})$ has $L_l(\mathcal{A}_{x,C}) = |(x, C)|$. Also for the empty expression () we may consider a single state ECA with no transition: $(\{q\}, \Sigma, \emptyset, \{q\}, \{q\})$. Then we adapt the usual constructions [HU] for union, concatenation and star:

Fix two ECA $\mathcal{A}_i = (Q_i, \Sigma, \delta_i, Q_0^i, F_i)$, $i \in [2]$ with disjoint sets of states $(Q_1 \cap Q_2 = \emptyset)$. Then the ECA $\mathcal{A}_\cup = (Q_1 \cup Q_2, \Sigma, \delta_1 \cup \delta_2, Q_0^1 \cup Q_0^2, F_1 \cup F_2)$ has $L_l(\mathcal{A}_\cup) = L_l(\mathcal{A}_1) \cup L_l(\mathcal{A}_2)$.

What for chop, we note that drawing ϵ-transitions from final states of $\mathcal{A}_1$ to initial states of $\mathcal{A}_2$ does not work here since LETSS accepted by the resulting automaton would have an extra time point in τ, corresponding to the ϵ-transition. Hence we first transform both automata such that initial states do no have incoming transitions and final states do not have outgoing transitions. Then build the automaton $\mathcal{A}_; = (Q, \Sigma, \delta, Q_0, F)$ where

- Q consists of nonfinal states in $\mathcal{A}_1$, noninitial states in $\mathcal{A}_2$ and pairs $(q, r) \in F_1 \times Q_0^2$;
- Q_0 consists of the initial and nonfinal states of $\mathcal{A}_1$ and, for each initial and final state s in $\mathcal{A}_1$, all pairs (s, r) with r initial in $\mathcal{A}_2$;
- similarily, F consists of final and noninitial locations of $\mathcal{A}_2$ and, for each final and initial state s in $\mathcal{A}_2$, all pairs (r, s) with r final in $\mathcal{A}_1$;
- δ consists of all transitions from δ_1 that do not involve some final state and all transitions from δ_2 not involving some initial state; moreover, each transition in $\mathcal{A}_1$ leading to some final state r gives rise to transitions leading to all pair states containing r and each transition in $\mathcal{A}_2$ starting in some initial state r gives rise to transitions starting in all pair states containing r. Formally:

$$\begin{aligned}\delta = \{(q, x, C, r) \mid (q, x, C, r) \in \delta_1 \text{ and } r \notin F_1\} \cup \\ \{(q, x, C, r) \mid (q, x, C, r) \in \delta_1 \text{ and } q \notin Q_0^2\} \cup \\ \{(q, x, C, (r, s)) \mid (q, x, C, r) \in \delta_1, r \in F_1 \text{ and } s \in Q_0^2\} \cup \\ \{((q, r), x, C, s) \mid (r, x, C, s) \in \delta_2, r \in Q_0^2 \text{ and } q \in F_1\}\end{aligned}$$

The resulting automaton has $L_l(\mathcal{A}_;) = L_l(\mathcal{A}_1); L_l(\mathcal{A}_2)$.

The star construction is similar to the above, in the sense that each pair consisting of a final state and an initial one gives a new state. Hence we start, say, with $\mathcal{A}_1$ above, which is assumed to satisfy the requirement that initial states do not have incoming transitions and final states do not have outgoing transitions. Then construct $\mathcal{A}_* = (Q_*, \Sigma, \delta_*, Q_0^1, F_*)$ where Q_0^1 is taken from $\mathcal{A}_1$ and

$$Q_* = (Q_1 \setminus F_1) \cup (F_1 \times Q_0^1) \qquad F_* = F_1 \times Q_0^1$$

$$\delta = \{(q,x,C,r) \mid (q,x,C,r) \in \delta_1 \text{ and } r \notin F_1\} \cup \\ \{(q,x,C,(r,s)) \mid (q,x,C,r) \in \delta_1, r \in F_1 \text{ and } s \in Q_0^1\} \cup \\ \{((q,r),x,C,s) \mid (r,x,C,s) \in \delta_1, r \in Q_0^1 \text{ and } q \in F_1\}$$

The resulting automaton has $L_l(\mathcal{A}_*) = (L_l(\mathcal{A}))^*$.

For the left-to-right inclusion, fix some ECA $\mathcal{A} = (Q, \Sigma, \delta, Q_0, F)$ with $Q = \{q_1, q_2, \ldots, q_n\}$ and denote $\mathcal{A}_{ij} = (Q, \Sigma, \delta, \{q_i\}, \{q_j\})$. Build as in [HU] the sequence of ϵ-regular expressions $(E_{ij}^k)_{1 \leq i,j,k \leq n}$ whose concrete semantics equals the set of LETSS which have a run in $\mathcal{A}_{ij}$ whose intermediate states (i.e. excepting q_i and q_j) have indices less than k:

$$E_{ij}^0 = \begin{cases} \sum\limits_{(q_i,x,C,q_j) \in \delta} (x,C) & \text{iff } i \neq j \\ \sum\limits_{(q_i,x,C,q_j) \in \delta} (x,C) + () & \text{iff } i = j \end{cases}$$

$$E_{ij}^k = E_{ij}^{k-1} + E_{ik}^{k-1}; (E_{kk}^{k-1})^*; E_{kj}^{k-1}$$

Then $|\sum\limits_{q_i \in Q_0} \sum\limits_{q_j \in F} E_{ij}^n| = L_l(\mathcal{A})$ □

Note that the proof of this theorem may also run as follows: we can define *symbolic semantics* of ϵ-regular expressions as words over the set of symbols $\Omega = \Sigma \times Constr(\Sigma)$ that are the classical semantics of the ϵ-regular expression, e.g. the symbolic semantics of $(a,C)^*$ consists of all the symbolic words with n concatenated symbols $(a,C)\ldots(a,C)$. These symbolic words are in fact star-free sum-free ϵ-regular expressions. We can do the same for ECA by defining the *symbolic language* accepted by the ECA as the set of words over Ω which are the concatenation of the labels on some run, or, equivalently, the set of words over Ω which are accepted by the ECA when it is regarded as a finite automaton over Ω. The classical Kleene theorem then says that the classes of symbolic semantics of ϵ-regular expressions and of symbolic languages of ECA are equal. Note how the empty expression plays the role of the empty symbolic word.

Then two more things are left to prove: the first is that the concrete semantics of ϵ-regular expressions equals the union of the concrete semantics of all symbolic words in the symbolic semantics of the ϵ-regular expression. The second is that the limited observation language of an ECA equals the union of the concrete semantics of all symbolic words in the symbolic language of the ECA. Both proofs are straightforward.

Theorem 2 (Abstract Kleene theorem). *The class of sets of TSS which are the abstract language of some ECA equals the class of sets of TSS which are the abstract semantics of some ϵ-regular expression.*

This is a corollary of Theorem 1 and Proposition 1:

Note that similarily to all the above we may aprove that ϵ-free ECA are equivalent to regular expressions.

5 Conclusions

We have described here a class of regular expressions that have the same expressive power as event-clock automata. The work can easily be extended to automata on infinite (extended) timed state sequences and ω-regular expressions as in the untimed case [Tho90] or the timed automata case [ACM97]. Our result relies on a semantics of concatenation of regular expressions as a *partial operation* on a suitable defined class of timed state sequences. We think this happens as event-clock automata are an interleaving model of distributed real-time systems.

Recently Bouyer and Petit [BP99] have proved a Kleene theorem for timed automata that does not require renaming, as in [ACM97]. The problem seems to come again from defining concatenation on real-time items. In [BP99] this is solved by defining *several* concatenations constrained by the set of clocks to be reset at the "chopping" point. It would be interesting to see if our approach (i.e. keeping a unique, but partial, concatenation while changing the items to be concatenated) can be applied to timed automata too.

References

[AD94] R. Alur and D.L. Dill. A theory of timed automata, *Theoretical Computer Science*, 126, 183-235, 1994.

[AFH94] R. Alur, L. Fix and T.A. Henzinger. A determinizable class of timed automata, *Theoretical Computer Science*, 211, 253-273, 1999.

[ACM97] E. Asarin, P. Caspi and O. Maler, A Kleene Theorem for Timed Automata, in G. Winskel (Ed.) *Proc. LICS'97*, 160-171, 1997.

[BP99] P. Bouyer and A. Petit, Decomposition and composition of timed automata, to appear in *Proc. of ICALP '99*, LNCS series, 1999.

[Di1] C. Dima. Real-time automata and their class of accepted languages, draft of a UNU/IIST report, available at `http://funinf.math.unibuc.ro/~cdima/work/expressive.ps.gz`.

[Di2] C. Dima. Complementation of real-time automata, submitted to *FCT&TCS'99*, available at `http://funinf.math.unibuc.ro/~cdima/work/mfcs.ps.gz`.

[Di3] C. Dima. Automata and Regular Expressions for Real-Time Languages, submitted to *AFL'99*, abstract available at `http://funinf.math.unibuc.ro/~cdima/work/abstract.ps.gz`.

[Di4] C. Dima. Removing ϵ-transitions from event-clock automata, in preparation, available at `http://funinf.math.unibuc.ro/~cdima/work/eps-elim.ps.gz`.

[HRS98] T.A. Henzinger, J.-F. Raskin and P.-Y. Schobbens. The regular real-time languages, in *Proceedings of the 25-th International Colloquium on Automata, Languages and Programming* LNCS series, Springer Verlag, 1998.

[HU] John E. Hopcroft and Jeffrey D. Ullman, *Introduction to Automata Theory, Languages and Computation*, Addison-Wesley/Narosa Publishing House, eighth edition, New Delhi, 1992.

[RS97] J.-F. Raskin and P.-Y. Schobbens. State-clock logic: a decidable real-time logic, in *Hybrid and Real-Time Systems*, LNCS 1201, 33-47, Springer Verlag, 1997.

[Tho90] W. Thomas. Automata on infinite objects, in J. van Leeuwen (Ed.), *Handbook of Theoretical Computer Science*, vol B, 133-191, Elsevier, Amsterdam, 1997.

Strong Iteration Lemmata for Regular, Linear, Context-Free, and Linear Indexed Languages

Pál Dömösi[1] and Manfred Kudlek[2]

[1] Institute of Mathematics and Informatics, Lajos Kossuth University
Debrecen, Egyetem tér 1, H-4032, Hungary
email : domosi@math.klte.hu

[2] Fachbereich Informatik, Universität Hamburg
D-22527 Hamburg, Vogt-Kölln-Str. 30, D-22527 Hamburg, Germany
email : kudlek@informatik.uni-hamburg.de

Abstract. New iteration lemmata are presented, generalizing most of the known iteration lemmata for regular, linear, context-free, and linear indexed languages.

1 Introduction

In [8] a new iteration lemma for context-free languages was presented, generalizing the Bader Moura Lemma ([2]). In this article is shown that it can be further generalized in the following directions :

(1) strengthening the lower bound of the number of distinguished positions of a word z to be linear in the number of excluded positions of z instead of being exponential,

(2) strengthening the upper bound of the number of distinguished positions of vwx in the same way.

Most of the known iteration lemmata ([2], [3], [14], [15]) then follow from specializations, namely to have no excluded positions, or to distinguish all positions.

Such strong iteration lemmata are also shown for regular and linear context-free (for short linear) languages. Using a characterization of linear indexed languages ([9]), also a new strong iteration lemma for that class is shown.

All notations not explicitly defined are standard, and may be found in [10], [12], [16].

A *linear indexed grammar* is a 5-tuple $G = (V, X, I, S, P)$, where V, X, I are finite pairwise disjoint sets, the set of *variables, terminals,* and *indices*, respectively, $S \in V$ is the *start variable*, and P is a finite set of pairs (Af, a), with $A \in V$, $f \in I \cup \{\lambda\}$, $a \in X^* \cup X^*VI^*X^*$, the set of *productions.* $(Af, a) \in P$ is denoted by $Af \to a$.

Let $\alpha = uAf_1f_2\cdots f_\ell v$ with $u, v \in X^*, A \in V, f_1 \in I \cup \{\lambda\}, f_2, \cdots, f_\ell \in I$, $\ell \geq 1$. If $Af_1 \to u'Bf_1' \cdots f_m'v' \in P$ with $u', v' \in X^*, B \in V, f_1', \cdots, f_m' \in I$, $m \geq 0$, then we set $\alpha \to \beta$ with $\beta = uu'Bf_1' \cdots f_m'f_2 \cdots f_\ell v'v$. Moreover, if $Af_1 \to u' \in P$ with $u' \in X^*$, then we set $\alpha \to \beta$ with $\beta = uu'f_2 \ldots f_\ell v$.

The *language generated by a linear indexed grammar* $G = (V, X, I, S, P)$ is the set $L(G) = \{w \mid w \in X^* \text{ and } S \overset{*}{\Rightarrow} w\}$, where $\overset{*}{\Rightarrow}$ denotes the reflexive and transitive closure of $\Rightarrow$. A language L is called *linear indexed* if $L = L(G)$ for some linear indexed grammar G.

Let $\delta(z)$ denote the number of *distinguished* positions in a word z, and $\epsilon(z)$ the number of *excluded* positions in z. Note that an excluded position might also be a distinguished one. Furthermore, let $\sigma(z)$ denote the number of *selected*, i.e. distinguished, but not excluded positions in z. Trivially, $\delta(z) \leq \sigma(z) + \epsilon(z)$.

The following theorems will be needed to prove the new iteration lemmata.

Theorem 1. *(Generalized Bader Moura Lemma, see [4], [8], [13]) Let L be a context-free language. Then there exists an integer $n = n(L) \geq 2$, depending only on the language L, such that for any $z \in L$ with $\delta(z) > n^{\epsilon(z)+1}$ there exist u, v, w, x, y such that $z = uvwxy$ with*

(1) $\epsilon(vx) = 0$ and
either $\delta(u) > 0, \delta(v) > 0, \delta(w) > 0$
or $\delta(w) > 0, \delta(x) > 0, \delta(y) > 0$,

(2) $\delta(vwx) \leq n^{\epsilon(w)+1}$

(3) $\forall i \geq 0 : uv^iwx^iy \in L$. □

An immediate consequence, by choosing $\epsilon(z) = 0$ in Theorem 1 is

Theorem 2. *Let L be a context-free language. There is an integer $n = n(L) \geq 2$, depending only on the language L, such that for any $z \in L$ with $\delta(z) > n$ there exist u, v, w, x, y such that $z = uvwxy$ with*

(1) either $\delta(u) > 0, \delta(v) > 0, \delta(w) > 0$
or $\delta(w) > 0, \delta(x) > 0, \delta(y) > 0$.

(2) $\delta(vwx) \leq n$

(3) $\forall i \geq 0 : uv^iwx^iy \in L$. □

Theorem 3. *(see [9]) If $L' \subseteq Y^*$ is a linear indexed language, then there exist an alphabet X, a context-free language $L \subseteq X^*$ and two homomorphisms $h_1, h_2 : X^* \rightarrow Y^*$ such that $L' = \{h_1(w)h_2(w)^R \mid w \in L\}$ holds where z^R denotes the mirror image of z.*

□

2 Results

The first new theorem generalizes the well known iteration lemma for regular languages to distinguished and excluded positions.

Theorem 4. *Let L be a regular language. Then there is an integer $n \geq 2$, depending only on L, such that for any $z \in L$ with $\delta(z) > n \cdot max(\epsilon(z), 1)$ there exist u, v, w such that $z = uvw$ with*

(1) $\epsilon(v) = 0$, $\delta(u) > 0$, $\delta(v) > 0$

(2) $\delta(uv) \leq n \cdot (\epsilon(u) + 1)$

(3) $\forall i \geq 0 : uv^iw \in L$.

Proof. Consider a type 3 grammar $G = (V, V_T, S, P)$ in normal form, generating $L = L(G)$. This means that all productions are of the forms $A \to Ba$ or $A \to a$ with $A \in V_N = V - V_T$ and $a \in V_T$. Then the derivation tree of any $z \in L$ has a comb-like structure, i.e. there is one main path with exactly one right child having a terminal label, except for the lowest node which also has a terminal label. The depth of this tree is exactly $|z|$.

Define a ***branch point*** to be a node on the main path whose right child is an excluded position.

Let a ***distinguished*** path be a subpath of the main path, where either the upper node is the root and the lower one a branch point, or the upper and lower ones are branch points, or the upper one is a branch point but not the root and the lower one the lowest (a leaf), and all intermediate nodes are no branch points. It is easy to see that there are at most $\epsilon(z) + 1$ distinguished paths.

Define $n = 2|V_N| + 5$.

If $\epsilon(z) = 0$ then there are more than $2|V_N| + 5$ distinguished positions in z. Let p be the ***main*** path. Then it consists of subpaths p_1, p_2, p_3 generating $\geq 1, |V_N| + 1, 1$ distinguished positions, respectively. Therefore there are two nodes on p_2 with the same label A, with $A \stackrel{*}{\Rightarrow} Av \stackrel{*}{\Rightarrow} uv$, $\delta(v) > 0$, $z = uvw$, and $\forall i \geq 0 : uv^iw \in L$. Since p_3 generates exactly 1 distinguished position it follows that $\delta(u) > 0$. Furthermore, $\delta(uv) \leq |V_N| + 2 \leq n$.

Let $\epsilon(z) > 0$. It is impossible that all distinguished paths generate at most $|V_N| + 2$ selected positions, since in that case it follows that

$\delta(z) \leq (\epsilon(z) + 1)(|V_N| + 2) + \epsilon(z) \leq (2|V_N| + 5) \cdot \epsilon(z)$.

Thus there is a distinguished path p generating more than $|V_N| + 1$ selected positions. These have at least $|V_N| + 2$ fathers on p. Divide again p into subpaths p_1, p_2, p_3 generating $\geq 1, |V_N| + 1, 1$ selected positions, respectively.

Then there are two such fathers on p_2 with the same label A, $A \stackrel{*}{\Rightarrow} Av \stackrel{*}{\Rightarrow} uv$, $\sigma(v) > 0$, $z = uvw$, and $\forall i \geq 0 : uv^iw \in L$. Since p_3 generates exactly 1 selected position it follows that $\sigma(u) > 0$, and trivially $\epsilon(v) = 0$. Choosing p to be the lowest of such paths implies

$\delta(uv) \leq (\epsilon(u) + 1) \cdot (|V_N| + 2) + \epsilon(u) \leq n \cdot (\epsilon(u) + 1)$.

□

The second new theorem states a similar result for linear languages, generalizing known iteration lemmata.

Theorem 5. *Let L be a linear language. Then there is an integer $n \geq 2$, depending only on L, such that for any $z \in L$ with $\delta(z) > n \cdot max(\epsilon(z), 1)$ there exist u, v, w, x, y such that $z = uvwxy$ with*

(1) $\epsilon(vx) = 0$ and
either $\delta(u) > 0$, $\delta(v) > 0$, $\delta(w) > 0$,
or $\delta(w) > 0$, $\delta(x) > 0$, $\delta(y) > 0$
(3) $\delta(uvxy) \leq n \cdot (\epsilon(uy) + 1)$
(3) $\forall i \geq 0 : uv^iwx^iy \in L$.

Proof. Consider a linear grammar $G = (V, V_T, S, P)$ in normal form, generating $L = L(G)$. This means that all productions are of the forms $A \to Ba$, $A \to aB$, or $A \to a$ with $A, B \in V_N = V - V_T$ and $a \in V_T$. Then the derivation tree of any $z \in L$ has a main path where the left or right children of nodes on it have terminal labels, except for the lowest node which also has a terminal label. The depth of that tree is exactly $|z|$.

Define a *branch point* to be a node on the main path whose left or right child is an excluded position.

Let a *distinguished* path be a subpath of the main path, where either the upper node is the root and the lower one a branch point, or the upper and lower ones are branch points, or the upper one is a branch point but not the root and the lower one the lowest (a leaf), and all intermediate nodes are no branch points. Again, there are at most $\epsilon(z) + 1$ distinguished paths.

Define $n = 4|V_N| + 7$.

If $\epsilon(z) = 0$ then there are more than $2|V_N| + 5$ distinguished positions in z. Let p be the *main* path. Then it consists of subpaths p_1, p_2, p_3, p_4 generating $1, |V_N| + 1, |V_N| + 1, \geq 1$ distinguished positions, respectively.

Case I. $p_2 p_3$ contains two nodes with identical labels A, $A \overset{*}{\Rightarrow} vAx \overset{*}{\Rightarrow} vwx$, $\delta(v) > 0$, $\delta(x) > 0$. Since $\delta(uy) > 0$, $\delta(w) > 0$ it follows that either $\delta(u) > 0$, $\delta(v) > 0$, $\delta(w) > 0$, or $\delta(w) > 0$, $\delta(x) > 0$, $\delta(y) > 0$, and $\forall i \geq 0 : uv^i wx^i y \in L$.

Case II. Let the paths p_i generate a_i (b_i) to the left (right), respectively. Note that p_4 always contributes $\delta(w) > 0$ since $\delta(a_4 b_4) > 0$.

Consider 6 subcases where L (R) denote that p_2, p_3 generate distinguished positions only on the left (right), respectively.

$RR : \delta(a_1 b_1) > 0$, $\delta(a_2) = 0$, $\delta(b_2) > 0$, $\delta(a_3) = 0$, $\delta(b_3) > 0$, $b_3 = c_3 x d_3$, $\delta(x) > 0$, and p_2 contributes $\delta(y) > 0$.

$RL1 : \delta(a_1) > 0$, $\delta(b_1) \geq 0$, $\delta(a_2) = 0$, $\delta(b_2) > 0$, $\delta(a_3) > 0$, $\delta(b_3) = 0$, $a_3 = c_3 v d_3$, $\delta(v) > 0$, and p_1 contributes $\delta(u) > 0$.

$RL2 : \delta(a_1) \geq 0$, $\delta(b_1) > 0$, $\delta(a_2) = 0$, $\delta(b_2) > 0$, $\delta(a_3) > 0$, $\delta(b_3) = 0$, $b_2 = c_2 x d_2$, $\delta(x) > 0$, and p_1 contributes $\delta(y) > 0$.

The other 3 subcases are symmetric, left and right, L and R, u and y, v and x, interchanged.

Then $\delta(uvxy) \leq 2|V_N| + 3 < n$.

Let $\epsilon(z) > 0$.

It is impossible that all distinguished paths generate at most $2|V_N|+3$ selected positions, since in that case

$\delta(z) \leq (\epsilon(z) + 1)(2|V_N| + 3) + \epsilon(z) \leq (4|V_N| + 7) \cdot \epsilon(z) = n \cdot \epsilon(z)$.

Thus there is a distinguished path p generating more than $2|V_N| + 3$ selected positions. These have at least $2|V_N|+4$ corresponding fathers on p. Divide again p into subpaths p_1, p_2, p_3, p_4 generating $1, |V_N|+1, |V_N|+1, \geq 1$ selected positions, respectively.

Case I. $p_2 p_3$ contains two nodes with identical labels A, $A \overset{*}{\Rightarrow} vAx \overset{*}{\Rightarrow} vwx$, $\sigma(v) > 0$, $\sigma(x) > 0$. Since $\sigma(uy) > 0$, $\sigma(w) > 0$ it follows that either $\sigma(u) > 0$, $\sigma(v) > 0$, $\sigma(w) > 0$, or $\sigma(w) > 0$, $\sigma(x) > 0$, $\sigma(y) > 0$, and $\forall i \geq 0 : uv^i wx^i y \in L$.

Case II. Let the paths p_i generate a_i (b_i) to the left (right), respectively. Note that p_4 always contributes $\delta(w) > 0$ since $\delta(a_4 b_4) > 0$.

Consider 6 subcases where L (R) denote that p_2, p_3 generate selected positions only on the left (right), respectively.

$RR : \sigma(a_1 b_1) > 0, \sigma(a_2) = 0, \sigma(b_2) > 0, \sigma(a_3) = 0, \sigma(b_3) > 0, b_3 = c_3 x d_3,$
$\sigma(x) > 0$, and p_2 contributes $\sigma(y) > 0$.

$RL1 : \sigma(a_1) > 0, \sigma(b_1) \geq 0, \sigma(a_2) = 0, \sigma(b_2) > 0, \sigma(a_3) > 0, \sigma(b_3) = 0,$
$a_3 = c_3 v d_3, \sigma(v) > 0$, and p_1 contributes $\sigma(u) > 0$.

$RL2 : \sigma(a_1) \geq 0, \sigma(b_1) > 0, \sigma(a_2) = 0, \sigma(b_2) > 0, \sigma(a_3) > 0, \sigma(b_3) = 0,$
$b_2 = c_2 x d_2, \sigma(x) > 0$, and p_1 contributes $\sigma(y) > 0$.

The other 3 subcases are symmetric, left and right, L and R, u and y, v and x, interchanged.

Choosing p to be the highest of such paths implies

$$\delta(uvxy) \leq (\epsilon(uy) + 1) \cdot (2|V_N| + 3) + \epsilon(uy) \leq (\epsilon(uy) + 1) \cdot (2|V_N| + 4)$$
$$\leq n \cdot (\epsilon(uy) + 1).$$

□

Theorem 6. *Let L be a context-free language. Then there exists an integer $n = n(L) \geq 2$, depending only on the language L, such that for any $z \in L$ with $\delta(z) > n \cdot max(\epsilon(z), 1)$ there exist u, v, w, x, y such that $z = uvwxy$ with*

(1) $\epsilon(vx) = 0$ and
either $\delta(u) > 0, \delta(v) > 0, \delta(w) > 0$
or $\delta(w) > 0, \delta(x) > 0, \delta(y) > 0$

(2) $\delta(vwx) \leq n \cdot (\epsilon(w) + 1)$

(3) $\forall i \geq 0 : uv^i wx^i y \in L$.

Proof. Let G be a context-free grammar in Chomsky normal form generating L, and consider a derivation tree T_z for z. Define a node to be a *branch point* if it has 2 children and both of them have excluded descendants. Define a node to be *free* if both children have no excluded descendants.

A (partial) path is called *distinguished* if

a) none of its intermediate nodes (i.e not initial and terminal) is a branch point, or it has no intermediate nodes.

b) the initial node is either a branch point or the root of the tree, and the terminal node is either a branch point or an excluded position.

From this definitions follows that T_z has exactly $\epsilon(z) - 1$ branch points, and that the number of distinguished paths is either $2 \cdot \epsilon(z) - 2$ if the root of T_z is a branch point, or $2 \cdot \epsilon(z) - 1$ if the root is not a branch point.

Now define $n = 2n' \cdot (2|V_N| + 3) + 1$ where n' is the constant from Theorem 2.

If $\epsilon(z) = 0$ the statement is just Theorem 2.

Let $\epsilon(z) > 0$. There are at most $2\epsilon(z) - 1$ distinguished paths, and more than $(n-1) \cdot \epsilon(z)$ selected positions. All these are generated by the free children of intermediate nodes on distinguished paths. It follows that at least one distinguished path p has to generate more than $\frac{1}{2}(n-1)$ selected positions, since otherwise $\frac{1}{2}(n-1) \cdot (2\epsilon(z) - 1) \leq (n-1) \cdot \epsilon(z)$. Distinguish two cases.

Case I. There exists a free child of p being the root of a binary tree generating more than n' selected positions. Then Theorem 2 can be applied yielding

$u''vwxy'$ with $\epsilon(vx) = 0$, either $\delta(u'') > 0, \delta(v) > 0, \delta(w) > 0$, or $\delta(w) > 0$, $\delta(x) > 0$, $\delta(y') > 0$, $\delta(vwx) \leq n' < n \cdot (\epsilon(w) + 1)$, and $\forall i \geq 0 : uv^iwx^iy \in L$ where $u = u'u''$, $y = y'y''$.

Case II. Each free child of p generates at most n' selected positions. To generate more than $\frac{1}{2}(n-1)$ selected positions there have to exist more than $2 \cdot |V_N| + 3$ free children of p, each generating at least one selected position, since otherwise at most $n' \cdot (2|V_N| + 3) = \frac{1}{2}(n-1)$ selected positions are generated by p.

Each of these free children has a father on p, and there are $\geq 2 \cdot |V_N| + 4$ such. p consists of subpaths p_1, p_2, p_3, p_4, generating $\geq 1, |V_N + 1|, |V_N| + 1, 1$ selected positions, respectively.

Thus both, p_2, p_3 contain $|V_N| + 1$ fathers of free children, each generating at least one selected position. Therefore on each, p_2, p_3 there exist two such fathers with identical labels A (B), respectively.

Let the paths p_i generate a_i (b_i) to the left (right), respectively. Note that p_4 always contributes $\delta(w) > 0$ since $\delta(a_4b_4) > 0$.

Case II.1. p_2p_3 contains 2 such fathers with identical labels A, and with $A \stackrel{*}{\Rightarrow} vAx$, $\sigma(v) > 0$, $\sigma(x) > 0$. Now, since $\sigma(a_1b_1) > 0$, it follows that either $\delta(u) > 0, \delta(v) > 0, \delta(w) > 0$, or $\delta(w) > 0$, $\delta(x) > 0, \delta(y) > 0$.

Case II.2. Consider 6 subcases where L (R) denote that p_2, p_3 generate selected positions only on the left (right), respectively.

$RR : \sigma(a_1b_1) > 0$, $\sigma(a_2) = 0$, $\sigma(b_2) > 0$, $\sigma(a_3) = 0$, $\sigma(b_3) > 0$, $b_3 = c_3xd_3$, $\sigma(x) > 0$, and p_2 contributes $\sigma(y) > 0$.

$RL1 : \sigma(a_1) > 0$, $\sigma(b_1) \geq 0$, $\sigma(a_2) = 0$, $\sigma(b_2) > 0$, $\sigma(a_3) > 0$, $\sigma(b_3) = 0$, $a_3 = c_3vd_3$, $\sigma(v) > 0$, and p_1 contributes $\sigma(u) > 0$.

$RL2 : \sigma(a_1) \geq 0$, $\sigma(b_1) > 0$, $\sigma(a_2) = 0$, $\sigma(b_2) > 0$, $\sigma(a_3) > 0$, $\sigma(b_3) = 0$, $b_2 = c_2xd_2$, $\sigma(x) > 0$, and p_1 contributes $\sigma(y) > 0$.

The other 3 subcases are symmetric, left and right, L and R, u and y, v and x, interchanged.

Choosing p to be nearest to a leaf, and on p the lower $2 \cdot |V_N| + 4$ such fathers, the children of which generate at least one selected, position, implies $\delta(vwx) < (2\epsilon(w) - 1) \cdot \frac{1}{2}(n-1) + \epsilon(w) \leq n \cdot \epsilon(w) \leq n \cdot (\epsilon(w) + 1)$.

□

Theorem 7. *Let L be a linear indexed language. There exists an integer $n \geq 2$ depending only on L such that, for any $z \in L$ with $\delta(z) > n \cdot (\epsilon(z) + 1)$ there exist $u_1, v_1, w_1, x_1, y_1, u_2, v_2, w_2, x_2, y_2$ such that*
$z = u_1v_1w_1x_1y_1y_2x_2w_2v_2u_2$ with

(1) $\epsilon(v_1x_1v_2x_2) = 0$ and
either $\delta(u_1u_2) > 0, \delta(v_1v_2) > 0, \delta(w_1w_2) > 0$,
or $\delta(w_1w_2) > 0, \delta(x_1x_2) > 0, \delta(y_1y_2) > 0$

(2) $\delta(v_jw_jx_j) \leq n \cdot (\epsilon(w_1w_2) + 1)$, $j = 1, 2$

(3) $\forall i \geq 0 : u_1{v_1}^iw_1{x_1}^iy_1y_2{x_2}^iw_2{v_2}^iu_2 \in L$

Proof. Consider a linear indexed language $L \subseteq Y^*$. By Theorem 3, there exist an alphabet X, a context-free language $L' \subseteq X^*$ and also two homomorphisms $h_1, h_2 : X^* \to Y^*$ such that $L = \{h_1(w)h_2(w)^R \mid w \in L'\}$.

Let $a = 2 \cdot max\{|h_i(x)| \mid i \in \{1,2\}, x \in X\}$ and $n = an'$ where n' is the constant for L' from Theorem 6. Note that $a > 0$ since otherwise $L = \{\lambda\}$, a trivial case.

Consider a word $z \in L$ with $\delta(z) > n \cdot (\epsilon(z) + 1)$. There is a word $p \in L'$ such that $z = h_1(p)h_2(p)^R$. Let $|p|$ be minimal, i.e. $z = h_1(p')h_2(p')^R, p' \in L'$ implies $|p| \leq |p'|$. Let $p = s_1 \cdots s_t \in X^+$, $s_1, \cdots, s_t \in X$.

Consider the words $z_1 = h_1(p)$ and $z_2 = h_2(p)$. Then $\epsilon(z) = \epsilon(z_1) + \epsilon(z_2)$. For $i \in \{1, \cdots, t\}$ exclude the ith position of $p = s_1 \cdots s_t$ if and only if one of the positions of $h_1(s_i)h_2(s_i)$ is excluded. Then $\epsilon(p) \leq \epsilon(z_1) + \epsilon(z_2) = \epsilon(z)$.

Furthermore, for $i \in \{1, \cdots, t\}$ distinguish the ith position of $p = s_1 \cdots s_t$ if and only if one of the positions of $h_1(x_i)h_2(x_i)$ is distinguished. This implies $a \cdot \delta(p) \geq \delta(z) > n \cdot max(\epsilon(z), 1) = (an') \cdot max(\epsilon(z), 1) \geq a \cdot n' \cdot max(\epsilon(p), 1)$, and this implies $\delta(p) > n' \cdot max(\epsilon(p), 1)$.

Then Theorem 6 can be applied to L' and $p \in L'$. Thus $p = uvwxy$ with either $\delta(u) > 0, \delta(v) > 0, \delta(w) > 0$, or $\delta(w) > 0, \delta(x) > 0, \delta(y) > 0$, $\epsilon(vx) = 0$, $\delta(vwx) \leq n \cdot (\epsilon(w) + 1)$, $\forall i \geq 0 : uv^iwx^iy \in L'$, and
$z = h_1(u)h_1(v)h_1(w)h_1(x)h_1(y)h_2(y)^R h_2(x)^R h_2(w)^R h_2(v)^R h_2(u)^R$.

Assume $h_1(v)h_1(x)h_2(x)h_2(v) = \lambda$. Now $\forall i \geq 0 : uv^iwx^iy \in L'$, and in particular $uwy \in L'$. Therefore $z = h_1(uwy)h_2(uwy)^R$. Since $|vx| > 0$, this is a contradiction to the minimality of $|p|$. Thus $|h_1(v)h_1(x)h_2(x)h_2(v)| > 0$.

Set $u_1 = h_1(u)$, $v_1 = h_1(v)$, $w_1 = h_1(w)$, $x_1 = h_1(x)$, $y_1 = h_1(y)$, and $y_2 = h_2(y)^R$, $x_2 = h_2(x)^R$, $w_2 = h_2(w)^R$, $v_2 = h_2(v)^R$, $u_2 = h_2(u)^R$. Then $|v_1x_1x_2v_2| > 0$. From $\epsilon(vx) = 0$ follows $\epsilon(v_1x_1x_2v_2) = 0$.

From $\delta(u) > 0, \delta(v) > 0, \delta(w) > 0$ follows that $\delta(u_1u_2) > 0$, $\delta(v_1v_2) > 0$, $\delta(w_1w_2) > 0$, and from $\delta(w) > 0, \delta(x) > 0, \delta(y) > 0$ follows that $\delta(w_1w_2) > 0$, $\delta(x_1x_2) > 0$, $\delta(y_1y_2) > 0$.

Theorem 6 gives $\delta(vwx) \leq (n') \cdot (\epsilon(w) + 1)$. Since $\epsilon(w) \leq \epsilon(w_1w_2)$ it follows that

$$\begin{aligned}\delta(v_jw_jx_j) &\leq a \cdot \delta(vwx) \leq a \cdot (n') \cdot (\epsilon(w) + 1) \leq (an') \cdot (\epsilon(w) + 1) \\ &\leq n \cdot (\epsilon(w) + 1) \leq n \cdot (\epsilon(w_1w_2) + 1)\end{aligned}$$

for $j = 1, 2$. □

References

1. Aho, A. V., *Indexed Grammars*. Journ. of ACM, **15** (1968), 647-671.
2. Bader, C., Moura, A., *A Generalization of Ogden's Lemma*. JACM **29**, no. 2, (1982), 404-407.
3. Bar-Hillel, Y., Perles, M., Shamir, E., *On Formal Properties of Simple Phrase Structure Grammars*. Zeitschrift für Phonetik, Sprachwissenschaft, und Kommunikationsforschung, **14** (1961), 143-172.
4. Berstel, J., Boasson, L., *Context-free Languages.*, in *Handbook of Theoretical Computer Sciences, Vol. B : Formal Models and Semantics*, van Leeuwen, J., ed., Elsevier/MIT, 1994, 60-102.
5. Dömösi, P., Duske, J., *Subword Membership Problem for Linear Indexed Languages*. Proc. Worksh. AMILP'95 (Algebraic Methods in Language Processing,

1995), Univ. of Twente, Enschede, The Netherlands, 6-8 Dec., A. Nijholt, G. Scollo, R. Steetskamp, eds., TWLT-11, Univ. of Twente, Enschede, 1995, 235-237.

6. Dömösi, P., Ito, M., *On Subwords of Languages*. RIMS Proceedings, Kyoto Univ., **910** (1995), 1-4.
7. Dömösi, P., Ito, M., *Characterization of Languages by Lengths of their Subwords*. Proc. Int. Conf. on Semigroups and their Related Topics, (Inst. of Math., Yunnan Univ., China,) Monograph Series, Springer-Verlag, Singapore, to appear.
8. Dömösi, P., Ito, M., Katsura, M, Nehaniv, C., *A New Pumping Property of Context-free Languages*. Combinatorics, Complexity and Logic (Proc. Int. Conf. DMTCS'96), ed. D.S. Bridges et al., Springer-Verlag, Singapore, 1996, 187-193 .
9. Duske, J., Parchmann, R., *Linear Indexed Grammars*. Theoretical Computer Science **32** (1984), 47-60.
10. Harrison, M. A., *Introduction to Formal Language Theory*. Addison-Wesley Publishing Company, Reading , Massachusetts, Menlo Park, California, London, Amsterdam, Don Mils, Ontario, Sidney, 1978.
11. Hayashi, T., *On Derivation Trees of Indexed Grammars - an Extension of the uvwxy - Theorem*. Publ. RIMS, Kyoto Univ., **9** (1973), 61-92.
12. Hopcroft, J. E., Ullmann, J. D., *Introduction to Automata Theory, Languages, and Computation*. Addison-Wesley, Reading , Massachusetts, Menlo Park, California, London, Amsterdam, Don Mils,Ontario, Sidney, 1979.
13. Horváth, S., *A Comparison of Iteration Conditions on Formal Languages*. In *Algebra, Combinatorics and Logic in Computer Science*, vol. II, pp. 453-464, *Colloquia Matematica Societatis János Bolyai*, **42**, North Holland 1986.
14. Nijholt, A., *An Annotated Bibliography of Pumping*. Bull. EATCS, **17** (June, 1982), 34-52.
15. Ogden, W., *A Helpful Result for Proving Inherent Ambiguity*. Math. Syst. Theory **2** (1968), 191-194
16. Salomaa, A., *Formal Languages*, Academic Press, New York, London, 1973.

Exponential Output Size of Top-Down Tree Transducers

Frank Drewes*

Department of Computer Science
University of Bremen
P.O. Box 33 04 40
D-28334 Bremen (Germany)
drewes@informatik.uni-bremen.de

Abstract. The exponential output size problem for top-down tree transducers asks whether the size of output trees grows exponentially in the size of input trees. In this paper the complexity of this problem is studied. It is shown to be NL-complete for total top-down tree transducers, but DEXPTIME-complete in general.

1 Introduction

Top-down tree transducers have been introduced in the late sixties by Rounds and Thatcher [Rou70,Tha70,Tha73] as a generalisation of finite-state transducers on strings. The main motivation was to provide a simple formal model of syntax-directed transformational grammars in mathematical linguistics and of syntax-directed translation in compiler construction (for the latter, see the recent book by Fülöp and Vogler [FV98]). Since these times it has turned out that top-down tree transducers are a useful tool for many other areas, too, and their properties and extensions have been studied by a variety of authors. For references see, e.g., [GS97,FV98].

As mentioned above, top-down tree transducers are a generalisation of finite-state string transducers (also called generalised sequential machines) to trees[1]. Like these, top-down tree transducers are one-way machines which process their input in one direction, using a finite number of states. However, while string transducers usually process their input from left to right, top-down tree transducers transform input trees to output trees from the root towards the leaves (which, of course, is the reason for calling them *top-down* tree transducers). Roughly speaking, the string case is the special case where the input and output trees are monadic trees (which can be viewed as "vertical strings").

---

* Partially supported by the EC TMR Network GETGRATS (General Theory of Graph Transformation Systems) and the ESPRIT Working Group APPLIGRAPH through the University of Bremen.

[1] In this context, a *tree* is a labelled, ordered tree whose labels are taken from a ranked alphabet (or signature), i.e., a term.

Although the generalisation is quite a direct one, the fact that trees instead of strings are considered makes a rather crucial difference in certain respects. This concerns, for example, closure properties which are known from the string case but do not carry over to the case of top-down tree transducers. For instance, an infinite hierarchy is obtained by considering compositions of top-down tree transducers (see [Eng82]). Another important difference is that, intuitively, the computations of top-down tree transducers are usually ramifying: when the topmost node of an input tree has been processed, the computation proceeds on all subtrees in parallel. In fact, there can even be some non-linearity in the sense that subtrees are copied and the copies are processed individually. One of the most distinct consequences of this fact is that, in contrast to the string case, the size of the output trees of a top-down tree transducer is not necessarily linear in the size of its input trees. As an example, consider the two rules $\gamma[g[x]] \to f[\gamma[x], \gamma[x]]$ and $\gamma[a] \to a$ (which should be considered as term rewrite rules in the usual way). Here, γ is a state and f, g, a are symbols of rank 2, 1, and 0, respectively. Without going into the details it should be clear that these rules transform the monadic tree $g[\cdots g[a] \cdots]$ of height n into a complete binary tree of the same height. Thus, the output size is exponential in the size of input trees. It follows directly from the definition of top-down tree transducers that an exponential size of output trees is the maximum growth they can achieve. However, it is as well possible to build a top-down tree transducer whose output size is given by a polynomial of degree k, for any given $k \in \mathbb{N}$. As a simple example, consider the rules

$$\gamma[g[x]] \to f[\gamma[x], \gamma'[x]], \quad \gamma[a] \to a, \quad \gamma'[g[x]] \to g[\gamma'[x]], \quad \gamma'[a] \to a.$$

Taking γ to be the initial state, an input tree $g[\cdots g[a] \cdots]$ of size $n+1$ is turned into the output tree $f[\cdots f[f[a, t_1], t_2], \ldots t_n]$, where each t_i is a tree $g[\cdots g[a] \cdots]$ of size i. In other words, the size of output trees is quadratic in the size of input trees. It is known from [AU71] and will turn out as a corollary in Section 4 that the output size of a top-down tree transducer is either bounded by a polynomial, or is exponential.

In this paper the complexity of the corresponding decision problem is studied: given top-down tree transducer td, is its output size $os_{td}(n)$ exponential in the size n of input trees? It turns out that this problem is efficiently solvable (namely NL-complete) for total top-down tree transducers, but is very hard (namely DEXPTIME-complete) in general. The NL- respectively DEXPTIME-hardness of the two variants is relatively easy to establish, but it is less obvious that these resources are indeed sufficient. In fact, a decision algorithm for the exponential output size problem is implicit in the results of [AU71] (using somewhat different notions), but complexity issues are not addressed in that paper.

Although the corresponding problem for bottom-up tree transducers will not be considered in this paper, it should be remarked that the same results can be proved for this case as well. The proofs are similar to some of the proofs given here, but are rather straight forward (and thus less interesting) because the computations of bottom-up tree transducers have a considerably simpler structure. In particular, the combinatorial results on trees developed in Section 4 are not required.

The paper is structured as follows. In the next section the basic notions and in Section 3 top-down tree transducers are recalled, and it is shown that the exponential output size problem can be reduced to the case of total deterministic string-to-tree transducers. In Section 4 a result is shown which will be used by the decision algorithms developed in Section 5, where the main result of this paper is presented. Finally, in Section 6 a short conclusion is given. Due to lack of space, a number of details in the proofs must be skipped. The complete proofs will be given in the full version of the paper.

2 Preliminaries

The sets of all natural numbers, including and excluding 0, are denoted by $\mathbb{N}$ and $\mathbb{N}_+$, respectively. For every $n \in \mathbb{N}$, $[n]$ denotes the set $\{1, \ldots, n\}$. The set of all finite sequence over a set A is denoted by A^*. The empty sequence is denoted by λ; the length of a sequence s is denoted by $|s|$. Concatenation of sequences is denoted by juxtaposition.

Like the length of a sequence, the cardinality of a set A is denoted by $|A|$. The canonical extensions of a function $f\colon A \to B$ to the power set of A and to A^* are denoted by f, too. Hence, $f(A') = \{f(a) \mid a \in A'\}$ for all $A' \subseteq A$, and $f(a_1 \cdots a_n) = f(a_1) \cdots f(a_n)$ for all $a_1, \ldots, a_n \in A$. The reflexive and transitive closure of a binary relation $r \subseteq A \times B$ is denoted by r^*. The domain of r, i.e., the set $\{a \in A \mid (a, b) \in r \text{ for some } b \in B\}$, is denoted by $dom(r)$.

A (finite, ordered) *unlabelled tree* is a finite prefix-closed subset T of $\mathbb{N}_+^*$. The elements of T are called its *nodes*. The *rank* of a node v in T is the number of distinct natural numbers i such that $vi \in T$. The rank of T is the maximum rank of its nodes. A *leaf* is a node of rank 0. A node u is a *descendant* of a node v if v is a proper prefix of u. Conversely, u is a *predecessor* of v if it is a proper prefix of v. The *subtree of T rooted at v* is the tree $\{v' \mid vv' \in T\}$. A *direct subtree of T* is a subtree of T rooted at v for some $v \in T \cap \mathbb{N}$. The *size* of T is $|T|$, its *height* is $ht(T) = \max\{|v| \mid v \in T\} - 1$, and its *width*, denoted by $wd(T)$, is the number of leaves in T.

A *labelled tree* is a mapping $t\colon T \to L$, where T is an unlabelled tree and L is a set of labels. The underlying unlabelled tree T is also denoted by $N(t)$ in this case. All notions and notations introduced for unlabelled trees above carry over to labelled trees in the obvious way. In the following, the attributes 'labelled' and 'unlabelled' will mostly be dropped when speaking about trees. As a general rule, unlabelled trees will be denoted by capital letters (usually T) whereas labelled trees will be denoted by lowercase letters (usually s and t).

For trees $t_1, \ldots, t_k$ and a label f, $f[t_1, \ldots, t_k]$ denotes the tree t such that $N(t) = \{\lambda\} \cup \bigcup_{i \in [k]} \{iv \mid v \in N(t_i)\}$, where $t(\lambda) = f$ and $t(iv) = t_i(v)$ for all $i \in [k]$ and $v \in N(t_i)$. The tree $f[]$ is usually denoted by f (which actually means that a single-node tree is identified with the label of that node).

A *signature* is a pair $(\Sigma, \mathit{rank}_\Sigma)$ consisting of a finite set Σ and a mapping $\mathit{rank}_\Sigma\colon \Sigma \to \mathbb{N}$ which assigns to every $f \in \Sigma$ its *rank*. For notational convenience, in the following we shall write Σ instead of $(\Sigma, \mathit{rank}_\Sigma)$. If Σ is clear from the context, writing $f^{(k)}$ for $f \in \Sigma$ is a shorthand for stating at the same time that $\mathit{rank}_\Sigma(f) = k$. A signature Σ is *monadic* if $\Sigma = \Sigma' \cup \{\epsilon^{(0)}\}$ for some signature Σ' all of whose symbols are of rank 1. For an arbitrary set S with $\epsilon \notin S$, $\mathit{mon}(S)$ denotes the monadic signature $\{f^{(1)} \mid f \in S\} \cup \{\epsilon^{(0)}\}$. A tree is called monadic if it has the form $f_1[\cdots f_n[\epsilon]\cdots]$ for some $f_1, \ldots, f_n$. (Notice that such a monadic tree could be identified with the string $f_1 \cdots f_n$.)

For a signature Σ and a set S of trees, $\Sigma(S)$ denotes the set of all trees $f[t_1, \ldots, t_k]$ such that $f^{(k)} \in \Sigma$ and $t_1, \ldots, t_k \in S$. Furthermore, $\mathrm{T}_\Sigma(S)$ denotes the set of *trees over* Σ *with subtrees in* S. It is the smallest set of trees such that $S \subseteq \mathrm{T}_\Sigma$ and, for all $f^{(k)} \in \Sigma$ and $t_1, \ldots, t_k \in \mathrm{T}_\Sigma(S)$, $f[t_1, \ldots, t_k] \in \mathrm{T}_\Sigma(S)$. The notation T_Σ is used as an abbreviation for $\mathrm{T}_\Sigma(\emptyset)$.

For the rest of this paper, let $X = \{x_1, x_2, \ldots\}$ be an indexed set of pairwise distinct *variables*. Variables are always viewed as symbols of rank 0. For every $n \in \mathbb{N}$, X_n denotes $\{x_1, \ldots, x_n\}$. The set of variables is assumed to be disjoint with all signatures under consideration. The variable x_1 is also denoted by x.

If t and $t_1, \ldots, t_n$ are trees, then $t[\![t_1, \ldots, t_n]\!]$ denotes the *substitution* of t_i for x_i in t ($i \in [n]$). More precisely, if $t = x_i$ for some $i \in [n]$ then $t[\![t_1, \ldots, t_n]\!] = t_i$ and if $t = f[s_1, \ldots, s_k]$ with $f \notin \{x_1, \ldots, x_n\}$ then

$$t[\![t_1, \ldots, t_n]\!] = f[s_1[\![t_1, \ldots, t_n]\!], \ldots, s_k[\![t_1, \ldots, t_n]\!]].$$

A *rewrite rule* is a pair $\rho = (l, r)$ of trees, called the *left-* and *right-hand side*, respectively, such that l contains every variable at most once and every variable in r occurs in l, too.[2] Consider some $n \in \mathbb{N}$ such that X_n contains all variables that occur in l. Then, ρ determines the binary relation $\to_\rho$ on trees such that $s \to_\rho t$ if s can be written as $s_0[\![l[\![s_1, \ldots, s_n]\!]]\!]$ for a tree s_0 which contains x_1 exactly once, and t equals $s_0[\![r[\![s_1, \ldots, s_n]\!]]\!]$. If R is a set of rewrite rules, $\to_R$ denotes the union of all $\to_\rho$ with $\rho \in R$.

3 Top-down tree transducers

Top-down tree transducers transform input trees into output trees in a top-down manner, using a restricted type of term rewrite rules.

Definition 1 (top-down tree transducer). *A* top-down tree transducer *is a tuple* $td = (\Sigma, \Sigma', \Gamma, R, \gamma_0)$ *such that*

- Σ *and* Σ' *are signatures, called the* input signature *and the* output signature,
- Γ *is a signature of* states *of rank 1 each, such that* Γ *is disjoint with* $\Sigma \cup \Sigma'$,
- $R \subseteq \Gamma(\Sigma(X)) \times \mathrm{T}_{\Sigma'}(\Gamma(X))$ *is a finite set of rewrite rules, and*
- $\gamma_0 \in \Gamma$ *is the* initial state.

[2] Notice that rewrite rules are always assumed to be "left-linear".

The top-down tree transduction computed by *td*, *which is also denoted by* *td*, *is the set of all pairs* $(s,t) \in T_\Sigma \times T_{\Sigma'}$ *such that* $\gamma_0[s] \rightarrow^*_{td} t$, *where* $\rightarrow_{td}$ *denotes the rewrite relation* $\rightarrow_R$.

In the following, for every top-down tree transducer $(\Sigma, \Sigma', \Gamma, R, \gamma_0)$ and every state $\gamma \in \Gamma$ the top-down tree transducer $(\Sigma, \Sigma', \Gamma, R, \gamma)$ is denoted by td_γ. As a convention, it is assumed that the variables in the left-hand side of a rule, read from left to right, are always $x_1, \ldots, x_k$ for some $k \in \mathbb{N}$. Thus, every rule of a top-down tree transducer $td = (\Sigma, \Sigma', \Gamma, R, \gamma_0)$ has the form

$$\gamma[f[x_1, \ldots, x_k]] \rightarrow t[\![\gamma_1[x_{i_1}], \ldots, \gamma_n[x_{i_n}]]\!],$$

where $\gamma, \gamma_1, \ldots, \gamma_n \in \Gamma$, $f^{(k)} \in \Sigma$, and $t \in T_{\Sigma'}(X_n)$ for some $k, n \in \mathbb{N}$. By convention, denoting a rule in this way means that t contains every variable in X_n exactly once. This convention carries over to the denotation of derivation steps: in $\gamma[f[t_1, \ldots, t_k]] \rightarrow_{td} t[\![\gamma_1[t_{i_1}], \ldots, \gamma_n[t_{i_n}]]\!]$ every $\gamma_j[t_{i_j}]$ is assumed to correspond to one particular occurrence of this subtree in $t[\![\gamma_1[t_{i_1}], \ldots, \gamma_n[t_{i_n}]]\!]$ (but notice that we may have $\gamma_j[t_{i_j}] = \gamma_{j'}[t_{i_{j'}}]$ for some distinct $j, j' \in [n]$, of course).

A rule of a top-down tree transducer is called a *γf-rule* if it has the form $\gamma[f[x_1, \ldots, x_k]] \rightarrow t$. Thus, a γf-rule is a rule that processes the input symbol f in state γ. A top-down tree transducer $td = (\Sigma, \Sigma', \Gamma, R, \gamma_0)$ is *total* (*deterministic*) if it contains at least one (at most one) γf-rule for every $\gamma \in \Gamma$ and $f \in \Sigma$. Clearly, if *td* is total then $td(t) \neq \emptyset$ for all $t \in T_\Sigma$, and if it is deterministic then it computes a partial function.

The *output size* of a top-down tree transducer *td* is given by $os_{td}(n) = \max\{|t| \mid (s,t) \in td \text{ and } |s| \leq n\}$ for all $n \in \mathbb{N}_+$.[3] Notice that os_{td} is a monotonic function. If there are $c \in \mathbb{R}$ and $n_0 \in \mathbb{N}$ such that $c > 1$ and $os_{td}(n) \geq c^n$ for all $n \geq n_0$, then os_{td} is *exponential*. The *exponential output size problem* is the problem to determine (on input *td*) whether os_{td} is exponential.

A (top-down) *string-to-tree transducer* is a top-down tree transducer $st = (\Sigma, \Sigma', \Gamma, R, \gamma_0)$ whose input signature Σ is monadic. For a derivation $\gamma[s] \rightarrow^*_{st} t$ with $\gamma \in \Gamma$, $s \in T_\Sigma$, and $t \in T_{\Sigma'}$ a corresponding *computation tree*, which is a tree with labels in Γ, is defined as follows. If the derivation has the form $\gamma[\epsilon] \rightarrow_{st} t$ then its computation tree is the tree γ. Otherwise, the derivation must have the form $\gamma[f[s']] \rightarrow_{st} t_0[\![\gamma_1[s'], \ldots, \gamma_k[s']]\!] \rightarrow^*_{st} t_0[\![t_1, \ldots, t_k]\!]$. In this case, its computation tree is $\gamma[t'_1, \ldots, t'_k]$, where t'_i is the computation tree of the i-th subderivation $\gamma_i[s'] \rightarrow^*_{st} t_i$, for $i \in [k]$. The set of all computation trees of derivations of *st* is denoted by *st-ct*.

In the case of string-to-tree transducers, instead of considering the size of output trees one may as well determine whether the number of leaves grows exponentially. This fact should be rather obvious, so the proof is skipped.

Lemma 2. *The output size of a string-to-tree transducer st is exponential if and only if there is some* $c \in \mathbb{R}$, $c > 1$, *such that st-ct contains an infinite number of trees t satisfying* $wd(t) \geq c^{ht(t)}$.

---

[3] As usual, $\max \emptyset$ is defined to be 0.

The next two lemmas are quite useful as they allow to restrict one's attention to the case of total deterministic string-to-tree transducers, which are considerably easier to deal with than general top-down tree transducers. First, every top-down tree transducer can be transformed into a deterministic one without affecting the output size too much.

Lemma 3. *For every top-down tree transducer td one can construct a deterministic top-down tree transducer td′ such that* $os_{td'}$ *is exponential if and only if* os_{td} *is exponential. The construction preserves totality and can be carried out on logarithmic space.*

The construction of *td′* mainly adds new unary symbols $\langle\gamma, i\rangle$ to the input signature and uses states of the form γ^i, where γ is a state of *td* and $i \in [m]$, m being the maximum number of rules with the same left-hand side in *td*. If a symbol $\langle\gamma, j\rangle$ or an input symbol f of *td* is encountered in state γ^i then *td′* switches to γ^j or applies the ith γf-rule of *td*, respectively. In this way the nondeterminism of *td* is shifted into the input signature.

Lemma 4. *For every top-down tree transducer td a total string-to-tree transducer st can be constructed, such that* os_{st} *is exponential if and only if* os_{td} *is exponential. The construction can be carried out on logarithmic space if td is total, and in exponential time otherwise.*

Proof. Let $td = (\Sigma, \Sigma', \Gamma, R, \gamma_0)$. By Lemma 3 it may be assumed that *td* is deterministic. Let ρ be the maximum rank of symbols in Σ, which is, without loss of generality, assumed to be at least 1. Intuitively, *st* is constructed in such a way that it reads the paths in the input trees of *td* and produces the corresponding part of the output tree.

Consider the total case first. Let q be the maximum number of occurrences of a variable x_i in a right-hand side of a rule in R. Then $st = (\Delta, \Delta', \Gamma, R', \gamma_0)$, where $\Delta = mon(\{f_i \mid f \in \Sigma, i \in [\rho]\})$, $\Delta' = \{h_n^{(n)} \mid n = 0, \ldots, q\}$, and R' is constructed as follows. For every rule $\gamma[f[x_1, \ldots, x_k]] \to t$ in R and every $i \in [\rho]$, if $\gamma_1[x_i], \ldots, \gamma_n[x_i]$ are the subtrees of t which are elements of $\Gamma(\{x_i\})$,[4] then R' contains the rule $\gamma[f_i[x]] \to h_n[\gamma_1[x], \ldots, \gamma_n[x]]$. Furthermore, for the sake of totality the (useless) rule $\gamma[\epsilon] \to h_0$ is added to R', for every $\gamma \in \Gamma$.

Clearly, totality is preserved by this construction, which can obviously be carried out on logarithmic space. Furthermore, by the definition of top-down tree transducers the rank of nodes in the input and output trees of *td* is bounded by a constant, and there is some $p \in \mathbb{N}$ such that $(s, s') \in td$ implies $ht(s') \leq p \cdot ht(s)$. These facts can be used to show that os_{st} is indeed exponential if and only if os_{td} is exponential, which yields the result for the total case.

Now, let us drop the assumption of totality. In this case it must be ensured that *st* does not produce a large output tree by processing some path in an

[4] More precisely, if $v_1, \ldots, v_n$ are the (pairwise distinct) nodes in $N(t)$ such that $t(v_1 1) = \cdots = t(v_n 1) = x_i$ then $\gamma_1[x_i], \ldots, \gamma_n[x_i]$ are the subtrees of t rooted at $v_1, \ldots, v_n$, respectively. Thus, a tree $\gamma'[x_i]$ occurs in this list as many times as in t.

input tree of *td* on which *td* would not yield any result. To see what this means, imagine a derivation of *td* on input s and consider the set $\{\gamma_1, \ldots, \gamma_m\}$ of states in which copies of a particular subtree s' of s are processed. Clearly, the derivation can produce an output tree only if s' lies in the intersection of the domains of $td_{\gamma_1}, \ldots, td_{\gamma_m}$. Therefore, *st* cannot simply use nondeterminism in order to process an arbitrary path of a tree in T_Σ because this could mean that it disregards such "dead ends" appearing elsewhere in the derivation of *td*. In *st* this problem will be solved by "cutting off" a derivation as soon as a configuration is reached for which there is no appropriate completion of the input path to an input tree of *td*. However, in order to implement this, one has to make use of more sophisticated states which keep track of the set of states of *td* in which it processes copies of a subtree of the input.

Let $st = (\Delta, \Delta', \Gamma', R'', \gamma_0')$, where Δ and Δ' are as above, Γ' is the set of all states of the form $\langle \gamma_1, \{\gamma_1, \ldots, \gamma_m\}\rangle$ with $\gamma_1, \ldots, \gamma_m \in \Gamma$, the initial state γ_0' is $\langle \gamma_0, \{\gamma_0\}\rangle$, and R'' is constructed as follows. Let $\langle \gamma_1, \{\gamma_1, \ldots, \gamma_m\}\rangle \in \Gamma'$ and $f \in \Sigma$. There are two cases which lead to different rules in R''.

1. For every $j \in [m]$ there is a $\gamma_j f$-rule $\gamma_j[f[x_1, \ldots, x_k]] \to t_j$ in R, and, for every $i \in [\rho]$, if Γ_i is the set of all states $\gamma \in \Gamma$ such that the subtree $\gamma[x_i]$ occurs in one of $t_1, \ldots, t_m$, then $\bigcap_{\gamma \in \Gamma_i} dom(td_\gamma) \neq \emptyset$. In this case, for every $i \in [\rho]$, R'' contains the rule
$$\langle \gamma_1, \{\gamma_1, \ldots, \gamma_m\}\rangle[f_i[x]] \to h_n[\langle \gamma_1', \Gamma_i\rangle[x], \ldots, \langle \gamma_n', \Gamma_i\rangle[x]],$$
where $\gamma_1'[x_i], \ldots, \gamma_n'[x_i]$ are the subtrees of t_1 which are elements of $\Gamma(\{x_i\})$.
2. Otherwise, the rule $\langle \gamma_1, \{\gamma_1, \ldots, \gamma_m\}\rangle[f_i[x]] \to h_0$ is in R'' for every $i \in [\rho]$.

As before, we also add to R'' all rules of the form $\langle \gamma_1, \{\gamma_1, \ldots, \gamma_m\}\rangle[\epsilon] \to h_0$.

By construction, *st* is total. Again, it can be shown that os_{st} is exponential if and only if os_{td} is. Therefore, it remains to argue that *st* can be constructed in exponential time. For this, it must be shown that it can be decided in exponential time whether $S = \bigcap_{\gamma \in \Gamma_0} dom(td_\gamma)$ is empty, for $\Gamma_0 \subseteq \Gamma$. However, it is well-known that S is a regular tree language which can be recognised by a tree automaton whose set of states is the power set of Γ. This tree automaton can be constructed easily in exponential time. Furthermore, the emptiness problem for regular tree languages can be decided in polynomial time in the size of a tree automaton defining the language (this is the emptiness problem for context-free string languages "in disguise"). Thus, altogether, the emptiness of S can be decided in exponential time, which completes the proof. □

4 The branching index of output trees

In this section it will be shown that, intuitively, trees of exponential width must necessarily contain trees with many ramifications on every path. In order to formalise this, the branching depth and the branching index of trees are defined.

Definition 5 (branching depth and branching index). *Let T be a tree. The* branching depth *of T is the smallest natural number b such that there is a leaf $v \in T$ which has exactly b distinct predecessors of rank ≥ 2. The* branching index *of T is the maximum branching depth of all trees $T' \subseteq T$.*

The lemma below states that the width of trees is polynomially bounded in their height if an upper bound is placed on their branching index.

Lemma 6. *Let S be a set of trees of bounded rank, such that there is a bound $b \in \mathbb{N}$ on the branching index of trees in S. Then, there is a polynomial p_b of degree b such that $wd(T) \leq p_b(ht(T))$ for all trees T.*

Proof. Let $r \in \mathbb{N}$ be a bound on the rank of trees in S and proceed by induction on b. A tree T of branching index 0 can at most have the rank 1, which implies $wd(T) = 1 = ht(T)^0$. Now consider some $b > 0$ and let T be a tree of branching index $\leq b$ having $k \leq r$ direct subtrees $T_1, \ldots, T_k$. If the branching index of one of $T_1, \ldots, T_k$ is greater than b or there are distinct $i, j \in [k]$ such that the branching index of both T_i and T_j is b, then it follows that the branching index of T is at least $b+1$. Therefore, at most one of the direct subtrees (T_1, say) has the branching index b and none of them has a larger branching index. According to the induction hypothesis, $T_2, \ldots, T_k$ satisfy $wd(T_i) \leq p_{b-1}(ht(T)-1)$. Therefore, $wd(T) \leq wd(T_1) + (r-1) \cdot p_{b-1}(ht(T)-1)$. Repeating the argument for T_1 until a tree of height 0 (and, therefore, of width 1) is reached, yields

$$\begin{aligned} wd(T) &\leq 1 + \textstyle\sum_{i=0}^{ht(T)-1} (r-1) \cdot p_{b-1}(i) \\ &\leq (r-1) \cdot ht(T) \cdot p_{b-1}(ht(T)) + 1, \end{aligned}$$

which is a polynomial in $ht(T)$ of degree b as p_{b-1} is one of degree $b-1$. □

Corollary 7. *Let S be an infinite set of trees of bounded rank and let $c \in \mathbb{R}$, $c > 1$. If $wd(T) \geq c^{ht(T)}$ for all $T \in S$, then there is no upper bound on the branching index of trees in S.*

It will now be proved that every set of labelled trees (with finitely many labels) of unbounded branching index contains a tree t such that there is a node v having two distinct descendants at the same height which carry the same label as v. Later on, this will be used in order to create a kind of pumping situation which characterises string-to-tree transducers having an exponential output size.

Lemma 8. *Let S be a set of trees with labels in a finite set L. If the branching index of trees in S is unbounded, then there exists a tree $t \in S$ which contains two distinct nodes v_0v and v_0v' such that $|v| = |v'|$ and $t(v_0v) = t(v_0) = t(v_0v')$.*

Proof. Consider some tree $t \in S$ whose branching index I is at least $\sum_{i=1}^{n} i$ for some $n \in \mathbb{N}$, and let $T \subseteq N(t)$ be a tree of branching depth I. Assume without loss of generality that the rank of T is at most 2. It is shown below that, if T does not contain two distinct nodes v_0v and v_0v' as required, then the set $t(T)$ of labels must exceed n (which proves the lemma by taking $n = |L|$).

For every node $w \in T$ let $\bar{t}(w)$ denote the set of labels of w and its predecessors, i.e., $\bar{t}(w) = \{t(w') \mid w = w'w'' \text{ for some } w'' \in \mathbb{N}_+^*\}$. Then the proof is finished as soon as it is shown that, for every $m \leq n$, there is a node $w \in T$ having at most $\sum_{i=1}^{m} i$ predecessors of rank 2, such that $|\bar{t}(w)| > m$. This is proved by induction on m. For $m = 0$ the assertion certainly holds, taking $w = \lambda$. For $m > 0$ consider some $v_0 \in T$ having at most $\sum_{i=1}^{m-1}$ predecessors of rank 2, such that $|\bar{t}(v_0)| \geq m$. By assumption, there should not exist two distinct descendants v_0v and v_0v' of v such that $|v| = |v'|$ and $t(v_0v), t(v_0v') \in \bar{t}(v_0)$. Now, choose some $v_0v \in T$ such that (a) there are m predecessors v_0v_1 of v_0v in T whose rank is 2 and (b) the length of v_0v is minimal (subject to requirement (a)). Notice that such a node v_0v exists because there are at most $\sum_{i=1}^{m-1} i$ predecessors of v_0 whose rank is 2, whereas every leaf of T has at least $\sum_{i=1}^{n}$ predecessors of rank 2, due to the branching depth of T. For the same reason, (b) implies that there is no leaf $v_0v' \in T$ such that $|v'| < |v|$. Therefore, the set $N = \{v_0v' \in T \mid |v'| = |v|\}$ satisfies $|N| > m$ (using (a)). Furthermore, again using (b), every node $v_0v' \in N$ has at most $m + \sum_{i=1}^{m-1} = \sum_{i=1}^{m}$ predecessors of rank 2. However, as pointed out above, there cannot be two distinct descendants v_0v' and v_0v'' of v_0 such that $t(v_0v') = t(v_0v'') \in \bar{t}(v_0)$. Therefore, there is at least one node $v_0v' \in N$ such that $t(v_0v') \notin \bar{t}(v_0)$, which means that $|\bar{t}(v_0v')| \geq \bar{t}(v_0) + 1 = m$, as asserted. □

The decision algorithm developed in the next section is based on the following theorem which characterises the class of total deterministic string-to-tree transducers with an exponential output size. In fact, the theorem could also be formulated for top-down tree transducers in general, but this would be technically more difficult.

Theorem 9. *The output size of a total deterministic string-to-tree transducer st is exponential if and only if there is some tree $t \in st\text{-}ct$ and there are distinct nodes $v_0v, v_0v' \in N(t)$ with $|v| = |v'|$, such that $t(v_0v) = t(v_0) = t(v_0v')$.*

Proof. Let $st = (\Sigma, \Sigma', \Gamma, R, \gamma_0)$. We shall consider both directions of the stated equivalence separately.

($\Rightarrow$) By Lemma 2, Corollary 7, and Lemma 8 there is some $t \in st\text{-}ct$ containing nodes v_0v and v_0v' such that $|v| = |v'|$ and $t(v_0v) = t(v_0) = t(v_0v')$, as required.

($\Leftarrow$) Consider some derivation $\gamma_0[s] \rightarrow_{st}^{*} s'$ whose computation tree t contains nodes v_0v and v_0v' of the type required. Then, s can be decomposed as $s = s_0[\![s_1[\![s_2]\!]]\!]$, where $ht(s_0) = |v_0|$ and $ht(s_1) = |v|$. Now, define $s_1^0 = x$ and $s_1^{i+1} = s_1^i[\![s_1]\!]$ for all $i \in \mathbb{N}$. Due to the assumption saying that st is total, it follows by a straightforward induction that there is a derivation $\gamma_0[s_0[\![s_1^i[\![s_2]\!]]\!]] \rightarrow_{st}^{*} s'^i$ for every $i \in \mathbb{N}$ (where $s'^i \in T_{\Sigma'}$) whose computation tree t^i satisfies $wd(t^i) \geq 2^i$. By Lemma 2 this means that the output size of st is exponential. □

As a by-product of the results in this section we get the following corollary (already known from [AU71]), which holds because the *only-if* direction of the proof of Theorem 9 could as well be proved using Lemma 6 instead of the weaker Corollary 7.

Corollary 10. *For every top-down tree transducer td, either* $os_{td} \in O(n^k)$ *for some* $k \in \mathbb{N}$*, or* os_{td} *is exponential.*

5 The main result

In this section the main result of the paper is proved: The exponential output size problem is NL-complete for total top-down tree transducers and DEXPTIME-complete in the general case. In order to prove that corresponding decision algorithms exist, it is shown that the problem can be reduced to a problem on graphs which is related to the well-known problem REACHABILITY (see [Pap94]).

In the following, a *graph* is a tuple $G = (V, E, \sigma, \tau, lab)$ consisting of a finite set V of nodes, a finite set E of edges, functions $\sigma: E \to V$ and $\tau: E \to V$ assigning to every edge its source and target node, respectively, and a labelling function $lab: E \to L$ assigning to every edge $e \in E$ a label $lab(e) \in L$, where L is some set of labels. (Notice that these graphs are multigraphs, i.e., there may be parallel edges carrying the same label.) For $v, v' \in V$ a vv'*-path* of length n in G is a sequence $e_1 \cdots e_n$ of edges such that $\sigma(e_1) = v$, $\tau(e_i) = \sigma(e_{i+1})$ for all $i \in [n-1]$, and $\tau(e_n) = v'$.

The following lemma is easy to prove by "guessing" appropriate paths.

Lemma 11. *Given a graph* $G = (V, E, \sigma, \tau, lab)$ *and nodes* $u, u' \in V$*, a nondeterministic Turing machine can determine on logarithmic space whether there is a* uu'*-path* p_0 *and there are distinct* $u'u'$*-paths* p *and* p' *such that* $lab(p) = lab(p')$*.*

It can now be shown that it is possible to determine nondeterministically on logarithmic space whether the output size of a total deterministic string-to-tree transducer is exponential.

Lemma 12. *The exponential output size problem for total deterministic string-to-tree transducers is in NL.*

Proof. By Theorem 9 it suffices to prove the following claim.

Claim. For a total deterministic string-to-tree transducer $st = (\Sigma, \Sigma', \Gamma, R, \gamma_0)$ it can be decided by a nondeterministic Turing machine on logarithmic space whether there is a tree $t \in st\text{-}ct$ and there are distinct nodes $v_0v, v_0v' \in N(t)$ with $|v| = |v'|$ such that $t(v_0v) = t(v_0) = t(v_0v')$.

In order to see how this can be done, define a graph $G = (\Gamma, E, \sigma, \tau, lab)$, as follows. For every state $\gamma \in \Gamma$, if R contains a rule $\gamma[f[x]] \to t$ then E contains an edge e_v for every $v \in N(t)$ with $t(v) \in \Gamma$, where $\sigma(e_v) = \gamma$, $\tau(e_v) = t(v)$, and $lab(e_v) = f$.

It should be clear that there is a tree $t \in st\text{-}ct$ containing nodes v_0v, v_0v' as stated in the claim if and only if there is some $\gamma \in \Gamma$ (namely $t(v_0)$) such that there is a $\gamma_0\gamma$-path and there are distinct $\gamma\gamma$-paths p, p' in G with $lab(p) = lab(p')$. By Lemma 11 this can be decided by a nondeterministic Turing machine on logarithmic space. □

Theorem 13. *The exponential output size problem is NL-complete for total top-down tree transducers and DEXPTIME-complete in general.*

Proof. By Lemma 4, Lemma 3, and Lemma 12 the problem is in NL for total top-down tree transducers (since logarithmic space reductions are closed under composition) and in DEXPTIME in general. It remains to prove NL-hardness and DEXPTIME-hardness, respectively.

In order to establish this for the total case, it is shown that the NL-complete problem REACHABILITY (given a graph G and two nodes v, v', is there a vv'-path in G?) can be reduced to the exponential output size problem for total top-down tree transducers. If V is the set of nodes of the input graph G, let $st = (\Sigma, \Sigma', \Gamma, R, \gamma_v)$ with $\Sigma = mon(V)$, $\Sigma' = \{f^{(2)}, \epsilon^{(0)}\}$, $\Gamma = \{\gamma_u \mid u \in V\}$, where R is defined as follows.

(1) For all $v, v' \in V$, R contains the rule $\gamma_u[u'[x]] \to \gamma_{u'}[x]$ if there is an edge from u to u' in G. If there is no such edge then R contains the rule $\gamma_u[u'[x]] \to \epsilon$.
(2) For all nodes $u \in V$, R contains the rule $\gamma_u[\epsilon] \to \epsilon$.
(3) In addition, R contains the rule $\gamma_{v'}[v[x]] \to f[\gamma_v[x], \gamma_v[x]]$.

Clearly, a work tape of logarithmic size is sufficient for a Turing machine to construct st. Furthermore, if G does not contain any vv'-path then the rule in (3), which is the only copying rule, will never be applied. Conversely, if there is a vv'-path $e_1 \cdots e_n$, let $s^0 = x$ and $s^{i+1} = v_1[\cdots v_n[v[s^i]] \cdots]$ for all $i \in \mathbb{N}$, where v_j is the target node of e_j for $j \in [n]$. By the rules in (1) and (3) $\gamma_v[s^1[x]] \to^*_{st} f[\gamma_v[x], \gamma_v[x]]$, which means that st contains all pairs $(s^i[\![\epsilon]\!], t^i)$, where t^i is a full binary tree of height i. Thus, the output size of st is exponential.

For the general case, Seidl [Sei94] showed that it is DEXPTIME-hard to decide whether $dom(td_1) \cap \cdots \cap dom(td_n) = \emptyset$ for top-down tree transducers $td_1, \ldots, td_n$ given as input. Suppose $td_i = (\Sigma_i, \Sigma'_i, \Gamma_i, R_i, \gamma_i)$ for $i \in [n]$ and assume without loss of generality that the sets of states are pairwise disjoint. Now, let $td = (\Sigma, \Sigma', \Gamma, R, \gamma_0)$ where the first four components are given by the union of the respective components of $td_1, \ldots, td_n$, except that new symbols $f^{(n+1)}$, $g^{(1)}$, and $\epsilon^{(0)}$ are added to Σ, $f^{(n+1)}$ and $\epsilon^{(0)}$ are added to Σ', states γ_0 and γ are added to Γ', and R contains the three additional rules

$$\begin{aligned}
\gamma_0[f[x_1, \ldots, x_{n+1}]] &\to f[\gamma_1[x_1], \ldots, \gamma_n[x_n], \gamma[x_{n+1}]],\\
\gamma[g[x]] &\to f[\gamma[x], \ldots, \gamma[x]], \text{ and}\\
\gamma[\epsilon] &\to \epsilon.
\end{aligned}$$

Clearly, the tree transduction td is empty if $dom(td_1) \cap \cdots \cap dom(td_n) = \emptyset$. Otherwise, choose some arbitrary $(s, t_1) \in td_1, \ldots, (s, t_n) \in td_n$ and define s^i ($i \in \mathbb{N}$) to be the tree $g[\cdots g[\epsilon] \cdots]$ of height i. Then there is a derivation $\gamma_0[f[s, \ldots, s, s^i]] \to^*_{td} f[t_1, \ldots, t_n, t^i]$ where t^i is a complete $(n+1)$-ary tree of height i over f and ϵ. Thus, the output size of td is exponential (except in the trivial case where $n = 0$), which completes the proof of the theorem. □

6 Conclusion

It was shown in this paper that the exponential output size problem is NL-complete for total top-down tree transducers and is DEXPTIME-complete for general top-down tree transducers. Intuitively, the reason for the huge complexity gap between these two variants is that, in the general case, solving the problem requires to solve the emptiness problem for top-down tree transductions, in addition. There are several directions for future research one could pursue. The complexity of the exponential output size problem for compositions of top-down tree transductions seems to be an interesting open problem. Another point is that, as mentioned in the introduction, for every $k \in \mathbb{N}$ one can construct top-down tree transductions whose output size is bounded from above by a polynomial of degree k (but not by a polynomial of degree $k-1$). In fact, by Corollary 10 the output size of a top-down tree transducer is bounded from above by a polynomial unless it is exponential. Thus, it may be interesting to search for efficient algorithms which determine, for a given top-down tree transducer td, the smallest natural number k such that $os_{td} \in O(n^k)$ (provided that such a k exists). Finally, there is the obvious question whether one can find natural classes of non-total top-down tree transducers for which the exponential output size problem is at least solvable on polynomial space.

Acknowledgement I thank J. Engelfriet for pointing out the related results in [AU71]. Furthermore, I would like to thank the anonymous referees for the useful comments they made.

References

[AU71] Alfred V. Aho and Jeffrey D. Ullman. Translations on a context free grammar. *Information and Control*, 19:439–475, 1971.

[Eng82] Joost Engelfriet. Three hierarchies of transducers. *Mathematical Systems Theory*, 15:95–125, 1982.

[FV98] Zoltán Fülöp and Heiko Vogler. *Syntax-Directed Semantics: Formal Models Based on Tree Transducers.* Springer, 1998.

[GS97] Ferenc Gécseg and Magnus Steinby. Tree languages. In G. Rozenberg and A. Salomaa, editors, *Handbook of Formal Languages. Vol. III: Beyond Words*, chapter 1, pages 1–68. Springer, 1997.

[Pap94] Christos H. Papadimitriou. *Computational Complexity.* Addison-Wesley, 1994.

[Rou70] William C. Rounds. Mappings and grammars on trees. *Mathematical Systems Theory*, 4:257–287, 1970.

[Sei94] Helmut Seidl. Haskell overloading is DEXPTIME-complete. *Information Processing Letters*, 52(2):57–60, 1994.

[Tha70] James W. Thatcher. Generalized^2 sequential machine maps. *Journal of Computer and System Sciences*, 4:339–367, 1970.

[Tha73] James W. Thatcher. Tree automata: an informal survey. In A.V. Aho, editor, *Currents in the Theory of Computing*, pages 143–172. Prentice Hall, 1973.

On Recognizable Languages in Divisibility Monoids

Manfred Droste and Dietrich Kuske

Institut für Algebra, Technische Universität Dresden, D-01062 Dresden,
{droste,kuske}@math.tu-dresden.de

Abstract. Kleene's theorem on recognizable languages in free monoids is considered to be of eminent importance in theoretical computer science. It has been generalized into various directions, including trace and rational monoids. Here, we investigate divisibility monoids which are defined by and capture algebraic properties sufficient to obtain a characterization of the recognizable languages by certain rational expressions as known from trace theory. The proofs rely on Ramsey's theorem, distributive lattice theory and on Hashigushi's rank function generalized to our divisibility monoids. We obtain Ochmański's theorem on recognizable languages in free partially commutative monoids as a consequence.

1 Introduction

In the literature, Kleene's theorem on recognizable languages of finite words has been generalized in several directions, e.g. to formal power series by Schützenberger [17], to infinite words by Büchi [5], and to rational monoids by Sakarovitch [16]. In all these cases, the notions of recognizability and of rationality where shown to coincide. In concurrency theory, several authors investigated recognizable languages in trace monoids (free partially commutative monoids) which generalize free monoids. It is known that here the recognizable languages only form a proper subclass of the rational languages, but a precise description of them using c-rational expressions could be given by Ochmański [13]. A further generalization of Kleene's and Ochmański's results to concurrency monoids was given in [8]. It is the goal of this paper to derive such a result for even more general monoids. At the same time, we obtain that well known combinatorial methods crucial in trace theory (like Levi's Lemma) are intimately related with algebraic properties (like distributivity) from classical lattice theory [3] or the theory of event structures [19].

Trace theory provides an important mathematical model for the sequential behavior of a parallel system in which the order of two independent actions is regarded as irrelevant. One considers pairs (T, I) where T is the set of actions, and I is a symmetric and irreflexive binary relation on T describing the independence of two actions. The trace monoid or free partially commutative monoid $\mathbb{M}(T, I)$ is then defined as the quotient $T^\star/{\sim}$ where $\sim$ is the congruence on the free monoid $T^\star$ generated by all pairs (ab, ba) with $(a, b) \in I$. For surveys on the many results obtained for trace monoids, we refer the reader to the collection [7].

An algebraic characterization of trace monoids was given by Duboc [10]. Here we use a lattice theoretically easy generalization of these algebraic conditions for the definition of divisibility monoids.

As for trace monoids, a divisibility monoid has a finite system of irreducible generators. They could be viewed as atomic transitions in a concurrent system. However, in comparison with trace monoids we allow much more general commutation possibilities for these generators. In our monoids it is possible, e.g., that $ab = cd$ or $ab = cc$ where a, b, c, d are four pairwise different irreducible generators. This would mean that the different sequential transformations ab and cd (cc, resp.) give rise to the same effect. It is clear that this is a much more general situation than in trace theory where $ab = cd$ implies $\{a, b\} = \{c, d\}$ (a, b, c, d generators as above) and even than in the situation of the automata with dynamic (situation dependent) independence of actions investigated in [8]. However, as for traces, we assume that any two sequential representations (i.e., products) by irreducible generators of a given monoid element have the same length. This is ensured by requiring that the divisibility monoid is cancellative and that the prefix (= left divisibility) relation satisfies natural distributivity laws well known from lattice theory (Birkhoff [3]). These classical distributivity laws suffice to deduce our results. Also, they enable us to develop and use a calculus of residuals similar to the one used e.g. in lambda calculus [2], term rewriting [4] and the models for concurrency considered in [14, 18].

In these divisibility monoids, we investigate closure properties of the class of recognizable languages under rational operations, analogously as in trace theory. To achieve this, we develop an extension of the notion of the *rank* of a language, which was already shown to be very useful in trace theory by Hashigushi [11], cf. [7, 6]. Under the assumption of a finiteness condition on the commutation behavior of the monoid elements, we can prove that the product of recognizable languages is again recognizable.

To deal with the iteration, analogously as in trace theory, we define when a monoid element is connected (intuitively, it cannot be split into disjoint components) using classical lattice-theoretic concepts. In trace theory, the iteration of a recognizable language consisting only of connected elements is again recognizable. We show (cf. Example 1) that, somewhat surprisingly, this fails in general in divisibility monoids. However, using the residuum operation mentioned above, we can define when a language is residually closed. Then we can show, using also Ramsey's Theorem, that the iteration of a recognizable residually closed language consisting only of connected elements is again recognizable. We call a language c-rational if it can be constructed from finite languages using the operations union, product and this restricted version of iteration. Thus, the closure properties indicated so far ensure that any c-rational language is recognizable.

Recall that an equation $ab = cd$ with irreducible generators a, b, c, d of M states that the different sequential executions ab and cd give rise to the same effect. If now $a \neq c$, the effect of a in the execution cd has to be resumed by that of d. Therefore, we consider the least equivalence on the irreducible generators of M identifying a and d that occur in an equation $ab = cd$ with $a \neq c$. Requiring that a and c are not equivalent whenever $ab = cd$ and $a \neq c$, we can prove the converse of the above result, i.e., we can show that any recognizable language is c-rational. With this requirement, our divisibility monoids are more similar to, but still more general than trace monoids. Our results can be summarized as follows (see the subsequent sections for the precise definitions)

Theorem 1. *Let $(M, \cdot, 1)$ be a labeled divisibility monoid with finite commutation behavior and $L \subseteq M$. Then L is recognizable iff L is c-rational.*

From these results, we obtain Ochmański's theorem for recognizable trace languages as an immediate consequence. Furthermore, a strengthening of the results from [8] for recognizable languages in concurrency monoids follows from our results (see the full paper [9]).

As the above examples and many others show, the class of divisibility monoids is much larger than the class of all concurrency monoids investigated in [8] which in turn is larger than the class of trace monoids.

The present divisibility monoids can hence be viewed as a general model for concurrent behaviors where it is still possible to describe recognizable sets of behaviors by certain rational expressions.

The complete proofs are contained in the full paper [9].

2 Preliminaries

Let $(M, \cdot, 1)$ be a monoid and $L \subseteq M$. A monoid morphism $\eta : M \to S$ into a finite monoid $(S, \cdot, 1)$ *recognizes* L if $\eta^{-1}\eta(L) = L$. The language L is *recognizable* if there exists a monoid morphism that recognizes L. For $x \in M$ let $x^{-1}L := \{y \in M \mid x \cdot y \in L\}$, the left quotient of L with respect to x. Then a classical result states that L is recognizable iff the set $\{x^{-1}L \mid x \in M\}$ is finite iff there is a finite M-automaton recognizing L.

Let $L, K \subseteq M$. Then $L \cdot K := \{l \cdot k \mid l \in L, k \in K\}$ is the product of L and K. By $\langle L \rangle$ we denote the submonoid of M generated by L, i.e., $\langle L \rangle = \{l_1 \cdot l_2 \cdot \ldots l_n \mid n \in \mathbb{N}, l_i \in L\}$. For a set T, $T^\star$ denotes the free monoid generated by T. Now let M be a free monoid and $L \subseteq M$. Then $\langle L \rangle$ is a subset of M while $L^\star$ is a set of words whose letters are elements of M. Classical formal language theory usually identifies the set $L^\star$ of words over L and the submonoid $\langle L \rangle$ of M generated by L. In this paper, we have to distinguish between them.

A language $L \subseteq M$ is *rational* if it can be constructed from the finite subsets of M by union, multiplication and iteration.

Now let T be a finite set and $L \subseteq M := T^\star$. By Kleene's Theorem, L is recognizable iff it is rational. In any monoid, the set of recognizable languages is closed under the usual set-theoretic operations, like complementation, intersection and difference.

For $x \in T^\star$, let $\alpha(x)$ denote the alphabet of x comprising all letters of T occurring in x. Then $L_B := \langle B \rangle \cap L \setminus (\bigcup_{A \subset B} \langle A \rangle)$ with $B \subseteq T$ is the set of elements x of L with $\alpha(x) = B$. If L is rational, the language L_B is rational, too. The language L is *monoalphabetic* if $L = L_B$ for some $B \subseteq T$. A language

$L \subseteq M$ is *monoalphabetic-rational* if it can be constructed from the finite subsets of M by union, multiplication and iteration where the iteration is applied to monoalphabetic languages, only. One can easily show that in a finitely generated free monoid any rational language is monoalphabetic-rational.

Let $(P, \leq)$ be a partially ordered set and $x \in P$. Then $\downarrow x$ comprises all elements dominated by x, i.e., $\downarrow x := \{y \in P \mid y \leq x\}$. If $A \subseteq P$, we write $A \leq x$ to denote that $A \subseteq \downarrow x$, i.e., that $a \leq x$ for all $a \in A$. The partially ordered set $(P, \leq)$ is a *lattice* if for any $x, y \in P$ the least upper bound $\sup(x, y) = x \vee y$ and the largest lower bound $\inf(x, y) = x \wedge y$ exist. The lattice $(P, \leq)$ is *distributive* if $x \wedge (y \vee z) = (x \wedge y) \vee (x \wedge z)$ for any $x, y, z \in P$. This is equivalent to $x \vee (y \wedge z) = (x \vee y) \wedge (x \vee z)$ for any $x, y, z \in P$. For properties of finite distributive lattices, we refer the reader to [3].

3 Divisibility monoids

In this section, we introduce divisibility monoids and investigate their basic properties.

Let $M = (M, \cdot, 1)$ be a monoid where $1 \in M$ is the unit element. We call M *cancellative* if $x \cdot y \cdot z = x \cdot y' \cdot z$ implies $y = y'$ for any $x, y, y', z \in M$. This in particular ensures that M does not contain a zero element and will be a very natural assumption (trivially satisfied in free monoids). For $x, y \in M$, *x is a left divisor of y* (denoted $x \leq y$) if there is $z \in M$ such that $x \cdot z = y$. In general, the relation $\leq$ is not antisymmetric, but we require this for a divisibility monoid.

Let $T := (M \setminus \{1\}) \setminus (M \setminus \{1\})^2$. The set T consists of those nonidentity elements of M that do not have a proper divisor, its elements are called *irreducible*. Note that T has to be contained in any set generating M.

Definition 1. *A monoid $(M, \cdot, 1)$ is called a* (left) divisibility monoid *provided the following hold*

1. *M is cancellative and its irreducible elements form a finite set of generators of M,*
2. *$(M, \leq)$ is a partial order such that any two elements $x, y \in M$ with an upper bound have a supremum, and*
3. *$(\downarrow m, \leq)$ is a distributive lattice for any $m \in M$.*

Since by condition 1 above a divisibility monoid $(M, \cdot, 1)$ is generated by the set T of its irreducible elements, there is a natural epimorphism from the free monoid $T^\star$ onto M. This epimorphism will be denoted by $[.]$.

Condition 2 is well known from domain theory and often regarded as "consistent completeness". It means that whenever two computations x and y from M allow a joint extension, there is a least such extension of them. In fact, the partial order $(M, \leq)$ can be seen as the compact elements of a Scott-domain. But $(M, \leq)$ is not necessarily a lattice since it may contain unbounded pairs of elements.

Using basic properties of distributive lattices, from conditions 1 and 3 one can infer that $\downarrow x$ is finite for any $x \in M$. It follows that any finite subset A

of M has an infimum in $(M, \leq)$, and if A has an upper bound, it also has a supremum. This supremum of A can be viewed as the least common multiple of A, whereas the infimum of A is the greatest common (left-)divisor of A. Observe that the distributivity required is a direct generalization of the triviality that in the multiplicative monoid $(\mathbb{N}, \cdot, 1)$ least common multiple and greatest common divisor distribute (i.e., $\gcd(x, \mathrm{lcm}(y, z)) = \mathrm{lcm}(\gcd(x, y), \gcd(x, z))$ for any x, y, z). In our general setting, the finiteness of $\downarrow x$ ensures that $(M, \leq)$ is even the set of compacts of a dI-domain. For the theory of dI-domains and their connection with lambda calculus we refer the reader to [1]. In particular we have $(x \vee y) \wedge z = (x \wedge z) \vee (y \wedge z)$ whenever the left hand side is defined.

Note that in left divisibility monoids the partial order is the prefix relation. Ordered monoids where the order relation is the intersection of the prefix and the suffix relation were investigated e.g. in [3] under the name "divisibility monoid". Since such monoids will not appear in this paper any more, we will simply speak of "divisibility monoids" as an abbreviation for "left divisibility monoid".

Next we show that for elements of a divisibility monoid a length can be defined in a natural way making the correspondence to computations even clearer: Let $x = x_1 x_2 \ldots x_n \in M$ with $x_i \in T$. Then $\{1, x_1, x_1 x_2, \ldots, x\}$ is a maximal chain in the finite distributive lattice $\downarrow x$. Since maximal chains in finite distributive lattices have the same size, any word u over T with $[u] = x$ has length n. Hence we can define the *length* of x to be $|x| = n$.

Divisibility monoids are defined algebraically, using classical notions from lattice theory. They can also be described combinatorially (and more similar to the original definition of trace monoids) using commutation conditions for their irreducible generators. A first step towards such a representation is provided by the following proposition.

Proposition 1. *Let M be a divisibility monoid and T the set of its irreducible elements. Let $\sim$ denote the least congruence on the free monoid T^* containing $\{(ab, cd) \mid a, b, c, d \in T$ and $a \cdot b = c \cdot d\}$. Then $\sim$ is the kernel of the natural epimorphism $[.] : T^* \to M$. In particular, $M \cong T^*/{\sim}$.*

On the other hand, there are sets of equations of the form $ab = cd$ such that $T^*/{\sim}$ is not a divisibility monoid. In [12], those sets of equations are described that give rise to divisibility monoids.

Let M be a divisibility monoid. Two elements x and y are *independent* (denoted by $x \parallel y$) if $x \wedge y = 1$ and $\{x, y\}$ is bounded above. Intuitively, this means that the computations x and y have no nontrivial joint past and are consistent. In this case the supremum $x \vee y$ exists in M. Since M is cancellative, there is a unique element z such that $y \cdot z = x \vee y$. This element z is called *the residuum of x after y* and denoted by $x \uparrow y$. Intuitively, $x \uparrow y$ denotes the computation that has to be performed after y in order to obtain the least common extension of x and y. Note that the residuum is defined for independent elements x and y only. Clearly, $x \uparrow y$ is defined iff $y \uparrow x$ is defined and in this case $x(y \uparrow x) = y(x \uparrow y) = x \vee y$.

Now assume M to be a trace monoid. Then two traces $x = [u]$ and $y = [v]$ in M are independent iff each letter occurring in u is independent from each letter occurring in v. This coincides with the usual definition of independence in trace theory. If x and y are independent, then it is known that $y \cdot x = x \cdot y = x \vee y$ and hence $x \uparrow y = x$ and similarly $y \uparrow x = y$.

Again, let M be an arbitrary divisibility monoid. Fixing $x \in M$, we define a unary partial function c_x from M to M with domain $\mathrm{dom}(c_x) := \{y \in M \mid x \parallel y\}$ by letting $c_x(y) := y \uparrow x$. The function c_x will be called the *commutation behavior of* x. In this paper, as usual, an equation $c_x(y) = c_z(y')$ means "$c_x(y)$ is defined iff $c_z(y')$ is defined and in this case they are equal". In other words, y is independent from x iff y' is independent from z and in this case $y \uparrow x = y' \uparrow z$.

Let $\mathbb{C}_M$ denote the set of all commutation behaviors of elements of M, i.e., $\mathbb{C}_M = \{c_x \mid x \in M\}$. Note that $\mathbb{C}_M$ is a set of partial functions from M to M that might be infinite. If $\mathbb{C}_M$ is actually finite, we say that M is a *divisibility monoid with finite commutation behavior*.

Let M again be a trace monoid. Recall that $y \uparrow x = y$ whenever $y \uparrow x$ is defined. Hence the commutation behavior c_x is the identity on its domain. This in particular implies that two traces have the same commutation behavior iff they have the same alphabet. Thus, if M is finitely generated, as a divisibility monoid it has finite commutation behavior.

The following lemma lists some properties of the commutation behaviors our proofs rely on.

Lemma 1. *Let* $(M, \cdot, 1)$ *be a divisibility monoid and* $x, x', y, z \in M$.

1. *The commutation behavior* c_x *is injective and length-preserving on its domain.*
2. $x \parallel yz$ *iff* $x \parallel y$ *and* $c_y(x) \parallel z$.
3. $c_{yz}(x) = c_z(c_y(x))$; *in other words* $x \uparrow (yz) = (x \uparrow y) \uparrow z$.
4. $c_x(yz) = c_x(y) \cdot c_{c_y(x)}(z)$; *equivalently* $yz \uparrow x = (y \uparrow x) \cdot (z \uparrow (x \uparrow y))$.
5. *If* $c_x = c_{x'}$ *and* $y \parallel x$ *then* $c_{c_y(x)} = c_{c_y(x')}$.

Note that the third statement of the lemma above in particular implies $c_z \circ c_y = c_{yz}$ where $\circ$ is the usual concatenation of partial functions. Hence $(\mathbb{C}_M, \circ, c_1)$ is a monoid, *the monoid of commutation behaviors of* M. The function $c : M \to \mathbb{C}_M : x \mapsto c_x$ is a monoid antihomomorphism. Thus, if M has finite commutation behavior, for any commutation behavior $c \in \mathbb{C}_M$, the set $\{x \in M \mid c_x = c\}$ of all elements of M with commutation behavior c is recognizable. This will be crucial for some proofs of our results. Unfortunately, we do not know whether actually each divisibility monoid has finite commutation behavior. This seems to be a difficult problem combining monoid theoretic, lattice theoretic and combinatorial concepts.

We will also need a lifting of the commutation behavior from a divisibility monoid M to the free monoid $T^\star$ which can be defined as follows. We define functions $d_u : T^\star \to T^\star$ for $u \in T^\star$ in such a way that equations like those from Lemma 1 hold: Recall that for $t \in T$ and $u \in T^\star$ with $[u] \parallel t$ we have $|t| = |c_{[u]}(t)|$ by Lemma 1 and therefore $c_{[u]}(t) \in T$. Hence $d_u(t) := c_{[u]}(t)$ (if $t \parallel [u]$) is a partial function mapping T to T. We extend it to a partial function from $T^\star$ to $T^\star$ by $d_u(tv) := d_u(t)d_{d_t(u)}(v)$. Then one gets properties similar to those listed in Lemma 1. In particular $d_u = d_v$ iff $c_{[u]} = c_{[v]}$ for any $u, v \in T^\star$.

Let $\mathbb{D}_M = \{d_u \mid u \in T^\star\}$ be the set of all commutation behaviors of words over T. Then $(\mathbb{D}_M, \circ, d_\varepsilon)$ is a monoid and $d : T^\star \to \mathbb{D}_M : u \mapsto d_u$ is a monoid antihomomorphism. Also, $d_u \mapsto c_{[u]}$ is a monoid isomorphism from $(\mathbb{D}_M, \circ, d_\varepsilon)$ to $(\mathbb{C}_M, \circ, c_1)$.

It is immediate that if $[u] \parallel [v]$ then $[v] = [w]$ implies $[d_u(v)] = c_{[u]}([v]) = c_{[u]}([w]) = [d_u(w)]$. The following lemma shows that not only the other implication holds as well but that even $\{d_u(w) \mid [v] = [w]\} = \{w' \mid [w'] = [d_u(v)]\}$. The proof relies on the fact that $(\downarrow x, \leq)$ is a distributive lattice and that projective intervals in distributive lattices are isomorphic.

Lemma 2. *Let $x \in M$, $u \in T^\star$ and $t_i \in T$ for $i = 1, 2, \ldots, n$ such that $c_{[u]}(x) = [t_1 t_2 \ldots t_n]$. Then there exist $s_i \in T$ for $i = 1, 2, \ldots, n$ such that $d_u(s_1 s_2 \ldots s_n) = t_1 t_2 \ldots t_n$. These elements s_i of T are unique.*

4 Commutation grids and the rank

In trace theory, the generalized Levi Lemma (cf. [6]) plays an important role. Here, we introduce a generalization to divisibility monoids using commutation grids. This enables us to obtain a concept of "rank" of a language in these monoids, similar to the one given by Hashigushi [11] for trace monoids. Let M be a divisibility monoid and $x, y \in M$. Recall that $c_x(y) = y{\uparrow}x$. Similarly, we define $v{\uparrow}u := d_u(v)$ whenever the latter is defined for $u, v \in T^\star$.

Definition 2. *For $0 \leq i \leq j \leq n$ let $x_j^i, y_i^j \in T^\star$. The tuple $(x_j^i, y_i^j)_{0 \leq i \leq j \leq n}$ is* a commutation grid *provided $x_j^i \parallel y_i^{j-1}$, $x_j^i{\uparrow}y_i^{j-1} = x_j^{i+1}$, and $y_i^{j-1}{\uparrow}x_j^i = y_i^j$ for any $0 \leq i < j \leq n$ (see Fig. 1).*

Lemma 3. *Let $z_0, z_1, \ldots, z_n, x, y \in T^\star$ with $[xy] = [z_0 z_1 \ldots z_n]$. Then there exists a commutation grid $(x_j^i, y_i^j)_{0 \leq i \leq j \leq n}$ such that $[x] = [x_0^0 x_1^0 \ldots x_n^0]$, $[y] = [y_0^n y_1^n \ldots y_n^n]$, and $[z_i] = [x_i^i y_i^i]$ for $i = 0, 1, \ldots, n$.*

Now we can introduce the notion of rank in the present context. Intuitively, it measures the amount of commutations of irreducible generators necessary to transform a product of two words into an equivalent word belonging to a given word language over T.

Definition 3. *Let $u, v \in T^\star$ and $X \subseteq T^\star$ such that $[uv] \in [X] := \{[w] \mid w \in X\}$. Let $\mathrm{rk}(u, v, X)$ denote the minimal integer n such that there exists a commutation grid $(u_j^i, v_i^j)_{0 \leq i \leq j \leq n}$ in $T^\star$ with $[u] = [u_0^0 u_1^0 \ldots u_n^0]$, $[v] = [v_0^n v_1^n \ldots v_n^n]$, and $u_0^0 v_0^0 u_1^1 v_1^1 \ldots u_n^n v_n^n \in X$.*

For $u, v \in T^\star$ and $X \subseteq T^\star$ with $[uv] \in [X] := \{[w] \mid w \in X\}$, one gets $\mathrm{rk}(u, v, X) \leq |uv|$. We define the *rank* $\mathrm{rk}(X)$ *of* X by

$$\mathrm{rk}(X) := \sup\{\mathrm{rk}(u, v, X) \mid u, v \in T^\star, [uv] \in [X]\} \in \mathbb{N} \cup \{\infty\}.$$

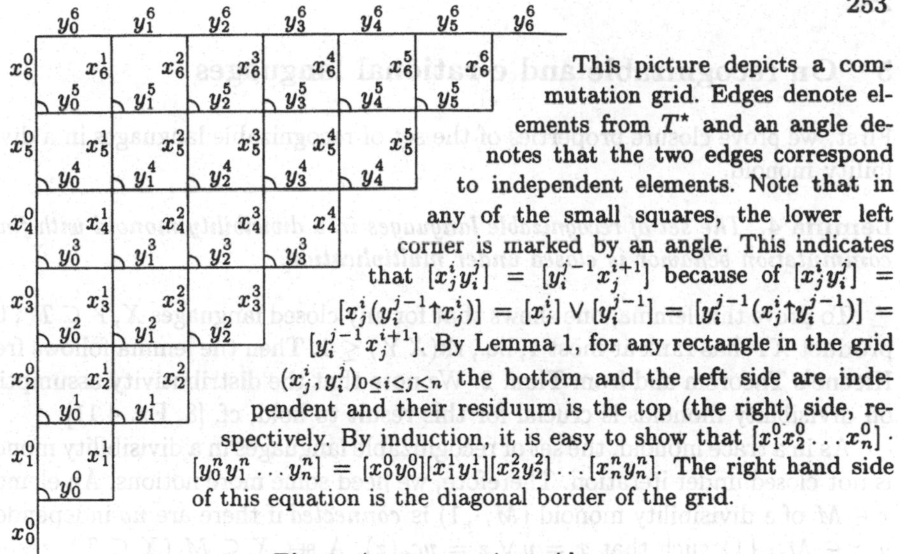

This picture depicts a commutation grid. Edges denote elements from T^* and an angle denotes that the two edges correspond to independent elements. Note that in any of the small squares, the lower left corner is marked by an angle. This indicates that $[x_j^i y_i^j] = [y_i^{j-1} x_j^{i+1}]$ because of $[x_j^i y_i^j] = [x_j^i(y_i^{j-1} \uparrow x_j^i)] = [x_j^i] \vee [y_i^{j-1}] = [y_i^{j-1}(x_j^i \uparrow y_i^{j-1})] = [y_i^{j-1} x_j^{i+1}]$. By Lemma 1, for any rectangle in the grid $(x_j^i, y_i^j)_{0 \le i \le j \le n}$ the bottom and the left side are independent and their residuum is the top (the right) side, respectively. By induction, it is easy to show that $[x_1^0 x_2^0 \ldots x_n^0] \cdot [y_0^n y_1^n \cdots y_n^n] = [x_0^0 y_0^0][x_1^1 y_1^1][x_2^2 y_2^2] \ldots [x_n^n y_n^n]$. The right hand side of this equation is the diagonal border of the grid.

Fig. 1. A commutation grid

A word language $X \subseteq T^*$ is *closed* if $[u] \in [X]$ implies $u \in X$ for any $u \in T^*$. Since $\text{rk}(u, v, X) = 0$ whenever $uv \in X$, the rank of a closed language equals 0.

We just note here that if M is a trace monoid then these notions coincide with the corresponding ones known from trace theory. Hence the following result generalizes [6, Thm. 3.2].

Theorem 2. *Let $(M, \cdot, 1)$ be a divisibility monoid with finite commutation behavior. Let $X \subseteq T^*$ be recognizable and $n := \text{rk}(X)$ be finite. Then $[X]$ is recognizable in M.*

Proof. Let η be a homomorphism into a finite monoid S recognizing X with $d_u = d_v$ whenever $\eta(u) = \eta(v)$. For $x \in M$ let $R(x)$ denote the subset

$$\{(\eta d(x_0), \eta d(x_1) \ldots \eta d(x_n))_{d \in \mathbb{D}_M} \mid x_0, x_1, \ldots, x_n \in T^* \text{ and } x = [x_0 x_1 \ldots x_n]\}$$

of $(S^{n+1})^{|\mathbb{D}_M|}$. Hence there are only finitely many sets $R(x)$. We show $R(x) = R(z) \Rightarrow x^{-1}[X] = z^{-1}[X]$, which implies that $[X]$ is recognizable.

So let $R(x) = R(z)$ and let $y \in x^{-1}[X]$. Since $\text{rk}(X) = n$, there exists a commutation grid $(u_j^i, v_i^j)_{0 \le i \le j \le n}$ such that $x = [u_0^0 u_1^0 \ldots u_n^0]$, $y = [v_0^n v_1^n \ldots v_n^n]$, and $u_0^0 v_0^0 u_1^1 v_1^1 \ldots u_n^n v_n^n \in X$. Then $(\eta d(u_0^0), \eta d(u_1^0) \ldots \eta d(u_n^0))_{d \in \mathbb{D}_M} \in R(x) = R(z)$. Hence there exist words $w_j^0 \in T^*$ with $\eta d(w_j^0) = \eta d(u_j^0)$ for each $0 \le j \le n$ and $d \in \mathbb{D}_M$, and $z = [w_0^0 w_1^0 \ldots w_n^0]$. Then $d_{w_j^0} = d_{u_j^0}$ implying the existence of a commutation grid $(w_j^i, v_i^j)_{0 \le i \le j \le n}$. Then one gets $zy = [w_0^0 v_0^0 w_1^1 v_1^1 \ldots w_n^n v_n^n] \in [X]$. Hence $y \in z^{-1}[X]$ and therefore $x^{-1}[X] = z^{-1}[X]$ as claimed above. □

5 On recognizable and c-rational languages

First, we prove closure properties of the set of recognizable languages in a divisibility monoid.

Lemma 4. *The set of recognizable languages in a divisibility monoid with finite commutation behavior is closed under multiplication.*

To prove this lemma, one shows that for any closed languages $X, Y \subseteq T^\star$, the product XY has rank at most 1, i.e., $\text{rk}(X\,Y) \leq 1$. Then the lemma follows from Kleene's Theorem and from Thm. 2. We note that the distributivity assumption on divisibility monoids is crucial for this result to hold, cf. [8, Ex. 4.11].

As in a trace monoid, the set of recognizable languages in a divisibility monoid is not closed under iteration. Therefore, we need some more notions: An element $x \in M$ of a divisibility monoid $(M, \cdot, 1)$ is *connected* if there are *no* independent $y, z \in M \setminus \{1\}$ such that $x = y \vee z = yc_y(z)$. A set $X \subseteq M$ ($X \subseteq T^\star$, respectively) is *connected* if all of its elements are connected ($[X] \subseteq M$ is connected, respectively). For trace monoids, this lattice theoretic definition is equivalent to the usual one via alphabets, and the iteration of a recognizable connected language is again recognizable. The following example shows that the latter is not the case for divisibility monoids.

Example 1. Let $T = \{a, b, c, d\}$ and let $\sim$ denote the least congruence on $T^\star$ with $ab \sim cd$ and $ba \sim dc$. Now we consider the monoid $M := T^\star/\!\sim$. Using the characterization from [12], one can show that M is a divisibility monoid. Moreover, it has finite commutation behavior. Since any irreducible element is trivially connected, $\{a, b\}$ is a recognizable connected language in M. Let L denote the iteration of this language in M, i.e., $L := \langle\{a, b\}\rangle \subseteq M$. To show that L is not recognizable, it suffices to prove that $X := \{w \in T^\star \mid [w] \in L\}$ is not recognizable in the free monoid $T^\star$. Note that X consists of those words that are equivalent to some word containing a's and b's, only. Clearly any such word has to contain the same number of c's and of d's. If X was recognizable, the language $Y = X \cap (ad)^\star(cb)^\star$ would be recognizable. We will derive a contradiction by showing $Y = \{(ad)^i(cb)^i \mid i \in \mathbb{N}\}$: By the observation above, $Y \subseteq \{(ad)^i(cb)^i \mid i \in \mathbb{N}\}$. Starting with $(ab)c \sim cdc \sim cba$, we obtain $(ab)^n c \sim c(ba)^n$ for any n. Thus $ad(ab)^n cb \sim adc(ba)^n b \sim aba(ba)^n b = (ab)^{n+2}$. Applying this equation to a word of the form $(ad)^i(cb)^i$ several times, one gets $(ad)^i(cb)^i \sim (ab)^{2i} \in X$ and therefore $Y = \{(ad)^i(cb)^i \mid i \in \mathbb{N}\}$. □

An analysis of this example leads to the following additional requirement on recognizable languages that we want to iterate: A language $X \subseteq T^\star$ is ***residually closed*** if it is closed under the application of d_u and d_u^{-1} for elements u of X (Note that in the example above $d_a^{-1}(b) = c \notin \{a, b\}$, i.e., this language is not residually closed.) A language $L \subseteq M$ is ***residually closed*** iff $\{w \in T^\star \mid [w] \in L\}$ is residually closed. Recall that in a trace monoid the commutation behaviors d_u are contained in the identity function on $T^\star$. Hence any trace language is residually closed.

Theorem 3. *Let $(M, \cdot, 1)$ be a divisibility monoid with finite commutation behavior. Let $X \subseteq T^\star$ be closed, connected, and residually closed. Then the rank* $\text{rk}(\langle X \rangle)$ *of the iteration of X is finite.*

Proof. Let $u, v \in T^\star$ and $x_0, x_1, \ldots, x_n \in X$ such that $[uv] = [x_0 x_1 \ldots x_n]$. One can show that there exists a commutation grid $(u_j^i, v_i^j)_{0 \leq i \leq j \leq n}$ in $T^\star$ such that $[u] = [u_0^0 u_1^0 \ldots u_n^0]$, $[v] = [v_0^n v_1^n \ldots v_n^n]$ and $[x_i] = [u_i^i v_i^i] \in [X]$.

Constructing a subgrid one shows that it is sufficient to consider the case $u_i^i \neq \varepsilon \neq v_i^i$ for all $0 \leq i \leq n$.

Now one can prove that there are no $1 \leq \alpha \leq \beta \leq \gamma \leq n$ with

$$d_{u_\alpha^0 u_{\alpha+1}^0 \ldots u_{\beta-1}^0} = d_{u_\beta^0 u_{\beta+1}^0 \ldots u_{\gamma-1}^0} = d_{u_\alpha^0 u_{\alpha+1}^0 \ldots u_{\gamma-1}^0}$$

since otherwise $[u_\alpha^\alpha v_\alpha^\alpha]$ would not be connected. Since $\mathbb{D}_M$ is finite, Ramsey's Theorem [15] bounds n and therefore the rank of $\langle X \rangle$. □

Using Kleene's Theorem and Thm. 2 one gets that the iteration of a connected, recognizable and residually closed language is recognizable.

A language $L \subseteq M$ is *c-rational* if it can be constructed from the finite subsets of M by union, multiplication and iteration where the iteration is applied to connected and residually closed languages, only. Since any element $x \in M$ has only finitely many prefixes, finite languages are recognizable. By Lemma 4 and Thm. 3, we get

Theorem 4. *Let $(M, \cdot, 1)$ be a divisibility monoid with finite commutation behavior. Let $L \subseteq M$ be c-rational. Then L is recognizable.*

Next we want to show the inverse implication of the theorem above. Let $(M, \cdot, 1)$ be a divisibility monoid, E a finite set and $\ell : T \to E$ a function. Then ℓ is a *labeling function* and (M, ℓ) is a *labeled divisibility monoid* if $\ell(s) = \ell(s \uparrow t)$ and $\ell(s) \neq \ell(t)$ for any $s, t \in T$ with $s \parallel t$. We note that the monoid M from Example 1 becomes a labeled divisibility monoid by putting $\ell(a) = \ell(d) = 0$ and $\ell(b) = \ell(c) = 1$. Thus, our main Thm. 5 holds for this monoid which is not a trace monoid.

Now let (M, ℓ) be a labeled divisibility monoid. The *label sequence* of a word $u_0 u_1 \ldots u_n \in T^\star$ is the word $\ell(u_0)\ell(u_1) \ldots \ell(u_n) \in E^\star$. We extend the mapping ℓ to words over T by $\ell(tw) = \{\ell(t)\} \cup \ell(w)$ and to elements of M by $\ell([u]) := \ell(u)$ for $u \in T^\star$. This latter is well defined by Prop. 1. Note that $\ell : M \to 2^E$ is a monoid homomorphism into the finite monoid $(2^E, \cup, \emptyset)$. One can show that $\ell(x) \cap \ell(y) = \emptyset$, $\ell(y) = \ell(y \uparrow x)$, and $\ell(x) \cup \ell(y) = \ell(x \vee y)$ for any $x, y \in M$ with $x \parallel y$.

A language $L \subseteq M$ is *monoalphabetic* if $\ell(x) = \ell(y)$ for any $x, y \in L$. It is an *mc-rational language* if it can be constructed from the finite subsets of M by union, multiplication and iteration where the iteration is applied to connected and monoalphabetic languages, only. Since, as we mentioned above, independent elements of M have disjoint label sets, any monoalphabetic language is residually closed. Hence mc-rational languages are c-rational.

Now let $\preceq$ be a linear order on the set E and let $x \in M$. The word $u \in T^*$ with $x = [u]$ is *the lexicographic normal form of* x (denoted $u = \text{lexNF}(x)$) if its label sequence is the least among all label sequences of words $v \in T^*$ with $x = [v]$. This lexicographic normal form is unique since two word $u, v \in T^*$ having the same label sequence with $[u] = [v]$ are equal. Let $\text{LNF} = \{\text{lexNF}(x) \mid x \in M\}$ denote the set of all words in T^* that are in lexicographic normal form. One can characterize the words from LNF similarly to trace theory. This characterization implies that LNF is recognizable in the free monoid T^*.

The crucial point in Ochmański's proof of the c-rationality of recognizable languages in trace monoids is that whenever a square of a word is in lexicographic normal form, it is actually connected. This does not hold any more for labeled divisibility monoids. But whenever a product of $|E| + 2$ words having the same set of labels is in lexicographic normal form, this product is connected.

We need another notation: For a set $A \subseteq E$ and $u \in T^*$ let $n_A(u)$ denote the number of occurrences of maximal factors w of u with $\ell(w) \subseteq A$ or $\ell(w) \cap A = \emptyset$. The number $n_A(u)$ is the number of blocks of elements of A and of $E \setminus A$ in the label sequence of u. Furthermore, we put $n_A(x) := n_A(\text{lexNF}(x))$ for $x \in M$.

Lemma 5. *Let $(M, \cdot, 1, \ell)$ be a labeled divisibility monoid, $x, y \in M$ and $x \parallel y$. Then $n_{\ell(x)}(x \vee y) \leq |E| + 1$.*

Lemma 6. *Let $X \subseteq T^*$ be a monoalphabetic language. Let $w \in X^{|E|+2} \cap \text{LNF}$. Then $[w]$ is connected.*

Proof. Let $n = |E| + 1$ and $x_i \in [X]$ with $[w] = x_0 x_1 \ldots x_n$. Furthermore assume $A = \ell(x_i)$ which is well defined since X is monoalphabetic. Now let $x, y \in M$ with $x \parallel y$ and $x \vee y = [w]$. Then $\ell(x) \cap \ell(y) = \emptyset$. If A contained an element from $\ell(x)$ and another one from $\ell(y)$, we would obtain $n_{\ell(x)}([w]) > n > |E| + 1$, contradicting Lemma 5. Hence $A \subseteq \ell(x)$ or $A \subseteq \ell(y)$. Now $\ell(x) \cup \ell(y) = \ell(x \vee y) = \ell(x_0 x_1 \ldots x_n) = A \subseteq \ell(x)$ implies $y = 1$. □

Now one can show that in a labeled divisibility monoid (with possibly infinite commutation behavior) any recognizable set is mc-rational. This proof follows the lines of the corresponding proof by Ochmański for traces using Lemma 6. Summarizing, we get the following theorem which in particular implies Thm. 1.

Theorem 5. *Let $(M, \cdot, 1)$ be a labeled divisibility monoid with finite commutation behavior and $L \subseteq M$. Then L is recognizable iff L is c-rational iff L is mc-rational.*

6 Open problems

Sakarovitch's and Ochmański's results are important generalizations of Kleene's Theorem to rational and to trace monoids, respectively; thus into "orthogonal" directions since any rational trace monoid is free. Our further extension of Ochmański's result is not "orthogonal" to Sakarovitch's approach any more (for instance $\{a, b, c, d\}^* / \langle ab = cd \rangle$ is both, a rational monoid and a divisibility monoid, but no free monoid). Hence our approach can be seen as a step towards a common generalization of Sakarovitch's and Ochmański's results.

References

1. R. M. Amadio and P.-L. Curien. *Domains and Lambda Calculi*. Cambridge Tracts in Theoretical Computer Science. 46. Cambridge: Cambridge University Press, 1998.
2. G. Berry and J.-J. Levy. Minimal and optimal computations of recursive programs. *J. ACM*, 26:148–175, 1979.
3. G. Birkhoff. *Lattice Theory*. Colloquium Publications vol. 25. American Mathematical Society, Providence, 1940; third edition, seventh printing from 1993.
4. G. Boudol. Computational semantics of term rewriting. In M. Nivat and J.C. Reynolds, editors, *Algebraic Methods in Semantics*, pages 169–236. Cambridge University Press, 1985.
5. J.R. Büchi. On a decision method in restricted second order arithmetics. In E. Nagel et al., editor, *Proc. Intern. Congress on Logic, Methodology and Philosophy of Science*, pages 1–11. Stanford University Press, Stanford, 1960.
6. V. Diekert and Y. Métivier. Partial commutation and traces. In G. Rozenberg and A. Salomaa, editors, *Handbook of Formal Languages*, volume 3. Springer, 1997.
7. V. Diekert and G. Rozenberg. *The Book of Traces*. World Scientific Publ. Co., 1995.
8. M. Droste. Recognizable languages in concurrency monoids. *Theoretical Comp. Science*, 150:77–109, 1995.
9. M. Droste and D. Kuske. On recognizable languages in left divisibility monoids. Technical Report MATH-AL-9-1998, TU Dresden, 1998.
10. C. Duboc. Commutations dans les monoïdes libres: un cadre théorique pour l'étude du parallelisme. Thèse, Faculté des Sciences de l'Université de Rouen, 1986.
11. K. Hashigushi. Recognizable closures and submonoids of free partially commutative monoids. *Theoretical Comp. Science*, 86:233–241, 1991.
12. D. Kuske. On rational and on left divisibility monoids. Technical Report MATH-AL-3-1999, TU Dresden, 1999.
13. E. Ochmański. Regular behaviour of concurrent systems. *Bull. Europ. Assoc. for Theor. Comp. Science*, 27:56–67, 1985.
14. P. Panangaden and E.W. Stark. Computations, residuals and the power of indeterminacy. In *Automata, Languages and Programming*, Lecture Notes in Comp. Science vol. 317, pages 439–454. Springer, 1988.
15. F.P. Ramsey. On a problem of formal logic. *Proc. London Math. Soc.*, 30:264–286, 1930.
16. J. Sakarovitch. Easy multiplications. I. The realm of Kleene's Theorem. *Information and Computation*, 74:173–197, 1987.
17. M.P. Schützenberger. On the definition of a family of automata. *Inf. Control*, 4:245–270, 1961.
18. E.W. Stark. Concurrent transition systems. *Theoretical Comp. Science*, 64:221–269, 1989.
19. G. Winskel. Event structures. In W. Brauer, W. Reisig, and G. Rozenberg, editors, *Petri nets: Applications and Relationships to Other Models of Concurrency*, Lecture Notes in Comp. Science vol. 255, pages 325–392. Springer, 1987.

Expressiveness of Point-to-Point versus Broadcast Communications

Cristian ENE[1,2], Traian MUNTEAN[1]

[1] University of Marseille
Laboratoire d'Informatique de Marseille (LIM-CNRS)
Parc Scientifique de Luminy - Case 925 - ESIL/ES2I;
F-13288 Marseille, France
muntean@lim.univ-mrs.fr, cene@esil.univ-mrs.fr

[2] on leave from University "Al. I. Cuza", Faculty of Computer Science,
6600 Iasi, Romania

Abstract. In this paper we address the problem of the expressive power of point-to-point communication to implement broadcast communication. We demonstrate that point-to-point communication as in *CCS* [M89] is "too asynchronous" to implement broadcast communication as in *CBS* [P95]. Milner's π-calculus [M91] is a calculus in which all communications are point-to-point. We introduce $b\pi$-calculus, using broadcast instead of rendez-vous primitive communication, as a variant of value-passing *CBS* in which communications are made on channels as in Hoare's *CSP* [H85] - and channels can be transmitted too as in π-calculus - but by a broadcast protocol: processes speak one at a time and are heard instantaneously by all others. In this paper, using the fact that π-calculus enjoys a certain interleaving property, whereas $b\pi$-calculus does not, we prove that there does not exist any uniform, parallel-preserving translation from $b\pi$-calculus into π-calculus, up to any "reasonable" equivalence. Using arguments similar to [P97], we also prove a separation result between *CBS* and *CCS*.

1 Introduction

Communication within processes is the main aspect of concurrency within distributed systems. One can specify basic communications from several points of view; primitives interactions can be, for instance, synchronous or asynchronous, associated to point-to-point or broadcast (one-to-many) message exchange protocols.

In theory (and in practice too), it arises naturally the question whether one mechanism can be "expressed" using the other (or, whether one can be implemented by the other). The first aspect (of synchrony/asynchrony) was, for instance, recently studied in [P97] in the framework of the π-calculus ([M91], [MPW92]). It was proved that there does not exist any uniform, parallel constructor-preserving, translation from π-calculus into the asynchronous π-calculus, up to any "reasonable" notion of equivalence.

In this paper, we address the question of the power of expressiveness of point-to-point communications versus broadcast communications. Indeed, we stress here that the broadcast is a natural manner of communication in concurrent and communicating systems found in many application fields. It has been also chosen as the hardware primitive exchange protocol for some networks, and in this case point-to-point message passing (when needed) is to be implemented on top of it. Broadcast can also be a natural primitive for many applications programming models (e.g. multimedia, data mining, etc.). Hence, it is natural to understand if a calculus for parallel/distributed computing based on basic broadcast communications ([EM99]) results in more powerful process calculi, or it is just a way of adding "syntactic sugar" to existing point-to-point based calculi. We chose Milner's π-calculus to study this problem, since it is recognised as one of the richest paradigm for concurrency introduced so far. In addition, the basic mechanism for communication in π-calculus is a point-to-point exchange event. When trying to express broadcast communication using point-to-point communication, what seems difficult, is how one can anticipate the size of the set of point-to-point communications needed when the number of potential receivers is "a priori" unknown. In addition, in π-calculus a system of two processes which exchange messages, can behave similarly in any context, while a broadcast communication is always "open" for the given environment. This is the intuition that we exploit in this paper to prove that broadcast communication cannot be "reasonably" simulated by using point-to-point communications of π-calculus.

The rest of the paper is as follows. In section 2 we briefly remind the bases of the π-calculus and then introduce the $b\pi$-calculus as a variant of a broadcast calculus (inspired from [P95]) together with some definitions concerning electoral systems. Section 3 presents the main result of the paper which proves the non-existence of any uniform encoding of $b\pi$-calculus into π-calculus. Section 4, discusses related works and presents future directions of research.

2 Preliminaries

In this section we briefly present the π-calculus, the $b\pi$-calculus (which is a variant of broadcast calculus), and then we introduce the notion of *electoral systems*.

2.1 The π-calculus

Let Ch_p be a countable set of *channels*. Then processes are defined in Table 1.

$$p ::= \quad nil \mid A\langle\tilde{x}\rangle \mid \Sigma_{i\in I}\alpha_i.p_i \mid p_1 \parallel p_2 \mid \nu xp \mid (rec\ A\langle\tilde{x}\rangle.p)\langle\tilde{y}\rangle$$

Table 1. Processes

where α_i belongs to the set of prefixes $\alpha ::= x(y) \mid \bar{x}y$.

Prefixes represent the basic actions of processes: $x(y)$ is the *input* of the name y on the channel x, and $\bar{x}y$ is the *output* of the name y on the channel

x. $\Sigma_{i\in I}\alpha_i.p_i$ represents guarded choice; when possible it begins by executing one of the atomic action α_i, and then its continuation is p_i. νxp is the creation of a new local channel x (whose initial scope is the process p). $p_1 \parallel p_2$ is the parallel composition of p_1 and p_2. $(recA\langle\tilde{x}\rangle.p)\langle\tilde{y}\rangle$ is a recursive process (allows to represent processes with infinite behaviour).

The operators νx and $y(x)$, are $x-$*binders*, i.e. in νxp and $y(x).p$, x appears bound, and $bn(p)$ represents the set of bound names of p. The free names of p are those that do not occur in the scope of any binder, and are denoted with $fn(p)$. The set of names of p is denoted with $n(p)$. The *alpha-conversion* is defined as usual.

In literature there have been defined relations among processes which relate processes which are "almost the same". Such a relation is a congruence, and allows to substitute a process by another congruent process, when needed.

Definition 1. *Structural congruence, denoted $\equiv$, is the smallest congruence over the set of processes which satisfies the conditions of Table 2.*

1 $p \equiv q$ if p and q are α-convertible
2 $(p \parallel q) \parallel r \equiv p \parallel (q \parallel r)$
3 $p \parallel q \equiv q \parallel p$
4 $p \parallel \nu xq \equiv \nu x(p \parallel q)$ if $x \notin fn(p)$

Table 2. Structural congruence

Definition 2. Actions, *ranged over α,β are given by the following syntax:*

$$\alpha \stackrel{def}{=} a(x) \mid \bar{a}x \mid \nu x\bar{a}x \mid \tau$$

where $a, x \in Ch_p$ and which reads as follows: an action is either a reception, a (possibly bound) output, or the silent action τ, denoting an uncontrollable transition.

We give an operational semantic of our calculus in terms of transitions over the set $\mathcal{P}_p$ of processes. Transitions are labeled by actions.

Definition 3. Transition system *The operational semantics of π-calculus is defined as a labeled transition system defined over the set $\mathcal{P}_p$ of processes. The judgement $p \xrightarrow{\alpha} p'$ means that that the process p is able to perform action α to evolve to p'. The operational semantics is given in Table 3.*

The semantic is an early one, i.e. the bound names of an input are instantiated as soon as possible, in the rule for input.

$$\frac{}{a(x).p\xrightarrow{a(z)}p[z/x]} \qquad \frac{}{\bar{a}x.p\xrightarrow{\bar{a}x}p} \qquad \frac{p_i\xrightarrow{\alpha}p'}{\Sigma_{i\in I}p_i\xrightarrow{\alpha}p'} \qquad \frac{p\xrightarrow{\bar{a}x}p'}{\nu xp\xrightarrow{\nu x\bar{a}x}p'}$$

$$\frac{p\xrightarrow{\alpha}p' \ \wedge\ x\notin n(\alpha)}{\nu xp\xrightarrow{\alpha}\nu xp'} \qquad \frac{p\xrightarrow{\alpha}p' \ \wedge\ bn(\alpha)\notin fn(q)}{p\|q\xrightarrow{\alpha}p'\|q} \qquad \frac{p\xrightarrow{\bar{a}x}p' \ \wedge\ q\xrightarrow{a(x)}q'}{p\|q\xrightarrow{\tau}p'\|q'}$$

$$\frac{p\xrightarrow{\nu x\bar{a}x}p' \ \wedge\ q\xrightarrow{a(x)}q'}{p\|q\xrightarrow{\tau}\nu x(p'\|q')} \qquad \frac{p[(rec\ A\langle\tilde{x}\rangle.p)/A,\tilde{y}/\tilde{x}]\xrightarrow{\alpha}p'}{(rec\ A\langle\tilde{x}\rangle.p)\langle\tilde{y}\rangle\xrightarrow{\alpha}p'} \qquad \frac{p\equiv p',\ p'\xrightarrow{\alpha}q',\ q'\equiv q}{p\xrightarrow{\alpha}q}$$

Table 3. Operational semantic of π-calculus

2.2 The $b\pi$-calculus

The $b\pi$-calculus is a process calculus in which broadcast is the fundamental communication paradigm. It is derived from broadcast calculus proposed by Prasad in [P95], and π-calculus. It differs from broadcast calculus, in that communications are made on channels (and transmitted values are channels too) which belong to a countable set Ch_b (like in π-calculus), and from π-calculus in the manner of use of channels: only for broadcast communications.

The syntax of *processes* is similar to that of π-calculus, as given in Table 1, and the meaning of process constructors is the same. Also, the set of *actions* (denoted $\mathcal{A}$), is defined similarly. The operational semantic (which is defined as a labeled transition system defined over the set $\mathcal{P}_b$ of $b\pi$-calculus processes) is different, and we describe it in Table 5. Before, we define similarly to [HR95], a relation $\longrightarrow\subseteq \mathcal{P}_b \times \mathcal{A}$ denoted $p \xrightarrow{\alpha:}$ (instead of $p \longrightarrow \alpha$) and which can be read "p discards the action α ".

$$\frac{\alpha{=}a(x)}{p\xrightarrow{\alpha:}} \qquad \frac{}{\bar{b}y.p\xrightarrow{\alpha:}} \qquad \frac{b{\neq}a}{b(x).p\xrightarrow{\bar{a}x:}}$$

$$\frac{\forall i{\in}I\ p_i\xrightarrow{\alpha:}}{\Sigma_{i\in I}p_i\xrightarrow{\alpha:}} \qquad \frac{p\xrightarrow{\bar{a}x:}}{p\xrightarrow{\nu x\bar{a}x:}} \qquad \frac{p\xrightarrow{\alpha:}}{\nu xp\xrightarrow{\alpha:}}$$

$$\frac{p\xrightarrow{\alpha:}\ \wedge\ q\xrightarrow{\alpha:}}{p\|q\xrightarrow{\alpha:}} \qquad \frac{p[(rec\ A\langle\tilde{x}\rangle.p)/A,\tilde{y}/\tilde{x}]\xrightarrow{\alpha:}}{(rec\ A\langle\tilde{x}\rangle.p)\langle\tilde{y}\rangle\xrightarrow{\alpha:}} \qquad \frac{p{\equiv}p'\ \wedge\ p'\xrightarrow{\alpha:}}{p\xrightarrow{\alpha:}}$$

Table 4 The "discard" relation

The operational semantic is given, like for π-calculus, via a transition system labeled by actions. Communication between processes is performed through unbuffered broadcasts. Comparing with π-calculus, outputs are non-blocking, i.e. there is no need of a receiving process. One of processes broadcasts an output and the remaining processes either receive or ignore the broadcast according to whether they are "listening" or not on the channel which serves as support of the output. A process which "listen" on a channel a, can not ignore any value send on this channel.

$$\frac{}{a(x).p \xrightarrow{a(z)} p[z/x]} \qquad \frac{}{\bar{a}x.p \xrightarrow{\bar{a}x} p} \qquad \frac{p_i \xrightarrow{\alpha} p'}{\Sigma_{i \in I} p_i \xrightarrow{\alpha} p'}$$

$$\frac{p \xrightarrow{\alpha} p' \ \wedge \ x \notin n(\alpha)}{\nu x p \xrightarrow{\alpha} \nu x p'} \qquad \frac{p \xrightarrow{\bar{a}x} p'}{\nu x p \xrightarrow{\nu x \bar{a} x} \nu x p'} \qquad \frac{p \xrightarrow{\bar{a}x} p'}{\nu a p \xrightarrow{\tau} \nu a p'}$$

$$\frac{p \xrightarrow{\nu x \bar{a} x} \nu x p' \ \wedge \ q \xrightarrow{a(x)} q'}{p \| q \xrightarrow{\nu x \bar{a} x} \nu x (p' \| q')} \qquad \frac{p \xrightarrow{\bar{a}x} p' \wedge q \xrightarrow{a(x)} q'}{p \| q \xrightarrow{\bar{a}x} p' \| q'} \qquad \frac{p \xrightarrow{a(x)} p' \ \wedge \ q \xrightarrow{a(x)} q'}{p \| q \xrightarrow{a(x)} p' \| q'}$$

$$\frac{p \xrightarrow{\alpha} p' \ \wedge \ q \xrightarrow{\alpha:} \ \wedge \ \ bn(\alpha) \notin fn(q)}{p \| q \xrightarrow{\alpha} p' \| q} \qquad \frac{p[(rec\ A\langle\tilde{x}\rangle.p)/A, \tilde{y}/\tilde{x}] \xrightarrow{\alpha} p'}{(rec\ A\langle\tilde{x}\rangle.p)\langle\tilde{y}\rangle \xrightarrow{\alpha} p'} \qquad \frac{p \equiv p',\ p' \xrightarrow{\alpha} q',\ q' \equiv q}{p \xrightarrow{\alpha} q}$$

Table 5. Operational semantic of $b\pi$-calculus

2.3 Electoral systems

In this subsection, we present the notion of an electoral system as given by Palamidessi in [P97]. All the notions given below hold both for the π-calculus and for the $b\pi$-calculus. We also use actions of the form $\bar{a}$ or a, when the names which are send or received do not matter.

A *cluster* is a system of parallel processes $P = P_1 \parallel P_2 \parallel \ldots \parallel P_n$. A *computation* C for the cluster is a (possibly ω-infinite) sequence of transitions[1]:

$$\begin{aligned} P_1 \parallel P_2 \parallel \ldots \parallel P_n \xrightarrow{\alpha_1} & P_1^1 \parallel P_2^1 \parallel \ldots \parallel P_n^1 \\ \xrightarrow{\alpha_2} & P_1^2 \parallel P_2^2 \parallel \ldots \parallel P_n^2 \\ & \vdots \\ \xrightarrow{\alpha_m} & P_1^m \parallel P_2^m \parallel \ldots \parallel P_n^m \\ (\xrightarrow{\alpha_{m+1}} & \ldots \quad) \end{aligned}$$

If $\tilde{\alpha} = \alpha_1.\alpha_2.\ldots.\alpha_m$, we will represent the computation C also by $C : P \stackrel{\tilde{\alpha}}{\Longrightarrow} P^m$ (or by $C : P \stackrel{\tilde{\alpha}}{\Longrightarrow} \ldots$ if C is infinite).

C' *extends* C if C: $P \stackrel{\tilde{\alpha}}{\Longrightarrow} P^m$, and there exists C'' : $P^m \stackrel{\tilde{\alpha}}{\Longrightarrow} P^{m+m'}$ or C'' : $P^m \stackrel{\tilde{\alpha}}{\Longrightarrow} \ldots$ such that $C' = CC''$, where the two occurrences of P^m are collapsed. The *projection* of C over P_i (C, P given as above), denoted $Proj(C, i)$, is defined as the "contribution" of P_i to the computation C.

We define the restriction of a sequence of actions $\tilde{\alpha}$ w.r.t. to a set of channels A (denoted $\tilde{\alpha}/A$) as being a word of the $(A \cup \bar{A})^*$, given as follows:

[1] As in [P97], for the sake of keeping notations simple, we suppose that each binder νx is pushed "to the top level" using repeatedly the rules for structural congruence. In addition, we do not represent explicitly the binders at top level and we suppose that the cluster will never perform a visible action on one of the names restricted by a binder

$$\tilde{\alpha}/A = \begin{cases} nil & \text{if } \tilde{\alpha} = nil, \\ \tilde{\beta}/A & \text{if } \tilde{\alpha} = \tau.\tilde{\beta}, \\ \tilde{\beta}/A & \text{if } \alpha = \bar{a}x.\tilde{\beta} \text{ or } \alpha = \nu x\bar{a}x.\tilde{\beta} \text{ or } \tilde{\alpha} = a(x).\tilde{\beta}, \text{ with } a \notin A, \\ \bar{a}.(\tilde{\beta}/A) & \text{if } \alpha = \bar{a}x.\tilde{\beta} \text{ or } \alpha = \nu x\bar{a}x.\tilde{\beta}, \text{ with } a \in A, \\ a.(\tilde{\beta}/A) & \text{if } \tilde{\alpha} = a(x).\tilde{\beta}, \text{ with } a \in A. \end{cases} \quad (1)$$

Like in [P97], for the definition of an electoral system, we assume that Ch_p and Ch_b contain the natural names $\mathbb{N}$, which will represent the identity of processes in a network. We shall use a definition slightly different from those used in [P97].

Definition 4. (Electoral system)
A cluster $P = P_1 \parallel P_2 \parallel \ldots \parallel P_n$ is an **electoral system** *if for every computation C of P, there exists an extension C' of C, and there exists $k \in \{1, \ldots, n\}$ (the "leader") such that for each $i \in \{1, \ldots, n\}$ the projection $Proj(C', i)$ contains exactly one output action of the form $\bar{k}$, and any trace of a P_i may contain at most one action of the form $\bar{l}$, with $l \in \{1, \ldots, n\}$.*

Note that for an electoral system, any infinite computation must contain already all the necessary output actions, because it cannot be strictly extended, and also, that for an electoral system, there exists always a finite computation which satisfies the requirements of Definition 4.

3 Encoding $b\pi$-calculus into π-calculus

In this section we prove the non-existence of an uniform encoding of the $b\pi$-calculus into π-calculus, under certain requirements on the encoding, which preserves "reasonable" semantics.

When translating a term from one calculus into another, we would like the translation be independent of the context, i.e. the encoding of t_1 in $C'[t_1]$ and in $C''[t_1]$ to be the same regardless of contexts C' and C''. This requires that the encoding is compositional. Concerning concurrent systems, it is reasonable to require at least that the parallel construction is preserved under the encoding, and more, that is exactly mapped in the parallel constructor, i.e. that

$$[\![P \parallel Q]\!] = [\![P]\!] \parallel [\![Q]\!] \quad (2)$$

Also, it seems reasonable to require that the encoding "behaves well" with renamings, i.e.

$$[\![\sigma(P)]\!] = \sigma([\![P]\!]) \quad (3)$$

We will call *uniform* an encoding which satisfies (2) and (3) and which in addition translates outputs (inputs) in the first calculus in related outputs (inputs) in the second calculus.

Definition 5. Uniform encoding[2]

An encoding $[\![\ ]\!] : \mathcal{P}_x \longrightarrow \mathcal{P}_y$ *(where* $(x,y) \in \{(p,b),(b,p)\}$*) is called uniform, if it satisfies the following conditions:*

1. $[\![P \parallel Q]\!] = [\![P]\!] \parallel [\![Q]\!]$,
2. $[\![\sigma(P)]\!] = \sigma([\![P]\!])$, *for any substitution* $\sigma : \mathbb{N} \longrightarrow \mathbb{N}$,
3. $(\exists \tilde{\alpha} : P \stackrel{\tilde{\alpha}}{\Longrightarrow} P'$ *if and only if* $\exists \tilde{\beta} : [\![P]\!] \stackrel{\tilde{\beta}}{\Longrightarrow} [\![P']\!])$, *where* $\tilde{\alpha}/\mathbb{N} = \tilde{\beta}/\mathbb{N}$,
4. *if* $[\![P]\!] \stackrel{\tilde{\beta}}{\Longrightarrow} Q$ *then* $\exists \tilde{\gamma} : [\![P]\!] \stackrel{\tilde{\beta}}{\Longrightarrow} Q \stackrel{\tilde{\gamma}}{\Longrightarrow} [\![P']\!]$.

In the same time, we should prefer that the encodings of two terms "equivalent" under a certain semantics in the first calculus, to be equivalent under a related semantics in the second calculus. So, in our case, following Palamidessi, "a semantic is << *reasonable* >> if it distinguishes two processes P and Q whenever in some computation of P the actions on certain intended channels are different from those of any computation of Q".

Lemma 1. *Any uniform encoding translates an electoral system P from $b\pi$-calculus into an electoral system $[\![P]\!]$ from π-calculus.*

Proof. Let $P = P_1 \parallel P_2 \parallel \ldots \parallel P_n$ be an electoral system. We shall prove that $R \stackrel{def}{=} [\![P]\!]$ is an electoral system.

We have

$$R \stackrel{def}{=} [\![P_1 \parallel P_2 \parallel \ldots \parallel P_n]\!] = [\![P_1]\!] \parallel [\![P_2]\!] \parallel \ldots \parallel [\![P_n]\!] = R_1 \parallel R_2 \parallel \ldots \parallel R_n$$

where $R_i \stackrel{def}{=} [\![P_i]\!]$.

Let $R_i \stackrel{\tilde{\beta^i}}{\Longrightarrow} Q_i$ be an arbitrary trace of R_i. From the Condition 4 of the Definition 5 we obtain that there exists a continuation

$$R_i = [\![P_i]\!] \stackrel{\tilde{\beta^i}}{\Longrightarrow} Q_i \stackrel{\tilde{\gamma^i}}{\Longrightarrow} R_i' = [\![P_i']\!]$$

Then, using the Condition 3 of the Definition 5, we obtain a corresponding trace of P_i

$$P_i \stackrel{\tilde{\alpha^i}}{\Longrightarrow} P_i'$$

such that $\tilde{\alpha^i}/\mathbb{N} = (\tilde{\beta^i}\tilde{\gamma^i})/\mathbb{N}$. Since P is an electoral system, α^i contains at most one action of the form $\bar{l}$, with $l \in \{1,\ldots,n\}$, and this is also true for the derivation $R_i \stackrel{\tilde{\beta^i}}{\Longrightarrow} Q_i$.

Let

$$D : R = R_1 \parallel R_2 \parallel \ldots \parallel R_n \stackrel{\tilde{\beta}}{\Longrightarrow} Q$$

[2] It is assumed implicitly in the definition that an uniform encoding "behaves well" with respect to restriction $[\![\nu a P]\!] = \nu L_a([\![P]\!])$ where L_a is the set of channels used to implements the channel a. Also, for the sake of simplicity, we do not mention explicitly the binders in the results or in the proofs.

be a computation. From the Condition 4 of the Definition 5 we obtain that there exists E, extension of D,

$$E : R = [\![P]\!] \overset{\tilde{\beta}}{\Longrightarrow} Q \overset{\tilde{\gamma}}{\Longrightarrow} R' = [\![P']\!]$$

Using the Condition 3 of the Definition 5, we obtain a corresponding computation

$$C : P \overset{\tilde{\alpha}}{\Longrightarrow} P'$$

such that $\tilde{\alpha}/\mathbb{N} = (\tilde{\beta}\tilde{\gamma})/\mathbb{N}$. Since P is an electoral system, there is an extension C' of C

$$C : P \overset{\tilde{\alpha}}{\Longrightarrow} P' \overset{\tilde{\alpha}'}{\Longrightarrow} P''$$

and there exists $k \in \{1, \ldots, n\}$ (the "leader") such that for each $i \in \{1, \ldots, n\}$ the projection $Proj(C', i)$ contains exactly one output action of the form $\bar{k}$. Using the Condition 3 of the Definition 5 we obtain the computation

$$D' : R = [\![P]\!] \overset{\tilde{\beta}}{\Longrightarrow} Q \overset{\tilde{\gamma}}{\Longrightarrow} [\![P']\!] \overset{\tilde{\beta}'}{\Longrightarrow} [\![P'']\!]$$

such that $(\tilde{\beta}\tilde{\gamma}\tilde{\beta}')/\mathbb{N} = (\tilde{\alpha}\tilde{\alpha}')/\mathbb{N}$. Because in $\tilde{\beta}\tilde{\gamma}\tilde{\beta}'$ there are n outputs of the form $\bar{k}$, with $k \in \{1, \ldots, n\}$, and since every component R_i can make at most one such action, it follows that for each $i \in \{1, \ldots, n\}$ the projection $Proj(D', i)$ contains exactly one output action of the form $\bar{k}$.□

Lemma 2. *Let* $P_i = \bar{a}i \parallel a(x).\bar{x}$.
For all $n \geq 2$, $P(n) = \prod_{i=1}^{n} P_i$ *is an electoral system.*

Proof. If

$$C : P(n) = P_1 \parallel P_2 \parallel \ldots \parallel P_n \overset{\alpha}{\longrightarrow} P_1^1 \parallel P_2^1 \parallel \ldots \parallel P_n^1 = P^1(n)$$

in a step, then there is $k \in \{1, \ldots, n\}$ such that $\alpha = \bar{a}k$, and $P_i^1 = \bar{k}$, $\forall i \in \{1, \ldots, n\}$. Then we can extend C to C':

$$C' : P(n) = P_1 \parallel \ldots \parallel P_k \parallel \ldots \parallel P_n \overset{\bar{a}k}{\longrightarrow} \bar{a}1 \parallel \bar{k} \parallel \ldots \parallel nil \parallel \bar{k} \parallel \ldots \parallel \bar{a}n \parallel \bar{k}$$

$$\overset{\bar{k}}{\longrightarrow} \bar{a}1 \parallel nil \parallel \ldots \parallel nil \parallel \bar{k} \parallel \ldots \parallel \bar{a}n \parallel \bar{k}$$

$$\vdots$$

$$\overset{\bar{k}}{\longrightarrow} \bar{a}1 \parallel nil \parallel \ldots \parallel nil \parallel nil \parallel \ldots \parallel \bar{a}n \parallel nil$$

and we can deal similarly with any other computation.□

Then, the main result of this paper is:

Theorem 1. *There exists no uniform encoding of the bπ-calculus into π-calculus preserving a reasonable semantics.*

Proof. Let suppose that there exists an uniform encoding $[\![\bullet]\!]$ of $b\pi$-calculus into π-calculus.

Let us denote $R_i \stackrel{def}{=} [\![P_i]\!]$, and $R(n) \stackrel{def}{=} R_1 \parallel R_2 \parallel \ldots \parallel R_n$, where P_i are the same as in the Lemma 2. We have

$$R(n) = R_1 \parallel R_2 \parallel \ldots \parallel R_n = [\![P_1]\!] \parallel [\![P_2]\!] \parallel \cdots \parallel [\![P_n]\!] =$$

$$[\![P_1 \parallel P_2 \parallel \ldots \parallel P_n]\!] = [\![P(n)]\!]$$

Since $\forall n \geq 2, P(n)$ is an electoral system, using the Lemma 1, we obtain that $\forall n \geq 2, R(n)$ is an electoral system too.

Let $m, m' \geq 2$, and let the renaming $\sigma : \mathbb{N} \longrightarrow \mathbb{N}, \sigma(i) = i + m'$, $\forall i \in \{1, \ldots, m\}$ and identity otherwise.

We have

$$R(m+m') = R_1 \parallel R_2 \parallel \ldots \parallel R_{m'} \parallel R_{m'+1} \parallel \ldots \parallel R_{m'+m} =$$

$$R_1 \parallel R_2 \parallel \ldots \parallel R_{m'} \parallel [\![P_{m'+1}]\!] \parallel \ldots \parallel [\![P_{m'+m}]\!] =$$

$$R_1 \parallel R_2 \parallel \ldots \parallel R_{m'} \parallel [\![\sigma(P_1)]\!] \parallel \ldots \parallel [\![\sigma(P_m)]\!] =$$

$$R_1 \parallel R_2 \parallel \ldots \parallel R_{m'} \parallel \sigma([\![P_1]\!]) \parallel \ldots \parallel \sigma([\![P_m]\!]) =$$

$$R_1 \parallel R_2 \parallel \ldots \parallel R_{m'} \parallel \sigma(R_1) \parallel \ldots \parallel \sigma(R_m) =$$

$$R(m') \parallel \sigma(R(m))$$

Since $R(m')$ is an electoral system, then there exists a computation $C_1 : R(m') \stackrel{\tilde{\alpha}}{\Longrightarrow} R^p(m')$ and a $k \in \{1, \ldots, m'\}$, such that for each $i \in \{1, \ldots, m'\}$ the projection $Proj(C_1, i)$ contains exactly one output action of the form $\bar{k}$. Similar, because $R(m)$ is an electoral system, then there exists a computation $C_2 : R(m) \stackrel{\tilde{\beta}}{\Longrightarrow} R^q(m)$ and a $k' \in \{1, \ldots, m\}$, such that for each $i \in \{1, \ldots, m\}$ the projection $Proj(C_2, i)$ contains exactly one output action of the form $\bar{k'}$. Hence, there exists a computation $C_3 : \sigma(R(m)) \stackrel{\sigma(\tilde{\beta})}{\Longrightarrow} \sigma(R^q(m))$ of $\sigma(R(m))$ and a $k' \in \{1, \ldots, m\}$, such that for each $i \in \{m'+1, \ldots, m'+m\}$ the projection $Proj(C_3, i)$ contains exactly one output action of the form $\overline{\sigma(k')}$.

Then we have the following computation

$$C_4 : R(m+m') = R(m') \parallel \sigma(R(m)) \stackrel{\tilde{\alpha}}{\Longrightarrow} R^p(m') \parallel \sigma(R(m))$$

$$\stackrel{\sigma(\tilde{\beta})}{\Longrightarrow} R^p(m') \parallel \sigma(R^q(m))$$

such that for each $i \in \{1, \ldots, m'\}$ the projection $Proj(C_4, i)$ contains exactly one output action of the form $\bar{k}$, and for each $i \in \{m'+1, \ldots, m'+m\}$ the projection $Proj(C_4, i)$ contains exactly one output action of the form $\overline{\sigma(k')}$. Since $\sigma(k') \in \{m'+1, \ldots, m'+m\}$, we have that $k \neq \sigma(k')$, and C_4 cannot be extended to a computation C as in the Definition 4, hence $R(m+m')$ cannot be an electoral system. □

If we call $b\pi_a$-calculus the asynchronous variant of $b\pi$-calculus (obtained from $b\pi$-calculus by forbidding output as prefix, asynchronous variant of π-calculus being obtained similarly from π-calculus), we obtain the following result:

Corollary 1. *There exists no uniform encoding of the $b\pi_a$-calculus into π-calculus preserving a reasonable semantics.*

Slightly changing the Definition 4, we can prove the following result:

Theorem 2. *There exists no uniform encoding of the $b\pi$-calculus into π-calculus which preserves weak bisimulation.*

Remark. Note that the condition 2 is essential to establish our result. We do not claim that there is no encoding at all; hence we cannot guarantee on the non-existence of a slighter encoding, which remains compositional, mapping $[\![P \parallel Q]\!]$ onto $C[[\![P]\!], [\![Q]\!]]$.

4 Conclusion and related work

The Theorem 1 corresponds to the separability result between synchronous and asynchronous π-calculus obtained by Palamidessi. She uses in her proof the fact that asynchronous π-calculus enjoys a certain kind of confluence. Hence, there are symmetric electoral systems in the π-calculus, whereas this does not hold in the asynchronous case, since whenever a first action occurs, all the other processes can execute their corresponding output action as well, and so on, in this way generating an infinite computation which never makes outputs on a special channel o (used for sending to the environment the leader).

In our paper we exploit another difference between the two calculus: while π-calculus enjoys an interleaving semantic (if $P \stackrel{\tilde{\alpha}}{\Longrightarrow} P'$ and $Q \stackrel{\tilde{\beta}}{\Longrightarrow} Q'$, then $P \parallel Q \stackrel{\tilde{\alpha}}{\Longrightarrow} P' \parallel Q \stackrel{\tilde{\beta}}{\Longrightarrow} P' \parallel Q'$), this does not hold for the $b\pi$-calculus ($P \stackrel{\tilde{\alpha}}{\Longrightarrow} P'$ does not imply $P \parallel Q \stackrel{\tilde{\alpha}}{\Longrightarrow} P' \parallel Q$).

The problem of encoding a broadcast calculus into π-calculus or CCS ([M89]) was already stated in [H93]. Holmer gives an encoding of CBS into $SCCS$ and he makes the conjecture that it is not possible to find a compositional translation from broadcast calculus to CCS: "CCS is << *too asynchronous* >> to interpret CBS in". His variant of broadcast calculus (CBS) is without value-passing. We can give a partial answer to his conjecture, by proving that value-passing CBS cannot be uniformly encoded in π-calculus and that CBS without value-passing cannot be uniformly encoded in CCS:

Proposition 1. *There exists no uniform encoding of value-passing CBS into π-calculus preserving a reasonable semantics.*

Hint of the proof. In newer variants of CBS, choice is also denoted by &, and outputs (inputs) are denoted differently ($v!P$ means a process which says v, and become P, while $x?Q$ is a process which listen, and once he heard v, it evolves to $Q[v/x]$).

Let $P_i = i! \,\&\, x?x!$. Then it is easy to remark that for all $n \geq 2$, $P(n) = \prod_{i=1}^{n} P_i$ is an electoral system and using arguments similar to Theorem 1, we obtain the result. □

Proposition 2. *There exists no uniform encoding of CBS without value-passing into CCS preserving a reasonable semantics.*

Hint of the proof. In *CBS* without value-passing, processes communicate by exchanging signals (by synchronisations). $v!P$ is a process which says v, and become P, while $v?Q$ is a process which listen, and once he heard v, it evolves to Q.

Let $P_i = a_i! \parallel \sum_{j=1}^{n} a_j?j!$. Then it is easy to remark that $P(n) = \prod_{i=1}^{n} P_i$ is a symmetric electoral system (for definition see Definition 3.1 and Definition 3.2 from [P97]). Combining this with a stronger version of the Theorem 5.2 from [P97] (which admits a similar proof for our definition of Electoral System), we obtain the result. □

Acknowledgements

We have gratefully appreciated the useful comments received from Catuscia Palamidessi and Gianluigi Zavattaro. The referees have also contributed by their remarks to improve the presentation of the paper.

References

[EM99] C. Ene and T. Muntean: A distributed calculus for nomadic processes. submitted, 1999.

[HR95] M. Hennessy and J. Rathke: Bisimulations for a Calculus of Broadcasting Systems. In I. Lee and S. Smolka, editors, *Proceedings of CONCUR 95*, Philadelphia, (1995) Lecture Notes in Computer Science **962**, Springer-Verlag 486–500.

[H85] C. A. R. Hoare: Communicating Sequential Processes. Prentice-Hall, 1985.

[H93] U. Holmer: Interpreting Broadcast Communication in SCCS. In *Proceedings of CONCUR 93*, (1993) Lecture Notes in Computer Science **715**, Springer-Verlag.

[M89] R. Milner: Communication and concurrency. Prentice-Hall, 1989.

[M91] R. Milner: The Polyadic π-Calculus: A Tutorial. In Friedrich L. Bauer and Wilfried Brauer and Helmut Schwichtenberg, editors, (1995) *Logic and Algebra of Specification* **94**. Available as Technical Report ECS-LFCS-91-180, University of Edinburgh, October 1991.

[MPW92] R. Milner and J. Parrow and D. Walker: A Calculus of Mobile Processes, Part I/II. Journal of Information and Computation **100** (1992) 1–77.

[P97] Catuscia Palamidessi: Comparing the Expressive Power of the Synchronous the Asynchronous π-calculus. In *Proceedings of POPL 1997* (ACM, Jan. 1997) 256–265.

[P95] K. V. S. Prasad: A Calculus of Broadcasting Systems. (1995) Science of Computer Programming **25**.

On Relative Loss Bounds in Generalized Linear Regression

Jürgen Forster

Universität Bochum, Germany
forster@lmi.ruhr-uni-bochum.de

Abstract. When relative loss bounds are considered, an on-line learning algorithm is compared to the performance of a class of off-line algorithms, called experts. In this paper we reconsider a result by Vovk, namely an upper bound on the on-line relative loss for linear regression with square loss – here the experts are linear functions. We give a shorter and simpler proof of Vovk's result and give a new motivation for the choice of the predictions of Vovk's learning algorithm. This is done by calculating the, in some sense, best prediction for the last trial of a sequence of trials when it is known that the outcome variable is bounded. We try to generalize these ideas to the case of generalized linear regression where the experts are neurons and give a formula for the "best" prediction for the last trial in this case, too. This prediction turns out to be essentially an integral over the "best" expert applied to the last instance. Predictions that are "optimal" in this sense might be good predictions for long sequences of trials as well.

1 Introduction

In the *on-line learning* protocols we consider here, a *learning algorithm* called *Learner* tries to predict real numbers in a sequence of trials. Real-world examples of applications for this protocol are weather or stockmarket predictions, or pattern recognition. This protocol can be seen as a game between Learner and an opponent, *Nature*. After Learner receives an instance x_t in the t–th trial, it makes a prediction $\hat{y}_t$ and Nature responds with the correct outcome y_t. Learner wants to keep the discrepancy between $\hat{y}_t$ and y_t as small as possible. This discrepancy is measured with a *loss function* L, and the total loss of Learner on a sequence of trials is the sum of the losses in each trial. One way to measure the quality of Learner's predictions is to compare the loss of Learner to that of a class of functions from the set of instances to the set of outcomes (such functions are called *experts*), i.e. to give *relative loss bounds*.

Like in [6], we consider the following protocol of interaction between Learner and Nature:

FOR $t = 1, 2, 3, \ldots, T$
  Nature chooses an instance $x_t \in \mathbb{R}^n$
  Learner chooses a prediction $\hat{y}_t \in \mathbb{R}$

Nature chooses an outcome $y_t \in \mathbb{R}$
END FOR

Learner does not necessarily know the number of trials T in advance. After t trials, the loss of Learner is

$$L_t(\text{Learner}) := \sum_{s=1}^{t}(y_s - \hat{y}_s)^2 \ . \tag{1}$$

We use the following notations: The vectors $x \in \mathbb{R}^n$ are column vectors. $x^\top$, the transposed vector of x, is a row vector. For $m, n \in \mathbb{N}$, $\mathbb{R}^{m\times n}$ is the set of real $m \times n$ matrices. The scalar product of $x, y \in \mathbb{R}^n$ is $x \cdot y = x^\top y = \sum_{i=1}^n x_i y_i$ and the 2-norm of x is $\|x\| = (x \cdot x)^{\frac{1}{2}}$. $I \in \mathbb{R}^{n\times n}$ is the $n \times n$ identity matrix. A matrix $A \in \mathbb{R}^{n\times n}$ is called positive semidefinite if $x^\top Ax \geq 0$ for all $x \in \mathbb{R}^n$ and it is called positive definite if $x^\top Ax > 0$ for all $x \in \mathbb{R}^n \setminus \{0\}$.

We search for strategies for Learner that ensure that its loss is not much larger than the loss of the, in some sense, best linear expert. A linear expert $w \in \mathbb{R}^n$ makes the prediction $w \cdot x$ on instance $x \in \mathbb{R}^n$ and its loss on the first t trials is

$$L_t(w) := \sum_{s=1}^{t}(y_s - w \cdot x_s)^2 \ . \tag{2}$$

For a fixed $a > 0$, we try to minimize

$$L_T(\text{Learner}) - \inf_{w\in\mathbb{R}^n} \left(a\|w\|^2 + L_T(w)\right) \ . \tag{3}$$

The term $a\|w\|^2$ gives the learner a start on expert w, according to the expert's "complexity" $\|w\|^2$.

Several prediction strategies for Learner have been considered. In [5], Kivinen and Warmuth gave relative loss bounds for the Exponentiated Gradient Algorithm. The ridge regression strategy was compared to a similar strategy, called AA^R, in [6] by Vovk. The best known upper bound on the relative loss holds for AA^R.

How can we get good strategies for Learner? It is not clear what the best prediction for Learner at trial t is. This is mainly because, when Learner gives $\hat{y}_t$, it does not know the number of trials T and it does not know anything about the future instances $x_{t+1}, \ldots, x_T$ and the future outcomes $y_t, \ldots, y_T$. But if it is known that there is some $Y \geq 0$ such that $y_T \in [-Y, Y]$, there is a prediction $\hat{y}_T$ for the last trial T that minimizes (3). This prediction is calculated in Theorem 1, it is essentially the prediction of AA^R.

If Learner makes in each trial t the prediction that would be optimal if $y_t \in [-Y, Y]$ is known and if trial t was known to be the last one of the sequence, there is an upper bound on (3) of the form $O(\ln T)$. This is proven in Theorem 3, which is a reproof of Theorem 1, [6]. Our new proof is shorter and simpler than Vovk's, which uses the Aggregating Algorithm (AA), "perfectly mixable" games and contains a lot of difficult calculations. An independent proof of Theorem 3 is given by Azoury and Warmuth in [1]. They discuss relative loss bounds in the context of density estimation using the exponential family of distributions.

It is possible to find other new on-line prediction strategies with the method of Theorem 1. As an example, in Sect. 3 we look at the case of generalized linear regression. Here the linear experts are replaced by neurons, i.e., we compare to functions $\mathbb{R}^n \ni x \mapsto \varphi(w \cdot x)$, $w \in \mathbb{R}^n$, where $\varphi : \mathbb{R} \to \mathbb{R}$ is a fixed differentiable and strictly increasing function. The matching loss is used to measure the loss of Learner and the losses of the experts (there is a motivation for using the matching loss in [4]). Under the assumption that the outcome variable is bounded, we can calculate the optimal prediction of Learner for the last trial of a sequence in this more general case, too.

2 Linear Regression with Square Loss

In this section we calculate the best prediction for Learner in the last trial when the outcome variable is bounded and we prove an upper bound on the relative loss for linear regression with square loss.

For a sequence $(x_1, y_1), \ldots, (x_t, y_t)$ of instances and outcomes it is easy to give a formula for the loss of the "best" linear expert. We use the following notation:

$$b_t := \sum_{s=1}^{t} y_s x_s \in \mathbb{R}^n \ , \tag{4}$$

$$A_t := aI + \sum_{s=1}^{t} x_s x_s^\top \in \mathbb{R}^{n \times n} \ . \tag{5}$$

Note that for $t = 0, 1, 2, \ldots$, A_t is a positive definite symmetric $n \times n$–matrix because aI is positive definite and $xx^\top$ is a positive semidefinite matrix for all $x \in \mathbb{R}^n$. (For each $y \in \mathbb{R}^n$: $y^\top (xx^\top) y = (x \cdot y)^2 \geq 0$.)

Lemma 1. *For all $t \geq 0$, function $f(w) := a\|w\|^2 + L_t(w)$ is minimal at a unique point, say w_t. Furthermore, w_t and $f(w_t)$ are given by*

$$w_t = A_t^{-1} b_t \quad \textit{and} \quad f(w_t) = \sum_{s=1}^{t} y_s^2 - b_t^\top A_t^{-1} b_t \ .$$

Proof. From

$$\begin{aligned} f(w) = a\|w\|^2 + L_t(w) &\overset{(2)}{=} a\|w\|^2 + \sum_{s=1}^{t} (y_s - w \cdot x_s)^2 \\ &= a w^\top w + \sum_{s=1}^{t} \left(y_s^2 - 2 y_s (w \cdot x_s) + (w^\top x_s)(x_s^\top w) \right) \end{aligned}$$

$$= \sum_{s=1}^{t} y_s^2 - 2\sum_{s=1}^{t} w \cdot (y_s x_s) + w^\top \left(aI + \sum_{s=1}^{t} x_s x_s^\top \right) w$$

$$\stackrel{(4),(5)}{=} \sum_{s=1}^{t} y_s^2 - 2w \cdot b_t + w^\top A_t w \ .$$

it follows that $\nabla f(w) = 2A_t w - 2b_t$, $Hf(w) = 2A_t$. Thus f is convex and it is minimal if $\nabla f(w) = 0$, i.e. for $w = A_t^{-1} b_t$. This shows that $w_t = A_t^{-1} b_t$ and we obtain

$$f(w_t) = f(A_t^{-1} b_t) = \sum_{s=1}^{t} y_s^2 - 2b_t^\top A_t^{-1} b_t + b_t^\top A_t^{-1} A_t A_t^{-1} b_t = \sum_{s=1}^{t} y_s^2 - b_t^\top A_t^{-1} b_t \ .$$

□

In [6], Vovk proposes the $\mathrm{AA^R}$ learning algorithm which makes the predictions

$$\hat{y}_t = b_{t-1}^\top A_t^{-1} x_t$$

and shows how these predictions can be computed in time $O(n^2)$ per trial. They are very similar to $w_t \cdot x_t = b_t^\top A_t^{-1} x_t$, the prediction of the expert w_t on the instance x_t. We will now show that $b_{T-1}^\top A_T^{-1} x_T$ is essentially the best prediction for the last trial T. For this, let

$$\mathrm{clip}(z, Z) := \begin{cases} -Z, & z \in (-\infty, -Z] \ , \\ z, & z \in [-Z, Z] \ , \\ Z, & z \in [Z, \infty) \ , \end{cases}$$

for $z, Z \in \mathbb{R}$, $Z \geq 0$. $\mathrm{clip}(z, Z)$ is the number in $[-Z, Z]$ that is closest to z.

Theorem 1. *If Learner knows that $y_T \in [-Y, Y]$, then the optimal prediction for the last trial T is*

$$\hat{y}_T = \mathrm{clip}(b_{T-1}^\top A_T^{-1} x_T, Y) \ .$$

Proof. Any $y_T \in [-Y, Y]$ can be chosen by Nature. Thus Learner should choose a $\hat{y}_T \in \mathbb{R}$ such that

$$\sup_{y_T \in [-Y,Y]} \left(L_T(\text{Learner}) - \inf_{w \in \mathbb{R}^n} \left(a\|w\|^2 + L_T(w) \right) \right)$$

$$\stackrel{(1),\ \text{Lemma 1}}{=} \sup_{y_T \in [-Y,Y]} \left(\sum_{t=1}^{T} (y_t - \hat{y}_t)^2 - \sum_{t=1}^{T} y_t^2 + b_T^\top A_T^{-1} b_T \right) .$$

is minimal. Because of

$$b_T^\top A_T^{-1} b_T \stackrel{(4)}{=} (b_{T-1} + y_T x_T)^\top A_T^{-1} (b_{T-1} + y_T x_T)$$
$$= b_{T-1}^\top A_T^{-1} b_{T-1} + 2y_T b_{T-1}^\top A_T^{-1} x_T + y_T^2 x_T^\top A_T^{-1} x_T$$

this expression is minimal if and only if (only terms that depend on y_T or on $\hat{y}_T$ are important)

$$\sup_{y_T\in[-Y,Y]} \left(-2y_T\hat{y}_T + \hat{y}_T^2 + 2y_T b_{T-1}^\top A_T^{-1} x_T + y_T^2 x_T^\top A_T^{-1} x_T\right)$$
$$= \sup_{y_T\in[-Y,Y]} \left(y_T^2(x_T^\top A_T^{-1} x_T) + 2y_T(b_{T-1}^\top A_T^{-1} x_T - \hat{y}_T) + \hat{y}_T^2\right)$$

is minimal. Since $x_T^\top A_T^{-1} x_T \geq 0$ (A_T^{-1} is positive definite), we only have to consider $y_T = -Y$ and $y_T = Y$. Thus we have to find a $\hat{y}_T$ for which

$$2Y|b_{T-1}^\top A_T^{-1} x_T - \hat{y}_T| + \hat{y}_T^2 \tag{6}$$

is minimal. For $\hat{y}_T \leq b_{T-1}^\top A_T^{-1} x_T$, (6) equals $2Y(b_{T-1}^\top A_T^{-1} x_T - \hat{y}_T) + \hat{y}_T^2$ and this is minimal at $\min(Y, b_{T-1}^\top A_T^{-1} x_T)$ on this domain. For $\hat{y}_T \geq b_{T-1}^\top A_T^{-1} x_T$, (6) equals $2Y(\hat{y}_T - b_{T-1}^\top A_T^{-1} x_T) + \hat{y}_T^2$ and this is minimal at $\max(-Y, b_{T-1}^\top A_T^{-1} x_T)$ on this domain. If $b_{T-1}^\top A_T^{-1} x_T \in [-Y, Y]$ this obviously implies the assertion. If $b_{T-1}^\top A_T^{-1} x_T \geq Y$, (6) is decreasing for $\hat{y}_T \leq Y$ and increasing for $\hat{y}_T \geq Y$. Thus (6) is minimal at Y. The case $b_{T-1}^\top A_T^{-1} x_T \leq -Y$ is similar. □

Theorem 1 does not show that $\hat{y}_t = \text{clip}(b_{t-1}^\top A_t^{-1} x_t, Y)$, $t = 1, \ldots, T$, are the optimal predictions over a sequence of T trials when it is known that $y_1, \ldots, y_T$ are bounded by Y, because the "best" expert on the first t trials might differ from the "best" expert on all T trials. But there are very good relative loss bounds for these predictions as we will see in Theorem 3.

A standard application of Theorem 1 results, when the learner knows in advance a global upper bound $Y \geq 0$ on all potential outcomes y_t (and thus in particular on the last outcome y_T). The next result shows that a learner without prior knowledge of such a global upper bound Y can use a sort of "empirical upper bound" without suffering much extra loss.

Theorem 2. *Let*

$$Y_0 := 0 \ , \qquad Y_t := \max_{s=1}^{t} |y_s| \ , \qquad t = 1, 2, 3, \ldots \ .$$

For all $p_t = p_t(x_1, \ldots, x_t, y_1, \ldots, y_{t-1}) \in \mathbb{R}$, $t = 1, 2, \ldots, T$, *with the predictions*

$$\hat{y}_t = \text{clip}(p_t, Y_{t-1})$$

the loss of Learner on a sequence of T *trials is at most by* Y_T^2 *larger than with the predictions* $\hat{y}_t = \text{clip}(p_t, Y)$ *for each* $Y \geq Y_{T-1}$.

Proof. We show that for all $t \in \{1, \ldots, T\}$:

$$(y_t - \text{clip}(p_t, Y_{t-1}))^2 \leq (y_t - \text{clip}(p_t, Y))^2 + (Y_t - Y_{t-1})^2 \ . \tag{7}$$

From this it would follow that the additional loss is at most

$$\sum_{t=1}^{T} (Y_t - Y_{t-1})^2 \leq \left(\sum_{t=1}^{T} (Y_t - Y_{t-1})\right)^2 = Y_T^2 \ .$$

Let $t \in \{1,\ldots,T\}$. If $|y_t| \le Y_{t-1}$, then p_t clipped by Y_{t-1} is closer to y_t than p_t clipped by $Y \ge Y_{t-1}$. Thus

$$(y_t - \mathrm{clip}(p_t, Y_{t-1}))^2 \le (y_t - \mathrm{clip}(p_t, Y))^2 \ .$$

So we can assume that $|y_t| > Y_{t-1}$. Then $|y_t| = Y_t$. If $\mathrm{clip}(p_t, Y_{t-1}) = \mathrm{clip}(p_t, Y)$, (7) is obvious. Assume that $\mathrm{clip}(p_t, Y_{t-1}) \ne \mathrm{clip}(p_t, Y)$. Thus $|\mathrm{clip}(p_t, Y_{t-1})| = Y_{t-1}$. If y_t and p_t have the same sign then

$$(y_t - \mathrm{clip}(p_t, Y_{t-1}))^2 = (Y_t - Y_{t-1})^2 \ .$$

Otherwise

$$\begin{aligned}(y_t - \mathrm{clip}(p_t, Y_{t-1}))^2 &= (Y_t + |\mathrm{clip}(p_t, Y_{t-1})|)^2 \\ &\le (Y_t + |\mathrm{clip}(p_t, Y)|)^2 = (y_t - \mathrm{clip}(p_t, Y))^2 \ .\end{aligned}$$

□

Even without clipping the $b_{t-1}^\top A_t^{-1} x_t$ (not clipping them means that the loss of Learner L_T(Learner) will only increase, but has the advantage that Learner does not need to know Y), there is a very good upper bound on the relative loss. To show this, we need the following lemma:

Lemma 2. *For all $t \ge 1$:*

$$A_{t-1}^{-1} - A_t^{-1} - A_t^{-1} x_t x_t^\top A_t^{-1} = (x_t^\top A_{t-1}^{-1} x_t) A_t^{-1} x_t x_t^\top A_t^{-1} \ .$$

Proof. From the equality $A_t - A_{t-1} \overset{(5)}{=} x_t x_t^\top$ we get

$$A_{t-1}^{-1} - A_t^{-1} = A_{t-1}^{-1} x_t x_t^\top A_t^{-1} \ , \tag{8}$$

$$A_{t-1}^{-1} - A_t^{-1} = A_t^{-1} x_t x_t^\top A_{t-1}^{-1} \ . \tag{9}$$

Thus

$$\begin{aligned}&A_{t-1}^{-1} - A_t^{-1} - A_t^{-1} x_t x_t^\top A_t^{-1} \\ &\overset{(8)}{=} (A_{t-1}^{-1} - A_t^{-1}) x_t x_t^\top A_t^{-1} \\ &\overset{(9)}{=} A_t^{-1} x_t x_t^\top A_{t-1}^{-1} x_t x_t^\top A_t^{-1} \ .\end{aligned}$$

□

Theorem 3 ([6], Theorem 1). *If Learner predicts with $\hat{y}_t = b_{t-1}^\top A_t^{-1} x_t$ for $1 \le t \le T$ and if the outcome variables $y_1, \ldots, y_T$ are bounded by $Y \ge 0$, then*

$$L_T(\text{Learner}) \le \inf_{w \in \mathbb{R}^n} \left(a\|w\|^2 + L_T(w) \right) + Y^2 \ln \left| \frac{1}{a} A_T \right| \ .$$

Proof. For $1 \le t \le T$ we have

$$\begin{aligned}
&(y_t - \hat{y}_t)^2 + \inf_{w \in \mathbb{R}^n}\left(a\|w\|^2 + L_{t-1}(w)\right) - \inf_{w \in \mathbb{R}^n}\left(a\|w\|^2 + L_t(w)\right) \\
&\overset{\text{Lemma 1}}{=} -2y_t\hat{y}_t + \hat{y}_t^2 - b_{t-1}^\top A_{t-1}^{-1} b_{t-1} + b_t^\top A_t^{-1} b_t \\
&\overset{(4)}{=} -2y_t b_{t-1}^\top A_t^{-1} x_t + b_{t-1}^\top A_t^{-1} x_t x_t^\top A_t^{-1} b_{t-1} - b_{t-1}^\top A_{t-1}^{-1} b_{t-1} \\
&\qquad + (b_{t-1} + y_t x_t)^\top A_t^{-1}(b_{t-1} + y_t x_t) \\
&= b_{t-1}^\top (A_t^{-1} x_t x_t^\top A_t^{-1} - A_{t-1}^{-1} + A_t^{-1}) b_{t-1} + y_t^2 x_t^\top A_t^{-1} x_t \\
&\overset{\text{Lemma 2}}{=} y_t^2 x_t^\top A_t^{-1} x_t - (x_t^\top A_{t-1}^{-1} x_t) b_{t-1}^\top A_t^{-1} x_t x_t^\top A_t^{-1} b_{t-1} \\
&= y_t^2 x_t^\top A_t^{-1} x_t - (x_t^\top A_{t-1}^{-1} x_t)\hat{y}_t^2 \\
&\le Y^2 x_t A_t^{-1} x_t \ .
\end{aligned}$$

Summing over $t \in \{1, \ldots, T\}$ and using (1) gives

$$L_T(\text{Learner}) - \inf_{w \in \mathbb{R}^n}\left(a\|w\|^2 + L_T(w)\right) \le Y^2 \sum_{t=1}^{T} x_t A_t^{-1} x_t \ .$$

Because of $\ln|\frac{1}{a}A_0| \overset{(5)}{=} 0$, it now suffices to show that for $t \in \{1, \ldots, T\}$:

$$x_t^\top A_t^{-1} x_t \le \ln \frac{|A_t|}{|A_{t-1}|} \ .$$

We first show that $x_t^\top A_t^{-1} x_t < 1$. This is trivial for $x_t = 0$. For $x_t \neq 0$, it is obtained from the following calculation:

$$\begin{aligned}
(x_t^\top A_t^{-1} x_t)^2 &= x_t^\top A_t^{-1} x_t x_t^\top A_t^{-1} x_t \overset{(5)}{=} x_t^\top A_t^{-1}(A_t - A_{t-1}) A_t^{-1} x_t \\
&= x_t^\top A_t^{-1} x_t - \underbrace{x_t^\top A_t^{-1} A_{t-1} A_t^{-1} x_t}_{>0} < x_t^\top A_t^{-1} x_t \ .
\end{aligned}$$

There is a symmetric, positive definite matrix $A \in \mathbb{R}^{n \times n}$ such that $A_t = AA$. Let $\xi := A^{-1} x_t$. Thus $x_t = A\xi$. From $\xi^\top \xi = x_t^\top A_t^{-1} x_t < 1$ we know that $I - \xi\xi^\top$ is positive definite and we get

$$|I - \xi\xi^\top| \le \prod_{i=1}^{n}(1 - \xi_i^2) \le \prod_{i=1}^{n} e^{-\xi_i^2} = e^{-\xi^\top \xi} \ , \tag{10}$$

where the first inequality holds because the determinant of a positive semidefinite matrix is bounded by the product of the entries on the diagonal of the matrix (e.g., see [2], Theorem 7 in Chapter 2). It follows that

$$x_t^\top A_t^{-1} x_t = \xi^\top \xi \overset{(10)}{\le} \ln \frac{1}{|I - \xi\xi^\top|} = \ln \frac{|AA|}{|AA - A\xi\xi^\top A|} \overset{(5)}{=} \ln \left|\frac{A_t}{A_{t-1}}\right| \ .$$

□

Vovk gives the following upper bounds on the term $\ln\left|\frac{1}{a}A_T\right|$ in [6]:

$$\ln\left|\frac{1}{a}A_T\right| \overset{(5)}{=} \ln\left|I + \frac{1}{a}\sum_{t=1}^{T} x_t x_t^\top\right|$$
$$\leq \sum_{i=1}^{n} \ln\left(1 + \frac{1}{a}\sum_{t=1}^{T} x_{t,i}^2\right) \leq n \ln\left(1 + \frac{TX^2}{a}\right) ,$$

where, for the first inequality, again [2], Chap. 2, Theorem 7 is used, and where we assume that $|x_{t,i}| \leq X$ for $t \in \{1,\ldots,T\}$, $i \in \{1,\ldots,n\}$.

3 Generalized Linear Regression

In generalized linear regression we consider the same protocol of interaction between Nature and Learner as before, but now expert $w \in \mathbb{R}^n$ makes the prediction $\varphi(w \cdot x)$ on an instance $x \in \mathbb{R}^n$, where $\varphi : \mathbb{R} \to \mathbb{R}$ is a strictly increasing, differentiable function. Now the losses are measured with the matching loss L_φ,

$$L_\varphi(y,\hat{y}) := \int_{\varphi^{-1}(y)}^{\varphi^{-1}(\hat{y})} (\varphi(\tau) - y)\, d\tau \tag{11}$$

$$= \int_{\varphi^{-1}(y)}^{\varphi^{-1}(\hat{y})} \varphi(\tau)\, d\tau + y\varphi^{-1}(y) - y\varphi^{-1}(\hat{y}) \tag{12}$$

for $y, \hat{y}$ in the range of φ (see Fig. 1). Examples of matching loss functions are the square loss ($\varphi = \mathrm{id}_{\mathbb{R}}$, $L_\varphi(y,\hat{y}) = \frac{1}{2}(y-\hat{y})^2$) and the entropic loss ($\varphi(z) = \frac{1}{1+e^{-z}}$, $L_\varphi(y,\hat{y}) = y \ln\frac{y}{\hat{y}} + (1-y)\ln\frac{1-y}{1-\hat{y}}$). Note that in the case of square loss ($\varphi = \mathrm{id}_{\mathbb{R}}$) all losses in this section differ by a factor of $\frac{1}{2}$ from those in Sect. 2. This was not harmonized because in Sect. 2 we wanted to use the same definitions as in [6], and in Sect. 3 we want to use the common definition of the matching loss.

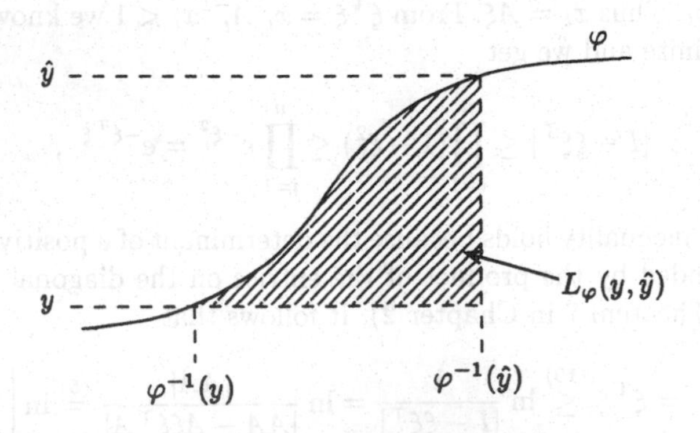

Fig. 1. The matching loss function L_φ

The loss of Learner and the loss of expert $w \in \mathbb{R}^n$ are now defined as

$$L_{\varphi,t}(\text{Learner}) = \sum_{s=1}^{t} L_\varphi(y_s, \hat{y}_s) \ , \qquad L_{\varphi,t}(w) = \sum_{s=1}^{t} L_\varphi(y_s, \varphi(w \cdot x_s)) \ .$$

Lemma 3. *For all $t \geq 0$, function $f(w) := \frac{a}{2}\|w\|^2 + L_{\varphi,t}(w)$ is minimal at a unique point, say w_t. w_t is differentiable in y_t and we have*

$$\frac{\partial w_t}{\partial y_t} \cdot x_t \geq 0 \ . \tag{13}$$

Let

$$\Phi(z) := -\frac{1}{2} z\varphi(z) + \int_0^z \varphi(\tau)\, d\tau \ , \qquad z \in \mathbb{R} \ ,$$

$$\Gamma_t := \sum_{s=1}^{t} \left(\frac{y_s}{2} w_t \cdot x_s - \Phi(w_t \cdot x_s) \right) .$$

Then

$$L_{\varphi,t}(\text{Learner}) - \inf_{w \in \mathbb{R}^n} \left(\frac{a}{2}\|w\|^2 + L_{\varphi,t}(w) \right)$$
$$= \sum_{s=1}^{t} \left(\int_0^{\varphi^{-1}(\hat{y}_s)} \varphi(\tau)\, d\tau - y_s \varphi^{-1}(\hat{y}_s) \right) + \Gamma_t \tag{14}$$

and

$$\frac{\partial \Gamma_t}{\partial y_t} = w_t \cdot x_t \ . \tag{15}$$

The proof of Lemma 3 is omitted here. We need one more small Lemma.

Lemma 4. *Let $f, g : \mathbb{R} \to \mathbb{R}$ be differentiable, convex functions with $f'(z) > g'(z)$ for all $z \in \mathbb{R}$. If there are numbers $Y^- \leq Y^+$, $Z \in \mathbb{R}$ such that f is minimal at Y^-, g is minimal at Y^+ and $f(Z) = g(Z)$, then $h := \max(f, g)$ is minimal at*

$$\text{clip}(Z, Y^-, Y^+) := \begin{cases} Y^- \ , & Z \in (-\infty, Y^-] \ , \\ Z \ , & Z \in [Y^-, Y^+] \ , \\ Y^+ \ , & Z \in [Y^+, \infty) \ . \end{cases}$$

Proof. Because of $f'(z) > g'(z)$ for $z \in \mathbb{R}$ and because of $f(Z) = g(Z)$, we have

$$h(z) = \begin{cases} g(z) \ , & z \leq Z \ , \\ f(z) \ , & z \geq Z \ . \end{cases}$$

If $Z \in [Y^-, Y^+]$, then h is decreasing on $(-\infty, Z]$ (because g is decreasing there) and h is increasing on $[Z, \infty)$ (because f is increasing there). Thus h is minimal at Z. If $Z < Y^-$ then h is decreasing on $(-\infty, Y^-]$ (because both f and g are decreasing there) and h is increasing on $[Y^-, \infty)$ (because f is increasing there). Thus h is minimal at Y^-. The case $Z > Y^+$ is very similar to the case $Z < Y^-$. □

Now we are ready to calculate the best prediction $\hat{y}_T$ for the last trial T when it is known that $y_T \in [Y^-, Y^+]$.

w_T and Γ_T are functions in $x_1, \ldots, x_T, y_1, \ldots, y_T$. If only $x_1, \ldots, x_T$ and $y_1, \ldots, y_{T-1}$ are known (which is the case when Learner makes the last prediction $\hat{y}_T$), we write $w_T = w_T(y_T)$, $\Gamma_T = \Gamma_T(y_T)$.

Theorem 4. *Let $Y^- < Y^+$ be in the range of φ. As a function of $\hat{y}_T$,*

$$\sup_{y_T \in [Y^-, Y^+]} \left(L_{\varphi,T}(\text{Learner}) - \inf_{w \in \mathbb{R}^n} \left(\frac{a}{2}\|w\|^2 + L_{\varphi,T}(w) \right) \right)$$

is minimal for

$$\hat{y}_T = \text{clip}\left(\varphi\left(\frac{1}{Y^+ - Y^-} \int_{Y^-}^{Y^+} w_T(y_T) \cdot x_T \, dy_T \right), Y^-, Y^+ \right) .$$

Proof. From (14) and (15) it follows that

$$\frac{\partial}{\partial y_T} \left(L_{\varphi,T}(\text{Learner}) - \inf_{w \in \mathbb{R}^n} \left(\frac{a}{2}\|w\|^2 + L_{\varphi,T}(w) \right) \right)$$
$$\overset{(15)}{=} w_T \cdot x_T - \varphi^{-1}(\hat{y}_T) ,$$
$$\frac{\partial}{\partial y_T} \frac{\partial}{\partial y_T} \left(L_{\varphi,T}(\text{Learner}) - \inf_{w \in \mathbb{R}^n} \left(\frac{a}{2}\|w\|^2 + L_{\varphi,T}(w) \right) \right) = \frac{\partial w_T}{\partial y_T} \cdot x_T \overset{(13)}{\geq} 0 .$$

Thus, for fixed $\hat{y}_T$, (14) (with $t = T$) is maximal for $y_T = Y^-$ or for $y_T = Y^+$. So we have to find a $\hat{y}_T$ such that

$$\int_0^{\varphi^{-1}(\hat{y}_T)} \varphi(\tau)\, d\tau + \max\left(-Y^- \varphi^{-1}(\hat{y}_T) + \Gamma_T(Y^-), -Y^+ \varphi^{-1}(\hat{y}_T) + \Gamma_T(Y^+) \right)$$
$$= \max\left(f(\varphi^{-1}(\hat{y}_T)), g(\varphi^{-1}(\hat{y}_T)) \right)$$

with

$$f(z) = \int_0^z \varphi(\tau)\, d\tau - Y^- z + \Gamma_T(Y^-), \qquad f'(z) = \varphi(z) - Y^- ,$$
$$g(z) = \int_0^z \varphi(\tau)\, d\tau - Y^+ z + \Gamma_T(Y^+) , \qquad g'(z) = \varphi(z) - Y^+ ,$$

is minimal. f and g are equal if $-Y^- \varphi^{-1}(\hat{y}_T) + \Gamma_T(Y^-) = -Y^+ \varphi^{-1}(\hat{y}_T) + \Gamma_T(Y^+)$, i.e., if

$$\varphi^{-1}(\hat{y}_T) = \frac{\Gamma_T(Y^+) - \Gamma_T(Y^-)}{Y^+ - Y^-} = \frac{1}{Y^+ - Y^-} \int_{Y^-}^{Y^+} \frac{\partial \Gamma_T}{\partial y_T}\, dy_T$$
$$\overset{(15)}{=} \frac{1}{Y^+ - Y^-} \int_{Y^-}^{Y^+} w_T \cdot x_T \, dy_T .$$

Because of Lemma 4 (f is minimal at $\varphi^{-1}(Y^-)$, g at $\varphi^{-1}(Y^+)$) this implies the assertion. □

Like in Theorem 2 it might be a good idea to use the predictions

$$\hat{y}_t = \mathrm{clip}\left(\varphi\left(\frac{1}{Y_{t-1}^+ - Y_{t-1}^-}\int_{Y_{t-1}^-}^{Y_{t-1}^+} w_t \cdot x_t \, dy_t\right), Y_{t-1}^-, Y_{t-1}^+\right)$$

where $Y_t^+ = \max_{s=1}^t y_s$, $Y_t^- = \min_{s=1}^t y_s$ when no bounds Y^-, Y^+ on the outcomes y_t are known.

4 Conclusion

We have shown that Vovk's prediction rule is essentially characterized by the property that it minimizes the maximal extra loss (compared to the best off-line expert) that might be suffered in the last trial (Theorem 1). For the sake of simplicity, let's call this the minmax-property. Note that instead of first inventing the learning rule and then proving Theorem 1, one could have gone the other direction. The calculus applied in the proof of Theorem 1 will then lead automatically to Vovk's rule. This is precisely the line of attack that we pursued in Section 3. There, we considered the (much more involved) generalized regression problem. In order to find a good *explicit* candidate prediction rule, we tried to find the rule *implicitly* given by the minmax-property. This finally has lead to the rule given in Theorem 4.

It is straightforward to ask whether this rule has (provably) good relative loss bounds. In other words, we need the analogue of Theorem 3 for the generalized regression problem. As yet, we can show that

$$L_\varphi(y_t, \hat{y}_t) + \inf_{w \in \mathbb{R}^n}\left(\frac{a}{2}\|w\|^2 + L_{\varphi,t-1}(w)\right) - \inf_{w \in \mathbb{R}^n}\left(\frac{a}{2}\|w\|^2 + L_{\varphi,t}(w)\right)$$
$$\leq \int_0^{\frac{\Gamma_t(Y^+) - \Gamma_t(Y^-)}{Y^+ - Y^-}} \varphi(\tau)\, d\tau + \frac{Y^+\Gamma_t(Y^-) - Y^-\Gamma_t(Y^+)}{Y^+ - Y^-} - \Gamma_{t-1} \qquad (16)$$

for all t. By summing over $t \in \{1, \ldots, T\}$ we get a bound on the relative loss over the sequence of T trials. Thus upper bounds on the terms on the right hand side of (16) would be very interesting.

5 Acknowledgements

The author is grateful to Hans Ulrich Simon and Manfred Warmuth for a lot of helpful suggestions.

References

1. Azoury, K., Warmuth, M.: Relative Loss Bounds for On-line Density Estimation with the Exponential Family of Distributions, to appear at the Fifteenth Conference on Uncertainty in Artificial Intelligence, UAI'99.

2. Beckenbach, E. F., Bellman, R.: Inequalities, Berlin: Springer, 1965.
3. Foster, D. P.: Prediction in the worst case, Annals of Statistics 19, 1084–1090.
4. Kivinen, J., Warmuth, M.: Relative Loss Bounds for Multidimensional Regression Problems. In Jordan, M., Kearns, M., Solla, S., editors, Advances in Neural Information Processing Systems 10 (NIPS 97), 287–293, MIT Press, Cambridge, MA, 1998.
5. Kivinen, J., Warmuth, M.: Additive versus exponentiated gradient updates for linear prediction, Information and Computation 132:1–64, 1997.
6. Vovk, V.: Competitive On-Line Linear Regression. Technical Report CSD-TR-97-13, Department of Computer Science, Royal Holloway, University of London, 1997.

Generalized P-Systems

Rudolf Freund

Institut für Computersprachen, Technische Universität Wien
Resselg. 3, A-1040 Wien, Austria
E-mail: rudi@logic.at

Abstract. We consider a variant of P-systems, a new model for computations using membrane structures and recently introduced by Gheorghe Păun. Using the membranes as a kind of filter for specific objects when transferring them into an inner compartment turns out to be a very powerful mechanism in combination with suitable rules to be applied within the membranes. The model of generalized P-systems, GP-systems for short, considered in this paper allows for the simulation of graph controlled grammars of arbitrary type based on productions working on single objects; for example, the general results we establish in this paper can immediately be applied to the graph controlled versions of context-free string grammars, n-dimensional #-context-free array grammars, and elementary graph grammars.

1 Introduction

One of the main ideas incorporated in the model of P-systems introduced in [7] is the membrane structure (for a chemical variant of this idea see [1]) consisting of membranes hierarchically embedded in the outermost *skin* membrane. Every membrane encloses a *region* possibly containing other membranes; the part delimited by the membrane labelled by k and its inner membranes is called *compartment* k. A region delimited by a membrane not only may enclose other membranes but also specific objects and operators, which in this paper are considered as multisets, as well as evolution rules, which in *generalized P-systems (GP-systems)* as introduced in this paper are evolution rules for the operators. Moreover, besides ground operators the most important kind of operators are transfer operators (simple rules of that kind are called travelling rules in [10]) allowing to transfer objects or operators (or even rules) either to the outer compartment or to an inner compartment delimited by a membrane of specific kind with also checking for some permitting and/or forbidding conditions on the objects to be transferred (in that way, the membranes act as a filter like in test tube systems, see [8]). In contrast to the original definition of P-systems we do not demand all objects to be affected in parallel by the rules; the proofs of the results established so far in various papers on P-systems, see [3], [6], and [9], show that only bounded parallelism is needed. Moreover, we also omit the feature of priority relations on the rules, because this feature can be captured in another way by using the transfer conditions in the transfer operators.

In the following section we shall give a general defintion of a grammar and then we define the notions of matrix grammars and graph controlled grammars in this general setting. In the third section we introduce our model of GP-systems, and in the fourth section we establish our main results showing that GP-systems allow for the simulation of graph controlled grammars of arbitrary type based on productions working on single objects; we also elaborate how these results can be used to show that graph controlled context-free string grammars, graph controlled n-dimensional #-context-free array grammars, and graph controlled elementary graph grammars can be simulated by GP-systems using the corresponding type of objects and underlying productions. An outlook to future research topics concludes the paper.

2 Definitions

First, we recall some basic notions from the theory of formal languages (for more details, the reader is referred to [2]).

For an alphabet V, by V^* we denote the free monoid generated by V under the operation of concatenation; the *empty string* is denoted by λ, and $V^* \setminus \{\lambda\}$ is denoted by V^+. Any subset of V^+ is called a *λ-free (string) language.*

A *(string) grammar* is a quadruple $G = (V_N, V_T, P, S)$, where V_N and V_T are finite sets of *non-terminal* and *terminal symbols,* respectively, with $V_N \cap V_T = \emptyset$, P is a finite set of *productions* $\alpha \to \beta$ with $\alpha \in V^+$ and $\beta \in V^*$, where $V = V_N \cup V_T$, and $S \in V_N$ is the *start symbol.* For $x, y \in V^*$ we say that *y is directly derivable from x in G*, denoted by $x \Longrightarrow_G y$, if and only if for some $\alpha \to \beta$ in P and $u, v \in V^*$ we get $x = u\alpha v$ *and* $y = u\beta v$. Denoting the reflexive and transitive closure of the derivation relation $\Longrightarrow_G$ by $\Longrightarrow_G^*$, the *(string) language generated by G* is $L(G) = \{w \in V_T^* \mid S \Longrightarrow_G^* w\}$. A production $\alpha \to \beta$ is called *context-free,* if $\alpha \in V_N$.

In order to prove our results in a general setting, we use the following general notion of a grammar:

A *grammar* is a quadruple $G = (B, B_T, P, A)$, where B and B_T are sets of *objects* and *terminal objects,* respectively, with $B_T \subseteq B$, P is a finite set of *productions,* and $A \in B$ is the axiom. A production p in P in general is a partial recursive relation $\subseteq B \times B$, where we also demand that the domain of p is recursive (i.e., given $w \in B$ it is decidable if there exists some $v \in B$ with $(w, v) \in p$) and, moreover, that the range for every w is finite, i.e., for any $w \in B$, $card(\{v \in B \mid (w, v) \in p\}) < \infty$. As for string grammars above, the productions in P induce a derivation relation $\Longrightarrow_G$ on the objects in B etc. The *language generated by G* is $L(G) = \{w \in B_T \mid A \Longrightarrow_G^* w\}$.

For example, a string grammar (V_N, V_T, P, S) in this general notion now is written as $\left((V_N \cup V_T)^*, V_T^*, P, S\right)$.

2.1 Control mechanisms

In the following, we give the necessary definitions of matrix and graph controlled grammars in our general setting. For detailed informations concerning these control mechanisms as well as many other interesting results about regulated rewriting in the theory of string languages, the reader is referred to [2].

A *matrix grammar* is a construct $G_M = (B, B_T, (M, F), A)$ where B and B_T are sets of *objects* and *terminal objects*, respectively, with $B_T \subseteq B$, $A \in B$ is the axiom, M is a finite set of matrices, $M = \{m_i \mid 1 \leq i \leq n\}$, where the matrices m_i are sequences of the form $m_i = (m_{i,1}, \ldots, m_{i,n_i})$, $n_i \geq 1$, $1 \leq i \leq n$, and the $m_{i,j}$, $1 \leq j \leq n_i$, $1 \leq i \leq n$, are productions over B, and F is a subset of $\bigcup_{1 \leq i \leq n,\, 1 \leq j \leq n_i} \{m_{i,j}\}$.

For $m_i = (m_{i,1}, \ldots, m_{i,n_i})$ and $v, w \in B$ we define $v \Longrightarrow_{m_i} w$ if and only if there are $w_0, w_1, \ldots, w_{n_i} \in B$ such that $w_0 = v$, $w_{n_i} = w$, and for each $j, 1 \leq j \leq n_i$,

- **either** w_j is the result of the application of $m_{i,j}$ to w_{j-1},
- **or** $m_{i,j}$ is not applicable to w_{j-1}, $w_j = w_{j-1}$, and $m_{i,j} \in F$.

The language generated by G_M is

$$L(G_M) = \{w \in B_T \mid A \Longrightarrow_{m_{i_1}} w_1 \ldots \Longrightarrow_{m_{i_k}} w_k = w,\; w_j \in B,\, m_{i_j} \in M \;\text{ for } 1 \leq j \leq k\}.$$

If $F = \emptyset$ then G_M is called a *matrix grammar without appearance checking.* G_M is said to be of type X if the corresponding underlying grammar $G = (B, B_T, P, A)$, where P exactly contains every production occuring in some matrix in M, is of type X.

A *graph controlled grammar* is a construct $G_C = (B, B_T, (R, L_{in}, L_{fin}), A)$; B and B_T are sets of *objects* and *terminal objects*, respectively, with $B_T \subseteq B$, $A \in B$ is the axiom; R is a finite set of rules r of the form $(l(r) : p(l(r)), \sigma(l(r)), \varphi(l(r)))$, where $l(r) \in Lab(G_C)$, $Lab(G_C)$ being a set of labels associated (in a one-to-one manner) to the rules r in R, $p(l(r))$ is a production over B, $\sigma(l(r)) \subseteq Lab(G_C)$ is the *success field* of the rule r, and $\varphi(l(r))$ is the *failure field* of the rule r; $L_{in} \subseteq Lab(G_C)$ is the set of initial labels, and $L_{fin} \subseteq Lab(G_C)$ is the set of final labels. For $r = (l(r) : p(l(r)), \sigma(l(r)), \varphi(l(r)))$ and $v, w \in B$ we define $(v, l(r)) \Longrightarrow_{G_C} (w, k)$ if and only if

- **either** $p(l(r))$ is applicable to v, the result of the application of the production $p(l(r))$ to v is w, and $k \in \sigma(l(r))$,
- **or** $p(l(r))$ is not applicable to v, $w = v$, and $k \in \varphi(l(r))$.

The language generated by G_C is

$$L(G_C) = \{w \in B_T \mid (A, l_0) \Longrightarrow_{G_C} (w_1, l_1) \Longrightarrow_{G_C} \ldots (w_k, l_k),\; k \geq 1,\; w_j \in B \text{ and } l_j \in Lab(G_C) \text{ for } 0 \leq j \leq k,\; w_k = w,\; l_0 \in L_{in},\; l_k \in L_{fin}\}.$$

If the failure fields $\varphi(q)$ are empty for all $q \in Lab$, then G_C is called a ***graph controlled grammar without appearance checking.*** G_C is said to be of type X if the corresponding underlying grammar $G = (B, B_T, P, A)$, where $P = \{p(q) \mid q \in Lab\}$ is of type X.

3 Generalized P-systems (GP-systems)

In this section we quite informally describe the model of generalized P-systems discussed in this paper. Only the features not captured by the original model of P-systems as described in [6] and [7] will be defined in more details.

The basic part of a (G)P-system is a *membrane structure* consisting of several membranes placed within one unique surrounding membrane, the so-called skin membrane. All the membranes can be labelled in a one-to-one manner by natural numbers; the outermost membrane (skin membrane) always is labelled by 0. In that way, a membrane structure can uniquely be described by a string of correctly matching parentheses, where each pair corresponds to a membrane. For example, the membrane structure depicted in Figure 1, which within the skin membrane contains two inner membranes labelled by 1 and 2, is described by $[_0[_1]_1[_2]_2]_0$. Figure 1 also shows that a membrane structure graphically can be represented by a Venn diagram, where two sets can either be disjoint or one set be the subset of the other one. In this representation, every membrane encloses a *region* possibly containing other membranes; the part delimited by the membrane labelled by k and its inner membranes is called *compartment* k in the following. The space outside the skin membrane is called *outer region*.

Informally, in [6] and [7] *P-systems* were defined as membrane structures containing multisets of objects in the compartments k as well as evolution rules for the objects. A priority relation on the evolution rules guarded the application of the evolution rules to the objects, which had to be affected in parallel (if possible according to the priority relation). The output was obtained in a designated compartment from a halting configuration (i.e., a configuration of the system where no rules can be applied any more).

A *generalized P-system (GP-system) of type* X is a construct G_P of the following form:

$$G_P = (B, B_T, P, A, \mu, I, O, R, f)$$

where

- (B, B_T, P, A) is a grammar of type X;
- μ is a membrane structure (with the membranes labelled by natural numbers $0, ..., n$);
- $I = (I_0, ..., I_n)$, where I_k is the initial contents of compartment k containing a (finite) multiset of objects from B as well as a (finite) multiset of operators from O and of rules from R; we shall assume $A \in I_0$ in the following;
- O is a finite set of operators (which will be described in detail below);
- R is a finite set of (evolution) rules of the form $(op_1, ..., op_k; op'_1, ..., op'_m)$ with $k \geq 1$ and $m \geq 0$, where op_1, ..., op_k, op'_1, ..., op'_m are operators from O;
- $f \in \{1, ..., n\}$ is the label of the final compartment; we shall always assume $I_f = \emptyset$ and $R_f = \emptyset$.

The main power of GP-systems lies in the operators, which can be of the following types:

- $P \subseteq O$, i.e., the productions working on the objects from B are operators;
- $O_0 \subseteq O$, where O_0 is a finite set of special symbols, which are called *ground operators*;
- $Tr_{in} \subseteq O$, where Tr_{in} is a finite set of transfer operators on objects from B of the form $(\tau_{in,k}, E, F)$, $1 \leq k \leq n$, $E \subseteq P$, $F \subseteq P$; the operator $(\tau_{in,k}, E, F)$ transfers an object w from B being in compartment m into compartment k provided
  1. region m contains membrane k,
  2. every production from E could be applied to w (hence, E is also called the permitting transfer condition),
  3. no production from F can be applied to w (hence, F is also called the forbidding transfer condition);
- $Tr_{out} \subseteq O$, where Tr_{out} is a finite set of transfer operators on objects from B of the form (τ_{out}, E, F), $1 \leq k \leq n$, $E \subseteq P$, $F \subseteq P$; the operator (τ_{out}, E, F) transfers an object w from B being in compartment m into compartment k provided
  1. region k contains membrane m,
  2. every production from E could be applied to w,
  3. no production from F can be applied to w;
- $Tr'_{in} \subseteq O$, where Tr'_{in} is a finite set of transfer operators working on operators from P, O_0, Tr_{in}, and Tr_{out} or even on rules from R; a transfer operator $\tau_{in,k}$ moves such an element in compartment m into compartment k provided region m contains membrane k;
- $Tr'_{out} \subseteq O$, where Tr'_{out} is a finite set of transfer operators working on operators from P, O_0, Tr_{in}, and Tr_{out} or even on rules from R; a transfer operator τ_{out} transfers such an element in compartment m into the surrounding compartment.

In sum, O is the disjoint union of P, O_0, and Tr, where Tr itself is the (disjoint) union of the sets of transfer operators Tr_{in}, Tr_{out}, Tr'_{in}, and Tr'_{out}. In the following we shall assume that the transfer operators in Tr'_{in}, and Tr'_{out} do not work on rules from R; hence, the distribution of the evolution rules is static and given by I. If in all transfer operators the permitting and the forbidding sets are empty, then G_P is called a *GP-system without transfer checking*.

A *computation* in G_P starts with the initial configuration with I_k being the contents of compartment k. A transition from one configuration to another one is performed by evaluating one evolution rule $(op_1, ..., op_k; op'_1, ..., op'_m)$ in some compartment k, which means that the operators op_1, ..., op_k, are applied to suitable elements in compartment k in the multiset sense (i.e., they are "consumed" by the usage of the rule; observe that ground operators have no arguments and are simply consumed in that way); thus we may obtain a new object by the application of a production and/or we may move elements out or into inner compartments by the corresponding transfer operators; yet we also obtain the operators op'_1, ..., op'_m (in the multiset sense) in compartment k.

The language generated by G_P is the set of all terminal objects $w \in B_T$ obtained in the terminal compartment f by some computation in G_P.

To give a first impression of the facilities offered by GP-systems we consider a simple example of a GP-system generating the non-context-free string language $L_1 = \{a^n b^n c^n \mid n \geq 1\}$:

Example 1. The string language L_1 can be generated by the regular matrix grammar with axiom (without ac)

$G_M = (\{A, B, C, a, b, c\}^*, \{a, b, c\}^*, (\{m, m'\}, \emptyset), ABC)$ with
$m = [p_1, p_2, p_3]$, $p_1 = A \to aA$, $p_2 = B \to bB$, $p_3 = C \to cC$,
$m' = [p'_1, p'_2, p'_3]$, $p'_1 = A \to a$, $p'_2 = B \to b$, $p'_3 = C \to c$.

Obviously, for any $n \geq 1$, from the axiom ABC we obtain $a^{n-1}Ab^{n-1}Bc^{n-1}C$ by applying $(n-1)$ times the matrix m and finally $a^n b^n c^n$ by once applying matrix m'. This regular matrix grammar with axiom can easily be simulated by a GP-system whose main ingredients are depicted in Figure 1:

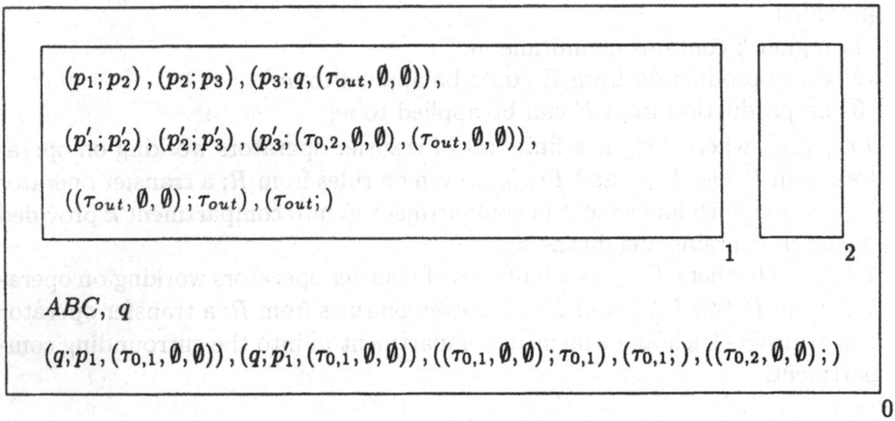

Membrane 1:
$(p_1; p_2), (p_2; p_3), (p_3; q, (\tau_{out}, \emptyset, \emptyset))$,
$(p'_1; p'_2), (p'_2; p'_3), (p'_3; (\tau_{0,2}, \emptyset, \emptyset), (\tau_{out}, \emptyset, \emptyset))$,
$((\tau_{out}, \emptyset, \emptyset); \tau_{out}), (\tau_{out};)$

Membrane 2 (empty)

Membrane 0:
ABC, q
$(q; p_1, (\tau_{0,1}, \emptyset, \emptyset)), (q; p'_1, (\tau_{0,1}, \emptyset, \emptyset)), ((\tau_{0,1}, \emptyset, \emptyset); \tau_{0,1}), (\tau_{0,1};), ((\tau_{0,2}, \emptyset, \emptyset);)$

Fig. 1. Membrane structure with rules and initial objects and operators

Within the skin membrane we start with the axiom ABC as the initial object and the ground operator q; we now can either choose $(q; p_1, (\tau_{0,1}, \emptyset, \emptyset))$ or $(q; p'_1, (\tau_{0,1}, \emptyset, \emptyset))$ from the rules available in the compartment delimited by the skin membrane labelled by 0. Using $(q; p_1, (\tau_{0,1}, \emptyset, \emptyset))$ yields the transfer operator $(\tau_{0,1}, \emptyset, \emptyset)$, which allows us to transfer the current string, in general being of the form $a^{n-1}Ab^{n-1}Bc^{n-1}C$ for some $n \geq 1$, into the compartment surrounded by the membrane labelled by 1. By applying $(\tau_{0,1}, \emptyset, \emptyset)$ we now gain the transfer operator $\tau_{0,1}$ which allows us to transfer the production p_1 into this compartment 1, too, by applying the rule $(\tau_{0,1};)$. In compartment 1, by using the rules $(p_1; p_2)$, $(p_2; p_3)$, and $(p_3; q, (\tau_{out}, \emptyset, \emptyset))$ sequentially we obtain the string $a^n Ab^n Bc^n C$ as well as the ground operator q and the transfer operator $(\tau_{out}, \emptyset, \emptyset)$, which then transfers the string out into the skin compartment through the rule $((\tau_{out}, \emptyset, \emptyset); \tau_{out})$, simultaneously yielding the tranfer operator τ_{out}, which finally transfers q, too, by the rule $(\tau_{out};)$.

In a similar way, after having applied $(q; p_1', (\tau_{0,1}, \emptyset, \emptyset))$ in the skin compartment, in the inner compartment 1 we obtain $a^n b^n c^n$ from $a^{n-1} A b^{n-1} B c^{n-1} C$ by sequentially using the rules $(p_1'; p_2')$, $(p_2'; p_3')$, and $(p_3'; (\tau_{0,2}, \emptyset, \emptyset), (\tau_{out}, \emptyset, \emptyset))$. Instead of the ground operator q now the transfer operator $(\tau_{0,2}, \emptyset, \emptyset)$ is generated and moved out into compartment 0, where according to the rule $((\tau_{0,2}, \emptyset, \emptyset);)$ it transfers the terminal string $a^n b^n c^n$ into the terminal compartment 2. □

The example given above already indicates how matrix grammars could be simulated by GP-systems. In general, the productions in the matrices need not all be different as in this example, hence, usually we will have to use different compartments for the application of different matrices. In the proof of Theorem 2 we will exploit this idea for showing how GP-systems can simulate graph controlled grammars when using only rules of the form $(op_1; op_2, ..., op_k)$ for some $k \geq 1$, which corresponds to some kind of context-freeness of rules, because op_1 is applied without checking for the applicability of other operators.

4 Results

The main result of this paper is to show how graph controlled grammars of arbitrary type can be simulated by GP-systems of the same type. The following result covers the case where also a kind of "context-sensitive" rules is taken into account.

Theorem 1. *Any graph controlled grammar of arbitrary type can be simulated by a GP-system of the same type with the simple membrane structure* $[_0[_1]_1[_2]_2]_0$.

Proof. Let $G_C = (B, B_T, (R, L_{in}, L_{fin}), A)$ be a graph controlled grammar of type X and $G = (B, B_T, P, A)$ be the corresponding underlying grammar of type X with $P = \{p(q) \mid q \in Lab\}$. The main ingredients of the GP-system G_P of type X generating the same language as G_C can be described in the following way (the complete formal description of G_P is obvious and therefore omitted):

- For each $q \in Lab$, in compartment 0 we take the following rules:
  For the success case, we use the rules $(q; q^{(1)}, (\tau_{in,1}, \emptyset, \emptyset))$, $((\tau_{in,1}, \emptyset, \emptyset); \tau_{in,1})$ and $(\tau_{in,1};)$ to transfer the current sentential form as well as the ground operator $q^{(1)}$, which represents the actual node in the control graph, into compartment 1.
  For the failure case, where $p(q)$ is not applicable to the current sentential form, we use $(q; q^*, (\tau_{in,1}, \emptyset, \{p(q)\}))$ as well as $((\tau_{in,1}, \emptyset, \{p(q)\}); \tau_{in,1})$ (and again $(\tau_{in,1};)$) for the transfer of the current sentential form and the ground operator q^*. The non-applicability of $p(q)$ is checked by the forbidding condition $\{p(q)\}$ in the transfer rule $(\tau_{in,1}, \emptyset, \{p(q)\})$.
- In compartment 1 we take the following rules for each $q \in Lab$:
  The rules $(q^{(1)}; q^{(2)}, p(q))$ and $(q^{(2)}, p(q); \alpha, (\tau_{out}, \emptyset, \emptyset))$ with $\alpha \in \sigma(q)$ guarantee the simulation of the application of the production $p(q)$ to the underlying sentential form as well as the proceeding in the control graph to a node

in the success field of q; the rules $((\tau_{out}, \emptyset, \emptyset); \tau_{out})$ and $(\tau_{out};)$ then transfer the result of the application of $p(q)$ as well as α into the skin compartment again.
In the failure case, we only have to proceed in the control graph to a node in the failure field of q, which is accomplished by one of the rules $(q^*; \beta, (\tau_{out}, \emptyset, \emptyset))$ for $\beta \in \varphi(q)$; the transfer back into compartment 0 again is performed by the rules $((\tau_{out}, \emptyset, \emptyset); \tau_{out})$ and $(\tau_{out};)$.

- The initial configuration starts with the axiom A and a special ground operator q_0, $q_0 \notin Lab$, in compartment 0. By using one of the rules $(q_0; q)$ for $q \in L_{in}$ we can proceed to a starting node in the control graph.
- For any $q \in L_{fin}$ we take the rule $(q; (\tau_{in,2}, \emptyset, \emptyset))$; the application of the rule $((\tau_{in,2}, \emptyset, \emptyset);)$ finally transfers a terminal object into the terminal compartment 2. □

As it is obvious from the proof given above, the use of forbidding transfer conditions in the operators moving the underlying objects is only needed for simulating the ac-case. In fact, we only need the skin compartment to simulate the derivations of a graph controlled grammar without ac: For every $q \in Lab$, we just take evolution rules of the form $(q, p(q); \alpha, p(\alpha))$ with $\alpha \in \sigma(q)$. Hence, we immediately infer the following result:

Corollary 1. *Any graph controlled grammar without ac of arbitrary type can be simulated by a GP-system without transfer checking of the same type with the simple membrane structure* $[_0[_1]_1]_0$.

When we only allow "context-free" rules of the form $(op_1; op_2, ..., op_k)$ for some $k \geq 1$, then we need a seperate compartment for every node of the control graph:

Theorem 2. *Any graph controlled grammar of arbitrary type with the control graph containing n nodes can be simulated by a GP-system of the same type with the membrane structure* $[_0[_1]_1 ... [_n]_n[_{n+1}]_{n+1}]_0$ *and rules of the forms* $(op_1;)$, $(op_1; op_2)$, *and* $(op_1; op_2, op_3)$.

Proof. The main ingredients of the GP-system G_P of type X generating the same language as the graph controlled grammar G_C, $G_C = (B, B_T, (R, L_{in}, L_{fin}), A)$, of type X with $card(Lab) = n$ are described as follows:

- For each $q \in Lab$, in compartment 0 we take the following rules:
For the success case, we use the rules $(q; q^{(1)}, (\tau_{in,q}, \emptyset, \emptyset))$, $((\tau_{in,q}, \emptyset, \emptyset); \tau_{in,q})$ as well as $(\tau_{in,q};)$ to transfer the current sentential form as well as the ground operator $q^{(1)}$ into the corresponding compartment q. For the failure case, we now use $(q; q^*, (\tau_{in,q}, \emptyset, \{p(q)\}))$ as well as $((\tau_{in,q}, \emptyset, \{p(q)\}); \tau_{in,q})$ (and again $(\tau_{in,q};)$) for the transfer of the current sentential form and the ground operator q^* into compartment q. The forbidding condition $\{p(q)\}$ in the transfer rule $(\tau_{in,q}, \emptyset, \{p(q)\})$ checks for the non-applicability of $p(q)$.

- For each $q \in Lab$, in compartment q we take the following rules: The rules $(q^{(1)}; p(q))$ and $(p(q); \alpha, (\tau_{out}, \emptyset, \emptyset))$ with $\alpha \in \sigma(q)$ allow for the simulation of the application of the production $p(q)$ to the underlying sentential form as well as to proceed to a ground operator α representing a node in the success field of q in the control graph; the rules $((\tau_{out}, \emptyset, \emptyset); \tau_{out})$ and $(\tau_{out};)$ then transfer the result of the application of $p(q)$ as well as α back into the skin compartment. The failure case again is accomplished by one of the rules $(q^*; \beta, (\tau_{out}, \emptyset, \emptyset))$ for $\beta \in \varphi(q)$ (as well as by the rules $((\tau_{out}, \emptyset, \emptyset); \tau_{out})$ and $(\tau_{out};)$).
- The axiom A and the special ground operator q_0, $q_0 \notin Lab$, constitute the initial configuration in compartment 0. The rules $(q_0; q)$ for $q \in L_{in}$ allow us to proceed to a starting node in the control graph.
- For any $q \in L_{fin}$, the rules $(q; (\tau_{in,n+1}, \emptyset, \emptyset))$ and $((\tau_{in,n+1}, \emptyset, \emptyset);)$ finally transfer a terminal object into the terminal compartment $n + 1$.

In contrast to the GP-system constructed in the preceding theorem, now the GP-system G_P can simulate the application of a production $p(q)$ at the node q in the control graph of G_C only by using a specific compartment q for each q.□

Corollary 2. *Any graph controlled grammar without ac of arbitrary type with the control graph containing n nodes can be simulated by a GP-system without transfer checking of the same type with the membrane structure* $[_0[_1]_1 \ldots [_n]_n[_{n+1}]_{n+1}]_0$, *and rules of the forms* $(op_1;)$, $(op_1; op_2)$, *and* $(op_1; op_2, op_3)$.

The general results proved above immediately apply for the string case, but as well for the objects being d-dimensional arrays and (directed) graphs, which will be elaborated in the following subsections. Moreover, in the ac-cases considered there, the graph controlled grammars can be constructed in such a way that only terminal objects from B_T have been derived when in a derivation a final node from L_{fin} is reached; hence, the results obtained in the terminal compartments of the GP-systems constructed in the proofs above are from B_T only, i.e., in this case no final intersection with B_T is necessary any more.

4.1 String languages

As shown in [2], context-free graph controlled string grammars can generate any recursively enumerable string language. The theorems proved above show that GP-systems using these context-free string productions can generate any recursively enumerable string language, too.

4.2 Array languages

Let Z denote the set of integers and let $d \in N$. Then a *d-dimensional array* $\mathcal{A}$ over an alphabet V is a function $\mathcal{A} : Z^d \to V \cup \{\#\}$, where $shape(\mathcal{A}) = \{v \in W \mid \mathcal{A}(v) \neq \#\}$ is finite and $\# \notin V$ is called the *background* or *blank symbol*. We usually shall write $\mathcal{A} = \{(v, \mathcal{A}(v)) \mid v \in shape(\mathcal{A})\}$. The set of all d-dimensional arrays over V is denoted by V^{*d}. The *empty array* in V^{*d} with empty shape is denoted by Λ_d. Moreover, we define $V^{+d} = V^{*d} \setminus \{\Lambda_d\}$. Any subset of V^{+d} is called a Λ-free d-dimensional array language.

The *translation* $\tau_v : Z^d \to Z^d$ is defined by $\tau_v(w) = w + v$ for all $w \in Z^d$, and for any array $\mathcal{A} \in V^{*d}$ we define $\tau_v(\mathcal{A})$, the corresponding d-dimensional array translated by v, by

$$(\tau_v(\mathcal{A}))(w) = \mathcal{A}(w - v) \quad \text{for all } w \in Z^d.$$

A *d-dimensional array production* p over V is a triple $(W, \mathcal{A}_1, \mathcal{A}_2)$, where $W \subseteq Z^d$ is a finite set and $\mathcal{A}_1$ and $\mathcal{A}_2$ are mappings from W to $V \cup \{\#\}$ such that *shape* $(\mathcal{A}_1) \neq \emptyset$; p is called *#-context-free*, if *card* (*shape* $(\mathcal{A}_1)) = 1$. We say that the array $\mathcal{C}_2 \in V^{*d}$ is *directly derivable* from the array $\mathcal{C}_1 \in V^{*d}$ by the d-dimensional array production $(W, \mathcal{A}_1, \mathcal{A}_2)$ if and only if there exists a vector $v \in Z^d$ such that $\mathcal{C}_1(w) = \mathcal{C}_2(w)$ for all $w \in Z^d \setminus \tau_v(W)$ as well as $\mathcal{C}_1(w) = \mathcal{A}_1(\tau_{-v}(w))$ and $\mathcal{C}_2(w) = \mathcal{A}_2(\tau_{-v}(w))$ for all $w \in \tau_v(W)$, i.e., the sub-array of $\mathcal{C}_1$ corresponding to $\mathcal{A}_1$ is replaced by $\mathcal{A}_2$, thus yielding $\mathcal{C}_2$.

Based on these definitions of d-dimensional array productions we can define d-dimensional array grammars, graph controlled d-dimensional array grammars etc. As was shown in [5], any recursively enumerable two-dimensional array language can even be generated by a graph controlled #-context-free two-dimensional array grammar without ac. Hence, Corollaries 1 and 2 apply showing that any recursively enumerable two-dimensional array language can even be generated by a GP-system without transfer checking using #-context-free two-dimensional array productions. The same result holds true for dimensions 1 and 3, too. For $d \geq 4$, we only know that graph controlled #-context-free d-dimensional array grammars can generate any recursively enumerable d-dimensional array language; hence, at least Theorems 1 and 2 are valid showing that recursively enumerable d-dimensional array languages can be generated by specific GP-systems using #-context-free d-dimensional array productions.

4.3 Graph languages

As shown in [4], any recursively enumerable graph language can be generated by a corresponding graph controlled graph grammar using only the following elementary graph productions:

1. add a new node with label K;
2. change the label of a node labelled by K to L;
3. delete a node with label K;
4. add a new edge labelled by a between two nodes labelled by K and M;
5. change the label a of an edge between two nodes labelled by K and M to b;
6. delete an edge labelled by a between two nodes labelled by K and M.

According to the theorems proved above, GP-systems using these elementary graph productions can generate any recursively enumerable graph language, too.

5 Summary and Future Research

The model of GP-systems investigated in the preceding sections has revealed great generative power with respect to arbitrary types of productions working on single objects. Yet among others, the following questions remained open:

- Is the hierarchy with respect to the number of inner membranes needed in the proof of Theorem 2 infinite or is there another proof showing that it collapses at a certain level? This question can be asked in general or for special types of grammars, e.g., for context-free string grammars.
- The effect of the permitting conditions in transfer operators corresponds with the permitting context condition in random context grammars (see [2]). Hence, in a similar way as in the proofs of Theorems 1 and 2 we can show that permitting random context grammars can be simulated by GP-systems with permitting transfer conditions of the same type.
- What is the generative power of GP-systems without ground operators and transfer operators working on them, i.e., when the only operators we allow are productions on objects and transfer operators to be applied to objects? Again we can ask this question in general or for special types of grammars. For example, in the string case (with at least context-free productions) the ground operators representing control variables can be encoded within the sentential forms; hence, at least in combination with the permitting and forbidding conditions in transfer operators moving objects GP-systems can reach the power of graph controlled string grammars again. A similar result can be established for graph controlled #-context-free d-dimensional array grammars and graph controlled elementary graph grammars, too. In all these cases, the proofs are much more complicated than the pure structural proofs of Theorems 1 and 2.

As already pointed out in [7], the idea of membrane structures offers a nearly unlimited variety of variants. Hence, this paper can be seen as a starting point for further investigations in this field of generalizing the idea of P-systems and the investigations of their modelling power. Let us mention just a few ideas to be considered in the future:

- In the proofs given above we have not made use of the possibilty to transfer productions working on the underlying objects or even transfer operators themselves as we did in Example 1, i.e., we only transferred ground operators and objects. Yet it might be interesting to use these features in general, too, and to investigate how the usage of other characteristic features can be reduced while still getting remarkable generative power.
- In this paper we have only considered the "sequential variant" of a GP-system, where in each derivation step only one evolution rule is evaluated. Yet we could also allow an arbitrary number of evolution rules to be evaluated in one step or we could even consider the forced "parallel variant" of a GP-system, where in each derivation step as many evolution rules as possible in all compartments have to be evaluated in parallel.

– Additional interesting features to be considered with GP-systems, which were already discussed for P-systems in [7], are the deletion and the generation of membranes.
– How can GP-systems using productions of a more complicated structure like splicing rules be defined in an adequate way? A splicing rule on strings over an alphabet V usually (e.g., see [8]) is written as $u_1\#v_1\$u_2\#v_2$ and its application to two strings over V of the form $x_1u_1v_1y_1$ and $x_2u_2v_2y_2$ yields the two strings $x_1u_1v_2y_2$ and $x_2u_2v_1y_1$, which are the result of cutting the two given strings at the sites u_1v_1 and u_2v_2 and immediately recombining the cut pieces in a crosswise way. Obviously, we can use splicing rules instead of other string productions by interpreting a splicing rule as a partial recursive relation $\subseteq V^* \times V^* \to V^* \times V^*$. Yet a suitable modelling of splicing systems (H-systems) and related systems is left to a forthcoming paper.

The formal investigation of the generative power of all these (and many other) variants of GP-systems and their complexity for simulating specific other generating devices remains for future research.

Acknowledgements

I gratefully acknowledge fruitful discussions with Grzegorz Rozenberg and especially with Gheorghe Păun, whose ideas inspired the model of GP-systems presented in this paper.

References

1. G. Berry and G. Boudol, *The chemical abstract machine,* Theoretical Computer Science 96 (1992), pp. 217-248.
2. J. Dassow and Gh. Păun, *Regulated Rewriting in Formal Language Theory* (Springer, Berlin, 1989).
3. J. Dassow and Gh. Păun, *On the power of membrane computing,* TUCS Research Report No. 217 (Dec. 1998).
4. R. Freund and B. Haberstroh, *Attributed elementary programmed graph grammars,* Proceedings 17th International Workshop on Graph-Theoretic Concepts in Computer Science (LNCS 570, Springer-Verlag, 1991), pp. 75–84.
5. R. Freund, *Control mechanisms on #-context-free array grammars,* Mathematical Aspects of Natural and Formal Languages (Gh. Păun, ed.), World Scientific, Singapore (1994), pp. 97-137.
6. Gh. Păun, *Computing with membranes,* TUCS Research Report No. 208 (Nov. 1998).
7. Gh. Păun, *Computing with membranes: an introduction,* Bulletin EATCS **67** (Febr. 1999), pp. 139-152.
8. Gh. Păun, G. Rozenberg, and A. Salomaa, *DNA Computing: New Computing Paradigms* (Springer-Verlag, Berlin, 1998).
9. Gh. Păun, G. Rozenberg, and A. Salomaa, *Membrane computing with external output,* TUCS Research Report No. 218 (Dec. 1998).
10. I. Petre, *A normal form for P-systems,* Bulletin EATCS **67** (Febr. 1999), pp. 165-172.

Optimal, Distributed Decision-Making: The Case of No Communication

Stavros Georgiades[1], *Marios Mavronicolas*[2], *and Paul Spirakis*[3]

[1] Department of Computer Science, University of Cyprus, Nicosia CY-1678, Cyprus. Email: `stavrosg@turing.cs.ucy.ac.cy`

[2] Department of Computer Science and Engineering, University of Connecticut, Storrs, CT 06269–3155. Part of the work of this author was performed while at Department of Computer Science, University of Cyprus, Nicosia CY-1678, Cyprus, and while at AT&T Labs – Research, as a visitor to the Special Year on Networks, DIMACS Center for Discrete Mathematics and Theoretical Computer Science. Email: `mavronic@engr.uconn.edu`

[3] Department of Computer Engineering and Informatics, University of Patras, Patras, Greece & Computer Technology Institute, Patras, Greece. Partially supported by the EU ESPRIT Long Term Research Project ALCOM-IT (Proj. # 20244), and by the Greek Ministry of Education. Email: `spirakis@cti.gr`

Abstract. We present a combinatorial framework for the study of a natural class of *distributed optimization problems* that involve *decision-making* by a collection of *n distributed agents* in the presence of incomplete information; such problems were originally considered in a *load balancing* setting by Papadimitriou and Yannakakis (*Proceedings of the 10th Annual ACM Symposium on Principles of Distributed Computing*, pp. 61–64, August 1991). For any given *decision protocol* and assuming *no communication* among the agents, our framework allows to obtain a combinatorial inclusion-exclusion expression for the probability that no "overflow" occurs, called the *winning probability*, in terms of the *volume* of some simple combinatorial polytope.

Within our general framework, we offer a complete resolution to the special cases of *oblivious algorithms*, for which agents do not "look at" their inputs, and *non-oblivious algorithms*, for which they do, of the general optimization problem. In either case, we derive optimality conditions in the form of combinatorial polynomial equations. For oblivious algorithms, we explicitly solve these equations to show that the optimal algorithm is simple and *uniform*, in the sense that agents need not "know" n. Most interestingly, we show that optimal non-oblivious algorithms must be *non-uniform:* we demonstrate that the optimality conditions admit different solutions for particular, different "small" values of n; however, these solutions improve in terms of the winning probability over the optimal, oblivious algorithm. Our results demonstrate an interesting trade-off between the amount of knowledge used by agents and uniformity for optimal, distributed decision-making with no communication.

1 Introduction

In a *distributed optimization problem,* each of n *distributed agents* receives a private *input,* communicates possibly with other agents to learn about their own inputs, and decides, based on this possibly partial knowledge, on an *output*; the task is to maximize a common objective function. Such problems were originally introduced by Papadimitriou and Yannakakis [9], in an effort to understand the crucial economic value of *information* [1] as a computational resource in a distributed system (see, also, [2, 4, 8, 10]). Intuitively, the more information available to agents, the better decisions they make, but naturally the more expensive the solution becomes due to the need for increased communication. Such natural trade-offs between communication cost and the quality of decision-making have been studied in the contexts of *communication complexity* [7] and *concurrency control* [6] as well.

Papadimitriou and Yannakakis [9] examined the special case of such distributed optimization problems where there are just *three* agents. More specifically, Papadimitriou and Yannakakis focused on a natural *load balancing* problem (see, e.g., [3, 5, 11], where each agent is presented with an input, and must decide on a binary output, representing one of two available "bins," each of capacity one; the input is assumed to be distributed uniformly in the unit interval $[0, 1]$. The load balancing property is modeled by requiring that no "overflow" occurs, namely that inputs dropped into each "bin" not exceed together its capacity. Papadimitriou and Yannakakis [9] pursued a comprehensive study of how the best possible probability, over the distribution of inputs, of "no overflow" depends on the amount of communication available to the agents. For each possible communication pattern, Papadimitriou and Yannakakis [9] discovered the corresponding optimal decision protocol to be unexpectedly sophisticated. The proof techniques of Papadimitriou and Yannakakis [9] were surprisingly complex, even for this seemingly simplest case, combining tools from nonlinear optimization with geometric and combinatorial arguments; these techniques have not been hoped to be conveniently extendible to instances of even this particular load balancing problem whose size exceeds three.

In this work, we introduce a novel combinatorial framework in order to enhance the study of general instances of distributed optimization problems of the kind considered by Papadimitriou and Yannakakis [9]. More specifically, we proceed to the general case of n agents, with each still receiving an input uniformly distributed over $[0, 1]$ and having to choose one out of two "bins"; however, in order to render the problem interesting, we make the technical assumption that the capacity of each "bin" is equal to δ, for some real number δ possibly greater than one, so as to compensate for the increase in the number of players. Papadimitriou and Yannakakis [9] focused on a specific kind of decision protocols by which each agent chooses a "bin" by comparing a "weighted average" of the inputs it "sees" against some "threshold" value; in contrast, our framework allows for the consideration of *general* decision protocols by which each agent decides by using *any* (computable) function of the inputs it "sees".

Our starting point is a combinatorial result that provides an explicit ***inclusion-exclusion*** formula [12, Section 2.1] for calculating the ***volume*** of any particular geometric polytope, in any given dimension, of some speficic form (Proposition 1). Roughly speaking, such polytopes are the intersection of a simplex in the positive quadrant with an orthogonal parallelepiped. An immediate implication of this result are inclusion-exclusion formulas for calculating the (conditional) probability of "no overflow" for a *single* "bin," as a function of the capacity δ and the number of inputs that are dropped into the "bin" (Lemmas 1 and 2).

In this work, we focus on the case where there is no communication among the agents, which we completely settle for the case of general n. Since communication comes at a cost, which it would be desirable to avoid, it is both natural and interesting to choose the case of no communication as an initial "testbed". We consider both *oblivious algorithms,* where players do not "look" at their inputs, and *non-oblivious algorithms,* where they do. For each case, we are interested in optimal algorithms.

We first consider oblivious algorithms. Our first major result is a combinatorial expression in the form of an inclusion-exclusion formula for the probability that "no overflow" occurs for either of the "bins" (Theorem 1). This formula incorporates a suitable inclusion-exclusion summation, over all possible input vectors, of the probabilities, induced by any particular decision algorithm, on the space of all possible decision vectors, as a function of the corresponding input vector. The coefficients of these probabilities in the summation are independent of any specific parameters of the algorithm, while they do depend on the input vector. A first implication of this expression is the reduction of the general problem of computing the probability that "no overflow" occurs to the problem of computing, given a particular decision algorithm, the probability distribution of the binary output vectors it yields. Most significantly, this expression contributes a methodology for the design of *optimal* decision algorithms "compatible" with any specific pattern of communication, and not just for the case of no communication that we particularly examine: one simply renders only those parameters of the decision algorithm that correspond to the possible communications, and computes values for these parameters that maximize the combinatorial expression as a function of these parameters. This is done by solving a certain system of *optimality conditions* (Corollary 2).

We demonstrate that our methodology for designing optimal algorithms for distributed decision-making is both effective and useful by applying it to the special case of no communication that we consider. We manage to settle down completely this case for oblivious algorithms. We exploit the underlying "symmetry" with respect to different agents in order to simplify the optimality conditions (by observing that all parameters satisfying them must be equal). This simplification reveals a beautiful combinatorial structure; more specifically, we discover that each optimality condition eventually amounts to zeroing a particular "symmetric" polynomial of a single variable. In turn, we explicitly solve these conditions to show that the best possible oblivious algorithm for the case of no communication is the very simple one by which each agent uses 1/2 as its "threshold" value; given that the optimal (non-oblivious) algorithms presented by Papadimitriou and Yannakakis for the special case where $n = 3$ are somehow unexpectedly sophisticated, it is perhaps surprising that such simple oblivious algorithm is indeed optimal for *all* values of n.

We next turn to non-oblivious algorithms, still for the case of no communication. In that case, we demonstrate that the optimality conditions do not admit a "constant" solution. Through a more sophisticated analysis, we are able to compute more complex expressions for the optimality conditions, which still allow exploitation of "symmetry". We consider the particular instances of the optimality conditions where $n = 3$ and $\delta = 1$ (considered by Papadimitriou and Yannakakis [9]), and $n = 4$ and $\delta = 4/3$. We discover that the optimal algorithms are different in each of these cases. However, they achieve larger winning probabilities than their oblivious counterparts. This shows that the improved performance of non-oblivious algorithms comes at the cost of sacrificing *uniformity*.

We believe that our work opens up the way for the design and analysis of algorithms for general instances of the problem of distributed decision-making in the presence of incomplete information. We envision that algorithms that are more complex, general communication patterns, and more realistic assumptions on the distribution of inputs, can all be treated in our combinatorial framework to yield optimal algorithms for distributed decision-making for these cases as well.

2 Framework and Preliminaries

Throughout, for any bit $b \in \{0,1\}$ and real number $\alpha \in [0,1]$, denote $\bar{b}$ the complement of b, and $\alpha^{(b)}$ to be α if $b = 1$, and $1 - \alpha$ if $b = 0$. For any binary vector $\mathbf{b}$, denote $|\mathbf{b}|$ the number of entries of $\mathbf{b}$ that are equal to one.

2.1 Model and Problem Definition

Consider a collection of n distributed agents $P_1, P_2, \ldots, P_n$, called *players,* where $n \geq 2$. Each player P_i receives an *input* x_i, which is the value of a random variable distributed uniformly over $[0,1]$; denote $\mathbf{x} = \langle x_1, x_2, \ldots, x_n \rangle^{\mathrm{T}}$ the *input vector.* Associated with each player P_i is a (local) *decision algorithm* A_i, that may be either deterministic or randomized, and "maps" the input x_i to P_i's *output* y_i. A *distributed decision algorithm* is a collection $\mathcal{A} = \langle A_1, A_2, \ldots, A_n \rangle$ of (local) decision algorithms, one for each player.

A *deterministic decision algorithm* is a function $A_i : [0,1] \rightarrow \{0,1\}$, that maps the input x_1 to P_i's *output* $y_i = A_i(x_i)$; denote $\mathbf{y}_{\mathcal{A}}(\mathbf{x}) = \langle A_1(x_1), A_2(x_2), \ldots,$ $A_n(x_n) \rangle^{\mathrm{T}}$ the *output vector of $\mathcal{A}$ on input vector* $\mathbf{x}$. A *deterministic, single-threshold decision algorithm* is a deterministic decision algorithm A_i that is a *single-threshold* function; that is,

$$A_i(x_i) = \begin{cases} 0, & x_i \leq a_i \\ 1, & x_i > a_i \end{cases},$$

where $0 \leq a_i \leq \infty$. Distributed, deterministic decision algorithms and distributed, deterministic single-threshold decision algorithms can be defined in the natural way.

Say that $\mathcal{A}$ is *randomized oblivious* if for each i, $1 \leq i \leq n$, A_i is a ***probability distribution*** on $\{0,1\}$; that is, $A_i(0)$ (resp., $A_i(1)$) is the probability that player P_i decides 0 (resp., 1). Denote $a_i = A_i(0)$. Thus, a ***distributed, randomized oblivious decision algorithm*** is a collection $\mathcal{A} = \langle a_i, a_2, \ldots, a_n\rangle$ is a collection of (local) randomized, oblivious decision algorithms, one for each player.

For each $b \in \{0,1\}$, define $S_b = \sum_{i:A_i(\mathbf{x})=b} x_i$; thus, S_b is the sum of the inputs of the players that "decide" b. For each parameter $\delta > 0$, we are interested in the event that neither S_0 nor S_1 exceeds δ; denote $\mathbf{Pr}_{\mathcal{A}}(S_0 \leq \delta \text{ and } S_1 \leq \delta)$ the probability, taken over all input vectors $\mathbf{x}$, that this event occurs. We wish to maximize $\mathbf{Pr}_{\mathcal{A}}(S_0 \leq \delta \text{ and } S_1 \leq \delta)$ over all protocols $\mathcal{A}$; any protocol $\mathcal{A}$ that maximizes $\mathbf{Pr}_{\mathcal{A}}(S_0 \leq \delta \text{ and } S_1 \leq \delta)$ is a corresponding *optimal* protocol.

2.2 Combinatorial Preliminaries

For any *polytope* $\mathbf{\Pi}$, denote $\mathsf{Vol}(\mathbf{\Pi})$ the *volume* of $\mathbf{\Pi}$. A cornerstone of our analysis is the following combinatorial result that calculates the volume of any particular polytope that has some specific form. Fix any integer $m \geq 2$. Consider any pair of vectors $\mathbf{a} = \langle \alpha_1, \alpha_2, \ldots, \alpha_m\rangle^{\mathrm{T}}$, and $\mathbf{b} = \langle \beta_1, \beta_2, \ldots, \beta_m\rangle^{\mathrm{T}}$, where for any l, $1 \leq l \leq m$, $0 \leq \alpha_l, \beta_l < \infty$. Define the m-dimensional polytope

$$\mathbf{\Pi}^{(m)}(\mathbf{a},\mathbf{b}) = \{\langle x_1, x_2, \ldots, x_m\rangle^T \in [0,\beta_1]\times[0,\beta_2]\times\ldots\times[0,\beta_m] \mid \sum_{l=1}^{m} \frac{x_l}{\alpha_l} \leq 1\}.$$

Thus, $\mathbf{\Pi}^{(m)}(\mathbf{a},\mathbf{b})$ is the intersection of the m-dimensional simplex

$$\mathbf{\Pi}^{(m)}(\mathbf{a}) = \{\langle x_1, x_2, \ldots, x_m\rangle^T \mid \sum_{l=1}^{m} \frac{x_l}{\alpha_l} \leq 1\},$$

with the m-dimensional orthogonal parallelepiped $[0,\beta_1] \times [0,\beta_2] \times \ldots \times [0,\beta_m]$; The vectors $\mathbf{a}$ and $\mathbf{b}$ determine the simplex and the orthogonal parallelepiped, respectively. We provide an explicit inclusion-exclusion formula for calculating the volume of $\mathbf{\Pi}^{(m)}(\mathbf{a},\mathbf{b})$.

Proposition 1.

$$\mathsf{Vol}(\mathbf{\Pi}^{(m)}(\mathbf{a},\mathbf{b})) =$$

$$V_\emptyset - \sum_{1\leq i\leq m} V_{\{i\}} + \sum_{1\leq i<j\leq m} V_{\{i,j\}} - \sum_{1\leq i<j<k\leq m} V_{\{i,j,k\}} + \cdots + (-1)^m V_{\{1,2,\ldots,m\}},$$

where

$$V_\emptyset = \frac{1}{m!}\prod_{l=1}^{m} \alpha_l,$$

and for each set of indices $\mathcal{I} \subseteq \{1, 2, \ldots, m\}$,

$$V_{\mathcal{I}} = \begin{cases} V_{\emptyset}(1 - \sum_{l \in \mathcal{I}} \beta_l \alpha_l^{-1})^m, & 1 > \sum_{l \in \mathcal{I}} \beta_l \alpha_l^{-1} \\ 0, & 1 \leq \sum_{l \in \mathcal{I}} \beta_l \alpha_l^{-1} \end{cases}.$$

2.3 Probabilistic Lemmas

In this section, we present two straightforward implications of Proposition 1 that will be used later.

Lemma 1. *Assume that for each* i, $1 \leq i \leq m$, x_i *is uniformly distributed over* $[0, \beta_i]$. *Then, for any parameter* $\delta > 0$,

$$\mathbf{Pr}(\sum_{i=1}^{m} x_i \leq \delta) = \frac{1}{m!} \frac{\sum_{I \subseteq \{1,2,\ldots,m\}, \sum_{l \in I} \beta_l < \delta} (-1)^{|I|} (\delta - \sum_{l \in I} \beta_l)^m}{\prod_{l=1}^{m} \beta_l}.$$

An immediate implication of Lemma 1 concerns the special case where for each i, $1 \leq i \leq n$, $\beta_i = 1$.

Corollary 1. *Assume that for each* i, $1 \leq i \leq m$, x_i *is uniformly distributed over* $[0, 1]$. *Then, for any parameter* $\delta > 0$,

$$\mathbf{Pr}(\sum_{i=1}^{m} x_i \leq \delta) = \frac{1}{m!} \sum_{0 \leq l \leq m, l < \delta} (-1)^l \binom{m}{l} (\delta - l)^m .$$

We also show:

Lemma 2. *Assume that for each* i, $1 \leq i \leq m$, x_i *is uniformly distributed over* $[\beta_i, 1]$. *Then, for any parameter* $\delta > 0$,

$$\mathbf{Pr}(\sum_{i=1}^{m} x_i \leq \delta) =$$

$$= 1 - \frac{1}{m!} \frac{\sum_{I \subseteq \{1,2,\ldots,m\}, |I| - \sum_{l \in I} \beta_l < m - \delta} (-1)^{|I|} (m - \delta - |I| + \sum_{l \in I} \beta_l)^m}{\prod_{l=1}^{m} (1 - \beta_l)}.$$

3 Oblivious Algorithms

3.1 The Winning Probability

We show:

Theorem 1. *Assume that* $\mathcal{A}$ *is any randomized oblivious algorithm. Then,*

$$\mathbf{Pr}_{\mathcal{A}}(S_0 \leq \delta \text{ and } S_1 \leq \delta) =$$

$$= \delta^n \sum_{\mathbf{b} \in \{0,1\}^n} (\frac{1}{|\mathbf{b}|!} \sum_{0 \leq l \leq |\mathbf{b}|, l < \delta} (-1)^l \binom{|\mathbf{b}|}{l} \left(1 - \frac{l}{\delta}\right)^{|\mathbf{b}|} \cdot$$

$$\frac{1}{(n - |\mathbf{b}|!)} \sum_{0 \leq l \leq n - |\mathbf{b}|, l < \delta} (-1)^l \binom{n - |\mathbf{b}|}{l} \left(1 - \frac{l}{\delta}\right)^{n - |\mathbf{b}|} \cdot \prod_{i=1}^{n} \alpha_i^{(b_i)}).$$

The proof of Theorem 1 relies on appropriately using Corollary 1. Theorem 1 immediately implies *necessary* conditions for any optimal protocol. These conditions are determined by simultaneously vanishing the partial derivatives with respect to all parameters of the algorithm.

Corollary 2 (Optimality conditions for oblivious algorithms). *Assume that $\mathcal{A}$ is an optimal, randomized oblivious algorithm. Then, for any index k,*

$$\sum_{\mathbf{b}\in\{0,1\}^n} \Big(\frac{1}{|\mathbf{b}|!} \sum_{0\leq l\leq |\mathbf{b}|, l<\delta} (-1)^l \binom{|\mathbf{b}|}{l} \left(1-\frac{l}{\delta}\right)^{|\mathbf{b}|} \cdot$$

$$\frac{1}{(n-|\mathbf{b}|)!} \sum_{0\leq l\leq n-|\mathbf{b}|, l<\delta} (-1)^l \binom{n-|\mathbf{b}|}{l} \left(1-\frac{l}{\delta}\right)^{n-|\mathbf{b}|} \cdot$$

$$\frac{\partial}{\partial \alpha_{kl}} \alpha_k^{(b_k)} \prod_{i=1, i\neq k}^{n} \alpha_i^{(b_i)}\Big)$$

$$= 0 .$$

3.2 The Optimal Oblivious Algorithm

For each i, set $1 \leq i \leq n$,

$$A_i(\mathbf{x}) = \begin{cases} 0, \; x_i \leq a_{ii} \\ 1, \; x_i > a_{ii} \end{cases};$$

it follows that $\mathbf{Pr}_{\mathcal{A}}(y_i = 0) = a_{ii}$ and $\mathbf{Pr}_{\mathcal{A}}(y_i = 1) = 1 - a_{ii}$. We show that the optimal winning probability is achieved by the very simple protocol for which, for each i, $1 \leq i \leq n$, $a_i = 1/2$.

Theorem 2. *Consider the oblivious case. Then,*

$$\max_{\mathcal{A}} \mathbf{Pr}_{\mathcal{A}}(S_0 \leq \delta \textit{ and } S_1 \leq \delta) = \frac{1}{n!}\left(\frac{\delta}{2}\right)^n \cdot$$

$$\sum_{r=0}^{n} \Big(\binom{n}{r}^2 \sum_{0\leq l\leq r, l<\delta} (-1)^l \binom{r}{l}\left(1-\frac{l}{\delta}\right)^r \cdot \sum_{\substack{0\leq l\leq n-r,\\ l<\delta}} (-1)^l \binom{n-r}{l}\left(1-\frac{l}{\delta}\right)^{n-r}\Big)$$

Proof. Take any optimal protocol $\mathcal{A}$. By Theorem 1,

$$\mathbf{Pr}_{\mathcal{A}}(S_0 \leq \delta \text{ and } S_1 \leq \delta)$$

$$= \delta^n \sum_{\mathbf{b}\in\{0,1\}^n} \Big(\frac{1}{|\mathbf{b}|!} \sum_{0\leq l\leq |\mathbf{b}|, l<\delta} (-1)^l \binom{|\mathbf{b}|}{l}\left(1-\frac{l}{\delta}\right)^{|\mathbf{b}|} \cdot$$

$$\frac{1}{(n-|\mathbf{b}|)!} \sum_{0\leq l\leq n-|\mathbf{b}|, l<\delta} (-1)^l \binom{n-|\mathbf{b}|}{l}\left(1-\frac{l}{\delta}\right)^{n-|\mathbf{b}|} \cdot \prod_{i=1}^{n} a_{ii}^{(b_i)}\Big).$$

Fix any index k, $1 \leq k \leq n$. By Corollary 2,

$$\sum_{\mathbf{b}\in\{0,1\}^n} \Big(\frac{1}{|\mathbf{b}|!} \sum_{0\leq l\leq|\mathbf{b}|,l<\delta} (-1)^l \binom{|\mathbf{b}|}{l} \left(1-\frac{l}{\delta}\right)^{|\mathbf{b}|} \cdot$$

$$\frac{1}{(n-|\mathbf{b}|)!} \sum_{0\leq l\leq n-|\mathbf{b}|,l<\delta} (-1)^l \binom{n-|\mathbf{b}|}{l} \left(1-\frac{l}{\delta}\right)^{n-|\mathbf{b}|} \cdot$$

$$\prod_{i=1,i\neq k}^{n} a_{ii}^{(b_i)} \frac{\partial}{\partial a_{kl}} \mathbf{Pr}_{\mathcal{A}}(y_k = b_k)\Big)$$

$$= 0,$$

so that

$$\sum_{\mathbf{b}\in\{0,1\}^n,b_k=1} \Big(\frac{1}{|\mathbf{b}|!} \sum_{0\leq l\leq|\mathbf{b}|,l<\delta} (-1)^l \binom{|\mathbf{b}|}{l} \left(1-\frac{l}{\delta}\right)^{|\mathbf{b}|} \cdot$$

$$\frac{1}{(n-|\mathbf{b}|)!} \sum_{0\leq l\leq n-|\mathbf{b}|,l<\delta} (-1)^l \binom{n-|\mathbf{b}|}{l} \left(1-\frac{l}{\delta}\right)^{n-|\mathbf{b}|} \cdot \prod_{i=1,i\neq k}^{n} a_{ii}^{(b_i)}\Big) -$$

$$\sum_{\mathbf{b}\in\{0,1\}^n,b_k=0} \Big(\frac{1}{|\mathbf{b}|!} \sum_{0\leq l\leq|\mathbf{b}|,l<\delta} (-1)^l \binom{|\mathbf{b}|}{l} \left(1-\frac{l}{\delta}\right)^{|\mathbf{b}|} \cdot$$

$$\frac{1}{(n-|\mathbf{b}|)!} \sum_{0\leq l\leq n-|\mathbf{b}|,l<\delta} (-1)^l \binom{n-|\mathbf{b}|}{l} \left(1-\frac{l}{\delta}\right)^{n-|\mathbf{b}|} \cdot \prod_{i=1,i\neq k}^{n} a_{ii}^{(b_i)}\Big)$$

$$= 0.$$

By symmetry of optimality conditions, it follows that $a_{11} = a_{22} = \ldots = a_{nn}$; denote α their common value. Clearly,

$$\sum_{\mathbf{b}\in\{0,1\}^n,b_k=1} \Big(\frac{1}{|\mathbf{b}|!} \sum_{0\leq l\leq|\mathbf{b}|,l<\delta} (-1)^l \binom{|\mathbf{b}|}{l} \left(1-\frac{l}{\delta}\right)^{|\mathbf{b}|} \cdot \frac{1}{(n-|\mathbf{b}|)!} \cdot$$

$$\sum_{0\leq l\leq n-|\mathbf{b}|,l<\delta} (-1)^l \binom{n-|\mathbf{b}|}{l} \left(1-\frac{l}{\delta}\right)^{n-|\mathbf{b}|} \cdot \alpha^{|\mathbf{b}|-1}(1-\alpha)^{n-1-(|\mathbf{b}|-1)}\Big) -$$

$$\sum_{\mathbf{b}\in\{0,1\}^n,b_k=0} \Big(\frac{1}{|\mathbf{b}|!} \sum_{0\leq l\leq|\mathbf{b}|,l<\delta} (-1)^l \binom{|\mathbf{b}|}{l} \left(1-\frac{l}{\delta}\right)^{|\mathbf{b}|} \cdot$$

$$\frac{1}{(n-|\mathbf{b}|)!} \sum_{0\leq l\leq n-|\mathbf{b}|,l<\delta} (-1)^l \binom{n-|\mathbf{b}|}{l} \left(1-\frac{l}{\delta}\right)^{n-|\mathbf{b}|} \cdot \alpha^{|\mathbf{b}|}(1-\alpha)^{n-1-|\mathbf{b}|}\Big)$$

$$= 0.$$

There are $\binom{n-1}{|\mathbf{b}|-1}$ vectors $\mathbf{b} \in \{0,1\}^n$ with $b_k = 1$; for any such vector, $1 \leq |\mathbf{b}| \leq n$. Similarly, there are $\binom{n-1}{|\mathbf{b}|}$ vectors $\mathbf{b} \in \{0,1\}^n$ with $b_k = 0$; for any such vector, $0 \leq |\mathbf{b}| \leq n-1$. It follows that

$$\sum_{|\mathbf{b}|=1}^{n} \binom{n-1}{|\mathbf{b}|-1} \Big(\frac{1}{|\mathbf{b}|!} \sum_{0 \leq l \leq |\mathbf{b}|, l<\delta} (-1)^l \binom{|\mathbf{b}|}{l} \left(1-\frac{l}{\delta}\right)^{|\mathbf{b}|} \cdot$$

$$\frac{1}{(n-|\mathbf{b}|)!} \sum_{0 \leq l \leq n-|\mathbf{b}|, l<\delta} (-1)^l \binom{n-|\mathbf{b}|}{l} \left(1-\frac{l}{\delta}\right)^{n-|\mathbf{b}|} \cdot \left(\frac{\alpha}{1-\alpha}\right)^{|\mathbf{b}|-1} \Big) -$$

$$\sum_{|\mathbf{b}|=0}^{n-1} \binom{n-1}{|\mathbf{b}|} \Big(\frac{1}{|\mathbf{b}|!} \sum_{0 \leq l \leq |\mathbf{b}|, l<\delta} (-1)^l \binom{|\mathbf{b}|}{l} \left(1-\frac{l}{\delta}\right)^{|\mathbf{b}|} \cdot$$

$$\frac{1}{(n-|\mathbf{b}|)!} \sum_{0 \leq l \leq n-|\mathbf{b}|, l<\delta} (-1)^l \binom{n-|\mathbf{b}|}{l} \left(1-\frac{l}{\delta}\right)^{n-|\mathbf{b}|} \cdot \left(\frac{\alpha}{1-\alpha}\right)^{|\mathbf{b}|} \Big)$$

$$= 0.$$

The left-hand side is a polynomial in $\alpha/(\alpha-1)$ of degree $n-1$. Consider any integer r, where $0 \leq r \leq n-1$; We show that the coefficients of $(\alpha/(\alpha-1))^r$ and $(\alpha/(\alpha-1))^{n-1-r}$ are the negative of each other.

Thus, the left-hand side is a symmetric polynomial of degree $n-1$. Moreover, we can establish along similar lines that for the case where n is odd, the coefficient of $(\alpha/(\alpha-1))^{(n-1)/2}$ is identically zero. This implies that 1 is the only one real root of this polynomial; setting $\alpha/(\alpha-1) = 1$ yields $\alpha = 1/2$, with corresponding optimal winning probability

$$\mathbf{Pr}_{\mathcal{A}}(S_0 \leq \delta \text{ and } S_1 \leq \delta)$$

$$= \delta^n \sum_{\mathbf{b} \in \{0,1\}^n} \Big(\frac{1}{|\mathbf{b}|!} \sum_{0 \leq l \leq |\mathbf{b}|, l<\delta} (-1)^l \binom{|\mathbf{b}|}{l} \left(1-\frac{l}{\delta}\right)^{|\mathbf{b}|} \cdot$$

$$\frac{1}{(n-|\mathbf{b}|)!} \sum_{0 \leq l \leq n-|\mathbf{b}|, l<\delta} (-1)^l \binom{n-|\mathbf{b}|}{l} \left(1-\frac{l}{\delta}\right)^{n-|\mathbf{b}|} \cdot \left(\frac{1}{2}\right)^n$$

$$= \frac{1}{n!} \left(\frac{\delta}{2}\right)^n \sum_{\mathbf{b} \in \{0,1\}^n} \Big(\binom{n}{|\mathbf{b}|} \sum_{0 \leq l \leq |\mathbf{b}|, l<\delta} (-1)^l \binom{|\mathbf{b}|}{l} \left(1-\frac{l}{\delta}\right)^{|\mathbf{b}|} \cdot$$

$$\sum_{0 \leq l \leq n-|\mathbf{b}|, l<\delta} (-1)^l \binom{n-|\mathbf{b}|}{l} \left(1-\frac{l}{\delta}\right)^{n-|\mathbf{b}|} \Big)$$

$$= \frac{1}{n!} \left(\frac{\delta}{2}\right)^n \sum_{r=0}^{n} \Big(\binom{n}{r}^2 \sum_{0 \leq l \leq r, l<\delta} (-1)^l \binom{r}{l} \left(1-\frac{l}{\delta}\right)^r \cdot$$

$$\sum_{0 \leq l \leq n-r, l<\delta} (-1)^l \binom{n-r}{l} \left(1-\frac{l}{\delta}\right)^{n-r} \Big),$$

as needed.

Theorem 2 implies that for any integer n, the optimal winning probability of an oblivious algorithm is computable in exponential time.

4 Non-oblivious Algorithms

4.1 The Winning Probability

Theorem 3. *Assume that $\mathcal{A}$ is any randomized non-oblivious algorithm. Then,*

$$\mathbf{Pr}_{\mathcal{A}}(S_0 \leq \delta \text{ and } S_1 \leq \delta)$$

$$= \sum_{\mathbf{b}\in\{0,1\}^n} \left(\frac{1}{(n-|\mathbf{b}|)!} \sum_{I\subseteq\{i:b_i=0\},\sum_{l\in I}\beta_l<\delta} (-1)^{|I|}\left(\delta - \sum_{l\in I}\beta_l\right)^{n-|\mathbf{b}|} \cdot\right.$$

$$\left.\left(\prod_{l=1}^{|\mathbf{b}|}(1-\beta_l) - \frac{1}{|\mathbf{b}|!} \sum_{\substack{I\subseteq\{i:b_i=1\},\\ |I|-\sum_{l\in I}\beta_l<|\mathbf{b}|-\delta}} (-1)^{|I|}\left(|\mathbf{b}| - \delta - |I| + \sum_{l\in I}\beta_l\right)^{|\mathbf{b}|}\right)\right).$$

4.2 Optimality Conditions

For non-oblivious algorithms, the analysis is more involved since it must take into account the conditional probabilities "created" by the knowledge of inputs by the agents. We show:

Theorem 4 (Optimality conditions for non-oblivious algorithms). *Assume that $\mathcal{A}$ is an optimal, randomized non-oblivious algorithm. Then, for any index k,*

$$\sum_{|\mathbf{b}|=0}^{n}\binom{n-1}{|\mathbf{b}|}\left(\frac{1}{(n-1-|\mathbf{b}|)!}\sum_{\substack{0\leq l\leq n-1-|\mathbf{b}|,\\ \delta-\beta l>0}}(-1)^l\binom{n-1-|\mathbf{b}|}{l}(\delta-\beta l)^{n-1-|\mathbf{b}|}\right)\cdot$$

$$\left(-(1-\beta)^{|\mathbf{b}|} - \frac{(|\mathbf{b}|+1)}{(|\mathbf{b}|+1)!}\sum_{\substack{1\leq l\leq|\mathbf{b}|+1,\\ |\mathbf{b}|+1-\delta-l+\beta l>0}}(-1)^l\binom{|\mathbf{b}|}{l-1}l\,(\mathbf{b}+1-\delta-l+\beta l)^{|\mathbf{b}|}\right)+$$

$$\sum_{|\mathbf{b}|=0}^{n-1}\binom{n-1}{|\mathbf{b}|}\left(\left((1-\beta)^{|\mathbf{b}|} - \left(\frac{1}{|\mathbf{b}|!}\sum_{\substack{0\leq l\leq|\mathbf{b}|,\\ |\mathbf{b}|-\delta-l+\beta l>0}}(-1)^l\binom{|\mathbf{b}|}{l}(|\mathbf{b}|-\delta-l+\beta l)^{|\mathbf{b}|}\right)\right)\right)\cdot$$

$$\frac{-(n-|\mathbf{b}|)}{(n-|\mathbf{b}|)!}\sum_{1\leq l\leq n-|\mathbf{b}|,\delta-\beta l>0}(-1)^l\binom{n-|\mathbf{b}|-1}{l-1}l\,(\delta-\beta l)^{n-|\mathbf{b}|-1})$$

$$=0\,.$$

Unfortunately, the conditions in Theorem 4 do not admit a uniform solution (independent of n). We discover that the solutions for $n=3$ and $n=4$ are different. The solution for $n = 3$ and $\delta = 1$ satisfies the polynomial equation $\beta^2 - 2\beta + 6/7 = 0$; the solution is calculated to be equal to $1 - \sqrt{1/7} = 0.622$, which is the threshold value conjectured by Papadimitriou and Yannakakis in [9] to imply optimality for the same case. On the other hand, the solution for $n = 4$ and $\delta = 4/3$ satisfies the polynomial equation $-(26/3)\beta^3 + (98/3)\beta^2 - (368/9)\beta + 416/27 = 0$; the solution is calculated to be equal to approximately 0.678.

References

1. K. Arrow, *The Economics of Information,* Harvard University Press, 1984.
2. G. Brightwell, T. J. Ott, and P. Winkler, "Target Shooting with Programmed Random Variables," *Proceedings of the 24th Annual ACM Symposium on Theory of Computing,* pp. 691–698, May 1992.
3. P. Fizzano, D. Karger, C. Stein and J. Wein, "Job Scheduling in Rings," *Proceedings of the 6th Annual ACM Symposium on Parallel Algorithms and Architectures,* pp. 210–219, June 1994.
4. X. Deng and C. H. Papadimitriou, "Competitive, Distributed Decision-Making," *Proceedings of the 12th IFIP Congress,* pp. 350–356, 1992.
5. J. M. Hayman, A. A. Lazar, and G. Pacifici, "Joint Scheduling and Admission Control for ATM-Based Switching Nodes," *Proceedings of the ACM SIGCOMM,* pp. 223–234, 1992.
6. P. C. Kanellakis and C. H. Papadimitriou, "The Complexity of Distributed Concurrency Control," *SIAM Journal on Computing,* Vol. 14, No. 1, pp. 52–75, February 1985.
7. E. Kushilevitz and N. Nisan, *Communication Complexity,* Cambridge University Press, Cambridge, 1996.
8. C. H. Papadimitriou, "Computational Aspects of Organization Theory," *Proceedings of the 4th Annual European Symposium on Algorithms,* September 1996.
9. C. H. Papadimitriou and M. Yannakakis, "On the Value of Information in Distributed Decision-Making," *Proceedings of the 10th Annual ACM Symposium on Principles of Distributed Computing,* pp. 61–64, August 1991.
10. C. H. Papadimitriou and M. Yannakakis, "Linear Programming Without the Matrix," *Proceedings of the 25th Annual ACM Symposium on Theory of Computing,* pp. 121–129, May 1993.
11. C. Polychronopoulos and D. Kuck, "Guided Self-Scheduling: A Practical Scheduling Scheme for Parallel Computers," *IEEE Transactions on Computers,* Vol. 12, pp. 1425–1439, 1987.
12. R. P. Stanley, *Enumerative Combinatorics: Volume 1,* The Wadsworth & Brooks/Cole Mathematics Series, 1986.

Generalized PCP Is Decidable for Marked Morphisms*

Vesa Halava[1], Tero Harju[2], and Mika Hirvensalo[3,4]**

[1] Turku Centre for Computer Science, Lemminkäisenkatu 14 A, 4th floor, FIN-20520, Turku, Finland, vehalava@cs.utu.fi
[2] Department of Mathematics, University of Turku, FIN-20014, Turku, Finland. harju@utu.fi
[3] Department of Mathematics, University of Turku, FIN-20014, Turku, Finland. mikhirve@cs.utu.fi
[4] Turku Centre for Computer Science

Abstract We prove that the generalized Post Correspondence Problem (GPCP) is decidable for marked morphisms. This result gives as a corollary a shorter proof for the decidability of the binary PCP, proved in 1982 by Ehrenfeucht, Karhumäki and Rozenberg.

1 Introduction

Let A and B be two finite alphabets and $h, g\colon A^* \to B^*$. The *Post Correspondence Problem*, PCP for short, is to determine if there exists a nonempty word $w \in A^*$ such that $h(w) = g(w)$. It was proved by Post [5] that this problem is undecidable. The PCP is one of the most useful problems for deriving other undecidability results.

Restricting the *instances* (h, g) of the PCP may make the problem decidable. For example, if we assume that the *size* of the instance, i.e. $|A|$, is at most two, then the PCP is decidable, see Ehrenfeucht, Karhumäki and Rozenberg [1]. On the other hand it is known that if $|A| \geq 7$, then the problem remains undecidable, see Matiyasevich and Sénizerques [4].

It was proved by Halava, Hirvensalo and de Wolf that the PCP is decidable if the morphisms are *marked* [2]. A morphism h is called marked, if $h(x)$ and $h(y)$ start with a different letter whenever $x, y \in \Sigma$ and $x \neq y$. In this paper we consider a modification of the PCP, called the *generalized PCP*, GPCP for short. An instance of the GPCP consists of two morphisms $h, g\colon A^* \to B^*$ and words $p_1, p_2, s_1, s_2 \in B^*$. The GPCP is to tell whether or not there exists a word $w \in A^*$ such that

$$p_1 h(w) s_1 = p_2 g(w) s_2.$$

We shall denote the instance of the GPCP by $((p_1, p_2), h, g, (s_1, s_2))$. Actually Ehrenfeucht, Karhumäki and Rozenberg proved in [1] that the GPCP for marked

---

* For more detailed proofs, see full paper: http://www.tucs.abo.fi/publications/techreports/TR283.html

** Supported by the Academy of Finland under grant 44087.

morphisms is decidable for binary alphabets A, and since every binary instance of the PCP is either periodic (see [3]) or can be reduced to an instance of the binary GPCP with marked morphisms, they had that the binary PCP is decidable. In this paper we shall prove the following theorem.

Theorem 1. *The GPCP with marked morphisms is decidable for any alphabet size.*

The decidability of the marked GPCP was mentioned as an open problem in [2]. As a general reference for the PCP and the GPCP we give [3].

Our proof of Theorem 1 uses the idea of reducing a problem instance to finitely many new instances such that one of these new instances has a solution if and only if the original one has. We will show that the iterative use of this reduction process will eventually give us the desired decision method.

We fix first some notations. The *empty word* is denoted by ε. A word $x \in A^*$ is said to be a *prefix* of $y \in A^*$, if there is $z \in A^*$ such that $y = xz$. This will be denoted by $x \leq y$. Similarly, a word $x \in A^*$ is said to be a *suffix* of $y \in A^*$, if there is $z \in A^*$ such that $y = zx$. This will be denoted by $x \preccurlyeq y$. We say that x and y are *comparable* if $x \leq y$ or $y \leq x$. Moreover, a word $x \in A^*$ is said to be a *factor* of $y \in A^*$, if there are words $z_1, z_2 \in A^*$ such that $y = z_1 x z_2$.

Finally we introduce a convention that will smoothen some concepts to be used later: If $h : A^* \to B^*$ is a marked morphism, then clearly $|A| \leq |B|$. Renaming the source alphabet A we can always assume even that $A \subseteq B$ and also that for each $a \in A$, $h(a)$ begins with a. However, notice carefully that for a morphism pair $h, g : A^* \to B^*$ given as an instance of the GPCP, we *cannot assume* that *both* $h(a)$ and $g(a)$ begin with a. That $h(a)$ begins with a will be a permanent premise hereafter.

2 Modified instances

Let $I = ((p_1, p_2), h, g, (s_1, s_2))$, where $h, g\colon A^* \to B^*$, be an instance of the marked GPCP. A word $w \in A^*$ is called a *solution* of I, if

$$p_1 h(w) s_1 = p_2 g(w) s_2.$$

A solution $w = w_0 v w_1$ is called *minimal* if for all $v \neq \varepsilon$, $w_0 w_1$ is not a solution.

The instances

$$I = ((p_1, p_2), h, g, (s_1, s_2)) \qquad (1)$$

can be reduced to instances, where $p_1 = \varepsilon$ or $p_2 = \varepsilon$ and $s_1 = \varepsilon$ or $s_2 = \varepsilon$, since to have a solution we must have $p_1 \leq p_2$ or $p_2 \leq p_1$, and $s_1 \preccurlyeq s_2$ or $s_2 \preccurlyeq s_1$.

We first modify the marked GPCP by requiring that the solutions begin with a new fixed letter #. For this, let # be a new letter. If in (1) $p_1 \neq \varepsilon$ or $p_2 \neq \varepsilon$ we extend the morphisms by defining $h(\#) = \#p_1$ and $g(\#) = \#p_2$. On the other hand, if $p_1 = \varepsilon = p_2$, we fix a letter $a_0 \in A$ and define $h(\#) = \#h(a_0)$ and $g(\#) = \#g(a_0)$.

In both cases, the extended morphisms $h, g\colon (A\cup\{\#\})^* \to (B\cup\{\#\})^*$ remain marked, and in the latter case, (1) has a solution that begins with a_0 if and only if the modified instance $((\varepsilon, \varepsilon), h, g, (s_1, s_2))$ has a solution in $\#A^*$. Clearly the marked GPCP is decidable if we can decide whether a solution exists for each $a_0 \in A$.

Therefore we can restrict to the instances of the form

$$(\#, h, g, (s_1, s_2)) \qquad (s_1 = \varepsilon \text{ or } s_2 = \varepsilon), \tag{2}$$

where the solutions w are required to satisfy $w \in \#(A \setminus \{\#\})^*$ and $h(w)s_1 = g(w)s_2$.

If an instance (2) has $s_1 = s_2 = \varepsilon$, then we have to check whether there is a solution of the marked PCP beginning with the letter $\#$, but this is decidable by the result of [2]. Therefore we shall assume that $s_1 s_2 \neq \varepsilon$. We have

Lemma 1. *The GPCP for marked morphisms is decidable if and only if it decidable for the instances $(\#, h, g, (s_1, s_2))$, where $s_1 = \varepsilon$ or $s_2 = \varepsilon$.*

3 Blocks and successors of instances

A *block* of an instance $I = (h, g)$ of the marked PCP is a pair $(u, v) \in A^+ \times A^+$ such that $h(u) = g(v)$ and for all nonempty prefixes $u_1 \leq u$, $v_1 \leq v$, $h(u_1) = g(v_1)$ implies $u_1 = u$ and $v_1 = v$. If there is no danger of confusion, we will also say that $h(u) = g(v)$ is a block. The words u and v are called *block words* (of h and g respectively). Letter $a \in A$ is a *block letter* if there is a block (u, v) such that $a \leq h(u), g(v)$ (now also $a \leq u$, since we assumed that $h(a)$ begins with a and h is marked). Accordingly, a block is a minimal nontrivial solution of the equation $h(x) = g(y)$.

Lemma 2. *Let (h, g) be an instance of the marked PCP for $h, g\colon A^* \to B^*$. Then for each letter $a \in A$, there exists at most one block (u, v) such that $a \leq u$. In particular, the instance (h, g) has at most $|A|$ blocks. Moreover, the blocks of (h, g) can be effectively found.*

The blocks can be constructed by constructing sequences (u_i, v_i) that always satisfy $h(u_i)s = g(v_i)$ or $h(u_i) = g(v_i)s$. In the first case, if there exists $u' \in A$ such that $h(u_i u')s' = g(v_i)$ or $h(u_i u') = g(v_i)s'$, then $(u_{i+1}, v_{i+1}) = (u_i u', v_i)$. In the second case, if there exists $v' \in A$ such that $h(u_i)s' = g(v_i v')$ or $h(u_i) = g(v_i v')s'$, then $(u_{i+1}, v_{i+1}) = (u_i, v_i v')$.

For two words $s_1, s_2 \in A^*$ with $s_1 = \varepsilon$ or $s_2 = \varepsilon$, a pair (u, v) is called an *end block* (or an (s_1, s_2)-*end block*, to be precise) if $h(u)s_1 = g(v)s_2$ and (u_1, v_1) is not a block for any $u_1 \leq u$ and $v_1 \leq v$. Let

$$E_a = \{(u, v) \mid (u, v) \text{ is an end block and } a \leq h(u) \text{ or } a \leq g(v)\}$$

be the set of all end blocks for the letter a.

Lemma 3. *Let $I = (\#, h, g, (s_1, s_2))$ be an instance of the marked GPCP, $s_1 = \varepsilon$ or $s_2 = \varepsilon$ and a a fixed letter. The set of end blocks E_a is a rational relation and can be effectively found. Moreover,*
(i) If a is a block letter, E_a is finite.
(ii) If E_a is infinite, then it is a union of a finite set and finite number of sets

$$\{(xu^k, yv^k w) \mid k \geq 0\} \text{ and } \{(xu^k w, yv^k) \mid k \geq 0\}$$

for some words u, v, x, y, w.

We shall call $(xu^k, yv^k w)$ and $(xu^k w, yv^k)$ in Lemma 3 (ii) *extendible end blocks.*

Let $I = (\#, h, g, (s_1, s_2))$ be an instance of the marked GPCP. For a solution $w \in A^*$, $h(w)s_1 = g(w)s_2$, of I,

$$w = u_1 u_2 \dots u_{k+1} = v_1 v_2 \dots v_{k+1}$$

is a *block decomposition* for w, if

1. (u_i, v_i) is a block for each $i = 1, 2, \dots, k$,
2. (u_{k+1}, v_{k+1}) is an (s_1, s_2)-end block.

Because the blocks are minimal solutions to $h(u_i) = g(v_i)$, it is easy to see that the following lemma holds.

Lemma 4. *Every solution $w \in A^*$ of I has a unique block decomposition.*

Note that, since the block decomposition of a solution may consist only of an end block, it is necessary to construct also the set $E_\#$.

Let $I = (h, g)$ be an instance of the marked PCP with $h, g \colon A^* \to B^*$ and

$$A' = \{a \in A \mid a \text{ is a block letter}\}. \tag{3}$$

We define the *successor* of I to be $I' = (h', g')$, where the morphisms h' and g' are from $(A')^*$ into A^* such that

$$h'(a) = u \quad \text{and } g'(a) = v, \tag{4}$$

where (u, v) is a block for the letter $a \in A'$.

Lemma 5. *Let $I = (h, g)$ be an instance of the marked PCP and $I' = (h', g')$ be its successor.*
(i) I' is an instance of the marked PCP.
(ii) I has a solution if and only if I' has.
(iii) $hh'(x) = gg'(x)$ for all $x \in (A')^*$.
(iv) If $a \leq h(a)$ for each a, then also $a \leq h'(a)$ for each a.

Proof. This is clear by the construction and by the definition of the blocks. □

The definition of a successor gives inductively a sequence of instances I_i, where $I_0 = I$ and $I_{i+1} = I'_i$. The decidability of the marked PCP in [2] was eventually based on the fact that the successor sequence defined above has only finitely many distinct instances. The authors of [2] used two measures for an instance I of the marked PCP, namely the size of the alphabet and the *suffix complexity*:

$$\sigma(I) = |\bigcup_{a\in A} \{x \mid x \text{ is a proper suffix of } g(a)\}| \\ + |\bigcup_{a\in A} \{x \mid x \text{ is a proper suffix of } h(a)\}|.$$

It is clear that for alphabet sizes of I' and I we have $|A'| \leq |A|$. That $\sigma(I') \leq \sigma(I)$ is not so straightforward. The next lemma can be found in [2].

Lemma 6. *If I is an instance of the marked PCP and I' is its successor then $\sigma(I') \leq \sigma(I)$.*

The previous lemma together with $|A'| \leq |A|$ yields the following result, see [2].

Lemma 7. *Let I be an instance of the marked PCP. Then there exist n_0 and d such that $I_{i+d} = I_i$ for all $i \geq n_0$. The numbers n_0 and d can be effectively found.*

The previous lemma means that after n_0 consecutive successors the instances begin to cycle: $I_{n_0}, \ldots, I_{n_0+d} = I_{n_0}, \ldots$. In this loop the alphabet size and the suffix complexity are constant.

Next we define the successors of the instances $I = (\#, h, g, (s_1, s_2))$ of the marked GPCP. Let (h', g') be the successor of (h, g) and let (u, v) be any end block of I. Then

$$I'(u, v) = (\#, h', g', (s'_1, s'_2))$$

is the *successor* of I w.r.t. (u, v), where (s'_1, s'_2) is defined as follows: If $v \preccurlyeq u$, then $s'_1 = uv^{-1}$ and $s'_2 = \varepsilon$ and if $u \preccurlyeq v$, then $s'_1 = \varepsilon$ and $s'_2 = vu^{-1}$. Otherwise $I'(u, v)$ is not defined. Note that for $(u, v) \in E_\#$ the successor $I'(u, v)$ is defined if and only if $u = v$, since $\#$ is a special symbol not in A. Moreover, $u = v$ is a solution of I.

Lemma 8. *An instance $I = (\#, h, g, (s_1, s_2))$ has a solution if and only if the successor $I'(u, v) = (\#, h', g', (s'_1, s'_2))$ has a solution for some end block (u, v). Moreover, each solution w to I can be written as $w = h'(w')u = g'(w')v$, where w' is a solution of I' and (u, v) an end block of I.*

Proof. Assume first that I has a solution w with the block decomposition

$$w = u_1 u_2 \ldots u_{k+1} = v_1 v_2 \ldots v_{k+1},$$

where (u_i, v_i) is a block for the letter a_i, for $1 \leq i \leq k$, and (u_{k+1}, v_{k+1}) is an end block. Clearly $u_{k+1} \preccurlyeq v_{k+1}$ or $v_{k+1} \preccurlyeq u_{k+1}$. If the first case holds, then $s'_1 = \varepsilon$, $s'_2 = v_{k+1}u_{k+1}^{-1}$ and $I'(u, v) = (\#, h', g', (s'_1, s'_2))$. Now

$$h'(a_1 \ldots a_k) = wu_{k+1}^{-1} = g'(a_1 \ldots a_k)s'_2,$$

i.e. $I'(u,v)$ has a solution $w' = a_1 \dots a_k$ and $w = h'(w')u_{k+1} = g'(w')v_{k+1}$. Case $v_{k+1} \preccurlyeq u_{k+1}$ is symmetric.

Assume then that $I'(u,v) = (\#, h', g', (s_1', s_2'))$ has a solution w'. Then also $h'(w')u = g'(w')v$ and by the definition of the end blocks and Lemma 5,

$$h(h'(w')u)s_1 = h(h'(w'))h(u)s_1 = g(g'(w'))g(v)s_2 = g(g'(w')v)s_2,$$

so $h'(w')u = g'(w')v$ is a solution of I. □

For $(u,v) \in E_\#$ in the previous theorem $w' = \varepsilon$. Clearly it is of no use to construct the successors of I for $(u,v) \in E_\#$ with $u = v$, since we already found a solution for I. We shall return to this in Lemma 11.

4 Reducing the extendible end blocks

By Lemma 8 we can reduce an instance I to its successors for all end blocks. The problem in this approach is that by Lemma 3, I potentially has infinitely many successors. We shall next show that also the extendible end blocks reduce to a finite number of successors.

Let $I = (\#, h, g, (s_1, s_2))$ and that assume that we have the successors

$$I'(xu^k w, yv^k) = I_k' \text{ for } k \geq 0 \tag{5}$$

with morphisms $h', g' \colon (A')^* \to A^*$ as defined for the successor of (h,g). Successors $I'(xu^k, yv^k w)$ are treated analogously.

Lemma 9. *Let I_k', for $k \geq 0$, be as in the above and $|u| = |v|$. Then there are only finitely many distinct successors I_k' and they can be effectively found.*

Proof. A successor with respect to $(xu^k w, yv^k)$ is defined only if either $yv^k \preccurlyeq xu^k w$ or $xu^k w \preccurlyeq yv^k$. In the first case, $s_1' = xu^k w (yv^k)^{-1}$ and $s_2' = \varepsilon$. Let ℓ be the least number such that $|yv^\ell| > |w|$ and $yv^\ell \preccurlyeq xu^\ell w$. If there is no such ℓ, then there is only finitely many possible k such that $yv^k \preccurlyeq xu^k w$. If $yv^k \preccurlyeq xu^k w$ for some $k > \ell$, then however $s_1' = xu^k w (yv^k)^{-1} = xu^\ell w (yv^\ell)^{-1}$ and $s_2' = \varepsilon$. The second case is similar. □

Lemma 10. *Assume that the successors I_k' are as in (5) and that $|u| \neq |v|$. Then there are only finitely many k such that I_k' can have a solution and these numbers k can be effectively found.*

Proof. Let $b, c \in A$ and assume that $b \leq xu^k$ and $c \leq yv^k$. Notice first that b (resp. c) is not a first letter of any image of h' (resp. g'), since it is a first letter of an extendible end block. To prove the claim it is sufficient to show that for all $f \in A'$ there are only finitely many k such that I_k' has an end block for f.

The end blocks can be effectively constructed for all f by constructing the same sequence $(u_i, v_i) \in (A')^* \times (A')^*$ as for blocks. If there is an end block for some k, then, for some i, necessarily

$$h'(u_i) = g'(v_i)z_i \quad \text{or} \quad g'(v_i) = h'(u_i)z_i,$$

where $c \leq z_i$ or $b \leq z_i$, respectively. If no such z_i exists before the sequence terminates, then there is no end block for f. Assume the first case, i.e. $c \leq z_i$, the other case is symmetric. Then the possible end block is of the form $h'(u_i u')xu^k w = g'(v_i)yv^k$ for some $u' \in A'$. This u' can be effectively found if it exists, by defining a sequence, which adds letters to u' and check whether $(h'(u_i u'))^{-1} g'(v_i)yv^k$ begins with b. This process is finite, since yv has only finitely many suffixes.

Once we have found the unique possible u', we can check by the lengths whether for some k $|h'(u_i u')xu^k w| = |g'(v_i)yv^k|$. Since $|u| \neq |v|$ such a k is unique. This proves the claim. □

Let

$$S' = \{(s_1', s_2') \mid (\#, h', g', (s_1', s_2')) \text{ is a successor of } I\}$$

be the set of the end words of the successor of I. We introduce a collective notation

$$\mathcal{I}' = (\#, h', g', S') \tag{6}$$

to stand for all the successors of I. By Lemma 8, I has a solution if and only if one of the successors (6) has, but by Lemmata 9 and 10, we can always assume that S' is a finite set. Thus we obtain a chain of sets of successors reducing the original instance $I_0 = (\#, h_0, g_0, (s_1, s_2))$ to its successors $\mathcal{I}_1 = (\#, h_1, g_1, S_1)$, then all these to get $\mathcal{I}_2 = (\#, h_2, g_2, S_2)$, etc. If eventually some successors $\mathcal{I}_i = (\#, h_i, g_i, S_i)$ have very simple morphisms (h_i, g_i) (i.e. $|A_i| = 1$ or the suffix complexity is zero), we can decide for each $(\#, h_i, g_i, S_i)$ if there is a solution or not. Thus we could also solve the original problem. Otherwise, we know by Lemma 7 that there is a number n_0 such that $(h_{i+d}, g_{i+d}) = (h_i, g_i)$ for each $i \geq n_0$, i.e. the morphisms start so cycle. Clearly to decide the marked GPCP it suffices to show how to solve these cycling instances. This is done in the next section.

In our solution we do not construct the successors for the end blocks in $E_\#$, but check at each step whether there is $(u, v) \in E_\#$ such that $u = v$.

Lemma 11. *We can effectively check whether there is an end block $(u, v) \in E_\#$ such that $u = v$.*

By a *successor sequence* we mean a sequence

$$(\#, h_0, g_0, (s_1^{(0)}, s_2^{(0)})), \ldots, (\#, h_i, g_i, (s_1^{(i)}, s_2^{(i)})), \ldots$$

of instances of the marked GPCP such that each $(\#, h_{i+1}, g_{i+1}, (s_1^{(i+1)}, s_2^{(i+1)}))$ is a successor of $(\#, h_i, g_i, (s_1^{(i)}, s_2^{(i)}))$.

To end this section, notice that if $\mathcal{I}_i = (\#, h_i, g_i, S_i)$ is the set of all ith members in the successor sequences, we can always assume that

(A) morphism pair (h_i, g_i) has a block for the letter # and
(B) $s_1 s_2 \neq \varepsilon$ for each $(s_1, s_2) \in S_i$.

For if the condition (A) does not hold, we know that no instance in $\mathcal{I}_i$ can have a solution beginning with # and if (B) is not satisfied by an instance, then that instance reduces to the marked PCP, which is decidable by [2].

5 Cycling instances

In this section we show how to treat the instances that begin to cycle, i.e. there exists a d such that for all successor sequences

$$(\#, h_0, g_0, (s_1^{(0)}, s_2^{(0)})), \ldots, (\#, h_i, g_i, (s_1^{(i)}, s_2^{(i)})), \ldots \tag{7}$$

$(h_i, g_i) = (h_{i+d}, g_{i+d})$ for all $i \geq 0$. We shall call such an instance I_0 a *loop instance*, and d is the *length of the loop*. Next lemma follows from [2].

Lemma 12. *The sequence $I_i = (h_i, g_i)$ has the following properties.*
(i) *The size of the alphabet is constant and $\sigma(I_i) = \sigma(I_0)$ for all $i \geq 0$.*
(ii) *An instance (h_0, g_0) of the marked PCP has a solution beginning with a if and only if $h_i(a) = a = g_i(a)$ for all $i \geq 0$.*

Notice that because the alphabet size does not decrease, there is a block for each letter $a \in A$. In particular, there cannot be extendible end blocks.

Corollary 1. *Assume that the instances cycle as in* (7) *and that a solution exists. Then we have two cases:*
(i) *If $h_0(\#) = \# = g_0(\#)$ then the minimal solution of I_0 is $\#w$, where w the initial letter a of w satisfies $h_0(a) \neq g_0(a)$. Hence also $h_i(a) \neq g_i(a)$ for all $i \geq 0$.*
(ii) *If $h_0(\#) \neq g_0(\#)$, then the minimal solution beginning with # does not have a solution of the PCP as a prefix.*

Hereafter we will assume that $h_0(\#) \neq g_0(\#)$, since the case (i) reduces to the following problem for each a such that $h_0(a) \neq g_0(a)$: Does the cycling instance I_0 have a solution beginning with a? But this is essentially the original problem, # replaced with a.

We would like to have some upper bound for the lengths of the new end blocks in the loop (7). We demonstrate that if a solution exists, there is a limit number L such that, if a solution exists, then the minimal solution is found in some sequence (7) shorter than L. Moreover, this limit can be effectively found, hence the main result follows.

In what follows, we assume that $I = (\#, h, g, (s_1, s_2))$ has a minimal solution beginning with # and which does not have a solution of the PCP as prefix. Then this minimal solution is unique, since the morphisms are marked. Consequently each I has a unique end block (u, v) in the block decomposition of the minimal solution. It follows that there exists a unique successor sequence $I_0, I_1, \ldots$ of instances such that

$$I_{i+1} = I_i(u_i, v_i). \tag{8}$$

and (u_i, v_i) is the end block of the minimal solution (beginning with $\#$) of I_i. This successor sequence is called ***the branch of minimal solutions.*** Note that we cannot determine, which is the end block of the minimal solution, but the desired limit will be obtained anyway.

Let $I_i = (\#, h_i, g_i, (s_1^{(i)}, s_2^{(i)}))$ be an instace of the branch of the minimal solutions and w_i the minimal solution of I_i. Recall that we permanently assume that $s_1^{(i)} s_2^{(i)} \neq \varepsilon$ and also that $g_0(\#) \neq h_0(\#)$, which implies that also $g_i(\#) \neq h_i(\#)$ for each i.

Lemma 13. *Let w_i be the minimal solution of I_i beginning with $\#$ and $(h_i, g_i) = (h_{i+d}, g_{i+d})$ for each i. Then $w_{i+d} \leq w_i$ but $w_{i+d} \neq w_i$ for each i.*

Proof. The instances

$$I_d = (\#, h_i, g_i, (s_1^{(i)}, s_2^{(i)})) \text{ and } I_{i+d} = (\#, h_i, g_i, (s_1^{(i+d)}, s_2^{(i+d)}))$$

share the marked morphisms, so clearly $w_i \leq w_{i+d}$ or $w_{i+d} \leq w_i$, since the minimal solutions cannot have a solution of PCP as a prefix (recall that $h_i(\#) \neq g_i(\#)$). If w is a minimal solution to some instance I, then by Lemma 8, there is a solution w' to the successor of I such that $w = h'(w')u = g'(w')v$. Notice that w and w' begin with the same letter. Since $s_1^{(i)} s_2^{(i)} \neq \varepsilon$, then also $uv \neq \varepsilon$ and consequently $|w| > |w'|$, because the morphisms are nonerasing. Hence $|w_{i+1}| + 1 \leq |w_i|$ Inductively $|w_{i+t}| + t \leq |w_i|$ for all t. This proves the claim. □

As a byproduct we obtain

Lemma 14. *If an instance occurs twice in a successor sequence, it has no solutions.*

Proof. By the proof of the previous lemma, the length of the minimal solution decreases strictly. □

An end block (u, v) of an instance $I = (\#, h, g, (s, \varepsilon))$ satisfies the equation

$$h(u)s = g(v).$$

If this is an end block of a solution, then necessarily $u = s'v$ or $v = s'u$ for some word s', and I', the successor of I has the end words (s', ε) or (ε, s'), respectively.

Lemma 15. *Let $I_i = (\#, h_i, g_i, (s_1^{(i)}, s_2^{(i)}))$ be the branch of the minimal solutions of a cycling instance with loop length d. Let also w_i be the minimal solution of I_i. Then $h_i(w_{i+d})s_1^{(i+d)} = g_i(w_{i+d})s_2^{(i+d)}$ is a prefix of $h_i(w_i)$ and $g_i(w_i)$.*

Proof. It is sufficient to prove the claim for $i = 0$. By Lemma 13, $w_d \leq w_0$, and so $h_0(w_d) \leq h_0(w_0)$ and $g_0(w_d) \leq g_0(w_0)$. We can prove that $|h_0(w_d)s_1^{(d)}| \leq |h_0(w_0)|$ and $|g_0(w_d)s_2^{(d)}| \leq |g_0(w_0)|$.

Assume by symmetry that $s_1^{(d)} \neq \varepsilon$. Then $g_0(w_d) = h_0(w_d)s_1^{(d)} \leq g_0(w_0)$, and, since $h_0(w_0)$ and $g_0(w_0)$ are comparable, the claim follows. Similarly we can prove that $|g_0(w_d)s_2^{(d)}| \leq |g_0(w_0)|$. □

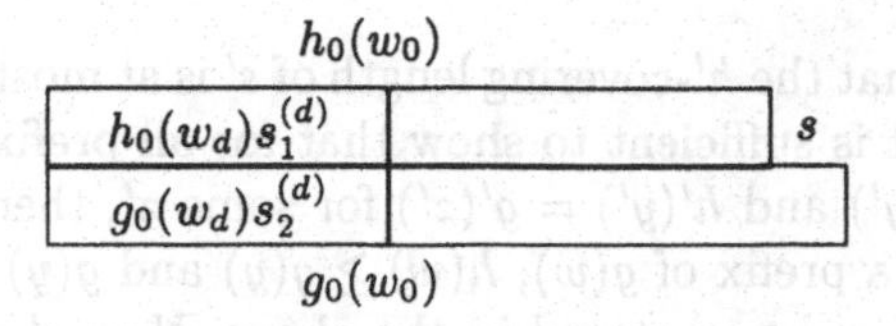

Figure1. Prefix property

The previous lemma is used in the proof our last lemma, which gives an upper bound for the size of the end blocks in the branch of the minimal solutions.

For an occurrence of a word u in $g(w)$, its *g-block covering* in a solution w of an instance $(\#, h, g, (s_1, s_2))$ is a word $z = g(v_1)g(v_2)\dots g(v_k)$ such that

1. $v_1 v_2 \dots v_k$ is a factor of w,
2. u is a factor of z,
3. u is not a factor of $g(v_2)\dots g(v_k)$ or $g(v_1)\dots g(v_{k-1})$,
4. for each i, $g(v_i) = h(u_i)$ is a block for morphism pair (h, g).

Note that a g-block covering for an occurrence of a factor u (in $g(w)$) is unique. Hence we can define the integer k to be the *g-covering length* of the occurrence of u (in w).

Lemma 16. *Let $I_i = (\#, h_i, g_i, (s_1^{(i)}, s_2^{(i)}))$ be the branch of minimal solutions of a cycling instance having loop length d. For all $i \geq d$*

(i) If $s_1^{(i)} \neq \varepsilon$, then the h_{i+1}-covering lengths of $s_1^{(i+1)}$ and $s_2^{(i+1)}$ are at most the g_i-covering length of $s_1^{(i)}$.

(ii) If $s_2^{(i)} \neq \varepsilon$ then the g_{i+1}-covering lengths of $s_1^{(i+1)}$ and $s_2^{(i+1)}$ are at most the h_i-covering length of $s_2^{(i)}$

Proof. We denote $I_i = I = (\#, h, g, (s, \varepsilon))$ We shall consider the proof of the case (i), the proof for the other case is similar. Assume that for the end block of the minimal solution w of I is (u, v). Then $h(u)s = g(v)$ and $u \preccurlyeq v$ or $u \preccurlyeq v$. We have two cases to consider.

(1) If $u = s'v$, then $|s'| \leq |u|$ and $s' \leq h'(a)$ for some letter a i.e. the h'-covering length is 1.

(2) Suppose $v = s'u$. Recall that for all $x \in A^*$, $hh'(x) = gg'(x)$, and therefore this is a catenation $|x|$ blocks in $I = (\#, h, g, (s, \varepsilon))$.

Consider the minimal solution w' of the successor $I' = (\#, h', g', (\varepsilon, s'))$, which is obtained from I by the end part $h(u)s = g(v)$, where $g(v) = g(s')g(u)$.

Then by Lemma 8, the (unique) corresponding solution of I equals wp, where $w = g'(w') = h'(w')s'$ is the solution for the instance $(\#, h, g, (\varepsilon, g(s')))$.

By Lemma 15, we can assume that the end words s and s' are covered by h and h', respectively. That is, s is a prefix of $h(x)$ and s' is a prefix of $g'(x')$ for some words x and x'.

It is clear that the g-covering length of the word $g(s')$ is at most that of the word s.

We show then that the h'-covering length of s' is at most the g-covering length of $g(s')$. For this, it is sufficient to show that for all prefixes $g'(y')$ of $g(w')$, for which $g'(w') \geq h'(y')$ and $h'(y') = g'(z')$ for some z', there corresponds a word y such that $g(y)$ is a prefix of $g(w)$, $h(w) \geq g(y)$ and $g(y) = h(z)$ for some z.

Let then y' and z' be as stated in the above. Now $z' = w'x'$ for some word x', where $x' \leq s'$. We have $g(g'(z')) = g(g'(w')g'(x')) \leq g(g'(w')s') = g(w)$. Also, $g(g'(z')) = h(h'(z')) = h(h'(w'))h(h'(x)) = h(w)h(h'(x))$, and therefore $h(w) \leq g(g'(z'))$. Since $g(g'(z')) = h(h'(z))$, the word $y = h'(z)$ satisfies the requirement, see Figure 2.

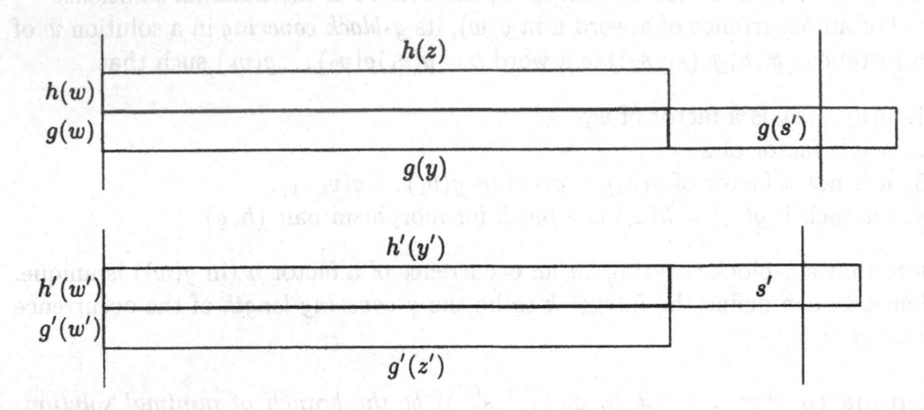

Figure2. Relation between $g(s')$ and s'

So we have proved that the h'-covering length of s' is at most the g-covering length of s. □

The previous lemma gives us a tool for recognizing instances which *are not* in the branch of minimal solutions. Let I_0 be a cycling instance with loop length d and consider *all* the instances $\mathcal{I}_d$ found by the first d reductions. If I_0 has a solution then there is a unique $I \in \mathcal{I}_d$ in the branch of the minimal solutions.

Let M be the maximal g- or h-covering length of all the end words s_1 and s_2 in $\mathcal{I}_d$. It now follows by Lemma 16 that in the branch of the minimal solutions the g_i or h_i-covering length is always less than or equal to M. For a sequence of cycling instances, the suffix complexity is constant $\sigma(I_0)$ and since the blocks of an instance I_i are the images of the successor I_{i+1}, the block length can never be more than $\sigma(I_0) + 1$. By the previous lemma we have

Corollary 2. *Let $I_0, \ldots, I_i, \ldots$ be the branch of the minimal solutions of a cycling instance with loop length d. For each $i \geq d$, the end words of I_i are not longer $M(\sigma(I) + 1)$.*

Now we are ready to prove the Theorem 1.

Proof of Theorem 1. It remains to be shown how to solve the marked GPCP for the cycling instances I_0. A cycling instance has the blocks for all the letters. In particular, there are no extendible end blocks and only finitely many successors. The successor relation naturally defines a tree $\mathcal{T}$ having I_0 as the root, all the successors of I_0 as the vertices and the pairs (I, I') as the edges.

The decision procedure is based on constructing $\mathcal{T}$ partially by first inserting the vertices having depth (the distance from the root) at most d and then computing the number M, the maximal covering length of the end words of instances at the depth d. For all vertices we check whether there is an end block $(u, v) \in E_\#$ such that $u = v$ as in Lemma 11. And for all vertices $I = (\#, h, g, (s_1, s_2))$ that have $s_1 s_2 = \varepsilon$, we can always decide if they have a solution or not. If some such vertex I has no solution, then I and all the successors of I can be removed. On the other hand, if some such I has a solution, then I_0 also has a solution and the procedure may stop.

For the vertices having depth greater than d, the (partial) construction of $\mathcal{T}$ is more specific: Only the successors $I = (\#, h, g, (s_1, s_2))$ that satisfy $|s_1 s_2| \leq M(\sigma(I_0) + 1)$ are inserted. By Corollary 2, the branch of minimal solutions is included in the partial construction.

But now there are only finitely many instances to be inserted, so each path (successor sequence) in the partially constructed $\mathcal{T}$ will eventually contain an instance twice, thus I_0 has no solution by Lemma 14, unless some vertex $I = (\#, h, g, (\varepsilon, \varepsilon))$ has or at some vertex $(u, u) \in E_\#$. □

In [1] Ehrenfeucht, Karhumäki and Rozenberg proved that the binary PCP, i.e. for the instances (h, g), where $h, g\colon A^* \to B^*$ and $|A| = 2$, is decidable if and only if the binary GPCP is decidable for marked morphisms. Therefore Theorem 1 has the following corollary.

Corollary 3. *The binary PCP is decidable.*

References

1. A. Ehrenfeucht, J. Karhumäki, and G. Rozenberg. *The (generalized) Post correspondence problem with lists consisting of two words is decidable.* Theoretical Computer Science, 21(2):119–144, 1982.
2. V. Halava, M. Hirvensalo and R. de Wolf. *Decidability and Undecidability of Marked PCP.* STACS'99 (C. Meinel and S. Tison, eds.), Lecture Notes in Comput. Sci, vol 1563, Springer-Verlag, 1999, pp. 207-216.
3. T. Harju and J. Karhumäki. *Morphisms.* In Handbook of Formal Languages, volume 1, 439–510. edited by G. Rozenberg and A. Salomaa, eds. Springer-Verlag, Berlin, 1997.
4. Y. Matiyasevich and G. Sénizergues. *Decision problems for semi-Thue systems with a few rules.* In Proceedings of the 11th IEEE Symposium on Logic in Computer Science, pages 523–531, 1996.
5. E. L. Post. *A variant of a recursively unsolvable problem.* Bulletin of the American Mathematical Society, 52:264–268, 1946.

On Time-Constructible Functions in One-Dimensional Cellular Automata

Chuzo Iwamoto, Tomonobu Hatsuyama, Kenichi Morita, and Katsunobu Imai

Hiroshima University, Higashi-Hiroshima, 739-8527 Japan
iwamoto@ke.sys.hiroshima-u.ac.jp

Abstract. In this paper, we investigate time-constructible functions in one-dimensional cellular automata (CA). It is shown that (i) if a function $t(n)$ is computable by an $O(t(n)-n)$-time Turing machine, then $t(n)$ is time-constructible by CA and (ii) if two functions are time-constructible by CA, then the sum, product, and exponential functions of them are time-constructible by CA. As an example for which time-constructible functions are required, we present a time-hierarchy theorem based on CA. It is shown that if $t_1(n)$ and $t_2(n)$ are time-constructible functions such that $\lim_{n\to\infty} \frac{t_1(n)}{t_2(n)} = 0$, then there is a language which can be recognized by a CA in $t_2(n)$ time but not by any CA in $t_1(n)$ time.

1 Introduction

One of the simplest models of parallel computation is cellular automata (CA). A CA is a one-dimensional array of identical finite-state automata, called *cells*, which are uniformly interconnected. Every cell operates synchronously at discrete time steps and changes its state depending on the previous states of itself and its neighbors.

Various algorithms have been designed on CA: For example, Cole [5] presented real-time recognition algorithms for concrete languages, including palindromes and the set of strings of the form ww. Korec [11,12] designed real-time algorithms for generating primes. Mazoyer and Terrier [14] considered signals and investigated constructibility of functions, such as i^k, k^i, $ki+\lfloor i^{1/2}\rfloor$, $i+\lfloor\log i\rfloor$, and $i!$, where $k\geq 0$ is an integer and $i=0,1,2,\cdots$. Also, Buchholz and Kutrib [2] investigated constructibility of functions in one-way CA, in which information is allowed to move in one direction. (More information on CA-algorithms may be found in [1,3,6,9,10,13,17–20].)

In this paper, we continue the study of constructible functions, started by Mazoyer and Terrier [14]. We show that if a function $t(n)$ is computable by an $O(t(n)-n)$-time Turing machine (TM), then $t(n)$ is constructible by CA. Since the set of computable functions by TMs is known to be very rich [8], most common functions are constructible by CA. The set of CA-constructible functions includes cn^r, $n+\log^r n$, $n\log^s n$, $n(\log\log n)^s$, etc., where $c\geq 1$, $r\geq 1$, and $s\geq 0$ are rational constants. Furthermore, we show that if functions $f(n)\geq$

n and $g(n) \geq n$ are constructible by CA, then $f(n) + g(n) - n \geq n$, $\lfloor f(n) \cdot g(n)/n \rfloor \geq n$, $g(n)^{f(n)}$, etc. are constructible by CA. (Of course, $f(n)+g(n) \geq 2n$ and $f(n) \cdot g(n) \geq n^2$ are also constructible.)

As an example for which constructible functions are required, we present a time-hierarchy theorem based on CA. It is shown that if $t_1(n)$ and $t_2(n)$ are constructible functions such that $\lim_{n\to\infty} \frac{t_1(n)}{t_2(n)} = 0$, then there is a language which can be recognized by a CA in $t_2(n)$ time but not by any CA in $t_1(n)$ time. Therefore, a slight increase in the growth rate of a time-function yields a new CA-based complexity class.

The first general investigation of constructibility of functions was given in [14]. The model in [14] is the so-called impulse cellular automaton, which is a semi-infinite array (with left boundary) of cells such that, at initial time, all cells are in the quiescent state except the leftmost cell is in a distinguished state. On this model, Mazoyer and Terrier presented signals which reach the leftmost cell at time $i^k, k^i, ki + \lfloor i^{1/2} \rfloor, i + \lfloor \log i \rfloor, i!$, etc., where $k \geq 0$ is an integer and $i = 0, 1, 2, \cdots$. Our model is essentially the same model as [14]; however, in order to present a hierarchy of languages recognized by CA, our model is defined as a string acceptor. In our model, the input string $a_1 a_2 \cdots a_n$ is fed serially to the leftmost cell. We consider constructibility of functions $t(n)$ in the sense that the leftmost cell falls into an accepting state at time $t(n)$.

A bounded CA, which is a finite array (delimited by special cell # at both ends) of n cells, is another common model for cellular acceptors. All results in this paper concerning constructible functions hold even if the model is a bounded CA. Buchholz and Kutrib [2] used a bounded CA and investigated constructibility of functions under the condition that information is allowed to move in one direction (called one-way CA, OCA). They showed that the set of OCA-constructible functions includes k, kn, $kn + \lfloor n^{1/2} \rfloor$, $n + \lfloor \log n \rfloor$, and $2n + \lfloor \log \log n \rfloor$, where k is an integer. They also showed that for any integer k, there is an OCA-constructible function in $\Theta(n^k)$. However, whether n^k $(k \geq 2)$ is OCA-constructible remains an open problem.

For the time-complexity class of n-time OCA, several separation results have been known. It was shown that one-letter languages $\{0^p \mid p \text{ is prime}\}$ [7] and $\{0^{2^i} \mid i \text{ is integer}\}$ [4] can be recognized by $O(n)$-time OCA but not by any n-time OCA. Terrier [16] showed that the class of languages accepted by n-time OCA is not closed under concatenation; in the proof, a language recognized by $O(n)$-time OCA but not by n-time OCA is presented.

In the following section, we give the definition of CA. Main theorems are summarized in Section 3. The proofs are given in Sections 4, 5, and 6.

2 Cellular Automata

A cellular automaton (CA) is a synchronous highly parallel string acceptor, consisting of a one-dimensional semi-infinite array (with left boundary) of identical finite-state automata, called *cells*, which are uniformly interconnected (Fig. 1).

A cellular automaton M is a 6-tuple $M = (Q, \Sigma, \#, \delta, q, Q_A)$, where

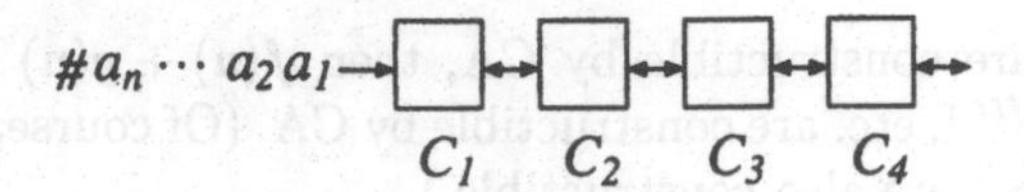

Fig. 1. Cellular automaton

(1) Q is the finite nonempty set of cell states,
(2) Σ is the finite input alphabet,
(3) # is the special boundary symbol not in Σ,
(4) $\delta : Q \cup \Sigma \cup \{\#\} \times Q \times Q \to Q$ is the local transition function,
(5) q is the quiescent state such that $\delta(q, q, q) = q$,
(6) Q_A is the accepting subset of Q.

C_i denotes the cell assigned to the integer $i \geq 1$. At step $t = 0$, the state of each cell is the quiescent state. The input string $a_1 a_2 \cdots a_n$, where $a_i \in \Sigma$, is fed serially to the leftmost cell C_1. The symbol a_i, $1 \leq i \leq n$, is received by the cell C_1 at step $i - 1$. After step $n - 1$, it receives the boundary symbol #. The leftmost cell C_1 is called the *accepting cell*. A parallel time-complexity measure $t(n)$ is introduced as the number of steps used to make the accepting cell fall into an accepting state on an input of length n. A function $t(n)$ is said to be *constructible* if, for each n, there is a CA whose accepting cell falls into an accepting state at step $t(n)$ on all inputs of length n.

3 Main Results

It is known that the set of functions computable by TMs is very rich. Thus, we first show the relation between TM-computability and CA-constructibility. Let $bin(n)$ denote the binary representation of the value n.

Theorem 1. *Suppose that $t(n)$ is an arbitrary function such that there is a TM which, given a string $bin(n)$ of length $\lceil \log n \rceil$, generates $bin(t(n))$ in time $O(t(n) - n)$. Then, the function $t(n)$ is constructible by CA.*

The proof of Theorem 1 is given in Section 4. This theorem implies that the set of functions constructible by CA includes functions computable by polynomial-time TMs.

Corollary 1. *Suppose that $t(n)$ is an arbitrary function such that (i) $t(n) \geq cn$ for some constant $c > 1$ and (ii) there is a TM which, given a string $bin(n)$ of length $\lceil \log n \rceil$, generates $bin(t(n))$ in time polynomial in $\log n$. Then, the function $t(n)$ is constructible by CA.*

The set of constructible functions includes cn^r, $n + \log^r n$, $n \log^s n$, $n(\log \log n)^s$, etc., where $c \geq 1$, $r \geq 1$, and $s \geq 0$ are rational constants. Furthermore, we show that if two functions are constructible by CA, then the sum, product, and exponential functions of them are constructible.

Theorem 2. *(i) If $n+f(n)$ and $n+g(n)$ are constructible, then $n+f(n)+g(n)$ and $n+f(n)\cdot g(n)$ are constructible. (ii) If $n\cdot f(n)$ and $n\cdot g(n)$ are constructible functions not bounded by $(1+o(1))n$, then $\lfloor n\cdot f(n)\cdot g(n)\rfloor$ is constructible. (iii) If $f(n)$ and $g(n)$ are constructible and k is an integer, then $k^{f(n)}$, $g(n)^k$, and $g(n)^{f(n)}$ are constructible.*

Remark 1. It is possible to show that *if $f(n)\geq n$ and $g(n)\geq n$ are constructible, then $f(n)+g(n)$ ($\geq 2n$) and $f(n)\cdot g(n)$ ($\geq n^2$) are constructible.* Constructible functions in (i) and (ii) of Theorem 2 include functions below $2n$ and n^2, respectively.

We prove Theorem 2 in Section 5. A typical example for which constructible functions are required is a time-hierarchy theorem. We show that a slight increase in the growth rate of a time-function yields a new CA-based complexity class.

Theorem 3. *Suppose that $t_1(n)$ and $t_2(n)$ are constructible functions such that $\lim_{n\to\infty}\frac{t_1(n)}{t_2(n)}=0$. Then, there is a language which can be recognized by a CA in $t_2(n)$ time but not by any CA in $t_1(n)$ time.*

The proof of Theorem 3 is given in Section 6.

4 Time Constructibilities of Functions

4.1 Data Structures

In this section, we investigate constructibility of functions in cellular automata. Our strategy is not based on signals. We compute the value of a function in binary, and terminate the machine at the time specified by the value. We use the following data structure.

Each cell of CA is divided into tracks. The states of the first and second tracks in the ith cell are denoted by c_i and b_i, respectively, where $c_i\in\{-1,0,1\}$ and $b_i\in\{0,1\}$. (If the ith cell is in the quiescent state, c_i and b_i are regarded as 0.) We represent a single value v using the first and second tracks. Configurations, say, $B=b_1b_2\cdots b_i\cdots$ and $C=c_1c_2\cdots c_i\cdots$, represent value v if

$$v=\sum_{i=1}^{\infty}b_i2^{i-1}+\sum_{i=1}^{\infty}c_i2^i.$$

Intuitively, each c_i plays a role as a carry bit. For example,

$$\begin{aligned}C&=0\,1\,0\,1\,0\,0\,0\,0\cdots\\ B&=1\,0\,1\,1\,0\,0\,0\,0\cdots\end{aligned}\tag{1}$$

represent value 33. The pair of tracks is called the *counter*. The counter is said to be *stable* if $c_i=0$ for all $i\geq 1$. The counter changes its configuration towards a stable one. That is, transition rules for the first and second tracks are as follows:

$$\text{(a) } \delta\begin{pmatrix} \$ * * \\ \$ 1 * \end{pmatrix} \to \begin{pmatrix} 0 \\ 1 \end{pmatrix}, \quad \text{(b) } \delta\begin{pmatrix} 1 * * \\ * 1 * \end{pmatrix} \to \begin{pmatrix} 1 \\ 0 \end{pmatrix}, \quad \text{(c) } \delta\begin{pmatrix} 0 * * \\ * 1 * \end{pmatrix} \to \begin{pmatrix} 0 \\ 1 \end{pmatrix},$$

$$\text{(d) } \delta\begin{pmatrix} \$ * * \\ \$ 0 * \end{pmatrix} \to \begin{pmatrix} 0 \\ 0 \end{pmatrix}, \quad \text{(e) } \delta\begin{pmatrix} 1 * * \\ * 0 * \end{pmatrix} \to \begin{pmatrix} 0 \\ 1 \end{pmatrix}, \quad \text{(f) } \delta\begin{pmatrix} 0 * * \\ * 0 * \end{pmatrix} \to \begin{pmatrix} 0 \\ 0 \end{pmatrix}.$$

Here, $\$ \in \Sigma \cup \{\#\}$ and $* \in \{0, 1\}$. For example, the above configuration (1) changes towards a stable one as follows:

$$\begin{matrix} \$\,010100 \\ \$\,101100 \end{matrix} \to \begin{matrix} \$\,001000 \\ \$\,100110 \end{matrix} \to \begin{matrix} \$\,000100 \\ \$\,100010 \end{matrix} \to \begin{matrix} \$\,000010 \\ \$\,100000 \end{matrix} \to \begin{matrix} \$\,000000 \\ \$\,100001 \end{matrix}$$

Each of the five pairs represents the same value 33.

4.2 Storing Value n into a Counter in n Steps

Lemma 1. *We can store the value n of the input length into the counter in n steps.*

Proof. Let M be a CA. Each cell of M is divided into three tracks. The first and second tracks are the counter. The third track is used for indicating the right-most 1 in the counter (which we need in Lemma 2). M changes its configuration according to the following rules:

$$\text{(a') } \delta\begin{pmatrix} \# * * \\ \# 1 * \end{pmatrix} \to \begin{pmatrix} 0 \\ 1 \end{pmatrix}, \quad \text{(b) } \delta\begin{pmatrix} 1 * * \\ * 1 * \end{pmatrix} \to \begin{pmatrix} 1 \\ 0 \end{pmatrix}, \quad \text{(c) } \delta\begin{pmatrix} 0 * * \\ * 1 * \end{pmatrix} \to \begin{pmatrix} 0 \\ 1 \end{pmatrix},$$

$$\text{(d') } \delta\begin{pmatrix} \# * * \\ \# 0 * \end{pmatrix} \to \begin{pmatrix} 0 \\ 0 \end{pmatrix}, \quad \text{(e) } \delta\begin{pmatrix} 1 * * \\ * 0 * \end{pmatrix} \to \begin{pmatrix} 0 \\ 1 \end{pmatrix}, \quad \text{(f) } \delta\begin{pmatrix} 0 * * \\ * 0 * \end{pmatrix} \to \begin{pmatrix} 0 \\ 0 \end{pmatrix},$$

$$\text{(g) } \delta\begin{pmatrix} a * * \\ a 1 * \end{pmatrix} \to \begin{pmatrix} 1 \\ 0 \end{pmatrix}, \quad \text{(h) } \delta\begin{pmatrix} a * * \\ a 0 * \end{pmatrix} \to \begin{pmatrix} 0 \\ 1 \end{pmatrix}.$$

Here, $a \in \Sigma$ and $* \in \{0, 1\}$. Note that rules (a) and (d) are replaced by (a'), (d'), (g), and (h).

We give an example for an input string $a_1 a_2 \cdots a_6 \in \Sigma^*$. At step 6, the value in the counter becomes 6. The configuration of the counter changes until it becomes a stable one (but the value 6 does not change).

$$\begin{matrix} a_1 000000 \cdots \\ a_1 000000 \cdots \end{matrix} \to \begin{matrix} a_2 000000 \cdots \\ a_2 100000 \cdots \end{matrix} \to \begin{matrix} a_3 100000 \cdots \\ a_3 000000 \cdots \end{matrix} \to \begin{matrix} a_4 000000 \cdots \\ a_4 110000 \cdots \end{matrix}$$

$$\to \begin{matrix} a_5 100000 \cdots \\ a_5 010000 \cdots \end{matrix} \to \begin{matrix} a_6 010000 \cdots \\ a_6 100000 \cdots \end{matrix} \to \begin{matrix} \#100000 \cdots \\ \#001000 \cdots \end{matrix} \to \begin{matrix} \#000000 \cdots \\ \#011000 \cdots \end{matrix}$$

■

Lemma 2. *Suppose the counter contains the value v. We can decrease the value in the counter one by one from v to 0 in v steps.*

Proof. We replace rules (a),(d) in Section 4.1 by (a"),(d"), and add rules (i),(j).

$$\text{(a")}\ \delta\begin{pmatrix}\$**\\ \$1*\end{pmatrix}\rightarrow\begin{pmatrix}0\\0\end{pmatrix},\quad \text{(b)}\ \delta\begin{pmatrix}1**\\ *1*\end{pmatrix}\rightarrow\begin{pmatrix}1\\0\end{pmatrix},\quad \text{(c)}\ \delta\begin{pmatrix}0**\\ *1*\end{pmatrix}\rightarrow\begin{pmatrix}0\\1\end{pmatrix},$$

$$\text{(d")}\ \delta\begin{pmatrix}\$**\\ \$0*\end{pmatrix}\rightarrow\begin{pmatrix}-1\\1\end{pmatrix},\quad \text{(e)}\ \delta\begin{pmatrix}1**\\ *0*\end{pmatrix}\rightarrow\begin{pmatrix}0\\1\end{pmatrix},\quad \text{(f)}\ \delta\begin{pmatrix}0**\\ *0*\end{pmatrix}\rightarrow\begin{pmatrix}0\\0\end{pmatrix},$$

$$\text{(i)}\ \delta\begin{pmatrix}-1**\\ *1*\end{pmatrix}\rightarrow\begin{pmatrix}0\\0\end{pmatrix},\quad \text{(j)}\ \delta\begin{pmatrix}-1**\\ *0*\end{pmatrix}\rightarrow\begin{pmatrix}-1\\1\end{pmatrix}.$$

Here, $\$ \in \Sigma \cup \{\#\}$ and $* \in \{-1, 0, 1\}$. The following example for $v = 12$ illustrates the proof. (Here, we use the symbol $\bar{1}$ instead of -1.) Recall that one of the tracks is used for indicating the rightmost 1 in the counter (see Lemma 1). When the rightmost 1 reaches the left end, we use rules (a),(d) instead of (a"),(d").

$$\begin{matrix}\$110000\\ \$011000\end{matrix}\rightarrow\begin{matrix}\$\bar{1}11000\\ \$100000\end{matrix}\rightarrow\begin{matrix}\$0\bar{1}0000\\ \$011100\end{matrix}\rightarrow\begin{matrix}\$\bar{1}00000\\ \$110100\end{matrix}\rightarrow\begin{matrix}\$000000\\ \$000100\end{matrix}$$

$$\rightarrow\begin{matrix}\$\bar{1}00000\\ \$100100\end{matrix}\rightarrow\begin{matrix}\$0\bar{1}0000\\ \$010100\end{matrix}\rightarrow\begin{matrix}\$\bar{1}0\bar{1}000\\ \$111100\end{matrix}\rightarrow\begin{matrix}\$000000\\ \$001000\end{matrix}\rightarrow\begin{matrix}\$\bar{1}00000\\ \$101000\end{matrix}$$

$$\rightarrow\begin{matrix}\$0\bar{1}0000\\ \$011000\end{matrix}\rightarrow\begin{matrix}\$\bar{1}00000\\ \$110000\end{matrix}\rightarrow\begin{matrix}\$000000\\ \$000000\end{matrix}$$

■

4.3 CA-Constructibility of Functions

Now we are ready to give the proof of Theorem 1. For simplicity, we first prove Corollary 1 in this section. Then, we extend the proof to the general case in Section 4.4.

Suppose that $t(n)$ is a function such that there is a TM which, given a string $bin(n)$ of length $\lceil \log n \rceil$, generates $bin(t(n))$ in time polynomial in $\log n$. We construct a CA M whose accepting cell falls into an accepting state at step $t(n)$.

For simplicity, we assume $t(n) \geq 2n$. The case where $t(n) = cn$ for $1 < c < 2$ is considered in Section 4.4. We divide $t(n)$ steps into $\lfloor t(n)/n \rfloor$ *stages* of n steps and the remaining x steps, where $t(n) = \lfloor t(n)/n \rfloor \cdot n + x$. Each cell of M is divided into tracks in order to use counters. In the first stage, M uses the algorithm given in Lemma 1; namely, M stores the value n into a counter in n steps. This counter is used for counting the n steps of the second stage. In the second stage, M also generates the values of $\lfloor t(n)/n \rfloor - 2$ and x in some tracks by simulating TMs computing $\lfloor t(n)/n \rfloor - 2$ and x. (Since $t(n)$ is computable in time polynomial in $\log n$, so are $\lfloor t(n)/n \rfloor - 2$ and x. These polylog-time procedures can be done in the n steps of the second stage because of the linear speed-up theorem [15].) Therefore, at step $2n$, M has values n, $\lfloor t(n)/n \rfloor - 2$, and x. M counts from n to 0, $\lfloor t(n)/n \rfloor - 2$ times (which consumes $n \times (\lfloor t(n)/n \rfloor - 2)$ steps). Finally, M counts x. Hence, the accepting cell can fall into an accepting state at step $t(n) = 2n + (\lfloor t(n)/n \rfloor - 2)n + x$.

In order to count n repeatedly, M uses three counters, say, CT_1, CT_2, and CT_3. CT_2 and CT_3 are used alternately; while CT_3 (resp. CT_2) is counting n steps, M copies the value n in CT_1 into CT_2 (resp. CT_3). This completes the proof of Corollary 1.

4.4 Extension to the General Case

In this section, we give the proof of Theorem 1. Suppose that $t(n)$ is a function such that there is a TM which, given a string $bin(n)$, generates $bin(t(n))$ in time $O(t(n) - n)$. We construct a CA M whose accepting cell falls into an accepting state at step $t(n)$. We divide $t(n)$ steps into three ***stages***. In the first stage, M stores the value n into counters in n steps as in Section 4.3.

The second stage is further divided into ***sub-stages***. In the first sub-stage, M executes the following (i) and (ii) simultaneously: (i) M generates the value $t(n)$ by simulating a TM computing $t(n)$. By assumption, this can be done in $(t(n) - n)/c$ steps for constant c. Then M computes $t(n) - n$ in time $O(\log n)$, which is bounded by $O(t(n) - n)$ because the input of the $O(t(n) - n)$-time TM computing $t(n)$ has length $\lceil \log n \rceil$. (ii) M counts the number, say, u_1, of steps of the first sub-stage. When the first sub-stage is finished (i.e., at step $n + u_1$), M has the value $t(n) - n$. In the second sub-stage, M computes $t(n) - n - u_1$ by simulating a TM, while M counts the number, say, u_2, of steps of the second sub-stage. Thus, at step $n + u_1 + u_2$ and M has the value $t(n) - n - u_1$. The value $t(n) - n - u_1$ can be computed in $O(\log u_1)$ steps because u_1 is represented in binary. Similarly, in the third sub-stage, M computes $t(n) - n - u_1 - u_2$ in $O(\log u_2)$ steps ($= O(\log \log u_1)$ steps), and counts u_3. Continuing this procedure until the number, say, u_l, of steps of the lth sub-stage is $u_l = 1$. At the end of the lth sub-stage (i.e., at step $n + u_1 + u_2 + \cdots + u_l$), M has the value $t(n) - n - u_1 - u_2 - \cdots - u_{l-1}$.

In the third stage, M uses the algorithm given in Lemma 2; the value $t(n) - n - u_1 - \cdots - u_{l-1}$ is decreased one by one. When the value becomes 1 (at step $t(n)$), M makes the accepting cell fall into an accepting state.

It should be noted that the value $t(n) - n - u_1 - \cdots - u_{l-1}$ is larger than 0 if the value $t(n) - n$ can be generated in time $u_1 = (t(n) - n)/c$ for a sufficiently large c. The remaining $u_2, u_3, \ldots, u_{l-1}$ are much smaller than u_1, since $u_i = O(\log u_{i-1})$ for $2 \leq i \leq l - 1$. This completes the proof of Theorem 1.

5 Constructibilities of $f + g$, $f \cdot g$, and g^f

5.1 Constructibility of $f(n) + g(n)$

Suppose that functions $n + f(n)$ and $n + g(n)$ are constructible. We construct a CA M whose accepting cell falls into an accepting state at step $n + f(n) + g(n)$. Without loss of generality, we assume $f(n) \leq g(n)$.

Each cell of M is divided into tracks. M executes the following (i) and (ii) simultaneously: (i) M simulates two CAs whose accepting cells fall into accepting states at step $n+g(n)$ and at step $n+f(n)$, respectively. (ii) At step n, M starts to count the number of steps. At step $n+f(n)$, the value becomes $f(n)$. At step $n+g(n)$, M starts to decrease the value $f(n)$ one by one. When the value becomes 0 (at step $n+f(n)+g(n)$), M makes the accepting cell fall into an accepting state.

5.2 Constructibility of $f(n) \cdot g(n)$

Construction from $n+f(n)$ and $n+g(n)$: Suppose that functions $n+f(n)$ and $n+g(n)$ are constructible by CA. We construct a CA M whose accepting cell falls into an accepting state at step $n+f(n) \cdot g(n)$. Since $n+f(n)$ and $n+g(n)$ are constructible, the values $f(n)$ and $g(n)$ are integers for all n. Without loss of generality, we assume $f(n) \leq g(n)$ and $f(n) \geq 1$.

From step 0 to step $n+g(n)$, M generates the values $f(n)$, $g(n)$, and $n+g(n)$ in counters in a manner similar to Section 5.1. M decreases the value from $g(n)$ to 0, $(f(n)-1)$ times. By this procedure, M consumes $g(n) \times (f(n)-1)$ steps, and thus it is now at step $n+f(n) \cdot g(n) = (n+g(n)) + g(n) \times (f(n)-1)$. M makes the accepting cell fall into an accepting state.

Construction from $n \cdot f(n)$ and $n \cdot g(n)$: Suppose that functions $n \cdot f(n)$ and $n \cdot g(n)$ are constructible by CA. We construct a CA M whose accepting cell falls into an accepting state at step $\lfloor n \cdot f(n) \cdot g(n) \rfloor$. Without loss of generality, we assume $f(n) \leq g(n)$. Since $n \cdot f(n)$ and $n \cdot g(n)$ are constructible functions, the values of them are integers for all n. However, the value $f(n)$ or $g(n)$ may not be an integer. Since $n \cdot f(n)$ is not bounded by $(1+o(1))n$, $f(n) \geq 1+\epsilon$ for some constant $\epsilon > 0$.

From step 0 to step $n \cdot g(n)$, M generates the values $n \cdot f(n)$ and $n \cdot g(n)$. M executes the following procedures (i) and (ii) simultaneously. (i) M counts the number, say, u_1, of steps required for (ii). (ii) As in Section 4.3, M computes the value $\lfloor (n^2 \cdot f(n) \cdot g(n))/n \rfloor - n \cdot g(n)$ (by simulating a TM). Thus, at step $n \cdot g(n) + u_1$, M has the value $\lfloor n \cdot f(n) \cdot g(n) \rfloor - n \cdot g(n)$. Then, M computes $\lfloor n \cdot f(n) \cdot g(n) \rfloor - n \cdot g(n) - u_1$, while M counts the number, say, u_2, of steps required for this procedure. Continuing this procedure until M has the value $\lfloor n \cdot f(n) \cdot g(n) \rfloor - n \cdot g(n) - u_1 - \cdots - u_{l-1}$ at step $n \cdot g(n) + u_1 + \cdots + u_l$, where $u_l = 1$. M decreases this value one by one. When the value becomes 1 (i.e., at step $\lfloor n \cdot f(n) \cdot g(n) \rfloor$), M makes the accepting cell fall into an accepting state.

It remains to show that $\lfloor n \cdot f(n) \cdot g(n) \rfloor - n \cdot g(n) - u_1 - \cdots - u_{l-1} > 0$. It is clear that $\lfloor n \cdot f(n) \cdot g(n) \rfloor - n \cdot g(n) \geq \epsilon n \cdot g(n) - 1$, since $f(n) \geq 1+\epsilon$. The value $\lfloor (n^2 \cdot f(n) \cdot g(n))/n \rfloor - n \cdot g(n)$ can be computed in $u_1 = O(\log^2(n \cdot g(n)))$, which is smaller than $\epsilon n \cdot g(n) - 1$. The remaining $u_2, u_3, \ldots, u_{l-1}$ are much smaller than u_1, since $u_i = O(\log u_{i-1})$ for $2 \leq i \leq l-1$.

5.3 Constructibility of $g(n)^{f(n)}$

Suppose that functions $f(n)$ and $g(n)$ are constructible by CA. We construct a CA M whose accepting cell falls into an accepting state at step $g(n)^{f(n)}$. (The proofs of constructibility of $k^{f(n)}$ and $g(n)^k$ are omitted, since they are analogous to $g(n)^{f(n)}$.)

M executes the following procedures (i) and (ii) simultaneously. (i) M counts the number, say, u_1, of steps required for (ii). (ii) M generates the values $f(n)$ and $g(n)$; then, M computes $g(n)^{f(n)}$ by simulating a TM working in $O(f(n)^2 \log^2 g(n))$ time. By this procedure, M has the value $g(n)^{f(n)}$ at step u_1. In the same technique as above, M has the value $g(n)^{f(n)} - u_1 - \cdots - u_{l-1}$ at step $u_1 + \cdots + u_l$, where $u_l = 1$. M decreases this value one by one. When the value becomes 1 (i.e., at step $g(n)^{f(n)}$), M makes the accepting cell fall into an accepting state.

6 Time Hierarchies of Cellular Automata

The proof is by diagonalization. We construct a $t_2(n)$-time CA M recognizing a language $L(t_1(n))$ which cannot be recognized by any $t_1(n)$-time CA. First of all, we fix the encoding rule of CA.

6.1 Encoding Rule of CA

All languages in this section are over $\{0,1\}$. We denote the states of CA by $q_1, q_2, \cdots$. For simplicity, we assume that q_1 is the unique accepting state and q_2 is the quiescent state. State q_i is encoded into string 10^i of length $i+1$. For example, we encode a transition rule, $\delta(q_3, q_5, q_2) = q_4$, into string 10001000001001000. The encoding of a CA, called the *encoding sequence*, is a concatenation of the encodings of the transition rules. The *encoding sequence* is followed by a sufficiently long string $y = 1100\cdots0$, called the *padding sequence*. The prefix 11 of the padding sequence indicates the boundary between encoding and padding sequences. Let $\psi(n)$ be a function defined as $\psi(n) = (\lfloor t_2(n)/t_1(n) \rfloor)^{1/4}$. Note that $\psi(n) \neq O(1)$. (The reason why we define such a function is given later.) The condition for y is $|x| \leq \psi(|xy|)$. For any encoding sequence x, there are an infinite number of strings xy such that $|x| \leq \psi(|xy|)$.

Let M_x denote the CA whose encoding sequence is x. If x is not a proper encoding sequence, we regard M_x as a CA accepting $\emptyset$. The language $L(t_1(n))$ is defined as $\{xy \mid M_x \text{ does not accept } xy \text{ within time } t_1(|xy|)\}$.

Lemma 3. *Any CA cannot accept the language $L(t_1(n))$ in time $t_1(n)$.*

Proof. Assume for contradiction that there exists a CA, say, M_x, which can accept $L(t_1(n))$ within $t_1(n)$ steps. Consider a string xy, where x is the encoding sequence of M_x and y is a sufficiently long padding sequence. If xy is given to M_x as an input, the following (i) or (ii) must be true: (i) M_x does not accept xy within $t_1(|xy|)$ steps. (ii) M_x accepts xy within $t_1(|xy|)$ steps. Suppose (i) is true. From the definition of $L(t_1(n))$, xy belongs to L, which contradicts the assumption and (i). Suppose (ii) is true. Again, from the definition of $L(t_1(n))$, xy does not belong to L, which contradicts the assumption and (ii). ∎

In the following section, we construct a CA M accepting $L(t_1(n))$ in time $t_2(n)$.

6.2 Constructing CA M Accepting $L(t_1(n))$

Since the linear speed-up theorem holds for CA [15], we construct an $O(t_2(n))$-time CA M. M accepts the input string if and only if the following conditions are met: (1) The tail of the input string is a padding sequence $y = 1100\cdots0$ (i.e., the input string can be written as xy for some $x \in \{0,1\}^*$). (2) x is a proper encoding sequence of some CA, say, M_x. (3) $|x| \le \psi(|xy|)$. (4) M_x does not accept xy in time $t_1(n)$.

In order to verify these conditions in parallel, each cell of M is divided into *tracks*. Verifying (1) in n steps is easy and is omitted. (2) Verifying whether the syntax of the encoding sequence is proper can be done by a single scan just as in finite automata. Therefore, M can verify conditions (1) and (2) in n steps. (3) The value of $\lfloor t_2(n)/t_1(n)\rfloor$ can be computed by counting how many times M can simulate a $t_1(n)$-time CA during $t_2(n)$ steps. The value of $\psi(n) = (\lfloor t_2(n)/t_1(n)\rfloor)^{1/4}$ can be computed in time polynomial in $\log\psi(n)$ by simulating a TM which computes the square root of a given value. If at least one of the three conditions is not satisfied, M simply rejects the input string. It remains to consider condition (4).

In order to verify whether M_x accepts xy within $t_1(n)$ steps, M simulates M_x on input xy. First of all, M stores the input string xy of length n into cells 1 through n. We denote the ith cell of M_x by s_i. Since M simulates $t_1(n)$ steps of M_x, M considers $t_1(n)$ cells of M_x. These cells are simulated by M using $|x|\cdot t_1(n)$ cells, where x is the encoding sequence of the input string. M divides $|x|\cdot t_1(n)$ cells into *blocks*, say, $B_1, B_2, \ldots$, each of length $|x|$. M's block B_i corresponds to M_x's cell s_i. Each block B_i is divided into two tracks in order to store x and the state of s_i. Therefore, every block has all transition rules of M_x. Generating such blocks in $|x|\cdot t_1(n)$ cells can be done in time $O(t_1(n)(\psi(n))^2)$. (Recall that $|x| \le \psi(n)$.) Thus, M can finish the above procedure in $t_1(n)(\psi(n))^3$ steps.

At step $t_1(n)(\psi(n))^3$, M starts to simulates M_x on input xy. M's computation is divided into *time-segments* each of length l. (l is fixed later.) A single step of M_x's cell s_i can be simulated by M's block B_i in $O((\psi(n))^2)$ steps, since each block has length $\psi(n)$. If l is larger than $c(\psi(n))^2$ for any large constant c, then M can simulate a single step of M_x in each time-segment. Therefore, l is defined as $l = (\psi(n))^3 + 2\psi(n)$. The reason for the additive $2\psi(n)$ is as follows. In order that every block starts each time-segment simultaneously, every block has a counter for counting $(\psi(n))^3$ steps (see Section 4.1 for a counter). At step $(\psi(n))^3$ in each time-segment, each block uses the firing squad synchronization algorithm so that every cell in the block simultaneously starts the simulation for the next time-segment, which requires additive $2\psi(n)$ steps.

Since $t_1(n)$ is a constructible function, M can make $t_1(n)$-step simulation of M_x. After the $t_1(n)$th time-segment, M accepts the input string xy if and only if M_x does not.

The time-complexity of the above simulation is bounded by $O(t_1(n)(\psi(n))^3)$, which is less than $t_2(n)$ because $\psi(n) \le (t_2(n)/t_1(n))^{1/4}$. Therefore, M can accept $L(t_1(n))$ in time $t_2(n)$. This completes the proof.

References

1. R. Balzer, An 8-state minimal time solution to the firing squad synchronization problem, *Inform. and Control*, **25** (1967) 22–42.
2. T. Buchholz and M. Kutrib, On time computability of functions in one-way cellular automata, *Acta Inform.*, **35** (1998) 329–352.
3. J.H. Chang, O.H. Ibarra, and A. Vergis, On the power of one-way communication, *J. ACM*, **35** 3 (1988) 697-726.
4. C. Choffrut and K. Culik II, On real-time cellular automata and trellis automata, *Acta Inform.*, **21** (1984) 393–407.
5. S.N. Cole, Real-time computation by n dimensional iterative arrays of finite-state machines, *IEEE Trans. on Computers*, **C-18** 4 (1969) 349–365.
6. C.R. Dyer, One-way bounded cellular automata, *Inform. and Control*, **44** (1980) 261–281.
7. P.C. Fischer, Generation of primes by a one-dimensional real-time iterative array, *J. ACM*, **12** 3 (1965) 388–394.
8. J.E. Hopcroft and J.D. Ullman, "Introduction to automata theory, languages and computation," Addison-Wesley, Reading, MA, 1979.
9. O.H. Ibarra and T. Jiang, On one-way cellular arrays, *SIAM J. Comput.*, **16** 6 (1987) 1135–1154.
10. O.H. Ibarra and T. Jiang, Relating the power of cellular arrays to their closure properties, *Theoret. Comput. Sci.*, **57** (1988) 225–235.
11. I. Korec, Real-time generation of primes by a one-dimensional cellular automaton with 11 states, in: *Proc. MFCS (LNCS1295)*, 1997, 358–367.
12. I. Korec, Real-time generation of primes by a one-dimensional cellular automaton with 9 states, in: *Proc. 2nd International Colloquium on Universal Machines and Computations*, 1998, 101–116.
13. J. Mazoyer, A 6-state minimal time solution to the firing squad synchronization problem, *Theoret. Comput. Sci.*, **50** (1987) 183–238.
14. J. Mazoyer and V. Terrier, Signals in one dimensional cellular automata, Research Report RR 94-50, Ecole Normale Supérieure de Lyon, 1994.
15. A.R. Smith III, Real-time recognition by one-dimensional cellular automata, *J. of Comput. and System Sci.*, **6** (1972) 233–253.
16. V. Terrier, On real time one-way cellular array, *Theoret. Comput. Sci.*, **141** (1995) 331–335.
17. H. Umeo, K. Morita, and K. Sugata, Deterministic one-way simulation of two-way real-time cellular automata and its related problems, *Inform. Process. Lett.*, **14** 4 (1982) 158–161.
18. R. Vollmar, Some remarks on pipeline processing by cellular automata, *Computers and Artificial Intelligence*, **6** 3 (1987) 263–278.
19. A. Waksman, An optimum solution to the firing squad synchronization problem, *Inform. and Control*, **9** (1980) 66–78.
20. J.B. Yunès, Seven state solutions to the firing squad synchronization problem, *Theoret. Comput. Sci.*, **127** 2 (1994) 313–332.

Dichotomy Theorem for the Generalized Unique Satisfiability Problem

Laurent Juban

LORIA (Université Henri Poincaré Nancy 1),
BP 239, 54506 Vandœuvre-lès-Nancy, France.
juban@loria.fr

Abstract. The unique satisfiability problem, that asks whether there exists a unique solution to a given propositional formula, was extensively studied in the recent years. This paper presents a dichotomy theorem for the unique satisfiability problem, partitioning the instances of the problem between the polynomial-time solvable and coNP-hard cases. We notice that the additional knowledge of a model makes this problem coNP-complete. We compare the polynomial cases of unique satisfiability to the polynomial cases of the usual satisfiability problem and show that they are incomparable. This difference between the polynomial cases is partially due to the necessity to apply parsimonious reductions among the unique satisfiability problems to preserve the number of solutions. In particular, we notice that the unique not-all-equal satisfiability problem, where we ask whether there is a unique model such that each clause has at least one true literal and one false literal, is solvable in polynomial time.

1 Introduction

The satisfiability problem SAT of a propositional formula in conjunctive normal form is a well-known NP-complete problem. Schaefer [Sch78] analyzed the generalized satisfiability problem, where each clause is represented by an arbitrary logical relation. He presented a Dichotomy Theorem for the generalized satisfiability problem, exhibiting conditions under which the problem is polynomial-time solvable, otherwise the problem is NP-complete. A similar dichotomy theorem was presented in [CH96] for the problem #SAT of counting the number of models (i.e., truth assignments) of a propositional formula. In particular, Creignou and Hermann show that if a decision satisfiability problem is intractable (NP-complete) then the corresponding counting satisfiability problem is also intractable (#P-complete).

The unique satisfiability problem UNIQUE SAT is defined as follows: given a propositional formula, is it true that it has a unique model (i.e., a unique satisfying truth assignment)? UNIQUE SAT is known to be coNP-hard [BG82], but it is not known whether it is in coNP. This problem is known to be only in DP, the class of languages equal to an intersection of two languages, one from NP and the other from coNP. UNIQUE SAT is therefore an intriguing problem from the point of view of collapsing complexity classes (see [Pap94, Chapter 17] or [CKR95]). We associate to UNIQUE SAT the problem ANOTHER SAT defined as follows: given a propositional formula ϕ and a model m of ϕ, is it true that there exists another model of ϕ different from m? It is clear that there is a certain relation between the two problems. If there is an instance of UNIQUE SAT that is true, then the corresponding instance of ANOTHER SAT must be false. Conversely, if an instance of ANOTHER SAT is true then the corresponding instance of UNIQUE SAT must be false. In this paper, we investigate the relation between UNIQUE SAT and ANOTHER SAT in terms of complexity. We study the polynomial-time solvable cases of both problems and compare it with the intractable cases. We also relate the intractable cases between the two problems.

Several polynomial-time solvable cases of UNIQUE SAT were studied in the literature. Minoux [Min92] noticed that for any subclass of polynomial-time solvable instance of the satisfiability problem with constants, UNIQUE SAT is also solvable in polynomial time. In particular, Hansen and Jaumard [HJ85] proposed a linear time UNIQUE SAT algorithm for 2SAT formulas, whereas several efficient UNIQUE SAT algorithms for Horn formulas were presented in the papers [Min92,BFS95,Pre93], ranging from quadratic to linear time. It would be interesting to know whether the four polynomial cases of SAT (namely Horn, anti-Horn, affine, and 2SAT formulas) are the only polynomial cases of UNIQUE SAT. Another interesting question is to know how the polynomial-time solvable cases of UNIQUE SAT and ANOTHER SAT relate to each other.

Both problems UNIQUE SAT and ANOTHER SAT involve some rudimentary counting. For this reason we cannot relate general unique satisfiability problems through ordinary polynomial many-one reductions. It is not enough to relate these problems through counting (sometimes also called weakly parsimonious) reductions (see [Pap94, Chapter 18], or [CH96] for a definition). Roughly speaking, a counting reduction R associates one solution of an input x to a constant number of solutions of the instance $R(x)$. Hence, in general, a counting reduction may reduce an instance x of a problem A with a unique solution to an instance $R(x)$ of a problem B with more solutions. Therefore we need to apply only reductions that exactly preserve the number of solutions between instances of the problems UNIQUE SAT and ANOTHER SAT, respectively. The number of solutions between instances is exactly preserved by the parsimonious reductions. Notice in this connection that it was not always possible to derive #P-hardness lower bounds for generalized satisfiability counting problems in [CH96] using only parsimonious reductions. Creignou and Hermann were obliged to apply weakly parsimonious reductions in the presence of the so-called complementive formulas. This indicates that complementive logical relations will be of special interest for the UNIQUE SAT and ANOTHER SAT problems.

2 Preliminaries

Let us recall some basic definitions and notions concerning complexity classes, reductions, and complete problems. More information can be found in the monographs [GJ79,Pap94]. Some parts of these preliminaries are taken from [Sch78] or [CH96] and are quoted only for self-containment of the paper.

We assume the knowledge of the following notions and notation. NP is the class of decision problems (languages) that can be solved in polynomial time by a *nondeterministic* Turing machine, coNP is the class of decision problems (languages) whose complements are in the class NP. For example, the problem SAT of deciding the satisfiability of a propositional formula in conjunctive normal form is in NP, whereas the UNSAT problem of deciding whether a propositional formula is unsatisfiable is in coNP.

Let A and B be two decision problems (languages). A polynomial-time many-one *reduction* from A to B is a polynomial-time computable function R from string to strings, such that for all inputs x the following holds: $x \in A$ if and only if $R(x) \in B$. For our purposes we will need ***parsimonious reductions*** that preserve the number of solutions. A polynomial reduction R from A to B is parsimonious if, for all $x \in A$, there is an equality between the number of solutions of x and $R(x)$. In particular, if there is a unique solution of the instance x of a problem A and there is a parsimonious reduction R from A to B then the instance $R(x)$ of the problem B has a unique solution, too.

Let C be a complexity class. A decision problem A is C-hard if for all problems $B \in$ C there exists a polynomial-time reduction from B to A. If in addition A is a member of C, then we say that the problem A is C-complete.

Let $S = \{R_1, \ldots, R_m\}$ be a finite set of logical relations. A logical relation is defined to be any subset of $\{0,1\}^k$ for some integer $k \geq 1$. An S-formula is any conjunction of clauses, each of the form $R_i(\boldsymbol{v})$, where $\boldsymbol{v}$ is a vector of not necessarily distinct variables. We overload the symbol R for a logical relation and the corresponding formula. The ***unique S-satisfiability problem*** UNIQUE SAT(S) is the problem of deciding whether a given S-formula has a unique model. The ***another S-model problem*** ANOTHER SAT(S) is the problem of deciding whether a given S-formula has another model different from a given model m. The problems UNIQUE SAT$_c(S)$ and ANOTHER SAT$_c(S)$ are the variations of UNIQUE SAT and ANOTHER SAT, respectively, where the Boolean constants are allowed to occur in the formulas (e.g., R(x,0,z) is allowed). The problems UNIQUE 3SAT and ANOTHER 3SAT are the versions of UNIQUE SAT and ANOTHER SAT, respectively, where every clause of the propositional formula contains three literals. The main result of our paper characterizes the complexity of UNIQUE SAT(S) and ANOTHER SAT(S) as properties of the logical relations in the set S.

If x is a variable, $\bar{x}$ denotes its negation. If ϕ is a formula, $\mathcal{V}ar(\phi)$ denotes the set of variables occurring in ϕ. We denote by $\mathrm{Sat}(\phi)$ the set of truth assignments (models) $m \colon \mathcal{V}ar(\phi) \to \{0,1\}$ that satisfy ϕ. We denote a model $m = (b_1, \ldots, b_n)$ as a string $b_1 \cdots b_n$ of its concatenated values. Let $m, m_1, m_2 \in \mathrm{Sat}(\phi)$ be models of the formula ϕ. We define the following four operations on models:

- $\bar{m}$ is defined by $\bar{m}(x) = 1$ iff $m(x) = 0$ and $\bar{m}(x) = 0$ otherwise,
- $m = m_1 \oplus m_2$: $m(x) = 1$ iff $m_1(x) \neq m_2(x)$ and $m(x) = 0$ otherwise,
- $m = m_1 \wedge m_2$: $m(x) = 1$ iff $m_1(x) = m_2(x) = 1$ and $m(x) = 0$ otherwise,
- $m = m_1 \vee m_2$: $m(x) = 0$ iff $m_1(x) = m_2(x) = 0$ and $m(x) = 1$ otherwise.

Two formulas ϕ and ψ are ***logically equivalent*** if and only if they have the same variable domains and their sets of models coincide. Two formulas ϕ and ψ are *quasi-equivalent* [CH96] if and only if there exists a bijection between the sets $\mathrm{Sat}(\phi)$ and $\mathrm{Sat}(\psi)$, such that each pair of models m and m' in the bijection coincides on the common variables of the formulas ϕ and ψ, i.e, such that $m(x) = m'(x)$ holds for every variable $x \in \mathcal{V}ar(\phi) \cap \mathcal{V}ar(\psi)$. As a consequence, the sets of models $\mathrm{Sat}(\phi)$ and $\mathrm{Sat}(\psi)$ have the same cardinality. The notion of quasi-equivalence is important in the presence of parsimonious and counting reductions, as it was shown in [CH96], since it preserves the number of solutions.

If ϕ is a formula, v is a variable, and l is a literal or a Boolean constant, then $\phi[l/v]$ denotes the formula obtained from ϕ by replacing each occurrence of v by l. If V is a set of variables, then $\phi[l/V]$ denotes the result of substituting l for every occurrence of each variable in V. We denote by $[\phi]$ the logical relation defined by the formula ϕ, when the variables are taken in lexicographic order. The relation 1-in-3 is the logical relation $\{001, 010, 100\}$.

The set of quasi-equivalent S-formulas with constants, $\mathcal{G}en(S)$, is the smallest set of formulas such that

- for all logical relations $R \in S$ and all vectors of variables $\boldsymbol{v}$, $R(\boldsymbol{v}) \in \mathcal{G}en(S)$,
- for all formulas $\phi, \psi \in \mathcal{G}en(S)$ and all variables x, y, the following formulas are all in $\mathcal{G}en(S)$: $\phi \wedge \psi$, $\phi[y/x]$, $\phi[0/x]$, $\phi[1/x]$, and
- if $\phi \in \mathcal{G}en(S)$ and ψ is quasi-equivalent to ϕ then also $\psi \in \mathcal{G}en(S)$.

Hence, $\mathcal{G}en(S)$ is the smallest set of quasi-equivalent S-formulas closed under conjunction, renaming, and substitution by a Boolean constant, whereas $\mathcal{G}en_b(S)$ is the smallest set of quasi-equivalent S-formulas closed under conjunction, renaming, and substitution by the Boolean constant b. The set of quasi-equivalent S-formulas without Boolean constants is denoted by $\mathcal{G}en_{nc}(S)$.

We define the set of all ***relations representable*** by quasi-equivalent S-formulas ***with Boolean constants*** as $\mathcal{R}ep(S) = \{[\phi] \mid \phi \in \mathcal{G}en(S)\}$ and the set of all ***relations representable*** by quasi-equivalent S-formulas ***without Boolean constants*** as $\mathcal{R}ep_{nc}(S) = \{[\phi] \mid \phi \in \mathcal{G}en_{nc}(S)\}$. $\mathcal{R}ep_b(S)$ is the set of all relations that are representable by quasi-equivalent S-formulas with the Boolean constant b only.

We adopt the usual syntactic characterization of logical relations and formulas. A Horn formula is a formula in conjunctive normal form with at most one positive literal per clause. Dually, an anti-Horn formula is a formula in conjunctive normal form with at most one negative literal per clause. A k-CNF formula, for a positive integer k, is a propositional formula in conjunctive normal form with k literals per clause. We say that a logical relation R is

- **0-valid** if $(0 \cdots 0) \in R$, **1-valid** if $(1 \cdots 1) \in R$;
- **Horn** if $R(\boldsymbol{v})$ is logically equivalent to a Horn formula, **anti-Horn** if $R(\boldsymbol{v})$ is logically equivalent to an anti-Horn formula;
- **affine** if the formula $R(\boldsymbol{v})$ is logically equivalent to a system of linear equations over the smallest Boolean ring $\mathbb{Z}_2$;
- **2SAT** if the formula $R(\boldsymbol{v})$ is logically equivalent to a 2-CNF formula;
- **complementive** if for every model $(a_1 \cdots a_n) \in R$ there exists the complementary model $(1 - a_1 \cdots 1 - a_n) \in R$.

3 General unique satisfiability problem

Theorem 1 (Dichotomy Theorem). *Let S be a finite set of logical relations. If S satisfies one of the conditions (1) to (6) below, then* ANOTHER SAT(S) *and* UNIQUE SAT(S) *are polynomial-time solvable. Otherwise,* ANOTHER SAT(S) *is NP-complete and* UNIQUE SAT(S) *is coNP-hard.*

1. *Every relation in S is 0-valid and 1-valid.*
2. *Every relation in S is complementive.*
3. *Every relation in S is Horn.*
4. *Every relation in S is anti-Horn.*
5. *Every relation in S is affine.*
6. *Every relation in S is 2SAT.*

Notice that UNIQUE SAT(S) becomes coNP-complete if either every relation in S is 0-valid or every relation in S is 1-valid, since then it can be expressed as the complement of the problem ANOTHER SAT(S) with the given model $0 \cdots 0$ or $1 \cdots 1$, respectively. This result can be generalized to an arbitrary model. The problem UNIQUE SAT(S) with the additional information that there exists a model is just the complement of the problem ANOTHER SAT(S), and therefore it is coNP-complete.

Notice also that the problem UNIQUE NOT-ALL-EQUAL SAT, asking whether there is a unique model, such that in no clause are all literals evaluated to the same Boolean constant (i.e., $(0 \cdots 0)$ and $(1 \cdots 1)$ are excluded), is polynomial-time solvable, since the relation $nae = \{001, 010, 011, 100, 101, 110\}$ is complementive. Indeed, if m is a model of a complementive formula ϕ then also the dual $\bar{m}$ is a model of ϕ. Hence, a complementive formula has never an odd number of models. On the other hand, recall that the satisfiability problem NOT-ALL-EQUAL SAT is NP-complete.

The rest of the paper is devoted to the proof of the Dichotomy Theorem for UNIQUE SAT(S) and ANOTHER SAT(S).

Proposition 1. *Let S be a finite set of logical relations. If S satisfies one of the conditions (3) to (6) of Theorem 1, then $\mathcal{R}ep_b(S)$ satisfies the same condition. Otherwise, $\mathcal{R}ep_b(S)$ is the set of all logical relations.*

The proof of this proposition requires several intermediate results.

First, we need a tool for detecting the polynomial cases. See [Sch78] or [CH96] for details.

Proposition 2. *Let R be a logical relation and let $\phi = R(v)$ be the corresponding formula. Then*

- *R is Horn iff $m_1, m_2 \in \mathrm{Sat}(\phi)$ implies $(m_1 \wedge m_2) \in \mathrm{Sat}(\phi)$;*
- *R is anti-Horn iff $m_1, m_2 \in \mathrm{Sat}(\phi)$ implies $(m_1 \vee m_2) \in \mathrm{Sat}(\phi)$;*
- *R is affine iff $m_1, m_2, m_3 \in \mathrm{Sat}(\phi)$ implies $(m_1 \oplus m_2 \oplus m_3) \in \mathrm{Sat}(\phi)$;*
- *R is 2SAT iff $m_1, m_2, m_3 \in \mathrm{Sat}(\phi)$ implies $(m_1 \vee m_2) \wedge (m_2 \vee m_3) \wedge (m_3 \vee m_1) \in \mathrm{Sat}(\phi)$.*

Lemma 1. *Let R be a logical relation. If R is not Horn then the set $\mathcal{R}ep_b(\{R\})$ contains the relations $[x \not\equiv y]$ or $[x \vee y]$ for each $b \in \{0,1\}$. If R is not anti-Horn then $\mathcal{R}ep_b(\{R\})$ contains the relations $[x \not\equiv y]$ or $[\bar{x} \vee \bar{y}]$ for each $b \in \{0,1\}$.*

Proof. We do the proof only for the case of R not being Horn and $b = 0$. The proof of the other cases is similar.

Let R be a logical relation which is not Horn and let $\phi = R(\boldsymbol{v})$ be the corresponding formula. We show that $\mathcal{R}ep_0(\{R\}) \cap \{[x \not\equiv y], [x \vee y]\} \neq \emptyset$. Following Proposition 2, there exist two models $m_1, m_2 \in \mathrm{Sat}(\phi)$ such that $(m_1 \wedge m_2) \notin \mathrm{Sat}(\phi)$. Moreover, we have that $m_1 \neq (0\cdots0)$ (since $(0\cdots0) \wedge m_2 = (0\cdots0) \in \mathrm{Sat}(\phi)$), $m_1 \neq (1\cdots1)$ (since $(1\cdots1) \wedge m_2 = m_2 \in \mathrm{Sat}(\phi)$), $m_1 \neq m_2$ (since it would imply $m_1 \wedge m_2 = m_1 \in \mathrm{Sat}(\phi)$), and there exists a variable $x \in \mathcal{V}ar(\phi)$ such that $m_1(x) = 1$ and $m_2(x) = 0$ (otherwise we get $m_1 \wedge m_2 = m_1 \in \mathrm{Sat}(\phi)$). Symmetrically, we have that $m_2 \neq (0\cdots0)$, $m_2 \neq (1\cdots1)$, and there exists a variable $y \in \mathcal{V}ar(\phi)$ such that $m_2(y) = 1$ and $m_1(y) = 0$.

Construct a new formula $\psi = \phi[0/V_0, x/V_x, y/V_y]$ where $V_0 = \{v \in \mathcal{V}ar(\phi) \mid m_1(v) = 0 \wedge m_2(v) = 0\}$, $V_x = \{v \in \mathcal{V}ar(\phi) \mid m_1(v) = 1\}$, and $V_y = \{v \in \mathcal{V}ar(\phi) \mid m_1(v) = 0 \wedge m_2(v) = 1\}$. The sets V_x and V_y are nonempty, hence the formula ψ contains both variables x and y. It is clear that V_0, V_x, and V_y are disjoint and $V_0 \cup V_x \cup V_y = \mathcal{V}ar(\phi)$. Clearly, $[\psi] \in \mathcal{R}ep_0(\{R\})$. The relation $[\psi]$ contains 01 and 10 but it does not contain 00. Hence, the relation $[\psi]$ is either $[\psi] = \{01, 10\} = [x \not\equiv y]$ or $[\psi] = \{01, 10, 11\} = [x \vee y]$, depending on whether $[\psi]$ contains 11 or not. □

Corollary 1. *If S contains some relation which is not Horn and some relation which is not anti-Horn then $\mathcal{R}ep_b(S)$ contains the relation $[x \not\equiv y]$.*

Proof. Let $R_1 \in S$ be a non-Horn relation. Then $\mathcal{R}ep_b(\{R_1\}) \cap \{[x \not\equiv y], [x \vee y]\} \neq \emptyset$ following Lemma 1. Let $R_2 \in S$ be a non anti-Horn relation. Similarly, $\mathcal{R}ep_b(\{R_2\}) \cap \{[x \not\equiv y], [\bar{x} \vee \bar{y}]\} \neq \emptyset$.

Assume that $[x \not\equiv y] \notin \mathcal{R}ep_b(S)$ holds. Then both relations $[x \vee y]$ and $[\bar{x} \vee \bar{y}]$ are included in $\mathcal{R}ep_b(S)$. Therefore $\mathcal{R}ep_b(S)$ contains also the relation $[(x \vee y) \wedge (\bar{x} \vee \bar{y})] = [x \not\equiv y]$, contradiction. Hence $\mathcal{R}ep_b(S)$ contains $[x \not\equiv y]$. □

Lemma 2 (Negated Substitution). *Let the relation $[x \not\equiv y]$ be included in $\mathcal{R}ep_b(S)$. If a formula ϕ belongs to $\mathcal{G}en_b(S)$ and u, v are variables, then the formula $\phi[\bar{u}/v]$ is contained in the set $\mathcal{G}en_b(S)$, too.*

Proof. By assumption, there exists a formula in $\mathcal{G}en_b(S)$ logically equivalent to $x \not\equiv y$, therefore we can construct the formula $\phi[u'/v] \wedge (u' \not\equiv u)$. The formulas $\phi[\bar{u}/v]$ and $\phi[u'/v] \wedge (u' \not\equiv u)$ are quasi-equivalent, when u' is a new variable not occurring in ϕ. □

Lemma 3 ([CH96]). *Let R be a non-affine relation. Then $\mathcal{R}ep(\{R, [x \not\equiv y]\})$ contains the relations $[x \vee y]$, $[\bar{x} \vee y]$, $[x \vee \bar{y}]$, and $[\bar{x} \vee \bar{y}]$.*

Let R be a b-valid and non-affine relation. Then the set $\mathcal{R}ep_b(\{R\})$ contains the relations $[x \vee y]$, $[\bar{x} \vee y]$, $[x \vee \bar{y}]$, and $[\bar{x} \vee \bar{y}]$.

Lemma 4. *Let R be a non-2SAT relation. Then the relation* 1-in-3 *is contained in* $\mathcal{R}ep_b(\{R, [x \not\equiv y], [x \vee y]\})$.

Proof. We do the proof only for $b = 0$, the proof for $b = 1$ is similar.

Let R be a non-2SAT logical relation and let $\phi = R(v)$ be the corresponding formula. Following Proposition 2, there exist three models $m_1, m_2, m_3 \in \mathrm{Sat}(\phi)$ such that $(m_1 \vee m_2) \wedge (m_2 \vee m_3) \wedge (m_3 \vee m_1) \notin \mathrm{Sat}(\phi)$. Let ϕ' be a formula constructed from ϕ by replacing each variable $x \in \mathcal{V}ar(\phi)$, such that $m_1(x) = 1$ holds, by its negation $\bar{x}$. From the Negated Substitution Lemma follows that $\phi' \in \mathcal{G}en(\{R, [x \not\equiv y]\})$. Let m'_2 and m'_3 be models of ϕ' corresponding to the models m_2 and m_3 of ϕ. For each $i = 2, 3$, if $m_1(x) = 0$ then $m'_i(x) = m_i(x)$ else $m'_i(x) = \bar{m}_i(x)$.

Let V_0, V_x, V_y, and V_z be the following sets of variables: $V_0 = \{v \in \mathcal{V}ar(\phi') \mid m'_2(v) = 0 \wedge m'_3(v) = 0\}$, $V_x = \{v \in \mathcal{V}ar(\phi') \mid m'_2(v) = 1 \wedge m'_3(v) = 1\}$, $V_y = \{v \in \mathcal{V}ar(\phi') \mid m'_2(v) = 0 \wedge m'_3(v) = 1\}$, and $V_z = \{v \in \mathcal{V}ar(\phi') \mid m'_2(v) = 1 \wedge m'_3(v) = 0\}$. Construct the formula $\psi = \phi'[0/V_0, x/V_x, y/V_y, z/V_z]$.

Note that the sets V_x, V_y, and V_z are nonempty. Therefore $[\psi]$ contains the models 000, 101, and 110, but it does not contain 100 since the original relation R is not 2SAT. Construct the formula $\omega = \psi[\bar{x}/x] \wedge (\bar{x} \vee \bar{y}) \wedge (\bar{y} \vee \bar{z}) \wedge (\bar{z} \vee \bar{x})$. From Negated Substitution Lemma follows that $\omega \in \mathcal{G}en_0(\{R, [x \not\equiv y], [x \vee y]\})$ and that $[\omega] = \{001, 010, 100\}$, i.e., $[\omega]$ is the relation 1-in-3. □

Lemma 5. $\mathcal{R}ep_b(\{\text{1-in-3}\})$ *is the set of all logical relations, for each* $b \in \{0, 1\}$.

Proof. Let $R(x, y, z)$ be the formula corresponding to the relation 1-in-3. Let $\phi_0 = R(x, u_1, u_4) \wedge R(y, u_2, u_4) \wedge R(u_1, u_2, u_5) \wedge R(u_3, u_4, u_6) \wedge R(z, u_3, 0)$, $\phi_1 = R(x, u_1, u_4) \wedge R(y, u_2, u_4) \wedge R(u_1, u_2, u_5) \wedge R(u_3, u_4, u_6) \wedge R(z, u_3, u_7) \wedge R(u_7, u_8, 1)$, $\psi_0 = R(x, y, 0)$, and $\psi_1 = R(x, y, u_1) \wedge R(u_1, u_2, 1)$. It is easy to verify that the formulas ϕ_0 and ϕ_1 are quasi-equivalent to $x \vee y \vee z$, and similarly that the formulas ψ_0 and ψ_1 are quasi-equivalent to $x \not\equiv y$.

There exists a parsimonious reductions from the satisfiability problem SAT of a propositional formula in conjunctive normal form to the satisfiability problem 3SAT of a propositional formula in conjunctive normal form with 3 literals per clause (see, e.g. [Koz92]). Hence, for each SAT formula ϕ there exists a quasi-equivalent 3-CNF formula ϕ'. Now, for each $i = 0, 1$, the formulas ϕ_i and ψ_i, using also the Negated Substitution Lemma, allow us to convert the 3-CNF formula ϕ' to a quasi-equivalent formula in $\mathcal{G}en_i(\{R\})$. Therefore, for every propositional formula ϕ we have that $[\phi] \in \mathcal{R}ep_i(\{R\})$, i.e., that $\mathcal{R}ep_i(\{R\})$ is the set of all logical relations. □

We are now able to prove Proposition 1. The proof is essentially the same as of Theorem 3.0 in [Sch78].

Proof of Proposition 1: We focus only on the case when S does not satisfy any of the conditions (3) to (6) of Theorem 1. The other cases are clear.

If S does not satisfy any of the conditions (3) to (6) then S contains a relation R_1 which is not Horn, a relation R_2 which is not anti-Horn, a relation R_3 which is not affine, and a relation R_4 which is not 2SAT. Corollary 1 implies that $[x \not\equiv y] \in \mathcal{R}ep_b(\{R_1, R_2\})$. From Lemma 3 follows that $[x \vee y] \in \mathcal{R}ep_b(\{R_1, R_2, R_3\})$. From Lemma 4 follows that the set $\mathcal{R}ep_b(\{R_1, R_2, R_3, R_4\})$ contains the relation 1-in-3. Therefore, by Lemma 5, $\mathcal{R}ep_b(\{R_1, R_2, R_3, R_4\})$ is the set of all logical relations. Hence, also $\mathcal{R}ep_b(S)$ is the set of all relations. □

Lemma 6 ([CH96]). *Let S be a nonempty finite set of logical relations. At least one of the following conditions holds:* (1) *Every relation in S is 0-valid.* (2) *Every relation in S is 1-valid.* (3) *$\mathcal{R}ep_{nc}(S)$ contains the relation $[\bar{x} \wedge y]$.* (4) *$\mathcal{R}ep_{nc}(S)$ contains the relation $[x \not\equiv y]$.*

Moreover, if $[\bar{x} \wedge y] \notin \mathcal{R}ep_{nc}(S)$ and $[x \not\equiv y] \in \mathcal{R}ep_{nc}(S)$ hold then every relation in S is complementive.

Proposition 3. *Let S be a finite set of logical relations. If the relations in S are neither all 0-valid, nor all 1-valid, nor all complementive, then there exists a parsimonious reduction from* UNIQUE SAT$_c(S)$ *to* UNIQUE SAT(S) *and from* ANOTHER SAT$_c(S)$ *to* ANOTHER SAT(S).

Proof. If the relations in S are neither all 0-valid, nor all 1-valid, nor all complementive, then $[\bar{x} \wedge y] \in \mathcal{R}ep_{nc}(S)$ or $[x \not\equiv y] \notin \mathcal{R}ep_{nc}(S)$ holds following Lemma 6. If $[\bar{x} \wedge y] \in \mathcal{R}ep_{nc}(S)$ holds then the proof is the same as in case 1 of Proposition 4.12 in [CH96]. If we have $[\bar{x} \wedge y] \notin \mathcal{R}ep_{nc}(S)$ then the relation $[x \not\equiv y]$ is contained in $\mathcal{R}ep_{nc}(S)$ following the first part of Lemma 6. But we have that $[x \not\equiv y] \notin \mathcal{R}ep_{nc}(S)$ since S contains a relation that is not complementive, following the second part of Lemma 6: contradiction. Hence $\mathcal{R}ep_{nc}(S)$ must contain the relation $[\bar{x} \wedge y]$. □

Theorem 2. *Let S be a finite set of logical relations. If S satisfies one of the conditions (3) to (6) of Theorem 1 then* UNIQUE SAT$_c(S)$ *and* ANOTHER SAT$_c(S)$ *are polynomial-time solvable. Otherwise,* ANOTHER SAT$_c(S)$ *is NP-complete and* UNIQUE SAT$_c(S)$ *is coNP-hard.*

Proof. If every relation in S is Horn then every S-formula is a Horn propositional formula. To compute a model of a Horn formula in polynomial time, apply exhaustively the unit resolution, followed by setting the unresolved variable to 0 (see [DG84] for details). Dually, if every relation in S is anti-Horn then we compute a model of such S-formula in polynomial time by exhaustive unit resolution, followed by setting the unresolved variables to 1. If every relation in S is affine then such S-formula is equivalent to a system of linear equations over the ring Z_2. Its solution can be found by Gaussian elimination in polynomial time. If every relation in S is 2SAT then a model of such S-formula can be found in polynomial time by the Davis-Putnam procedure.

Let $\phi(x_1, \ldots, x_n)$ be an S-formula. If S satisfies one of the polynomial conditions, we compute a model m of ϕ in the case of UNIQUE SAT in polynomial time by one of the previous methods. In the case of ANOTHER SAT the model m is already given. We can decide whether there is another model by the following polynomial-time algorithm:

$i \leftarrow 0$; *another* $\leftarrow$ *false*;
while $\neg$*another* $\wedge (i < n)$ **do** $i \leftarrow i + 1$; *another* $\leftarrow$ *sat*$(\phi[\bar{m}(x_i)/x_i])$ **od**

The call *sat*$(\phi[\bar{m}(x_i)/x_i])$ means that we instantiate in the formula ϕ the variable x_i by the dual value of $m(x_i)$ and test whether this instance is satisfiable. The satisfiability of $\phi[\bar{m}(x_i)/x_i]$ can be computed in polynomial time since S satisfies one of the polynomial conditions. If *another* = *false* (there are no other models) then return *false* in the case of ANOTHER SAT and *true* for UNIQUE SAT. Otherwise, return *true* for ANOTHER SAT and *false* for UNIQUE SAT.

If S does not satisfy any of the polynomial conditions then we show that there exists a parsimonious reduction from UNIQUE 3SAT to UNIQUE SAT$_c(S)$, and from ANOTHER 3SAT to ANOTHER SAT$_c(S)$. Indeed, consider the relations $R_0 = [x \vee y \vee z]$, $R_1 = [\bar{x} \vee y \vee z]$, $R_2 = [\bar{x} \vee \bar{y} \vee z]$, and $R_3 = [\bar{x} \vee \bar{y} \vee \bar{z}]$. Let $\phi_i(x, y, z)$ be a formula in $\mathcal{G}en_b(S)$ quasi-equivalent to $R_i(x, y, z)$, for $i = 0, 1, 2, 3$. Such formulas exist by Proposition 1. Let ψ be a 3-CNF formula. Construct the formula ψ' by replacing each clause of ψ by a corresponding formula ϕ_i. This reduction is parsimonious.

For proving both lower bounds, we use the same construction as in [BG82]. Let $\alpha(x_1, \ldots x_n)$ be a 3-CNF formula. Construct the formula $\beta(x_0, x_1, \ldots x_n) = (x_0 \wedge x_1 \wedge \cdots \wedge x_n) \vee (\bar{x}_0 \wedge \alpha(x_1, \ldots, x_n))$. Transform β to conjunctive normal form (there is no exponential blow-up in this case), getting a 4-CNF formula. Transform β to a quasi-equivalent 3-CNF formula β' (see [Koz92]). It is clear that β has a unique model, namely $1 \cdots 1$, iff the formula α is unsatisfiable. If α represents an instance of UNSAT, the problem of unsatisfiability of a propositional 3-CNF formula that is coNP-complete, then this reduction proves the coNP-hardness of UNIQUE SAT$_c(S)$. If α represents an instance of 3SAT, the satisfiability problem of a propositional 3-CNF formula that is NP-complete, then this reduction proves the NP-hardness of ANOTHER SAT$_c(S)$. To prove membership of ANOTHER SAT$_c(S)$ in NP, guess an assignment m' different from m and check in polynomial time if m' satisfies the formula ϕ. □

We have assembled now all the necessary tools to prove Theorem 1.

Proof of Theorem 1: If every relation in S is 0-valid and 1-valid then every S-formula without constants has at least two models: $0 \cdots 0$ and $1 \cdots 1$. Hence, the solution of this instance for UNIQUE SAT and ANOTHER SAT is trivial.

Let every relation in S be complementive and let ϕ be an S-formula without constants. Following the definition of a complementive relation, if m is a model of ϕ then also its dual $\bar{m}$ is a model of ϕ. Hence, for this case UNIQUE SAT is alway false and ANOTHER SAT is always true. The rest of the polynomial cases is decided by the same algorithm as in Theorem 2.

Assume that S does not satisfy any of the polynomial conditions (1) to (6), i.e., that S contains a relation that is not both 0-valid and 1-valid, a relation that is not complementive, a relation R_1 that is not Horn, a relation R_2 that is not anti-Horn, a relation R_3 that is not affine, and a relation R_4 that is not 2SAT. There are two cases to analyze.

Case 1: If S contains a relation which is not 0-valid and a relation which is not 1-valid, then there exists a parsimonious reduction from UNIQUE SAT$_c(S)$ to UNIQUE SAT(S) and from ANOTHER SAT$_c(S)$ to ANOTHER SAT(S) following Proposition 3. Since UNIQUE SAT$_c(S)$ is coNP-hard and ANOTHER SAT$_c(S)$ is NP-complete following Theorem 2, this proves that UNIQUE SAT(S) is coNP-hard and ANOTHER SAT(S) is NP-hard.

Case 2: If every relation in S is either 0-valid or 1-valid, but not both, and not complementive, then we have that $[x \not\equiv y] \in \mathcal{R}ep_b(\{R_1, R_2\})$ following Corollary 1, $[x \vee y] \in \mathcal{R}ep_b(\{R_3\})$ following Lemma 3, and the set $\mathcal{R}ep_b(\{R_1, R_2, R_3, R_4\})$ contains the logical relation 1-in-3 following Lemma 4. Let $R(x, y, z)$ be the formula representing the relation 1-in-3. If there exists a formula ϕ containing the Boolean constant b, construct the formula ϕ' by replacing the constant b by a new variable x_b. Create a new formula ψ as follows. If $b = 0$ then let $\psi = \phi' \wedge R(x_b, x_b, u)$, otherwise if $b = 1$ then let $\psi = \phi' \wedge R(x_b, u, u)$, where u is a new variable. It is clear that the formulas ϕ and ψ are quasi-equivalent. Hence, there exists a parsimonious reduction from UNIQUE SAT$_c(S)$ to UNIQUE SAT(S) and from ANOTHER SAT$_c(S)$ to ANOTHER SAT(S).

In both cases, since UNIQUE SAT$_c(S)$ is coNP-hard and ANOTHER SAT$_c(S)$ is NP-complete following Theorem 2, this proves that UNIQUE SAT(S) is coNP-hard and ANOTHER SAT(S) is NP-hard. Membership of ANOTHER SAT(S) in NP is proved as in Theorem 2, hence ANOTHER SAT(S) is NP-complete. □

4 Concluding remarks

The main result of the paper is a Dichotomy Theorem for the UNIQUE SAT and ANOTHER SAT problems (Theorem 1), showing that every instance of both problems is either solvable in polynomial time or it is coNP-hard, respectively NP-complete. We noticed that both considered problems have the same polynomial-time solvable instances. Moreover, we showed that the additional knowledge of the existence of a model pushes the problem UNIQUE SAT from the difference class DP down to coNP, making it coNP-complete. Compare it with the result in [VV86] that UNIQUE SAT is DP-complete under randomized reductions.

We also proved Minoux's claim (see [Min92]) that the SAT and UNIQUE SAT problems have the same solvable in polynomial time instances in the presence of propositional formulas with Boolean constants (Theorem 2). On the other hand, if we consider formulas without constants, the solvable in polynomial time instances of SAT and UNIQUE SAT are incomparable. There are solvable in polynomial time SAT instances with corresponding coNP-complete UNIQUE SAT instances (0-valid or 1-valid), whereas there are NP-complete SAT instances with corresponding solvable in polynomial time UNIQUE SAT instances (complementive). Among the latter we find the problem NOT-ALL-EQUAL SAT for which the satisfiability is NP-complete, but which has never a unique model.

Acknowledgment: I thank Miki Hermann for many valuable comments on the previous versions of the paper.

References

[BFS95] K. A. Berman, J. Franco, and J. S. Schlipf. Unique satisfiability of Horn sets can be solved in nearly linear time. *Discrete Applied Mathematics*, 60(1-3):77–91, 1995.

[BG82] A. Blass and Y. Gurevich. On the unique satisfiability problem. *Information and Control*, 55(1-3):80–88, 1982.

[CH96] N. Creignou and M. Hermann. Complexity of generalized satisfiability counting problems. *Information and Computation*, 125(1):1–12, 1996.

[CKR95] R. Chang, J. Kadin, and P. Rohatgi. On unique satisfiability and the threshold behavior of randomized reductions. *Journal of Computer and System Science*, 50(3):359–373, 1995.

[DG84] W. F. Dowling and J. H. Gallier. Linear-time algorithms for testing the satisfiability of propositional Horn formulae. *Journal of Logic Programming*, 1(3):267–284, 1984.

[GJ79] M. R. Garey and D. S. Johnson. *Computers and intractability: A guide to the theory of NP-completeness*. W.H. Freeman and Co, 1979.

[HJ85] P. Hansen and B. Jaumard. Uniquely solvable quadratic Boolean equations. *Discrete Applied Mathematics*, 12(2):147–154, 1985.

[Koz92] D. C. Kozen. *The design and analysis of algorithms*, chapter 26: Counting problems and #P, pages 138–143. Springer-Verlag, 1992.

[Min92] M. Minoux. The unique Horn-satisfiability problem and quadratic Boolean equations. *Annals of Mathematics and Artificial Intelligence*, 6(1-3):253–266, 1992.

[Pap94] C. H. Papadimitriou. *Computational complexity*. Addison-Wesley, 1994.

[Pre93] D. Pretolani. A linear time algorithm for unique Horn satisfiability. *Information Processing Letters*, 48(2):61–66, 1993.

[Sch78] T. J. Schaefer. The complexity of satisfiability problems. In *Proceedings 10th Symposium on Theory of Computing (STOC'78), San Diego (California, USA)*, pages 216–226, 1978.

[VV86] L. G. Valiant and V. V. Vazirani. NP is as easy as detecting unique solutions. *Theoretical Computer Science*, 47(1):85–93, 1986.

A General Categorical Connection between Local Event Structures and Local Traces

H.C.M. KLEIJN[1], R. MORIN[2], and B. ROZOY[2]

[1] LIACS, Leiden University, P.O. Box 9512, 2300 RA Leiden, The Netherlands
[2] L.R.I., Bât. 490, Université de Paris Sud, 91405 Orsay Cedex, France
kleijn@wi.leidenuniv.nl, morin@lri.lri.fr, rozoy@lri.lri.fr

Abstract. Local event structures and local traces are generalizations of the classical prime event structures and Mazurkiewicz' traces in which independence is no longer a global binary property. We consider the problem of lifting the categorical connection between prime event structures and Mazurkiewicz' traces to this more general setting. Using a generic approach it is shown how certain subcategories of local event structures and local trace languages can be related. Moreover, every coreflection between subcategories generalizing the connection between prime event structures and Mazurkiewicz' traces fits into this approach.

Introduction

The traces introduced by Mazurkiewicz [5] and Winskel's prime event structures [6] are well-known abstract models for describing the behavior of concurrent systems, in particular 1-safe Petri nets. Whereas traces can be used to describe the non-conflicting sequential executions of the system together with an equivalence relation induced by an independence relation over the actions, a prime event structure provides explicit information on the relationships between events in terms of a partial ordering and a binary conflict relation. Despite the fact that the prime event structure model is more abstract than the trace model, they are closely related [8]. In particular, a coreflection between the categories of prime event structures and of Mazurkiewicz trace languages has been established [10]. This categorical approach allows not only to compare the models as objects, but also to compare their behavioral aspects (see, e.g., [1, 7]).

When an abstract model is meant to represent the behavior of dynamic systems, it is reasonable to extend it to a category by equipping it with behavior preserving morphisms to capture a notion of simulation. Categories can be compared by functors which relate objects to objects and morphisms to morphisms, and thus preserve the dynamic behavior of the systems. An adjunction between categories consists of two functors, one in each direction, that fit together in a particular way. This is a formal way to express that one model is more abstract than another, as it allows to canonically represent objects from the more concrete model that are mapped to the same object in the more abstract model. If in addition going from an object in the more abstract model to an object in the more concrete model and then back using the functors, leads to an isomorphic object, the adjunction is called a coreflection. Thus establishing a coreflection between two categories proves a very strong relationship, as it shows that the functor to the abstract model gives a faithful description of the concrete model. For more technical details on categories, functors, adjunctions and coreflections, the reader is referred to [9].

Local traces and local event structures were introduced in [2, 3] to lift the semantical theory of 1-safe Petri nets to the level of more general Petri nets in which concurrency and conflict are not structural properties, but depend on the current marking (the state of the system). Therefore a local independence relation describes which sets of actions may occur concurrently after a given execution of the system, while local event structures require a concurrency axiom local to their configurations. Both the local trace semantics and the local event structure semantics of Petri nets are respectively proper conservative extensions of the trace semantics and the prime event structure semantics of 1-safe Petri nets. In [4] these extended models have been considered as independent notions without their connection to Petri nets and some direct links between the two classes have been established. Due to the locally defined concurrency of events, every local event structure defines in a natural way a local trace language. *To associate a local event structure to a local trace language is however more complicated, since this requires the identification of events.* Following the classical approach, events are viewed as equivalence classes of prime intervals of the local trace language. Given an equivalence relation over prime intervals which satisfies certain elementary conditions, a local trace language directly defines a local event structure. The least equivalence satisfying these conditions is called Projectivity. It corresponds to the equivalence used to relate Mazurkiewicz' trace languages to prime event structures [6, 8, 10]. Projectivity however is too fine an equivalence to properly represent the concurrency of a local independence relation in its associated local event structure. To achieve this, more prime intervals need to be identified. It is shown that History, the equivalence corresponding to the relation used in [3] to associate a local event structure to a Petri net, is the least generalization of Projectivity which allows to extend the connection between Mazurkiewicz' trace languages and prime event structures to the more general setting of local trace languages and local event structures.

Whereas in [4] also other examples of equivalences of prime intervals are mentioned, it is only for History that a coreflection is established between subcategories of local event structures and local event structures. It is left open whether more general equivalences might lead to categorical connections between larger subclasses. This paper gives an affirmative answer to that question. Rather than pursuing individual examples of suitable equivalence relations it focusses on a generic approach. First, we recall the straightforward and intuitive representation of local event structures as local trace languages. This map easily extends to a functor which however admits *no* right-adjoint (Th. 2.4) except for restricted subcategories of each model [10, 4]. Next we introduce a notion of *punctuation*, based on equivalences of prime intervals, which admits Projectivity and History as particular examples; our main result is that any punctuation determines a coreflection between some associated subcategories of local event structures and local trace languages (Th. 2.12). Moreover we show that any generalization of the coreflection between prime event structures and trace languages may be obtained by this generic coreflection (Th. 3.4). Finally, we briefly indicate some possible implications of this research for the theory of Petri nets.

1 Basic Notions and Results

Preliminaries. We will use the following notations: for any (possibly infinite) alphabet Σ, and any words $u \in \Sigma^*$, $v \in \Sigma^*$, we write $u \leq v$ if u is a prefix of v, i.e. there is $z \in \Sigma^*$ such that $u.z = v$; the empty word is denoted by ε. We write $|u|_a$ for the number of occurrences of $a \in \Sigma$ in $u \in \Sigma^*$ and $\wp_f(\Sigma)$ denotes the set of finite subsets of Σ; for any $p \in \wp_f(\Sigma)$, $\mathrm{Lin}(p) = \{u \in p^* \mid \forall a \in p, |u|_a = 1\}$ is the set of linearisations of p. Finally, if $\lambda : \Sigma \rightharpoonup \Sigma'$ is a partial function from Σ to Σ', we also write $\lambda : \Sigma^* \to \Sigma'^*$ and $\lambda : \wp_f(\Sigma) \to \wp_f(\Sigma')$ to denote the naturally associated monoid morphisms.

Local Trace Languages. Local traces are a generalization of the classical Mazurkiewicz' traces since they are based on an independence relation which is left-context dependent and which specifies sets of independent actions rather than pairs.

DEFINITION 1.1. *A* local independence relation *on Σ is a non-empty subset I of $\Sigma^* \times \wp_f(\Sigma)$. The* (local) trace equivalence $\sim$ *induced by I is the least equivalence on Σ^* such that*
TE_1*: $\forall u, u' \in \Sigma^*, \forall a \in \Sigma, u \sim u' \Rightarrow u.a \sim u'.a$;*
TE_2*: $\forall (u,p) \in I, \forall p' \subseteq p, \forall v_1, v_2 \in \mathrm{Lin}(p'), u.v_1 \sim u.v_2$.*
A (local) trace *is an $\sim$-equivalence class $[u]$ of a word $u \in \Sigma^*$.*

By TE_1 local trace equivalences are right-congruences. TE_2 asserts that for every subset of actions which are independent after a sequence u, all sequences obtained by executing first u and then in an arbitrary order the actions from this subset, are equivalent. Note also that local trace equivalences are Parikh equivalences: $u \sim u' \Rightarrow \forall a \in \Sigma, |u|_a = |u'|_a$.

A local independence relation can be thought of as a representation of the behavior of a concurrent system. It provides information on possible sequential observations as well as information on their equivalence. Thus every local independence relation defines a prefix-closed language of sequential observations and a set of traces, the equivalence classes of these sequential observations. As observed in [4], the assumptions in these definitions can be translated into explicit additional conditions on the local independence relation without affecting the resulting sets of observations and traces. A local independence relation satisfying these additional conditions is called complete and is a maximal representative among local independence relations defining the same sequential observations and traces. In this paper we directly define local trace languages as combinations of a language (of sequences) and a complete local independence relation.

DEFINITION 1.2. *A* local trace language *(LTL) over Σ is a structure $\mathcal{L} = (\Sigma, I, L)$ where $L \subseteq \Sigma^\star$ and I is a local independence relation on Σ such that*

LTL_1: $(u,p) \in I \wedge p' \subseteq p \Rightarrow (u,p') \in I$;
LTL_2: $(u,p) \in I \wedge p' \subseteq p \wedge v \in \mathrm{Lin}(p') \Rightarrow (u.v, p \setminus p') \in I$;
LTL_3: $u \sim u' \wedge (u,p) \in I \Rightarrow (u',p) \in I$;
LTL_4: $(u.a, \emptyset) \in I \Rightarrow (u, \{a\}) \in I$;
LTL_5: $u \in L \Leftrightarrow (u, \emptyset) \in I$.

The requirements LTL_1 through LTL_4 make the local independence relation complete. LTL_1 makes explicit what TE_2 from Def. 1.1 guarantees for the trace equivalence: if a set of actions p can be executed concurrently after u, then so can any subset of p; moreover, following LTL_2, the step p can be split into a sequential execution v and a concurrent step of the remaining actions. LTL_3 states that after two equivalent sequences the independency and thus unorderedness of actions is the same; it corresponds to the right-congruence property TE_1 from Definition 1.1. LTL_4 guarantees that whenever $u.a$ is a sequential execution, then action a is allowed as a step after u. Finally, LTL_5 ensures that L is precisely the set of sequential observations associated to I in [4]: from LTL_4 and LTL_1 it follows that L is prefix-closed; moreover by LTL_3 we know that L is closed under the trace equivalence induced by I.

Local Event Structures. A local event structure is a family of configurations equipped with an enabling relation that specifies locally the possible concurrency of events.

DEFINITION 1.3. *A* local event structure *(LES) is a triple $\mathcal{E} = (E, C, \vdash)$ where E is a set of events, $C \subseteq \wp_f(E)$ is a set of finite subsets of events called configurations and $\vdash \subseteq C \times \wp_f(E)$ is an enabling relation such that*

LES_1: $(\emptyset \vdash \emptyset) \wedge (\forall e \in E, \exists c \in C, e \in c)$;
LES_2: $\forall c \in C$: $c \neq \emptyset \Rightarrow \exists e \in c, c \setminus \{e\} \vdash \{e\}$;
LES_3: $\forall c \in C, \forall p \in \wp_f(E)$: $c \vdash p \Rightarrow c \cap p = \emptyset$;
LES_4: $\forall c \in C, \forall p \in \wp_f(E), \forall p' \subseteq p$: $c \vdash p \Rightarrow (c \vdash p' \wedge c \cup p' \vdash p \setminus p')$.

LES_1 guarantees that the empty set is always a configuration and that the enabling relation is never empty. Also by LES_1, each event occurs in at least one configuration. LES_2 ensures that every non-empty configuration can be reached from the (initial) empty configuration. LES_3 implies that each event occurs at most once and by LES_4 each concurrent set can be split arbitrarily into subsets of concurrent events.

To each local event structure $\mathcal{E}$ a set of (finite) sequential observations can be associated which we call the paths of $\mathcal{E}$; formally, $\mathrm{Paths}(\mathcal{E}) = \{e_1...e_n \in E^\star \mid \forall i \in [1,n], \{e_1,...,e_{i-1}\} \vdash \{e_i\}\}$. As shown in [3], an event appears at most once along a path and each path u leads to a unique configuration $\mathrm{Cfg}(u)$ defined by $\mathrm{Cfg}(u) = \{e \mid |u|_e = 1\}$.

To associate a local trace language to a local event structure we use the map defined in [4] which translates the enabling relation in a natural way into a local independence relation. The local trace language thus obtained faithfully represents the concurrency between events.

DEFINITION 1.4. *Let $\mathcal{E} = (E, C, \vdash)$ be a local event structure. The local trace language* $\mathsf{ltl}(\mathcal{E})$ *associated to* $\mathcal{E}$ *is* $\mathsf{ltl}(\mathcal{E}) = (E, I, \text{Paths}(\mathcal{E}))$ *where* $I = \{(u,p) \in \Sigma^\star \times \wp_f(\Sigma) | \ u \in \text{Paths}(\mathcal{E})$ *and* $\text{Cfg}(u) \vdash p\}$.

Equivalences of Prime Intervals. In order to associate a local event structure to a given local trace language $\mathcal{L} = (\Sigma, I, L)$ one has to define events. The event structure should properly reflect the sequential behavior and the independencies represented by the local trace language. Thus we want events to represent occurrences of actions. For the prefix-closed language L, occurrences of actions correspond to *prime intervals* of the partial order $(L, \leq)$; these may be defined as the pairs $(u, a) \in \Sigma^\star \times \Sigma$ such that $u.a \in L$; we write $\Pr(\mathcal{L})$ for the set of prime intervals of $\mathcal{L}$. It may however be the case that different occurrences of an action (hence different prime intervals) should correspond to the same event: for instance, if $(u, \{a, b\}) \in I$ then actions a and b may occur simultaneously after u; an observer cannot distinguish the occurrence of a after u from the occurrence of a after $u.b$; thus (u, a) and $(u.b, a)$ must be identified as the same occurrence of a. Furthermore if u and u' are equivalent (in the same local trace), then the prime intervals (u, a) and (u', a) should not be distinguished either. For these reasons we need an equivalence relation over the prime intervals of $\mathcal{L}$ when associating a local event structure to $\mathcal{L}$. The requirements which any such equivalence should satisfy are defined below.

DEFINITION 1.5. *Let $\mathcal{L} = (\Sigma, I, L)$ be a local trace language; an* equivalence of prime intervals *of $\mathcal{L}$ is an equivalence $\asymp_\mathcal{L}$ over* $\Pr(\mathcal{L})$ *which satisfies*

Ind: $(u, \{a, b\}) \in I \wedge \ a \neq b \Rightarrow (u, a) \asymp_\mathcal{L} (u.b, a)$ *[Independence]*
Cfl: $(u, a) \in \Pr(\mathcal{L}) \wedge (u', a) \in \Pr(\mathcal{L}) \wedge u \sim u' \Rightarrow (u, a) \asymp_\mathcal{L} (u', a)$ *[Confluence]*
Lab: $(u, a) \asymp_\mathcal{L} (v, b) \Rightarrow a = b$ *[Labeling]*
Occ: $u.a \leq v.a \wedge \ (u, a) \asymp_\mathcal{L} (v, a) \Rightarrow u = v$ *[Occurrence Separation]*

Ind and Cfl specify which prime intervals should definitely be identified whereas Lab and Occ limit rationally the allowed identifications: Lab ensures that equivalent prime intervals correspond to occurrences of the same action and by Occ no execution sequence allows more than one occurrence of the same event. Given an equivalence of prime intervals of a local trace language, we can define a local event structure.

DEFINITION 1.6. *Let $\mathcal{L} = (\Sigma, I, L)$ and let $\asymp_\mathcal{L}$ be an equivalence of prime intervals of $\mathcal{L}$. For any word $u \in L$, the set of events in u is $Eve_{\asymp_\mathcal{L}}(u) = \{\langle v, b\rangle_\mathcal{L} \mid v.b \leq u\}$, where $\langle v, b\rangle_\mathcal{L}$ denotes the $\asymp_\mathcal{L}$-class of (v, b). The local event structure $\mathsf{les}_{\asymp_\mathcal{L}}(\mathcal{L})$ is the triple $(E, C, \vdash)$ where $C = \{Eve_{\asymp_\mathcal{L}}(u) \mid u \in L\}$, $E = \cup C$, and*

$$c \vdash \{e_1, ..., e_n\} \Leftrightarrow \begin{cases} \exists u \in \Sigma^\star, \exists a_1, ..., a_n \in \Sigma, (u, \{a_1, ..., a_n\}) \in I \\ \wedge \ Eve_{\asymp_\mathcal{L}}(u) = c \wedge \ \forall i \in [1, n], e_i = \langle u, a_i\rangle_\mathcal{L} \end{cases}$$

This definition is essentially the same as Definition 2.3 of [4], the only difference being the representation of the independence relation through a local trace language. The proof that $\mathsf{les}_{\asymp_\mathcal{L}}(\mathcal{L})$ is indeed a local event structure is completely analogous to the proof in [4].

2 Generic Categorical Connection

In this section the generic approach towards relating local event structures and local trace languages is outlined. By equipping both classes of objects with suitable behavior-preserving morphisms we obtain the categories $\mathbb{LES}$ and $\mathbb{LTL}$.

Categories $\mathbb{LES}$ and $\mathbb{LTL}$. First, we provide local trace languages with morphisms which preserve sequential executions and independencies.

DEFINITION 2.1. *A (local trace) morphism λ from $\mathcal{L} = (\Sigma, I, L)$ to $\mathcal{L}' = (\Sigma', I', L')$ is a partial function $\lambda : \Sigma \rightharpoonup \Sigma'$ such that*
- $\forall (u,p) \in I$, $(\lambda(u), \lambda(p)) \in I'$;
- $\forall (u, \{a, b\}) \in I$: $a \neq b \wedge$ $\lambda(a)$ *and* $\lambda(b)$ *are both defined* $\Rightarrow \lambda(a) \neq \lambda(b)$.

We denote by $\mathbb{LTL}$ the category of LTL provided with these morphisms.

Note that, due to Axiom LTL_5 of Def. 1.2, $\lambda(L) \subseteq L'$; moreover if u_1 and u_2 are trace equivalent according to I then $\lambda(u_1)$ and $\lambda(u_2)$ are trace equivalent according to I'. Note finally that if two distinct actions a and b are independent after u and if $\lambda(a)$ and $\lambda(b)$ are both defined then they should be independent after $\lambda(u)$ in order to respect concurrency: therefore in this case we require that $\lambda(a) \neq \lambda(b)$. Morphisms of local event structures preserve the enabling relation and hence the concurrency between events.

DEFINITION 2.2. *A LES morphism η from $\mathcal{E} = (E, C, \vdash)$ to $\mathcal{E}' = (E', C', \vdash')$ is a partial function $\eta : E \rightharpoonup E'$ such that $\forall c \in C$, $\forall p \in \wp_f(E)$: $c \vdash p \Rightarrow \eta(c) \vdash' \eta(p)$. We denote by $\mathbb{LES}$ the category of LES provided with these morphisms.*

We should stress here that Winskel's prime event structures form a full subcategory of local event structures, as detailed in [3]; moreover, Mazurkiewicz' trace languages of [10] can easily be seen as a full subcategory of $\mathbb{LTL}$.

Impossibility Result. In ltl we already have a map from the objects of $\mathbb{LES}$ to those of $\mathbb{LTL}$. This map can be extended to a functor from $\mathbb{LES}$ to $\mathbb{LTL}$, because each local event structure morphism $\eta : E \rightharpoonup E'$ from $\mathcal{E}$ to $\mathcal{E}'$ induces a local trace morphism $\mathsf{ltl}(\eta) : E \rightharpoonup E'$ from $\mathsf{ltl}(\mathcal{E})$ to $\mathsf{ltl}(\mathcal{E}')$ defined by $\mathsf{ltl}(\eta) = \eta$.

LEMMA 2.3. ltl *is a functor from $\mathbb{LES}$ to $\mathbb{LTL}$.*

This is the expected intuitive translation which extends very simply the classical functor from prime event structures to trace languages [10]. On the other hand, there are in general various ways to map a local trace language to a local event structure, depending on the chosen equivalence of prime intervals. Given a family of such equivalences (one for each local trace language) one could attempt to lift the corresponding family of mappings to a functor from $\mathbb{LTL}$ to $\mathbb{LES}$. As we will see in the examples to follow, for certain families of equivalences of prime intervals this can indeed be done. Yet, none of these functors can act as a right-adjoint to ltl in an adjunction between $\mathbb{LES}$ and $\mathbb{LTL}$. In fact we even have

THEOREM 2.4. *There is no adjunction between $\mathbb{LES}$ and $\mathbb{LTL}$ with ltl as the left-adjoint.*

This result directly follows from a result (Th. 3.4) given in Section 3.

Punctuation. Given the impossibility of an adjunction between $\mathbb{LES}$ and $\mathbb{LTL}$ with $\mathfrak{ltl}$ as the left-adjoint, we now investigate the relationships between subcategories. We are particularly interested in having $\mathfrak{ltl}$ as the left-adjoint and a functor based on prime interval respecting maps as the right-adjoint. Using a generic approach which abstracts from concrete equivalences of prime intervals, it is possible to identify conditions on objects and arrows in the categories which lead to subcategories between which coreflections of the desired nature do exist.

Equivalences of prime intervals are essential to separate occurrences of the same action which are perceived as different events (see Axiom Occ of Def. 1.5). Thus the choice of such an equivalence is a *semantic issue.* In analogy with the use of punctuation in a text to influence its meaning, we call the choice of an equivalence $\asymp_{\mathcal{L}}$ for each local trace language $\mathcal{L}$ a *punctuation.* A natural condition on a punctuation is that it should respect isomorphisms between local trace languages, as otherwise behaviorally equivalent local trace languages could be given different semantics.

DEFINITION 2.5. *A punctuation is a family of equivalences $\pi = (\asymp_{\mathcal{L}})_{\mathcal{L}\in\mathbb{LTL}}$ such that each $\asymp_{\mathcal{L}}$ is an equivalence of prime intervals of $\mathcal{L}$ and for any isomorphism $\lambda : \mathcal{L} \to \mathcal{L}'$ of $\mathbb{LTL}$: $(u,a) \asymp_{\mathcal{L}} (v,b) \Rightarrow (\lambda(u),\lambda(a)) \asymp_{\mathcal{L}'} (\lambda(v),\lambda(b))$.*

Thus a punctuation π determines a representation of each local trace language $\mathcal{L}$ as a local event structure $\mathfrak{les}_{\asymp_{\mathcal{L}}}(\mathcal{L})$; hence it provides a translation $\mathfrak{les}_\pi$ from local trace languages to local event structures for which $\mathcal{L}$ maps to $\mathfrak{les}_{\asymp_{\mathcal{L}}}(\mathcal{L})$.

Before extending this map to a functor, we restrict the arrows of $\mathbb{LTL}$ to those local trace morphisms that respect the choice of events prescribed by π.

DEFINITION 2.6. *Let $\pi = (\asymp_{\mathcal{L}})_{\mathcal{L}\in\mathbb{LTL}}$ be a punctuation. A morphism $\lambda : \mathcal{L} \to \mathcal{L}'$ is π-stable if $(u,a) \asymp_{\mathcal{L}} (v,a) \wedge \ \lambda(a)$ is defined $\Rightarrow (\lambda(u),\lambda(a)) \asymp_{\mathcal{L}'} (\lambda(v),\lambda(a))$.*

Note that by Def. 2.5, local trace isomorphisms are always stable. Each π-stable local trace morphism λ from $\mathcal{L}$ to $\mathcal{L}'$ induces a local event structure morphism $\mathfrak{les}_\pi(\lambda)$ from $\mathfrak{les}_\pi(\mathcal{L})$ to $\mathfrak{les}_\pi(\mathcal{L}')$ defined by $\mathfrak{les}_\pi(\lambda)(\langle u,a\rangle_{\mathcal{L}}) = \langle\lambda(u),\lambda(a)\rangle_{\mathcal{L}'}$. Thus $\mathfrak{les}_\pi(\lambda)$ can be extended to a functor between the subcategory of $\mathbb{LTL}$ with only π-stable morphisms and the category $\mathbb{LES}$.

LEMMA 2.7. *Let $\pi = (\asymp_{\mathcal{L}})_{\mathcal{L}\in\mathbb{LTL}}$ be a punctuation; $\mathfrak{les}_\pi$ is a functor from the category of local trace languages with π-stable morphisms to $\mathbb{LES}$.*

Thus, with this restriction to π-stable morphisms, we get a morphism preserving way back from local trace languages to local event structures. As the following examples will demonstrate, for some particular punctuations π, this restriction is not a real restriction as all local trace morphisms are π-stable.

Examples. The simplest punctuation is called *Projectivity*; it consists of the equivalences $\pi^p = (\asymp^p_{\mathcal{L}})_{\mathcal{L}\in\mathbb{LTL}}$ such that each $\asymp^p_{\mathcal{L}}$ is the least equivalence over $\mathrm{Pr}(\mathcal{L})$ satisfying the conditions Ind and Cfl of Definition 1.5. Thus each $\asymp^p_{\mathcal{L}}$ identifies only those prime intervals that should definitely be identified as explained

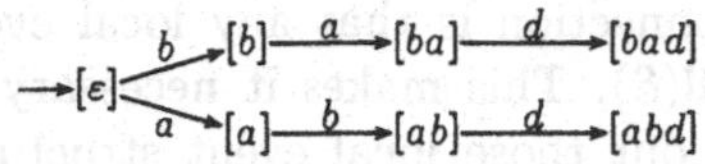

FIG. 1. $\mathcal{L}_1$

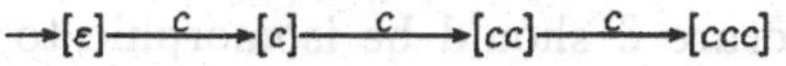

FIG. 2. $\mathcal{L}_2$

just before Definition 1.5. This rule of identification was used in [8] to connect prime event structures and Mazurkiewicz traces, and initially in [6] for the connection with 1-safe Petri nets. Note that each $\asymp^p_{\mathcal{L}}$ also satisfies Lab and Occ. So it is the least equivalence of prime intervals of $\mathcal{L}$.

The punctuation *History* π^h corresponds to the rule of identification introduced in [3] in order to connect Petri nets with local event structures. In [4] it has been formally defined by $\pi^h = (\asymp^h_{\mathcal{L}})_{\mathcal{L}\in\mathbb{LTL}}$ such that each $\asymp^h_{\mathcal{L}}$ is the least equivalence over $\mathrm{Pr}(\mathcal{L})$ satisfying Ind and Cjc (Conjunction) defined by:
Cjc : $(u,a) \in \mathrm{Pr}(\mathcal{L}) \wedge (u',a) \in \mathrm{Pr}(\mathcal{L}) \wedge \mathrm{Eve}_{\asymp_{\mathcal{L}}}(u) = \mathrm{Eve}_{\asymp_{\mathcal{L}}}(u') \Rightarrow (u,a) \asymp_{\mathcal{L}} (u',a)$
Informally speaking, Conjunction requires the identification of two occurrences of an action a, whenever the histories u and u' preceding these occurrences describe the same events and thus will lead to the same configuration in the local event structure to be associated to $\mathcal{L}$. Since every $\asymp^h_{\mathcal{L}}$ is an equivalence of prime intervals, it follows that $\asymp^p_{\mathcal{L}} \subseteq \asymp^h_{\mathcal{L}}$ for all local trace languages $\mathcal{L}$. In general this inclusion is strict: History leads to more identifications of prime intervals than Projectivity (for an example in connection with Petri nets see [4]).

The punctuation $\pi^q = (\asymp^q_{\mathcal{L}})_{\mathcal{L}\in\mathbb{LTL}}$ consists of the equivalences $\asymp^q_{\mathcal{L}}$ such that each $\asymp^q_{\mathcal{L}}$ is the least equivalence over $\mathrm{Pr}(\mathcal{L})$ which satisfies Ind and the following condition: $[(u,a) \in \mathrm{Pr}(\mathcal{L}) \wedge (u',a) \in \mathrm{Pr}(\mathcal{L}) \wedge \forall x \in \Sigma, |u|_x = |u'|_x] \Rightarrow (u,a) \asymp_{\mathcal{L}} (u',a)$. Thus in the history preceding an occurrence of action only the number of occurrences of each action is relevant. This resembles the situation in Petri nets where firing sequences lead to the same marking whenever in these sequences, for each transition its number of occurrences is the same. It is easy to see that for each local trace language $\asymp^h_{\mathcal{L}} \subseteq \asymp^q_{\mathcal{L}}$. In general, History identifies less prime intervals than π^q does. Consider for example the local trace language $\mathcal{L}_1$ depicted in Fig. 1. Here we have $(ab,d) \not\asymp^h_{\mathcal{L}_1} (ba,d)$ but $(ab,d) \asymp^q_{\mathcal{L}_1} (ba,d)$.

Note that each of the punctuations π^p, π^h, and π^q has the property that all morphisms are stable w.r.t. it. However this property does not hold for all punctuations. Consider, e.g., the *Counting* punctuation $\pi^c = (\asymp^c_{\mathcal{L}})_{\mathcal{L}\in\mathbb{LTL}}$ defined by $(u,a) \asymp^c_{\mathcal{L}} (v,b)$ iff $a = b \wedge |u|_a = |v|_b$. Thus in Counting two occurrences of an action a are identified, if a has occurred the same number of times in the histories preceding these occurrences. As an example of a local trace morphism that is not π^c-stable consider the morphism λ from $\mathcal{L}_1$ to $\mathcal{L}_2$ (depicted in Fig. 1 and 2) which is defined by $\lambda(a) = \lambda(b) = \lambda(d) = c$. This morphism is not π^c-stable because $(b,a) \asymp^c_{\mathcal{L}_1} (\varepsilon,a)$ but $(c,c) \not\asymp^c_{\mathcal{L}_2} (\varepsilon,c)$.

Main Result. In the rest of this section, we consider a fixed punctuation $\pi = (\asymp_{\mathcal{L}})_{\mathcal{L}\in\mathbb{LTL}}$. We establish that the functors ltl and les$_\pi$ form a coreflection when restricted to certain subcategories of $\mathbb{LES}$ and $\mathbb{LTL}$, determined by π.

First, one essential property of such a connection is that any local event structure $\mathcal{E}$ should be isomorphic to $\mathsf{les}_\pi \circ \mathsf{ltl}(\mathcal{E})$. This makes it necessary to cut down on the objects of $\mathbb{LES}$. We single out those local event structures which have the property that for each event of $\mathcal{E}$ all its occurrences in $\mathsf{ltl}(\mathcal{E})$ are equivalent.

DEFINITION 2.8. *A local event structure* $\mathcal{E}$ *is* π-singular *if* $\forall u_1.e, u_2.e \in \mathrm{Paths}(\mathcal{E})$, $(u_1, e) \asymp_{\mathsf{ltl}(\mathcal{E})} (u_2, e)$. $\mathbb{LES}_\pi$ *is the full subcategory of* π*-singular LES.*

As the next proposition shows, the π-singular local event structures have the desired property; moreover, as far as *finitely branching* local event structures are concerned, π-singularity is an optimal restriction.

PROPOSITION 2.9. *Let* $\mathcal{E} = (E, C, \vdash)$ *be a local event structure.*

1. *If* $\mathcal{E}$ *is* π*-singular then* $\mathcal{E}$ *is isomorphic to* $\mathsf{les}_\pi \circ \mathsf{ltl}(\mathcal{E})$*;*
2. *If* $\mathcal{E}$ *is isomorphic to* $\mathsf{les}_\pi \circ \mathsf{ltl}(\mathcal{E})$ *and if* $\forall c \in C$, $\mathrm{Card}\{e \in E \mid c \vdash \{e\}\}$ *is finite then* $\mathcal{E}$ *is* π*-singular.*

Thus now it is necessary to define the subclass of local trace languages which are mapped by les_π to π-singular local event structures.

DEFINITION 2.10. *A local trace language* $\mathcal{L}$ *is* π-adequate *if* $\mathsf{les}_\pi(\mathcal{L})$ *is* π*-singular.*

Despite these restrictions, it may however still be the case that a π-adequate local trace language $\mathcal{L}$ and its associated local event structure $\mathsf{les}_\pi(\mathcal{L})$ are not close enough. Motivated by a result from [4], we focus on the local trace languages $\mathcal{L} = (\Sigma, I, L)$ such that the equivalence $\asymp_\mathcal{L}$ satisfies Cjc and Sym (Symmetry):

Cjc : $(u, a) \in \mathrm{Pr}(\mathcal{L}) \wedge (u', a) \in \mathrm{Pr}(\mathcal{L}) \wedge \mathrm{Eve}_{\asymp_\mathcal{L}}(u) = \mathrm{Eve}_{\asymp_\mathcal{L}}(u') \Rightarrow (u, a) \asymp_\mathcal{L} (u', a)$

Sym : $(u, p) \in I \wedge u' \in L \wedge \mathrm{Eve}_{\asymp_\mathcal{L}}(u) = \mathrm{Eve}_{\asymp_\mathcal{L}}(u') \Rightarrow (u', p) \in I$.

Conjunction and Symmetry are necessary consequences of the requirement that the translation les_π should respect the sequential observations and the concurrency of a local trace language $\mathcal{L}$. Thus $\mathcal{L}$ should be such that whenever two sequences u and u' define the same configuration of $\mathsf{les}_\pi(\mathcal{L})$, then after u and u' one cannot distinguish through $\asymp_\mathcal{L}$ occurrences of the same action a (Cjc) and the same subsets of actions are independent (Sym). Together Conjunction and Symmetry guarantee that the behaviors described by $\mathcal{L}$ and $\mathsf{les}_\pi(\mathcal{L})$ are bisimilar. Here they provide sufficient conditions for a coreflection.

DEFINITION 2.11. $\mathbb{LTL}_\pi$ *is the subcategory of* $\mathbb{LTL}$ *whose arrows are the* π*-stable morphisms and whose objects are the* π*-adequate local trace languages that satisfy* Cjc *and* Sym *w.r.t. the punctuation* π*.*

We are now ready to state our main result:

THEOREM 2.12. *The functor* $\mathsf{les}_\pi : \mathbb{LTL}_\pi \to \mathbb{LES}_\pi$ *is a right-adjoint of* $\mathsf{ltl} : \mathbb{LES}_\pi \to \mathbb{LTL}_\pi$ *which forms a coreflection.*

EXAMPLE 2.13. Consider the punctuation History π^h defined earlier in the examples. By Theorem 2.12, $\mathbb{LES}_{\pi^h}$ is coreflective in $\mathbb{LTL}_{\pi^h}$, which was also proved in [4]. There however, this categorical result was proved for History alone and there was no need to introduce the notions of π-adequacy and π-stability. The following observations show how that directly obtained result fits into the generic approach of this paper. First, we note that all equivalences of History satisfy Conjunction by definition. Moreover, it is not difficult to check that all local trace languages $\mathcal{L}$ for which $\asymp^h_{\mathcal{L}}$ satisfies Symmetry are π^h-adequate. Hence the objects of $\mathbb{LTL}_{\pi^h}$ are all local trace languages $\mathcal{L}$ for which $\asymp^h_{\mathcal{L}}$ satisfies Sym and the restriction to π^h-adequate languages is void. Furthermore, as observed before, all local trace morphisms are π^h-stable. Consequently, $\mathbb{LTL}_{\pi^h}$ is the full subcategory of local trace languages satisfying Sym w.r.t. History.

In general though these restrictions may be real. Consider, e.g., the punctuation π^q which was also defined in the earlier examples. The local trace language $\mathcal{L}_1$ of Fig. 1 satisfies both Conjunction and Symmetry, but $\mathcal{L}_1$ is not π^q-adequate: in the local event structure $\mathsf{les}_{\pi^q}(\mathcal{L}_1)$ we have paths $\langle \varepsilon, a\rangle.\langle a, b\rangle.\langle ab, d\rangle$ and $\langle \varepsilon, b\rangle.\langle b, a\rangle.\langle ba, d\rangle$ with the property that the four events $\langle \varepsilon, a\rangle$, $\langle a, b\rangle$, $\langle \varepsilon, b\rangle$, and $\langle b, a\rangle$ are all distinct while $\langle ab, d\rangle = \langle ba, d\rangle$ are the same event (due to $\asymp^q_{\mathcal{L}_1}$). Thus in the local trace language of $\mathsf{les}_{\pi^q}(\mathcal{L}_1)$, the prime intervals $(\langle \varepsilon, a\rangle\langle a, b\rangle, \langle ab, d\rangle)$ and $(\langle \varepsilon, b\rangle\langle b, a\rangle, \langle ba, d\rangle)$ will not be identified even though $\langle ab, d\rangle$ and $\langle ba, d\rangle$ are the same.

The significance of Theorem 2.12 as the main result of this paper lies in its being optimal. As detailed in the next section, $\mathbb{LES}_\pi$ and $\mathbb{LTL}_\pi$ are the largest subcategories of $\mathbb{LES}$ and $\mathbb{LTL}$ for which a coreflection with ltl and les_π is possible. Moreover, any extension of the coreflection between prime event structures and Mazurkiewicz' trace languages is an instantiation of Theorem 2.12 with an appropriate punctuation π.

3 Optimal Aspects

Theorem 2.12 gives conditions on subcategories of $\mathbb{LES}$ and $\mathbb{LTL}$ in terms of the chosen family π of equivalences which guarantee that ltl and les_π form a coreflection between those subcategories. These conditions are now reconsidered. First of all, in $\mathbb{LTL}_\pi$ no other local trace morphisms can be allowed than those which never map equivalent prime intervals to non-equivalent prime intervals. Without this restriction to π-stable morphisms it would not be possible to extend the map les_π to a functor mapping morphisms to corresponding local event structure morphisms (Lemma 2.7). Secondly, if ltl and les_π are to form a coreflection, then the restriction to π-singular local event structures is necessary, at least for finitely branching local event structures (Proposition 2.9), and consequently the local trace languages have to be restricted to the π-adequate ones. Finally, the resulting subcategory of $\mathbb{LTL}$ consisting of the π-adequate local trace languages and the π-stable morphisms between them is even further restricted to its full subcategory $\mathbb{LTL}_\pi$ by the requirement that the equivalences should satisfy Conjunction and Symmetry. (Note that this is a condition on the local trace languages once π is given.) As the following proposition shows, this restriction cannot be weakened, at least not for *finitely branching* local trace languages.

PROPOSITION 3.1. *Let* $\mathbb{LTL}'$ *be a subcategory of* $\mathbb{LTL}$ *whose arrows are the* π*-stable morphisms between its objects and let* $\mathcal{L} = (\Sigma, I, L)$ *be a local trace language of* $\mathbb{LTL}'$ *such that* $\forall u \in L$, $\text{Card}\{a \in \Sigma \mid u.a \in L\}$ *is finite. If* $\mathfrak{ltl} : \mathbb{LES}_\pi \to \mathbb{LTL}'$ *admits* $\mathfrak{les}_\pi$ *as a right-adjoint then* $\mathcal{L}$ *satisfies* Cjc *and* Sym *w.r.t. the punctuation* π.

Finitely branching in local trace languages is similar to finitely branching in local event structures. In both cases only a finite number of actions are enabled at any stage of the execution. From the above observations it follows that Theorem 2.12 is optimal in the sense that each of the conditions in the definitions of $\mathbb{LES}_\pi$ and $\mathbb{LTL}_\pi$ is necessary if $\mathfrak{ltl}$ and $\mathfrak{les}_\pi$ are to form a coreflection. Thus $\mathbb{LES}_\pi$ and $\mathbb{LTL}_\pi$ are the largest subcategories of $\mathbb{LES}$ and $\mathbb{LTL}$ for which a coreflection with $\mathfrak{ltl}$ and $\mathfrak{les}_\pi$ is possible. In fact, the technique of associating a local event structure to a local trace language using an equivalence of prime intervals is a very general one. As explained below any possible extension of the coreflection from prime event structures to Mazurkiewicz' trace languages may be obtained with an appropriate punctuation. Our main technical lemma is the following.

LEMMA 3.2. *Let* $\mathbb{LES}'$ *be a full subcategory of* $\mathbb{LES}$ *which includes the prime event structures and* $\mathbb{LTL}'$ *be a full subcategory of* $\mathbb{LTL}$. *If the functor* $\mathfrak{ltl} : \mathbb{LES}' \to \mathbb{LTL}'$ *admits a right adjoint* $\mathfrak{les}' : \mathbb{LTL}' \to \mathbb{LES}'$ *then there exists a punctuation* π *such that*

A_1*:* $\mathbb{LES}' \subseteq \mathbb{LES}_\pi$ *and* $\mathbb{LTL}' \subseteq \mathbb{LTL}_\pi$*;*

A_2*:* $\forall \mathcal{L} \in \mathbb{LTL}'$, $\mathfrak{les}_\pi(\mathcal{L})$ *is isomorphic to* $\mathfrak{les}'(\mathcal{L})$.

Thus, any method $\mathfrak{les}'$ to represent local trace languages by local event structures is (modulo a renaming of the equivalence classes) based on a punctuation. One can even show that such a family of equivalences of prime intervals is unique.

LEMMA 3.3. *Let* $\mathbb{LES}'$ *and* $\mathbb{LTL}'$ *be as in Lemma 3.2 and let the functor* $\mathfrak{les}'$: $\mathbb{LTL}' \to \mathbb{LES}'$ *be a right-adjoint to* $\mathfrak{ltl}$. *If* $\pi^1 = (\asymp^1_{\mathcal{L}})_{\mathcal{L} \in \mathbb{LTL}}$ *and* $\pi^2 = (\asymp^2_{\mathcal{L}})_{\mathcal{L} \in \mathbb{LTL}}$ *are two punctuations satisfying* A_1 *and* A_2, *then for each local trace language* $\mathcal{L}$ *of* $\mathbb{LTL}'$, $\asymp^1_{\mathcal{L}} = \asymp^2_{\mathcal{L}}$.

The technical Lemma 3.2 can be translated into the following informal converse of Theorem 2.12.

THEOREM 3.4. *Any adjunction between full subcategories of* $\mathbb{LES}$ *and* $\mathbb{LTL}$ *with* $\mathfrak{ltl}$ *as left-adjoint may be obtained by Theorem 2.12 with a unique family of equivalences of prime intervals — as soon as the local event structures considered include the prime event structures.*

We should note here that this result is the basis of a proof of Th. 2.4 as follows. Assume that $\mathfrak{ltl} : \mathbb{LES} \to \mathbb{LTL}$ admits a right-adjoint. Then, by Th. 3.4, there is a punctuation π such that $\mathbb{LES} = \mathbb{LES}_\pi$ and $\mathbb{LTL} = \mathbb{LTL}_\pi$. Now it is easy to prove that $\pi = \pi^c$ the Counting punctuation. Thus $\mathbb{LTL} = \mathbb{LTL}_{\pi^c}$, a contradiction since as observed above not all local trace morphisms are π^c-stable.

Our main Theorem 2.12 presents a generic framework to extend the classical connection between Winskel's prime event structures and Mazurkiewicz' trace languages. First of all, Projectivity π^p induces a coreflection from $\mathbb{LES}_{\pi^p}$ to $\mathbb{LTL}_{\pi^p}$ which (strictly) include prime event structures and Mazurkiewicz' trace languages respectively. Furthermore, any other punctuation brings a similar extension of the classical connection. However, it seems reasonable to restrict ourself to connections between *full* subcategories of each model in order to respect the structure of the original categories. Then Theorem 3.4 asserts that any such connection is an instance of the generic coreflection of Th. 2.12 — as soon as the translation from local event structures to local trace languages is the obvious functor $\mathbb{ltl}$.

Conclusion

To conclude the paper, we now briefly consider how the theory presented here relates to the event structure semantics of Petri nets. Recall first that the firing sequences of a 1-safe net determine Mazurkiewicz' traces and thus a prime event structure [10]. Similarly, as established in [2], each Place/Transition net is naturally associated to a local trace language; this latter can then be represented by a local event structure with the help of an appropriate punctuation. Applying this method to History leads precisely to the semantics introduced in [3]. Now different punctuations would lead to different event structure semantics that should be studied more precisely in the future. Furthermore the complete generic connection presented here should be also useful to propose semantics for generalized Petri nets with capacities, read arcs or inhibitor arcs.

Acknowledgment. Thanks to the constructive criticism of some anonymous referees on a previous version of this paper, we could improve the presentation of our results.

References

1. Degano, P., Gorrieri, R., Vigna, S.: *On Relating some Models for Concurrency.* TAPSOFT, LNCS **668** (1993) 15–30
2. Hoogers, P.W., Kleijn, H.C.M., Thiagarajan, P.S.: *A Trace Semantics for Petri Nets.* Information and Computation **117** (1995) 98–114
3. Hoogers, P.W., Kleijn, H.C.M., Thiagarajan, P.S.: *An Event Structure Semantics for General Petri Nets.* Theoretical Computer Science **153** (1996) 129–170
4. Kleijn, H.C.M., Morin, R., Rozoy, B.: *Event Structures for Local Traces.* Electronic Notes in Theoretical Computer Science **16-2** (1998) – 16 pages
5. Mazurkiewicz, A.: *Trace Theory.* Advanced Course on Petri Nets, Bad Honnef, Germany, LNCS **254** (1987) 269–324
6. Nielsen, M., Plotkin, G., Winskel, G.: *Petri nets, events structures and domains, Part I.* Theoretical Computer Science **13** (1981) 85–108
7. Nielsen, M., Sassone, V., Winskel, G.: *Relationships between Models of Concurrency.* LNCS **803** (1994) 425–475
8. Rozoy, B., Thiagarajan, P.S.: *Trace monoids and event structures.* Theoretical Computer Science **91** (1991) 285–313
9. Pierce, B.C.: *Category Theory for Computer Scientists.* (The MIT Press, 1991)
10. Winskel, G., Nielsen, M.: *Models for Concurrency.* In *Handbook of Logic in Computer Science, Vol. 4: Semantic Modelling,* S. Abramsky, D.M. Gabbay and T.S.E. Maibaum, eds (Oxford University Press, Oxford, 1995) 1–148

Correct Translation of Mutually Recursive Function Systems into TOL Collage Grammars*

Renate Klempien-Hinrichs, Hans-Jörg Kreowski, and Stefan Taubenberger

University of Bremen, Department of Computer Science,
P.O. Box 33 04 40, D-28334 Bremen, Germany
{rena,kreo,taube}@informatik.uni-bremen.de

Abstract. In this paper, mutually recursive function systems, picture-generating devices known in the area of fractal geometry, are translated into TOL collage grammars. The translation works in such a way that the infinite sequence of pictures which is specified by a mutually recursive function system through the Hutchinson operator contains exactly those pictures which belong to the language generated by the corresponding TOL collage grammar. In other words, the translation is correct.

1 Introduction

In this paper, we relate a picture-generating device from the area of fractal geometry with a grammatical approach to picture generation. More precisely, we construct a translation of mutually recursive function systems into TOL collage grammars and prove the correctness of this translation, meaning that every input system generates the same pictures as the resulting grammar.

Mutually recursive function systems are a generalization of the much better known and frequently used iterated function systems. An iterated function system (see, e.g., Hutchinson [Hut81] and Barnsley [Bar88]) consists of a finite set of transformations on $\mathbb{R}^d$. If all transformations are applied to a picture, i.e., a set of points, then the union of the results is again a picture. This construction of a picture from a picture is called Hutchinson operator. It can be repeated ad infinitum. The resulting infinite picture sequence converges if all transformations of the iterated function system are contractions and the initial picture is a compact set. The limit is called attractor and can be shown to be the fixed point of the Hutchinson operator. This means that the attractor is the union of its own transformed copies. In the case of affine transformations, this phenomenon is called self-affinity (and self-similarity if all transformations are similarity transformations). If one of the transformations is the identity, the

* This research was partially supported by the EC TMR Network GETGRATS (General Theory of Graph Transformation Systems), the ESPRIT Basic Research Working Group APPLIGRAPH (Applications of Graph Transformation), and the *Deutsche Forschungsgemeinschaft* (DFG) under grant no. Kr-964/6-1.

attractor still exists, but depends on the initial picture. A function system of this type is called iterated function system with condensation.

A mutually recursive function system, as it is called by Culik II and Dube [CD93a], is not just an unstructured set of transformations, but may be seen as a directed graph the edges of which are labelled with transformations. In this case, the Hutchinson operator gets a vector of pictures – one picture for each node –, applies the transformation of each edge to the picture of the source node, returns the result to the target node, and overlays the resulting pictures at each of the nodes. Therefore, the repetition of the Hutchinson operator yields an infinite sequence of vectors of pictures, where the picture sequence of one distinguished display node is considered as the semantics of the system. Peitgen, Jürgens, and Saupe [PJS92] call such function systems *networked multiple reduction copy machines* and *hierarchical iterated function systems*. Moreover, various other denominations are used in the literature they refer to.

Collage grammars (see, e.g., Drewes, Habel, Kreowski, and Taubenberger [DK99, HKT93]) are rule-based, syntactic picture-generating devices, whose generating mechanism is based on hyperedge replacement as known from the area of graph transformation (see, e.g., Drewes, Habel, and Kreowski [DHK97, Hab92]).

A collage consists of a set of parts and a sequence of pin points. A part may be an arbitrary set of points (in a Euclidean space) – usually taken from some standard set of geometric objects like line segments, circles, triangles, polygons, polyhedra, etc. that have simple finite descriptions and are easy to deal with on graphical user interfaces. The pin points are used to paste collages into collages. A collage represents a picture by the overlay of its parts.

To generate collages from collages by the application of productions, they are decorated with hyperedges in intermediate steps. A hyperedge is a labelled item with an ordered number of tentacles each of which is attached to a point. A hyperedge serves as a place holder. It may eventually be replaced by a decorated collage, provided that there is a transformation of the pin points into the attachment points of the hyperedge. This kind of hyperedge replacement establishes the rewrite steps of a collage grammar if the label of the replaced hyperedge and the replacing decorated collage form a production. The grammar is a TOL collage grammar if it provides a set of tables, i.e., a set of sets of productions, and in each derivation step all hyperedges are replaced simultaneously, using productions from only one of the tables. A TOL collage grammar generates a set of collages in the usual way of language generation.

While mutually recursive function systems generate pictures from pictures in a bottom-up fashion, TOL collage grammars derive pictures top-down by refining and eventually terminating nonterminal hyperedges within intermediate pictures. Nevertheless, both picture-generating processes are related to each other in a meaningful way, as our main result shows.

The paper is organized in the following way. The basic notions and notations of collage grammars and mutually recursive function systems are recalled in Sections 2 and 3. The translation is constructed and proved correct in Section 4. Finally, Section 5 contains some concluding remarks.

2 TOL collage grammars

In this section, the basic notions and notations concerning collages and TOL collage grammars (cf. [DK99, DKK99]) are recalled and illustrated by two examples.

For a set A, A^* denotes the set of all finite sequences over A, including the empty sequence λ; $\wp(A)$ denotes its powerset; and $A \setminus B$ denotes the complement of a set B in A. For a function $f: A \to B$, the canonical extensions of f to $\wp(A)$ and to A^* are denoted by f as well, i.e., $f(S) = \{f(a) \mid a \in S\}$ for $S \subseteq A$ and $f(a_1 \cdots a_n) = f(a_1) \cdots f(a_n)$ for $a_1, \ldots, a_n \in A$.

As usual, for a binary relation $\Longrightarrow \subseteq S \times S$ one may write $s \Longrightarrow s'$ instead of $(s, s') \in \Longrightarrow$ and $s_0 \Longrightarrow^n s_n$ or $s_0 \Longrightarrow^* s_n$ to abbreviate $s_0 \Longrightarrow s_1 \Longrightarrow \cdots \Longrightarrow s_n$.

The sets of natural numbers and real numbers are denoted by $\mathbb{N}$ and $\mathbb{R}$, respectively. Familiarity with the basic notions of Euclidean geometry is assumed (see, e.g., Coxeter [Cox89]). $\mathbb{R}^d$ denotes the Euclidean space of dimension d for some $d \geq 1$, which is equipped with the usual Euclidean distance function $dist: \mathbb{R}^d \times \mathbb{R}^d \to \mathbb{R}$.

A collage consists of a set of geometric objects, called parts, and a sequence of so-called pin points. To generate sets of collages, they are decorated with hyperedges. A hyperedge has a label and an ordered finite set of tentacles, each of which is attached to a point. A handle – a collage consisting of a single hyperedge – is particularly useful as initial collage. Each collage specifies a picture, called induced pattern, by the overlay of all its parts. All these notions are defined more precisely as follows.

A *collage* (in $\mathbb{R}^d$) is a pair $(PART, pin)$, where $PART \subseteq \wp(\mathbb{R}^d)$ is a finite set of so-called *parts* and $pin \in (\mathbb{R}^d)^*$ is the sequence of *pin points*. $\mathcal{C}$ denotes the class of all collages. The overlay of the set of all parts of a collage C yields the *induced pattern* $pattern(C) = \bigcup_{part \in PART_C} part$.

Let N be a set of *labels* and let, for each $A \in N$, $pin_A \in (\mathbb{R}^d)^*$ be a fixed sequence of pin points associated with A. A *(hyperedge-)decorated collage* (over N) is a construct $C = (PART, E, att, lab, pin)$, where $(PART, pin)$ is a collage, E is a finite set of *hyperedges*, $att: E \to (\mathbb{R}^d)^*$ is a mapping, called the *attachment*, and $lab: E \to N$ is a mapping, called the *labelling*, such that for each hyperedge $e \in E$ there is a unique affine transformation $a(e): \mathbb{R}^d \to \mathbb{R}^d$ satisfying $att(e) = a(e)(pin_{lab(e)})$. The class of all decorated collages over N is denoted by $\mathcal{C}(N)$, and $PART_C$, E_C, att_C, lab_C, and pin_C denote the components of $C \in \mathcal{C}(N)$.

A collage can be seen as a decorated collage C where $E_C = \emptyset$. In this sense, $\mathcal{C} \subseteq \mathcal{C}(N)$. Thus, we may drop the components E_C, att_C, and lab_C if $E_C = \emptyset$. Furthermore, we will use the term *collage* also in the case of decorated collages as long as this is not a source of misunderstandings.

Example 1 (collage). Collages will be depicted as shown in Figure 1. Pin points and attachment points of hyperedges are drawn as small bullets. The pin points are numbered according to their order in the sequence of pin points. A hyperedge is represented as a square carrying its label inside, together with numbered lines

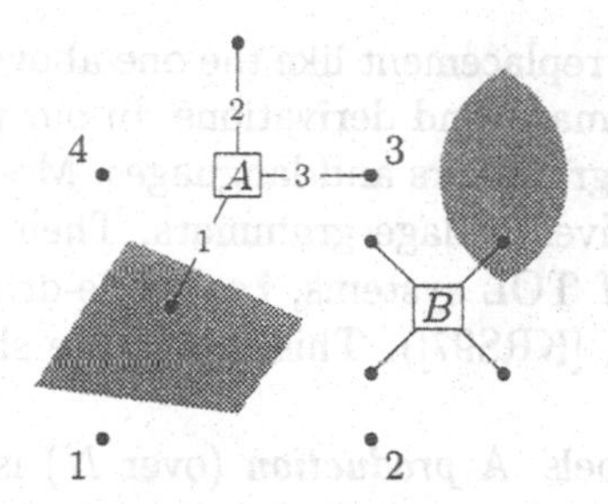

Fig. 1. A collage

indicating the attachment. If the sequence of attachment points is obtained from the sequence of pin points by scaling and translation only, then the numbers are dropped (see the B-labelled hyperedge). ◁

Let A be a label in N. The ***handle induced by A*** is the decorated collage $A^\bullet = (\emptyset, \{e\}, att, lab, pin_A)$ with $lab(e) = A$ and $att(e) = pin_A$. (Recall that pin_A denotes the sequence of pin points associated with the label A.)

Removing a set $B \subseteq E_C$ of hyperedges from a collage yields the collage $C - B = (PART_C, E_C \setminus B, att, lab, pin_C)$, where att and lab are the restrictions of att_C and lab_C to $E_C \setminus B$, respectively.

The addition of a collage C' to a collage C is defined by $C + C' = (PART_C \cup PART_{C'}, E_C \uplus E_{C'}, att, lab, pin_C)$, where $\uplus$ denotes the disjoint union of sets,

$$att(e) = \begin{cases} att_C(e) & \text{for } e \in E_C \\ att_{C'}(e) & \text{for } e \in E_{C'}, \end{cases} \quad \text{and} \quad lab(e) = \begin{cases} lab_C(e) & \text{for } e \in E_C \\ lab_{C'}(e) & \text{for } e \in E_{C'}. \end{cases}$$

As an abbreviation, one may write $C + \sum_{i=1}^{n} C_i$ for $C + C_1 + \cdots + C_n$.

The transformation of a collage C by a transformation $a \colon \mathbb{R}^d \to \mathbb{R}^d$ is given by $a(C) = (a(PART_C), E_C, att, lab_C, a(pin_C))$, where $att(e) = a(att_C(e))$ for all $e \in E_C$.

Hyperedges in decorated collages serve as place holders for (decorated) collages. Hence, the key construction is the replacement of a hyperedge in a decorated collage with a (decorated) collage. While a hyperedge is attached to some points, a (decorated) collage is equipped with a number of pin points. If there is an affine transformation which maps the pin points to the attached points of the hyperedge, the transformed (decorated) collage may replace the hyperedge.

Let $C \in \mathcal{C}(N)$ be a collage and let $B \subseteq E_C$. Furthermore, let $repl \colon B \to \mathcal{C}(N)$ be a mapping such that, for all $e \in B$, there is a unique affine transformation $a(e)$ which satisfies $att_C(e) = a(e)(pin_{repl(e)})$. Then the ***replacement of B in C through repl*** yields the decorated collage $C[repl]$ constructed by

1. removing the hyperedges in B from C,
2. transforming $repl(e)$ by $a(e)$ for all $e \in B$, and
3. adding the transformed collages $a(e)(repl(e))$ to $C - B$.

Thus, to be more precise, $C[repl] = (C - B) + \sum_{e \in B} a(e)(repl(e))$.

Whenever a notion of replacement like the one above is given, one may easily define productions, grammars, and derivations. In our particular case this leads to the notions of collage grammars and languages. More specifically, we will directly introduce table-driven collage grammars. Their definition is inspired by the well-known notion of TOL systems, i.e., table-driven context-free Lindenmayer systems (see, e.g., [KRS97]). This is why we shall call these grammars TOL collage grammars.

Let N be a set of labels. A *production* (over N) is a pair $p = (A, R)$ with $A \in N$ and $R \in \mathcal{C}(N)$ such that pin_A is the pin point sequence of R. A is called the *left-hand side* of p and R is its *right-hand side*, denoted by $lhs(p)$ and $rhs(p)$, respectively. A production $p = (A, R)$ is also denoted by $A ::=_p R$ or simply $A ::= R$.

Let $C \in \mathcal{C}(N)$, $B \subseteq E_C$, and let P be a set of productions over N. We call $b\colon B \to P$ a *base* on B in C if, for all $e \in B$, $lab_C(e) = lhs(b(e))$. As, moreover, $pin_{lhs(b(e))} = pin_{rhs(b(e))}$, there is a unique affine transformation $a(e)$ which satisfies the equation $att_C(e) = a(e)(pin_{rhs(b(e))})$; for all $e \in B$. Thus we can say that C *directly derives* $C' \in \mathcal{C}(N)$ *through* b if $C' = C[repl]$, where $repl(e) = rhs(b(e))$ for all $e \in B$. A direct derivation is denoted by $C \Longrightarrow_P C'$. A sequence $C \Longrightarrow_P^* C'$ of direct derivations is called a *derivation* from C to C'.

Definition 2 (***TOL collage grammar***)**.** A *TOL collage grammar* is a system $G = (N, T, Z)$, where N is a finite set of *nonterminal labels* (or *nonterminals*, for short), T is a finite set of *tables* with each table $P \in T$ being a finite set of productions over N such that $\{lhs(p) \mid p \in P\} = N$, and a decorated collage $Z \in \mathcal{C}(N)$ is the *axiom*.

For collages $C, C' \in \mathcal{C}(N)$ we write $C \Longrightarrow_T C'$ if C derives C' through a base $b\colon E_C \to P$ for some $P \in T$. The collage language generated by G is

$$L(G) = \{C \in \mathcal{C} \mid Z \overset{*}{\underset{T}{\Longrightarrow}} C\}. \qquad \diamond$$

One may write $C \Longrightarrow C'$ for $C \Longrightarrow_T C'$ if T is understood. Notice that every direct derivation is required to replace *all* hyperedges of a collage in parallel. This is essential because otherwise the division of the set of productions into tables would be meaningless. Notice also that every table contains at least one production for each left-hand side $A \in N$, which implies that every table can be applied at any step of a derivation.

Examples 3. The first example is the TOL collage grammar $G_{\text{HILBERT}} = (\{A, B\}, \{\{p_1, p_2\}, \{p_1', p_2'\}\}, B^\bullet)$, where p_1, p_2, p_1', and p_2' are as shown in Figure 2 and $pin_B = pin_{rhs(p_2)}$. This grammar generates all regular approximations of the Hilbert curve, of which some are shown in Figure 3.

The second example is the collage grammar $G_{\text{SPECKLES}} = (\{A, B, C\}, \{T_1, T_2\}, A^\bullet)$, where $T_1 = \{p_1, p_2, p_3\}$ and $T_2 = \{p_1', p_2', p_3'\}$ contain productions as shown in Figure 4 and $pin_A = pin_{rhs(p_1)}$. The patterns of some of the collages in the language L_{SPECKLES} are depicted in Figure 5. $\triangleleft$

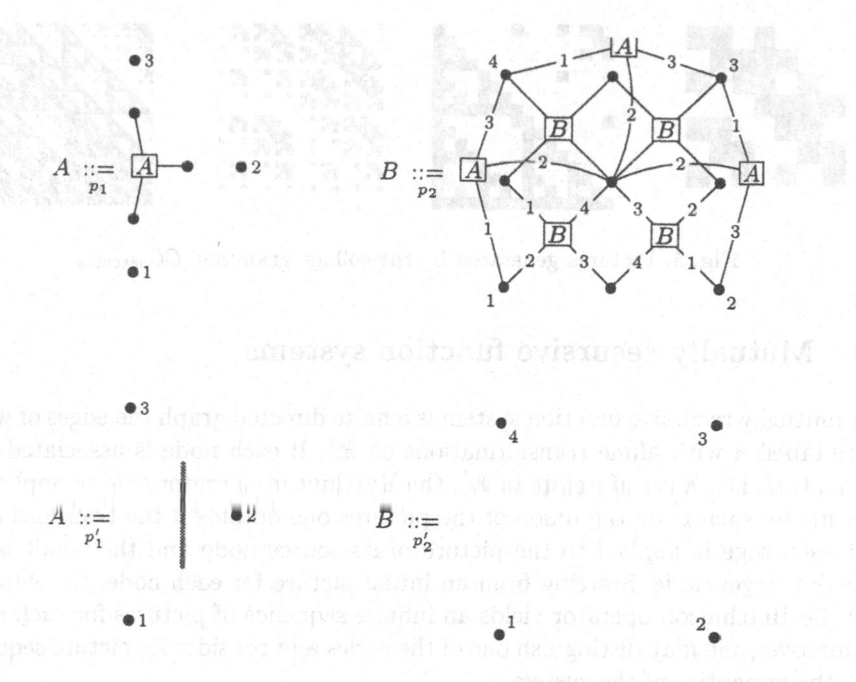

Fig. 2. The productions of the TOL collage grammar CG_{HILBERT} in Example 3

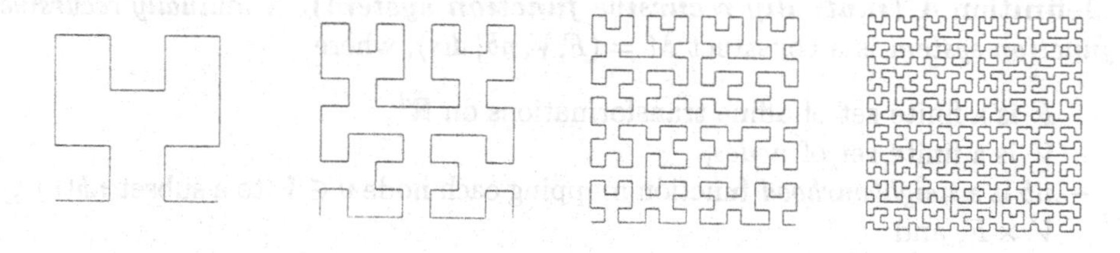

Fig. 3. Pictures generated by the collage grammar CG_{HILBERT}

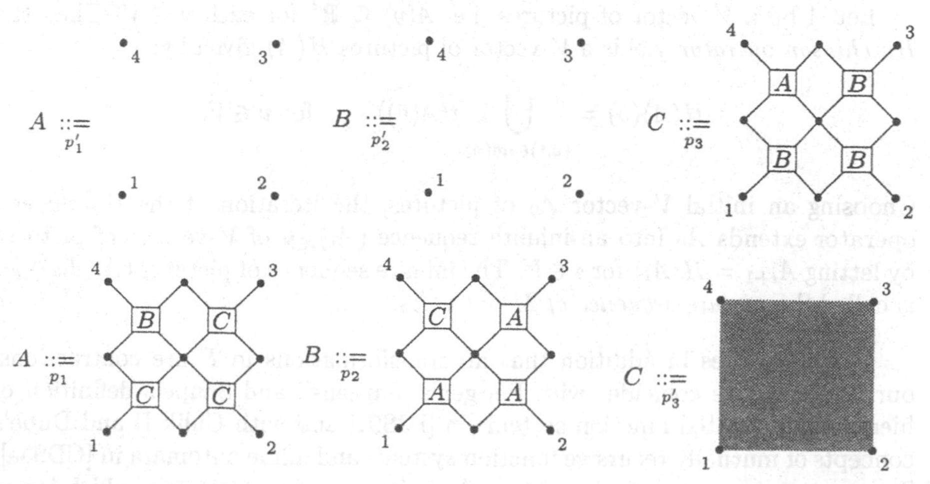

Fig. 4. The productions of the TOL collage grammar CG_{SPECKLES} in Example 3

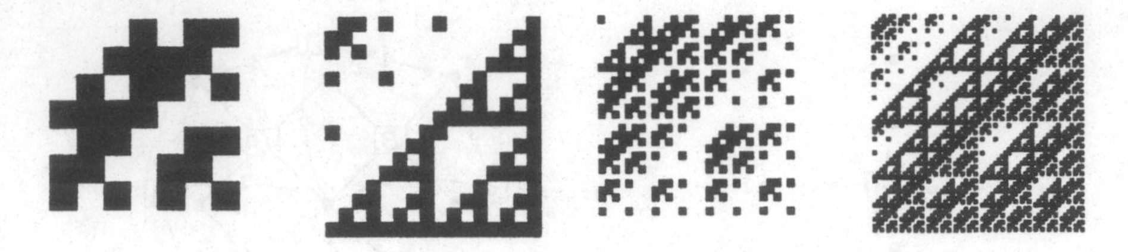

Fig. 5. Pictures generated by the collage grammar CG_{SPECKLES}

3 Mutually recursive function systems

A mutually recursive function system is a finite directed graph the edges of which are labelled with affine transformations on $\mathbb{R}^d$. If each node is associated with a picture, i.e., a set of points in $\mathbb{R}^d$, the Hutchinson operator can be applied. It yields for each node the union of the pictures one obtains if the transformation of each edge is applied to the picture of its source node and the result is put to the target node. Starting from an initial picture for each node, the iteration of the Hutchinson operator yields an infinite sequence of pictures for each node. Moreover, one may distinguish one of the nodes and consider its picture sequence as the semantics of the system.

Definition 4 (*mutually recursive function system*). A *mutually recursive function system* is a construct $M = (F, V, nb, dis)$, where

- F is a finite set of affine transformations on $\mathbb{R}^d$,
- V is a finite set of *nodes,*
- nb is a *neighbourhood* function mapping each node $v \in V$ to a subset $nb(v) \subseteq V \times F$, and
- $dis \in V$ is the *display node.*

Let A be a V-vector of pictures, i.e. $A(v) \subseteq \mathbb{R}^d$ for each $v \in V$. Then the *Hutchinson operator* yields a V-vector of pictures $H(A)$ given by:

$$H(A)(v) = \bigcup_{(\bar{v},t)\in nb(v)} t(A(\bar{v})) \qquad \text{for } v \in V.$$

Choosing an initial V-vector A_0 of pictures, the iteration of the Hutchinson operator extends A_0 into an infinite sequence $(A_i)_{i\in\mathbb{N}}$ of V-vectors of pictures by letting $A_{i+1} = H(A_i)$ for $i \in \mathbb{N}$. The infinite sequence of pictures $(A_i(dis))_{i\in\mathbb{N}}$ is called the *picture sequence of M* w.r.t. A_0. ◇

If one requires in addition that all transformations in F are contractions, our notion above coincides with Peitgen's, Jürgens', and Saupe's definition of hierarchical iterated function systems in [PJS92] and with Culik II and Dube's concepts of mutually recursive function systems and affine automata in [CD93a]. The contraction property guarantees the existence of an attractor, which is not needed in this paper (cf. the end of this section).

The graph structure underlying a mutually recursive function system is given in terms of a neighbourhood function rather than a set of edges because the Hutchinson operator is more easily defined in this way. However, a neighbourhood function nb represents the set of edges $E = \{(\bar{v}, t, v) \mid (\bar{v}, t) \in nb(v)\}$ where the first component is the source node, the second the label, and the third the target node. Conversely, a set of edges $E \subseteq V \times F \times V$ induces the neighbourhood function given by $nb(v) = \{(\bar{v}, t) \mid (\bar{v}, t, v) \in E\}$ for all $v \in V$.

Examples 5. The first example is the mutually recursive function system HILBERT which approximates the Hilbert curve. It consists of two nodes and eight edges forming the graph

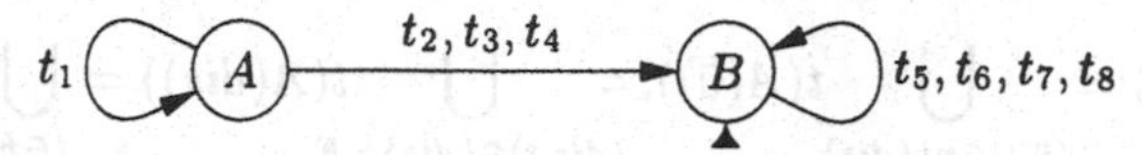

where the display node B is indicated by ▲ and parallel edges are represented by a single edge labelled with the list of the original labels. The eight affine transformations are defined according to the following table giving, for each transformation, the scaling factor, the rotation angle in degrees (for counter-clockwise rotation), and the translation vector:

	scaling factor	rotation angle	translation vector
t_1	1/2	0	(0,0)
t_2	1	0	(0,0)
t_3	1	270	(2,2)
t_4	1	180	(4,0)
t_5	1/2	0	(0,1)
t_6	1/2	0	(2,1)
t_7	1/2	270	(1,0)
t_8	1/2	90	(3,−2)

In Figure 3, the fourth to seventh picture of the picture sequence of HILBERT w.r.t. H_0 are depicted, where the vector H_0 comprises as initial picture for node A the straight line with the end points $(1, -1)$ and $(1, 1)$, and as initial picture for node B the empty set of points. The i-th iteration of the Hutchinson operator yields at the node A the straight line with end points $(2^{-i}, -2^{-i})$ and $(2^{-i}, 2^{-i})$, for $i \in \mathbb{N}$.

The second example SPECKLES is given by the graph:

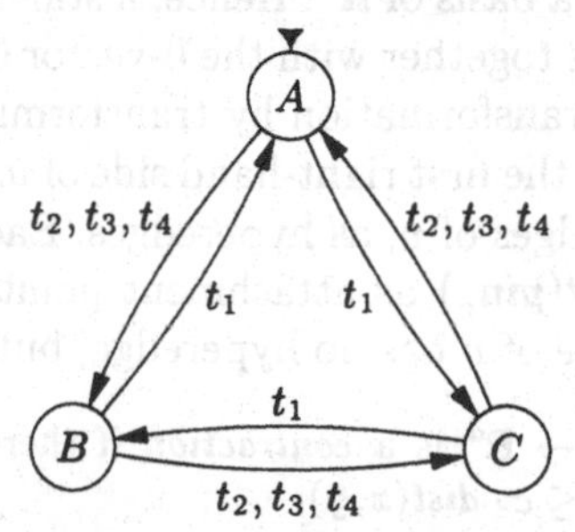

The four affine transformations have all the scaling factor 1/2. In addition, t_1 translates by (0,1), t_3 by (1,0), and t_4 by (1,1).

If one chooses as initial pictures of the nodes A and B the empty set and as initial picture for the node C the square with the corners (0,0), (2,0), (2,2), and (0,2), forming the vector S_0, one obtains the pictures in Figure 5 as the fifth to eighth picture of the picture sequence of SPECKLES w.r.t. S_0. ◁

If a mutually recursive function system $M = (F, V, nb, dis)$ has only one node, i.e. $V = \{dis\}$, and maximum neighbourhood, i.e. $nb(dis) = \{dis\} \times F$, then M is uniquely determined by the set F. In this case, the Hutchinson operator just applies all transformations to the given picture and overlays the results:

$$H(A)(dis) = \bigcup_{(\bar{v},t)\in nb(dis)} t(A(\bar{v})) = \bigcup_{(dis,t)\in\{dis\}\times F} t(A(dis)) = \bigcup_{t\in F} t(A(dis))$$

In other words, such a mutually recursive function system coincides with the well-known notion of an iterated function system. If, moreover, the set of transformations contains the identity, the case of iterated function systems with condensation is covered.

Under certain conditions, mutually recursive function systems allow the definition of fractal pictures. If the compositions of the transformations along all simple cycles in the underlying graph of the mutually recursive function system are contractions[1], the infinite sequence of V-vectors of pictures turns out to be a Cauchy sequence provided that the initial pictures are compact sets. Hence it has a limit, which is often called the *attractor* of the mutually recursive function system and can be shown to be the fixed point of the Hutchinson operator.

4 Correct translation

We are now able to present the correct translation of mutually recursive function systems together with initial vectors of pictures into TOL collage grammars. The nodes of an input system become the nonterminal labels with the display node as axiom. For each such label v, there are two rules with v as left-hand side. The required pin points can be chosen arbitrarily; but to guarantee correctness, the pin points of v must uniquely identify affine transformations, i.e., $t_1(pin_v) = t_2(pin_v)$ implies $t_1 = t_2$. This means that there are at least $d+1$ distinct pin points and d of them form a basis of $\mathbb{R}^d$. Hence, a standard choice of pin points is the set of the d unit vectors together with the 0-vector $\mathbb{O}$. The pin point condition allows the encoding of a transformation by transforming the pin points. This is used in the construction of the first right-hand side of v. It has the neighbourhood $nb(v)$, i.e., the incoming edges of v, as hyperedges. Each such hyperedge $(\bar{v}, t) \in nb(v)$ has $\bar{v}$ as label and $t(pin_v)$ as attachment points, encoding t in this way. The second right-hand side of v has no hyperedge, but the initial picture of v as

[1] A transformation $A\colon \mathbb{R}^d \to \mathbb{R}^d$ is a *contraction* if there is $c < 1$ such that for all $x, y \in \mathbb{R}^d$, $dist(t(x), t(y)) \le c \cdot dist(x, y)$.

only part. All rules of the latter type belong to one table, all rules of the former type to another.

Construction 6. Let $M = (F, V, nb, dis)$ be a mutually recursive function system and A_0 an initial V-vector of pictures, i.e., $A_0(v) \in \mathbb{R}^d$ for $v \in V$. Moreover, let $\boldsymbol{pin}$ be an arbitrary choice of pin points for each $v \in V$, i.e., $pin_v \in (\mathbb{R}^d)^*$, subject to the condition that $t_1(pin_v) = t_2(pin_v)$ implies $t_1 = t_2$ for all affine transformations t_1 and t_2.

Then the translation of M, A_0, and $\boldsymbol{pin}$ yields the TOL collage grammar $CG(M, A_0, pin) = (V, \{P_1, P_2\}, dis^\bullet)$, where $P_1 = \{(v, R_1(v)) \mid v \in V\}$ with $R_1(v) = (\emptyset, nb(v), att_v, lab_v, pin_v)$ and att_v and lab_v defined by $att_v((\bar{v}, t)) = t(pin_{\bar{v}})$ and $lab_v((\bar{v}, t)) = \bar{v}$ for all $(\bar{v}, t) \in nb(v)$, and $P_2 = \{(v, R_2(v)) \mid v \in V\}$ with $R_2(v) = (\{A_0(v)\}, \emptyset, \emptyset, \emptyset, pin_v)$.

The pin points do not have to be part of the input of the translator because there is a standard choice given by $\mathbb{O}u_1 \cdots u_d$ with $\mathbb{O} = (0, \ldots, 0)$ and $u_i = (\underbrace{0, \ldots, 0}_{i-1}, 1, \underbrace{0, \ldots, 0}_{d-i})$ for $i = 1, \ldots, d$. □

Examples 7. Consider the mutually recursive function system HILBERT and the vector H_0 of initial pictures given in Examples 5. If one chooses $pin_A = (0, -2)(2, 0)(0, 2)$ as pin points of A and $pin_B = (0, -2)(4, -2)(4, 2)(0, 2)$ as pin points of B, then the translation yields the TOL collage grammar CG_{HILBERT} of Examples 3, i.e., $CG(\text{HILBERT}, H_0, pin) = CG_{\text{HILBERT}}$.

Analogously, the translation of SPECKLES and the vector S_0 of Examples 5 yields CG_{SPECKLES} of Examples 3 if the pin points are chosen as $pin_A = pin_B = pin_C = (0, 0)(2, 0)(2, 2)(0, 2)$. ◁

The TOL collage grammars resulting from the translation have a special structure. In each of the two tables, there is exactly one rule for every label, and all right-hand sides of the second table are terminal. Hence, every derivation applies the first table for some steps and ends possibly with an application of the second table. Moreover, each derivation step is uniquely determined by the choice of the table. In particular, for each v there is exactly one terminal collage $C_{v,n}$ which is derived from the handle of v through n applications of the first table followed by an application of the second table. If one applies the Context-Freeness Lemma of collage grammars to such a derivation for $n+1$, one obtains a characterization of $C_{v,n+1}$ in terms of $R_1(v)$ and $C_{\bar{v},n}$ for all labels $\bar{v}$. Altogether, these considerations lead to the following observations.

Observations 8. Let M, A_0, $\boldsymbol{pin}$ and $CG(M, A_0, pin)$ be as in the construction above. Then the following hold.

1. For $v \in V$ and $n \in \mathbb{N}$, there is exactly one derivation of the form
$$v^\bullet \underset{P_1}{\overset{n}{\Longrightarrow}} \bar{C}_{v,n} \underset{P_2}{\Longrightarrow} C_{v,n}.$$
2. Let $v^\bullet \Longrightarrow^* C$ be a derivation with $v \in V$ and $C \in \mathcal{C}$. Then there is $n \in \mathbb{N}$ with $C_{v,n} = C$.

3. In particular, $L(CG(M, A_0, pin)) = \{C_{dis,n} \mid n \in \mathbb{N}\}$.
4. For $v \in V$ and $n \in \mathbb{N}$, we have $C_{v,n+1} = R_1(v)[repl]$ with $repl\colon nb(v) \to C_V$ given by $repl((\bar{v}, t)) = C_{\bar{v},n}$ for all $(\bar{v}, t) \in nb(v)$ (with $t(pin_{\bar{v}}) = att_v((\bar{v}, t))$).
5. In particular,

$$PART_{C_{v,n+1}} = \bigcup_{(\bar{v},t)\in nb(v)} t(PART_{C_{\bar{v},n}}).$$ □

These observations allow to prove the correctness of the translations, i.e., that the derivation process of the collage grammar simulates the picture generation process given by the Hutchinson operator.

Theorem 9. *Let M, A_0, pin, and $CG(M, A_0, pin)$ be as in the construction above. Let $(A_i)_{i\in\mathbb{N}}$ be the infinite sequence of V-vectors of pictures specified by M and A_0 through the Hutchinson operator. Let $C_{v,n}$ for $v \in V$ and $n \in \mathbb{N}$ be as in the observations above.*

Then $pattern(C_{v,n}) = A_n(v)$ for all $v \in V$ and $n \in \mathbb{N}$.

In particular, $pattern(L(CG(M, A_0, pin))) = \{A_n(dis) \mid n \in \mathbb{N}\}$.

Proof (by induction on n). If $n = 0$, then $pattern(C_{v,0}) = pattern(R_2(v)) = pattern(\{A_0(v)\}) = A_0(v)$.

Let the statement hold for n. Using the observations above, the induction hypothesis, and some properties of sets of points and affine transformations, we obtain for $n + 1$:

$$\begin{aligned}
pattern(C_{v,n+1}) &= \{x \in \mathbb{R}^d \mid x \in part,\ part \in PART_{C_{v,n+1}}\} \\
&= \{x \in \mathbb{R}^d \mid x \in part,\ part \in t(PART_{C_{\bar{v},n}}),\ (\bar{v}, t) \in nb(v)\} \\
&= \{x \in \mathbb{R}^d \mid x \in t(part'),\ part' \in PART_{C_{\bar{v},n}},\ (\bar{v}, t) \in nb(v)\} \\
&= \{x \in \mathbb{R}^d \mid x \in t(pattern(C_{\bar{v},n})),\ (\bar{v}, t) \in nb(v)\} \\
&= \{x \in \mathbb{R}^d \mid x \in t(A_n(\bar{v})),\ (\bar{v}, t) \in nb(v)\} = A_{n+1}(v)
\end{aligned}$$

Finally, the proof is completed by the following specialization:

$$\begin{aligned}
pattern(L(CG(M, A_0, pin))) &= pattern(\{C_{dis,n} \mid n \in \mathbb{N}\}) \\
&= \{pattern(C_{dis,n}) \mid n \in \mathbb{N}\} \\
&= \{A_n(dis) \mid n \in \mathbb{N}\}.
\end{aligned}$$ □

5 Conclusion

We have presented a correct translation of mutually recursive function systems into TOL collage grammars. Clearly, this sheds some light on the relationship between two picture generation methods which are inspired by different areas: fractal geometry and formal language theory. What is gained in addition? First of all, as iterated and mutually recursive function systems are frequently used in visualization and animation, the translation provides a wealth of examples of TOL collage grammars – and of context-free collage grammars as well by ignoring the table structure. Moreover, one may try to carry over concepts and results known for iterated and mutually recursive function systems to TOL collage grammars, and vice versa. Working out such synergy effects will be subject of future research. The results do not seem to be obvious because the two approaches have quite different focusi.

While collage grammars and mutually recursive function systems are studied in this paper, Culik II and Dube [CD93b] have related mutually recursive function systems with L-systems – and there are other candidates for comparison like chain-code picture languages. Further research is expected to lead to a unified theory of picture-generating devices.

Acknowledgement. We would like to thank Frank Drewes for many enlightening discussions on the subjects of this paper and the anonymous referees for their helpful hints.

References

[Bar88] Michael Barnsley. *Fractals Everywhere.* Academic Press, Boston, 1988.

[CD93a] Karel Culik II and Simant Dube. Affine automata and related techniques for generation of complex images. *Theoretical Computer Science*, 116:373–398, 1993.

[CD93b] Karel Culik II and Simant Dube. L-systems and mutually recursive function systems. *Acta Informatica*, 30:279–302, 1993.

[Cox89] H.S.M. Coxeter. *Introduction to Geometry.* John Wiley & Sons, New York, Wiley Classics Library, second edition, 1989.

[DHK97] Frank Drewes, Annegret Habel, and Hans-Jörg Kreowski. Hyperedge replacement graph grammars. In G. Rozenberg, editor, *Handbook of Graph Grammars and Computing by Graph Transformation. Vol. 1: Foundations*, chapter 2, pages 95–162. World Scientific, 1997.

[DK99] Frank Drewes and Hans-Jörg Kreowski. Picture generation by collage grammars. In H. Ehrig, G. Engels, H.-J. Kreowski, and G. Rozenberg, editors, *Handbook of Graph Grammars and Computing by Graph Transformation*, volume 2. World Scientific, 1999. To appear.

[DKK99] Frank Drewes, Renate Klempien-Hinrichs, and Hans-Jörg Kreowski. Table-driven and context-sensitive collage languages. In *Proc. Developments in Language Theory '99*, 1999. To appear.

[Hab92] Annegret Habel. *Hyperedge Replacement: Grammars and Languages*, volume 643 of *Lecture Notes in Computer Science*. Springer, 1992.

[HKT93] Annegret Habel, Hans-Jörg Kreowski, and Stefan Taubenberger. Collages and patterns generated by hyperedge replacement. *Languages of Design*, 1:125–145, 1993.

[Hut81] John E. Hutchinson. Fractals and self similarity. *Indiana University Mathematics Journal*, 30:713–747, 1981.

[KRS97] Lila Kari, Grzegorz Rozenberg, and Arto Salomaa. L systems. In G. Rozenberg and A. Salomaa, editors, *Handbook of Formal Languages. Vol. I: Word, Language, Grammar*, chapter 5, pages 253–328. Springer, 1997.

[PJS92] Heinz-Otto Peitgen, Hartmut Jürgens, and Dietmar Saupe. *Chaos and Fractals. New Frontiers of Science.* Springer, New York, 1992.

Synchronized Product of Linear Bounded Machines

Teodor Knapik and Étienne Payet

IREMIA, Université de La Réunion,
BP 7151, 97715 Saint Denis Messag. Cedex 9, France
{knapik,epayet}@univ-reunion.fr

Abstract. This paper introduces a class of graphs associated to linear bounded machines. It is shown that this class is closed, up to observational equivalence, under synchronized product. The first-order theory of these graphs is investegated and shown to be undecidable. The latter result extends to any logic in which the existence of sinks may be stated.

1 Introduction

Finite transition systems together with their synchronized product define a simple and elegant theoretical framework for specification and verification of systems of communicating processes. This framework is known as the Arnold–Nivat approach [2, 16]. A number of equivalent approaches as *e.g.* CCS [14] or Meije [3] and decision procedures for various logics (see *e.g.* [8] and [11]) have provided grounds for the model checking (see *e.g.* [12] or [13]). In spite of encouraging time–complexity results in this area, the approaches based on finite transition systems encounter space–complexity problems. To face up these problems, many compression–like techniques, as *e.g.* binary decision diagrams, have been developed [4].

The problem of storage space for a representation of a process may be overcome using (possibly) infinite transition systems. Among these, the best known are the pushdown transition systems *viz* the transition graphs of pushdown machines. Since the result of [15] about the decidability of the monadic second–order logic of these graphs, more general families of graphs that enjoy this decidability property have been discovered in terms of several descriptions (see [7] and [5]).

Although the latter approaches provide an increased expressive power, they did not give rise to an important development of the "infinite model checking" theory and practice. In the authors' opinion, this is due to the fact that interacting processes cannot be described within these approaches, because the classes of graphs of [5], [7] and [15] are not closed under the synchronized product.

The classes of finite (resp. pushdown) transition systems are naturally related to rational (resp. context–free) languages. Both classes have been already investegated, the former more deeply than the latter. But almost nothing is known about graphs related to the next level of the Chomsky hierarchy, namely the context–sensitive languages. The present paper goes into this direction with the emphasis on the synchronized product.

We consider a multi-tape linear bounded machine with a single, read-only input tape and we define the transition graphs of such devices. We study two transformations on these machines. The first-one is similar to the usual simulation of a multi-tape Turing machine by a single-tape one. The second-one consists in a construction of a multi-tape machine that behaves like several communicating single-tape machines. In both cases, we show that the transformations preserve observational equivalence of associated graphs (this is the main difference with the usual treatment where isomorphisms are considered.) The composition of both transformations allows to establish that the class of graphs of linear bounded machines is closed, up to observational equivalence, under synchronized product.

Unfortunately, as established in the paper, the first-order theory of the graphs of linear bounded machines is not decidable.

2 Preliminaries

Throughout this paper, the empty word is written ε and, if $n \in \mathbb{N}$, $[n]$ stands for the set $\{1, \ldots, n\}$ (with $[0] = \varnothing$).

2.1 Rooted Graphs, Their Synchronized Product and Their ϵ-Equivalence

A simple directed edge-labelled graph $\mathcal{G}$ (or more simply a ***graph***) over C is a set of *edges*, *i.e.* a subset of $D \times C \times D$ where D is an arbitrary set, the elements of which are called the *vertices* of $\mathcal{G}$ and C is an alphabet possibly extended with the empty word. Given d and d' in D, an edge from d to d' labelled by $c \in C$ is written $d \overset{c}{\to} d'$. Thus, $\overset{c}{\to}$ is a binary relation on D for each $c \in C$. A (finite) ***path*** in $\mathcal{G}$ from d to d' is a sequence of edges of the form $d_0 \overset{c_1}{\to} d_1, \ldots, d_{n-1} \overset{c_n}{\to} d_n$ such that $d_0 = d$ and $d_n = d'$. The word $w = c_1 \ldots c_n$ is then the ***label*** of the path. In this case, we may write $d \overset{w}{\to} d'$. We shall constantly consider graphs, the vertices of which are all accessible from some distinguished vertex. Thus, a graph $\mathcal{G}$ is said to be ***rooted on a vertex*** d if there exists a path from d to each vertex of $\mathcal{G}$. The maximal subgraph of $\mathcal{G}$ that is rooted on a vertex e is written $\mathcal{G}[e]$.

Synchronized Product of Rooted Graphs. The synchronized product of graphs has been introduced by Arnold and Nivat [2,16]. It is an essential part of the semantic of interacting processes. For more material, the reader may refer to [1]. We introduce here a definition that is a variant of Arnold and Nivat's one. Indeed, in the scope of this paper, we need a product that takes as entry some rooted graphs and returns a rooted graph as well. Given n alphabets $C_1, \ldots, C_n$ possibly extended with ε, a ***synchronization constraint*** $\mathcal{C}$ over $C_1, \ldots, C_n$ is a subset of $\prod_{i \in [n]} C_i$. Let $\mathcal{G}_1[d_1], \ldots, \mathcal{G}_n[d_n]$ be some rooted graphs over $C_1, \ldots, C_n$. The ***synchronized product*** of $\mathcal{G}_1[d_1], \ldots, \mathcal{G}_n[d_n]$ with respect to $\mathcal{C}$, written $\prod^{\mathcal{C}}_{i \in [n]} \mathcal{G}_i[d_i]$, is the graph $\mathcal{G}[(d_1, \ldots, d_n)]$ where

$$\mathcal{G} = \{ e \overset{c}{\to} e' \mid c \in \mathcal{C} \text{ and } \forall i \in [n],\ e_i \overset{c_i}{\to} e'_i \in \mathcal{G}_i[d_i] \} .$$

In this definition, e_i (resp. c_i, resp. e'_i) stands for the ith coordinate of tuple e (resp. c, resp. e').

ε-Equivalence of Rooted Graphs. In the following definition, $\overset{\varepsilon}{\dashrightarrow}$ stands for the reflexive–transitive closure of $\overset{\varepsilon}{\rightarrow}$ and $\overset{a}{\dashrightarrow} = \overset{\varepsilon}{\dashrightarrow} \circ \overset{a}{\rightarrow} \circ \overset{\varepsilon}{\dashrightarrow}$ for any a in alphabet Σ. Let $\mathcal{G}_1[e_1]$ and $\mathcal{G}_2[e_2]$ be two rooted graphs over $\Sigma \cup \{\varepsilon\}$ with sets of vertices respectively D_1 and D_2. These graphs are said to be *ε-equivalent* if there exists a relation $\leftrightsquigarrow \subseteq D_1 \times D_2$ such that:

1. $\mathrm{Dom}(\leftrightsquigarrow) = D_1$ and $\mathrm{Ran}(\leftrightsquigarrow) = D_2$;
2. $e_1 \leftrightsquigarrow e_2$;
3. for each a in Σ, each path $d_1 \overset{a}{\dashrightarrow} d'_1$ in $\mathcal{G}_1[e_1]$ and each vertex d_2 of $\mathcal{G}_2[e_2]$ such that $d_1 \leftrightsquigarrow d_2$, there exists a vertex d'_2 in $\mathcal{G}_2[e_2]$ such that $d'_1 \leftrightsquigarrow d'_2$ and $d_2 \overset{a}{\dashrightarrow} d'_2$ is a path of $\mathcal{G}_2[e_2]$;
4. for each a in Σ, each path $d_2 \overset{a}{\dashrightarrow} d'_2$ in $\mathcal{G}_2[e_2]$ and each vertex d_1 of $\mathcal{G}_1[e_1]$ such that $d_1 \leftrightsquigarrow d_2$, there exists a vertex d'_1 in $\mathcal{G}_1[e_1]$ such that $d'_1 \leftrightsquigarrow d'_2$ and $d_1 \overset{a}{\dashrightarrow} d'_1$ is a path of $\mathcal{G}_1[e_1]$.

It should be noted that ε-equivalence is an observational equivalence (also called weak bisimulation) as defined by Milner in [14] if one considers ε to be a nonobservable event.

2.2 Linear Bounded Machines and Their Graphs

We use a definition of a Linear Bounded Machine[1] (*LBM* for short) that is slightly different from standard ones and has a flavour of a *Chaitin computer* [6]. Our motivation does not rely on any languages theory aspect. Actually, we are interested in LBM's as an approach for modelling process behaviour that is defined as a graph associated to an LBM. For that matter, we need LBM's such that the motion of the input tape head is one–way, from the left to the right. When this head moves from a cell c to the right neighboor of c, the machine *reads* the content of c. Moreover, the work tapes are infinite to the left and to the right and, in addition to the usual moves (left and right), each work tape head may stay at its place.

Formally, a k work tape LBM $\mathcal{L}$ is a tuple $(Q, \Sigma, \Gamma_1, \ldots, \Gamma_k, \delta, q_0)$ where Q is the finite set of *states*, Σ is the *input alphabet*, $\Gamma_1, \ldots, \Gamma_k$ are the *work tape alphabets*, q_0 is the *initial state*, and δ is the set of *transitions*:

$$\delta \subseteq Q \times \Sigma \cup \{\varepsilon\} \times \Gamma_1 \times \cdots \times \Gamma_k \times \Gamma_1 \times \{\blacktriangleleft, \blacktriangleright, \blacksquare\} \times \cdots \times \Gamma_k \times \{\blacktriangleleft, \blacktriangleright, \blacksquare\} \times Q$$

where $\blacktriangleleft$ (resp. $\blacktriangleright$ and $\blacksquare$) symbolizes a move to the left (resp. a move to the right and no move) and an input ε represents the special case where the input tape head of $\mathcal{L}$ does not move and does not read any character. We assume that the blank character, written $\Box$, belongs to each Γ_i.

[1] A linear bounded machine is a linear bounded automaton with no final state.

An *internal configuration* $(\mu_1 q\nu_1, \ldots, \mu_k q\nu_k)$ of $\mathcal{L}$ is an element of the set $\Gamma_1{}^*.Q.\Gamma_1{}^* \times \cdots \times \Gamma_k{}^*.Q.\Gamma_k{}^*$. This encodes the description of $\mathcal{L}$ at a time as follows: q is the current state, for all i in $[k]$, $\mu_i\nu_i$ is the content of the ith work tape from the leftmost nonblank character to the rightmost nonblank character and for all i in $[k]$, the head of the ith work tape is reading ν_i's first character or $\Box$ if ν_i is empty. Note that for all i in $[k]$, both μ_i or ν_i may contain $\Box$. An internal configuration, every coordinate of which equals q_0, is called *initial configuration* of $\mathcal{L}$.

To every LBM $\mathcal{L} = (Q, \Sigma, \Gamma_1, \ldots, \Gamma_k, \delta, q_0)$ we associate the graph $\mathcal{G}_{\mathcal{L}}[\iota]$ where ι is the initial configuration of $\mathcal{L}$ and $\mathcal{G}_{\mathcal{L}}$ is defined as follows. The vertices of $\mathcal{G}_{\mathcal{L}}$ are all the internal configurations of $\mathcal{L}$ and the labels of $\mathcal{G}_{\mathcal{L}}$ belong to the set $\Sigma \cup \{\varepsilon\}$. Moreover, $(\mu_1 q\nu_1, \ldots, \mu_k q\nu_k) \xrightarrow{c} (\mu'_1 q'\nu'_1, \ldots, \mu'_k q'\nu'_k)$ is an edge of $\mathcal{G}_{\mathcal{L}}$ if and only if for each $i \in [k]$, there exist $X_i, Y_i \in \Gamma_i$ and $\blacklozenge_i \in \{\blacktriangleleft, \blacktriangleright, \blacksquare\}$ such that $(q, c, X_1, \ldots, X_k, Y_1, \blacklozenge_1, \ldots, Y_k, \blacklozenge_k, q') \in \delta$ and either $\blacklozenge_i = \blacktriangleleft$ and one of the following holds:

- $\exists \alpha_i, \beta_i \in \Gamma_i{}^*, \exists Z_i \in \Gamma_i, \mu_i = \alpha_i Z_i, \nu_i = X_i\beta_i, \mu'_i = \alpha_i$ and, if $Z_iY_i\beta_i \notin \{\Box\}^*$, then $\nu'_i = Z_iY_i\beta_i$, else $\nu'_i = \varepsilon$;
- $\mu_i = \varepsilon$, $\exists \alpha_i \in \Gamma_i{}^*$, $\nu_i = X_i\alpha_i$, $\mu'_i = \varepsilon$ and, if $Y_i\alpha_i \notin \{\Box\}^*$, then $\nu'_i = \Box Y_i\alpha_i$, else $\nu'_i = \varepsilon$;
- $X_i = \Box$, $\mu_i = \varepsilon$, $\nu_i = \varepsilon$, $\mu'_i = \varepsilon$ and, if $Y_i \neq \Box$, then $\nu'_i = \Box Y_i$, else $\nu'_i = \varepsilon$;

or $\blacklozenge_i = \blacktriangleright$ and one of the following holds:

- $\exists \alpha_i \in \Gamma_i{}^*$, $\nu_i = X_i\alpha_i$, $\nu'_i = \alpha_i$ and, if $\mu_iY_i \notin \{\Box\}^*$, then $\mu'_i = \mu_iY_i$ else $\mu'_i = \varepsilon$;
- $X_i = \Box$, $\nu_i = \varepsilon$, $\nu'_i = \varepsilon$, and if $\mu_iY_i \notin \{\Box\}^*$, then $\mu'_i = \mu_iY_i$, else $\mu'_i = \varepsilon$;

or $\blacklozenge_i = \blacksquare$ and one of the following holds:

- $\exists \alpha_i \in \Gamma_i{}^*$, $\nu_i = X_i\alpha_i$, $\mu'_i = \mu_i$ and, if $Y_i\alpha_i \notin \{\Box\}^*$, then $\nu'_i = Y_i\alpha_i$, else $\nu'_i = \varepsilon$;
- $X_i = \Box$, $\nu_i = \varepsilon$, $\mu'_i = \mu_i$ and, if $Y_i \neq \Box$, then $\nu'_i = Y_i$, else $\nu'_i = \varepsilon$.

We say that two LBM's are *ε-equivalent* if the associated graphs are so.

3 Multi-work Tape LBM's

In this section, it is established that every LBM with k work tapes is ε-equivalent to an LBM with one work tape. This is done by providing a construction of the one work tape LBM from the k work tapes one.

It is important to note that this result is not necessarily a consequence of the fact that the languages recognized by the linear bounded automata with n work tapes are the same as those recognized by the linear bounded automata with m work tapes, for all n and m in $\mathbb{N}$ (a linear bounded automaton is an LBM provided with a set of final states). When two kinds of devices accept the same family of languages then the classes of graphs generated by both devices need not to be the same up to observational equivalence. For instance, the graphs of pushdown automata are not, in general, ε-equivalent to the graphs of realtime pushdown automata whereas both kinds of automata accept exactly the family of context-free languages.

Let $\mathcal{L} = (Q, \Sigma, \Gamma_1, \ldots, \Gamma_k, \delta, q_0)$ be a k work tape LBM. The one work tape LBM $\mathcal{L}'$, that is ε-equivalent to $\mathcal{L}$, is constructed in the following way. On one hand, the work tape inscription of $\mathcal{L}'$ consists of the concatenation of the inscriptions of all the work tapes of $\mathcal{L}$ separated by ***delimiters***. On the other hand, the motion of the k work tape heads of $\mathcal{L}$ is simulated by the single work tape head of $\mathcal{L}'$ by means of ***head marks***.

More precisely, we introduce the following new characters: for each i in $[k+1]$, a delimiter, written $\%_i$, and, for each work tape character X, a corresponding head mark, written $\dot{X}$. In the sequel, for all i in $[k]$, H_i denotes the set $\{\dot{X} \mid X \in \Gamma_i\}$. An internal configuration $(\mu_1 q \nu_1, \ldots, \mu_k q \nu_k)$ of $\mathcal{L}$ is simulated by an inscription $\%_1\ \alpha_1 \dot{X}_1 \beta_1\ \%_2\ \ldots\ \%_k\ \alpha_k \dot{X}_k \beta_k\ \%_{k+1}$ on the work tape of $\mathcal{L}'$. This inscription is such that, for each i in $[k]$, $\nu_i \neq \varepsilon \Rightarrow (\exists \gamma_i \in \{\Box\}^*,\ \nu_i \gamma_i = X_i \beta_i)$, $\nu_i = \varepsilon \Rightarrow (X_i = \Box \text{ and } \beta_i \in \{\Box\}^*)$ and $\exists \gamma_i \in \{\Box\}^*,\ \gamma_i \mu_i = \alpha_i$.

The LBM $\mathcal{L}'$ simulates $\mathcal{L}$ in the following way. First, in order to simulate the initial configuration of $\mathcal{L}$, $\mathcal{L}'$ copies the word $\%_1\ \dot{\Box}\ \%_2\ \ldots\ \%_k\ \dot{\Box}\ \%_{k+1}$ on its work tape. Let δ_{copy} denote the set of transitions of $\mathcal{L}'$ performing this copy, Q_{copy} denote the set of states involved in δ_{copy} and suppose that after these operations, $\mathcal{L}'$ switches to state q_0. A computation for $\mathcal{L}$ consists in overprinting a character Y_i on a character X_i on each work tape i and possibly moving the head of i to the left or to the right. This is simulated by $\mathcal{L}'$ in the following way. If $\mathcal{L}$ does not move the head of i, then $\mathcal{L}'$ overprints $\dot{Y}_i$ on $\dot{X}_i$. If $\mathcal{L}$ moves the head of i to the left (resp. right), then $\mathcal{L}'$ overprints Y_i on $\dot{X}_i$ and, if Z is the character written to the left (resp. right) of Y_i, it overprints $\dot{Z}$ on Z. Overprintings on each ith portion of the work tape of $\mathcal{L}'$ (*i.e.* the part of this tape corresponding to the ith work tape of $\mathcal{L}$) that are due to a transition $t \in \delta$ are performed by means of state $q_{t,i}$. Moreover, in order to perform each overprinting due to t, the work tape head of $\mathcal{L}'$ must be able to move from the ith portion of the work tape to the next one to the right (if it exists); this is done by means of the state $m_{t,i}$; when the last overprinting is done on the kth portion, the work tape head comes back to the first portion of the tape by means of the state $m_{t,k}$.

Whenever $\mathcal{L}$ is placing the head of i before the beginning (resp. after the end) of the inscription, $\mathcal{L}'$ has to insert $\dot{\Box}$ to the right of $\%_i$ (resp. to the left of $\%_{i+1}$). For that matter, $\mathcal{L}'$ shifts to the left (resp. to the right) the portion of its inscription from $\%_1$ to $\%_i$ (resp. from $\%_{i+1}$ to $\%_{k+1}$) and then writes $\Box$ in the right cell. These operations are performed by a set of transitions denoted by δ_{shift} and they start from $l_{t,i} \in Q_{shift}$ (resp. $r_{t,i+1} \in Q_{shift}$) where Q_{shift} is the set of states involved in δ_{shift}. We suppose that after shifting the portion of its tape and writting $\Box$, $\mathcal{L}'$ switches to $m_{t,i}$.

Finally, take it that $\mathcal{L}' = (Q', \Sigma, \Gamma', \delta_{copy} \cup \delta_{shift} \cup \delta', q_0')$ where Q' is the set $Q \cup \{q_{t,i} \mid t \in \delta \text{ and } i \in [k]\} \cup \{m_{t,i} \mid t \in \delta \text{ and } i \in [k]\} \cup Q_{copy} \cup Q_{shift}$, Γ' is the set $\Gamma_1 \cup \cdots \cup \Gamma_k \cup \{\%_1, \ldots, \%_{k+1}\} \cup \mathrm{H}_1 \cup \cdots \cup \mathrm{H}_k$ and δ' is constructed this way: $t = (q, c, X_1, \ldots, X_k, Y_1, \blacklozenge_1, Y_2, \blacklozenge_2, \ldots, Y_k, \blacklozenge_k, q') \in \delta$ if and only if the following set is part of δ':

$$\{(q,c,\dot{X}_1,Y_1,\blacklozenge_1,q_{t,1})\} \cup \{(q_{t,i},\varepsilon,X,\dot{X},\blacktriangleright,m_{t,i}) \mid i \in [k-1],\ X \in \Gamma_i\} \cup$$

$$\{(q_{t,i},\varepsilon,\%_i,\dot{\%}_i,\blacksquare,l_{t,i}) \mid i \in [k]\} \cup \{(q_{t,i},\varepsilon,\%_{i+i},\dot{\%}_{i+1},\blacksquare,r_{t,i+1}) \mid i \in [k]\} \cup$$

$$\{(q_{t,k},\varepsilon,X,\dot{X},\blacksquare,m_{t,k}) \mid X \in \Gamma_k\} \cup$$

$$\{(m_{t,i},\varepsilon,X,X,\blacktriangleright,m_{t,i}) \mid i \in [k-1],\ X \in \Gamma_i \cup \{\%_{i+1}\} \cup \Gamma_{i+1}\} \cup$$

$$\{(m_{t,i},\varepsilon,\dot{X}_{i+1},Y_{i+1},\blacklozenge_{i+1},q_{t,i+1}) \mid i \in [k-1]\} \cup$$

$$\{(m_{t,k},\varepsilon,X,X,\blacktriangleleft,m_{t,k}) \mid X \in \textstyle\bigcup_{i\in[k]} \Gamma_i \cup \{\%_1,\ldots,\%_k\} \cup \bigcup_{i\in[k]\smallsetminus\{1\}} \mathrm{H}_i\} \cup$$

$$\{(m_{t,k},\varepsilon,\dot{X},\dot{X},\blacksquare,q') \mid X \in \Gamma_1\}\,.$$

Consequently, the internal configurations of $\mathcal{L}'$ are the elements of

$$\%_1.\Gamma_1{}^*.Q'.\Gamma_1{}^*.\mathrm{H}_1.\Gamma_1{}^*.\%_2.\Gamma_2{}^*.\mathrm{H}_2.\Gamma_2{}^*.\%_3\ \ldots \%_k\Gamma_k{}^*.\mathrm{H}_k.\Gamma_k{}^*.\%_{k+1} \cup \ldots$$

(the work tape head may be located anywhere, so the dots mean "the same with Q' anywhere else").

Obviously, $\mathcal{L}'$ is an LBM because the number of cells that are used on its work tape is a linear function of the size of the input word; this number is $(k+1)+\sum_{i\in[k]} S_i(n)$ where $k+1$ stands for the number of cells containing a delimiter, n is the size of the input word and the S_i are the k linear functions in the size of the input word of $\mathcal{L}$. Note that the number of states of $\mathcal{L}'$ is in $O(k.|\delta|)$ and the number of transitions of $\mathcal{L}'$ is in $O(|\delta|.\sum_{i\in[k]}|\Gamma_i|)$.

In view of the definition of $\mathcal{L}'$, the following proposition is straightforward.

Proposition 3.1. *Let ι and ι' denote the initial configuration of $\mathcal{L}$ and $\mathcal{L}'$ respectively. Then, $\mathcal{G}_{\mathcal{L}}[\iota]$ and $\mathcal{G}_{\mathcal{L}'}[\iota']$ are ε-equivalent.*

4 Synchronized Product of LBM's

We consider n LBM $\mathcal{L}_i$, $i \in [n]$. Suppose that for all i in $[n]$, $\mathcal{L}_i$ has k_i work tapes, ι_i denotes the initial configuration of $\mathcal{L}_i$ and $\mathcal{L}_i = (Q_i,\Sigma_i,\Gamma_{i,1},\ldots,\Gamma_{i,k_i},\delta_i,q_{0i})$.

Let $C \subseteq \prod_{i\in[n]}(\Sigma_i \cup \{\varepsilon\})$ be a synchronization constraint. The LBM *composed* according to C is the $\sum_{i\in[n]} k_i$ work tape LBM $\mathcal{L}$ defined as follows: the set of states is $\prod_{i\in[n]} Q_i$, the input alphabet is C, the work tape alphabets are all the $\Gamma_{i,j}$ for $i \in [n]$ and $j \in [k_i]$, the initial state is $(q_{01},\ldots,q_{0n})$, and the set of transitions consists of tuples

$$\Big(q,a,(X_{1,j})_{j\in[k_1]}\,,\ldots,(X_{n,j})_{j\in[k_n]}\,,(Y_{1,j},\blacklozenge_{1,j})_{j\in[k_1]}\,,\ldots,(Y_{n,j},\blacklozenge_{n,j})_{j\in[k_n]}\,,q'\Big)$$

such that $a \in C$ and, for each $i \in [n]$, $(q_i,a_i,(X_{i,j})_{j\in[k_i]}\,,(Y_{i,j},\blacklozenge_{i,j})_{j\in[k_i]}\,,q'_i) \in \delta_i$. Here, for each $i \in [n]$, $(X_{i,j})_{j\in[k_i]}$ stands for the sequence $X_{i,1}\,,\ldots,X_{i,k_i}$ and

$(Y_{i,j}, \blacklozenge_{i,j})_{j\in[k_i]}$ stands for the sequence $Y_{i,1}, \blacklozenge_{i,1}, \ldots, Y_{i,k_i}, \blacklozenge_{i,k_i}$. Moreover, q_i (resp. a_i, resp. q'_i) stands for the ith coordinate of tuple q (resp. a, resp. q'). Note that the number of states of $\mathcal{L}$ is equal to $\prod_{i\in[n]} |Q_i|$ and that the number of transitions of $\mathcal{L}$ is at most equal to $\prod_{i\in[n]} |\delta_i|$.

Proposition 4.1. *Let ι denote the initial configuration of $\mathcal{L}$. Then, graphs $\mathcal{G}_{\mathcal{L}}[\iota]$ and $\prod^{C}_{i\in[n]} \mathcal{G}_{\mathcal{L}_i}[\iota_i]$ are isomorphic.*

Proof. Let ϕ be the one–to–one correspondence

$$\prod_{i\in[n]} \Big(\prod_{j\in[k_i]} \Gamma_{i,j}{}^*.Q_i.\Gamma_{i,j}{}^* \Big) \longrightarrow \prod_{i\in[n],j\in[k_i]} \Gamma_{i,j}{}^*.\Big(\prod_{l\in[n]} Q_l\Big).\Gamma_{i,j}{}^* \quad \text{such that}$$

$$\phi\Big((\mu_{1,1}q_1\nu_{1,1}, \ldots, \mu_{1,k_1}q_1\nu_{1,k_1}), \ldots, (\mu_{n,1}q_n\nu_{n,1}, \ldots, \mu_{n,k_n}q_n\nu_{n,k_n})\Big)$$
$$= \left(\mu_{1,1}(q_1, \ldots, q_n)\nu_{1,1}, \ldots, \mu_{n,k_n}(q_1, \ldots, q_n)\nu_{n,k_n}\right).$$

The roots $\iota_1, \ldots, \iota_n$ and ι are such that $\phi(\iota_1, \ldots, \iota_n) = \iota$. Moreover, according to the definition of transitions of $\mathcal{L}$, it is clear that $d \xrightarrow{a} d'$ is an edge of $\prod^{C}_{i\in[n]} \mathcal{G}_{\mathcal{L}_i}[\iota_i]$ if and only if $\phi(d) \xrightarrow{a} \phi(d')$ is an edge of $\mathcal{G}_{\mathcal{L}}[\iota]$. □

In view of the above, the following corollary is obvious.

Corollary 4.2. *The class of graphs of LBM's is closed, up to isomorphism, under synchronized product.*

The next corollary is obtained as a composition of Proposition 3.1 and Proposition 4.1.

Corollary 4.3. *The class of graphs of single work tape LBM's is closed, up to ε-equivalence, under synchronized product.*

5 An Example

We consider a small portion of a railway network that is composed of three stations S_0, S_1 and S_2 linked together by a single track:

Train crossing is allowed only at station S_1 which has two platforms. Note that a train arriving to S_1 from S_0 (resp. S_2) may go back to S_0 (resp. S_2). We model the behaviour of this railway portion by means of a synchronized product of LBM's.

This portion of the railway network may be seen as a *composed process* in the sense that its *behaviour* is the result of the *parallel* working of station S_1 and portions S_0S_1 and S_1S_2 of the track. Note that this composed process, unlike its composing parts, is not a pushdown one *i.e.* its behaviour cannot be modelled by a pushdown transition graph. Indeed, suppose that left–to–right motions are distinguished from right–to–left motions. In the left–to–right direction, the departure of a train from S_0 is represented by d_0, the arrival of a train at S_1 is represented by a_1, the departure of a train from S_1 is represented by d_1' and the arrival of a train at S_2 is represented by a_2. Notations for the other direction are depicted above. Then, if $\mathcal{G}$ is a rooted graph modelling the behaviour of the whole portion, the set Lang($\mathcal{G}$) of the labels of the paths from the root to any other vertex looks like

$$\Big\{w \in \{a_0, d_0, a_1, d_1, a_1', d_1', a_2, d_2\}^* \;\Big|\; \forall u, v \in \{a_0, d_0, a_1, d_1, a_1', d_1', a_2, d_2\}^*, \\ uv = w \Rightarrow (|u|_{a_1} \leq |u|_{d_0}, |u|_{a_2} \leq |u|_{d_1'}, |u|_{a_1'} \leq |u|_{d_2}, |u|_{a_0} \leq |u|_{d_1}, \\ |u|_{d_1} \leq |u|_{a_1} + |u|_{a_1'} \text{ and } |u|_{d_1'} \leq |u|_{a_1} + |u|_{a_1'})\Big\}.$$

Obviously, Lang($\mathcal{G}$) is context–sensitive and one can establish, using Ogden's Lemma, that Lang($\mathcal{G}$) is not context–free. But, if $\mathcal{G}$ was the graph of a pushdown process, then Lang($\mathcal{G}$) would be a context–free language [5]. Thus, $\mathcal{G}$ is not the graph of a pushdown process.

In order to model the behaviour of the portion of a railway network, we need the following construction. We associate to any LBM $\mathcal{L} = (Q, \Sigma, \Gamma_1, \ldots, \Gamma_k, \delta, q_0)$ the LBM $\mathcal{L}^{\mathbf{1}} = (Q, \Sigma, \Gamma_1, \ldots, \Gamma_k, \delta \cup \delta^{\mathbf{1}}, q_0)$ where

$$\delta^{\mathbf{1}} = \{(q, \varepsilon, X_1, \ldots, X_k, X_1, \blacksquare, \ldots, X_k, \blacksquare, q) \mid q \in Q \text{ and } \forall i \in [k],\ X_i \in \Gamma_i\}$$

Observe that $\mathcal{G}_{\mathcal{L}^{\mathbf{1}}}$ differs from $\mathcal{G}_{\mathcal{L}}$ only by ε-loops added to each vertex by $\delta^{\mathbf{1}}$.

Let us now specify each part of this railway portion using LBM's. The behaviour of the track between S_0 and S_1 can be modelled by the machine $\mathcal{L}_0^{\mathbf{1}}$ where $\mathcal{L}_0 = \big(\{q_0, q_{d_0}, q_{a_1}, q_{d_1}, q_{a_0}\}, \{a_0, d_0, a_1, d_1\}, \{\Box, d_0, d_1\}, \delta_0, q_0\big)$ with

$$\delta_0 = \left\{\begin{array}{l} (q_0, d_0, \Box, d_0, \blacksquare, q_{d_0}), (q_0, d_1, \Box, d_1, \blacksquare, q_{d_1}), (q_{d_0}, d_0, d_0, d_0, \blacktriangleright, p_{d_0}), \\ (q_{d_0}, \varepsilon, \Box, \Box, \blacksquare, q_0), (q_{d_0}, a_1, d_0, \Box, \blacktriangleleft, q_{d_0}), (p_{d_0}, \varepsilon, \Box, d_0, \blacksquare, q_{d_0}), \\ (q_{d_1}, d_1, d_1, d_1, \blacktriangleright, p_{d_1}), (q_{d_1}, \varepsilon, \Box, \Box, \blacksquare, q_0), (q_{d_1}, a_0, d_1, \Box, \blacktriangleleft, q_{d_1}), \\ (p_{d_1}, \varepsilon, \Box, d_1, \blacksquare, q_{d_1}) \end{array}\right\}.$$

The graph of $\mathcal{L}_0$ may be depicted as follows:

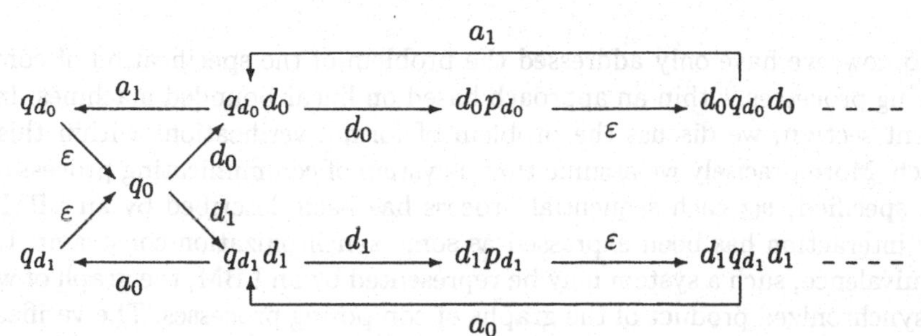

Notice that Lang($\mathcal{G}_{\mathcal{L}_0^1}$), the set of the labels of the paths in $\mathcal{G}_{\mathcal{L}_0^1}$ from q_0 to any other vertex, is L^* where

$$L = \{w \in \{d_0, a_1\}^* \mid \forall u, v \in \{d_0, a_1\}^* (uv = w \Rightarrow |u|_{a_1} \leq |u|_{d_0})\} \cup \{w \in \{d_1, a_0\}^* \mid \forall u, v \in \{d_1, a_0\}^* (uv = w \Rightarrow |u|_{a_0} \leq |u|_{d_1})\} .$$

This means that there cannot be more train arrivals than train departures in each direction on this portion of the track and that any train arriving at S_1 may go back to S_0. The behaviour of the track between S_1 and S_2 can be modelled by the LBM $\mathcal{L}_2^1$, the input alphabet of which is $\{a_1', d_1', a_2, d_2\}$, defined in the same way as $\mathcal{L}_0^1$.

Station S_1 is composed of two platforms. One of them is modelled by the LBM $\mathcal{L}_{11}^1 = (\{q_a, q_d\}, \{a, d\}, \{\square\}, \{(q_a, a, \square, \square, \blacktriangleright, q_d), (q_d, d, \square, \square, \blacktriangleleft, q_a)\} \cup \delta_{11}^1, q_a)$ which is such that Lang($\mathcal{G}_{\mathcal{L}_{11}^1}$) $= (ad)^*$. The other one is modelled by the LBM $\mathcal{L}_{12}^1$, the input alphabet of which is $\{a', d'\}$, which is defined in the same way as $\mathcal{L}_{12}^1$. Letters a, d, a' and d' have the same meaning as $a_0 \dots$ above.

The whole portion of the railway network can now be modelled by the synchronized product of $\mathcal{L}_0^1$, $\mathcal{L}_{11}^1$, $\mathcal{L}_{12}^1$ and $\mathcal{L}_2^1$ with respect to the synchronization constraint $\mathcal{C}$ described as follows: any arrival to S_1 from a track corresponds to an arrival on a platform and any departure from a platform corresponds to a departure from S_1 to a track. Let Σ_0, Σ_{11}, Σ_{12} and Σ_2 denote the input alphabet of $\mathcal{L}_0^1$, $\mathcal{L}_{11}^1$, $\mathcal{L}_{12}^1$ and $\mathcal{L}_2^1$ respectively. Then, constraint $\mathcal{C}$ is the set of the tuples $(c_0, c_{11}, c_{12}, c_2) \in \Sigma_0 \cup \{\varepsilon\} \times \Sigma_{11} \cup \{\varepsilon\} \times \Sigma_{12} \cup \{\varepsilon\} \times \Sigma_2 \cup \{\varepsilon\}$ such that

$$(c_0 = a_1 \Leftrightarrow (c_{11} = a \text{ or } c_{12} = a')),\ (c_2 = a_1' \Leftrightarrow (c_{11} = a \text{ or } c_{12} = a')),$$
$$(c_0 = d_1 \Leftrightarrow (c_{11} = d \text{ or } c_{12} = d')) \text{ and } (c_2 = d_1' \Leftrightarrow (c_{11} = d \text{ or } c_{12} = d')) .$$

We do not give a complete description of the machine that models the whole portion of the network because it is quite a big machine (to have an idea, one may compute its number of states which is equal to $|Q_0| \times |Q_{11}| \times |Q_{12}| \times |Q_2| = 100$ where Q_0, Q_{11}, Q_{12} and Q_2 stand for the set of states of $\mathcal{L}_0^1$, $\mathcal{L}_{11}^1$, $\mathcal{L}_{12}^1$ and $\mathcal{L}_2^1$ respectively).

6 First-Order Logic on Graphs of Linear Bounded Machines

Up to now, we have only addressed the problem of the specification of communicating processes within an approach based on linear bounded machines. In the present section, we discuss the problem of formal verifications within this approach. More precisely we assume that a system of communicating processes has been specified, *viz* each sequential process has been described by an LBM and their interaction has been expressed as some synchronization constraint. Up to ε-equivalence, such a system may be represented by an LBM, the graph of which is a synchronized product of the graphs of composing processes. The verification problem consists then in checking the truth of a formula of some logic on the resulting graph considered as a model–theoretic structure.

Concerning the verification problem, we claim that in the area of LBM specifications, even for rather weak logics, one should not expect algorithmic solutions but rather semi–algorithmic ones. More precisely we establish that the first–order theory of the graphs of LBM's is not recursive (even not recursively enumerable).

Formally, the *first-order theory* of a rooted graph $\mathcal{G} \subseteq D \times C \times D$ where $C = \Sigma \cup \{\varepsilon\}$, is defined as follows. The variables form an infinite countable set X and are interpreted as vertices of $\mathcal{G}$. The binary predicates $\mathbf{s}_c$ for each $c \in C$, $=$ and the unary predicate $\mathbf{r}$ allow to build atomic formulae. These predicates are interpreted on $\mathcal{G}$ resp. as $\xrightarrow{c}$, the identity relation on D and the singleton $\{e\}$, where e stands for the root of $\mathcal{G}$. Using classical connectives and quantifiers, from atomic formulae that are of the form $\mathbf{s}_c(x, y)$, $x = y$ or $\mathbf{r}(x)$, first order formulae are constructed in the usual way. The set of valid sentences on $\mathcal{G}$, *i.e.* valid formulae on $\mathcal{G}$ with no free variable, is called *the first-order theory of* $\mathcal{G}$.

Example 6.1. The sentence $\forall x \, \exists y \, \bigvee_{c \in C} \mathbf{s}_c(x, y)$ belongs to the first–order theory of a graph $\mathcal{G}$ if and only if $\mathcal{G}$ has no sink.

The indecidability result that we have to establish involves linear bounded automata (LBA for short). Formally an LBA $\mathcal{L}_f$ is a single–work–tape[2] LBM $\mathcal{L} = (Q, \Sigma, \Gamma, \delta, q_0)$ with a distinguished state $f \in Q$, called the *final state.* The language accepted by $\mathcal{L}_f$, written Lang($\mathcal{L}_f$), is the set of labels of all paths in the graph $\mathcal{G}_{\mathcal{L}}[\iota]$ from the initial configuration ι to $\mu f \nu$ for some $\mu, \nu \in \Gamma^*$. Two LBA's $\mathcal{L}_f$ and $\mathcal{L}'_{f'}$ are *equivalent* if Lang($\mathcal{L}_f$) = Lang($\mathcal{L}'_{f'}$).

The emptiness problem for LBA's is the following decision problem.

Instance: An LBA $\mathcal{L}_f$.

Question: Lang($\mathcal{L}_f$) $= \varnothing$?

It is well known that this problem is not recursively enumerable. Using this fact, we can establish the following.

Theorem 6.2. *The problem*

Instance: *An LBM $\mathcal{L}$ and a first-order sentence φ.*

Question: *Does φ belong to the first-order theory of $\mathcal{G}_{\mathcal{L}}$?*

is not recursively enumerable.

Proof. We show that the emptiness problem for LBA's is many–one reducible to the problem of the statement.

Let $\mathcal{L}_f = (Q, \Sigma, \Gamma, \delta, q_0, f)$ be an LBA. Let $\mathcal{L}'_f = (Q', \Sigma, \Gamma, \delta', q_0, f)$ be an LBA equivalent to $\mathcal{L}_f$ and satisfying the following properties:

(1) $\delta' \cap (\{f\} \times C \times \Gamma \times \Gamma \times \{\blacktriangleleft, \blacktriangleright, \blacksquare\} \times Q') = \varnothing$,

(2) $\delta' \cap (\{q\} \times C \times \{X\} \times \Gamma \times \{\blacktriangleleft, \blacktriangleright, \blacksquare\} \times Q') \neq \varnothing$ for all $(q, X) \in (Q \setminus \{f\}) \times \Gamma$.

Such an LBA may readily be constructed. Concerning (1), for each rule of δ that has the form $(q_1, c_1, X_1, Y_1, \blacklozenge_1, f)$, we add a rule $(q_1, c_1, X_1, Y_1, \blacklozenge_1, q')$ where

[2] In fact an LBA can be multitape but for the purpose of the paper single–work–tape LBA's suffice.

$q' \notin Q$ is a new state and each rule $(f, c_2, X_2, Y_2, \blacklozenge_2, q_2) \in \delta$ is replaced by the rule $(q', c_2; X_2, Y_2, \blacklozenge_2, q_2)$. Concerning (2), we add a new state $p \notin Q$, the rules $(p, \varepsilon, X, X, \blacksquare, p)$ for all $X \in \Gamma$ and for each $(q, X) \in (Q \setminus \{f\}) \times \Gamma$ such that $\delta \cap (\{q\} \times \Sigma \times \{X\} \times \Gamma \times \{\blacktriangleleft, \blacktriangleright, \blacksquare\} \times Q) = \varnothing$, we add the rule $(q, \varepsilon, X, X, \blacksquare, p)$.

Now $\mathrm{Lang}(\mathcal{L}_f) = \mathrm{Lang}(\mathcal{L}'_f)$ and the following holds: a vertex $\mu q \nu$ of $\mathcal{G}_{\mathcal{L}'_f}$ is a sink if and only if $q = f$. Consequently $\mathrm{Lang}(\mathcal{L}_f) = \varnothing$ if and only if the sentence of Example 6.1 belongs to the first-order theory of $\mathcal{G}_{\mathcal{L}'_f}$. □

Taking into account the fact that the property of being a sink is expressible in the Hennessy–Milner logic [9], the following corollary may be derived from above proof.

Corollary 6.3. *The Hennessy–Milner logic is not semi-decidable on the graphs of LBM's.*

We close this section by the following remark. If the existence of sinks is expressible within a logic, then it cannot exist a complete formal system for checking the truth of the formulae of the logic on graphs of arbitrary LBM's.

7 Conclusion

We have defined transition graphs associated to a peculiar kind of linear bounded machines that read their input performing all computations on work tapes. The closure under synchronized product of the family of graphs thus defined has been established up to observational equivalence (considering ε–transitions as non observable) using two transformations. We hope that both transformations may be improved using some speedup and tape compression techniques so as to preserve bisimulation (or even isomorphism) instead of observational equivalence.

As a consequence of the closure result and similarly to the Arnold–Nivat approach, the linear bounded machines provide a uniform framework for the specification of communicating processes. Moreover the expressive power within this framework seems to be very satisfactory. However, this has a counterpart in the undecidability result. We have established that the first–order theory of the graphs of linear bounded machines is not recursively enumerable. This result extends to any logic in which the existence of sinks may be stated. It may be observed that this is one of the weakest safety properties that one should be able to express within a logic usable for verification purposes, since it corresponds to the existence of deadlocks. In spite of this negative result, we believe that some semi–decision techniques adequate for the graphs of linear bounded machines may be developed for various logics. An elementary example of this kind may be found in [10].

The transition graph of linear bounded machine has been defined as the maximal subgraph of the configuration graph that is accessible from the initial configuration. When this accessibility requirement is dropped, we have a transition graph, the vertices of which are all configurations. Since our undecidability result was related to the reachability problem, more precisely to the language of an LBA, the emptiness problem for LBA does not lead to a similar result in the latter case. Is the first-order theory of such graphs still undecidable ? Currently, we do not know the answer to this question.

References

1. A. ARNOLD. *Finite Transition Systems.* Prentice Hall Int., 1994.
2. A. ARNOLD and M. NIVAT. Comportements de processus. In *Colloque AFCET "Les mathématiques de l'Informatique"*, pages 35–68, 1982.
3. G. BOUDOL. Notes of algebraic calculi of processes. In *Logics and Models of Concurrent Systems*, volume F-13 of *NATO ASI series*, pages 261–303. 1985.
4. R. BRYANT. Binary decision diagrams and beyond: Enabling technologies for formal verification. In *Proceedings of the International Conference on Computer Aided Design, ICCAD'95*, 1995.
5. D. CAUCAL. On infinite transition graphs having a decidable monadic second-order theory. In F. M. auf der Heide and B. Monien, editors, *23th International Colloquium on Automata Languages and Programming*, LNCS 1099, pages 194-205, 1996.
6. G. CHAITIN. A Theory of Program Size Formally Identical to Information Theory. *J. Assoc. Compt. Mach.*, (22):329–340, 1975.
7. B. COURCELLE. The monadic second-order logic of graphs, II: Infinite graphs of bounded width. *Mathematical System Theory*, 21:187–221, 1989.
8. A. EMERSON. Temporal and modal logic. In J. van Leeuwen, editor, *Formal Models and Semantics*, volume B of *Handbook of Theoretical Computer Science*, pages 997-1072. Elsevier, 1990.
9. M. HENNESSY and R. MILNER. Algebraic laws for nondeterminism and concurrency. *J. ACM*, 32:137–162, 1985.
10. T. KNAPIK. Domains of word-functions and Thue specifications. Technical Report INF/96/11/05/a, IREMIA, Université de La Réunion, 1997.
11. D. KOZEN and J. TIURYN. Logics of programs. In J. van Leeuwen, editor, *Formal Models and Semantics*, volume B of *Handbook of Theoretical Computer Science*, pages 789–840. Elsevier, 1990.
12. R. P. KURSHAN. *Computer-aided Verification of Coordinating Processes.* Princeton University Press, 1994.
13. K. MCMILLAN. *Symbolic Model Checking.* Kluwer, 1993.
14. R. MILNER. *Calculus of Communicating Systems.* LNCS 82. Springer Verlag, 1980.
15. D. E. MULLER and P. E. SCHUPP. The theory of ends, pushdown automata and second-order logic. *Theoretical Comput. Sci.*, 37:51-75, 1985.
16. M. NIVAT. Sur la synchronisation des processus. *Revue technique Thomson-CSF*, (11):899–919, 1979.

On Maximal Repetitions in Words *

Roman Kolpakov[1] and Gregory Kucherov[2]

[1] French-Russian Institute for Informatics and Applied Mathematics, Moscow University, 119899 Moscow, Russia, e-mail: roman@vertex.inria.msu.ru
[2] LORIA/INRIA-Lorraine, 615, rue du Jardin Botanique, B.P. 101, 54602 Villers-lès-Nancy France, e-mail: kucherov@loria.fr

Abstract. A (fractional) repetition in a word w is a subword with the period of at most half of the subword length. We study maximal repetitions occurring in w, that is those for which any extended subword of w has a bigger period. The set of such repetitions represents in a compact way all repetitions in w.
We first study maximal repetitions in Fibonacci words – we count their exact number, and estimate the sum of their exponents. These quantities turn out to be linearly-bounded in the length of the word. We then prove that the maximal number of maximal repetitions in general words (on arbitrary alphabet) of length n is linearly-bounded in n, and we mention some applications and consequences of this result.

1 Introduction

Repetitions (called also periodicities) play a fundamental role in many topics of word combinatorics, formal language theory and applications. Several notions of repetition has been used in the literature. In its simplest form, a repetition is a word of the form uu, commonly called a *square*. A natural generalization is to consider, instead of squares, arbitrary powers, that is words of the form $u^n = \underbrace{uu\ldots u}_{n}$ for $n \geq 2$. We call such repetitions *integer repetitions* (or *integer powers*). If a word is not an integer repetition, it is called *primitive*. Integer repetitions can be further generalized to *fractional repetitions*, that is words of the form $w = u^n v$, where $n \geq 2$ and v is a proper prefix of u. u is called a *root* of w. If u is primitive, quantity $n + \frac{|v|}{|u|}$ is called the *exponent* of w, and $|u|$ is the *period* of w. Considering repetitions with fractional exponent may turn to be very useful and may provide a deeper insight of combinatorial properties of words [Dej72,Lot83,MP92,MRS95,CS96,JP99].

* The work has been done during the first author's visit of LORIA/INRIA-Lorraine supported by a grant from the French Ministry of Public Education and Research. The first author has been also in part supported by the Russian Foundation of Fundamental Research, under grant 96-01-01068, and by the Russian Federal Programme "Integration", under grant 473. The work has been done within a joint project of the French-Russian A.M.Liapunov Institut of Applied Mathematics and Informatics at Moscow University

Depending on the problem, the difference between the above three notions of repetition may not be relevant (for example if one wants to check whether a word is repetition-free) but, as will be seen below, may be important. Besides, if one wants to find (or to count) all repetitions in a word, it must be specified whether all *distinct* repetitions are looked for (that is, their position in the word is not relevant) or all the occurrences of (possibly syntactically equal) repetitions. In this paper we will be mainly concerned with the latter case, and we will sometimes say ***positioned repetitions*** to underline this meaning.

When one considers (integer or fractional) repetitions in a word, it is natural to consider "maximal" ones, that is those which cannot be further extended to the right/left to a bigger repetition with the same period. However, the definition of maximality differs depending on whether integer or fractional repetitions are considered. In case of integer repetitions, this amounts to those repetitions u^k, $k \geq 2$, which are not followed or preceeded by another occurrence of u. In case of fractional repetitions, a maximal repetition is a subword $u^n v$ (v a prefix of u, $n \geq 2$) which cannot be extended *by one letter* to the right or to the left without changing (increasing) the period. For example, the subword 10101 in the word $w = 1011010110110$ is a maximal fractional repetition (with period 2), while the subword 1010 is not. Another maximal fractional repetitions of w are prefix 10110101101 (period 5), suffix 10110110 (period 3), prefix 101101 (period 3), and the three occurrences of 11 (period 1).

In this paper we study maximal positioned fractional repetitions that, for the sake of shortness, we will call simply *maximal repetitions*.[1]. Maximal repetitions are important objects as they encode, in a most compact way, all repetitions in the word. For example, if we know all maximal repetitions in a word, we can easily obtain all squares in this word, with both primitive and non-primitive roots.

The question "How many repetitions can a word contain?" is interesting from both theoretical and applicative perspective. However, one must specify carefully which repetitions are counted.

A word of length n contains $O(n \log n)$ positioned primitively-rooted squares. This follows, in particular, from Lemma 10 of [CR95] which asserts that a word cannot contain in its prefixes more than $\log_\phi n$ primitive-rooted squares which immediately implies the $n \log_\phi n$ upper bound (ϕ is the golden ratio). On the other hand, in [Cro81] it was shown that Fibonacci words contain $\Omega(n \log n)$ positioned squares. Since all squares in Fibonacci words are primitively-rooted, this proves that $O(n \log n)$ is the asymptotically tight bound. A formula for the exact number of squares in Fibonacci words has been obtained in [FS99]. Note that in contrast, the number of distinct squares in Fibonacci words and in general, the maximal number of distinct squares in general words (over an arbitrary alphabet) is linear in the length [FS99,FS98].

The situation is different if only distinct squares are counted. In [FS99], it is shown that the k-th Fibonacci word f_k contains $2(|f_{k-2}|-1) = 2(2-\phi)|f_k|+o(1)$

[1] maximal repetitions have been called *runs* in [IMS97], *maximal periodicities* in [Mai89], and *m-repetitions* in [KK98]

distinct squares (ϕ is the golden ratio). The number of distinct squares in general words of length n is bounded by $2n$ (for an arbitrary alphabet), that was shown in [FS98] using a result from [CR95]. It is conjectured that this number is actually smaller than n, at least for the binary alphabet. Thus, in contrast to positioned squares, the maximal number of distinct squares is linear.

In [Cro81], Crochemore studies positioned primitively-rooted maximal integer powers, that is those subwords u^k, $k \geq 2$, which are not followed or preceeded by another occurrence of u. Similar to positioned squares, the maximal number of such repetitions is $\Theta(n \log n)$. The lower bound easily follows from the $\Omega(n \log n)$ bound for positioned squares in Fibonacci words, as Fibonacci words don't contain 4-powers, and an occurrence of a 3-power is an extension of two square occurrences. Therefore, the number of maximal integer powers in Fibonacci words is at least half the number of positioned squares, and is then $\Theta(n \log n)$.

What happens if we count the number of maximal repetitions instead of integer powers or just squares? Note that a word can contain much less maximal repetitions than maximal integer powers: e.g. if v is a square-free word over $\{a, b, c\}$, then word $v\#v\#v$ contains $|v| + 1$ (maximal) integer powers but only one maximal repetition. What is the maximal number of maximal repetitions in a word?

In the first part of the paper, we study maximal repetitions in Fibonacci words. The results of [IMS97] imply that Fibonacci words contain a linear number of maximal repetitions, with respect to the length of the word. This is showed, however, in an indirect way by presenting a linear-time algorithm which enumerates all maximal repetitions in a Fibonacci word. In this paper we first obtain directly the exact number of maximal repetitions in Fibonacci words, which is equal to $2|f_{k-2}| - 3$. Incidentally (or maybe not?), a Fibonacci word contains one less maximal repetitions than distinct squares.

We also estimate the sum of exponents of all maximal repetitions in a Fibonacci word. It is known ([MP92]) that Fibonacci words contain no subword of exponent greater than $2 + \phi$ but contain subwords of exponent greater than $2 + \phi - \varepsilon$ for every $\varepsilon > 0$. Therefore, from our previous result, the sum of exponents of all maximal repetitions is bounded from above by $(2+\phi)(2|f_{k-2}|-3) = 2(2-\phi)(2+\phi)|f_k| + o(1) = 2(3-\phi)|f_k| + o(1) \approx 2.764|f_k| + o(1)$. We could not obtain the exact formula for the sum of exponents, but we give a good estimation of it showing that this number is bounded asymptotically between $1.922 \cdot |f_k|$ and $1.926 \cdot |f_k|$.

Fibonacci words are known to contain "many" repetitions, and the fact that in Fibonacci words there is a linear number of maximal repetitions, rises the question if this is true for general words. We confirm this conjecture and prove that a word of length n over an arbitrary alphabet contains $O(n)$ maximal repetitions. The result is both of theoretical and practical interest. From the theoretical point of view, it contrasts to the above results about the $O(n \log n)$ number of positioned squares or integer repetitions, and shows that maximal repetitions are indeed a compact (linear) representation of all repetitions in a word. In particular, this answers the open question rised in [IMS97] whether all repetitions can be encoded in a linear-size structure and in particular, whether the number of maximal repetitions is linearly-bounded.

From the practical point of view, this result allows us to derive a linear-time algorithm of enumerating all maximal repetitions in a word. This algorithm, which is a modification of Main's algorithm [Mai89], will be briefly commented in the end of this paper, but will be presented in full details in an accompanying paper.

2 Definitions and Basic Results

Consider a word $w = a_1 \ldots a_n$. Any word $a_i \ldots a_j$ for $1 \leq i \leq j \leq n$, which we denote $w[i..j]$, is a *subword* of w. A position in w is an integer number between 0 and n. Each position π in w defines a decomposition $w = w_1 w_2$ where $|w_1| = \pi$. The position of letter a_i in w is $(i-1)$. We say that subword $v = w[i..j]$ *crosses* a position π in w, if $i \leq \pi < j$.

If w is a subword of u^n for some natural n, $|u|$ is called a *period* of w, and word u is a *root* of w. Clearly, p is a period of $w = a_1 \ldots a_n$ iff $a_i = a_{i+p}$ whenever $1 \leq i, i+p \leq n$. Another equivalent definition is (see [Lot83]): p is a period of $w = a_1 \ldots a_n$ iff $w[1..n-p] = w[p+1..n]$. The last definition shows that each word w has the minimal period that we will denote $p(w)$ and call often simply *the* period of w. The ratio $\frac{|w|}{p(w)}$ is called the *exponent* of w and denoted $e(w)$. Clearly, a root u of w such that $|u| = p(w)$, is *primitive*, that is u cannot be written as v^n for $n \geq 2$. Following [Lot83, Chapter 8], we call the roots u with $|u| = p(w)$ *cyclic roots*.

Consider $w = a_1 \ldots a_n$. A *repetition* in w is any subword $r = w[i..j]$ with $e(r) \geq 2$. A *maximal repetition* in w is a repetition $r = w[i..j]$ such that

(i) if $i > 1$, then $p(w[i-1..j]) > p(w[i..j])$,
(ii) if $j < n$, then $p(w[i..j+1]) > p(w[i..j])$.

In other words, a maximal repetition is a repetition $r = w[i..j]$ such that no subword of w which contains r as a proper subword has the same minimal period as r. Note that any repetition in a word can be extended to a unique maximal repetition. For example, the repetition 1010 in word $w = 1011010110110$ extends to the maximal repetition 10101 obtained by one letter extension to the right.

A basic result about periods is the Fine and Wilf's theorem (see [Lot83]):

Theorem 1 (Fine and Wilf). *If w has periods p_1, p_2, and $|w| \geq p_1 + p_2 - gcd(p_1, p_2)$, then $gcd(p_1, p_2)$ is also a period of w.*

The following Lemma states some useful facts about maximal repetitions.

Lemma 1. *(i) Two distinct maximal repetitions with the same period p cannot have an overlap of length greater than or equal to p,*
(ii) Two maximal repetitions with minimal periods p_1, p_2, $p_1 \neq p_2$, cannot have an overlap of length greater than or equal to $(p_1 + p_2 - gcd(p_1, p_2)) \leq 2\max\{p_1, p_2\}$.

Proof. Part (i) is easily proved by analyzing relative positions of two repetitions of period p and showing that if they intersect on at least p letters, at least one of them is not maximal. Part (ii) is a consequence of Fine and Wilf's theorem. If the intersection is at least $(p_1 + p_2 - gcd(p_1, p_2))$ long, then at least one of the cyclic roots of the two repetitions is not primitive, which is a contradiction.

3 Maximal Repetitions in Fibonacci Words

Fibonacci words are binary words defined recursively by $f_0 = 0$, $f_1 = 1$, $f_n = f_{n-1}f_{n-2}$ for $n \geq 2$. The length of f_n, denoted F_n, is the n-th Fibonacci number. Fibonacci words have numerous interesting combinatorial properties and often provide a good example to test conjectures and analyse algorithms on words (cf [IMS97]).

As it was noted in Introduction, Fibonacci word f_n contains $\Theta(F_n \log F_n)$ squares all of which are primitively-rooted. In [FS99], the exact number of squares in Fibonacci words has been obtained, which is asymptotically $\frac{2}{5}(3 - \phi)nF_n + O(F_n)$. Since general words of length n contain $O(n \log n)$ primitively-rooted squares [CR95], Fibonacci words contain asymptotically maximal number of primitively-rooted squares (at least up to a multiplicative constant).

In this section, we first count the exact number of maximal repetitions in Fibonacci words. Let R_n be the number of maximal repetitions in f_n. We prove the following

Theorem 2. *For all $n \geq 4$, $R_n = 2F_{n-2} - 3$.*

We follow the general proof scheme used in [FS99] for counting the number of positioned squares. Consider the decomposition $f_n = f_{n-1}f_{n-2}$ and call the position between f_{n-1} and f_{n-2} the *boundary*. Clearly, the maximal repetitions in f_n are divided into those which lie entirely in f_{n-1} or f_{n-2} and those which cross the boundary, that is intersect with f_{n-1} (call this intersection the left part) and with f_{n-2} (right part). We call the latter *crossing* repetitions. Note first that the left part and the right part of a crossing repetition cannot be both of exponent ≥ 2, since Fibonacci words don't have subwords of exponent 4. If either the left or the right part is of exponent ≥ 2, then the crossing repetition is an extension of a maximal repetition of respectively f_{n-1} or f_{n-2}. This implies that the only new crossing repetitions of f_n that should be counted are those that don't have their right and left part of exponent ≥ 2. Denote $c(n)$ the number of such crossing repetitions that we will call *composed* maximal repetitions of f_n. Then

$$R_n = R_{n-1} + R_{n-2} + c(n). \tag{1}$$

The following argument gives the solution.

Lemma 2. *For all $n \geq 8$, $c(n) = c(n-2)$.*

Consider the representation

$$f_n = f_{n-1}|f_{n-2} = f_{n-2}f_{n-3}|f_{n-3}f_{n-4} = f_{n-2}[f_{n-3}|f_{n-4}]f_{n-5}f_{n-4} \tag{2}$$

where | denotes the boundary, $n \geq 5$, and square brackets delimit the occurrence of f_{n-2} with the same boundary as for the whole word f_n. It is known that every repetition in Fibonacci words has the period F_k for some k (this is mentioned in [FS99] as a "folklore" result, proved in [Séé85]). Since $F_{n-3} > F_{n-4} > 2F_{n-6}$, it follows from (2) that if a composed maximal repetition of f_n has the period F_k for $k \leq n-6$, then it is also a composed maximal repetition of f_{n-2} and therefore is counted in $c(n-2)$. Vice versa, every composed maximal repetition of f_{n-2} with period F_k for $k \leq n-6$, is also a composed maximal repetition of f_n. We now examine the maximal repetitions of f_n with periods F_{n-2}, F_{n-3}, F_{n-4}, F_{n-5} which cross the boundary.

Crossing repetitions with period F_{n-2}. The last term of (2) shows that square $(f_{n-2})^2$ is a prefix of f_n that crosses the boundary. As $F_{n-1} < 2F_{n-2}$, the corresponding maximal repetition does not have a square in its left or right part and therefore is composed for f_n. Since $F_{n-2} > F_n/3$, any two maximal repetitions of f_n with period F_{n-2} intersect by more than F_{n-2} letters. By Lemma 1(i), this shows that f_n has only one maximal repetition with period F_{n-2}. Trivially, the maximal repetition under consideration is not a maximal repetition of f_{n-2}.

Crossing repetitions with period F_{n-3}. From the decomposition $f_n = f_{n-2}f_{n-3}|f_{n-3}f_{n-4}$ (see (2)), there is a square $(f_{n-3})^2$ with the root length F_{n-3} crossing the boundary. The corresponding maximal repetition does not extend to the left of the left occurrence of f_{n-3}, as the last letters of f_{n-3} and f_{n-2} are different (the last letters of f_i's alternate). Therefore, this maximal repetition does not have a square in its left or right part, and thus is composed for f_n. As this maximal repetition has a period both on the left and on the right of the boundary, it is the only maximal repetition with period F_{n-3} crossing the boundary (see Lemma 1(i)). Again, from length considerations, it is not an maximal repetition of f_{n-2}.

Crossing repetitions with period F_{n-4}. As $f_n = f_{n-2}[f_{n-3}|f_{n-4}]f_{n-5}f_{n-4} = f_{n-3}f_{n-4}[f_{n-4}f_{n-5}|f_{n-5}f_{n-6}]f_{n-5}f_{n-4} = f_{n-3}f_{n-4}[f_{n-4}\underbrace{f_{n-5}|f_{n-6}f_{n-7}}_{f_{n-4}}f_{n-6}]$ $f_{n-5}f_{n-4}$ for $n \geq 7$, this reveals a maximal repetition of period F_{n-4} which crosses the boundary. However, this is not a composed maximal repetition of f_n, as it has a square on the left of the boundary. On the other hand, the restriction of this maximal repetition to f_{n-2} (subword in square brackets) *is* a composed maximal repetition for f_{n-2}.

It can be shown that this is the only maximal repetition of period F_{n-4} crossing the boundary. (There is another one which touches the boundary from the right, but does not extend to the left of it.) In conclusion, there is one composed maximal repetition of period F_{n-4} in f_{n-2} and no such maximal repetition in f_n.

Crossing repetitions with period F_{n-5}. Rewrite $f_n = f_{n-2}[f_{n-4}f_{n-5}|f_{n-5}f_{n-6}]f_{n-5}f_{n-4}$ which shows that there is a square of root length F_{n-5} crossing the boundary. Since the boundary is the center of this square, the latter corresponds to the only maximal repetition with period F_{n-5}

crossing the boundary. However, this maximal repetition is not a composed maximal repetition for f_n, as it has a square in its right part, as shown by the following transformation: $f_n = f_{n-2}[f_{n-4}f_{n-5}|f_{n-5}f_{n-6}]f_{n-6}f_{n-7}f_{n-4} = f_{n-2}[f_{n-4}f_{n-5}|f_{n-5}\underbrace{f_{n-6}]f_{n-7}}_{f_{n-5}}f_{n-8}f_{n-7}f_{n-4}$ for $n \geq 8$. On the other hand, the restriction of this maximal repetition to f_{n-2} (subword in square brackets) *is* a composed maximal repetition for f_{n-2}. Thus, there is one composed maximal repetition of the period F_{n-5} in f_{n-2} and no such maximal repetition in f_n.

In conclusion, two new composed maximal repetitions arise in f_n in comparison to f_{n-2}, but two composed maximal repetitions of f_{n-2} are no more composed in f_n, as they extend in f_n to form a square in its right or left part. This shows that $c(n) = c(n-2)$ for $n \geq 8$ and proves the Lemma.

A direct counting shows that $R_0 = 0$, $R_1 = 0$, $R_2 = 0$, $R_3 = 0$, $R_4 = 1$, $R_5 = 3$, $R_6 = 7$, $R_7 = 13$. Therefore, $c(3) = 0$, $c(4) = 1$, $c(5) = 2$, $c(6) = 3$, $c(7) = 3$. Since $c(n) = c(n-2)$ for all $n \geq 8$, then $c(n) = 3$ for all $n \geq 6$. We then have the recurrence relation $R_n = R_{n-1} + R_{n-2} + 3$ for $n \geq 6$ with boundary conditions $R_4 = 1$, $R_5 = 3$. Resolving it, we get $R_n = 2F_{n-2} - 3$ for $n \geq 4$. Theorem 2 is proved.

Thus, in contrast to squares, the number of maximal repetitions in Fibonacci words is linear. Using the same approach, we now estimate the sum of exponents of all maximal repetitions in f_n. A direct consequence of Theorem 2 and the fact that Fibonacci words don't contain exponents greater than $(2+\phi)$ [MP92], is that the sum of exponents is no greater, asymptotically, than $2(3-\phi)|f_k| \approx 2.764 \cdot |f_k|$. We now obtain a more precise estimation.

Denote $SR(n)$ the sum of exponents of all maximal repetitions in Fibonacci word f_n. We prove the following estimation for $SR(n)$.

Theorem 3. $SR(n) = C \cdot |f_n| + o(1)$, *where* $1.922 \leq C \leq 1.926$.

Similarly to (1), we write the recurrent relation

$$SR(n) = SR(n-1) + SR(n-2) + cx(n), \tag{3}$$

where $cx(n)$ is the sum of exponents of those left and right parts of crossing repetitions, which have the exponent smaller than 2. (If the exponent of the left or right part is 2 or more, it is counted in $SR(n-1)$ or $SR(n-2)$ respectively.) As before, the goal is to reduce $cx(n)$ to $cx(n-2)$, and a similar argument shows that for all crossing repetitions with the period F_k for $k \leq n-6$, nothing has to be done, as they occur completely inside f_{n-2} (see (2)) and are counted in $cx(n-2)$. As for Theorem 2, it remains to analyse repetitions with periods F_{n-2}, F_{n-3}, F_{n-4}, F_{n-5}.

The crossing repetition with period F_{n-2} is composed (both its left and right part is of exponent < 2), its length can be shown to be $F_n - 2 = F_{n-1} + F_{n-2} - 2$, and the exponent $\frac{F_{n-1}+F_{n-2}-2}{F_{n-2}}$. The crossing repetition with period F_{n-3} is also composed, of the length $2F_{n-3} + F_{n-4} = F_{n-2} + F_{n-3}$, and of the exponent

$\frac{F_{n-2}+F_{n-3}}{F_{n-3}}$. Let us turn to the crossing repetition with period F_{n-4}. Recall that it extends a repetition present in f_{n-2}. Its right part is of exponent < 2, and is inside f_{n-2}, therefore it is already counted in $cx(n-2)$, and it does not have to be added. Its left part is of exponent ≥ 2, and does not have to be counted in $cx(n)$. However, a part of it which is in f_{n-2} (namely $f_{n-4}f_{n-5}$), is of exponent < 2, and therefore has been counted in $cx(n-2)$. We then have to substract $\frac{F_{n-4}+F_{n-5}}{F_{n-4}} = \frac{F_{n-3}}{F_{n-4}}$. Similarly, the crossing repetition with the period F_{n-5} has the left part which is already counted in $cx(n-2)$, and the right part which should not be counted, but the part of it of exponent $\frac{F_{n-5}+F_{n-6}}{F_{n-5}} = \frac{F_{n-4}}{F_{n-5}}$ has been counted in $cx(n-2)$ and should be substracted. Putting everything together, we obtain the recurrence

$$cx(n) = cx(n-2) + 2 - 2/F_{n-2} + F_{n-1}/F_{n-2} + F_{n-2}/F_{n-3} - F_{n-3}/F_{n-4} - F_{n-4}/F_{n-5}, \quad (4)$$

for $n \geq 8$. Transforming further this expression, we obtain

$$cx(n) = n-1-2(1/F_{n-2}+1/F_{n-4}+\ldots+1/F_4+1/F_2)+F_{n-1}/F_{n-2}+F_{n-2}/F_{n-3}$$

for even $n \geq 8$, and

$$cx(n) = n+1/2-2(1/F_{n-2}+1/F_{n-4}+\ldots+1/F_3+1/F_1)+F_{n-1}/F_{n-2}+F_{n-2}/F_{n-3}$$

for odd $n \geq 9$. To join the cases, we rewrite (3) into

$$SR(n) = 2SR(n-2) + SR(n-3) + cx(n) + cx(n-1) = 2SR(n-2) + SR(n-3) + 2n - 3/2 - 2(\sum_{j=1}^{n-2} 1/F_j) + F_{n-1}/F_{n-2} + 2F_{n-2}/F_{n-3} + F_{n-3}/F_{n-4}.$$

The following estimation can be obtained using some elementary consideration.

$$-2(\sum_{j=1}^{n-1} 1/F_j) + F_n/F_{n-1} + 2F_{n-1}/F_{n-2} + F_{n-2}/F_{n-3} < 2,$$

for $n \geq 8$. We omit the proof. Using this estimation, we get that for all $n \geq 9$,

$$SR(n) \leq 2SR(n-2) + SR(n-3) + 2n + 1/2.$$

Solving this recurrence with initial conditions $SR(4) = 2, SR(5) = 6.5, SR(6) = 15\frac{11}{30}, SR(7) = 29\frac{27}{40}, SR(8) = 53\frac{142}{195}$, we obtain that

$$SR(n) \leq \frac{33}{520}(-1)^{n+1} + \frac{1}{\sqrt{5}}(40\frac{47}{130} - 25\frac{281}{1560}\bar{\phi})\phi^{n-6} + \frac{1}{\sqrt{5}}(40\frac{47}{130} - 25\frac{281}{1560}\phi)\bar{\phi}^{n-6} - n - \frac{15}{4} < \frac{1}{\sqrt{5}}(40\frac{47}{130} - 25\frac{281}{1560}\bar{\phi})\phi^{n-6} \approx 1.926 \cdot |f_n|.$$

The lower bound can be obtained as follows. A direct calculation gives the values $SR(23) = 1.922328 \cdot |f_{23}|, SR(24) = 1.922520 \cdot |f_{24}|$. Then using the obvious inequality $SR(n) \geq SR(n-1) + SR(n-2)$, we get $SR(n) \geq 1.922328 \cdot |f_n|$. Theorem 3 is proved.

4 Maximal Number of Maximal Repetitions in a Word

Since Fibonacci words contain "many" repetitions, Theorem 2 suggests the following question: Is it true that general words contain only a linear number of maximal repetitions? We answer this question affirmatively. We prove that the maximal number of maximal repetitions in words of length n is a linear function on n, regardless of the underlying alphabet. Denote by $Rep(n)$ the maximal number of maximal repetitions in words of length n (the alphabet is not fixed).

Theorem 4. *$Rep(n) = O(n)$.*

The proof of Theorem 4 is rather technical and cannot be given here because of space limitations. Actually, we prove that there exist absolute positive constants C_1, C_2 such that

$$Rep(n) \leq C_1 n - C_2 \sqrt{n} \log n$$

For the proof we refer the reader to [KK98].

5 Applications, Generalizations, Open Questions

In this concluding section we mention an important algorithmic application of Theorem 4, discuss its possible generalization, and formulate several related open questions.

An important application of Theorem 4 is that it allows to derive an algorithm which finds all maximal repetitions in a word in time linear in the length of the word.

The problem of searching for repetitions in a string (or testing if a string contains repetitions) has been studied since early 80's. Let us first survey known results. In early 80's, Slisenko [Sli83] claimed a linear (real-time) algorithm for finding all *distinct* maximal repetitions in a word. Independently, Crochemore [Cro83] described a simple and elegant linear algorithm for finding square in a word (and thus checking if a word is repetition-free). The algorithm was based on a special factorization of the word, called s-factorizarion (f-factorization in [CR94]). Another linear algorithm for checking whether a word contains a square was proposed in [ML85].

If one wants to explicitely list all squares (or integer powers) occurring in a word, there is no hope to do it in linear time, as their number may be of order $n \log n$. Several algorithms have been proposed in order to find all repetitions in time $O(n \log n)$. In 1981, Crochemore [Cro81] proposed an $O(n \log n)$ algorithm for finding all occurrences of primitively-rooted maximal integer powers in a word. Using a suffix tree technique, Apostolico and Preparata [AP83] described an $O(n \log n)$ algorithm for finding all positioned *right-maximal* fractional repetitions. Finally, Main and Lorentz [ML84] proposed another algorithm which actually finds all maximal repetitions in $O(n \log n)$ time. In 1989, using Crochemore's s-factorization, Main [Mai89] proposed a *linear-time* algorithm which finds all *leftmost* occurrences of distinct maximal repetitions in a word.

As far as other related works are concerned, Kosaraju [Kos94] describes an $O(n)$ algorithm which, given a word, finds for each position the shortest square starting at this position. He also claims a generalization which finds all primitively-rooted squares in time $O(n+S)$ where S is the number of such squares. In [SG98a], Stoye and Gusfield proposed several algorithms that are based on a unified suffix tree framework. Their results are based on an algorithm which finds in time $O(n \log n)$ all "branching tandem repeats". In our terminology, branching tandem repeats are (not necessarily primitively-rooted) square suffixes of maximal repetitions. In a very recent paper, Stoye and Gusfield [SG98b] proposed a different approach, combining s-factorization (called Lempel-Ziv factorization in the paper) and suffix tree techniques. The goal achieved is to find, in linear time, a representative of each *distinct* square. The feasibility of this task is supported by the result of [FS98] mentioned in Introduction. The approach allows also to solve some other problems, e.g. to achieve the results claimed in [Kos94].

However, so far it has been an open question whether a *linear* algorithm for finding *all* maximal repetitions exists. In the concluding section of [Mai89], Main speculates that such an algorithm might exist. The same question is raised in [IMS97]. However, there has been no evidence in support of this conjecture as the number of maximal repetitions has not been known to be linear. Theorem 4 provides this argument. Using Theorem 4, it can be shown that Main's algorithm can be modified in order to find all maximal repetitions in linear time. This allows also to solve other related problems, e.g. to output all squares in a word in time $O(n+S)$, where S is the output size (cf [Kos94,SG98b]). The algorithm will be described in an accompaining paper. An interested reader may consult [KK98].

The results of this paper suggest an interesting question: Can Theorem 3 asserting the linearity of the sum of exponents of the maximal repetitions in Fibonacci words be also generalized to general words? Putting in direct terms, is the sum of exponents of maximal repetitions in a word also bounded linearly in the length of the word?

This conjecture is somewhat related to the hypothesis suggested in [SG98a] about the linearity of the maximal number of "branching tandem repeats" in a word. Branching tandem repeats are squares uu (not necessarily primitively-rooted) which are not followed by the first letter of u. To relate this to maximal repetitions, branching tandem repeats are suffixes of the maximal repetitions of length $2kp(r)$, where r is the corresponding maximal repetition and $k \geq 1$. The linearity of the maximal number of branching tandem repeats is stronger than our Theorem 4, as there are at least as many branching tandem repeats as maximal repetitions (each branching tandem repeat corresponds to a maximal repetition but one maximal repetition may contain several branching tandem repeats).

If the maximal sum of exponents of all maximal repetitions in a word were proved also linearly bounded, this would imply both our Theorem 4 and the conjecture of [SG98a], and also shed some light on some facts we will mention below. Both authors of this paper strongly believe that this hypothesis is true.

This is supported by computer experiments which show that in binary words that realize the maximal number of maximal repetitions, maximal repetitions are all of small exponent, typically not bigger than 3. This phenomenon is also illustrated by Fibonacci words, which contain "many" maximal repetitions, all of which are of exponent smaller than $2+\phi \approx 3.618$. The above hypothesis would shed light on this fact.

Let us make some other remarks about our results.

The main drawback of our proof of Theorem 4 is that it does not allow to extract a "reasonable" constant factor in the linear bound. It remains an open question if a simpler proof can be found which would imply a constant factor. We conjecture that for the binary alphabet this constant factor is equal to 1, which is supported by computer experiments.

Concerning counting results of Section 3, we note that Fibonacci words don't realize the maximal number of maximal repetitions among the binary words. For example, for length 21 this number is 15 (realized, e.g., by word 000101001011010010100) while Fibonacci word f_7 of length 21 contains 13 maximal repetitions.

While the number of maximal repetitions in Fibonacci words is one less than the number of distinct squares, computer experiments show that the maximal number of maximal repetitions in binary words of length n is apparently slightly bigger than the maximal number of distinct squares. In spite of this closeness between the number of maximal repetitions and that of distinct squares, there is no apparent connection between them. It is possible to conceive words with a big number of maximal repetitions and small number of distinct squares. For example, the result of [FS95] implies that there exist words with only three distinct squares but with unbounded number of maximal repetitions. Still, we are wondering if the fact that the number of maximal repetitions in Fibonacci words is one less than the number of distinct squares is a simple coincidence or it has some combinatorial explanation.

References

[AP83] A. Apostolico and F.P. Preparata. Optimal off-line detection of repetitions in a string. *Theoretical Computer Science*, 22(3):297–315, 1983.

[CR94] M. Crochemore and W. Rytter. *Text algorithms*. Oxford University Press, 1994.

[CR95] M. Crochemore and W. Rytter. Squares, cubes, and time-space efficient string searching. *Algorithmica*, 13:405–425, 1995.

[Cro81] M. Crochemore. An optimal algorithm for computing the repetitions in a word. *Information Processing Letters*, 12:244–250, 1981.

[Cro83] M. Crochemore. Recherche linéaire d'un carré dans un mot. *Comptes Rendus Acad. Sci. Paris Sér. I Math.*, 296:781–784, 1983.

[CS96] J.D. Currie and R.O. Shelton. Cantor sets and Dejean's conjecture. *Journal of Automata, Languages and Combinatorics*, 1(2):113–128, 1996.

[Dej72] F. Dejean. Sur un théorème de Thue. *J. Combinatorial Th. (A)*, 13:90–99, 1972.

[FS95] A.S. Fraenkel and J. Simpson. How many squares must a binary sequence contain? *Electronic Journal of Combinatorics*, 2(R2):9pp, 1995. `http://www.combinatorics.org/Journal/journalhome.html`.

[FS98] A.S. Fraenkel and J. Simpson. How many squares can a string contain? *J. Combinatorial Theory (Ser. A)*, 82:112–120, 1998.

[FS99] A.S. Fraenkel and J. Simpson. The exact number of squares in Fibonacci words. *Theoretical Computer Science*, 218(1):83–94, 1999.

[IMS97] C.S. Iliopoulos, D. Moore, and W.F. Smyth. A characterization of the squares in a Fibonacci string. *Theoretical Computer Science*, 172:281–291, 1997.

[JP99] J. Justin and G. Pirillo. Fractional powers in Sturmian words. Technical Report LIAFA 99/01, Laboratoire d'Informatique Algorithmique: Fondements et Applications (LIAFA), 1999.

[KK98] R. Kolpakov and G. Kucherov. Maximal repetitions in words or how to find all squares in linear time. Rapport Interne LORIA 98-R-227, Laboratoire Lorrain de Recherche en Informatique et ses Applications, 1998. available from URL: `http://www.loria.fr/~kucherov/res_activ.html`.

[Kos94] S. R. Kosaraju. Computation of squares in string. In M. Crochemore and D. Gusfield, editors, *Proceedings of the 5th Annual Symposium on Combinatorial Pattern Matching*, number 807 in Lecture Notes in Computer Science, pages 146–150. Springer Verlag, 1994.

[Lot83] M. Lothaire. *Combinatorics on Words*, volume 17 of *Encyclopedia of Mathematics and Its Applications*. Addison Wesley, 1983.

[Mai89] M. G. Main. Detecting leftmost maximal periodicities. *Discrete Applied Mathematics*, 25:145–153, 1989.

[ML84] M.G. Main and R.J. Lorentz. An $O(n \log n)$ algorithm for finding all repetitions in a string. *Journal of Algorithms*, 5(3):422–432, 1984.

[ML85] M.G. Main and R.J. Lorentz. Linear time recognition of square free strings. In A. Apostolico and Z. Galil, editors, *Combinatorial Algorithms on Words*, volume 12 of *NATO Advanced Science Institutes, Series F*, pages 272–278. Springer Verlag, 1985.

[MP92] F. Mignosi and G. Pirillo. Repetitions in the Fibonacci infinite word. *RAIRO Theoretical Informatics and Applications*, 26(3):199–204, 1992.

[MRS95] F. Mignosi, A. Restivo, and S. Salemi. A periodicity theorem on words and applications. In *Proceedings of the 20th International Symposium on Mathematical Foundations of Computer Science (MFCS)*, volume 969 of *Lecture Notes in Computer Science*, pages 337–348. Springer Verlag, 1995.

[Sée85] P. Séébold. Propriétés combinatoires des mots infinis engendrés par certains morphismes. Rapport 85-16, LITP, Paris, 1985.

[SG98a] J. Stoye and D. Gusfield. Simple and flexible detection of contiguous repeats using a suffix tree. In M. Farach-Colton, editor, *Proceedings of the 9th Annual Symposium on Combinatorial Pattern Matching*, number 1448 in Lecture Notes in Computer Science, pages 140–152. Springer Verlag, 1998.

[SG98b] J. Stoye and D. Gusfield. Linear time algorithms for finding and representing all the tandem repeats in a string. Technical Report CSE-98-4, Computer Science Department, University of California, Davis, 1998.

[Sli83] A.O. Slisenko. Detection of periodicities and string matching in real time. *Journal of Soviet Mathematics*, 22:1316–1386, 1983.

Axiomatization of the Coherence Property for Categories of Symmetries

Dorel Lucanu

"A.I. Cuza" University
Faculty of Computer Science
Berthelot 16, 6600 Iaşi, Romania
e-mail: dlucanu@infoiasi.ro

Abstract. Given an equational theory (Σ, E), a relaxed (Σ, E)-system is a category S enriched with a Σ-algebra structure on both objects and arrows such that a natural isomorphism $\alpha_S : t_S \Rightarrow t'_S$, called natural symmetry, exists for each $t =_E t'$. A symmetry is an instance of a natural symmetry. A category of symmetries, which includes only symmetries, is a free object in the category of relaxed (Σ, E)-systems. The coherence property states that the diagrams in a category of symmetries are commutative. In this paper we present a method for expressing the coherence property in an axiomatic way.

1 Introduction

(Σ, E)-systems are categories enriched with a Σ-algebra structure on both objects and arrows such that the functors induced by terms satisfy the equations E. The satisfaction uses the strict interpretation in the sense that an equation $t = t'$ is satisfied by a system S iff the functors t_S and t'_S are equal. These categories were introduced in [7] to define models for rewriting logic. In [5] a relaxed version for these systems was defined, where the equations E are relaxedly interpreted, that is an equation $t = t'$ is satisfied by a system S iff there exists a natural isomorphism $t_S \cong t'_S$, called natural symmetry. A relaxation is partial iff a part of equations are always strictly interpreted. A category of symmetries is a free object in the category of relaxed (Σ, E)-systems, where (Σ, E) is a given equational theory. The arrows in such a category are instances of natural symmetries and are called symmetries. A well known example is that of the monoidal categories: the signature is $\Sigma = \{+, 0\}$ and the set of equations E consists of the associativity, left and right units. In such a category the three equations are interpreted by natural isomorphisms [6]. A category of symmetries is coherent if all diagrams are commutative. The commutativity of diagrams in the category of symmetries implies the commutativity of certain diagrams in any (Σ, E)-system [5]. Sometimes the coherence is partial in the sense that only the diagrams of a certain sort are commutative. For example, this is the case of symmetric strict monoidal categories.

The axiomatic definition of the free $\mathcal{R}$-systems allowed as the well known algebraic specification language OBJ3 to be extended such that it handles specifications of concurrent systems. The most representatives extensions in this direction are Maude [8] and CafeOBJ [3]. In these languages we can describe concurrent systems using rewrite specifications whose standard semantics is the initial semantics. The problem becomes more complicated if we use coherent relaxed models for rewrite specifications because these models include also symmetries. In [5] an axiomatic construction for the free relaxed models is given. But the construction given there assumes the subcategory of symmetries already defined. So that a full axiomatic construction of the free relaxed model requires an axiomatization of the symmetries.

In this paper we investigate the possibilities to obtain complete axiomatizations for categories of symmetries. The key point consists in associating a rewrite theory $\mathcal{R}(\Sigma, E)$ with the equational specification by turning the equations into rewrite rules. The elegant construction of the free $\mathcal{R}$-groupoid given in [7] provides already an axiomatization of the free (Σ, E)-system (the non-coherent category of symmetries). The problem of finding axioms which expresses the commutativity of the diagrams still remains to be solved. We show that if equations E, viewed as rewrite rules, form a convergent (confluent and terminating) rewriting system then these axioms are obtained by computing all critical pairs. Each confluent rewriting generated by a critical pair produce an equation. The set of all equations obtained in this way forms a specification of the commutative diagrams. The method can be generalized to the case when E is convergent modulo a theory T.

The paper has four sections. Section 2 presents the basic notions and notations from rewrite logic used in the paper. Section 3 is devoted to the axiomatization of the symmetries and includes the main results. It also includes representative examples which show that the proposed method is simple and efficient. The last section includes conclusions and directions for further work. Due to the space limitation, the proofs are omitted but they (and more details and examples) can be found in [4].

2 Preliminaries

2.1 Rewriting logic

A **(unconditional labelled) rewrite specification** $\mathcal{R}$ is a 4-tuple $\mathcal{R} = (\Sigma, E, L, R)$ where Σ is a signature, E is a set of Σ-equations, L is a set called the set of **labels**, and R is a set of labelled rewrite rules written as $r : t(\overline{x}) \to t'(\overline{x})$. When the set L of labels is derived from the context we shall omit to write it. The models for rewrite signatures are (small) categories where both classes of objects and arrows have an algebraic structure over the rewrite signature. We can define these categories in a general framework. Let E be a set of Σ-equations. A (strict) (Σ, E)**-system** is a category $\mathcal{S}$ such that:

(i) for any $n \in \omega$ and $f \in \mathcal{S}_n$, there exists a functor $f_{\mathcal{S}} : \mathcal{S}^n \to \mathcal{S}$, and
(ii) the induced functors $t_{\mathcal{S}}$ and $t'_{\mathcal{S}}$ are equal, for any terms $t(\overline{x})$ and $t'(\overline{x})$ with $E \models t(\overline{x}) = t'(\overline{x})$.

A (Σ, E)-**homomorphism** between two (Σ, E)-systems is a functor which preserves the operations in Σ. We denote by $\mathbf{Sys}(\Sigma, E)$ the category of (Σ, E)-systems.

Given a rewrite specification $\mathcal{R} = \langle \Sigma, E, L, R \rangle$, a (strict) $\mathcal{R}$-**system** ($\mathcal{R}$-model) $\mathcal{S}$ is a (Σ, E)-system $\mathcal{S}$ such that, for each rewrite rule $r : t(\overline{x}) \to t'(\overline{x})$ in R, there exists a natural transformation $r_{\mathcal{S}} : t_{\mathcal{S}} \Rightarrow t'_{\mathcal{S}}$. We say that a $\mathcal{R}$-system $\mathcal{S}$ **satisfies** the sequent $[t(\overline{x})]_E \to [t'(\overline{x})]_E$ iff there exists a natural transformation $\tau : t_{\mathcal{S}} \Rightarrow t'_{\mathcal{S}}$. A **sequent** $[t(\overline{x})]_E \to [t'(\overline{x})]_E$ is a **semantical consequence** of the rewrite specification $\mathcal{R}$, written $\mathcal{R} \models [t(\overline{x})]_E \to [t'(\overline{x})]_E$, iff it is satisfied by all $\mathcal{R}$-systems. An $\mathcal{R}$-**homomorphism** $F : \mathcal{S} \to \mathcal{S}'$ between two $\mathcal{R}$-systems is a (Σ, E)-homomorphism $F : \mathcal{S} \to \mathcal{S}'$ which preserves the natural transformations corresponding to the rules in $\mathcal{R}$. We denote by $\mathcal{R}$-**Sys** the category of (strict) $\mathcal{R}$-systems.

The free model $\mathcal{T}_{\mathcal{R}}(X)$ in $\mathcal{R}$-**Sys** has as arrows those defined by the following inference rules:

(I1) *Identities.* For each $[t] \in T_{\Sigma,E}(X)$,

$$\overline{[t] : [t] \to [t]}$$

(I2) *Σ-structure.* For each $f \in \Sigma_n, n \in \omega$,

$$\frac{\alpha_1 : [t_1] \to [t'_1], \ldots, \alpha_n : [t_n] \to [t'_n]}{f(\alpha_1, \ldots, \alpha_n) : [f(t_1, \ldots, t_n)] \to [f(t'_1, \ldots, t'_n)]}$$

(I3) *Replacement.* For each rule $r : [t(\overline{x})] \to [t'(\overline{x})]$ in R,

$$\frac{\alpha_1 : [w_1] \to [w'_1], \ldots, \alpha_n : [w_n] \to [w'_n]}{r(\overline{\alpha}) : [t(\overline{w}/\overline{x})] \to [t'(\overline{w}'/\overline{x})]}$$

(I4) *Composition.*

$$\frac{\alpha_1 : [t_1] \to [t_2], \alpha_2 : [t_2] \to [t_3]}{\alpha_1; \alpha_2 : [t_1] \to [t_3]}$$

modulo the axioms given by:

(A1) *Category.*
- Associativity. For all α, β, γ,
  $$(\alpha; \beta); \gamma = \alpha; (\beta; \gamma).$$
- Identities. For each $\alpha : [t] \to [t']$
  $$\alpha; [t'] = \alpha, \qquad [t]; \alpha = \alpha.$$

(A2) *Functoriality of the Σ-algebraic structure.* For each $f \in \Sigma_n, n \in \omega$, all $\alpha_1, \ldots, \alpha_n, \beta_1, \ldots, \beta_n$,

$$f(\alpha_1; \beta_1, \ldots, \alpha_n; \beta_n) = f(\alpha_1, \ldots, \alpha_n); f(\beta_1, \ldots, \beta_n).$$

(A3) *Axioms in E.* For all $t(\overline{x}) =_E t'(\overline{x})$, $\overline{\alpha} = (\alpha_1 : [u_1] \to [v_1], \ldots, \alpha_n : [u_n] \to [v_n])$

$$t(\alpha_1, \ldots, \alpha_n) = t'(\alpha_1, \ldots, \alpha_n)$$

(A4) *Decomposition.* For each rule $r : [t(\overline{x})] \to [t'(\overline{x})]$ in R,

$$\frac{\alpha_1 : [w_1] \to [w'_1], \ldots, \alpha_n : [w_n] \to [w'_n]}{r(\overline{\alpha}) = r(\overline{[w]}); t'(\overline{\alpha})}$$

(A5) *Exchange.* For each rule $r : [t(\overline{x})] \to [t'(\overline{x})]$ in R,

$$\frac{\alpha_1 : [w_1] \to [w'_1], \ldots, \alpha_n : [w_n] \to [w'_n]}{r(\overline{[w]}); t'(\overline{\alpha}) = t(\overline{\alpha}); r(\overline{[w']})}$$

The $\mathcal{R}$-system $\mathcal{T}_{\mathcal{R}} = \mathcal{T}_{\mathcal{R}}(\emptyset)$ is initial in **$\mathcal{R}$-Sys**. A rewriting logic $\mathcal{R}$ is an **equational logic** (modulo E) iff the "conservative condition"

$$\mathcal{R} \models [t] \to [t'] \text{ iff } \mathcal{R} \models [t'] \to [t]$$

is satisfied. A model for an equational logic is an $\mathcal{R}$-system whose category structure is a groupoid and is called **$\mathcal{R}$-groupoid**. We denote by **$\mathcal{R}$-Grpd** the (sub)category of $\mathcal{R}$-groupoids. The free system $\mathcal{T}_{\overleftrightarrow{\mathcal{R}}}(X)$ for an equational logic $\mathcal{R}$ is obtained by the inference rules I1-4 plus the rule

(I5) *Inversion.*

$$\frac{\alpha : [t] \to [t']}{\alpha^{-1} : [t'] \to [t]}$$

modulo the axioms A1-5 plus the axiom

(A6) *Inverse.* For any $\alpha : [t] \to [t']$ in $\mathcal{T}_{\overleftrightarrow{\mathcal{R}}}(X)$,

$$\alpha; \alpha^{-1} = [t], \qquad \alpha^{-1}; \alpha = [t'].$$

2.2 Relaxed (Σ, E)-systems

In a relaxed model the equations are preserved up to a (canonical) isomorphism. A relaxed (Σ, E)-system is coherent if for each equation e the isomorphism which preserves e is unique. We will see that the coherence property is equivalent to the commutativity of the diagrams of some sort.

Definition 1. *Let (Σ, E) be an equational presentation.*
1. We say that the category $\mathcal{S}$ is a **relaxed (Σ, E)-system** *iff:*

(i) for any $n \in \omega$ and $f \in \Sigma_n$ there exists a functor $f_{\mathcal{S}} : \mathcal{S}^n \to \mathcal{S}$, and
(ii) for each pair $t(\overline{x}) =_E t'(\overline{x})$, there exists a natural isomorphism $t_{\mathcal{S}} \cong t'_{\mathcal{S}}$, called **natural symmetry**.

A **symmetry** *in $\mathcal{S}$ is an instance of a natural symmetry.*
2. A **relaxed (Σ, E)-homomorphism** *$F : \mathcal{S} \to \mathcal{S}'$ between two relaxed (Σ, E)-systems is a (Σ, E)-homomorphism which preserves the symmetries.*

An equivalent way to define relaxed (Σ, E)-systems is as follows. We associate with (Σ, E) the rewrite specification

$$\mathcal{R}(\Sigma, E) = (\Sigma, \emptyset, \{\mathrm{SYM}^{u,v} : u \to v \mid u = v \text{ in } E\}).$$

A relaxed (Σ, E)-system is now an $\mathcal{R}(\Sigma, E)$-system $\mathcal{S}$ where the natural transformation $t_{\mathcal{S}} \Rightarrow t'_{\mathcal{S}}$ is an isomorphism, for each $t = t'$ in E. We denote this natural isomorphism by $\mathrm{SYM}_{\mathcal{S}}(t, t')$ or by $\mathrm{SYM}^{t,t'}$ if the system $\mathcal{S}$ is understood from the context. It is easy to see that if $t =_E t'$ then there exists a natural isomorphism $t_{\mathcal{S}} \cong t'_{\mathcal{S}}$. Unfortunately, this isomorphism is not always unique. In the next subsection we give a condition under that it is unique. However, we use the notation $\mathrm{SYM}_{\mathcal{S}}(t, t')$ and for pairs $t =_E t'$ in the sense that it denotes an arbitrarily chosen natural isomorphism $t_S \cong t'_S$. Note that a relaxed (Σ, E)-system $\mathcal{S}$ is not necessarily an $\mathcal{R}(\Sigma, E)$-groupoid because only the symmetries are invertible in $\mathcal{S}$.

We denote by $\mathbf{RSys}(\Sigma, E)$ the category of relaxed (Σ, E)-systems. Since every strict (Σ, E)-system is also a relaxed one, where the symmetries are identities, we have the full subcategory inclusion $\mathbf{Sys}(\Sigma, E) \hookrightarrow \mathbf{RSys}(\Sigma, E)$.

We show now a way to get initial and free objects in $\mathbf{RSys}(\Sigma, E)$. We consider again the rewrite theory $\mathcal{R}(\Sigma, E)$. Every $\mathcal{R}(\Sigma, E)$-system $\mathcal{S}$ can be transformed into a relaxed (Σ, E)-system $\mathcal{S}^r$. $\mathcal{S}^r$ is the quotient of the groupoid $\mathcal{T}_{\underset{\mathcal{R}(\Sigma,E)}{\longleftrightarrow}}(\mathcal{S})$ modulo the congruence $\equiv$ defined as follows:

- for each $t =_E t'$ there is a $\alpha_S : t_S \Rightarrow t'_S$ in $\mathcal{S}$ and $\alpha : t \Rightarrow t'$ in $\mathcal{T}_{\mathcal{R}(\Sigma,E)}(\mathcal{S}) \hookrightarrow \mathcal{T}_{\underset{\mathcal{R}(\Sigma,E)}{\longleftrightarrow}}(\mathcal{S})$ and we set $\alpha_S \equiv \alpha$; if there is α_S^{-1} then we also set $\alpha_S^{-1} \equiv \alpha^{-1}$.

If $\mathcal{S}$ is a $\mathcal{R}(\Sigma, E)$-system which is a relaxed (Σ, E)-system, too, then $\mathcal{S}^r \cong \mathcal{S}$. It follows that the inclusion $\mathbf{RSys}(\Sigma, E) \hookrightarrow \mathcal{R}(\Sigma, E)$-$\mathbf{Sys}$ is reflective. The $\mathcal{R}(\Sigma, E)$-groupoid $\mathcal{T}_{\underset{\mathcal{R}(\Sigma,E)}{\longleftrightarrow}}(X)$ is freely generated by X in $\mathbf{RSys}(\Sigma, E)$ and it is the image by the reflector functor of the free $\mathcal{R}(\Sigma, E)$-system $\mathcal{T}_{\mathcal{R}(\Sigma,E)}(X)$. We prefer to denote the free relaxed system $\mathcal{T}_{\underset{\mathcal{R}(\Sigma,E)}{\longleftrightarrow}}(X)$ by $Sym^*_{\Sigma,E}(X)$. Because all arrows in $Sym^*_{\Sigma,E}(X)$ are symmetries or identities, we call it the **category of symmetries** (corresponding to (Σ, E)).

2.3 Coherent relaxed (Σ, E)-systems

Because an equation $t =_E t'$ may have different equational deductions it follows that we can have more than one natural symmetries corresponding to $t =_E t'$ in a relaxed system $\mathcal{S}$. Each deduction gives rise, by composition, to a possible distinct natural symmetry. In this subsection we define the subclass of relaxed (Σ, E)-systems in which these natural symmetries are identified.

Definition 2. *Let (Σ, E) be an equational specification.*
1. A **relaxed (Σ, E)-system** *$\mathcal{S}$ is* **coherent** *iff all diagrams involving only symmetries are commutative.*
2. A **coherent relaxed (Σ, E)-homomorphism** *$F : \mathcal{S} \to \mathcal{S}'$ between two coherent relaxed (Σ, E)-systems is a homomorphism of (Σ, E)-systems.*

If we denote by C the set of equations corresponding to the commutative diagrams of symmetries, then the quotient of a relaxed (Σ, E)-system modulo the congruence generated by C is a coherent system. See [5] for a more precise definition of the diagrams involved by the coherence property.

Example 1. Here we consider the case of monoidal categories. The signature Σ consists of a constant 0 and the binary operator $+$, and E consists of the axioms:

$$\begin{array}{ll} (A(+))\; x + (y + z) = (x + y) + z & (t_1 = t'_1) \\ (L(+))\; 0 + x = x & (t_2 = t'_2) \\ (R(+))\; x + 0 = x & (t_3 = t'_3) \end{array} \quad (1)$$

The rewrite theory $\mathcal{R}(\Sigma, E)$ is:

$$\begin{array}{l} \mathrm{SYM}^{t_1,t'_1} : x + (y+z) \to (x+y) + z \\ \mathrm{SYM}^{t_2,t'_2} : 0 + x \to x \\ \mathrm{SYM}^{t_3,t'_3} : x + 0 \to x \end{array} \quad (2)$$

In order a (Σ, E)-system S to be a monoidal category, i.e., a coherent relaxed (Σ, E)-system, the diagrams of the following sort:

$$\begin{array}{ccccc} x+(y+(z+u)) & \xrightarrow{\mathrm{SYM}^{t_1,t'_1}_{x,y,z+u}} & (x+y)+(z+u) & \xrightarrow{\mathrm{SYM}^{t_1,t'_1}_{x+y,z,u}} & ((x+y)+z)+u \\ \Big\downarrow {\scriptstyle x+\mathrm{SYM}^{t_1,t'_1}_{y,z,u}} & & & & \Big\uparrow {\scriptstyle \mathrm{SYM}^{t_1,t'_1}_{x,y,z}+u} \\ x+((y+z)+u) & & \xrightarrow[\mathrm{SYM}^{t_1,t'_1}_{x,y+z,u}]{} & & (x+(y+z))+u \end{array}$$

are necessarily commutative [6]. The axiom corresponding to this diagram is

$$\mathrm{SYM}^{t_1,t'_1}_{x,y,z+u};\mathrm{SYM}^{t_1,t'_1}_{x+y,z,u} = x + \mathrm{SYM}^{t_1,t'_1}_{y,z,u};\mathrm{SYM}^{t_1,t'_1}_{x,y+z,u};\mathrm{SYM}^{t_1,t'_1}_{x,y,z} + u$$

This example is studied in detail in subsection 3.1.

We denote by $\mathbf{CRSys}(\Sigma, E)$ the category of coherent relaxed (Σ, E)-systems. The inclusion $\mathbf{CRSys}(\Sigma, E) \hookrightarrow \mathbf{RSys}(\Sigma, E)$ is reflective and hence the inclusion $\mathbf{CRSys}(\Sigma, E) \hookrightarrow \mathcal{R}(\Sigma, E)$-**Sys** is reflective. It follows that the free coherent relaxed (Σ, E)-system on X, denoted also by $Sym^*_{\Sigma,E}(X)$, is the image by the reflector functor of the free $\mathcal{R}(\Sigma, E)$-system $\mathcal{T}_{\mathcal{R}(\Sigma,E)}(X)$. It is easy to see that the free coherent system $Sym^*_{\Sigma,E}(X)$ is isomorphic to the quotient of the free $\mathcal{R}(\Sigma, E)$-groupoid $\mathcal{T}_{\overleftrightarrow{\mathcal{R}(\Sigma,E)}}(X)$ modulo the equations corresponding to the commutative diagrams.

3 Axiomatization of categories of symmetries

The axiomatic construction of the free strict systems is generalized in [5] in order to obtain free relaxed models. But this construction assumes the subcategory of symmetries already defined. In this section we deal with the axiomatization of the free system $Sym^*_{\Sigma,E}(X)$ in $\mathbf{CRSys}(\Sigma, E)$. The definition of the $\mathcal{R}(\Sigma, E)$-groupoid $\mathcal{T}_{\overleftrightarrow{\mathcal{R}(\Sigma,E)}}(X)$ already provides a partial axiomatization. In order to obtain a complete axiomatization we have to add axioms which specify the commutativity of the diagrams. We show that if E is a confluent and terminating theory then these axioms can be computed using the critical pairs. The method is then extended to the case when E is confluent and terminating modulo a congruence.

3.1 Confluent and terminating theories

Assume that E, viewed as a term rewriting system where the equations are oriented from left to right, is confluent and terminating. We first show that, if we have a set of equations C which specifies the commutativity of the diagrams in $\mathcal{T}_{\mathcal{R}(\Sigma,E)}(X)$, then the equations C specify the commutativity of the diagrams in $\mathcal{T}_{\overleftrightarrow{\mathcal{R}(\Sigma,E)}}(X)$. We then show how we can get the equations C which specifies the commutativity of the diagrams in $\mathcal{T}_{\mathcal{R}(\Sigma,E)}(X)$.

Theorem 1. *Let C be a set of equations such that the quotient of $\mathcal{T}_{\mathcal{R}(\Sigma,E)}(X)$ modulo C is coherent. Then the quotient of $\mathcal{T}_{\overleftrightarrow{\mathcal{R}(\Sigma,E)}}(X)$ modulo C is coherent.*

The axioms corresponding to the commutative diagrams in $\mathcal{T}_{\mathcal{R}(\Sigma,E)}(X)$ can be found by computing the critical pairs. Let (s,t) be a critical pair of $\mathcal{R}(\Sigma,E)$. Then there exists u such that $s \leftarrow_E u \rightarrow_E t$. We denote by $\alpha_1 : u \rightarrow s$ and $\alpha_2 : u \rightarrow t$ the corresponding arrows in $\mathcal{T}_{\mathcal{R}(\Sigma,E)}(X)$. Because $\mathcal{R}(\Sigma,E)$ is confluent and terminating it follows that there exists v such that $s \overset{*}{\rightarrow}_E v \overset{*}{\leftarrow}_E t$. Let $\alpha_3 : s \rightarrow v$ and $\alpha_4 : t \rightarrow v$ be the corresponding arrows in $\mathcal{T}_{\mathcal{R}(\Sigma,E)}(X)$. For each such a tuple $(\alpha_1, \alpha_2, \alpha_3, \alpha_4)$ we add to C the equation $\alpha_1;\alpha_3 = \alpha_2;\alpha_4$.

Theorem 2. *The set C of equations computed as above specifies the commutativity of the diagrams in $\mathcal{T}_{\mathcal{R}(\Sigma,E)}(X)$.*

Example: monoidal categories Recall that the equational specification for the monoidal categories is given in (1) and that the rewrite theory (2) is confluent and terminating. For the sake of simplicity, we use the more usual notations:

$$\begin{aligned}
\alpha_{x,y,z} &= \mathrm{SYM}^{t_1,t_1'}_{x,y,z} : x + (y+z) \rightarrow (x+y) + z, \\
\lambda_x &= \mathrm{SYM}^{t_2,t_2'}_{x} : 0 + x \rightarrow x, \\
\rho_x &= \mathrm{SYM}^{t_3,t_3'}_{x} : x + 0 \rightarrow x.
\end{aligned}$$

All rewritings generated by the critical pairs are:

$$\begin{aligned}
& x + (y + (z+u)) \rightarrow x + ((y+z)+u) \rightarrow (x + (y+z)+u) \\
& \qquad\qquad\qquad\quad \rightarrow ((x+y)+z)+u \\
& x + (y + (z+u)) \rightarrow (x+y) + (z+u) \rightarrow ((x+y)+z)+u \\
& x + (y+0) \rightarrow x + y \\
& x + (y+0) \rightarrow (x+y)+0 \rightarrow x+y \\
& x + (0+y) \rightarrow x+y \\
& x + (0+y) \rightarrow (x+0)+y \rightarrow x+y \\
& 0 + (x+y) \rightarrow x+y \\
& 0 + (x+y) \rightarrow (0+x)+y \rightarrow x+y \\
& 0 + 0 \rightarrow 0
\end{aligned}$$

From these rewritings we can deduce the following equations, corresponding to the commutative diagrams in monoidal categories:

$(AA)\ \ (x+\alpha_{y,z,u});\alpha_{x,y+z,u};(\alpha_{x,y,z}+u)=\alpha_{x,y,z+u};\alpha_{x+y,z,u}$
$(AR)\ \ x+\rho_y=\alpha_{x,y,0};\rho_{x+y}$
$(AL1)\ x+\lambda_y=\alpha_{x,0,y};(\rho_x+b)$
$(AL2)\ \lambda_{x+y}=\alpha_{0,x,y};(\lambda_x+y)$
$(LR)\ \ \lambda_0=\rho_0$

In order to have a complete axiomatization for $Sym^*_{\Sigma,E}(X)$ we have to add the equations coming from the definition of the strict $\mathcal{R}(\Sigma,E)$-groupoid $\mathcal{T}_{\underset{\mathcal{R}(\Sigma,E)}{\longleftrightarrow}}(X)$:

- associativity of ";":
  $(A(;))\ \beta_1;(\beta_2;\beta_3)=(\beta_1;\beta_2);\beta_3,$
- identities of ";":
  $(L(;))\ w;\beta=\beta \quad (R(;))\ \beta;w'=\beta$
  if $\beta: w\rightarrow w'$,
- preservation of composition:
  $(PC)\quad (\beta_1;\beta_2)+(\beta_1';\beta_2')=(\beta_1+\beta_1');(\beta_2+\beta_2'),$
- decomposition:
  $(DE1)\quad \alpha(\beta_1,\beta_2,\beta_3)=\alpha_{w1,w2,w3};((\beta_1+\beta_2)+\beta_3)$
  $(DE2)\quad \lambda(\beta_1)=\lambda_{w1};\beta_1$
  $(DE3)\quad \rho(\beta_1)=\rho_{w1};\beta_1$
  if $\beta_1: w1\rightarrow w1', \beta_2: w2\rightarrow w2', \beta_3: w3\rightarrow w3'$, and $\alpha(\beta_1,\beta_2,\beta_3): w1+(w2+w3)\rightarrow(w1'+w2')+w3'$,
- exchange:
  $(EX1)\quad (\beta_1+(\beta_2+\beta_3));\alpha_{w1',w2',w3'}=\alpha_{w1,w2,w3};((\beta_1+\beta_2)+\beta_3)$
  $(EX2)\quad (0+\beta_1);\lambda_{w1'}=\lambda_{w1};\beta_1$
  $(EX3)\quad (\beta_1+0);\rho_{w1'}=\rho_{w1};\beta_1$
  if $\beta_1: w1\rightarrow w1', \beta_2: w2\rightarrow w2', \beta_3: w3\rightarrow w3'$,
- inverse:
  $(INV(\alpha))\ \alpha_{x,y,z};\alpha^{-1}_{x,y,z}=x+(y+z),\quad \alpha^{-1}_{x,y,z};\alpha_{x,y,z})=(x+y)+z,$
  $(INV(\lambda))\ \lambda_x;\lambda_x^{-1}=0+x,\quad \lambda_x^{-1};\lambda_x=x,$
  $(INV(\rho))\ \rho_x;\rho_x^{-1}=x+0,\quad \rho_x^{-1};\rho_x=x,$
  if $\beta_1: w1\rightarrow w1'$ and $\beta_2: w2\rightarrow w2'$.

Remark 1. The rôle of the decomposition laws is to represent the simultaneous rewritings as sequential rewritings and are very important in modeling the concurrent calculus. From the algebraic specification point of view, the operations like $\alpha(\beta_1,\beta_2,\beta_3)$, where β_i are symmetries, are derived operators. Thus, these axioms can be omitted in the axiomatization of the symmetries.

Often in practice it is more convenient to work only with the axiomatization of the coherent system $\mathcal{T}_{\mathcal{R}(\Sigma,E)}$.

3.2 Confluence and termination modulo a congruence

We suppose now that the equational presentation is $(\Sigma, E \cup T)$ and E (viewed as a term rewriting system) is confluent and terminating modulo T. We assume further that there exists a set of equations $C(T)$ corresponding to the commutative diagrams in $\mathcal{T}_{\underset{\mathcal{R}(\Sigma,T)}{\longleftrightarrow}}(X)$. We denote by $\mathcal{R}(\Sigma, E/T)$ the rewrite theory

$$(\Sigma, T, \{\mathrm{SYM}^{[t]_T,[t']_T} : [t]_T \to [t']_T \mid t = t' \text{ in } E\}).$$

By the hypothesis, $\mathcal{R}(\Sigma, E/T)$ is confluent and terminating. We can get the set of equations specifying the commutativity of diagrams in $\mathcal{T}_{\mathcal{R}(\Sigma,E/T)}(X)$ in the same way as in subsection 3.1. We denote this set by $C(E/T)$. The subcategory $\mathcal{R}(\Sigma, E/T)$**-Sys** is reflective in $\mathcal{R}(\Sigma, E \cup T)$**-Sys** and the system $\mathcal{T}_{\mathcal{R}(\Sigma,E/T)}(X)$ is the quotient of $\mathcal{T}_{\mathcal{R}(\Sigma,E\cup T)}(X)$ modulo the congruence generated by T [5]. It follows that the set of equations $C(T) \cup C(E/T)$ specifies the commutativity of a part of diagrams in $\mathcal{T}_{\mathcal{R}(\Sigma,E\cup T)}(X)$. It remains to be solved the diagrams which include ;-compositions of arrows in $\mathcal{T}_{\mathcal{R}(\Sigma,E/T)}(X)$ with arrows in $\mathcal{T}_{\mathcal{R}(\Sigma,T)}(X)$. But a reasoning similar to that in the previous subsection shows that the set of these equations can be obtained by computing the (E,T)-critical pairs. Recall that each such a critical pair is confluent [2].

Example: symmetric monoidal categories The signature is the same as for the monoidal categories but the equations are the following ones:

$$(A(+))\ x + (y + z) = (x + y) + z, \qquad (L(+))\ 0 + x = x,$$
$$(R(+))\ x + 0 = x, \qquad (C(+))\ x + y = y + x.$$

We consider the following rewrite system E:

$$\alpha_{x,y,z} : x + (y + z) \to (x + y) + z,$$
$$\lambda_x : 0 + x = x,$$
$$\rho_x : x + 0 \to x$$

modulo $T = \{C(+)\}$. We denote by $\gamma_{x,y}$ the symmetry $\gamma_{x,y} : x + y \cong y + x$.

The axioms produced by the (E,E)-critical pairs are obtained in the same manner as for the monoidal categories:

$$(AA)\ \ (x + \alpha_{y,z,u}); \alpha_{x,y+z,u}; (\alpha_{x,y,z} + u) = \alpha_{x,y,z+u}; \alpha_{x+y,z,u}$$
$$(AL1)\ x + \lambda_y = \alpha_{x,0,y}; (\gamma_{x,0} + y); (\lambda_x + b)$$
$$(AL2)\ \lambda_{x+y} = \alpha_{0,x,y}; (\lambda_x + y)$$
$$(AR)\ \ x + \rho_y = \alpha_{x,y,0}; \rho_{x+y}$$
$$(LR)\ \ \lambda_0 = \rho_0$$

The (E,T)-critical pairs are:

$$x + (y + z) \to (x + y) + z = z + (x + y) \to (z + x) + y,$$
$$x + (y + z) = x + (z + y) \to (x + z) + y = (z + x) + y,$$
$$x + (y + z) \to (x + y) + z = (y + x) + z = z + (y + x) \to (z + y) + x,$$
$$x + (y + z) = x + (z + y) = (z + y) + x,$$

$x+0 \to x,$
$x+0 = 0+x \to x,$
$0+x \to x,$
$0+x = x+0 \to x,$

and produce the following equations:

$(AC1)\ \alpha_{x,y,z};\gamma_{x+y,z};\alpha_{z,x,y} = (x+\gamma_{y,z});\alpha_{x,z,y};(\gamma_{x,z}+y),$
$(AC2)\ \alpha_{x,y,z};(\gamma_{x,y}+z);\gamma_{y+x,z};\alpha_{z,y,x} = (x+\gamma_{y,z});\gamma_{x,y+z},$
$(RC)\quad \rho_x = \gamma_{x,0};\lambda_x,$
$(LC)\quad \lambda_x = \gamma_{0,x};\rho_x,$

The axioms which specify the commutativity of diagrams in $\mathcal{T}_{\mathcal{R}(\Sigma,T)}^{\longleftrightarrow}(X)$:

$(C1)\ \gamma_{x,0} = x$
$(C2)\ \gamma_{x,y};\gamma_{y,x} = x+y$ (equivalent to $\gamma_{x,y}^{-1} = \gamma_{y,x}$),
$(C3)\ \gamma_{w_1,w_2};(\beta_2+\beta_1) = (\beta_1+\beta_2);\gamma_{w_1',w_2'}$ (exchange),
if $\beta_i : w_i \to w_i'$ $i = 1,2$.

All these axioms together with $A(;)$, $L(;)$, $R(;)$, PC, $DE1{-}3$, $EX1{-}3$, and INV form a complete axiomatization for symmetric monoidal categories. Note that these axioms are not independent. For example, LC can be deduced from RC, $C1$, and $A(;)$.

Recall that the symmetrical monoidal categories are not full coherent relaxed (Σ, E)-systems because not all symmetry diagrams commute. A typical example is the following one:

$$a+a \xrightarrow[a+a]{\mathrm{SYM}_{a,a}^{x+y,y+x}} a+a$$

We have no $\mathrm{SYM}_{a,a}^{x+y,y+x} = a+a$. This property is crucial in the definition of the concatenable processes for Petri nets [10]. Nevertheless, this example is representative for our method because the class of commutative diagrams in $\mathcal{T}_{\mathcal{R}(\Sigma,T)}^{\longleftrightarrow}(X)$ is an input for the procedure we propose here.

3.3 Partial relaxation

The definitions in section 2.2 can have different nuances. For example, we may be interested in the category of relaxed models where the natural isomorphism corresponding to the associativity is always the equality of functors. We call these models strictly associative. In a similar way, we can have strictly commutative models, strictly commutative and associative models, and so on. Generally, if E' is the subset of the equations strictly interpreted, i.e., if $t = t'$ in E' then the functors $t_{\mathcal{S}}$ and $t'_{\mathcal{S}}$ are equals in all systems $\mathcal{S}$, and E'' is the subset of the equations relaxedly interpreted, i.e., if $t = t'$ in E'' then there exists a natural symmetry $t_{\mathcal{S}} \cong t'_{\mathcal{S}}$ in each system $\mathcal{S}$, then we write **relaxed (Σ, E', E'')-system** for relaxed (Σ, E)-system. We denote by **RSys**(Σ, E', E''), **CRSys**(Σ, E', E''), and $\mathbf{CRSys}_C(\Sigma, E', E'')$, the corresponding full subcategories of (coherent)/(C-coherent) (Σ, E', E'')-systems. $\mathbf{RSys}(\Sigma, E', E'')$ is reflective in $\mathbf{RSys}(\Sigma, E' \cup E'')$ and the category of symmetries in $\mathbf{RSys}(\Sigma, E', E'')$ is the quotient of the category of symmetries in $\mathbf{RSys}(\Sigma, E' \cup E'')$ modulo a congruence [5]. This congruence coincides with $=_{E'}$ on objects and identifies the E'-symmetries with the identities.

There is also a direct axiomatic way to define the category of symmetries in $\mathbf{RSys}(\Sigma, E', E'')$. By the second theorem of isomorphism, there exists a congruence $=_{E/E'}$ such that $t =_E t'$ iff $[t]_{E'} =_{E/E'} [t']_{E'}$. In fact, because $E = E' \cup E''$, $=_{E/E'}$ coincides with the congruence generated by E'' over $T_{\Sigma,E'}(X)$. Consider the rewrite specification

$$\mathcal{R}(\Sigma, \pi) = (\Sigma, E', \{\mathrm{SYM}^{[t]_{E'},[t']_{E'}} : [t]_{E'} \to [t']_{E'} \mid t = t' \text{ in } E'\}).$$

The category of symmetries in $\mathbf{RSys}(\Sigma, E', E'')$, $Sym^*_{\Sigma,E',E''}(X)$, is the strict groupoid $\mathcal{T}_{\underset{\mathcal{R}(\Sigma,\pi)}{\longleftrightarrow}}(X)$.

The axioms corresponding to the commutative diagrams for the coherent category $Sym^*_{\Sigma,E',E''}(X)$ are obtained in two steps: we first compute the axioms corresponding to commutative diagrams in $Sym^*_{\Sigma,E' \cup E''}(X)$, as in subsection 3.1, afterwards we replace in each equations any occurrence of a term t (denoting the identity) by $[t]_{E'}$ and any occurrence of a symmetry $\mathrm{SYM}^{t,t'}_{\overline{w}}$ with $t =_{E'} t'$ by the identity $[t(\overline{w})]_{E'}$.

Example: symmetric strict monoidal categories A symmetric strict monoidal category is a symmetric monoidal category where all symmetries corresponding to $A(+), L(+)$ and $R(+)$ are identities. The axiomatization for symmetric strict monoidal categories is obtained from the axioms for symmetric monoidal categories by replacing the symmetries $\alpha_{x,y,z}$, λ_x, and ρ_x by identities. For example, $AC1$ becomes

$$(AC) \quad \gamma_{x+y,z} = (x + \gamma_{y,z}); (\gamma_{x,z} + y)$$

and $AC2$ becomes a consequence of AC. Also, we have to add the axioms which are A3-instances for the associativity and unit.

The result axiomatization is the same with that given in [1, 9]:

- associativity of ";":
  $(A(;)) \quad \beta_1; (\beta_2; \beta_3) = (\beta_1; \beta_2); \beta_3,$
- identities of ";":
  $(L(;)) \ w; \beta = \beta \quad (R(;)) \ \beta; w' = \beta$
  if $\beta : w \to w'$,
- preservation of composition:
  $(PC) \quad (\beta_1; \beta_2) + (\beta'_1; \beta'_2) = (\beta_1 + \beta'_1); (\beta_2 + \beta'_2),$
- non-relaxed axioms in E:
  $(A(+)) \ \beta_1 + (\beta_2 + \beta_3) = (\beta_1 + \beta_2) + \beta_3,$
  $(U(+)) \ \beta + 0 = \beta,$
- all exchange axioms become trivial identities,

– commutativity of diagrams:

$(AC)\ \gamma_{x+y,z} = (x + \gamma_{y,z}); (\gamma_{x,z} + y),$ $\quad (C1)\ \gamma_{x,0} = x,$

$(C3)\ \gamma_{w_1,w_2}; (\beta_2 + \beta_1) = (\beta_1 + \beta_2); \gamma_{w'_1,w'_2},$ $\quad (C2)\ \gamma_{x,y}; \gamma_{y,x} = x + y,$

if $\beta_i : w_i \to w'_i\ i = 1, 2.$

We note that symmetrical strict monoidal categories are not full coherent because the symmetrical monoidal categories are not.

4 Conclusions and further work

Until now, the problem of finding axiomatizations for categories enriched with an algebraic structure was studied only on particular cases [1, 10]. In this paper we presented a method which can be applied to a large class of such categories. The method presented here is based on the computation of the critical pairs. A problem which arises is that if the completion procedures [2] can be adapted such that they can compute the axioms corresponding to the coherence property.

The use of categories as models for rewrite specifications increases their practical utility. By axiomatizing these categories, we have the possibility to program them in specification languages as Maude or CafeOBJ. Further work will be focused on this aspect.

References

1. V.E. Căzănescu and Gh. Ştefănescu. Classes of finite relations as initial abstract data types I. *Discrete Mathematics*, 90:233–265, 1991.
2. N. Dershowitz and J.-P. Jouannaud. Rewrite systems. In *Handbook of Theoretical Computer Science*. Elsevier Science, 1990.
3. Răzvan Diaconescu and Kokichi Futatsugi. *CafeOBJ Report: The Language, Proof Techniques, and Methodologies for Object-Oriented Algebraic Specification*, volume 6 of *AMAST Series in Computing*. World Scientific, 1998.
4. D. Lucanu. On the axiomatization of the category of symmetries. Technical report TR-98-03, University "Al.I.Cuza" of Iaşi, Computer Science Department, December 1998. `http://www.infoiasi.ro/~dlucanu/reports.html`.
5. D. Lucanu. Relaxed models for concurrent rewriting logic. Internal report, University "Al.I.Cuza" of Iaşi, Computer Science Department, November 1998. URL: `http://www.infoiasi.ro/~dlucanu/`.
6. S. MacLane. *Category theory for working mathematician*. Springer Verlag Berlin, 1971.
7. J. Meseguer. Conditional rewriting logic as unified model of concurrency. *Theoretical Computer Science*, 96:73–155, 1992.
8. J. Meseguer. A logical theory of concurrent objects and its realization in the Maude language. In Gul Agha, Peter Wegner, and Akinori Yonezawa, editors, *Research Directions in Concurrent Object-Oriented Programming*. The MIT Press, 1993.
9. V. Sassone. On the category of Petri net computation. In *TAPSOFT'95, LNCS*, pages 334–348. Springer Verlag, 1995.
10. V. Sassone. An axiomatization of the algebra of Petri net concatenable processes. *to appear in Theoretical Computer Science*, 1997.

Sewing Grammars *

Carlos Martin-Vide[1] and Alexandru Mateescu[2]

[1] Research Group on Mathematical Linguistics and Language Engineering (GRLMC) Universitat Rovira i Virgili, 43005 Tarragona, Spain, email: cmv@astor.urv.es
[2] Turku Centre for Computer Science (TUCS), 20520 Turku, Finland and Faculty of Mathematics, University of Bucharest, Romania, email: mateescu@utu.fi

Abstract. Sewing grammars are very simple grammars, still able to define families of mildly context-sensitive languages. These grammars are inspired from Marcus contextual grammars and simple matrix grammars. We consider various families of sewing grammars. Some of them lead to very special families of languages such that each such family is a mildly context-sensitive family of languages, and, moreover, most of the fundamental problems, like the equivalence problem, the inclusion problem, etc., are decidable.

1 Introduction

The need of *mildly context-sensitive families of languages* was emphasized in connection with linguistics, see [3] and [7]. As far as we know, no systematic investigation of grammars or other devices that define such families of languages has been done. The aim of our paper is to introduce very simple grammars that can define mildly context-sensitive families of languages. Our paper is just a first step in order to have a systematic view of those grammars that define mildly context-sensitive families of languages.

A mildly context-sensitive family of languages should contain the most significant languages that occur in the study of natural languages. Languages in such a family must be semilinear languages, and, moreover, they should be computationally feasible, i.e., the membership problem for languages in such a family must be solvable in deterministic polynomial time complexity.

It is well known that the hierarchy of Chomsky does not contain such a family. Whereas the family of context-free languages has good computational properties, it does not contain some important languages that appear in the study of natural languages. The family of context-sensitive languages contains all important languages that occur in the study of natural languages, but no algorithm in deterministic polynomial time is knew for its membership problem.

Remark 1. By a ***mildly context-sensitive family of languages*** we mean a family $\mathcal{L}$ of languages such that the following conditions are fulfilled:

* This work has been partially supported by the Project 137358 of the Academy of Finland.

(i) each language in $\mathcal{L}$ is semilinear,
(ii) for each language in $\mathcal{L}$ the membership problem is solvable in deterministic polynomial time, and
(iii) $\mathcal{L}$ contains the following three non-context-free languages:
- multiple agreements: $L_1 = \{a^ib^ic^i \mid i \geq 0\}$,
- crossed agreements: $L_2 = \{a^ib^jc^id^j \mid i,j \geq 0\}$, and
- duplication: $L_3 = \{ww \mid w \in \{a,b\}^*\}$.

Note that in the literature some authors consider that such a family contains all context-free languages, and/or some other non-context-free languages: ***k multiple agreements:*** $L_1' = \{a_1^ia_2^i \dots a_k^i \mid i \geq 0\}$ where $k \geq 3$, ***marked duplication:*** $L_3' = \{wcw \mid w \in \{a,b\}^*\}$. In the sequel we will consider these variants, too

The paper is organized as it follows. Firstly, we introduce the basic type of sewing grammar and we show that the corresponding family of languages is a mildly context-sensitive family of languages. This type of grammars is extended to a more general type of grammars. We investigate pumping lemmata for the languages defined by these grammars as well as closure properties of these families of languages.

Next we define some families of languages that are almost mildly context-sensitive, and, moreover, for languages in these families the problems of equivalence, of inclusion, etc., are decidable problems.

Finally, we discuss some other extended models as well as further topics of research.

Now we recall some terminology and definitions that we will use in this paper.

Let Σ be an alphabet and let Σ^* be the free monoid generated by Σ with the identity denoted by λ. The free semigroup generated by Σ is $\Sigma^+ = \Sigma^* - \{\lambda\}$. Elements in Σ^* (Σ^+) are referred to as *words* (*nonempty words*). λ is the *empty word.* A *context* is a pair of words, i.e., (u,v), where $u,v \in \Sigma^*$.

The families of regular, linear, context-free, context-sensitive and recursively enumerable languages are denoted by REG, LIN, CF, CS and RE, respectively.

Assume that $\Sigma = \{a_1, a_2, \dots, a_k\}$. The *Parikh mapping*, denoted by Ψ, is:

$$\Psi : \Sigma^* \longrightarrow N^k,$$

$$\Psi(w) = (|w|_{a_1}, |w|_{a_2}, \dots, |w|_{a_k}).$$

If L is a language, then the *Parikh set* of L is defined by

$$\Psi(L) = \{\Psi(w) \mid w \in L\}.$$

A *linear set* is a set $M \subseteq N^k$ such that $M = \{v_0 + \sum_{i=1}^m v_ix_i \mid x_i \in N\}$, for some $v_0, v_1, \dots, v_m$ in N^k. A *semilinear set* is a finite union of linear sets and a *semilinear language* is a language L such that $\Psi(L)$ is a semilinear set.

In the sequel we recall the definition of a simple matrix grammar.

Definition 1. *A simple matrix grammar of degree n (see [2]) is an ordered system $G = (N_1, \dots, N_n, \Sigma, P, S)$ where $N_i, 1 \leq i \leq n$, are pairwise disjoint* *alphabets of nonterminals, Σ is a terminal alphabet, S is the start symbol,*

$$S \notin \Sigma \cup \bigcup_{i=1}^{n} N_i,$$

and P is a finite set of n-dimensional vectors of rules, $(r_1, \ldots, r_n)$, where each rule r_i is a context-free rule over the alphabet $N_i \cup \Sigma$ such that for all pairs of rules, $r_i : A_i \longrightarrow x_i$, $r_j : A_j \longrightarrow x_j$ it follows that $|x_i|_{N_i} = |x_j|_{N_j}$, $1 \le i, j \le n$.

Moreover, P contains rules of the form $(S \to u)$, with $u \in \Sigma^*$ and also rules of the form $(S \to A_1 A_2 \ldots A_n)$, where $A_i \in N_i$, $1 \le i \le n$.

Let G be a simple matrix grammar of degree n. G defines a relation of *direct derivation* as follows:

$$S \Rightarrow_G v \text{ iff } (S \to v) \in P$$

and

$$u_1 X_1 u_1' \ldots u_n X_n u_n' \Rightarrow_G u_1 v_1 u_1' \ldots u_n v_n u_n' \text{ iff } (X_1 \to v_1, \ldots, X_n \to v_n) \in P,$$

where $u_j \in \Sigma^*$, $u_j' \in (\Sigma \cup \bigcup_{i=1}^{n} N_i)^*, j = 1, \ldots, n, X_i \in N_i, i = 1, \ldots, n$.

The *derivation* relation induced by G, denoted $\Rightarrow_G^*$, is the reflexive and transitive closure of $\Rightarrow_G$.

The *language generated* by G is: $L(G) = \{w \in \Sigma^* \mid S \Rightarrow_G^* w\}$.

In this paper the so called regular (linear) simple matrix grammars are of a special interest.

Definition 2. *A regular (linear) simple matrix grammar of degree n, where $n \ge 1$, is a simple matrix grammar of degree n, $G = (N_1, \ldots, N_n, \Sigma, P, S)$ such that all the rules occurring as components in the n-dimensional vectors from P, excepting the rules starting with S, are Chomsky regular (linear) rules.*

Contextual grammars were firstly considered in [4] with the aim to model some natural aspects from descriptive linguistics like for instance the acceptance of a word (construction) only in certain contexts. For a detailed presentation of this topic, the reader is referred to the recent monograph [6].

Definition 3. *A Marcus simple contextual grammar is an ordered system $G = (\Sigma, B, C)$, where Σ is the alphabet of G, B is a finite subset of Σ^*, the base of G, and C is a finite set of contexts, i.e., a finite set of pairs of words over Σ.*

Let $G = (\Sigma, B, C)$ be a Marcus simple contextual grammar. The *direct derivation relation* with respect to G is a binary relation between words over Σ, denoted $\Rightarrow_G$, or $\Rightarrow$, if G is understood from context. By definition, $x \Rightarrow_G y$, where $x, y \in \Sigma^*$ iff $y = uxv$ for some $(u, v) \in C$. The *derivation relation* with respect to G, denoted $\Rightarrow_G^*$, or $\Rightarrow^*$, if G is understood from context, is the reflexive and transitive closure of the relation $\Rightarrow_G$.

Definition 4. *Let $G = (\Sigma, B, C)$ be a Marcus simple contextual grammar. The language generated by G, denoted $L(G)$, is defined as:*

$$L(G) = \{y \in \Sigma^* \mid \text{ there exists } x \in B, \text{ such that } x \Rightarrow_G^* y\}.$$

The reader is referred to [8] or [7] for the basic notions of formal languages we use in the sequel and to [1] and [7] for interrelations between linguistics and formal languages.

2 Sewing grammars: the basic model

Here we introduce the basic model of *sewing grammars*, we give some examples and we show some properties of these grammars.

Definition 5. *A sewing grammar is a construct $G = (\Sigma, B, C, n, f)$, where Σ is an alphabet, $n \geq 1$ is an integer called the degree of G, $B \subset (\Sigma^*)^n$, B finite, is the base of G, $C \subset (\Sigma^*)^n$, C finite, is the set of contexts (or rules) of G and f is a recursive function, $f : (\Sigma^*)^n \longrightarrow \Sigma^*$, called the zipper function of G.*

Using the above notations, a sewing grammar G defines a relation of *direct derivation*, denoted $\Longrightarrow_G$ or $\Longrightarrow$, between elements in $(\Sigma^*)^n$. By definition

$$(x_1, x_2, \ldots, x_n) \Longrightarrow_G (y_1, y_2, \ldots y_n)$$

iff there exists $(z_1, z_2, \ldots, z_n) \in C$, such that $y_i = x_i z_i$, $1 \leq i \leq n$.

The reflexive and transitive closure of $\Longrightarrow_G$ or $\Longrightarrow$ is denoted by $\Longrightarrow_G^*$ or $\Longrightarrow^*$ and called the relation of *derivation* defined by G.

The *n-ary language* defined by G, denoted by $nL(G)$, is by definition:

$$nL(G) = \{(x_1, x_2, \ldots, x_n) \in (\Sigma^*)^n \mid (u_1, u_2, \ldots, u_n) \Longrightarrow_G^* (x_1, x_2, \ldots, x_n),$$
$$\text{for some } (u_1, u_2, \ldots, u_n) \in B\}.$$

Definition 6. *Let $G = (\Sigma, B, C, n, f)$ be a sewing grammar. The language defined by G is:*

$$L(G) = \{f(x_1, x_2, \ldots, x_n) \mid (x_1, x_2, \ldots, x_n) \in nL(G)\}.$$

Therefore the language defined by a sewing grammar $G = (\Sigma, B, C, n, f)$ is the set of all words obtained by applying the zipper function f to the n-tuples from the n-ary language $nL(G)$.

Notation. We denote by $\mathcal{SW}_n(f)$ the family of all languages generated by sewing grammars of degree n and with the zipper function f.

Remark 2. In this paper we mainly consider that the zipper function f is the catenation function of arity n, denoted cat_n, i.e., the function, $cat_n : (\Sigma^*)^n \longrightarrow \Sigma^*$, $cat_n(u_1, u_2, \ldots, u_n) = u_1 u_2 \ldots u_n$.

We will drop the indice f whenewer this is the case. For instance $\mathcal{SW}_n$ denotes the family $\mathcal{SW}_n(f)$, where f is the catenation function.

Moreover, note that for $n < m$, it follows that $\mathcal{SW}_n \subset \mathcal{SW}_m$.

Theorem 1. *Let Σ be an alphabet.*
(i) The languages $\emptyset$, Σ^*, $F \subset \Sigma^*$, F finite are in SW_n for every $n \geq 1$.
(ii) The language of multiple agreements, $L_1 = \{a^i b^i c^i \mid i \geq 0\}$, is in SW_n for every $n \geq 3$.
(iii) The language of crossed agreements, $L_2 = \{a^i b^j c^i d^j \mid i, j \geq 0\}$ is in SW_n for every $n \geq 4$.
(iv) The language of duplication, $L_3 = \{ww \mid w \in \{a,b\}^*\}$, is in SW_n for every $n \geq 2$.

Proof. (*i*) The language $\emptyset$ is generated by a sewing grammar of degree 1 with $B = \emptyset$. The language Σ^* is generated by a sewing grammar of degree 1 with $B = \{(\lambda)\}$ and the contexts $C = \{(a) \mid a \in \Sigma\}$. Finally, a finite language F is generated by a sewing grammar of degree 1 with $B = \{(u) \mid u \in F\}$ and the contexts $C = \{(\lambda)\}$.

Therefore all these languages are in SW_1 and, using Remark 2 it follows that they are in all families SW_n, with $n \geq 1$.

(*ii*) The language L_1 is generated by the sewing grammar

$$G_1 = (\{a, b, c\}, B, C, 3, cat_3),$$

where $B = \{(\lambda, \lambda, \lambda)\}$ and $C = \{(a, b, c)\}$.

(*iii*) The language L_2 is generated by the sewing grammar

$$G_2 = (\{a, b, c, d\}, B, C, 4, cat_4),$$

where $B = \{(\lambda, \lambda, \lambda, \lambda)\}$ and $C = \{[(a, \lambda, c, \lambda)], [(\lambda, b, \lambda, d)]\}$.

(*iv*) Finally, the language L_3 is generated by the sewing grammar

$$G_3 = (\{a, b, \}, B, C, 2, cat_2),$$

where $B = \{(\lambda, \lambda)\}$ and $C = \{(a, a), (b, b)\}$.

Therefore, using Remark 2 we conclude the proof. □

Comment. Concerning the language:

k-multiple agreements: $L_1' = \{a_1^i a_2^i \dots a_k^i \mid i \geq 0\}$ where $k \geq 3$, one can easily prove, using the method from Theorem 1, that L_1' is in SW_n, for $n \geq k$.

As in Theorem 1, it can be proved that the language:

marked duplication: $L_3' = \{wcw \mid w \in \{a,b\}^*\}$.
is in SW_2 (replace in the grammar associated to L_3, the set B with $B = \{(\lambda, c)\}$.

Theorem 2. *Each language in SW_n, where $n \geq 1$ is a semilinear language.*

Proof. Let L be a language in SW_n, $n \geq 1$. Let $G = (\Sigma, B, C, n, cat_n)$ be a sewing grammar of degree n such that $L(G) = L$. Define the Chomsky regular grammar $G' = (\{S\}, \Sigma, S, \pi)$, where

$$\pi = \{S \longrightarrow u_1 u_2 v_2 \dots u_n S \mid [(u_1, u_2, \dots u_n)] \in C\} \cup$$

$$\cup\{S \longrightarrow x_1x_2\ldots x_n \mid (x_1,x_2,\ldots x_n) \in B\}.$$

Let L' be the language $L(G')$. One can easily see that L and L' are letter equivalent, i.e., $\Psi(L) = \Psi(L')$. Since L' is a regular language, it follows that L' is a semilinear language. Therefore, also L is a semilinear language. □

The *membership problem* consists in the following problem: given a language $L \subseteq \Sigma^*$ (defined by a certain type of grammar, automaton, etc.) and a word $w \in \Sigma^*$ to decide by an algorithm whether or not w is in L. The existence of such an algorithm as well as its complexity are very important from the practical point of view. The next theorem shows that the membership problem for languages in $\mathcal{SW}_p$, $p \geq 1$ is in $\mathcal{P}$, the class of all deterministic polynomial time complexity problems.

Theorem 3. *For every $p \geq 1$ and for every $L \in \mathcal{SW}_p$ the membership problem is solvable in deterministic polynomial time.*

From Theorem 1, Theorem 2 and Theorem 3 we obtain the following:

Theorem 4. *For every integer $n \geq 4$, the family $\mathcal{SW}_n$ is a mildly context-sensitive family of languages.*

Now we prove some pumping lemmata for languages in $\mathcal{SW}_n$, where $n \geq 1$.

Let $G = (\Sigma, B, C, n, cat_n)$ be a sewing grammar of degree n, $n \geq 1$. Let $x = (x_1, x_2, \ldots, x_n)$ be a vector from B. The *length* of x, denoted $|x|$, is by definition, $|x| = |x_1| + |x_2| + \ldots + |x_n|$. Similarly, the *length of a context* $c \in C$, $c = (u_1, u_2, \ldots, u_n)$ is by definition $|c| = |u_1| + |u_2| + \ldots + |u_n|$. Note that for every $c \in C$, $|c| > 0$, since we can assume that G does not contain the completely empty context.

Theorem 5. *Let $L \subseteq \Sigma^*$ be a language in $\mathcal{SW}_n$, $n \geq 1$. There exist two integers $m \geq 1$ and $k \geq 1$ such that:*

(i) (pumping an arbitrary context) If $w \in L$ such that $|w| > m$, then w has a decomposition $w = x_1u_1x_2u_2\ldots x_nu_nx_{n+1}$, with $0 < |u_1|+|u_2|+\ldots+|u_n| \leq k$, such that for all $i \geq 0$, the following words are in L:

$$w_i = x_1u_1^ix_2u_2^ix_3\ldots x_nu_n^ix_{n+1}.$$

(ii) (pumping an innermost context) If $w \in L$ such that $|w| > m$, then w has a decomposition $w = x_1u_1y_1x_2u_2y_2x_3\ldots x_nu_ny_nx_{n+1}$, with $0 < |u_1| + |u_2| + \ldots + |u_n| \leq k$, and $|y_1| + |y_2| + \ldots + |y_n| \leq m$, such that for all $i \geq 0$, the following words are in L:

$$w_i = x_1u_1^iy_1x_2u_2^iy_2x_3\ldots x_nu_n^iy_nx_{n+1}.$$

(iii) (pumping an outermost context) If $w \in L$ such that $|w| > m$, then w has a decomposition $w = u_1y_1u_2y_2\ldots u_ny_n$, with $0 < |u_1| + |u_2| + \ldots + |u_n| \leq k$, such that for all $i \geq 0$, the following words are in L:

$$w_i = u_1^iy_1u_2^iy_2\ldots u_n^iy_n.$$

(iv) *(pumping all occurring contexts) If $w \in L$ such that $|w| > m$, then w has a decomposition $w = u_1 y_1 u_2 y_2 \ldots u_n y_n$, with $0 < |y_1| + |y_2| + \ldots + |y_n| \leq m$, such that for all $i \geq 0$, the following words are in L:*

$$w_i = u_1^i y_1 u_2^i y_2 \ldots u_n^i y_n.$$

(v) *(interchanging contexts) If $w, w' \in L$ such that $|w| > m$ and $|w'| > m$, then w and w' have decomposition: $w = x_1 u_1 x_2 u_2 \ldots x_n u_n x_{n+1}$, and $w' = x_1' u_1' x_2' u_2' \ldots x_n' u_n' x_{n+1}'$, with $0 < |u_1| + |u_2| + \ldots + |u_n| \leq k$, and with $0 < |u_1'| + |u_2'| + \ldots + |u_n'| \leq k$, such that also the following two words, z and z', are in L,*

$$z = x_1 u_1' x_2 u_2' \ldots x_n u_n' x_{n+1}, \text{ and}$$

$$z' = x_1' u_1 x_2' u_2 \ldots x_n' u_n x_{n+1}'.$$

Theorem 5 can be used to show that certain languages are not in a family $\mathcal{SW}_n$, $n \geq 1$.

Theorem 6. *If $m, n \geq 1$ such that $m < n$, then $\mathcal{SW}_m \subset \mathcal{SW}_n$ and the inclusion is strict.*

Proof. Clearly, $\mathcal{SW}_m \subseteq \mathcal{SW}_n$. see Remark 2. It remains to show that the inclusion is strict. Consider the language:

$$L = \{a_1^i a_2^i \ldots a_n^i \mid i \geq 0\}.$$

Let $G = (\Sigma, B, C, n, cat_n)$ be the following sewing grammar, where:
$\Sigma = \{a_1, a_2, \ldots, a_n\}$, $B = \{(\underbrace{\lambda, \lambda, \ldots, \lambda}_{n})\}$, and $C = \{(a_1, a_2, \ldots, a_n)\}$.

It is easy to see that $L(G) = L$ and hence $L \in \mathcal{SW}_n$. On the other hand L is not in $\mathcal{SM}_m$ since L does not satisfy the condition from Theorem 5 (*i*). Therefore $\mathcal{SW}_m$ is strictly included in $\mathcal{SW}_n$. □

Combining Theorem 4 and Theorem 6 we obtain:

Theorem 7. *The families $(\mathcal{SW}_n)_{n \geq 4}$ define an infinite hierarchy of mildly context-sensitive languages.*

3 Sewing families of languages and decidable problems

In this section we introduce some special families of sewing languages. Each such a family is almost a mildly context-sensitive family of languages, and, moreover, each such family has good decidability properties.

We start by introducing a special type of sewing grammar. If n is an integer, $n \geq 1$, then $[n]$ denotes the set $\{1, 2, \ldots, n\}$. An n-function is a function $g : [n] \longrightarrow N$, where N denotes the set of all positive integers. The length of an n-function g, denoted $|g|$, is defined as $|g| = g(1) + g(2) + \ldots + g(n)$.

Definition 7. *Let $n \geq 1$ be a fixed integer, let g be an n-function and let $k \geq 1$ be a fixed integer. A sewing grammar $G = (\Sigma, B, C, n, cat_n)$ is of type (g, k) iff for all $c = (c_1, c_2, \ldots, c_n) \in C$, $|c_i| = g(i)$, $i = 1, 2, \ldots, n$ and for all $b = (b_1, b_2, \ldots, b_n) \in B$, $b_i = \lambda$ for all $i \neq k$ and $|b_k| \leq |g|$.*

Notation. Let $n \geq 1$ be an integer, let g be an n-function and let $1 \leq k \leq n$ be an integer. We denote by $\mathcal{SW}_{n,g,k}$ the following family of languages:

$$\mathcal{SW}_{n,g,k} = \{L \mid \text{ there exists a sewing grammar } G = (\Sigma, B, C, n, cat_n)$$

$$\text{of type } (g,k) \text{ such that } L(G) = L\}.$$

Remark 3. Assume that $n \geq 1$ is a fixed integer. Let g be an n-function and let $1 \leq k \leq n$ be an integer. Consider an alphabet Σ and let Σ_1 and Σ_2 be the following two alphabets:

$$\Sigma_1 = \{[\alpha] \mid |\alpha| = |g|, \alpha \in \Sigma^*\}$$

and

$$\Sigma_2 = \{[\beta] \mid |\beta| < |g|, \beta \in \Sigma^*\}$$

Note that for each $w \in \Sigma^*$ there exist and are unique two integers $p, r \geq 0$ such that

$$|w| = p|g| + r \text{ and } 0 \leq r < |g|.$$

Moreover, notice that for w does exist a unique decomposition:

$$w = w_1 w_2 \ldots w_k \beta w_{k+1} \ldots w_n,$$

such that for all $i = 1, 2, \ldots, n$, $|w_i| = pg(i)$ and $|\beta| = r$. Hence, each w_i, $1 \leq i \leq n$ is the catenation of p words from Σ^*, $w_i = w_i^{(1)} w_i^{(2)} \ldots w_i^{(p)}$ with $w_i^{(j)} \in \Sigma^*$, $1 \leq j \leq p$, such that $|w_i^{(j)}| = g(i)$, where $1 \leq j \leq p$.

Now define the function: $\varphi_g^{n,k} : \Sigma^* \longrightarrow \Sigma_1^* \Sigma_2$, such that,

$$\varphi_g^{n,k}(w) = [w_1^{(1)} w_2^{(1)} \ldots w_n^{(1)}][w_1^{(2)} w_2^{(2)} \ldots w_n^{(2)}] \ldots [w_1^{(p)} w_2^{(p)} \ldots w_n^{(p)}][\beta].$$

One can easily prove the following:

Proposition 1. *The function $\varphi_g^{n,k}$ is a bijective function.*

Next two results show the importance of the function $\varphi_g^{n,k}$.

Proposition 2. *If $G = (\Sigma, B, C, n, cat_n)$ is a sewing grammar of type (g, k), then the language $\varphi_g^{n,k}(L(G))$ is a regular language.*

Proof. We define a regular grammar $G' = (\Sigma, \{S\}, S, P)$ where S is a new symbol. The set of productions P is defined as follows:

$$P = \{S \longrightarrow [u_1 u_2 \dots u_n]S \mid (u_1, u_2, \dots, u_n) \in C\} \cup$$
$$\cup \{S \longrightarrow [v_1 v_2 \dots v_n] \mid (v_1, v_2, \dots, v_n) \in B\}.$$

One can easily prove that $L(G') = \varphi_g^{n,k}(L(G))$. Hence, $\varphi_g^{n,k}(L(G))$ is a regular language. □

As a consequence of this result we show that each family $\mathcal{SW}_{n.g,k}$ has good decidability properties.

Theorem 8. *For each family $\mathcal{SW}_{n,g,k}$ the following problems are decidable:*
(*i*) *the equivalence problem* ($L_1 = L_2$?).
(*ii*) *the inclusion problem* ($L_1 \subseteq L_2$?).
(*iii*) *the completeness problem* ($L = \Sigma^*$?).

Proof. (*i*) Clearly, for a given sewing grammar G of degree n and of type (g, k) one can effectively find a regular grammar G' such that

$$L(G') = \varphi_g^{n,k}(L(G)).$$

Since the function $\varphi_g^{n,k}$ is bijective, it follows that two sewing grammars G_1 and G_2 of degree n and of type (g, k) are equivalent if and only if

$$\varphi_g^{n,k}(L(G_1)) = \varphi_g^{n,k}(L(G_2)).$$

Note that the above equality is an equality between regular languages and thus it is decidable if this equality is true or not.

(*ii*) Similarly, the inclusion $L(G_1) \subseteq L(G_2)$ if and only if

$$\varphi_g^{n,k}(L(G_1)) \subseteq \varphi_g^{n,k}(L(G_2)).$$

Note that the above inclusion is an inclusion between regular languages and hence it is decidable if this inclusion is true or not.

(*iii*) Let G be a sewing grammar of degree n and of type (g, k). One can easily verify that

$$L(G) = \Sigma^* \text{ if and only if } \varphi_g^{n,k}(L(G)) = \Sigma_1^* \Sigma_2.$$

Again, the second equality is decidable since $\varphi_g^{n,k}(L(G))$ is a regular language. □

Remark 4. Other problems, for instance the membership problem, the emptiness problem, the finiteness problem are decidable problems for each family $\mathcal{SW}_n$.

Remark 5. For every $n \geq 4$, each family $\mathcal{SW}_{n,g,k}$ is almost a mildly context-sensitive family of languages, i.e., each such family satisfies all conditions to be a mildly context-sensitive family of languages, except that the language of crossed agreements is not in such a family.

A rather long combinatorial argument can be used to show that the language of crossed agreements is not in any such family $\mathcal{SW}_{n,g,k}$.

4 Isometric zipper functions

Let n be a fixed natural number, $n \geq 1$, and let $\mathcal{S}_n$ be the group of all permutations of degree n. Let $\mathcal{B}_n = \{0,1\}^n$ be the set of all Boolean n-dimensional vectors. We define the multiplication of two vectors from $\mathcal{B}_n$ componentwise using the rules: $0.0 = 1.1 = 1$ and $0.1 = 1.0 = 0$.

For every $p \in \mathcal{S}_n$ and for every $k = (k_1, \ldots, k_n) \in \mathcal{B}_n$, define the catenative function $cat_{p,k} : (\Sigma^*)^n \to \Sigma^*$ by $cat_{p,k}(u_1, \ldots, u_n) = k_1(u_{p(1)}) \ldots k_n(u_{p(n)})$, where $0(v) = mi(v)$ and $1(v) = v$.

Let $\mathcal{G}_n$ be the set $\{cat_{p,k} \mid p \in \mathcal{S}_n, k \in \mathcal{B}_n\}$. Define a binary operation on $\mathcal{G}_n$, denoted ".",

$$(cat_{p,k}).(cat_{r,j}) = cat_{pr,kj}.$$

It is easy to see that $(\mathcal{G}_n, .)$ is a group of order $2^n n!$ and, moreover, that $\mathcal{G}_n$ is isomorphic with the group of all isometries (symmetries) of the n-dimensional hypercube.

If L is a subset of $(\Sigma^*)^n$, then there are $2^n n!$ functions which may transform L into a language over Σ^*.

Definition 8. *An isometric zipper function of degree n is a function $cat_{p,k}$, where $p \in \mathcal{S}_n$ and $k \in \mathcal{B}_n$.*

Notation. We denote by $\mathcal{SW}_n(cat_{p,k})$ the family of those languages generated by sewing grammars of degree n using the zipper function $cat_{p,k}$.

Theorem 9. *Each family $\mathcal{SW}_n(cat_{p,k})$ is a mildly context-sensitive family of languages, where $n \geq 4$, $p \in \mathcal{S}_n$ and $k \in \mathcal{B}_n$.*

Note that all pumping conditions from Theorem 5 can be established for each family $\mathcal{SW}_n(cat_{p,k})$ with some adequate modifications.

Remark 6. Note that although for a fixed $n \geq 1$ there are $2^n n!$ different zipper functions the number of different families $\mathcal{SW}_n(cat_{p,k})$ is much smaller. For instance, one can easily see that for every $p, q \in \mathcal{S}_n$ it follows that $\mathcal{SW}_n(cat_{p,k}) = \mathcal{SW}_n(cat_{q,k})$.

However this is not true if someone change the Boolean vector k.

It is an open problem how many different families $\mathcal{SW}_n(cat_{p,k})$ there are for a given number n.

5 Comparison with other families of languages

In this section we investigate the interrelations between the families $\mathcal{SW}_n$, $n \geq 1$ and the families of languages in the Chomsky hierarchy as well as with families of simple matrix languages.

Notation. Let $\psi_g^{n,k}$ be the function

$$\psi_g^{n,k} : \Sigma_1^* \Sigma_2 \longrightarrow \Sigma^*$$

such that $\psi_g^{n,k}$ is the inverse of the function $\varphi_g^{n,k}$.

Theorem 10. *Let $L \subseteq \Sigma_1^* \Sigma_2$ be a regular language that can be generated by a regular grammar with only one nonterminal and with only one terminal production. Then the language $\psi_g^{n,k}(L)$ is a sewing language of degree n and of type (g,k).*

Proposition 3. *A language L is in $\mathcal{SW}_n$ if and only if L can be generated by a regular matrix grammar of degree n having only one nonterminal.*

Comments. From the above Theorem one can find alternative proofs for the Theorems 1, 2, 3 and 6.

Proposition 4. *Assume that $n = 2$ and let $p \in S_n$ be a permutation. Consider the Boolean vectors $b = (1,0)$ and $d = (0,1)$. The following equalities are true:*

$$\mathcal{SW}_n(cat_{p,b}) = \mathcal{SW}_n(cat_{p,d}) = \mathcal{SM}.$$

where, we recall that $\mathcal{SM}$ is the family of all Marcus simple contextual languages.

Remark 7. The families $\mathcal{SW}_n(cat_{p,k})$, $n > 1$, are not comparable with any of the families REG, LIN and CF. The reason is that the language $a^* \cup b^*$ is not contained in any of the families $\mathcal{SW}_n(cat_{p,k})$, whereas each of the families $\mathcal{SW}_n(cat_{p,k})$, where $n > 1$ contains non-context-free languages.

6 Conclusion

Sewing grammars provide a very simple generative device able to define classes of mildly context-sensitive languages. A modern trend in linguistics referred to as "minimality" requires very simple models in order to capture most of the facts that occur in natural languages. We hope that sewing grammars will be used as a tool in linguistics. Also, sewing grammars and sewing languages are suitable for other investigations in formal languages and combinatorics of words.

References

[1] Delany, P. and Landow, G. P. (eds); *Hypermedia and Literary Studies*, The MIT Press, 1991.

[2] Ibarra, O.; "Simple matrix languages", Inform. and Control, 17 (1970) 359-394.

[3] Joshi, A.K., Vijay-Shanker, K. and Weir, D.; "The convergence of mildly context-sensitive grammatical formalisms", in *Foundations Issues in Natural Language Processing*, Sells, P., Shieber, S. and Wasow, T. (eds.), MIT Press, 1991.

[4] Marcus, S.; "Contextual Grammars", Rev. Roum. Math. Pures et Appl., 14, 10 (1969) 1525-1534.

[5] Martin-Vide, C. and Mateescu, A.; "Contextual Grammars with Trajectories", to appear.

[6] Păun, G.; *Marcus Contextual Grammars*, Kluwer Acad. Publ., 1997.

[7] Rozenberg, G. and Salomaa, A. (eds.); *Handbook of Formal Languages*, Vol. 1-3, Springer, Berlin, New York, 1997.

[8] Salomaa, A.; *Formal Languages*, Academic Press, New York, London, 1973.

State and Transition Complexity of Watson-Crick Finite Automata

Andrei Păun and Mihaela Păun

Department of Computer Science
University of Western Ontario
London, Ontario, Canada N6A 5B7
E-mail: apaun@csd.uwo.ca

Abstract. We consider the number of states and the number of transitions in Watson-Crick finite (non-deterministic) automata as descriptional complexity measures. The succinctness of recognizing regular languages by Watson-Crick (arbitrary or 1-limited) automata in comparison with non-deterministic finite automata is investigated, as well as decidability and computability questions. Major differences are found between finite automata and Watson-Crick finite automata from both these points of view.

1 Introduction

Watson-Crick finite automata are language recognizing devices similar to finite automata, recently introduced in DNA computing area, [4]. They use a double-stranded tape, whose strands are separately scanned by read-only heads controlled by a common state; the symbols placed in corresponding cells of the two strands are linked by a complementarity relation. Several variants were investigated in [4], [5]; see a comprehensive presentation in Chapter 5 from [9]. It is known that restricted variants of Watson-Crick automata characterize the recursively enumerable languages modulo morphic images.

The complexity of recognizing languages by Watson-Crick finite automata was not yet investigated. We contribute here to filling in this gap, by considering two descriptional complexity measures: the number of states and the number of transitions. We mainly investigate two problems: (1) how efficient the Watson-Crick automata are in comparison with finite automata (when recognizing regular languages), and (2) decidability and computability questions usual in the descriptional complexity area, [6].

We find that there are regular languages which need arbitrarily many states or transitions in finite automata which recognize them, but require only a bounded number of states or transitions when they are recognized by Watson-Crick automata. In what concerns the second problem, we prove that many questions about Watson-Crick automata are undecidable (while they are known to be easily decidable for finite automata).

These results show that the use of a double-stranded tape is quite powerful, a fact already observed in many places in DNA computing.

We only partially solve another basic problem in descriptional complexity, that of non-triviality of the considered measures for Watson-Crick automata (the existence of languages of an arbitrary complexity). In particular, we completely leave open the connectedness problem (whether or not for each natural number greater than a given threshold there is a language of the complexity equal to that number).

In the last section we briefly investigate another complexity measure, the maximal distance between the two heads of a Watson-Crick automaton reached when recognizing the strings of a language.

2 Definitions

We only give here some notations and the definitions of Watson-Crick automata. For formal language elements we refer to [11]; in particular, we use [13] for details about finite automata.

We write a (non-deterministic) finite automaton in the form $A = (K, V, s_0, F, P)$, where K is the set of states, V is the alphabet, s_0 is the initial state, F is the set of final states, and P is the set of transition rules, of the form $sa \to s'$, $s, s' \in K, a \in V$ (in state s one reads the symbol a and one passes to state s'). Thus, the recognized language is defined by $L(A) = \{x \in V^* \mid s_0x \Longrightarrow^* s_f$, for some $s_f \in F\}$. (V^* is the free monoid generated by V; the empty string is denoted by λ and $V^* - \{\lambda\}$ is denoted by V^+.)

The family of regular languages is denoted by REG.

Let us now consider an alphabet V and a "complementarity" relation on V (like the Watson-Crick complementarity relation among the four DNA nucleotides), $\rho \subseteq V \times V$, which is symmetric. Denote

$$WK_\rho(V) = \{\begin{bmatrix} a \\ b \end{bmatrix} \mid a, b \in V, (a, b) \in \rho\}^*.$$

The set $WK_\rho(V)$ is called the *Watson-Crick domain* associated to V and ρ. The elements $\begin{bmatrix} a_1 \\ b_1 \end{bmatrix}\begin{bmatrix} a_2 \\ b_2 \end{bmatrix} \dots \begin{bmatrix} a_n \\ b_n \end{bmatrix} \in WK_\rho(V)$ are also written in the form $\begin{bmatrix} w_1 \\ w_2 \end{bmatrix}$, for $w_1 = a_1a_2 \dots a_n$, and $w_2 = b_1b_2 \dots b_n$. We call such elements $\begin{bmatrix} w_1 \\ w_2 \end{bmatrix} \in WK_\rho(V)$ *molecules*. According to the usual way of representing DNA molecules as double-stranded sequences, we also write the product monoid (V^*, V^*) in the form $\begin{pmatrix} V^* \\ V^* \end{pmatrix}$ and its elements in the form $\begin{pmatrix} x \\ y \end{pmatrix}$.

A *Watson-Crick finite automaton* (in short, a WK automaton) is a construct

$$A = (K, V, \rho, s_0, F, P),$$

where K and V are disjoint alphabets, $\rho \subseteq V \times V$ is a symmetric relation, $s_0 \in K$, $F \subseteq K$, and P is a finite set of transition rules of the form $s\begin{pmatrix} x \\ y \end{pmatrix} \to s'$, for $s, s' \in K$, $x, y \in V^*$.

The elements of K are called *states*, V is the (input) alphabet, ρ is a complementarity relation on V, s_0 is the initial state, F is the set of final states. The interpretation of a rule $s\begin{pmatrix} x \\ y \end{pmatrix} \to s'$ is: in the state s, the automaton passes over x in the upper level strand and over y in the lower level strand of a double-stranded sequence, and enters the state s'. Intuitively, a WK automaton looks as suggested in Figure 1.

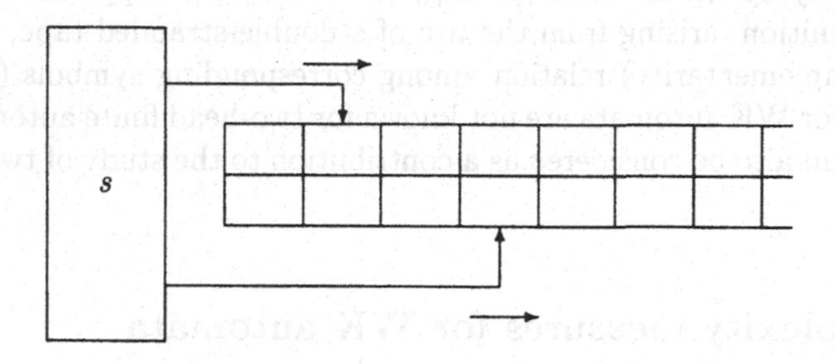

Fig. 1. A representation of a WK automaton.

For $\begin{pmatrix} u_1 \\ u_2 \end{pmatrix}, \begin{pmatrix} w_1 \\ w_2 \end{pmatrix} \in \begin{pmatrix} V^* \\ V^* \end{pmatrix}$ and $s, s' \in K$, we write

$$s\begin{pmatrix} u_1 \\ u_2 \end{pmatrix}\begin{pmatrix} w_1 \\ w_2 \end{pmatrix} \Longrightarrow s'\begin{pmatrix} w_1 \\ w_2 \end{pmatrix} \text{ iff } s\begin{pmatrix} u_1 \\ u_2 \end{pmatrix} \to s' \in P.$$

We denote by $\Longrightarrow^*$ the reflexive and transitive closure of the relation $\Longrightarrow$. The language recognized by a WK automaton is

$$L(A) = \{w_1 \in V^* \mid s_0 \begin{bmatrix} w_1 \\ w_2 \end{bmatrix} \Longrightarrow^* s_f, \text{ for some } w_2 \in V^* \text{ and } s_f \in F\}.$$

We emphasize the important fact that we start from molecules (elements of $WK_\rho(V)$) and we stop when the two strands are completely parsed and we reach a final state.

In [4], [9] one also defines other languages associated with a WK automaton, but we do not consider them here.

One can see that a WK automaton is a finite automaton with a double-stranded tape (and two read heads, one for each strand of the tape).

A WK automaton is said to be in the *1-normal form* if, for each transition rule $s\begin{pmatrix} x \\ y \end{pmatrix} \to s'$, we have $|xy| = 1$ ($|w|$ is the length of w).

It is proved in [9] that for each WK automaton A there is an automaton A' in the 1-normal form, such that $L(A) = L(A')$.

Convention. When comparing the languages recognized by two automata, the empty string is ignored.

In the same way as we pass from WK automata in the 1-normal form to arbitrary WK automata (we call them *block* WK automata), we can pass from usual finite automata to finite automata with transitions of the form $sx \to s'$: in state s we read the string x and pass to state s'. We call such an automaton a *block* finite automaton.

As it is already mentioned in [9], the WK automata are equivalent in power to the two-head finite automata, [7], [8], [10]. Still, the WK automata are motivated not only by the DNA computing, but also by the supplementary freedom in their definition, arising from the use of a double-stranded tape, with a symmetric (complementarity) relation among corresponding symbols (the variants considered for WK automata are not known for two-head finite automata). Thus, our paper can also be considered as a contribution to the study of two-head finite automata.

3 Complexity measures for WK automata

We denote by NFA_1, WK_1 the sets of nondeterministic finite automata and of Watson-Crick finite automata in the 1-normal form; by NFA_b, WK_b we denote the set of block automata of these types, respectively. Clearly, $NFA_1 \subseteq NFA_b$ and $WK_1 \subseteq WK_b$; because each finite automaton can be simulated in a natural way by a Watson-Crick automaton (scan the upper strand as in the finite automaton and scan any symbol in the lower strand in any state), we also consider that $NFA_1 \subseteq WK_1$ and $NFA_b \subseteq WK_b$.

These observations raise the question of comparing the complexity of describing a given regular language by finite automata and by WK automata. The basic complexity measure of finite automata is the number of states, which was investigated in many papers, see [13] and its bibliography, [1], [12], etc. Another parameter which estimates the size of a nondeterministic finite automaton is the number of transition rules. Up to now, none of these measures was considered for WK automata.

Formally, if K and P are the sets of states and of transition rules of a given finite automaton A or a WK automaton A, then we denote

$$State(A) = card(K),$$
$$Trans(A) = card(P).$$

We extend these measures in the usual way to languages; for instance,

$$State_{NFA_1}(L) = \min\{State(A) \mid L = L(A), A \in NFA_1\}.$$

In this way we get $State_X(L), Trans_X(L)$ for $X \in \{NFA_1, NFA_b, WK_1, WK_b\}$.

Many natural problems are to be investigated for such measures. The following three sections are devoted to three classes of such problems.

4 Decidability questions

Several decidability problems for descriptional complexity measures are systematized in [6]. We consider here only a few of them.

Theorem 1. (i) *None of the mappings* $M_X(L(A))$, *for* $M \in \{State, Trans\}$, $X \in \{WK_1, WK_b\}$, *can be computed algorithmically.*

(ii) *Given an integer* n, *it is not decidable whether or not* $M_X(L(A)) = n$, *for any* $M \in \{State, Trans\}, X \in \{WK_1, WK_b\}$, *and an arbitrary* $A \in X$.

(iii) *There is no algorithm able to construct* A_0 *such that* $L(A_0) = L(A)$ *and* $M(M_0) = M_X(L(A))$, *for an arbitrary* $A \in X$, $X \in \{WK_1, WK_b\}$.

Proof. Consider an instance of the Post Correspondence Problem (in short, PCP) over the alphabet $\{a, b\}$, $x = (x_1, x_2, \dots, x_n), y = (y_1, y_2, \dots, y_n)$; thus, $x_i, y_i \in \{a, b\}^+, 1 \le i \le n$. We construct the WK automaton

$$A = (\{s_0, s_1, s_2, s_3\}, \{a, b, c\}, \{(a, a), (b, b), (c, c)\}, s_0, \{s_1, s_3\}, P),$$

$$P = \{s_0 \binom{a}{a} \to s_1,\ s_1 \binom{a}{a} \to s_1,\ s_0 \binom{c}{c} \to s_2,\ s_2 \binom{\lambda}{c} \to s_3\}$$

$$\cup \{s_2 \binom{x_i}{y_i} \to s_2,\ s_2 \binom{x_i c}{y_i} \to s_2 \mid 1 \le i \le n\}.$$

It is easy to see that using the states s_0, s_1 we can recognize any string of the form a^m, $m \ge 1$. If we use first the transition $s_0 \binom{c}{c} \to s_2$, then we have to finish with the transition $s_2 \binom{\lambda}{c} \to s_3$ (this is the only terminal state accessible from s_2). This means that the molecule we recognize is of the form $\begin{bmatrix} cwc \\ cwc \end{bmatrix}$, with $w \in \{a, b\}^*$. In order to obtain such a molecule, before using the transition $s_2 \binom{\lambda}{c} \to s_3$ we have to use one time only a transition of the form $s_2 \binom{x_i c}{y_i} \to s_2$; before this transition, we can use any number of times transitions of the form $s_2 \binom{x_i}{y_i} \to s_2, 1 \le i \le n$. Consequently, $w = x_{i_1} x_{i_2} \dots x_{i_k} = y_{i_1} y_{i_2} \dots y_{i_k}$ (and, because we use a transition of the form $s_2 \binom{x_i c}{y_i} \to s_2$, we have $k \ge 1$), that is w corresponds to a solution to the PCP for x, y.

In conclusion,

$$L(A) = \begin{cases} a^+, & \text{if PCP}(x, y) \text{ has no solution,} \\ a^+ \cup L, & \text{otherwise,} \end{cases}$$

where L is an infinite subset of $\{c\}\{a, b\}^+\{c\}$.

If $L(A) = a^+$, then we obviously have

$$State_{WK_1}(L(A)) = 1 = State_{WK_b}(L(A)),$$
$$Trans_{WK_b}(L(A)) = 1,$$
$$Trans_{WK_1}(L(A)) = 2.$$

However, if $L(A) \neq a^+$, then:

- $State_{WK_b} \geq 2$: from Lemma 5.11 in [9] we know that if $State_{WK_b}(L) = 1$, then $L = L^+$; because such an equality is not true for our language, it follows that we need at least two states in order to recognize it.
- $Trans_{WK_b}(L(A)) \geq 2$: we need at least one transition passing over strings composed of the symbol a only (in order to recognize elements of a^+) and at least one transition also passing over strings containing symbols b and c, needed for recognizing strings of the form cwc, with $w \in \{a,b\}^+$.
- $Trans_{WK_1}(L(A)) \geq 6$: we need transitions of the form $s\begin{pmatrix}\alpha\\ \lambda\end{pmatrix} \to s'$ and $s\begin{pmatrix}\lambda\\ \alpha\end{pmatrix} \to s'$ for all $\alpha \in \{a,b,c\}$.

Consequently, $M_X(L_1) < M_X(L_2)$ in all cases, for $L_1 = a^+$ and $L_2 = a^+ \cup L$ as above, that is, $State_{WK_1}(L(A)) = 1 = State_{WK_b}(L(A)), Trans_{WK_b}(L(A)) = 1$ and $Trans_{WK_1}(L(A)) = 2$ if and only if PCP(x,y) has no solution, which is not decidable. This proves both points (i) and (ii). Because both parameters are trivially computable for automata, also point (iii) follows. □

5 Succinctness questions

In general, when having two sets of generative (or recognizing) mechanisms, $\mathcal{G}_1, \mathcal{G}_2$ such that $\mathcal{G}_1 \subseteq \mathcal{G}_2$, and a complexity measure M defined on $\mathcal{G}_2$ and extended in the natural way to languages, we have $M_{\mathcal{G}_1}(L) \geq M_{\mathcal{G}_2}(L)$, for all languages generated (recognized) by devices in $\mathcal{G}_1$. Stronger forms of this relation can be considered (see, for instance, [2]). We write:

$\mathcal{G}_1 \overset{1}{>} \mathcal{G}_2(M)$ iff there is L such that $M_{\mathcal{G}_1}(L) > M_{\mathcal{G}_2}(L)$,

$\mathcal{G}_1 \overset{2}{>} \mathcal{G}_2(M)$ iff for each n there is L_n such that $M_{\mathcal{G}_1}(L_n) - M_{\mathcal{G}_2}(L_n) \geq n$,

$\mathcal{G}_1 \overset{3}{>} \mathcal{G}_2(M)$ iff there are $L_n, n \geq 1$, such that $\lim_{n\to\infty} \dfrac{M_{\mathcal{G}_1}(L_n)}{M_{\mathcal{G}_2}(L_n)} = \infty$,

$\mathcal{G}_1 \overset{4}{>} \mathcal{G}_2(M)$ iff there are $L_n, n \geq 1$, such that $M_{\mathcal{G}_1}(L_n) \geq n$ and $M_{\mathcal{G}_2}(L) \leq k$, for a given constant k.

Clearly, $\overset{i}{>}$ implies $\overset{i-1}{>}$ for each $i = 2,3,4$.

We are now going to compare the finite automata and the WK automata from the points of view of measures $State$ and $Trans$, looking for relations $\overset{i}{>}$ as above with as large i as possible.

For some integer $k \geq 1$, let us consider the language

$$L_k = \{x^n \mid n \geq 1, x \in \{a,b\}^k\}.$$

Lemma 1. $State_{WK_b}(L_k) \leq 3$.

Proof. The language L_k can be recognized by the WK automaton

$$A = (\{s_0, s_1, s_2\}, \{a, b\}, \{(a, a), (b, b)\}, s_0, \{s_2\}, P),$$

$$P = \{s_0 \binom{x}{\lambda} \to s_1,\ s_1 \binom{x}{x} \to s_1,\ s_1 \binom{\lambda}{x} \to s_2 \mid x \in \{a, b\}^k\}.$$

Indeed, when parsing a molecule $\begin{bmatrix} x_1 x_2 \dots x_n \\ y_1 y_2 \dots y_n \end{bmatrix}$, for some $x_i, y_i \in \{a, b\}^k, 1 \leq i \leq n, n \geq 1$ (strings whose length is not a multiple of k cannot be recognized), we must have $x_2 = y_1, x_3 = y_2, \dots, x_n = y_{n-1}$ (from the form of the transition rules) and $x_i = y_i, 1 \leq i \leq n$ (from the complementarity relation). Therefore, $x_i = x_{i+1}, 1 \leq i \leq n-1$, that is the recognized string is in L_k. Conversely, each string in L_k can easily be seen that it can be recognized by our automaton. □

Lemma 2. $State_{NFA_b}(L_k) \geq 2^k + 1$.

Proof. The set $\{a, b\}^k$ contains 2^k strings; let us denote them, in any given order, by $x_1, \dots, x_{2^k}$.

Let $A = (K, \{a, b\}, s_0, F, P)$ be a block finite automaton recognizing the language L_k. Because, for each $i, 1 \leq i \leq 2^k$, we have $x_i^+ \in L_k$, there is a state $s_i \in K$ such that there is a cycle in the transition graph of A recognizing a string $w_i \neq \lambda$, that is $s_i w_i \Longrightarrow^* s_i$. This means that also $s_i w_i^n \Longrightarrow^* s_i$ is possible, for all $n \geq 1$. This implies that $w_i^+ \subseteq Sub(x_i^+)$ ($Sub(L)$ is the set of substrings of strings in L).

Assume now that there are $i \neq j, 1 \leq i, j \leq 2^k$, such that $s_i = s_j$. The following two parsings are possible in A:

$$s_0 y_i w_i^p z_i \Longrightarrow^* s_i w_i^p z_i \Longrightarrow^* s_i z_i \Longrightarrow^* s_{fi},$$
$$s_0 y_j w_j^q z_j \Longrightarrow^* s_i w_j^q z_j \Longrightarrow^* s_i z_j \Longrightarrow^* s_{fj},$$

such that $y_i w_i^p z_i = x_i^g$ and $y_j w_j^q z_j = x_j^h$, for some $y_i, z_i, y_j, z_j \in \{a, b\}^*$, with $s_{fi}, s_{fj} \in F$. This is true for all $p, q \geq 1$, so we may suppose that g, h are arbitrarily large. However, also the following parsing is possible in A:

$$s_0 y_i w_i^p w_j^q z_j \Longrightarrow^* s_i w_i^p w_j^q z_j \Longrightarrow^* s_i w_j^q z_j \Longrightarrow^* s_i z_j \Longrightarrow^* s_{fj}.$$

Because p and q can be arbitrary, the string $y_i w_i^p w_j^q z_j$ is of the form $x_i^l w x_j^t$, where l and t can be arbitrarily large. Such a string is in L_k if and only if $i = j$, which is a contradiction. Consequently, K contains 2^k different states $s_i, 1 \leq i \leq 2^k$.

None of these states can be the initial state of A. If s_i is the initial state, then the following parsings are possible:

$$s_i w_i^p z_i \Longrightarrow^* s_i z_i \Longrightarrow^* s_{fi},$$
$$s_i y_j w_j^q z_j \Longrightarrow^* s_i w_j^q z_j \Longrightarrow^* s_i z_j \Longrightarrow^* s_{fj},$$

such that $w_i^p z_i = x_i^g$ and $y_j w_j^q z_j = x_j^h$, for some $z_i, y_j, z_j \in \{a, b\}^*$, with $s_{fi}, s_{fj} \in F$. As above, we can mix the two parsings, and get a parsing for the string

$w_i^p y_j w_t^q z_j = x_i^l w x_j^t$, with arbitrarily large l and t, which implies the contradictory equality $i = j$.

In conclusion, $card(K) \geq 2^k + 1$, that is, $State_{NFA_b}(L_k) \geq 2^k + 1$. □

Actually, in the previous lemma we have equality, because the following block finite automaton clearly recognizes the language L_k:

$$A = (\{s_0\} \cup \{s_i \mid 1 \leq i \leq 2^k\}, \{a, b\}, s_0, \{s_i \mid 1 \leq i \leq 2^k\}, P),$$
$$P = \{s_0 x_i \to s_i,\ s_i x_i \to s_i \mid 1 \leq i \leq 2^k\}.$$

Combining the previous lemmas, we get:

Theorem 2. $NFA_b \overset{4}{>} WK_b(State)$ *and* $NFA_1 \overset{4}{>} WK_b(State)$.

Lemma 3. $State_{WK_1}(L_k) \leq 4k + 1$.

Proof. Using the same idea as in the proof of Lemma 1, we can see that the following WK automaton recognizes the language L_k:

$$A = (K, \{a, b\}, \{(a, a), (b, b)\}, s_0, \{s_{k+1}\}, P),$$

where

$$\begin{aligned}
K = \{&s_i \mid 0 \leq i \leq k+1\} \\
&\cup \{s_{a,i}, s_{b,i}, s_{ab,i} \mid 1 \leq i \leq k-1\} \cup \{s_{a,k}, s_{b,k}\},
\end{aligned}$$

$$\begin{aligned}
P = \{&s_i \binom{a}{\lambda} \to s_{i+1},\ s_i \binom{b}{\lambda} \to s_{i+1} \mid 0 \leq i \leq k-1\} \\
\cup \{&s_k \binom{a}{\lambda} \to s_{a,1},\ s_k \binom{b}{\lambda} \to s_{b,1}\} \\
\cup \{&s_{a,i} \binom{\lambda}{a} \to s_{ab,i},\ s_{b,i} \binom{\lambda}{b} \to s_{ab,i}, \\
&s_{ab,i} \binom{a}{\lambda} \to s_{a,i+1},\ s_{ab,i} \binom{b}{\lambda} \to s_{b,i+1} \mid 1 \leq i \leq k-1\} \\
\cup \{&s_{a,k} \binom{\lambda}{a} \to s_k,\ s_{b,k} \binom{\lambda}{b} \to s_k,\ s_k \binom{\lambda}{a} \to s_{k+1}, \\
&s_k \binom{\lambda}{b} \to s_{k+1},\ s_{k+1} \binom{\lambda}{a} \to s_{k+1},\ s_{k+1} \binom{\lambda}{b} \to s_{k+1}\}.
\end{aligned}$$

(On the path from s_0 to s_k one recognizes an arbitrary string $x \in \{a, b\}^k$ in the upper strand, on the cycles from s_k to s_k one recognizes arbitrary strings of length k in both strands, but shifted with k positions; because of the shift and of the complementarity relation, the string recognized when passing from s_0 to s_k should be repeated, that is, we get a string in L_k.)

One can see that $card(K) = 4k + 1$. □

Combining Lemmas 2 and 3, we get:

Theorem 3. $NFA_1 \overset{3}{>} WK_1(State)$.

Note that we cannot also compare NFA_b with WK_1, because we do not have the inclusion $NFA_b \subseteq WK_1$.

Theorem 4. $NFA_1 \overset{3}{>} WK_1(Trans)$, $NFA_1 \overset{4}{>} WK_b(Trans)$, *and* $NFA_b \overset{3}{>} WK_b(Trans)$.

Proof. If a finite automaton has n reachable states, then it has at least n transitions; thus, $Trans_{NFA_b}(L_k) \geq 2^k + 1$. However, from the construction in the proof of Lemma 3 we see that $Trans_{WK_1}(L_k) \leq 6k + 4$. Thus, the first and the third relations in the theorem follow.

For the second relation in the theorem we consider the language $(a^k)^*$, for some $k \geq 1$. It is easy to see that a finite automaton with transitions of the form $sa \to s'$ needs at least k transitions (and at least k states) in order to recognize this language, while a block WK automaton with one state and one transition can recognize it ($s_0 \begin{pmatrix} a^k \\ a^k \end{pmatrix} \to s_0$ is enough). Consequently, the relation $NFA_1 \overset{4}{>} WK_b(Trans)$ follows. □

We do not know which of the relations $\overset{3}{>}$ in Theorems 3 and 4 can be replaced by the stronger relation $\overset{4}{>}$.

It is interesting to note that even on the one-letter alphabet (in this case, the complementarity relation is unique and, at the first sight, looks useless) the WK automata are more efficient than finite automata. Let us consider the singleton language

$$L'_k = \{a^{2k}\}.$$

Obviously, $State_{NFA_1}(L'_k) = 2k + 1$, but we can recognize this language with the WK automaton

$$A = (\{s_i \mid 0 \leq i \leq k\} \cup \{s_a, s'_a\}, \{a\}, \{(a,a)\}, s_0, \{s_k\}, P),$$
$$P = \{s_0 \begin{pmatrix} a \\ \lambda \end{pmatrix} \to s_a,\ s_a \begin{pmatrix} \lambda \\ a \end{pmatrix} \to s'_a,\ s'_a \begin{pmatrix} \lambda \\ a \end{pmatrix} \to s_0\}$$
$$\cup \{s_i \begin{pmatrix} a \\ \lambda \end{pmatrix} \to s_{i+1} \mid 0 \leq i \leq k-1\}.$$

The cycle (s_0, s_a, s'_a, s_0) introduces one symbol in the upper strand and two in the lower one, while the path from s_0 to s_k introduces k symbols in the upper strand. This means that we can repeat the cycle exactly k times, otherwise the numbers of symbols in the two strands are not equal and the initial molecule is not a complete one. Consequently, we have only one correct parsing, recognizing the string a^{2k}.

Thus, on the one-letter alphabet we have at least relations of the form $\overset{2}{>}$ among NFA_1 and WK_1 for both measures $State$ and $Trans$.

6 Connectedness/Triviality questions

A measure M of descriptional complexity (of languages generated or recognized by devices in a given class $\mathcal{G}$) is said to be *non-trivial* if for each n there is a language L_n such that $M_{\mathcal{G}}(L_n) \geq n$, and *connected* if there is n_0 such that for each $n \geq n_0$ there is L_n with $M_{\mathcal{G}}(L_n) = n$.

Theorem 5. *Both measures State and Trans are non-trivial for WK automata in the 1-normal form.*

Proof. Consider the alphabet $V = \{a, b\}$ and assume that all languages over V which can be recognized by WK automata in the 1-normal form can be recognized by such automata with at most n states. The number of automata using at most n states and two symbols is finite, so they can recognize only a finite number of languages. This is contradictory, because the WK automata can recognize an infinite number of languages (for instance, all regular languages) over V. Similarly for the number of transitions (if we have at most n transitions, then we have at most $n + 1$ states). □

The connectivity problem (as well as the triviality problem for block WK automata) remains *open*. In view of the results in the previous sections, we expect some difficulties in solving this problem. One further evidence to this assertion is provided by the following observation.

Consider an alphabet V and its barred version, $\bar{V} = \{\bar{a} \mid a \in V\}$. Define the morphisms $h, \bar{h}$ on $V \cup \bar{V}$ by $h(a) = a, h(\bar{a}) = \lambda$ and $\bar{h}(a) = \lambda, \bar{h}(\bar{a}) = a$, for all $a \in V$. The *twin-shuffle* language over V is defined by

$$TS_V = \{x \in (V \cup \bar{V})^* \mid h(x) = \bar{h}(x)\}.$$

In [3] it is proved that each recursively enumerable language L can be written in the form $L = g(TS_{\{0,1\}})$, where g is a deterministic gsm mapping. The "universal" language $TS_{\{0,1\}}$ can be recognized by a block WK automaton with only one state:

$$A = (\{s_0\}, \{0, 1, \bar{0}, \bar{1}\}, \{(0,0), (1,1), (\bar{0}, \bar{0}), (\bar{1}, \bar{1})\}, s_0, \{s_0\},$$
$$\{s_0 \binom{0}{\lambda} \to s_0,\ s_0 \binom{1}{\lambda} \to s_0,\ s_0 \binom{\bar{0}}{0} \to s_0,$$
$$s_0 \binom{\bar{1}}{1} \to s_0,\ s_0 \binom{\lambda}{\bar{0}} \to s_0,\ s_0 \binom{\lambda}{\bar{1}} \to s_0\}).$$

All occurrences of symbols 0, 1 in the upper strand (the morphism $\bar{h}$ erases them) and all the barred symbols in the lower strand are ignored (the morphism h erases them). However, with each barred symbol in the upper strand its non-barred variant is associated in the lower strand, such that the order of paired symbols is the same in the two strands. This corresponds to the checking of the equality $h(x) = \bar{h}(x)$.

We can conclude that the power of small Watson-Crick automata is impressive.

7 Final remarks

We have considered here two natural measures of complexity for (nondeterministic) Watson-Crick automata, the number of states and the number of transitions. The main result of the paper is that these automata can recognize regular languages in a significantly more efficient manner than finite automata.

These two complexity measures are classic in the case of finite automata. We can also consider a measure which is specific to Watson-Crick automata (which, however, is not a descriptional complexity parameter, but a dynamical one, defined on computations): the maximal distance between the two read heads of the automaton during the recognition of a string.

Let $A = (K, V, \rho, s_0, F, P)$ be a WK automaton and Δ: $s_0 \begin{bmatrix} w_1 \\ w_2 \end{bmatrix} \Longrightarrow^* s_f, s_f \in F$, be a computation with respect to A. We define

$$Dist(\Delta) = \max\{||x_1| - |x_2|| \mid \Delta : s_0 \begin{bmatrix} w_1 \\ w_2 \end{bmatrix} \Longrightarrow^* s \begin{pmatrix} x_1 \\ x_2 \end{pmatrix} \Longrightarrow^* s_f,$$
$$x_1, x_2 \text{ are suffixes of } w_1, w_2, \text{ respectively}\}.$$

For $w \in L(A)$, we put

$$Dist(w, A) = \min\{Dist(\Delta) \mid \Delta : s_0 \begin{bmatrix} w \\ w' \end{bmatrix} \Longrightarrow^* s_f, s_f \in F\}$$

and we get

$$Dist(A) = \sup\{Dist(w, A) \mid w \in L(A)\}.$$

Then, for a language L, we define

$$Dist_{WK_\alpha}(L) = \inf\{Dist(A) \mid L = L(A), A \in WK_\alpha\}, \alpha \in \{1, b\}.$$

It is easy to see that we have

Theorem 6. $REG = \{L \mid Dist_{WK_b}(L) < \infty\}$.

(If the two heads of a WK automaton are always at a bounded distance, then a "window" of a bounded length plus the state can control the work of the automaton without using two heads, hence a finite memory and a usual single-stranded tape suffice. Conversely, it is obvious that for each regular language L we have $Dist_{WK_1}(L) = 1$.)

Moreover, we have the following result (contrast it with the triviality of the corresponding problem for measures *State* and *Trans*):

Theorem 7. *The mapping $Dist(A)$ is not algorithmically computable.*

Proof. For the WK automaton A in the proof of Theorem 1, we have

$$Dist(A) = \begin{cases} 0, & \text{PCP}(x, y) \text{ has no solution,} \\ 1, & \text{otherwise,} \end{cases}$$

which implies that we cannot compute $Dist(A)$ for this particular WK automaton (associated with the instance x, y of the Post Correspondence Problem). □

The measure *Dist* deserves a closer investigation.

Returning to the equivalence of WK automata with two-head finite automata, each two-head automaton can be considered a WK automaton and, conversely, for each WK automaton $A = (K, V, \rho, s_0, F, P)$ in the 1-normal form we can construct an equivalent two-head automaton $A' = (K, V, s_0, F, P')$, with

$$P' = \{s(a, \lambda) \to s' \mid s\binom{a}{\lambda} \to s' \in P\}$$
$$\cup \{s(\lambda, c) \to s' \mid s\binom{\lambda}{a} \to s' \in P, (c, a) \in \rho\}.$$

Therefore, the state complexity is preserved, but not the transition complexity. The question whether or not WK automata are more efficient than two-head finite automata from the point of view of the measure *Trans* remains *open*.

Note. Work supported by Grants OGP0041630 and OGP0007877 of the Natural Sciences and Engineering Research Council of Canada.

References

1. C. Câmpeanu, N. Sântean, S. Yu, Minimal cover-automata for finite languages, *International Workshop on Implementing Automata, WIA 98*, Rouen, 1998, 32–42.
2. J. Dassow, Gh. Păun, *Regulated Rewriting in Formal Language Theory*, Springer-Verlag, Berlin, 1989.
3. J. Engelfriet, G. Rozenberg, Fixed point languages, equality languages, and representations of recursively enumerable languages, *Journal of the ACM*, 27 (1980), 499–518.
4. R. Freund, Gh. Păun, G. Rozenberg, A. Salomaa, Watson-Crick finite automata, *Proc. of the Third Annual DIMACS Symp. on DNA Based Computers*, Philadelphia, 1997, 305–317.
5. R. Freund, Gh. Păun, G. Rozenberg, A. Salomaa, Watson-Crick automata, *Technical Report 97-13*, Dept. of Computer Sci., Leiden Univ., 1997.
6. J. Gruska, Descriptional complexity of context-free languages, *Proc. MFCS '73*, High Tatras, 1973, 71–83.
7. J. Hromkovic, One-way multihead deterministic finite automata, *Acta Informatica*, 19, 4 (1983), 377–384.
8. O. H. Ibarra, C. E. Kim, On 3-head versus 2-head finite automata, *Inform. Control*, 4 (1975), 193–200.
9. Gh. Păun, G. Rozenberg, A. Salomaa, *DNA Computing. New Computing Paradigms*, Springer-Verlag, Heidelberg, 1998.
10. A. L. Rosenberg, On multihead finite automata, *IBM J. R. and D.*, 10 (1966), 388–394.
11. G. Rozenberg, A. Salomaa, eds., *Handbook of Formal Languages*, Springer-Verlag, Heidelberg, 1997.
12. K. Salomaa, S. Yu, Q. Zhuang, The state complexities of some basic operations on regular languages, *Theoretical Computer Sci.*, 125 (1994), 315-328.
13. S. Yu, Regular languages, Chapter 2 in vol. 1 of [11], 41–110.

A Confluence Result for a Typed λ-Calculus of Exception Handling with Fixed-Point

Catherine Pilière

LORIA
Campus Scientifique - B.P. 239
54506 Vandœuvre-lès-Nancy Cedex – FRANCE
piliere@loria.fr

Abstract. We prove the confluence of a λ-calculus of exception handling whose typing system and evaluation rules are initially based on classical logic through the Curry-Howard isomorphism and to which we have added a general fixed-point operator.

1 Introduction

Philippe de Groote proposed in [3] a computational interpretation of classical logic through a simply typed λ-calculus which features an exceptions handling mechanism inspired by the ML language.

This calculus possesses several interesting properties (strong normalisation, subject reduction...), the main of them being given by its typing system which ensures that every raised exception is eventually handled whenever the whole term is correctly typed.

Our goal consists in seeing if this interpretation can provide a realistic system of exceptions handling, with emphasis on the study of its behaviour in the presence of a general fixed-point combinator. We here establish that the rule for fixed-point preserves the confluence of the calculus, even if it implies the lost of the strong normalisation.

Prawitz [8] studied the confluence properties of natural deduction for intuitionistic logic. The typing and evaluation rules of λ_{exn} present two major differences. Firstly, they are initially based on classical logic whose constructive aspects are less natural than the ones of intuitionistic logic. Secondly, the addition of a general fixed-point operator implies that we leave the framework of logic and penetrate deeper the one of functional programming by increasing the expressive power of the language.

From a programming point of view, a lambda-term models the set constituted by the program and the data it is applied to. It follows that if a lambda-calculus, that is, an evaluation process of lambda-terms, is confluent, then every program which terminates has a unique result. It then becomes simple to define an operational semantics for the reduction system in question by considering the normal form of a term, provided it does exist, as the result of its evaluation.

We first present the calculus and discuss in which sense it satisfies the Curry-Howard isomorphism. The second part is devoted to the proof of confluence of its evaluation rules.

2 Definition of the λ_{exn}-Calculus

Definition 1. *The types of the calculus are given by the following grammar:*

$$\mathcal{T} ::= \iota \mid exn \mid \mathcal{T} \to \mathcal{T}$$

where ι and exn stand for the base types, exn being the type of exceptional values.

λ_{exn} features an exception handling mechanism by means of exception variables y which act as datatype constructors: these exceptional variables are of functional type, say $\tau \to exn$ and then, when applied to a term of type τ, return exceptions. An exception acts like all the terms of base type but may also be raised under the form of packets, which are then propagated and possibly handled.

The packet $(\mathcal{R}aise M)$ is represented by the term $(\mathcal{R}M)$, the exception declaration *let exception* $y : \alpha \to exn$ *in* M *handle* $(yx) \Rightarrow N$ being denoted by $\langle y \cdot M | x \cdot N \rangle$.
Multiple declarations such as $\langle y_1 \cdot \langle y_2 \cdot \ldots \langle y_n \cdot M | x \cdot N_n \rangle \ldots \rangle | x \cdot N_1 \rangle$ are abbreviated by $\langle \overrightarrow{y} \cdot M | x \cdot \overrightarrow{N} \rangle$.
The calculus represents all recursive functions by way of the binding operator μ. Hence, the recursive function solution of the equation $f = \lambda x \cdot M(f, x)$ is denoted by the term $\mu f \cdot \lambda x \cdot M$.

Definition 2. *The syntax of the expressions of the calculus is the following:*

$$E ::= \bar{n} \mid x \mid y \mid \lambda x \cdot E \mid (EE) \mid (\mathcal{R}E) \mid \langle y \cdot E | x \cdot E \rangle \mid \mu f \cdot \lambda x \cdot E$$

The set $FV(T)$ of free variables of a term T is defined as usual. In particular, the free occurences of y in M and x in N are bound in $\langle y \cdot M | x \cdot N \rangle$ and similarly, the free occurences of f in $\lambda x \cdot M$ are bound in $\mu f \cdot \lambda x \cdot M$.

Definition 3. *Define a typing environment to be a function that assigns a type to every variable. Let Γ stand for such an environment. The expressions of the language are typed in the following manner:*

$$\Gamma \vdash \bar{n} : \iota$$

$$\Gamma \vdash x : \Gamma(x)$$

$$\frac{\Gamma, x:\alpha \vdash M:\beta}{\Gamma \vdash \lambda x \cdot M : \alpha \to \beta}$$

$$\frac{\Gamma \vdash M:\alpha \to \beta \qquad \Gamma \vdash N:\alpha}{\Gamma \vdash MN:\beta}$$

$$\frac{\Gamma \vdash M : exn}{\Gamma \vdash (\mathcal{R}M):\alpha}$$

$$\frac{\Gamma, y:\alpha \to exn \vdash M:\beta \qquad \Gamma, x:\alpha \vdash N:\beta}{\Gamma \vdash \langle y \cdot M | x \cdot N\rangle : \beta}$$

$$\frac{\Gamma, x:\alpha, f:\alpha \to \beta \vdash M:\beta}{\Gamma \vdash \mu f \cdot \lambda x \cdot M : \alpha \to \beta}$$

If exn is seen as the absurdity type $false$, then the type system above, provided that we forget the rule for fixed-point, corresponds to classical logic[1] through the Curry-Howard isomorphism. Now, a natural question arises: what is the meaning of the last rule? Another way this rule can be expressed is the following:

$$\frac{\Gamma, f:\alpha \to \beta \vdash \lambda x \cdot M : \alpha \to \beta}{\Gamma \vdash \mu f \cdot \lambda x \cdot M : \alpha \to \beta},$$

which appears as a very unexpected logical deduction: $(A \Rightarrow A) \vdash A$ ...

But, we can also see it as a rough approximation for the Noetherian induction:

$$\frac{\forall k\ ([\forall i < k\ A(i)] \Rightarrow\ A(k))}{\forall k\ A(k)},$$

where "<" denotes a *well-founded order.*

However, it is clear that with the rule for fixed-point, the typing system of λ_{exn} does no longer fit with classical logic. So, by adding such a recursion operator to the calculus, we certainly increase its expressiveness, but also loose interesting properties.

λ_{exn} is based on the higher order functional language ML. For that reason, we are interested in a call-by-value evaluation process. Therefore, the β-reduction will be performed only if the argument belongs to a particular set $\mathcal{V}$: the *set of values.*

Definition 4. *The set $\mathcal{V}$ of values is defined by:*

$$\mathcal{V} ::= \bar{n} \mid x \mid y \mid \lambda x \cdot M \mid (y\mathcal{V}).$$

In the following V,W (with possible subscript) will stand for members of $\mathcal{V}$.

[1] The typing rule for $\langle y \cdot M | x \cdot N\rangle$ matches the elimination of the disjonction in the particular case of *the excluded-middle.*

The reduction rules of the calculus are given by the table below [2]

β_V	$: (\lambda x \cdot M)V \rightarrow_v M[V/x]$
Raise$_{left}$	$: V(\mathcal{R}M) \rightarrow_v (\mathcal{R}M)$
Raise$_{right}$	$: (\mathcal{R}M)N \rightarrow_v (\mathcal{R}M)$
Raise$_{idem}$	$: (\mathcal{R}(\mathcal{R}M)) \rightarrow_v (\mathcal{R}M)$
Handle$_{simp}$	$: \langle y \cdot M \| x \cdot N \rangle \rightarrow_v M \quad if\ y \notin FV(M)$
Handle$_{Raise}$	$: \langle \overrightarrow{y} \cdot (\mathcal{R} y_i V) \| x \cdot \overrightarrow{N} \rangle \rightarrow_v \langle \overrightarrow{y} \cdot N_i[V/x] \| x \cdot \overrightarrow{N} \rangle$
Handle$_{left}$	$: V\langle y \cdot M \| x \cdot N \rangle \rightarrow_v \langle y \cdot VM \| x \cdot VN \rangle$
Handle$_{right}$	$: \langle y \cdot M \| x \cdot N \rangle O \rightarrow_v \langle y \cdot MO \| x \cdot NO \rangle$
Raise$_{Handle}$	$: (\mathcal{R}\langle y \cdot M \| x \cdot N \rangle) \rightarrow_v \langle y \cdot (\mathcal{R}M) \| x \cdot (\mathcal{R}N) \rangle$
Fix	$: \mu f \cdot \lambda x \cdot M \rightarrow_v \lambda x \cdot M[\mu f \cdot \lambda x \cdot M/f]$

Now, if we forget for a while the last rule and look at the table above from a logical point of view, the rules appear to be no more than proof reduction rules in natural deduction for classical logic [6] and have been proved to be strongly normalizing [3]. But of course, with our general fixed-point operator, these results become no longer available.

3 Confluence of λ_{exn}

As we obviously loose the strong normalisation property by adding the above rule for recursion, it would not be sufficient to establish that λ_{exn} is locally confluent [2]. Another way such a proof could be carried out is given by the Hindley-Rosen lemma [1]. It allows to conclude that the confluence (or equivalently, the Church-Rosser property) holds for a complex system if the rules it is made of possess themselves the property and behave together in a particular way:

Lemma 1 (Hindley-Rosen). *Let R_1 and R_2 be two binary relations such that:*

- *R_1 and R_2 possess the Church-Rosser property,*

---

[2] In Philippe de Groote's paper, the rule for **Handle$_{simp}$** was originaly written as $\langle y \cdot V | x \cdot N \rangle \rightarrow_v M \quad if\ y \notin FV(V)$. But this leads to divergence as was shown in [4] (take for example $z\langle y \cdot V | x \cdot N \rangle$: as zV is not itself a value, **Handle$_{simp}$** cannot be applied to $\langle y \cdot zV | x \cdot zN \rangle$). Hence, in order to achieve confluence, the rule has to be changed. Even if this last modification is not necessary, we also slightly modify the three rules **Raise$_{left}$**, **Raise$_{right}$**, **Raise$_{idem}$** to make them more general: contrarily to [3], they apply here to terms instead of values.

– R_1 *and* R_2 *commute* [3]

Then the system $R_1 \cup R_2$ *possesses the Church-Rosser property too.*

Because all the rules of λ_{exn} are well known or can be easely proved to be confluent ([1, 5]), let us focus on the interactions generated by their regrouping. If the commutation of each couple of rules can be established, the confluence of the calculus will follow immediately as a consequence of the lemma.

It can be easely checked that, in most cases, the rules behave in a very pleasant way, as the critical pairs which sometimes occur are solvable, *i.e.*, do not lead to divergence [2]. Nevertheless, serious difficulties appear with the three couples of rules $(H/Raise, H_{right})$, $(H/Raise, H_{left})$ and $(H/Raise, R/Handle)$: the situation for $(H/Raise, H_{left})$ is described by the figure 1 ($|\overrightarrow{y}|$ stands for the length of $\overrightarrow{y}$).

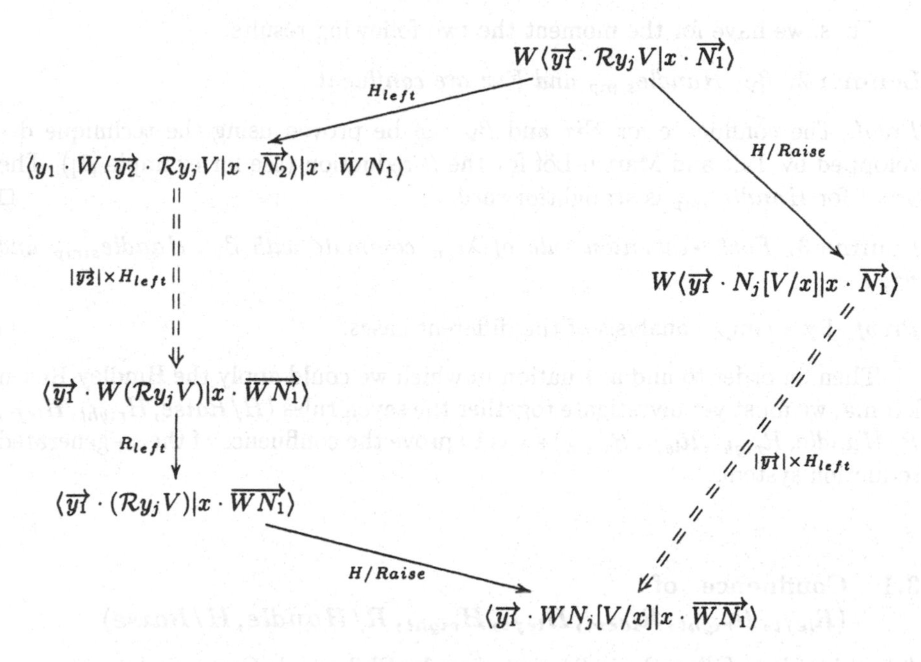

Fig. 1.

We here remark that the closure of the diamond for $(H/Raise, H_{left})$ can only be achieved in several steps, and especially by means of the rule R_{left}. Now,

[3] Two different binary relations $\overset{R_1}{\to}$ and $\overset{R_2}{\to}$ commute if the following holds:

$$\forall M_1, M_2, M_3 :\ M_1 \overset{R_1}{\to} M_2,\ M_1 \overset{R_2}{\to} M_3 \implies \exists M_4 \text{ such that } M_2 \overset{R_2}{\to} M_4,\ M_3 \overset{R_1}{\to} M_4.$$

whenever it seems possible to solve the inner problem by modifying the system so as to allow the rule H_{left} to deal with multiple exceptions declarations, clearly the creation of the redex $W(\mathcal{R}y_j V)$ which needs to be reduced using a sponger rule, is definitively undesirable...

As the couples $(H/Raise, H_{right})$ and $(H/Raise, R/Handle)$ impose in the same way the presence of the rules R_{right} and R_{idem}, the three systems $(H/Raise, H_{right})$, $(H/Raise, H_{left})$ and $(H/Raise, R/Handle)$ are obviously not confluent. It then becomes necessary to see if the systems $(H/Raise, H_{right}, R_{right})$, $(H/Raise, H_{left}, R_{left})$ and $(H/Raise, R/Handle, R_{idem})$ are confluent or not. If the confluence could be proved in the three cases, the idea would then be to use the commutation of each of the three rules which belongs to each of these subsystems with the remainder rules of λ_{exn}. But the problem is precisely that, to take only this example, the rule $H/Raise$ of the subsystem $(H/Raise, H_{left}, R_{left})$ do not commute with the rule H_{right} of the subsystem $(H/Raise, H_{right}, R_{right})$...

Thus, we have for the moment the two following results:

Lemma 2. *$\beta_{\mathcal{V}}$, $Handle_{simp}$ and Fix are confluent.*

Proof. The confluence for Fix and $\beta_{\mathcal{V}}$ can be proved using the technique developped by Tait and Martin-Löf for the β-reduction (see for example [1]). The proof for $Handle_{simp}$ is straightforward. □

Lemma 3. *Each evaluation rule of λ_{exn} commute with $\beta_{\mathcal{V}}$, $Handle_{simp}$ and Fix.*

Proof. By a simple analysis of the different cases. □

Then, in order to find a situation in which we could apply the Hindley-Rosen lemma, we must yet investigate together the seven rules $(H/Raise, H_{right}, H_{left}, R/Handle, R_{right}, R_{left}, R_{idem})$ so as to prove the confluence of the so-generated reduction system.

3.1 Confluence of $(R_{left}, R_{right}, R_{idem}, H_{left}, H_{right}, R/Handle, H/Raise)$

A first idea is to follow the indication given by Philippe de Groote in [3] and to use the previously mentioned method of Tait and Martin-Löf. We then aim at proving the confluence of a relation $\triangleright$ (to be defined) which possesses the same transitive closure than the union of $(R_{left}, R_{right}, R_{idem}, H_{left}, H_{right}, R/Handle, H/Raise)$ and which permits to solve the two problems pointed out by Figure 1:

- the rules H_{right}, H_{left} and $R/Handle$ have to be applied for each exception declaration enclosed in the term, that is, one step is required for each of the components of $\overrightarrow{y}$,

– It becomes then necessary to use one of the rules R_{right}, R_{left} or R_{idem} in order to regain a position allowing the application of $H/Raise$.

The inner problem can be given a solution provided we suppose to have an infinitely denumerable set of rules at our disposal for each of the three rules in question: thus, it becomes possible to cross in only one step all the exception declarations enclosed in a term, whatever the number of them is.

The second problem is due to the fact that, if we look at the case presented by Figure 1 (the two other triples work in a similar way, because of the symmetry of the rules), we may observe that the term $\langle \overrightarrow{y_1} \cdot W(\mathcal{R}y_jV)|x \cdot \overrightarrow{WN_1}\rangle$ is not a redex for $H/Raise$. Yet, it becomes such a redex if we reduce its subterm $W(\mathcal{R}y_jV)$ to $(\mathcal{R}y_jV)$ using the rule R_{left}. For this reason, the technique of Tait and Martin-Löf appears to be unsuitable for proving the confluence of the system. However, according to the method proposed by Aczel and Klop, this difficulty vanishes. Indeed, Aczel-Klop is an extension of Tait-Martin-Löf: both methods consist in proving the confluence of a relation whose transitive closure equals the one of the relation to be investigated, but there is a difference between them concerning the choice of this relation. In Tait-Martin-Löf, only the redexes which are present in the original term can be reduced at the same time. In Aczel-Klop, as soon as the reduction of the subterms confers to the whole term the status of being a redex, the reduction of these subterms followed by the reduction of the so generated redex may be seen as forming just one reduction step for the original term [7].

Before giving the definition of the relation $\triangleright$, a last observation is necessary. λ_{exn} being an exception handling language, all its terms may contain exception declarations, even the handlers themselves. We thus explicitely allow stacks of exception declarations at handlers-level, this in turn giving the possibility of stacks of exception raising to appear during the evaluation.

Now, as the rules H_{right}, H_{left} and $R/Handle$ must be applicable to terms containing multiple exception declarations, the rules R_{right}, R_{left} and R_{idem} must be applicable to terms containing multiple exception raising. In order to simplify the notation for such terms, we set up a new definition:

Definition 5. *The expression* $\Re\{T\}$ *is recursively defined as follows:*

$$\Re\{T\} \triangleq \mathcal{R}M \mid \langle \overrightarrow{y} \cdot \mathcal{R}M|x \cdot \overrightarrow{\Re\{N\}}\rangle$$

where $\{T\}$ *stands for the list built by adding* $M_{\overrightarrow{y},x}$ *to the list* $\{N\}$.

Remark 1. If $\Re\{T\} \triangleq \mathcal{R}M$, then we simply have $T = M$.

The relation $\triangleright$ can be described by structural induction in the following manner:

Definition 6.

$$\frac{M \triangleright V \quad N \triangleright \Re\{A\}}{MN \triangleright \Re\{A\}} R_l$$

$$\frac{M \triangleright \Re\{A\} \quad N \triangleright N'}{MN \triangleright \Re\{A\}} R_r$$

$$\frac{M \triangleright \Re\{A\}}{\mathcal{R}M \triangleright \Re\{A\}} R_i$$

$$\frac{M \triangleright V \quad N \triangleright \langle \overrightarrow{y} \cdot A | x \cdot \overrightarrow{B} \rangle}{MN \triangleright \langle \overrightarrow{y} \cdot VA | x \cdot \overrightarrow{VB} \rangle} H_l$$

$$\frac{M \triangleright \langle \overrightarrow{y} \cdot A | x \cdot \overrightarrow{B} \rangle \quad N \triangleright N'}{MN \triangleright \langle \overrightarrow{y} \cdot AN' | x \cdot \overrightarrow{BN'} \rangle} H_r$$

$$\frac{M \triangleright \langle \overrightarrow{y} \cdot A | x \cdot \overrightarrow{B} \rangle}{(\mathcal{R}M) \triangleright \langle \overrightarrow{y} \cdot (\mathcal{R}A) | x \cdot \overrightarrow{(\mathcal{R}B)} \rangle} R/H$$

$$\frac{M \triangleright (\mathcal{R}(y_j\, V)) \quad N \triangleright N'}{\langle \overrightarrow{y} \cdot M | x \cdot \overrightarrow{N} \rangle \triangleright \langle \overrightarrow{y} \cdot N'_j[V/x] | x \cdot \overrightarrow{N'} \rangle} H/R$$

The relation $\triangleright$ is reflexive and closed by the formation rules of terms. This property is expressed by the congruence rules:

$$M \triangleright M$$

$$\frac{M \triangleright M'}{\lambda x \cdot M \triangleright \lambda x \cdot M'}$$

$$\frac{M \triangleright M' \quad N \triangleright N'}{MN \triangleright M'N'}$$

$$\frac{M \triangleright M'}{(\mathcal{R}M) \triangleright (\mathcal{R}M')}$$

$$\frac{M \triangleright M' \quad N \triangleright N'}{\langle \overrightarrow{y} \cdot M | x \cdot \overrightarrow{N} \rangle \triangleright \langle \overrightarrow{y} \cdot M' | x \cdot \overrightarrow{N'} \rangle}$$

$$\frac{M \triangleright M'}{\mu f \cdot \lambda x \cdot M \triangleright \mu f \cdot \lambda x \cdot M'}$$

Remark 2. In the above statement concerning the rule H/R, $\overrightarrow{y}$ possesses at least one component y_j. On the contrary, in the case of the rules H_l, H_r and R/H, it seems possible for the vectors of exceptions declaration to be empty, and then for the corresponding handlers $x|\overrightarrow{\cdots}$ to be empty too. But in this case, these three rules become:

$$\frac{M \triangleright V \quad N \triangleright A}{MN \triangleright VA} H_l \qquad \frac{M \triangleright A \quad N \triangleright N'}{MN \triangleright AN'} H_r \qquad \frac{M \triangleright A}{(\mathcal{R}M) \triangleright (\mathcal{R}A)} R/H$$

hence appearing as particular cases of the congruence rules given previously. As a consequence, we will from now on consider that $\overrightarrow{y}$ stands for a vector with at least one component, in other words, that $\langle \overrightarrow{y} \cdot A|x \cdot \overrightarrow{B}\rangle$ contains a least one exception declaration.

Proposition 1. *Let $\rightarrow_\Sigma$ be the reduction system built from the rules $R_{left}, R_{left}, R_{idem}, H_{left}, H_{right}, R/Handle, H/Raise$. Then the relations $\rightarrow_\Sigma$ and $\triangleright$ have the same transitive closure.*

Proof. 1. $M \triangleright M' \Longrightarrow \exists n\ M \xrightarrow{n}_\Sigma M'$ (by induction on the definition of $\triangleright$)
2. $M \rightarrow_\Sigma M' \Longrightarrow M \triangleright M'$ (by structural induction on M)

□

Lemma 4.

$$\lambda x \cdot T \triangleright M \Longrightarrow M \equiv \lambda x \cdot T_1 : T \triangleright T_1 \ (\mathit{congruence})$$

$$ST \triangleright M \Longrightarrow \begin{cases} M \equiv S_1 T_1 : S \triangleright S_1, T \triangleright T_1 \ (\mathit{congruence}), \\ M \equiv \Re\{A\} : S \triangleright \Re\{A\}, T \triangleright T_1 \ (R_r) \\ M \equiv \Re\{A\} : S \triangleright V, T \triangleright \Re\{A\} \ (R_l) \\ M \equiv \langle \overrightarrow{y} \cdot AT_1|x \cdot \overrightarrow{BT_1}\rangle : S \triangleright \langle \overrightarrow{y} \cdot A|x \cdot \overrightarrow{B}\rangle, T \triangleright T_1 (H_r) \\ M \equiv \langle \overrightarrow{y} \cdot VA|x \cdot \overrightarrow{VB}\rangle : S \triangleright V, T \triangleright \langle \overrightarrow{y} \cdot A|x \cdot \overrightarrow{B}\rangle \ (H_l) \end{cases}$$

$$(\mathcal{R}T) \triangleright M \Longrightarrow \begin{cases} M \equiv (\mathcal{R}T_1) : T \triangleright T_1 \ (\mathit{congruence}) \\ M \equiv \langle \overrightarrow{y} \cdot (\mathcal{R}A)|x \cdot \overrightarrow{(\mathcal{R}B)}\rangle : T \triangleright \langle \overrightarrow{y} \cdot A|x \cdot \overrightarrow{B}\rangle \ (R/H) \\ M \equiv \Re\{A\} : T \triangleright \Re\{A\} \ (R_i) \end{cases}$$

$$\langle \overrightarrow{y} \cdot S|x \cdot \overrightarrow{T}\rangle \triangleright M \Longrightarrow \begin{cases} M \equiv \langle \overrightarrow{y} \cdot S_1|x \cdot \overrightarrow{T_1}\rangle : S \triangleright S_1, \overrightarrow{T} \triangleright \overrightarrow{T_1} \ (\mathit{congruence}) \\ M \equiv \langle \overrightarrow{y} \cdot T_{j1}[V/x]|x \cdot \overrightarrow{T_1}\rangle : S \triangleright (\mathcal{R}y_j\, V), \overrightarrow{T} \triangleright \overrightarrow{T_1} (H/R) \end{cases}$$

$$\mu f \cdot \lambda x \cdot T \triangleright M \Longrightarrow M \equiv \mu f \cdot \lambda x \cdot T_1 : T \triangleright T_1 \ (\mathit{congruence})$$

Proof. By induction on the definition of $\triangleright$. □

Corollary 1. $\forall M, \{A\} :\ \Re\{A\} \triangleright M \implies \exists\{B\}$ *such that* $M \equiv \Re\{B\}$.

Proof. (By structural induction on $\Re\{A\}$)

$\boldsymbol{\Re\{A\} \equiv \mathcal{R}A}$ From Lemma 4, three cases are possible:

1. $M \equiv \mathcal{R}A' \triangleq \Re\{A'\}$ (congruence)
2. $M \equiv \langle \overrightarrow{y} \cdot \mathcal{R}B|x \cdot \overrightarrow{\mathcal{R}C}\rangle \triangleq \Re\{B_{\overrightarrow{y},x}, C\}$ (rule R/H)
3. $M \equiv \Re\{B\}$ (rule R_i)

$\boldsymbol{\Re\{A\} \equiv \langle \overrightarrow{y} \cdot \mathcal{R}B|x \cdot \overrightarrow{\Re\{C\}}\rangle}$ From Lemma 4, two cases are possible for M:

M has been obtained from $\langle \overrightarrow{y} \cdot \mathcal{R}B|x \cdot \overrightarrow{\Re\{C\}}\rangle$ by H/R
We then have $M = \langle \overrightarrow{y} \cdot (\Re\{C\})'_j[V/x]|x \cdot \overrightarrow{(\Re\{C\})'}\rangle$ with $\mathcal{R}B \rhd \mathcal{R}(y_j V)$ and $\overrightarrow{(\Re\{C\})} \rhd \overrightarrow{(\Re\{C\})'}$. By induction hypothesis, $\exists D$ such that $\overrightarrow{(\Re\{C\})'} \equiv \overrightarrow{\Re\{D\}}$. Then $(\Re\{C\})'_j[V/x] \equiv \Re\{D_j\}[V/x]$. So we have:

$$
\begin{aligned}
M &\equiv \langle \overrightarrow{y} \cdot \Re\{D_j\}[V/x]|x \cdot \overrightarrow{\Re\{D\}}\rangle \\
&\triangleq \langle \overrightarrow{y} \cdot \langle \overrightarrow{y1} \cdot \mathcal{R}E[V/x]|x_1 \cdot \overrightarrow{\Re\{F[V/x]\}}\rangle|x \cdot \overrightarrow{\Re\{D\}}\rangle \quad (D_j = \{E_{\overrightarrow{y1},x_1}, F\}) \\
&\triangleq \langle \overrightarrow{y}, \overrightarrow{y1} \cdot \mathcal{R}E[V/x]|x \cdot \overrightarrow{\Re\{D\}}, x_1 \cdot \overrightarrow{\Re\{F[V/x]\}}\rangle \\
&\equiv \langle \overrightarrow{y2} \cdot \mathcal{R}E[V/x]|x_2 \cdot \overrightarrow{\Re\{D\}}[x_2/x], \overrightarrow{\Re\{F[V/x]\}}[x_2/x_1]\rangle \quad (x_2 \text{ is new}) \\
&\equiv \langle \overrightarrow{y2} \cdot \mathcal{R}G|x_2 \cdot \overrightarrow{\Re\{H\}}\rangle \\
&\triangleq \Re\{I\} \quad (\text{where } I = \{G_{\overrightarrow{y2},x_2}, H\}).
\end{aligned}
$$

M has been obtained from $\langle \overrightarrow{y} \cdot \mathcal{R}B|x \cdot \overrightarrow{\Re\{C\}}\rangle$ by congruence
Then $M \equiv \langle \overrightarrow{y} \cdot (\mathcal{R}B)'|x \cdot \overrightarrow{(\Re\{C\})'}\rangle$ with $\mathcal{R}B \rhd (\mathcal{R}B)'$ and $\overrightarrow{(\Re\{C\})} \rhd \overrightarrow{(\Re\{C\})'}$. For $\mathcal{R}B$, the base case holds. Thus, $\exists D$ such that $(\mathcal{R}B)' \equiv \Re\{D\}$.
By induction hypothesis, $\exists\{E\}$ such that $\overrightarrow{(\Re\{C\})'} \equiv \overrightarrow{\Re\{E\}}$. Then:

$$
\begin{aligned}
M &\equiv \langle \overrightarrow{y} \cdot \Re\{D\}|x \cdot \overrightarrow{\Re\{E\}}\rangle \\
&\triangleq \langle \overrightarrow{y}, \overrightarrow{y1} \cdot \mathcal{R}F|x \cdot \overrightarrow{\Re\{E\}}, x_1 \cdot \overrightarrow{\Re\{G\}}\rangle \quad (D = \{F_{\overrightarrow{y1},x_1}, G\}) \\
&\equiv \langle \overrightarrow{y2} \cdot \mathcal{R}F|x_2 \cdot \overrightarrow{\Re\{E\}}[x_2/x], \overrightarrow{\Re\{G\}}[x_2/x_1]\rangle \quad (x_2 \text{ new variable}) \\
&\equiv \langle \overrightarrow{y2} \cdot \mathcal{R}F|x_2 \cdot \overrightarrow{\Re\{H\}}\rangle \\
&\triangleq \Re\{I\} \quad (\text{where } I = \{F_{\overrightarrow{y2},x_2}, H\}).
\end{aligned}
$$

□

Lemma 5. *$\mathcal{V}$ and its complementary set are closed under the relation $\rhd$.*

Proof. (By induction on the definition of $\rhd$) □

Lemma 6 (Substitution). $M \rhd M_1, N \rhd N_1 \implies M[N/x] \rhd M_1[N_1/x]$

Proof. (By induction on the definition of $\rhd$) □

Proposition 2 (Confluence of $\rhd$). *For all M, M_1, M_2 such that $M \rhd M_1$ and $M \rhd M_2$, there exists M_3 such that $M_1 \rhd M_3$ and $M_2 \rhd M_3$.*

Proof. The proof is carried out by induction on the derivation of $M \rhd M_1$. As the whole proof can seem somehow tedious because we are dealing with numerous cases and because the rules R_r, R_l and R_i (and in the same way, their associated rules H_r, H_l and R/H) work rather closely, we here present only one of them.

$(M \rhd M_1) \equiv (PQ \rhd P_1Q_1)$ with $P \rhd P_1$ and $Q \rhd Q_1$
By Lemma 4, five cases are possible for M_2, the first of them matching the one presented by Figure 1 above:

$\boldsymbol{M \triangleright M_2}$ **by** $\boldsymbol{H_l}$: we then have $M_2 \equiv \langle \overrightarrow{y} \cdot VA|x \cdot \overrightarrow{VB}\rangle$ with the hypothesis $\begin{cases} P \triangleright V \\ Q \triangleright \langle \overrightarrow{y} \cdot A|x \cdot \overrightarrow{B}\rangle \end{cases}$

By Lemma 5 and induction hypothesis on P and Q: $\exists V_1$ such that $\begin{cases} P_1 \triangleright V_1 \\ V \triangleright V_1 \end{cases}$ and $\exists Q_2$ such that $\begin{cases} Q_1 \triangleright Q_2 \\ \langle \overrightarrow{y} \cdot A|x \cdot \overrightarrow{B}\rangle \triangleright Q_2. \end{cases}$

By Lemma 4, two cases are then possible for Q_2:

$\boldsymbol{\langle \overrightarrow{y} \cdot A|x \cdot \overrightarrow{B}\rangle \triangleright Q_2}$ **by congruence:**

$$Q_2 \equiv \langle \overrightarrow{y} \cdot A_1|x \cdot \overrightarrow{B_1}\rangle \text{ with } \begin{cases} A \triangleright A_1 \\ \overrightarrow{B} \triangleright \overrightarrow{B_1}. \end{cases}$$

By congruence, we get $\langle \overrightarrow{y} \cdot VA|x \cdot \overrightarrow{VB}\rangle \triangleright \langle \overrightarrow{y} \cdot V_1A_1|x \cdot \overrightarrow{V_1B_1}\rangle$

As $P_1 \triangleright V_1$ and $Q_1 \triangleright \langle \overrightarrow{y} \cdot A_1|x \cdot \overrightarrow{B_1}\rangle$, it comes
$P_1Q_1 \triangleright \langle \overrightarrow{y} \cdot V_1A_1|x \cdot \overrightarrow{V_1B_1}\rangle$ by H_l.

$\boldsymbol{\langle \overrightarrow{y} \cdot A|x \cdot \overrightarrow{B}\rangle \triangleright Q_2}$ **by H/R:**

$$Q_2 \equiv \langle \overrightarrow{y} \cdot B_{1j}[W/x]|x \cdot \overrightarrow{B_1}\rangle \text{ with } \begin{cases} A \triangleright (\mathcal{R}y_j\, W) \\ \overrightarrow{B} \triangleright \overrightarrow{B_1}. \end{cases}$$

By R_l, we get: $VA \triangleright (\mathcal{R}y_j\, W)$ and by congruence: $\overrightarrow{VB} \triangleright \overrightarrow{V_1B_1}$.

Then, by H/R $\langle \overrightarrow{y} \cdot VA|x \cdot \overrightarrow{VB}\rangle \triangleright \langle \overrightarrow{y} \cdot (V_1B_1)_j[W/x]|x \cdot \overrightarrow{V_1B_1}\rangle$.

On the other hand, from $Q_1 \triangleright \langle \overrightarrow{y} \cdot B_{1j}[W/x]|x \cdot \overrightarrow{B_1}\rangle$ and $P_1 \triangleright V_1$, we conclude by H_l that: $P_1Q_1 \triangleright \langle \overrightarrow{y} \cdot V_1B_{1j}[W/x]|x \cdot \overrightarrow{V_1B_1}\rangle$, *i.e.*, $P_1Q_1 \triangleright \langle \overrightarrow{y} \cdot (V_1B_1)_j[W/x]|x \cdot \overrightarrow{V_1B_1}\rangle$ as $x \notin FV(V_1)$

$\boldsymbol{M \triangleright M_2}$ **by** $\boldsymbol{R_l}$ Thus we have $M_2 \equiv \Re\{A\}$ with the hypothesis $P \triangleright V$, $Q \triangleright \Re\{A\}$.
By induction hypothesis, $\exists P_3$ such that $P_1 \triangleright P_3$ and $V \triangleright P_3$ and $\exists Q_3$ such that $Q_1 \triangleright Q_3$ and $\Re\{A\} \triangleright Q_3$.
From the corollary 1: $\exists\{B\}$ such that $Q_3 \equiv \Re\{B\}$.
As we know from Lemma 5, that P_3 is a value, we finally get $M_1 \equiv P_1Q_1 \triangleright \Re\{B\}$ by R_l.

We do not detail the three remaining cases (M_2 obtained by H_r, R_r or congruence) which are similar or easy to check.

□

Theorem 1. *The* λ_{exn}*-calculus is confluent.*

Proof. By Lemmas 1, 2, 3 and Proposition 2. □

4 Conclusion and Further Works

The previous result can be considered from two points of view. If we focus on the computational side, as it follows from the confluence property that the normal form of a term, if any, is unique, λ_{exn} can now be given in a simple way an operational semantics. The operational behaviour of the calculus will in turn have to be investigated in order to define precisely what is the contribution of λ_{exn} to the field of functional programming. In the same way, it would be useful to see if other type constructors, especially pairs and sums, can in turn be added with benefit to the language. From a computer science point of view, this would in particular allow the pattern-matching, whose interest is clear.

But the logical side of the calculus may be of interest too. Indeed, we know that all the evaluation rules, excepted the rule for *Fix*, correspond to normalisation rules for classical natural deduction. Therefore, one may wonder whether a form of ***subformula property*** [5, 6] holds for λ_{exn}. If such a property can be caracterized, the addition of new rules for disjunction will need some precautions.

5 Acknowledgments

I am very indebted to Philippe de Groote for (definitely) helpful discussions.

References

[1] H.P. BARENDREGT. *The Lambda Calculus.* Studies in Logic and the Foundations of Mathematics. Elsevier Science, 1984.
[2] R.V. BOOK and F. OTTO. *String-rewriting systems.* Texts and Monographs in Computer Science. Springer-Verlag, 1993.
[3] Ph. DE GROOTE. A simple calculus of exception handling. *Lectures Notes in Computer Science*, 902:201–215, 1995.
[4] K. FUJITA. Calculus of classical proofs 1. *Lectures Notes in Computer Science*, 1345:321–335, 1997.
[5] J.H. GALLIER. On the correspondence between proofs and λ-terms. *Cahiers du Centre de Logique (Université Catholique de Louvain)*, 8:55–138, 1995.
[6] J.Y. GIRARD, Y. LAFONT, and P. TAYLOR. *Proofs and Types*, volume 7 of *Cambridge Tracts in Computer Science.* Cambridge University Press, 1989.
[7] J.W. KLOP, V. VAN OOSTROM, and F. VAN RAAMSDONK. Combinatory reduction systems: introduction and survey. *Theorical Computer Science*, 121:279–308, 1993.
[8] D. PRAWITZ. *Natural deduction: a proof-theoretical study.* Almquist and Wiksell, Stockholm, 1965.

δ-Uniformly Decidable Sets and Turing Machines

Adriana Popovici[1] and Dan Popovici[2]

[1] Department of Computer Science, West University of Timişoara, Bd. V. Pârvan nr. 4, 1900 Timişoara, Romania, E-mail: apopovic@info.uvt.ro
[2] Department of Mathematics, West University of Timişoara, Bd. V. Pârvan nr. 4, 1900 Timişoara, Romania, E-mail:danp@gaspar2.uvt.ro

Abstract. We give a characterization of the archimedean fields in which nontrivial δ-uniform decidable sets exist. More exactly, after we introduce a notion of Turing closure of an archimedean field we prove that such a field posseses nontrivial δ-uniformly decidable sets if and only if it is not Turing closed. Moreover, if a function is δ-uniformly computable on a Turing closed field then it is rational over each of the connected components induced on the halting set by the reals. Finally, given a field which is not Turing closed, we obtain as a consequence that there exists a δ-uniform machine computing a total function which is not rational.

1 Definitions and Notations

Following Turing [Tur], it is possible to study computations of Turing machines whose input tapes contain representations of real numbers (and that are allowed to output similar representations, as well): this approach is known as *Type 2 recursion theory* [Wei].

In a paper of Boldi and Vigna [BV] the authors introduce a version of the BSS model of Blum, Shub and Smale [BSS], called a δ-uniform machine, in which exact tests are not allowed. In other words, a δ-uniform machine can only decide whether two numbers are very close, but cannot decide whether they are truly equal or not. There is a strict relation between δ-uniform computability and Type 2 recursion theory. More exactly, for any archimedean field the halting sets of δ-uniform BSS machines with coefficients in $\mathbb{T}$ (the field of Turing computable reals) or $\mathbb{Q}$ are exactly the halting sets of Type 2 Turing machines [BV]. Thus, the restriction of δ-uniformity reduces the full power of the BSS model, making it closer to Turing machines.

A *finite dimensional BSS machine* M over the archimedean field A consists of three spaces: the input space $I = A^l$, the output space $O = A^m$ and the state space $S = A^n$, together with a finite node set $N = \{1, 2, \ldots, p\}$ divided into four subsets: input, computation, branch and output:

1. node 1 is the only *input node* (with fan-in 0 and fan-out 1); there is a linear function with integer coefficients $i(-)$ mapping I to S;
2. node p is the only *output node* (with fan-out 0); there is a linear function with integer coefficients $o(-)$ mapping S to O;
3. $q \notin \{1,p\}$ can be a *computation node* (with fan-out 1); there is a rational function $r_q : S \to S$;
4. $q \notin \{1,p\}$ can be a *branching node* (with fan-out 2 and its two succesors are denoted by $\beta^-(q)$ and $\beta^+(q)$; branching on $-$ or $+$ will depend upon whether or not the first coordinate of the state space is negative).

M induces a *computing endomorphism* $e : N \times S \to N \times S$ by the relations:

$$e(1,x) = (\beta(1),x),\ e(p,x) = (p,x),$$
$$e(q,x) = (\beta(q), r_q(x)),\ \text{if } q \text{ is a computation node},$$

and

$$e(q,x) = \begin{cases} (\beta^-(q),x), & x_1 < 0 \\ (\beta^+(q),x), & x_1 \geq 0 \end{cases},\ \text{if } q \text{ is a branching node} \tag{1}$$

(if q has fan-out 1 we denoted by $\beta(q)$ the next node in the graph after q).

If the *computation* of M under input a (that is the orbit generated by the computing endomorphism starting from $(1, i(a))$) reaches a fixed point of the form (p,b) with $b \in S$ then we say that the machine *halted* obtaining a correspondence φ_M between the input a and the output $o(b)$. φ_M is called the *partial function computed by the machine M*. The sets of all inputs on which the machine M halts is denoted by Ω_M and called the *halting set of M*.

Formally, given a BSS machine M and a $\delta \geq 0$ (called a *threshold*), we define the δ-computing endomorphism much as above, but substituting the test case (1) as follows:

$$e(q,x) = \begin{cases} (\beta^-(q),x), & \text{if } x_1 < -\delta \\ (\beta^+(q),x), & \text{if } x_1 \geq -\delta \end{cases} \tag{2}$$

(if q is a branching node).

The induced δ-halting set is denoted by Ω^δ_M and the δ-computed function by φ^δ_M.

Definition 1. *A BSS machine is δ-uniform if $\Omega_M^\delta = \Omega_M$ and $\varphi_M^\delta = \varphi_M$ for all $\delta \in (0,1)$.*

A set which is the halting set of some BSS machine is called *semi-decidable.* The set is called *decidable* if, moreover, its complement is also semi-decidable. A partial function is *computable* if it is computed by some BSS machine. A set X is *semi-decidable relative to* Y if $X \cap Y$ is semi-decidable; it is *decidable relative to* Y if both $X \cap Y$ and $\mathcal{C}_X \cap Y$ are semi-decidable (we denoted by $\mathcal{C}_X$ the complement of the set X).

Similarly, if we consider a δ-uniform BSS machine we obtain respectively, the notions: *δ-uniformly semi-decidable, δ-uniformly decidable* and *δ-uniformly computable.* As observed by Boldi and Vigna, every δ–uniformly semi-decidable set is open.

The following theorem [BV] gives an equivalence between δ-uniform and Type 2 decidability.

Theorem 1. *Let $X \subseteq A^m$. Then X is δ-uniformly semi-decidable by a machine M with coefficients $\alpha_1, \ldots, \alpha_r$ if and only if there exists a Type 2 Turing machine M' with $m+r$ input tapes such that for all $(x_1, \ldots x_m) \in A^m$,*

$$(x_1, \ldots x_m) \in X \Leftrightarrow M' \textit{ halts on input } (x_1, \ldots, x_m, \alpha_1, \ldots, \alpha_r)\,.$$

2 Some Topological and Algebraic Preliminaries

The basic notions of topology quoted bellow can be found, for example, in [Gaa]. The results stated here are necessary in order to prove the connection between δ-uniform computability and Turing closed fields.

An open set is called *regular* if it is the interior of its own closure.

Given a connected topological space T and a dense subspace $D \subseteq T$, for every set $U \subseteq D$ open in D we define $\tilde{U} = (U \cup \mathcal{C}_D)^\circ$.

Proposition 1. *Let T be a connected topological space and $D \subseteq T$ a dense subspace. If $U, V \subseteq D$ are open in D then*

1. *$U = \tilde{U} \cap D$ and it is a dense subset of $\tilde{U}$; if moreover C is a component of $\tilde{U}$ then $\widetilde{C \cap D} = C$;*
2. *if $U \cap V = \emptyset$ then $\tilde{U} \cap \tilde{V} = \emptyset$; if moreover $U \cup V = D$ then $\tilde{U}$ and $\tilde{V}$ are regular.*

Theorem 2. *Let T be a topological space and U, V two disjoint regular open sets with $U \cup V$ dense in T. Then $\partial U = \partial V = \mathcal{C}_{U \cup V}$ (∂U denotes the boundary of U). If moreover T is connected and U and V are nonempty then the boundary is nonempty.*

For the following basic algebra notions we can mention [Lan] as a general reference.

Let $k \subseteq K$ be a field extension and $M \subseteq K$. We denote by $k(M)$ the intersection of all fields containing M and k. This field extension is said to be

- *finite* if $K = k(M)$ for some finite M;
- *simple* if $K = k(x)$, where x is an element of K called *primitive element;*
- *algebraic* if every $x \in K$ is algebraic over k; if θ is algebraic over k then there is a surjective homomorphism $v : k[X] \to k[\theta]$. Because $k[X]$ is principal, the ideal $\ker v$ is generated by an irreductible polynomial $p \in k[X]$ (assumed monic) called the *minimum polynomial of* θ. Consequently, by the theorem of isomorphism
$$k[X]/\langle p\rangle \cong k[\theta] = k(\theta); \tag{3}$$
- *separable* if it is algebraic and the minimum polynomial of each $x \in K$ has only simple roots.

The *primitive element theorem* states that

Theorem 3. *Every finite and separable field extension is simple.*

Let us note that every finite field extension is algebraic. Furthermore, if k has characteristic 0 then k is *perfect,* that is every algebraic field extension is separable. Consequently, by the previous theorem, if k has characteristic 0 and the field extension $k \subseteq K$ is finite then it is simple.

Consider a $\delta-$uniform BSS machine over the archimedean field A with coefficients $a_1, \dots, a_r$. Using a renumerotation (if necessary) we take $s \leq r$ as the minimum natural number such that the extension $\mathbb{Q}(a_1, \dots, a_s) \subseteq \mathbb{Q}(a_1, \dots, a_r)$ is algebraic. It follows that $\mathbb{Q}(a_1, \dots, a_s)$ is isomorphic with $\mathbb{Q}(X_1, \dots, X_s)$, the field of all rational functions with s arguments and coefficients in $\mathbb{Q}$. Moreover, because the above extension is finite and $\mathbb{R}$ has characteristic 0, by the primitive element theorem it is simple, that is there is an $\theta \in \mathbb{Q}(a_1, \dots, a_r)$ such that $\mathbb{Q}(a_1, \dots, a_r) = \mathbb{Q}(a_1, \dots, a_s)(\theta)$. By (3) there is an irreductible polynomial p such that

$$\mathbb{Q}(X_1, \dots, X_s)[X]/\langle p\rangle \cong \mathbb{Q}(a_1, \dots, a_r).$$

Consequently all field operations of $\mathbb{Q}(a_1, \dots, a_r)$ can be performed simbolically in $\mathbb{Q}(X_1, \dots, X_s)[X]/\langle p\rangle$ as well as equality tests. So we can emulate the computation of M with a machine M' that also keeps track of the intermediate results of the computation of M.

3 Main Results

Since every ordered archimedean field is isomorphic to an ordered subfield of $\mathbb{R}$ [Wae] we can regard an archimedean field A as an ordered subfield of $\mathbb{R}$.

Definition 2. *Let $F \supseteq A$ be a field extension. An element $f \in F$ is said to be Turing over A if there are $m \in \mathbb{N}$, $w \in A^m$ and a Turing machine M (with m input tapes) such that $M(w) = f$. If every element of F is Turing over A, then F is said to be a Turing extension of A. A is Turing closed if it does not have any proper Turing extension. The Turing closure of A is the intersection of all Turing closed fields containing A.*

Proposition 2. *The Turing closure of A is given exactly by the set T of all reals that are Turing over A.*

Proof. We just have to show that T is a Turing closed field. If $t, t' \in T$ then there are $w \in A^m$, $w' \in A^{m'}$ and Turing machines M, M' such that $M(w) = t$ and $M'(w') = t'$. But then there is a machine M'' with $m + m'$ input tapes that on input (w, w') computes internally t, t' and writes the sum $t + t'$ on the output tape. Analogously for the other operations. Turing closedness can be easily shown by a suitable composition of Turing machines. □

We are now in the position to prove that

Theorem 4. *The following conditions are equivalent:*

1. *there is an $m > 0$, an open subset $Z \subseteq A^m$ such that $\tilde{Z}$ is connected, and a nonempty proper subset $X \subset Z$ which is δ-uniformly decidable relative to Z;*
2. *there is a nonempty set $X \subset (0,1) \cap A$ which is δ-uniformly decidable relative to $(0,1) \cap A$;*
3. *there is an $\alpha \in \mathbb{R} \setminus A$ which is Turing over A;*
4. *there is an $\alpha \in \mathbb{R} \setminus A$ such that $\{x \in A \mid x < \alpha\}$ is δ-uniformly decidable*

(in order to avoid ambiguities we choose to use the symbol tilde only for $T = \mathbb{R}^m$ and $D = A^m$.)

Proof. (1.) ⇒ (2.) Let $X' = Z \setminus X$. By applying Proposition 3 with $T = \tilde{Z}$ and $D = Z$, we have that $Y = [X \cup (\tilde{Z} \setminus Z)]^\circ$ and $Y' = [X' \cup (\tilde{Z} \setminus Z)]^\circ$ are regular disjoint subsets of $\tilde{Z}$. Because $Z = X \cup X' \subset Y \cup Y' \subset \tilde{Z}$ and Z is dense in $\tilde{Z}$ their union is also dense in $\tilde{Z}$. Thus, we can apply Theorem 4, and consider a point y in $\partial Y = \partial Y'$ (the boundary in $\tilde{Z}$). Since $y \in \tilde{Z}$ and $\tilde{Z}$ is open there is an open ball $B \ni y$ entirely contained in $\tilde{Z}$. X being dense in Y and $B \cap Y$ open in Y the intersection $B \cap X = B \cap Y \cap X$ is non-void. We take $x \in B \cap X$ and analogously $x' \in B \cap X'$, and consider the path f connecting x to x', parameterized by $f(t) = t(x'-x)+x$, where $t \in [0,1] \subseteq \mathbb{R}$: observe that $t \in A$ iff $f(t) \in A^m$ and $f|A : A \to A^m$ is bijective.

Note that $f^{-1}(X) \cap (0,1)$ and $f^{-1}(X') \cap (0,1)$ are complementary in $(0,1) \cap A$. Indeed, if for some $t \in (0,1) \cap A$, $f(t)$ is not in X then, since $f(t) \in B \subset \tilde{Z}$, $f(t)$ is in $A^m \cap \tilde{Z} = Z$, that is $f(t) \in X'$. Let M be a δ-uniform BSS machine semi-deciding $f^{-1}(X) \cap (0,1)$. The same is true for $f^{-1}(X') \cap (0,1)$. Moreover, both $f^{-1}(X) \cap (0,1)$ and $f^{-1}(X') \cap (0,1)$ are nonempty; otherwise, if we suppose that $f^{-1}(X) \cap (0,1) = \emptyset$ then every neighbourhood of 0 in A contains points of $f^{-1}(X')$; thus, every neighbourhood of x in Z contains points of X'. But X is open and so contains a neighbourhood of x contradicting the fact that X and X' are disjoint.

(2.) $\Rightarrow$ (3.) We define a Turing machine working as follows: given a dyadic interval (l, r) containing some points of both X and $X' = (0,1) \cap A \setminus X$, and initially set to $(0,1)$, we find the minimum $k > 0$ such that the set of $2^k - 1$ dyadics of the form $l + i(r-l)/2^k$, for $0 < i < 2^k$, intersects both X and X' (in order to decide membership to X and X' we use Theorem 2); note that this minimization is terminating because the numbers of this form for all k are dense in (l, r), and thus must intersect both X and X', which contain open neighbourhoods (i.e., intervals) in $(l, r) \cap A$. Then, we find the first j such that $l + j(r-l)/2^k \in X$ and $l + (j+1)(r-l)/2^k \in X'$ (we exchange the rôle of X and X' if such a j does not exists), and restart the process on the interval $(l + j(r - l)/2^k, l + (j+1)(r-l)/2^k)$, which certainly contains points of both X and X' (because they are open), and whose length is at most $| r - l | /2$. The sequence of intervals thus defined cannot converge to a point of A (for example if $\alpha \in X$ all the above intervals, excepting a finite number, are contained in X, since X is open); hence, it converges to some number $\alpha \in \mathbb{R} \setminus A$, whose signed binary digits can be increasingly output each time a new subinterval is found.

(3.) $\Rightarrow$ (4.) Take the Turing machine M writing α and emulate it with a δ-uniform machine M'. Then, for every input a generate α with enough precision in order to decide whether $a < \alpha$ or $a > \alpha$ (the case $\alpha = a$ being impossible).

(4.) $\Rightarrow$ (1.) Take $m = 1$, $Z = A$ and $X = \{x \in A | x < \alpha\}$. □

The main application of the previous theorem is the following

Theorem 5. *Let A be an archimedean field. There are nontrivial $\delta-$uniformly decidable subsets of A^m iff A is not Turing closed.*

In particular, there are no nontrivial decidable subsets of $\mathbb{T}^m$ or $\mathbb{R}^m$. We now prove some restrictions about the functions computed over Turing closed fields:

Theorem 6. *Let M be a $\delta-$uniform machine, and C a component of Ω_M. If A is Turing closed, then $\varphi_{M|C \cap A^m}$ is a rational function.*

Proof. Let f_a be the rational function of the input computed by M on input a, $B = C \cap A^m$, and suppose $\varphi_{M|B}$ is not the restriction of a rational function. This implies that for some rational function g the sets $X = \{a \in B \mid f_a = g\}$ and $B \setminus X$ are both nonempty. Note that $\tilde{B} = C$ is connected by Proposition 3, and that X (hence $B \setminus X$) is $\delta-$uniformly decidable relative to B. Indeed, consider $E = \mathbb{Q}(a_1, \ldots, a_r) \subseteq A$, the extension of $\mathbb{Q}$ generated by the coefficients of M. By the primitive element theorem we can recode all constants appearing in the program of M as elements of $\mathbb{Q}(X_1, \ldots, X_s)[X]/\langle p \rangle$, where $\langle p \rangle$ is the principal ideal generated by a certain irreducible polynomial in $\mathbb{Q}(X_1, \ldots, X_s)[X]$ and $s \leq r$ (as explained in Section 2). We emulate the computation of M with a machine M' that also keeps track of the intermediate results of the computation of M under the form of polynomials (the variables now being the input) with coefficients in $\mathbb{Q}(X_1, \ldots, X_s)[X]/\langle p \rangle$; when M stops, the rational function computed can be tested exactly against g (also g can be coded, since its coefficients belong to E). By Theorem 8 (1) $\Leftrightarrow$ (3), A is not Turing closed. □

This implies, in particular, that the only total functions that are $\delta-$uniformly computable on a Turing closed field are the rational functions. Moreover, Theorem 10 gives also a necessary condition, as explained by the following

Theorem 7. *Let A be a field which is not Turing closed. Then, there exists a $\delta-$uniform machine M computing a total function which is not rational.*

Proof. We know, from Theorem 8, that there is some $\alpha \notin A$ such that $H = \{x \in A \mid x < \alpha\}$ is $\delta-$uniformly decidable. Then the characteristic function $\chi_H : A \to \{0, 1\}$ (which is clearly not rational) is computable.

□

References

[BSS] Blum, L., Shub, M., Smale, S.: On a theory of computation and complexity over the real numbers: NP-completeness, recursive functions and universal machines. Bull. Amer. Math. Soc. (N.S.) **21** (1989) 1–46

[BV] Boldi, P., Vigna, S.: δ–uniform BSS machines. J. Complexity **14** (1998) 234–256

[Gaa] Gaal, S.A.: Point set topology. Pure and Applied Mathematics, Academic Press, Boston (1964)

[Lan] Lang, S.: Algebra. Addison-Wesley (1971)

[Tur] Turing, A.M.: On computable numbers with an application to the Entscheidungsproblem. Proc. London Math. Soc. **42** (1936) 230–265. A correction **43** 544–546

[Wae] van der Waerden, B.L.: Algebra. Springer-Verlag, Berlin, eighth edition (1971)

[Wei] Weihrauch, K.: Type 2 recursion theory. Theoret. Comput. Sci. **38** (1985) 17–33

A Parallel Context-Free Derivation Hierarchy *

Klaus Reinhardt

Wilhelm-Schickhard Institut für Informatik, Universität Tübingen
Sand 13, D-72076 Tübingen, Germany
e-mail: reinhard@informatik.uni-tuebingen.de

Abstract. We consider the number of parallel derivation steps as complexity measure for context-free languages and show that a strict and dense hierarchy is obtained between logarithmic and linear (arbitrary) tree height. We hereby improve a result of Gabarro. Furthermore we give a non-regular languagc with logarithmic tree height disproving a conjecture of Culik and Maurer. As a new method we use counter-representations, where the successor relation can be handled as the complement of context-free languages.

1 Introduction

Like in complexity theory, where different measures for costs for recognizing languages are used, Book [Boo71] considered the costs of derivations of words in languages by grammars. Correspondences between the number of derivation steps and Turing machine computations were established by Igarashi in [Iga77].

The height of a derivation-tree corresponds to the minimal number of parallel derivation steps using a context-free grammar, where we allow an arbitrary number of variables to be replaced simultaneously. Correspondingly [Gab84] considered the space needed on the store of a push-down automaton and obtained a strict hierarchy of classes with $n^{1/q}$ space.

Heights of derivation trees generated by context-free grammars with regular parallel control languages were considered in [KPU79]. Brandenburg used the height of syntactical graphs[1] as complexity measure and showed equivalences to complexity classes [Bra81].

Why are context-free languages with sub-linear derivation tree height interesting? One reason is that their recognition can be parallelized efficiently as we will describe in Section 3. Unfortunately most context-free languages do not have this property. For some of them like $\{a^n b^n \mid n \in \mathcal{N}\}$ this was shown by Culik and Maurer [CM78] who also showed that regular languages have

logarithmic tree height. Furthermore, it was conjectured [Bra81,CM78] that context-free languages with logarithmic tree height are regular. We will disprove this conjecture in Section 6.

We will give a quite general criterion for languages not to have sub-linear derivation tree height by Theorem 2 in Section 5 using bounded languages.

* This research has been partially supported by the DFG Project La 618/3-2 KOMET.

[1] Generalizations of derivation trees for arbitrary grammars

On the other hand there are non-regular context-free languages which do not fit into this criterion [Boa97]: Consider the infinite word $w = baba^2ba^3b.....a^nb...$, which is a sequence of unary encodings of increasing numbers. Let the language L be the set of finite words which are not prefixes of w. L is context-free (the idea is that there exists a block of letters a with an exponent different from the number of letters b sitting at the left of the block). According to [Gol72] (see also [Gol76]) any bounded language in the full AFL generated by L is regular. Indeed this language has $\sqrt{n}$ tree height as we show in Section 6.

In section 7 we generalize this method to various kinds of representations of counters and define languages as sets of words which are not prefix of an infinite word being the sequence of such counter-representations (because of [AFG87] all these languages have to be inherently ambiguous). Furthermore we give a separation of tree height classes; this means that for every 'reasonable' function f between logarithmic and linear, we can construct a context-free language with $f(n)$ but no less tree height. Thus we have a strict and dense hierarchy.

In section 2 we show that derivation tree height corresponds to pushdown complexity and therefore our result improves [Gab84].

We consider context-free languages, which can be generated by a context-free grammar such that every word in the language has a derivation tree of height $f(|x|)$:

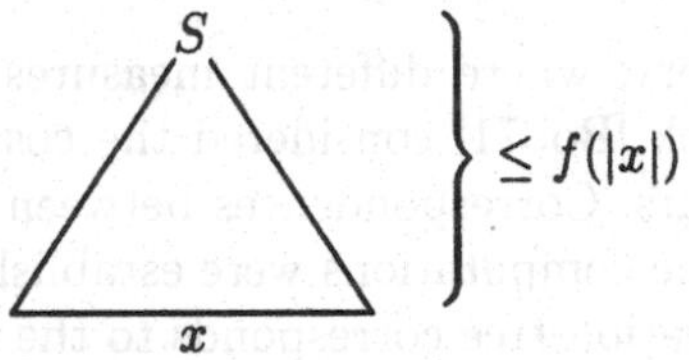

A main difference to E0L and similar systems is that in a parallel derivation step not all variables have to be replaced (this explains why we need not write $\underset{\|G}{\overset{\leq f(|x|)}{\Longrightarrow}}$ instead of $\underset{\|G}{\overset{f(|x|)}{\Longrightarrow}}$ in the following definition).

Definition 1. *Let $G = (V, \Sigma, P, S)$ be a context-free grammar. A* parallel derivation step *is defined by $\alpha_1 A_1 \alpha_2 A_2 ... \alpha_k A_k \alpha_{k+1} \underset{\|G}{\Longrightarrow} \alpha_1 r_1 \alpha_2 r_2 ... \alpha_k r_k \alpha_{k+1}$ with $A_i \to r_i \in P$ and $\alpha_i \in (V \cup \Sigma)^*$ for all $0 \leq i \leq k$. $\underset{\|G}{\overset{f(|x|)}{\Longrightarrow}}$ denotes $f(|x|)$ parallel derivation steps in sequence. $CFLth(f(n)) := \{L \subseteq \Sigma^* | \exists G, L = L(G), \forall x \in L$ $S \underset{\|G}{\overset{f(|x|)}{\Longrightarrow}} x\}$.*

Remark: It is easy to see that height can be compressed by a constant factor c by using $P' := \{A \to r \mid A \in V, A \underset{\|G}{\overset{\leq c}{\Longrightarrow}} r\}$. For that reason throughout the paper we need not care about additive and multiplicative constants. Observe that this destroys Chomsky normal form, which makes the O-notation for multiplicative constants necessary in this case.

Proposition 1. *[CM78] REG$\subseteq$ CFLth$(\log n)$.*

Clearly only finite languages can be generated with sub-logarithmic tree height.

2 The connection to pushdown complexity

Definition 2. *[Gab84] A language L has* pushdown complexity $f(n)$ *if L is recognized by a pushdown automaton such that every $w \in L$ has an accepting computation with pushdown space $O(f(|w|))$.*

By standard construction [HU79], every language in CFLth($f(n)$) has pushdown complexity $f(n)$. But in the other direction the standard construction in [HU79] leads to CFLth(lin) for any (even constant) pushdown complexity.

Lemma 1. *A language L with pushdown complexity $f(n)$ is in CFLth($f(n) + \log n$).*

Sketch of proof: Take the grammar $G = (V, \Sigma, P, S)$ obtained by the standard construction in [HU79] for the given pushdown automaton and build the equivalent grammar $G' = (V', \Sigma, P', S)$ with $V' = V \cup \{A_B | A, B \in V\}$ and $P' = P \cup \{A \mapsto A_A A, A_A \mapsto A_A A_A | A \in V\} \cup \{C_A \mapsto \alpha B_A | C \mapsto \alpha B \in P\} \cup \{C_A \mapsto \alpha | C \mapsto \alpha A \in P\}$. Words produced in a right recursion $A \overset{*}{\underset{G'}{\Rightarrow}} w_1 A \overset{*}{\underset{G'}{\Rightarrow}} w_1 w_2 A \overset{*}{\underset{G'}{\Rightarrow}} \ldots \overset{*}{\underset{G'}{\Rightarrow}} w_1 w_2 \ldots w_k A$ can now be produced in any balanced way via $A_A^k A$. Only parts of finite length and left recursions caused by the height on the pushdown store can not be balanced.

Corollary 1. *For $f(n) \geq \log n$ pushdown complexity and derivation tree height are the same and thus Theorem 1 in Section 4 and Theorem 3 in Section 7 are an improvement of Theorem 6 in [Gab84].*

3 The connection to parallel recognition

With the method in [GR88], context-free languages can be recognized by a CRCW-PRAM in $O(log(n))$ steps. But this method needs n^6 processors, which makes it very inefficient. On the other hand the CYK-algorithm [Kas65] allows an easy parallelization on a CRCW-PRAM in linear time with n^3 processors[2]. This idea can be used to recognize languages in CFLth($f(n)$) in time $O(f(n))$ with n^3 processors[3]. This means that the derivation tree height corresponds to the running time of the following parallel algorithm. For every production $A \rightarrow BC$ and every infix uv in the input word, a processor writes a 1 to the position for $A \overset{*}{\Rightarrow} uv$ if the positions for $B \overset{*}{\Rightarrow} u$ and $C \overset{*}{\Rightarrow} v$ have a 1. If $w \in L \in$CFLth($f(n)$), then $S \overset{*}{\Rightarrow} w$ will have a 1 after $O(f(n))$ steps.

---

[2] A more clever algorithm works on a CROW-PRAM in linear time with n^2 processors.

[3] In most cases an improvement to n^2 processors is possible.

4 Closure properties

Theorem 1. *CFLth(f) is an AFL and closed under reversal and substitution by other languages in CFLth(f) not containing* λ.

Proof. For the closure properties union, product, $*$, intersection with regular languages, reversal and substitution by other languages in CFLth(f) not containing λ we can just take the standard construction in [Ber79] for context-free languages and observe that the tree-height is only increased by an constant factor.

To build the grammar for $h(L)$ for $L \in$CFLth(f) and a non-erasing homomorphism h, we simply replace every terminal letter a by $h(a)$.

(An erasing homomorphism h could make the word significantly shorter.)

For the inverse homomorphism, a slight modification to the proof on page 31 in [Ber79] is necessary. The problem is that there may be letters z with $h(z) = \lambda$ occurring in $h^{-1}(w)$, which are produced in [Ber79] by a linear part of the grammar, which could increase the tree height to linear. To prevent this, we need to insert the production $\omega \to \omega\omega$ to P' in [Ber79] in order to be able to produce letters z with $h(z) = \lambda$ in binary trees which consume only logarithmic tree height. Furthermore we observe that the erasing homomorphism used for the construction of the inverse homomorphism in [Ber79] can decrease the length of the word only by a constant factor.

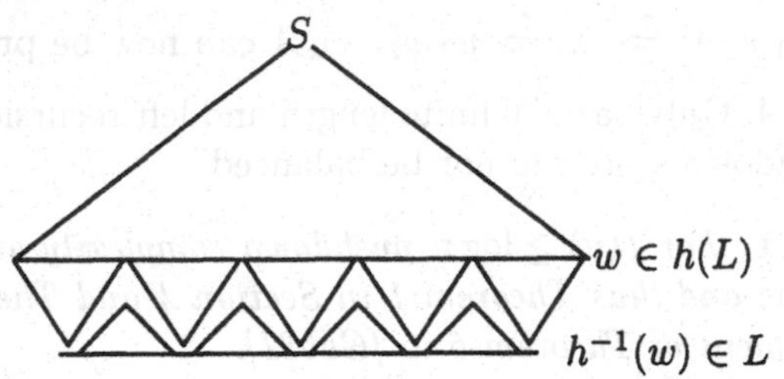

Open problem: Is CFLth(f) closed under erasing homomorphisms?

5 Bounded languages have linear derivation tree height

Definition 3. *[Gin66]*
A language L is bounded *if $L \subseteq w_1^* w_2^* ... w_m^*$ for $w_1, ..., w_m \in \Sigma^*$, $m \geq 1$.*

Theorem 2. *No bounded non regular language is in CFLth($o(n)$).*

Proof. We first start to prove the theorem for 2-bounded[4] languages. Since CFLth($o(n)$) is an AFL we may assume w.l.o.g. $L \subset a^*b^*$. Let G be a context-free grammar for L. Consider all variables $A, B, ...$ which are produced on the left or right side in the path from S to the border between a's and b's. The words producible by such a variable are in a sub-language $\subseteq a^* \cup b^*$ which is regular. Let

[4] $L \subseteq w_1^* w_2^*$ for $w_1, w_2 \in \Sigma^*$

s be the product of of all lengths of minimal loops[5] of the minimal deterministic automaton for all this languages. If L is not regular then there must be $k, k' \leq s$ such that $L' = L \cap a^k(a^s)^*b^{k'}(b^s)^*$ is not regular. Thus the semi-linear Parikh-image of L' must have a period incrementing a's and b's simultaneously. If L and thus $L' \in$CFLth($o(n)$), there must be infinite sub-languages on both sides. Then there is a word $w \in L'$ with an infinite sub-language on both sides, which means $(a^s)^*w(b^s)^* \subseteq L'$, thus L' is regular in contradiction to the assumption.

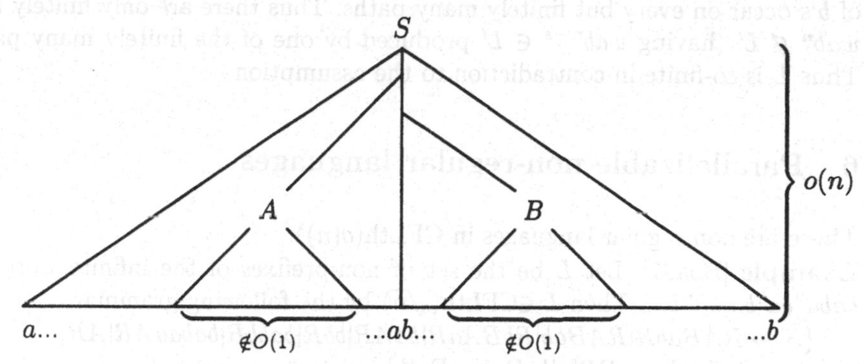

Now consider an m-bounded language $L \subset a_1^*a_2^*...a_m^*$ and construct L' analogously. If the semi-linear Parikh-image of L' has a period incrementing all a_i's simultaneously and if L and thus $L' \in$CFLth($o(n)$), there must be a word in L' with an infinite sub-language in every block, which means $(a_1^s)^*a_1^{c_1}(a_2^s)^*a_2^{c_2}...(a_m^s)^*a_m^{c_m} \subseteq L'$. Otherwise we have $(a_1^s)^*a_1^{c_1}(a_2^s)^*a_2^{c_2}...(a_m^s)^*a_m^{c_m} \cap L' = \emptyset$. In both cases the intersection with one of the languages $a_1^*...a_i^c...a_m^*$ must have the non-regular part, which means that the AFL generated by L and thus L' contains a $m-1$-bounded non-regular language, which leads to a contradiction by induction.

Corollary 2. *Every context-free language L generating an AFL containing a bounded non regular language is not in CFLth($o(n)$).*

If we knew that every AFL generated by a non-regular unambiguous context-free language contains a non-regular bounded language, we could proof the following:

Conjecture Every unambiguous context-free non-regular language L is not in CFLth($o(n)$).

For a more general separation result as Theorem 2 we need the following definition:

Definition 4. *Let Σ be an alphabet and $a \neq b$ two symbols. A language L is* $f(n)$ tail-bounded *if for every $w \in \Sigma^*$ and $n \in \mathcal{N}$ with $wab^n \notin L$ it holds* $n = f(|wa|)$.

Lemma 2. *No $f(n)$ tail-bounded non regular language is in CFLth($o(f(n))$)*

[5] A loop of the automaton reading the same input symbol, where every state occurs only once

Proof. Assume by the contrary G to be a context-free grammar w.l.o.g. for a language $L \subseteq \Sigma^*ab^*$. Consider all variables which are produced on the right side in the path from S to a in the derivation tree. The words producible by such a variable form regular sub-languages in b^*. Let s be the product of all lengths of minimal loops of the minimal deterministic automaton for all these languages. If L is not regular then there must be a $k \leq s$ such that $L' = L \cap \Sigma^*ab^k(b^s)^*$ is not regular. If L and thus $L' \in$CFLth($o(n)$), there must be an infinite sub-language of b's occur on every but finitely many paths. Thus there are only finitely many $wab^n \notin L'$ (having $wab^{n-s} \in L'$ produced by one of the finitely many paths). Thus L is co-finite in contradiction to the assumption.

6 Parallelizable non-regular languages

There are non-regular languages in CFLth($o(n)$):
Example [Boa97]: Let L be the set of non-prefixes of the infinite word $w = baba^2ba^3b.....a^nb....$ Then $L \in$CFLth($\sqrt{n}$) by the following grammar:

$\{S \to RABabR|RABab|RbBAaR|baAR|bbR|babbR|babaaAR|AR,$
$A \to AA|a, R \to RR|a|b, B \to aBa|b\}$

The variable R produces any string with logarithmic tree height. The variable B produces $\{a^nba^n \mid n \in \mathcal{N}\}$ thus $ABab$ produces $\{a^nba^mb \mid n \geq m \in \mathcal{N}\}$, which can not occur in w and $bBAa$ produces $\{ba^nba^m \mid m > n+1 \in \mathcal{N}\}$, which also can not occur in w. Thus a word in L can be produced making use of the first 'error' in respect to w, which is illustrated in the following picture:

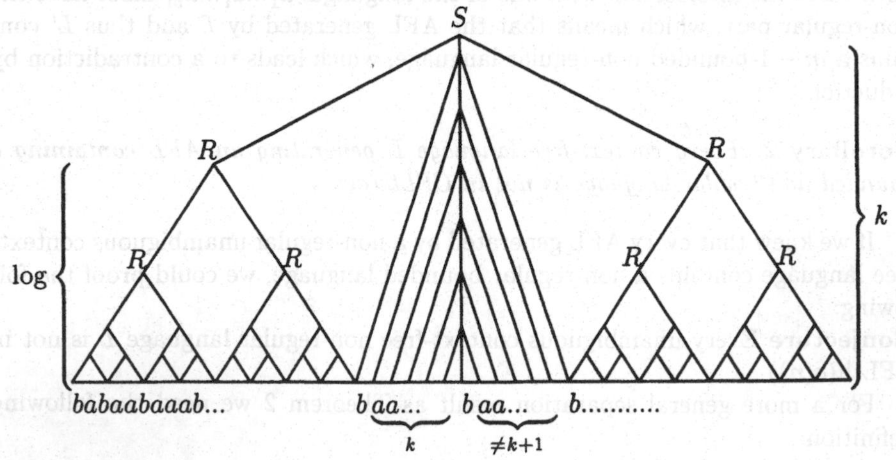

For a word in L we consider the derivation using the first position where a block has length k and the following block has not the length $k+1$. Thus we can estimate the length n of the word by $n \geq \sum_{i=1}^{k} i$ which means $k \in O(\sqrt{n})$.

We now improve this by constructing an example with the smallest possible derivation tree height and hereby disprove the conjecture in [Bra81,CM78]. There are non-regular languages in CFLth(log):

Example: Let L be the set of non-prefixes of the infinite word
$w = b0a1b10a11b100a101b110a111b...a1111000001b100010000a...bb_kab_{k+1}^Rb...$,
where $b_k \in \{0,1\}^*$ is the binary representation of k. It holds $L \in$CFLth(log(n)) by a grammar which produces words containing an 'error'-part showing that the word is not prefix of w; this may be some part ab_kbv^Ra or bb_k^Ravb with $v \neq b_{k+1}$ or some other syntactic 'error'. Since the length of the binary representations grow only logarithmically until an 'error' occurs, the tree-height is also logarithmic like the following picture shows:

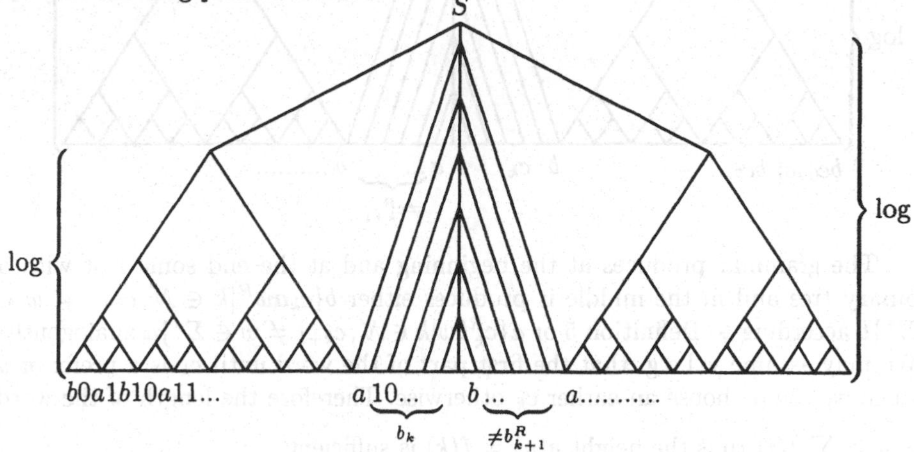

7 Strictness and denseness of the hierarchy

As an important tool we use a generalization of the representation of a counter to define context-free complement constructible functions.

Definition 5. *A function* $f : \mathcal{N} \to \mathcal{N}$ *is called* context-free complement constructible *(ccc), if there is an infinite sequence* $c_0, c_1, ...$ *of words called* context-free complement construction *over an alphabet* Σ_f *with* $|c_k| = f(k)$*, such that the complement of* $\{c_k\$c_{k+1}^R | k \in \mathcal{N}\}$ *is context-free.*

The following property of ccc functions will help us to establish the strictness and denseness of the hierarchy:

Theorem 3. *For all functions* $g : \mathcal{N} \to \mathcal{N}$ *with* $\log \leq g \leq lin$ *such that there is a ccc function* $f : \mathcal{N} \to \mathcal{N}$ *with* $g(\sum_{i=1}^k f(i)) = f(k)$ *for all* k*, there are languages in CFLth(g(n))\CFLth(o(g(n))).*

Proof. Let L be the set of non-prefixes of the infinite word
$u = bc_0ac_1^Rbc_2ac_3^R...ac_{n-1}^Rbc_na...$, with (c_i) being the context-free complement construction for f with $a, b \notin \Sigma_f$. Then $L \in$CFLth($g(n)$) by a grammar which works as the picture shows:

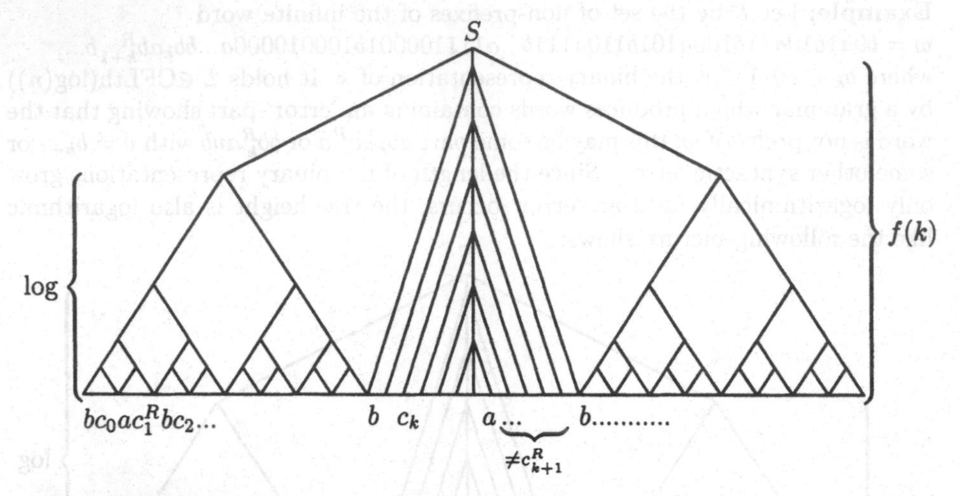

The grammar produces at the beginning and at the end some rest with a binary tree and in the middle it produces either $b\{c_k aw^R | k \in \mathcal{N}, c_{k+1} \neq w \in \Sigma^*\}b$ according to Definition 5 or $a\{c_k^R bw | k \in \mathcal{N}, c_{k+1} \neq w \in \Sigma^*\}a$ analogously. We may assume w.l.o.g. that the first part of the word until c_k is a prefix of u since we could choose an earlier c_k otherwise. Therefore the length of the word is $n \geq \sum_{i=1}^{k} f(i)$ thus the height $g(n) = f(k)$ is sufficient.

Let $L' := Lb^* \cup \overline{(\Sigma_f \cup \{a,b\})^* ab^*} \cup (\Sigma_f \cup \{a,b\})^* b\{vab^n \mid v \in \Sigma_f^*, n \neq |v|\}$. Every word $wab^n \in (\Sigma_f \cup \{a,b\})^* ab^*$ is in Lb^* unless wa is a prefix of u; in this case wab^n is in L' iff the length of the last counter representation in w is not equal to n, which can be tested within tree-height $g(|wa|)$ thus $L' \in$CFLth$(g(n))$. The only possibility for wab^n not to be in L' is $n = g'(|wa|)$ for a $g' \in \Theta(g)$, which means L' is $g'(n)$ tail-bounded. Because of Lemma 2, L' can not be in CFLth$(o(g'(n)))$ =CFLth$(o(g(n)))$.

Lemma 3. *1, 2^n and log are ccc. If f and g are ccc, then also $f+g$ and $f*g$ and the functions e, h with $e(n) = \sum_{i=1}^{n} f(i)$ and $h^{-1}(n) = \sum_{i=1}^{n} f^{-1}(i)$ are ccc.*

Proof. Let $c_{1,i} = 1$, $c_{2^n,i} = 1^{2^n}$ and $c_{\log,i} = bin(i)$. Assume w.l.o.g. $\Sigma_f \cap \Sigma_g = \emptyset$ and $d \notin \Sigma_f \cup \Sigma_g$.

Let $\Sigma_{f+g} = \Sigma_f \cup \Sigma_g$, $c_{f+g,i} := c_{f,i}c_{g,i}$. The complement of $\{c_{f+g,k}\$c_{f+g,k+1}^R \mid n \in \mathcal{N}\}$ is $\{c_{f+g,k}aw^R | k \in \mathcal{N}, c_{f+g,k+1} \neq w \in \Sigma^*\}b$ and can be generated by generating at least one counter wrongly.

Let $\Sigma_{f*g} := \Sigma_f \times \Sigma_g$ with the canonical projections π_1 and π_2 and $c_{f*g,i} := w$ with $\pi_1(w) = c_{f,i}^{|c_{g,i}|}$ and $\pi_2(w) = h(c_{g,i})$ with $h(a) := a^{|c_{f,i}|}$ and the complement of $\{c_{f*g,k}\$c_{f*g,k+1}^R | n \in \mathcal{N}\}$ can be generated analogously.

Let $\Sigma_e = \Sigma_h = \Sigma_f \cup \{d\}$ $c_{e,i} := d^{\sum_{j=1}^{i-1} |c_{f,j}|} c_{f,i}$ and the complement of $\{c_{e,k}\$c_{e,k+1}^R | n \in \mathcal{N}\}$ can be generated by either generating a wrong $c_{f,i+1}$ or not adding $|c_{f,i}|$ correctly to the number of d's.

The last transformation enables building roots of functions (log is fixed point of the transformation); the idea to construct the counter is to keep a size $h(i)$ and simulate the counter $c_{f,j}$ for f inside by $c_{h,i} = d^k c_{f,j}$, where the rest is filled with d's. This is done until the size is not anymore sufficient, then the space is incremented an the simulated counter for f is started from 0 again. Formally this is $c_{h,i+1} := \begin{cases} d^{|c_{h,i}|+1-|c_{f,0}|} c_{f,0} \text{ if } f(j+1) > h(i) \\ d^{k+|c_{f,j}|-|c_{f,j+1}|} c_{f,j+1} \text{ else} \end{cases}$.

Lemma 3 shows that all polynomials with rational coefficient and its multiplications with poly-logarithms are ccc. For f ranging from logarithmic to very big polynomials, the function g with $g(\sum_{i=1}^{k} f(i)) = f(k)$ ranges from logarithmic towards linear. If $g(n)$ is a polynomial $n^{p/q}$ with $q > p$ then a ccc function $f(n) \in \Theta(n^{p/(q-p)})$ fulfills the condition $g(\sum_{i=1}^{k} f(i)) = f(k)$ and Theorem 3 can be applied. If $g(n)$ is a polylogarithmic function $\log^{p/q} n$, we do not know how to find an appropriate function f to apply Theorem 3 directly. However, if we set $f(n) = \log^{p/q} n$, the obtained function $g(n)$ can be estimated by $\log^{(p/q)-\varepsilon} n < g(n) < \log^{p/q} n$ for every $\varepsilon > 0$ thus Theorem 3 can separate any CFLth($\log^r$) from CFLth($\log^{r'}$) with $r < r'$. Since the ccc functions f are dense, also the obtained functions g are dense between between log and lin; therefor we get:

Conclusion: The parallel context-free derivation hierarchy is strict and dense between CFLth(log) and CFLth(lin)=CFL.

Remark: It is easy to see (but difficult to express formally) that any nontrivial decision problem, where it has to be decided for a given grammar for a language L in CFLth($g'(n)$) whether L is in CFLth($g(n)$), is undecidable (provided corresponding ccc function f, f' exist) since counter representations could also contain configurations of Turing machines and the behavior of the length can change if an accepting configuration is reached.

Acknowledgment: We thank H. Fernau and K.-J. Lange for helpful remarks.

References

[AFG87] J.-M. Autebert, P. Flajolet, and J. Gabarró. Prefixes of infinite words and ambiguous context-free languages. *Information Processing Letters*, 25:211–216, 1987.

[Ber79] J. Berstel. *Transductions and context-free languages.* Teubner Studienbücher, Stuttgart, 1979.

[Boa97] L. Boasson. personal communication. 1997.

[Boo71] R. V. Book. Time-bounded grammars and their languages. *Journal of Computer and System Sciences*, 5(4):397–429, August 1971.

[Bra81] F. J. Brandenburg. On the height of syntactical graphs. In Peter Deussen, editor, *Proceedings of the 5th GI-Conference on Theoretical Computer Science*, volume 104 of *LNCS*, pages 13–21, Karlsruhe, FRG, March 1981. Springer.

[CM78] K. Culik II and H.A. Maurer. On the height of derivation trees. *Forschungsbericht Nr. 18, Inst. für Informationsverarbeitung TU Graz*, 1978.

[Gab84] J. Gabarró. Pushdown space complexity and related full-AFLs. In *Symposium of Theoretical Aspects of Computer Science*, volume 166 of *LNCS*, pages 250–259, Paris, France, 11–13 April 1984. Springer.

[Gin66] S. Ginsburg. *The Mathematical Theory of Context-Free Languages*. McGraw-Hill, New York, 1966.

[Gol72] J. Goldstine. Substitution and bounded languages. *Journal of Computer and System Sciences*, 6(1):9–29, February 1972.

[Gol76] J. Goldstine. Bounded AFLs. *Journal of Computer and System Sciences*, 12(3):399–419, June 1976.

[GR88] A. Gibbons and W. Rytter. *Efficient Parallel Algorithms*. Cambridge University Press, 1988.

[HU79] J. E. Hopcroft and J. D. Ullman. *Introduction to Automata Theory, Languages and Computation*. Addison-Wesley, 1979.

[Iga77] Y. Igarashi. General properties of derivational complexity. *Acta Inf.*, 8(3):267–283, 1977.

[Kas65] T. Kasami. An efficient recognition and syntax algorithm for context-free languages. Scientific Report AFCRL-65-758, Air Force Cambridge Research Laboratory, Bedford MA, 1965.

[KPU79] W. Kuich, H. Prodinger, and F. J. Urbanek. On the height of derivation trees. In H. A. Maurer, editor, *Automata, Languages and Programming, 6th Colloquium*, volume 71 of *Lecture Notes in Computer Science*, pages 370–384, Graz, Austria, 16–20 July 1979. Springer-Verlag.

Generalized Synchronization Languages

Isabelle Ryl, Yves Roos, and Mireille Clerbout

C.N.R.S. U.R.A. 369, L.I.F.L. Université de Lille I, Bât. M3, Cité Scientifique
59655 Villeneuve d'Ascq Cedex, FRANCE

Abstract. Generalized synchronization languages are a model used to describe the behaviors of distributed applications whose synchronization constraints are expressed by generalized synchronization expressions — an extension of synchronization expressions. Generalized synchronization languages were conjectured by Salomaa and Yu to be characterized by a semi-commutation. We show that this semi-commutation characterizes the images of generalized synchronization languages by a morphism-like class of rational functions.

Topics. Automata and formal languages, theory of parallel and distributed computation.

1 Introduction

Generalized synchronization languages, introduced in [10], are regular languages which correspond to generalized synchronization expressions, an extension of the synchronization expressions introduced by Govindarajan, Guo, Yu and Wang in [6] within the framework of the *ParC* project. These expressions allow a programmer to express minimal synchronization constraints of a program in a distributed context. A synchronization language can be seen as the set of correct executions of a distributed application where each action is split in two atomic actions, its start and its termination. In this sense, synchronization languages take place in interleaving semantics (see [11] for a comparison between interleaving semantics and non-interleaving semantics) with split of actions [12].

Salomaa and Yu conjectured in [10] that the family of generalized synchronization languages coincides with the family of regular st-languages, closed under a particular semi-commutation function named θ. Such a characterization may improve the space efficiency of the implementation of synchronization expressions. Moreover, as it is possible to decide whether a regular language is closed under a semi-commutation, the use of θ would give a decidable characterization of generalized synchronization languages. At last, we have to point out that generalized synchronization languages belong to a family for which the closure under θ is computable using Mtivier's semi-algorithm introduced in [4].

Salomaa and Yu have shown in [10] that their conjecture holds for finite languages but it seems to be very difficult to extend this result to the general case and, in fact, we think that this conjecture does not hold. Nevertheless, we show that the semi-commutation given by Salomaa and Yu allows to characterize the family of the images of generalized synchronization languages by a morphism-like class of rational functions — called st-morphisms. This result establishes a strong link between generalized synchronization languages and languages closed under θ. Moreover, it reduces the conjecture of Salomaa and Yu to the question: "are the generalized synchronization languages closed under st-morphisms ?"

2 Preliminaries

In the following, we shall denote by alph(u) the alphabet of a word u and by Π_Y the *projection* onto the sub-alphabet Y, i.e. the morphism defined by: for each letter x, if $x \in Y$ then $\Pi_Y(x) = x$, else $\Pi_Y(x) = \varepsilon$, where ε denotes the empty word.

The *shuffle* of two words u and v belonging to Σ^* is

$$u \sqcup\!\sqcup\, v = \{u_1v_1u_2v_2...u_nv_n \mid u_i \in \Sigma^*, v_i \in \Sigma^*, u = u_1u_2...u_n, v = v_1v_2...v_n\}$$

We denote by $\oplus$ the exclusive "or".

We just recall the definition of semi-commutations and partial commutations:

Definition 1. *A semi-commutation relation defined over an alphabet Σ is an irreflexive relation included in $\Sigma \times \Sigma$. With such a relation θ, we associate a rewriting system, that we will denote by S_θ, defined by: $S_\theta = \{xy \longrightarrow yx \mid (x,y) \in \theta\}$. When the semi-commutation is symmetrical, it is named partial commutation and the associated rewriting system is defined by $S_\theta = \{xy \longleftrightarrow yx \mid (x,y) \in \theta\}$ (for more details, see [3]).*

For a semi-commutation or a partial commutation θ defined over an alphabet Σ, we denote by $\mathrm{f}_\theta(u)$ the set of words which can be obtained applying rules of S_θ to the word u of Σ^* and we say that $\mathrm{f}_\theta(u)$ is the closure under θ of u. We extend this definition to languages: $\forall L \subseteq \Sigma^*, \mathrm{f}_\theta(L) = \bigcup_{u \in L} \mathrm{f}_\theta(u)$.

3 Generalized Synchronization Languages

Generalized synchronization expressions allow a programmer to express the synchronization constraints his distributed application has to respect. The statements are tagged and, during the execution, a statement can be executed immediately if it satisfies the constraints described by the expression, if it does not, the execution is delayed.

A generalized synchronization expression may be:

- a statement tag or ε for no action,
- if e_1 and e_2 are synchronization expressions:
  - $(e_1 \rightarrow e_2)$ which imposes that the execution of e_2 starts only after the end of the execution of e_1,
  - $(e_1 \mid e_2)$ which specifies that either e_1 or e_2 can be executed but not both,

- $(e_1 \parallel e_2)$ which allows the executions of e_1 and e_2 to overlap. Autoconcurrence is not allowed, so expressions like $a \parallel a$ means $a \to a$,
- $(e_1 \& e_2)$ which imposes that the execution satisfies both expressions e_1 and e_2,
- (e_1^*) which allows the execution of e_1 to be repeated an arbitrary number of times.

With each generalized synchronization expression, we associate a language whose words represent the possible executions with respect to the expression. Hence, the words of the languages are traces in which each action is represented by two instantaneous events, its start and its termination. So, from an expression e over Σ, we construct $L(e) \subseteq (\Sigma_s \cup \Sigma_t)^*$ which is an st-language:

Definition 2. *Let Σ be a finite alphabet. The alphabets Σ_s and Σ_t are defined by the relation:*

$$(a \in \Sigma) \Leftrightarrow (a_s \in \Sigma_s) \Leftrightarrow (a_t \in \Sigma_t).$$

A word $u \in (\Sigma_s \cup \Sigma_t)^*$ is an st-word if and only if for each $x \in \Sigma$, $\Pi_{\{x_s, x_t\}}(u) \in (x_s x_t)^*$. We extend this definition in a canonical way to languages. We denote by ST_Σ the set of st-words over the alphabet $\Sigma_s \cup \Sigma_t$. An st-primitive word is a non-empty word which has no proper left st-factor. The st-primitive factors of an st-word u are the st-primitive words $u_1, \ldots, u_n$ such that $u = u_1 \ldots u_n$. An st-word u is a sequence *if u belongs to $(\bigcup_{x\in\Sigma} x_s x_t)^*$.*

Definition 3. *Let Σ be an alphabet of actions (or tags). The generalized synchronization language $L(e) \subseteq (\Sigma_s \cup \Sigma_t)^*$ associated with an expression e over Σ is inductively defined by:*

- $L(\varepsilon) = \varepsilon$,
- *for each action a, $L(a) = a_s a_t$,*
- *if $e = e_1 \to e_2$ then $L(e) = L(e_1).L(e_2)$,*
- *if $e = e_1 \mid e_2$ then $L(e) = L(e_1) \cup L(e_2)$,*
- *if $e = e_1 \& e_2$ then $L(e) = L(e_1) \cap L(e_2)$,*
- *if $e = e_1 \parallel e_2$ then $L(e) = (L(e_1) \sqcup\!\sqcup L(e_2)) \cap \mathrm{ST}_\Sigma$,*
- *if $e = e_1^*$ then $L(e) = (L(e_1))^*$.*

We denote by $\mathrm{SL_G}$ the family of generalized synchronization languages.

By construction, generalized synchronization languages are clearly regular languages. Moreover, as we use an intersection after computing the shuffle product, generalized synchronization languages are st-languages. Note also that any regular set of sequences is clearly a generalized synchronization language associated with a generalized synchronization expression using regular operations $\to$, $\mid$ and $\star$.

Salomaa and Yu [10] have defined the semi-commutation θ in order to characterize generalized synchronization languages and they have shown that the closure under θ_Σ of any regular st-language over Σ is regular.

Definition 4. *Let* Σ *be an alphabet of actions. The semi-commutation* θ_Σ *is defined by:*

$$\theta_\Sigma = \bigcup_{x \neq y} \{(x_s, y_s), (x_t, y_t), (x_s, y_t)\}.$$

We denote by R_θ *the family of regular st-languages closed under* θ.

Example 1. Let $\Sigma = \{a, b\}$ and $u = a_s b_s a_t b_t$. For example, the word $b_s a_s a_t b_t$ belongs to $f_{\theta_\Sigma}(u)$. It is clear that a left factor $a_s b_s$ means that a and b are executed in parallel so, it is allowed to start b before a: we can start with $b_s a_s$. The word $a_s a_t b_s b_t$ also belongs to $f_{\theta_\Sigma}(u)$. Intuitively, we are allowed to use a rule $b_s a_t \longrightarrow a_t b_s$ because it is not forbidden to execute a and b in sequence when it is allowed to execute them concurrently.

Theorem 1 (Salomaa and Yu [10]). *Any generalized synchronization language over* $\Sigma_s \cup \Sigma_t$ *is closed under* θ_Σ.

Conjecture 1 (Salomaa and Yu [10]). An arbitrary regular st-language over $\Sigma_s \cup \Sigma_t$ closed under θ_Σ is a generalized synchronization language.

This conjecture is shown to be true in the case of finite languages in [10] but, in the conclusion of their paper Salomaa and Yu say that the proof of this conjecture may be very difficult because of the fact that, even for a simple language, the associated generalized synchronization expression have little structural resemblance to the language (for example $f_{\theta\{a,b\}}(a_s(b_s a_t a_s b_t)^* a_t)$ is associated with $((a \to b)^* \to a) \parallel (a \parallel b)^*)$.

We believe this conjecture does not hold and we work on the following possible counter-example: $f_{\theta\{a,b\}}(a_s b_s (b_t b_s (a_t a_s)^+ b_t b_s)^* a_t b_t)$. All the same, we will show that the family of the images of generalized synchronization languages by a subclass of rational functions called st-morphisms has the good properties of closure.

Definition 5. *Let* Σ *and* X *be two alphabets of actions. A strictly alphabetical morphism from* Σ^* *into* X^* *is called action morphism. With an action morphism* φ *from* Σ^* *into* X^**, we associate a morphism* $\bar{\varphi}$ *from* $(\Sigma_s \cup \Sigma_t)^*$ *into* $(X_s \cup X_t)^*$ *by* $\forall a \in \Sigma^*, \bar{\varphi}(a_s) = \bar{\varphi}(a)_s$ *and* $\bar{\varphi}(a_t) = \bar{\varphi}(a)_t$. *When there will be no ambiguity, we shall denote by* φ *in the sequel the action morphism* φ *and the corresponding morphism* $\bar{\varphi}$. *With each action morphism* φ *from* Σ^* *into* X^**, we associate a* morphism-like *rational function* $\hat{\varphi}$ *:*

$$\hat{\varphi} = \{(u, \varphi(u)) \mid u \in ST_\Sigma \text{ and } \varphi(u) \in ST_X\}.$$

Note that $\hat{\varphi}$ *is equal to* $(\cap ST_X) \circ \varphi \circ (\cap ST_\Sigma)$. *We name these functions* st-morphisms, *and we denote by* Φ_{st} *the st-morphisms family.*

The aim of this paper is to show the equality $R_\theta = \Phi_{st}(SL_G)$. In order to obtain the inclusion $R_\theta \subseteq \Phi_{st}(SL_G)$, we will define a coloring such that each regular st-language closed under θ is the image by an st-morphism of a regular colored st-language closed under a partial commutation. After that, using properties of partial commutations, asynchronous automata and some results about mixed product, we will show that the colored st-language is the image of a generalized synchronization language by an st-morphism. The proofs of these results are not immediate and use a lot of tools. For lack of space, we will only explain the reasoning and give ideas of the proofs.

4 The Coloring

The aim of this section is to get the first step of our construction: we will show that we can use a coloring and a partial commutation to compute the closure under θ. We will first show the idea of the coloring with an example. We have to keep in mind that, when we consider two st-words u_1 and u_2 defined over an alphabet of actions Σ, we have $\mathrm{f}_{\theta_\Sigma}(u_1.u_2) = \mathrm{f}_{\theta_\Sigma}(u_1).\mathrm{f}_{\theta_\Sigma}(u_2)$ (see [10]). So, let us consider an alphabet of two actions $\Sigma = \{a, b\}$ and a word $u = a_s b_s a_t b_t b_s b_t$. We have:

$$\mathrm{f}_{\theta_\Sigma}(u) = \{a_s b_s a_t b_t, b_s a_s a_t b_t, a_s b_s b_t a_t, b_s a_s b_t a_t, a_s a_t b_s b_t, b_s b_t a_s a_t\}.\{b_s b_t\}.$$

We will use "colors" (here some subscripts) in order to mark some dependent actions. First, as the letters of two consecutive st-primitive factors can never be mixed, we color alternatively the st-primitive factors with two disjoint sets of colors and we forbid actions whose colors do not belong to the same set to commute. Second, in an st-primitive factor, some actions are in sequence so, we will bring out some sequences with colors and actions which have the same color will not be allowed to commute. For example, the word $v = a_{1s} b_{2s} a_{1t} b_{2t} b_{5s} b_{5t}$ is a coloring of u which satisfies these conditions. So now, instead of the semi-commutation, we can use the partial commutation:

$$\varrho = \{(a_{1s}, b_{2s}), (a_{1s}, b_{2t}), (a_{1t}, b_{2s}), (a_{1t}, b_{2t})\},$$

and we obtain:

$$\begin{aligned}\mathrm{f}_\varrho(v) = \{\, & a_{1s} b_{2s} a_{1t} b_{2t}, b_{2s} a_{1s} a_{1t} b_{2t}, a_{1s} b_{2s} b_{2t} a_{1t}, \\ & b_{2s} a_{1s} b_{2t} a_{1t}, a_{1s} a_{1t} b_{2s} b_{2t}, b_{2s} b_{2t} a_{1s} a_{1t}\}.\{b_{5s} b_{5t}\}.\end{aligned}$$

With a partial commutation, we loose the non-symmetry. The use of the coloring makes up for this lost of power because the coloring contains some informations about dependence of actions. If the morphism φ removes colors, it is easy to see that $\mathrm{f}_{\theta_\Sigma}(u) = \varphi(\mathrm{f}_\varrho(v))$.

This simple idea can be extended to the general case. To color a word u, it suffices to add to each of its letters a set of colors. The set associated with an occurrence of letter is the union of the sets it would have received in the coloring of each projection over an alphabet of two actions of the word u.

Definition 6. *Let Σ be an alphabet of actions. With each pair of distinct actions a and b of Σ, we associate two disjoint sets of three colors denoted by $\mathrm{C}_{\{a,b\}}$ and*

$\bar{\mathrm{C}}_{\{a,b\}}$ *such that* $\mathrm{C}_{\{a,b\}}$, $\bar{\mathrm{C}}_{\{a,b\}}$, $\mathrm{C}_{\{c,d\}}$ *and* $\bar{\mathrm{C}}_{\{c,d\}}$ *are pairwise disjoint when* $\{a,b\} \neq \{c,d\}$. *Let* Σ_c *be the colored alphabet corresponding with* Σ:

$$\Sigma_c = \{(x,E) \mid x \in \Sigma, E \subseteq \bigcup_{a \neq x} (\mathrm{C}_{\{a,x\}} \cup \bar{\mathrm{C}}_{\{a,x\}}),$$

$$\forall a \neq x, E \cap \mathrm{C}_{\{a,x\}} = \emptyset \oplus E \cap \bar{\mathrm{C}}_{\{a,x\}} = \emptyset\}.$$

Notation 2 *In order to simplify the use of letters of* $\Sigma_{cs} \cup \Sigma_{ct}$, *we will write* (a_s, E) *instead of* $(a,E)_s$ *and, for each action* a *of* Σ, *we denote*

$$\alpha(a_s) = \alpha(a_t) = \alpha((a_s,E)) = \alpha((a_t,E)) = a.$$

To come back to the initial alphabet, we have to "remove" colors, this can be done using a strictly alphabetical morphism.

Definition 7. *Let* Σ *be an alphabet of actions. The strictly alphabetical morphism* φ_Σ *which removes colors is defined by:*

$$\begin{aligned} \varphi_\Sigma \; : (\Sigma_{cs} \cup \Sigma_{ct})^* &\longrightarrow (\Sigma_s \cup \Sigma_t)^* \\ (x,E) &\longmapsto x. \end{aligned}$$

Now, we are able to define the coloring that we will use until the end of this section. As we have already said, this coloring is the merging of the colorings of projections over sub-alphabets of two actions so, we will first define the coloring of st-primitive words over sub-alphabets of two actions.

Definition 8. *Let* $\Sigma = \{a,b\}$ *be an alphabet of actions. The language* P_{ab}, *given by Figure 1 contains all the st-primitive words over* Σ *colored with the set* $\mathrm{C}_{\{a,b\}}$. *The language* $\bar{\mathrm{P}}_{\{a,b\}}$ *is obtained in the same way using* $\bar{\mathrm{C}}_{\{a,b\}}$.

Let us remark some essential properties of this coloring. Each of them can be easily verified on the automaton. These properties are given for P_{ab} but they also hold for $\bar{\mathrm{P}}_{ab}$:

Property 1.

a - The language P_{ab} is an st-language and $\varphi_{\{a,b\}}(\mathrm{P}_{ab})$ is the set of st-primitive words over $\{a,b\}$.

b - For a given color, the projection of P_{ab} onto the sub-alphabet containing all the letters which possess this color is a sequence.

c - Let (a_t, E) and (b_s, F) be two letters of alph(P_{ab}). If u is an st-word of P_{ab} in the form $u = u_1(a_t,E)u_2(b_s,F)u_3$ with $E \cap F = \emptyset$ then, u_2 contains a sub-word $(a_s,G)(a_t,G)$ with $G \cap F \neq \emptyset$ or a sub-word $(b_s,G)(b_t,G)$ with $G \cap E \neq \emptyset$.

With the two languages we have presented, we can define the rational function we use to color the words of a language according to the colorings of their projections over sub-alphabets of two actions.

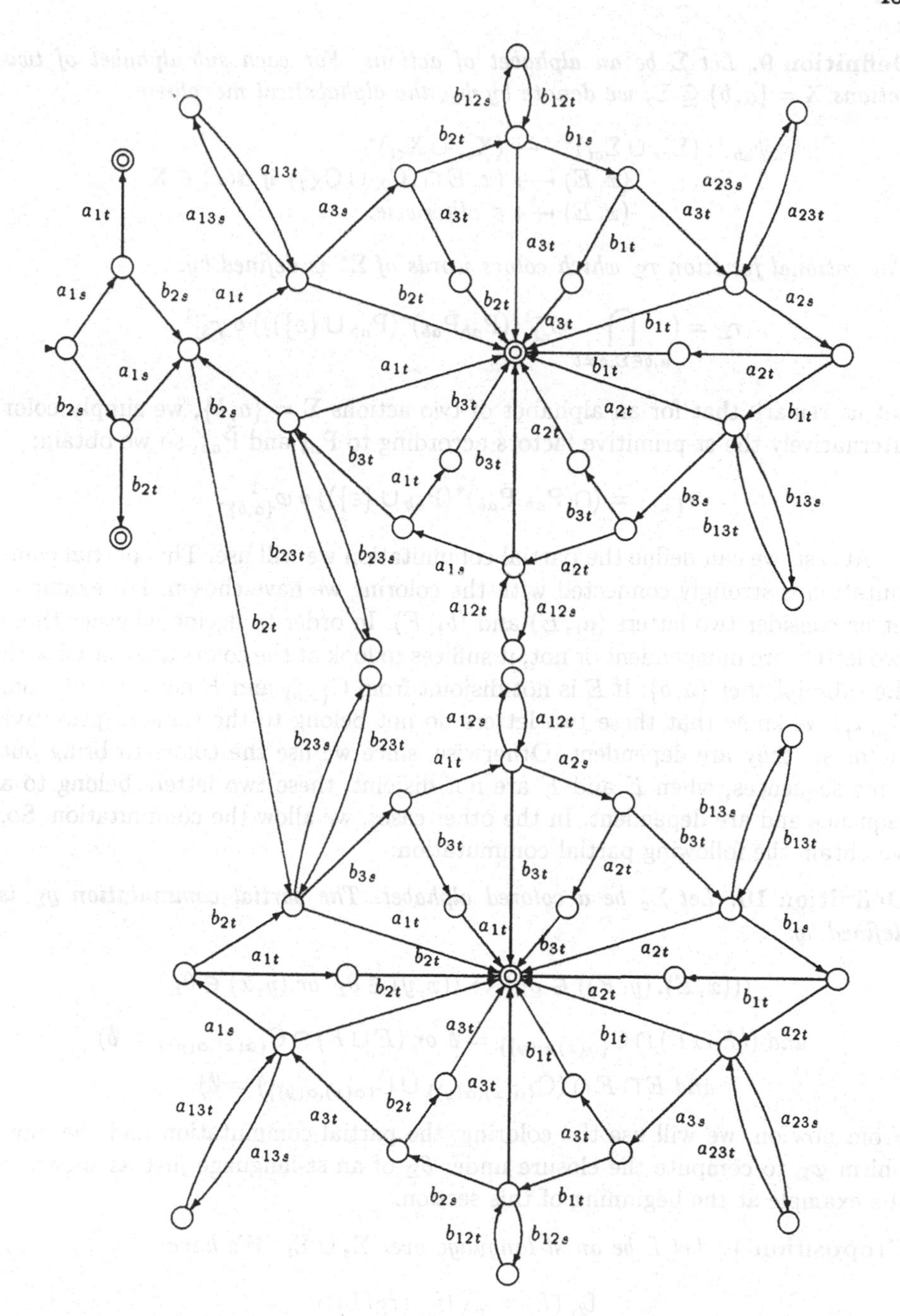

The subscripts 1, 2 and 3 represent the colors of the set $C_{\{a,b\}}$. The language recognized by this automaton is called P_{ab}. It suffices to replace the colors 1, 2 and 3 by respectively by 4, 5 and 6 (representing the colors of the set $\bar{C}_{\{a,b\}}$) to obtain $\bar{P}_{ab}$.

Fig. 1. The language P_{ab}

Definition 9. *Let Σ be an alphabet of actions. For each sub-alphabet of two actions $\mathrm{X} = \{a,b\} \subseteq \Sigma$, we denote by ψ_{ab} the alphabetical morphism:*

$$\begin{aligned}\psi_{ab}\ :(\Sigma_{cs}\cup\Sigma_{ct})^* &\longrightarrow (\mathrm{X}_{cs}\cup\mathrm{X}_{ct})^*\\ (x,E) &\longmapsto (x, E\cap(\mathrm{C}_\mathrm{X}\cup\bar{\mathrm{C}}_\mathrm{X}))\ \textit{if}\ \alpha(x)\in\mathrm{X}\\ (x,E) &\longmapsto \varepsilon\ \textit{otherwise}.\end{aligned}$$

The rational function τ_Σ which colors words of Σ^* is defined by:

$$\tau_\Sigma = \Big(\bigcap_{a,b\in\Sigma,a\neq b} \psi_{ab}^{-1}((\mathrm{P}_{ab}\bar{\mathrm{P}}_{ab})^*(\mathrm{P}_{ab}\cup\{\varepsilon\}))\Big)\circ\varphi_\Sigma^{-1}.$$

Let us remark that for an alphabet of two actions $\Sigma = \{a,b\}$, we simply color alternatively the st-primitive factors according to P_{ab} and $\bar{\mathrm{P}}_{ab}$, so we obtain:

$$\tau_{\{a,b\}} = (\cap(\mathrm{P}_{ab}.\bar{\mathrm{P}}_{ab})^*(\mathrm{P}_{ab}\cup\{\varepsilon\}))\circ\varphi_{\{a,b\}}^{-1}.$$

At last, we can define the partial commutation we will use. This partial commutation is strongly connected with the coloring we have chosen. For example, let us consider two letters (a_t, E) and (b_s, F). In order to decide whether these two letters are independent or not, it suffices to look at the colors associated with the sub-alphabet $\{a,b\}$. If E is not disjoint from $\mathrm{C}_{\{a,b\}}$ and F not disjoint from $\bar{\mathrm{C}}_{\{a,b\}}$, we know that these two letters do not belong to the same st-primitive factor so, they are dependent. Otherwise, since we use the colors to bring out some sequences, when E and F are not disjoint, these two letters belong to a sequence and are dependent. In the other cases, we allow the commutation. So, we obtain the following partial commutation:

Definition 10. *Let Σ_c be a colored alphabet. The partial commutation ϱ_Σ is defined by:*

$$(((x,E),(y,F))\in\varrho_\Sigma) \Leftrightarrow ((x,y)\in\theta_\Sigma\ \textit{or}\ (y,x)\in\theta_\Sigma)$$

$$\textit{and}\ ((E\cup F)\cap\mathrm{C}_{\{\alpha(x),\alpha(y)\}}=\emptyset\ \textit{or}\ (E\cup F)\cap\bar{\mathrm{C}}_{\{\alpha(x),\alpha(y)\}}=\emptyset)$$

$$\textit{and}\ E\cap F\cap(\mathrm{C}_{\{\alpha(x),\alpha(y)\}}\cup\bar{\mathrm{C}}_{\{\alpha(x),\alpha(y)\}})=\emptyset).$$

From now on, we will use the coloring, the partial commutation and the morphism φ_Σ to compute the closure under θ_Σ of an st-language just as shown in the example at the beginning of this section.

Proposition 1. *Let L be an st-language over $\Sigma_s\cup\Sigma_t$. We have:*

$$\mathrm{f}_{\theta_\Sigma}(L) = \varphi_\Sigma(\mathrm{f}_{\varrho_\Sigma}(\tau_\Sigma(L))).$$

The proof of this proposition contains two steps. First, we show the result over an alphabet of two actions and, more precisely, we can restrict us to st-primitive words. In this case, we use an induction on the length of the derivation and the properties of τ_Σ. Second, we use the Projection Lemma for partial commutation [3] and the notion of image of a semi-commutation by an alphabetical substitution [8].

5 Relation between R_θ and $\Phi_{\mathrm{st}}(\mathrm{SL_G})$

The second step of the construction consists in showing that, for a regular st-language L, the language $\mathrm{f}_{\varrho\Sigma}(\tau_\Sigma(L))$ belongs to $\Phi_{\mathrm{st}}(\mathrm{SL_G})$. In order to do this, we will use asynchronous automata since, fortunately, we only obtain regular languages:

Lemma 1. *Let L be a regular st-language. The language $\mathrm{f}_\varrho(\tau(L))$ is a regular language.*

The proof of this lemma is a direct consequence of Métivier's Theorem [7] and of the properties of the coloring.

As we have regular languages, we will be able to use asynchronous automata [13] and mixed product. The definition of mixed product we will recall is the one given by Duboc in [5], this definition is a little bit different from the first one introduced by De Simone [2].

Definition 11 (Duboc [5]). *Let X and Σ be two alphabets and L and M be two languages respectively over X^* and Σ^*. The mixed product of L and M in X and Σ is the set $L \sqcap M$ defined by:*

$$L \sqcap M = \{u \in (\mathrm{X} \cup \Sigma)^* \mid \Pi_{\mathrm{X}}(u) \in L \text{ and } \Pi_\Sigma(u) \in M\}.$$

Clearly, the mixed product is associative. We can easily show the following lemma using intersection and shuffle product to compute mixed product:

Lemma 2. *The mixed product in $\Sigma_{1s} \cup \Sigma_{1t}, \ldots, \Sigma_{ns} \cup \Sigma_{nt}$ of n generalized synchronization languages is a generalized synchronization language.*

Our main result is based on the link between asynchronous automata and mixed product established by Duboc. We first recall the definition of this class of automata. Each state of a deterministic asynchronous automaton (DAA), is a vector of local components. Each letter can only read and modify a subset of the components which are its associated components.

Definition 12 (Zielonka [13]). *A DAA – deterministic asynchronous automaton – $\mathcal{A}$ over the alphabets $(\Sigma_1, \ldots, \Sigma_n)$ is a tuple*

$$\mathcal{A} = ((\Sigma_1, \ldots, \Sigma_n), \bigotimes_{1 \le i \le n} Q_i, \delta, q_I, F)$$

such that:

- *$(\Sigma_1, \ldots, \Sigma_n)$ is a n-tuple of alphabets, $\Sigma = \bigcup_{1 \le i \le n} \Sigma_i$ is the alphabet of $\mathcal{A}$,*
- *the global states set $\bigotimes_{1 \le i \le n} Q_i$ is a direct product of local states sets Q_i for $1 \le i \le n$,*
- *q_I is the (global) initial state and F the set of (global) final states;*
- *the domain of a letter $a \in \Sigma$ is $I_a = \{1 \le i \le n \mid a \in \Sigma_i\}$,*

– *the transition function δ is defined by the partial functions. For each letter a, the partial function δ_a is a set of transitions labelled by a:*

$$\delta_a \subseteq (\bigotimes_{i \in I_a} Q_i) \times a \times (\bigotimes_{i \in I_a} Q_i),$$

and for each $q \in \bigotimes_{i \in I_a} Q_i$, there exists at most one $q' \in \bigotimes_{i \in I_a} Q_i$ such that $(q, a, q') \in \delta_a$. The function δ is defined by: $((q_1, \ldots, q_n), a, (q'_1, \ldots, q'_n)) \in \delta$ if and only if the partial transition $(((q_i)_{i \in I_a}), a, ((q'_i)_{i \in I_a}))$ belongs to δ_a and for each $i \notin I_a$, $q_i = q'_i$.

Two letters which are associated with disjoint subsets of components are independent. So, each DAA leads to a partial commutation and, by construction, a DAA recognizes a language closed under this partial commutation. Now, we will briefly explain our construction.

Proposition 2. *Let L be an st-language over the alphabet of actions Σ. Let $\Sigma_1, \ldots, \Sigma_n$ included in Σ be such that for each i, $\Pi_{\Sigma_{is} \cup \Sigma_{it}}(L)$ is a set of sequences. If L is recognized by a DAA over $(\Sigma_{1s} \cup \Sigma_{1t}), \ldots, (\Sigma_{ns} \cup \Sigma_{nt})$ then L is the image by an st-morphism of a generalized synchronization language.*

Sketch of proof 1 *The proof of this proposition is based on a result due to Duboc [5]: each language recognized by a DAA is the image by a strictly alphabetical morphism of a language recognized by a deterministic loosly cooperating automaton (DLCA). A DLCA is a DAA with a stronger condition: the function δ is defined by the local transition functions $\delta_i \subseteq Q_i \times \Sigma_i \times Q_i$, for each index i. A transition $((q_1, \ldots, q_n), x, (q'_1, \ldots, q'_n))$ belongs to δ if and only if for each $i \notin I_x$, $q_i = q'_i$ and if for each $i \in I_x$, $(q_i, x, q'_i) \in \delta_i$.*

Duboc gives a method to build a DLCA corresponding to a given DAA, we will use the same idea but we cannot use the same construction because it does not suit to st-languages.

Let us denote by $\mathcal{A} = ((\Sigma_1, \ldots, \Sigma_n), \bigotimes_{1 \le i \le n} Q_i, \delta, q_I, F)$, a DAA recognizing L whose states are both accessible and coaccessible. Since L is an st-language and since the languages defined by the projections over the components are sequences, we know that between an "a_s" and the corresponding "a_t", no letter can have an effect on the state components seen by a. So, for each letter a of Σ, for each transition of δ_{a_t}, we have such a scheme:

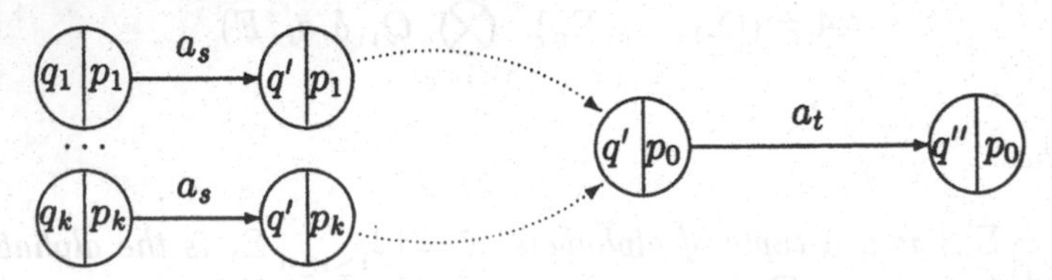

Now, we will convert $\mathcal{A}$ into a DLCA $\mathcal{A}'$ recognizing an st-language whose image by an st-morphism is L. We will rename the transitions in order that each label appears only on one transition of the partial transition function (this implies

that we have a DLCA) but also in order to get an st-language. For each transition labelled by a termination of action (in the previous scheme a_t), we name the action according to the states it depends on (here we obtain new actions $a_{(q_1,q',q'')}, \ldots, a_{(q_k,q',q'')}$) and we introduce new local states in the domain of this action (here $n_1, \ldots, n_k$). We obtain:

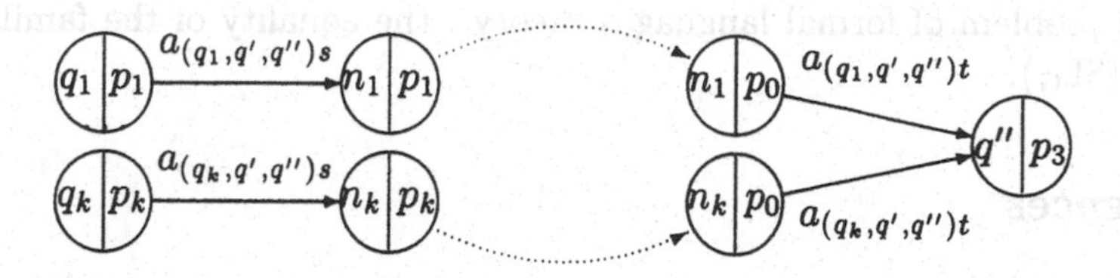

The automaton $\mathcal{A}'$ is a DLCA over new alphabets $(X_1, \ldots, X_n)$. Moreover, since the first partial state q_1 and the last partial state q'' are not changed, it is easy to see that L is the image by an action morphism of the language M recognized by $\mathcal{A}'$ and, since L and M are st-languages, L is the image of M by an st-morphism. According to Duboc [5], the language M, recognized by $\mathcal{A}'$, is the union of mixed products of regular languages, each of them being included in the projection of M onto a sub-alphabet X_i. By definition, for each i, $\Pi_{\Sigma_{is} \cup \Sigma_{it}}(L)$ is a set of sequences and the property is preserved by the construction: the projections of M over the alphabets $X_{1s} \cup X_{1t}, \ldots, X_{ns} \cup X_{nt}$ are sequences so they belong to SL_G. Since SL_G is closed under union and mixed product (Lemma 2), M belongs to SL_G.

Proposition 3. *The family of regular st-languages closed under θ is equal to the family of the images by st-morphisms of all the generalized synchronization languages : $R_\theta = \Phi_{st}(SL_G)$.*

Proof. The inclusion of $\Phi_{st}(SL_G)$ in R_θ has been shown in [9]. Let L be a regular st-language over $\Sigma_s \cup \Sigma_t$ closed under θ_Σ. According to Proposition 1, we have the equality $L = \varphi_\Sigma(f_{\varrho_\Sigma}(\tau_\Sigma(L)))$. Furthermore, by construction, $\varphi_\Sigma(f_{\varrho_\Sigma}(\tau_\Sigma(L)))$ is equal to $\hat{\varphi}_\Sigma(f_{\varrho_\Sigma}(\tau_\Sigma(L)))$. From Lemma 1, $f_{\varrho_\Sigma}(\tau_\Sigma(L))$ is regular, according to Zielonka's theorem [13], the language $f_{\varrho_\Sigma}(\tau_\Sigma(L))$ is recognized by a DAA on maximal cliques and according to [1], this DAA can be converted in a DAA on a covering by cliques that we choose. The coloring is built such that if we consider the cliques $(\Sigma_{1s} \cup \Sigma_{1t}), \ldots, (\Sigma_{ns} \cup \Sigma_{nt})$ defined by:

$$\forall 1 \leq i \leq n, \Sigma_i = \{(x, E), (y, F) \mid \alpha(x) = \alpha(y) \text{ or }$$
$$(E \cap C_{\{x,y\}} = \emptyset \oplus F \cap C_{\{x,y\}} = \emptyset) \text{ or } E \cap F \cap (C_{\{x,y\}} \cup \bar{C}_{\{x,y\}}) \neq \emptyset\},$$

we obtain a covering by cliques of the alphabet of $\tau_\Sigma(L)$ such that the projection of $f_{\varrho_\Sigma}(\tau_\Sigma(L))$ over each clique is a sequence. According to Proposition 2, $f_{\varrho_\Sigma}(\tau_\Sigma(L))$ is the image by an st-morphism of a generalized synchronization language. Since the composition of two st-morphisms is an st-morphism [9], L belongs to $\Phi_{st}(SL_G)$.

6 Conclusion

The family $\Phi_{st}(\mathrm{SL_G}) = \mathrm{R}_\theta$ has the properties of closure that we wanted and it seems reasonable, in a practical point of view to accept the renaming of actions. So, this family seems to be the good one.

The conjecture of Salomaa and Yu is still open but it is now reduced to a classical problem of formal languages theory : the equality of the families $\mathrm{SL_G}$ and $\Phi_{st}(\mathrm{SL_G})$.

References

1. Cori, R., Sopena, E., Latteux, M., and Roos, Y. 2-asynchronous automata. *Theoretical Computer Science 61* (1988), 93–102.
2. De Simone, R. Langages infinitaires et produit de mixage. *Theoretical Computer Science 31* (1984), 83–100.
3. Diekert, V., and Rozenberg, G., Eds. *The Book of Traces*. World Scientific, Singapore, 1995.
4. Diekert, V., Y. Métivier, Partial Commutation and Traces. *Handbook of Formal Languages* (1997), 457–533.
5. Duboc, C. *Commutations dans les Monoïdes libres : un Cadre Théorique pour l'Étude du Parallélisme.* PhD thesis, Université de Rouen, 1986.
6. Govindarajan, R., Guo, L., Yu, S., and Wang, P. ParC project: Practical constructs for parallel programming languages. In *Proc. IEEE 15th Annual Internationnal Computer Software & Applications Conference* (1991), pp. 183–189.
7. Métivier, Y. Contribution à l'étude des monoïdes de commutations. Thèse d'état, Université de Bordeaux I, 1987.
8. Roos, Y. *Contribution à l'Étude des Fonctions de Commutation Partielle.* PhD thesis, Université des Sciences et Technologies de Lille, Lille, France, 1989.
9. Ryl, I., Roos, Y., and Clerbout, M. About synchronization languages. In *Proc. MFCS'98* (Brno, Czech Republic, 1998), *LNCS* 1450, Springer-Verlag, Berlin, pp. 533–542.
10. Salomaa, K., and Yu, S. Synchronization expressions with extended join operation. *Theoretical Computer Science 207* (1998), 73–88.
11. Sassone, V., Nielsen, M., and Winskel, G. Models for concurrency: Towards a classification. *Theoretical Computer Science 170*, 1-2 (1996), 297–348.
12. van Glabbeck, R. J., and Vaandrager, F. The difference between splitting in n and $n+1$. *Information and Computation 136*, 2 (1997), 109–142.
13. Zielonka, W. Notes on finite asynchronous automata. *R.A.I.R.O. — Theoretical Informatics and Applications 21*, 2 (1987), 99–135.

A Generalization of Dijkstra's Calculus to Typed Program Specifications

Klaus-Dieter Schewe[1], Bernhard Thalheim[2]

[1] Technical University of Clausthal, Dept. of Computer Science
Julius-Albert-Str. 4, D-38678 Clausthal-Zellerfeld, Germany
schewe@informatik.tu-clausthal.de

[2] Brandenburgian Technical University at Cottbus, Dept. of Computer Science
Karl-Marx-Str. 17, D-03044 Cottbus, Germany
thalheim@informatik.tu-cottbus.de

Abstract. Dijkstra's predicate transformer calculus in its extended form gives an axiomatic semantics to program specifications including partiality and recursion. However, even the classical theory is based on infinitary first order logic which is needed to guarantee the existence of predicate transformers for weakest (liberal) preconditions. This theory can be generalized to higher-order intuitionistic logic.

Such logics can be interpreted in topoi. Then each topos E canonically corresponds to a definitionally complete theory T such that E is equivalent to the topos $\mathbb{E}(T)$ of definable types over T. Furthermore, each model of T in an arbitrary topos F canonically corresponds to a logical morphism $\mathbb{E}(T) \rightarrow F$.

This correspondence enables the definition of a type specification discipline with a semantics based on topoi such that the predicate transformers in the associated logic give an axiomatic semantics for typed program specifications.

1 Motivation

The semantics of programs and program specifications can be defined axiomatically in Dijkstra's calculus [6, 16]. Given a program (specification) S two predicate transformers $wlp(S)$ and $wp(S)$, i.e. mappings from formulae to formulae of a given logic $\mathcal{L}$, are associated with S with the following informal meaning:

- $wlp(S)(\mathcal{R})$ characterizes those initial states such that all terminating executions of S will reach a final state characterized by $\mathcal{R}$, and
- $wp(S)(\mathcal{R})$ characterizes those initial states such that all executions of S terminate and will reach a final state characterized by $\mathcal{R}$.

It has been shown in [16] that partial programs that are not defined on all initial states and recursive programs are comprised by this kind of semantics definition. Moreover, it has been shown that any pair of predicate transformers satisfying the *pairing condition* and *universal conjunctivity* corresponds to some program specification. Predicate transformers are usefull for proving properties of program specification [5].

Dijkstra's calculus is usually introduced in companion with guarded commands, but its application area is much more general. However, the existence of predicate transformers in $\mathcal{L}$ has to be guaranteed in any case. The classical theory uses infinitary first-order logic[1]. If we assume that the logic is interpreted in a structure St satisfying the domain closure property, then for each state σ, i.e. a variable assignment especially for the finitely many program variables used in the specification S, a *characterizing predicate* $\mathcal{P}_\sigma$ exists, i.e. we have $\models_{(St,\sigma)} \mathcal{P}_\tau$ iff $\tau = \sigma$ holds. From this the existence proof can be derived quite easily.

A disadvantage of the classical theory is that program variables as variables in $\mathcal{L}$ are untyped. The simplest generalization would be to use a many-sorted infinitary logic, but in this case a type would be just a sort in $\mathcal{L}$ interpreted by a set. This is not in accordance with any established approach to type theory in computer science [7, 9, 15, 23]. In particular, it can be hardly combined with approaches to type theory based on λ-calculi [4, 15, 17].

Fortunately, $\mathcal{L}^\omega_{\omega\infty}$ is not the only logic that assures the existence of predicate transformers as a basis for axiomatic semantics. An alternative is higher-order intuitionistic logic [8] or infinitary coherent logic [12]. The existence proof for predicate transformers in these logics is analogous to the classical case.

The main advantage is the close connection of these logics to the theory of topoi [2, 3, 8, 10, 11, 14]. Here, we consider the higher-order logic of Fourman and Scott [8, 22]. Each theory T of such a logic defines a topos $\mathbb{E}(T)$ of definable types. Conversely, each topos E defines a higher-order language $\mathcal{L}(E)$ and a canonical definitionally complete theory $T(E)$ with $E \cong \mathbb{E}(T(E))$. In particular, when given a topos E, the semantics of typed program specifications can be defined by predicate transformers in $\mathcal{L}(E)$. These results will be briefly presented in the next sections.

The theory can be applied to formal specifications with types. In this case proof obligations for static and dynamic consistency and the theory of consistency enforcement [21] can be generalized [18]. Another application considers the extension of object oriented database theory [19] to the case of dynamics in the data, in which case the polymorphic λ-calculus is chosen as the underlying type system [20].

Throughout the text, we assume some familiarity with basic notions of category theory [1, 13]. Furthermore, topos theory [2, 11, 14] and its connection to logic [3, 8, 10] must also be presupposed.

2 The Classical Calculus

This section gives a brief review of Dijkstra's classical calculus [6, 16]. Assume that S is a program specification and that X is the finite set of variables occurring

[1] It is also possible to use first order logic for arithmetic, but this alternative will not be considered here

in S. We usually call X a *state space*. If D is a set of values, then a *state* is simply a variable assignment $\sigma : X \to D$. Let Σ be the set of all such states.

Then the meaning of S can be given by a subset $\Delta(S) \subseteq \Sigma \times \Sigma \cup \{\infty\}$, where $(\sigma, \tau) \in \Delta(S)$ means that starting S in the initial state σ, may lead to the final state τ and ∞ represents non-termination. This description does not depend on the style of the specification S. Of course, this trivial semantics description comprises non-determinism and partiality.

Now consider an infinitary logic $\mathcal{L}$ with an equality predicate $=$. Regard formulae $\mathcal{R}$ with free variables in X. These are called X-predicates. Let $\mathcal{F}(X)$ be the set of all X-predicates. Let $St = (D, \omega)$ be a fixed structure for the interpretation of $\mathcal{L}$ with semantic domain D and assume that St satisfies the domain closure property, i.e. for each $d \in D$ there is some closed term $t \in \mathcal{T}(\mathcal{L})$ with $\omega(t) = d$. Obviously, a state σ is sufficient to interpret an X-predicate. Write $\models_\sigma \mathcal{R}$ if interpreting $\mathcal{R}$ in state σ yields *true*. Now define two mappings $wlp(S)$ and $wp(S)$ on equivalence classes of X-predicates.

$$\models_\sigma wlp(S)(\mathcal{R}) \text{ iff } (\sigma, \tau) \in \Delta(S) \wedge \tau \neq \infty \Rightarrow \models_\tau \mathcal{R} \quad \text{and} \tag{1}$$

$$\models_\sigma wp(S)(\mathcal{R}) \text{ iff } (\sigma, \tau) \in \Delta(S) \Rightarrow \tau \neq \infty \wedge \models_\tau \mathcal{R}\,. \tag{2}$$

we call $w(l)p(S)(\mathcal{R})$ the *weakest (liberal) precondition* of S for the postcondition $\mathcal{R}$. Note that this definition precisely formalizes the informal meaning of $wlp(S)$ and $wp(S)$. Moreover, the predicate transformers are uniquely determined by $\Delta(S)$ up to equivalence.

Theorem 1. *For a given program specification S the predicate transformers $wlp(S)$ and $wp(S)$ exist. Moreover, they satisfy*

$$wp(S)(\mathcal{R}) \Leftrightarrow wlp(S)(\mathcal{R}) \wedge wp(S)(true) \qquad \textit{(pairing condition)} \tag{3}$$

and

$$wlp(S)(\bigwedge_{i \in I} \mathcal{R}_i) \Leftrightarrow \bigwedge_{i \in I} wlp(S)(\mathcal{R}_i) \qquad \textit{(universal conjunctivity)}\,. \tag{4}$$

The following *inversion theorem* shows that universal conjunctivity and the pairing condition already suffice to find a specification S with corresponding predicate transformers. For this recall that the dual f^* of a predicate transformer f is defined as $f^*(\mathcal{R}) = \neg f(\neg \mathcal{R})$.

Theorem 2. *Let flp and fp be predicate transformers satisfying (3) and (4) in place of $wlp(S)$ and $wp(S)$. Then for a program specification S with*

$$\Delta(S) = \{(\sigma, \tau) \mid \models_\sigma flp^*(\mathcal{P}_\tau)\} \cup \{(\sigma, \infty) \mid \models_\sigma fp^*(false)\}$$

$wlp(S)(\mathcal{R}) \Leftrightarrow flp(\mathcal{R})$ and $wp(S)(\mathcal{R}) \Leftrightarrow fp(\mathcal{R})$.

In [16] recursion has been investigated with respect to the order $\sqsubseteq$ defined by $S \sqsubseteq T$ iff $wlp(T)(\mathcal{R}) \Rightarrow wlp(S)(\mathcal{R})$ and $wp(S)(\mathcal{R}) \Rightarrow wp(T)(\mathcal{R})$ hold for all

X-predicates $\mathcal{R}$. Therefore, for monotonic f with respect to $\sqsubseteq$ the program specification $T = \mu S.f(S)$ can be defined as a least fixpoint and $wlp(T)$ (resp. $wp(T)$) is defined by conjunction (disjunction).

It is well known how to define the axiomatic semantics of *guarded commands* using predicate transformers [16].

3 Type Systems, Topoi and Categorical Logic

Since we are interested in axiomatic semantics for typed programs, we start with a brief look at type systems and their semantics. First consider a rather simple type system with set semantics, i.e.

$$t \quad := \quad b \mid x \mid (a_1 : t_1, \ldots, a_n : t_n) \mid \{t\} \quad . \tag{5}$$

Here b denotes some collection of base types, x represents some type variable, and $(\cdot)$ and $\{\cdot\}$ are constructors for record- and set-types.

Next consider a function-type constructor. This leads to the typed λ-calculus with semantics in cartesian closed categories, i.e.

$$t \quad := \quad b \mid x \mid (a_1 : t_1, \ldots, a_n : t_n) \mid t_1 \rightarrow t_2 \mid \quad . \tag{6}$$

Finally, a sophisticated type system is given by the following version of the polymorphic λ-calculus with semantics defined in the effective topos, i.e.

$$t \quad := \quad b \mid x \mid t_1 \times \ldots \times t_n \mid t_1 \rightarrow t_2 \mid \Pi x.t \quad . \tag{7}$$

Π denotes impredicative polymorphic abstraction with x running over all types [15, 17]. As topoi are cartesian-closed and SET is a very simple topos, we may claim that the semantics of any reasonable type system can be defined in some suitable topos.

Definition 1. *A category E is a* topos *iff*

(i) E is finitely complete, i.e. finite products and equalizers always exist,
(ii) E is finitely cocomplete, i.e. finite coproducts and coequalizers always exist,
(iii) E is cartesian closed and
(iv) E has a subobject classifier, *i.e. a* truth object Ω *and a morphism* $true : \mathbb{1} \rightarrow \Omega$, *where $\mathbb{1}$ denotes a terminal object in E, such that pullbacks along true exist and for each monomorphism $f : A \rightarrow B$ in E there is a* classifying morphism $cl(f) : B \rightarrow \Omega$ *making f and $triv : A \rightarrow \mathbb{1}$ the pullback of $cl(f)$ and true.*

Now consider a higher-order intuitionistic logic $\mathcal{L}$ on the basis of Fourman-Scott languages [8, 22]. Recall that such a language consists of

- two sets *Sort* and *Const* of sorts and constants,
- a *power sort map* $[\cdot] : \bigcup_{n \in \mathbb{N}} Sort^n \rightarrow Sort$ written $(A_1, \ldots, A_n) \mapsto [A_1, \ldots, A_n]$,

– a family of countable sets $\{Var_s\}_{s \in Sort}$ indexed by the sorts and
– a map $\# : Const \to Sort$ assigning to each constant its sort.

We also use $Var = \bigcup_{s \in Sort} Var_s$ to refer to the set of all variables. Then for a given variable $x \in Var$ we write $\#x$ to refer to the sort of x. Moreover, we use $\mho = []$ as an abbreviation for the empty power sort which will be regarded as consisting of truth values.

Then the terms $\mathcal{T}_s(\mathcal{L})$ of sort $s \in Sort$ in $\mathcal{L}$ are constructed as the smallest set such that each variable x of sort s, each constant c with $\#c = s$ and $\mathbf{I}x.\varphi$ for each variable x with $\#x = s$ and each formula φ belong to $\mathcal{T}_s(\mathcal{L})$.

Informally $\mathbf{I}x.\varphi$ means "the unique x that satisfies φ". However, such an x may not exist. More generally, the problem is how to handle possibly empty domains. The intuitionistic logic used here deals with this problem by introducing a formal "existence predicate" $\mathbf{E}$, where $\mathbf{E}\tau$ means that τ exists. This is formalized by distinguishing domains $\tilde{A}$ of "possible elements" and to let $\mathbf{E}$ pick out the subdomains of actual elements. Then bound variables will range only over actual elements. Therefore, this version of higher order intuitionistic logic is also called "the logic of partial elements".

The introduction of an existence predicate also influences the *equality* predicate $=$ which is considered as a property of actual elements. In order to compare also possible elements an *equivalence* predicate $\equiv$ is introduced. Non-existing elements are all considered to be equivalent. Since then equality can be defined in terms of the equivalence and the existence predicates, only $\equiv$ is taken as a primitive in the logic. Then the *formulae* of $\mathcal{L}$ build the smallest set $\mathcal{F}(\mathcal{L})$ such that

– $\mathbf{E}\tau$ for each term $\tau \in \mathcal{T}(\mathcal{L})$,
– $\tau \equiv \sigma$ for terms σ, τ of the same sort s,
– $\tau(\sigma_1, \dots, \sigma_n)$ for terms $\sigma_i \in \mathcal{T}_{s_i}(\mathcal{L})$ and $\tau \in \mathcal{T}_{[s_1,\dots,s_n]}(\mathcal{L})$,
– $\varphi \wedge \psi$ for formulae φ and ψ,
– $\varphi \Rightarrow \psi$ for formulae φ and ψ and
– $\forall x.\varphi$ for variables $x \in Var$ and formulae φ

belong to $\mathcal{F}(\mathcal{L})$. The definition of axioms and rules is omitted here (see [8]). They define the derivation operator $\vdash$ and from this the definition of a *theory* T of $\mathcal{L}$ is standard. For the interpretation of $\mathcal{L}$ in a topos E we also refer to [8].

Given a theory T of $\mathcal{L}$, we have a *canonical interpretation* $\mathcal{I}(T)$ in the *topos* $\mathbb{E}(T)$ *of definable types*. A *type* A is a term of the form $\mathbf{I}y :: [s].\forall x :: s.(\varphi \Leftrightarrow y(x))$. A *relation* f from s to t is a term of the form $\mathbf{I}z :: [s,t].\forall x :: s, y :: t.(\varphi \Leftrightarrow z(x,y))$. A type A or a relation f is said to be *definable* iff the defining formula is closed.

As a more convenient notation write $A = \{x :: s \mid \varphi\}$ for a type A defined by the formula φ. For a term τ of sort s we then get the formula $\tau \in A$. For a variable x with $\#x = s$ we may use the quantifiers $\forall x \in A$ and $\exists x \in A$.

For a relation f we may use the notation $f^{\#}(\tau)$ for $\mathbf{I}y :: t.f(\tau, y)$ for $\tau \in \mathcal{T}_s(\mathcal{L})$ even if do not know whether f is the graph of a function. Furthermore, we use *functional abstraction* writing $\lambda x :: s.\sigma$ as an abbreviation for $\mathbf{I}z :: [s,t].\forall x :: s, y :: t.(y = \sigma \Leftrightarrow z(x,y))$.

Then two relations f, g from type A to type B are *equivalent* with respect to the theory T iff $T \vdash \forall x :: s.(f^{\#}(x) \equiv g^{\#}(x))$ holds.

Definition 2. *Let T be a theory over $\mathcal{L}$. The topos $\mathbb{E}(T)$ of* definable types *and* definable total functions *has as objects the definable types of $\mathcal{L}$ and as morphisms from A to B equivalence classes of definable relations from A to B such that $T \vdash \forall x \in A.f^{\#}(x) \in B$. For $f \in Hom(A,B)$ and $g \in Hom(B,C)$ the composition $g \circ f \in Hom(A,C)$ is defined by $\lambda x \in A.(g^{\#}(f^{\#}(x)))$.*

Then the canonical interpretation $\mathcal{I}(T)$ in the topos $\mathbb{E}(T)$ defined as follows: Each sort s is interpreted by the type $A_s = \{x :: s \mid x = x\}$. For a constant c with $\#c = s$ regard the morphism

$$f = \lambda z :: \mho.c : \{z :: \mho \mid z() \wedge \mathbf{E}c\} \to A_s \ ,$$

defined on a subobject of $\mathbb{1}$. Then let c be interpreted by the unique morphism $f : \mathbb{1} \to \tilde{A}_s$ defined by the partial morphism classifier η_{A_s}.

4 Predicate Transformers in Higher-Order Intuitionistic Logic

In order to find the analogon to the previous section we consider again a program specification S with state space X, but now assume that the variables in X are typed, i.e. for each $x \in X$ there is a sort s of $\mathcal{L}$ with $x :: s$, equivalently $X = \varsigma$ and $X\varsigma$ replaces the state space Σ.

As in the classical case we now assume that $\Delta(S)$ is a subobject of $(X\varsigma \times X\varsigma) \oplus X\varsigma$. In a topos coproducts are pullback stable, hence $\Delta(S) \cong \Delta'(S) \oplus \Sigma_0(S)$ with a subobject $\Delta'(S)$ of $X\varsigma \times X\varsigma$ and a subobject $\Sigma_0(S)$ of $X\varsigma$.

Definition of $wlp(S)$. For a formula $\mathcal{R}$ with free variables in ς take the subobject X' of $\Delta'(S)$ classified by

$$\Delta'(S) \hookrightarrow X\varsigma \times X\varsigma \xrightarrow{\pi_2} X\varsigma \xrightarrow{\mathcal{I}_\varsigma(\mathcal{R})} \Omega \ .$$

Then the formula $wlp(S)(\mathcal{R})$ is defined such that X' is the pullback of $\Delta'(S) \hookrightarrow X\varsigma \times X\varsigma$ and $X'' \times X\varsigma \hookrightarrow X\varsigma \times X\varsigma$, where X'' is the subobject of $X\varsigma$ classified by $\mathcal{I}_\varsigma(wlp(S)(\mathcal{R}))$.

Definition of $wp(S)$. For a formula $\mathcal{R}$ with free variables in ς take the subobject X_1 of $\Delta'(S)$ classified by

$$\Delta'(S) \hookrightarrow X\varsigma \times X\varsigma \xrightarrow{\pi_2} X\varsigma \xrightarrow{\mathcal{I}_\varsigma(\mathcal{R})} \Omega \ ,$$

the subobject X_2 of $X\varsigma$ classified by

$$X\varsigma \xrightarrow{cl(\Sigma_0(S))} \Omega \xrightarrow{\neg} \Omega$$

and let X' be the pullback of X_1 and $X_2 \times X_\varsigma$.

Then the formula $wp(S)(\mathcal{R})$ is defined such that X' is the pullback of $\Delta'(S) \hookrightarrow X_\varsigma \times X_\varsigma$ and $X'' \times X_\varsigma \hookrightarrow X_\varsigma \times X_\varsigma$, where X'' is the subobject of X_ς classified by $\mathcal{I}_\varsigma(wp(S)(\mathcal{R}))$.

Theorem 3. *If $\mathcal{I}$ is the canonical interpretation of $\mathcal{L}$ in $\mathbb{E}(T)$, then for a given program specification S the predicate transformers $wlp(S)$ and $wp(S)$ exist and are unique up to equivalence. Moreover, they satisfy (with $\#x_i = \#\sigma_i$)*

$$wp(S)(\mathcal{R}) \Leftrightarrow wlp(S)(\mathcal{R}) \wedge wp(S)(true) \qquad \textit{(pairing condition)} \tag{8}$$

and

$$wlp(S)(\forall x_1, \ldots, x_n.\tau(\sigma_1, \ldots, \sigma_n)) \Leftrightarrow \forall x_1, \ldots, x_n.wlp(S)(\tau(\sigma_1, \ldots, \sigma_n)) . \tag{9}$$

By abuse of notation we continue to call property (9) *universal conjunctivity*.

Proof (sketch).

In both cases X' is defined as a subobject of $X_\varsigma \times X_\varsigma$, hence $X' \xrightarrow{\pi_1} X_\varsigma$ factors through some subobject X_0 of X_ς. Let $f : X_\varsigma \to \Omega$ classify this subobject.

Since we consider $E = \mathbb{E}(T)$, this morphism f is equivalent to

$$\lambda x_1, \ldots, x_n.\mathbf{I}y :: \mho.(y() \Leftrightarrow \varphi(\mathbf{I}z.x_1(z), \ldots, \mathbf{I}z.x_n(z)))$$

for some description-free formula φ. On the other hand, for the canonical interpretation this term is just $\mathcal{I}_\varsigma(\varphi)$ [8]. Hence the stated existence of predicate transformers. The uniqueness is obvious.

Since *true* is an abbreviation for $\mathbf{E}(\mathbf{I}y :: \mho.y())$ we get $X_1 = \Delta'(S)$ and hence $X'' = X_2$, which implies the pairing condition.

For the universal conjunctivity use direct calculation. □

The full proof of Theorem 3 was given in [18, Theorem 4.2.11, p.136 ff.].

We also have an analog of the inversion theorem, but only up to double negation. For this let $f^*(\mathcal{R}) = \neg f(\neg\mathcal{R})$ denote the *conjugate predicate transformer* for any predicate transformer f.

Theorem 4. *Let flp and fp be predicate transformers on $\mathcal{L}$ satisfying (8) and (9) in place of $wlp(S)$ and $wp(S)$. Then with respect to the canonical interpretation in a topos $\mathbb{E}(T)$ define*

$$\Delta(S) = \coprod_{c_1, \ldots, c_n} \Delta_{c_1, \ldots, c_n} \oplus \Sigma_0(S) \ ,$$

where the coproduct ranges over constants $c_1, \ldots, c_n$, $\Sigma_0(S)$ is the subobject of X_ς classified by $\mathcal{I}_\varsigma(fp^*(false))$ and $\Delta_{c_1, \ldots, c_n}$ is the subobject of $X_\varsigma \times X_\varsigma$ classified by

$$X_\varsigma \times X_\varsigma \xrightarrow{\mathcal{I}_\varsigma(flp^*(x_1=c_1 \wedge \ldots \wedge x_n=c_n)) \times \mathcal{I}_\varsigma(x_1=c_1 \wedge \ldots \wedge x_n=c_n)} \Omega \times \Omega \xrightarrow{\wedge} \Omega .$$

Then we have $wlp(S)^*(\mathcal{R}) \Leftrightarrow flp^*(\mathcal{R})$ and $wp^*(S)(\mathcal{R}) \Leftrightarrow fp^*(\mathcal{R})$.

Proof (sketch).
We have $\mathcal{R} \Leftrightarrow \forall x_1, \ldots, x_n.\tau(c_1, \ldots, c_n)$ with $\tau = \mathbf{I}z.(z(x_1, \ldots, x_n) \Leftrightarrow \neg\mathcal{R})$. From this we calculate $\mathcal{I}_\varsigma(wlp(S)(\tau(x_1, \ldots, x_n))) = \mathcal{I}_\varsigma(flp(\tau(x_1, \ldots, x_n)))$, hence $wlp(S)(\mathcal{R}) \Leftrightarrow flp(\mathcal{R})$ because of the universal conjunctivity property.

Moreover, $\mathcal{I}_\varsigma(wp(S)(true)) = \neg \circ cl(\Sigma_0(S)) = \neg \circ \neg\mathcal{I}_\varsigma(fp(true))$. From this we conclude $wp(S)(true) \Leftrightarrow fp(true)$. Applying the pairing condition completes the proof. □

The full proof of Theorem 4 was given in [18, Theorem 4.2.16, p.143 ff.].

5 Dual Predicate Transformers

In the classical theory the conjugate predicate transformers $wlp(S)^*$ and $wp(S)^*$ already determine the predicate transformers $wlp(S)$ and $wp(S)$. This can not be expected in the non-classical case, since $\neg\neg\mathcal{R}$ is in general not equivalent to $\mathcal{R}$. However, the meaning of the conjugate predicate transformers in the classical theory can be informally characterized as follows:

- $wlp(S)^*(\mathcal{R})$ characterizes those initial states such that there exists a terminating execution of S which reaches a final state characterized by $\mathcal{R}$, and
- $wp(S)^*(\mathcal{R})$ characterizes those initial states such that there exists an execution of S which either fails to terminate or reaches a final state characterized by $\mathcal{R}$.

Therefore, we can also start from this meaning and define ***dual predicate transformers*** $\overline{wlp}(S)$ and $\overline{wp}(S)$ in the non-classical case. Then we have to prove an existence result and derive a ***dual pairing condition*** and a ***universal disjunctivity*** property.

As in case of $wlp(S)$, $wp(S)$ we start with a relational semantics of a ***state transition*** S defined by a subobject $\Delta(S)$ of the coproduct $(X_\varsigma \times X_\varsigma) \oplus X_\varsigma$. We may write $\Delta(S) \cong \Delta'(S) \oplus \Sigma_0(S)$ with a subobject $\Delta'(S)$ of $X_\varsigma \times X_\varsigma$ and a subobject $\Sigma_0(S)$ of X_ς.

Definition of $\overline{wlp}(S)$. For a ς-predicate $\mathcal{R}$ take the subobject X' of $\Delta'(S)$ classified by

$$\Delta'(S) \hookrightarrow X_\varsigma \times X_\varsigma \xrightarrow{\pi_2} X_\varsigma \xrightarrow{\mathcal{I}_\varsigma(\mathcal{R})} \Omega$$

and let the formula $\mathcal{I}_{(\varsigma,\varsigma)}(\varphi)$ be the classifier of $X' \hookrightarrow X_\varsigma \times X_\varsigma$. Then define

$$\begin{aligned}\overline{wlp}(S)(\mathcal{R}) &\Leftrightarrow \forall z :: \mho.(\forall y_1, \ldots, y_n.(\varphi \Rightarrow z()) \Rightarrow z()) \\ &\Leftrightarrow \exists y_1, \ldots, y_n.\varphi \,.\end{aligned}$$

Definition of $\overline{wp}(S)$. For a ς-predicate $\mathcal{R}$ define

$$\overline{wp}(S)(\mathcal{R}) \Leftrightarrow \overline{wlp}(S)(\mathcal{R}) \vee \varphi_0 ,$$

where $\mathcal{I}_\varsigma(\varphi_0)$ classifies $\Sigma_0(S) \hookrightarrow X_\varsigma$.

The predicate transformers $\overline{wlp}(S)$ and $\overline{wp}(S)$ will be called the *dual predicate transformers*of the state transition S. Now we can prove an existence and uniqueness theorem with the desired properties.

Theorem 5. *Let $\mathcal{I}$ be the canonical interpretation of $\mathcal{L}$ in $\mathbb{E}(T)$. Then for a given program specification S the predicate transformers $\overline{wlp}(S)$ and $\overline{wp}(S)$ exist and are unique up to equivalence. Moreover, they satisfy (with $\#y = [\#x_1, \ldots, \#x_n]$)*

$$\overline{wp}(S)(\mathcal{R}) \Leftrightarrow \overline{wlp}(S)(\mathcal{R}) \vee \overline{wp}(S)(false) \tag{10}$$

and

$$\overline{wlp}(S)(\exists y.\tau(y) \wedge y(x_1, \ldots, x_n)) \Leftrightarrow \exists y.\tau(y) \wedge \overline{wlp}(S)(y(x_1, \ldots, x_n)) . \tag{11}$$

(10) will be called the *dual pairing condition* and (11) *universal disjunctivity* property.

Proof (sketch).

Since X' is defined as a subobject of $X_\varsigma \times X_\varsigma$, we may take its classifier $f : X_\varsigma \times X_\varsigma \to \Omega$. Since we consider $E = \mathbb{E}(T)$, the morphism f is equivalent to

$$\lambda x_1, \ldots, x_n, y_1, \ldots, y_n.\mathbf{I}y :: \mho.$$
$$(y() \Leftrightarrow \varphi(\mathbf{I}z.x_1(z), \ldots, \mathbf{I}z.x_n(z), \mathbf{I}z.y_1(z), \ldots, \mathbf{I}z.y_n(z)))$$

for some description-free formula φ. On the other hand we know that for the canonical interpretation this term is just $\mathcal{I}_{(\varsigma,\varsigma)}(\varphi)$. Hence the existence of the predicate transformer $\overline{wlp}(S)$. The existence of $\overline{wp}(S)$ is shown analogously, the uniqueness is obvious.

For the proof of the dual pairing condition it is sufficient to show $\overline{wp}(S)(false) \Leftrightarrow \varphi_0$ with $\mathcal{I}_\varsigma(\varphi_0) = cl(\Sigma_0(S))$, which can be achieved by direct computation.

For the universal disjunctivity (11) write

$$\overline{wlp}(S)(\exists y.\tau(y) \wedge y(x_1, \ldots, x_n)) \Leftrightarrow \exists y_1, \ldots, y_n.\varphi$$

and $\overline{wlp}(S)(y(x_1, \ldots, x_n)) \Leftrightarrow \exists y_1, \ldots, y_n.\psi$. Then the required result follows from $\models \varphi \Leftrightarrow (\exists y.\tau(y) \wedge \psi)$. □

The full proof of Theorem 5 was given in [18, Theorem 4.2.13, p.140 ff.].

Next we may also ask whether we can get back $\overline{wlp}(S)$ and $\overline{wp}(S)$ from predicate transformers $\overline{flp}$ and $\overline{fp}$ satisfying the dual pairing condition and the universal disjunctivity property. In contrast to Theorem 4 we can achieve a stronger result.

Theorem 6. *Let $\overline{flp}$ and $\overline{fp}$ be predicate transformers on $\mathcal{L}$ satisfying (10) and (11) in place of $\overline{wlp}(S)$ and $\overline{wp}(S)$. Then with respect to the canonical interpretation in a topos $\mathbb{E}(T)$ define*

$$\Delta(S) = \coprod_{c_1,\ldots,c_n} \Delta_{c_1,\ldots,c_n} \oplus \Sigma_0(S)\,,$$

where the coproduct ranges over all constants $c_1, \ldots, c_n$ with $\#c_i = \#x_i$, $\Sigma_0(S)$ is the subobject of X_ς classified by $\mathcal{I}_\varsigma(\overline{fp}(false))$ and $\Delta_{c_1,\ldots,c_n}$ is the subobject of $X_\varsigma \times X_\varsigma$ classified by

$$X_\varsigma \times X_\varsigma \overset{\mathcal{I}_\varsigma(\overline{flp}(x_1=c_1\wedge\ldots\wedge x_n=c_n))\times\mathcal{I}_\varsigma(x_1=c_1\wedge\ldots\wedge x_n=c_n)}{\longrightarrow} \Omega\times\Omega \overset{\wedge}{\longrightarrow} \Omega\,.$$

Then we have $\overline{wlp}(S)(\mathcal{R}) \Leftrightarrow \overline{flp}(\mathcal{R})$ and $\overline{wp}(S)(\mathcal{R}) \Leftrightarrow \overline{fp}(\mathcal{R})$.

Proof (sketch).
$\Delta(S)$ is a well-defined subobject of $X_\varsigma \times X_\varsigma$, hence defines a state transition S. Let

$$\Delta'(S) = \coprod_{c_1,\ldots,c_n} \Delta_{c_1,\ldots,c_n}\,.$$

First write $\mathcal{R} \Leftrightarrow \exists y.(\tau(y) \wedge y(x_1,\ldots,x_n))$ with

$$\begin{aligned}\tau \;=\; & \mathbf{I}z.\forall y.(z(y) \Leftrightarrow \exists y_1,\ldots,y_n.\tau'(y_1,\ldots,y_n)\wedge \\ & (\forall x_1,\ldots,x_n.y(x_1,\ldots,x_n) \Leftrightarrow x_1 = y_1 \wedge \ldots \wedge x_n = y_n))\end{aligned}$$

and $\tau' = \mathbf{I}z.(\forall x_1,\ldots,x_n.z(x_1,\ldots,x_n) \Leftrightarrow \mathcal{R})$.

Then we may apply the universal disjunctivity property for $\overline{flp}$ and compute $\overline{flp}(\mathcal{R}) \quad\Leftrightarrow\quad \exists y_1,\ldots y_n.\psi$ with

$$\psi \Leftrightarrow (\tau'(y_1,\ldots,y_n) \wedge \overline{flp}(y(x_1,\ldots,x_n)))\,.$$

Now let Y' be the subobject of $X_\varsigma \times X_\varsigma$ classified by $\mathcal{I}_{(\varsigma,\varsigma)}(\psi)$ and X' the pullback of $\Delta'(S) \overset{i}{\hookrightarrow} X_\varsigma \times X_\varsigma$ and $Y' \hookrightarrow X_\varsigma \times X_\varsigma$.

We show $\mathcal{I}_{(\varsigma,\varsigma)}(\psi) \circ i = \mathcal{I}_\varsigma(\mathcal{R}) \circ \pi_2 \circ i$. Consequently, $X' \hookrightarrow \Delta'(S)$ is classified by $\mathcal{I}_\varsigma(\mathcal{R}) \circ \pi_2 \circ i$ and $X' \cong Y'$.

If $\mathcal{I}_{(\varsigma,\varsigma)}(\varphi)$ classifies $X' \to X_\varsigma \times X_\varsigma$, then this implies $\varphi \Leftrightarrow \psi$.

By definition we have $\overline{wlp}(S)(\mathcal{R}) \Leftrightarrow \exists y_1,\ldots,y_n.\varphi$, which completes the proof of $\overline{wlp}(S) = \overline{flp}$.

For $\overline{wp}(S)$ we simply exploit the dual pairing condition for $\overline{flp}$ and $\overline{fp}$. □

The full proof of Theorem 6 was given in [18, Theorem 4.2.18, p.145 ff.].

6 Conclusion

Dijkstra's original predicate transformer calculus is well known as one approach to axiomatic semantics. The major goal is to reason about program specifications in logical terms and to prove properties such as termination, determinacy, consistency, etc. Unfortunately, it is untyped in contradiction to the many results in the area of strongly typed programs.

Therefore, we asked for a generalization of Dijkstra's predicate transformer calculus to typed program specifications. First we argued that any reasonable type system defines its semantics in some suitable topos. More precisely, we could even restrict to recursive realizability topoi such as the effective topos or the recursive topos. Hence type semantics canonically defines a higher-order intuitionistic logic.

This logic associated with a topos is taken as a cornerstone for the work in this paper. Starting from the intensional meaning of predicate transformers we found the general characterization of $wlp(S)$ and $wp(S)$ using category- theoretical arguments. We could also derive properties that generalize the well known characteristics in the classical theory, i.e. the pairing condition and universal conjunctivity. This in turn allowed to establish a weak form of the inversion theorem, where programs are characterized uniquely up to double negation of their predicate transformers. In the same way we were also able to generalize dual predicate transformers and prove a second stronger inversion theorem.

The whole work is part of a larger project on specification theory with topos semantics. In that theory it is also possible to generalize consistency proof obligations and shift the theory of consistency enforcement to a level, where it coexists with type theory.

References

1. M. Barr, C. Wells: *Category Theory for Computing Science*, Prentice-Hall 1990
2. M. Barr, C. Wells: *Toposes, Triples and Theories*, Springer Grundlehren der mathematischen Wissenschaften 278, 1985
3. A. Boileau, A. Joyal: *La Logique des Topos*, Journal of Symbolic Logic, vol. 46 (1), 1981, 6-16
4. K. B. Bruce, A. R. Meyer: *The Semantics of Second Order Polymorphic Lambda Calculus*, in G. Kahn, D. B. MacQueen, G. Plotkin (Eds.): *Semantics of Data Types*, Springer LNCS 173, 1984, pp. 131-144
5. P. Cousot: *Methods and Logics for Proving Programs*, in J. van Leeuwen (Ed.): *The Handbook of Theoretical Computer Science*, vol B: "Formal Models and Semantics", Elsevier, 1990, pp. 841-993
6. E. W. Dijkstra, C. S. Scholten: *Predicate Calculus and Program Semantics*, Springer Texts and Monographs in Computer Science, 1989
7. H. Ehrig und B. Mahr: *Fundamentals of Algebraic Specification 1*, Springer EATCS Monographs, vol. 6, 1985
8. M. P. Fourman: *The Logic of Topoi*, in J. Barwise (Ed.): *Handbook of Mathematical Logic*, North-Holland Studies in Logic, vol. 90, 1977, pp. 1053-1090
9. J. A. Goguen: *Types as Theories*, Oxford University, 1990

10. R. Goldblatt: *Topoi – The Categorial Analysis of Logic*, North-Holland, Studies in Logic, vol. 98, 1984
11. P. Johnstone: *Topos Theory*, Academic Press, 1977
12. A. Kock, G. Reyes: *Doctrines in Categorial Logic*, in J. Barwise (Ed.): *Handbook of Mathematical Logic*, North-Holland Studies in Logic, vol. 90, 1977, pp. 283-313
13. S. Mac Lane: *Categories for the Working Mathematician*, Springer GTM, vol. 5, 1972
14. S. Mac Lane, I. Moerdijk: *Sheaves in Geometry and Logic – A First Introduction to Topos Theory*, Springer Universitext, 1992
15. J. C. Mitchell: *Type Systems for Programming Languages*, in J. van Leeuwen (Ed.): *The Handbook of Theoretical Computer Science*, vol B: "Formal Models and Semantics", Elsevier, 1990, pp. 365-458
16. G. Nelson: *A Generalization of Dijkstra's Calculus*, ACM TOPLAS, vol. 11 (4), 1989, pp. 517-561
17. J. C. Reynolds: *Polymorphism is not Set-Theoretic*, in G. Kahn, D. B. MacQueen, G. Plotkin (Eds.): *Semantics of Data Types*, Springer LNCS 173, 1984, pp. 145-156
18. K.-D. Schewe. *Specification of Data-Intensive Application Systems*. Habilitation Thesis. BTU Cottbus 1995. available from `http://www.in.tu-clausthal.de/~schewe/public/habil.ps.gz`
19. K.-D. Schewe, B. Thalheim. Fundamental Concepts of Object Oriented Databases. *Acta Cybernetica* vol. 11(4):49-84, 1993.
20. K.-D. Schewe. Fundamentals of object oriented database modelling. Интеллектуальны Системы (Intelligent Systems). Moskau 1996.
21. K.-D. Schewe, B. Thalheim. Towards a Theory of Consistency Enforcement. *Acta Informatica* vol. 36(4): 97-141, 1999.
22. D. S. Scott: *Identity and Existence in Intuitionistic Logic*, in M. P. Fourman, C. J. Mulvey, D. S. Scott: *Applications of Sheaves*, Springer LNM, vol. 753, 1979, pp. 660-698
23. M. Wirsing: *Algebraic Specification*, in J. van Leeuwen (Ed.): *The Handbook of Theoretical Computer Science*, vol B: "Formal Models and Semantics", Elsevier, 1990, pp. 675-788

Homomorphisms and Concurrent Term Rewriting

Franck Seynhaeve, Sophie Tison, and Marc Tommasi

LIFL, Bât M3, Université Lille 1, F59655 Villeneuve d'Ascq cedex, France
seynhaev,tison,tommasi@lifl.fr

Abstract. In this paper we study applications of relations based on rewrite systems to regular tree languages. For instance, we want to deal with decidability problems of the form "$\mathcal{R}el(L_1) \subseteq L_2$" where L_1, L_2 are regular tree languages and $\mathcal{R}el$ can be either IO, OI, parallel, or one step rewriting for a given rewrite system. Our method somehow standardizes previous ones because it reveals conditions $\mathcal{R}el$ must fulfill to preserve recognizability for the language $\mathcal{R}el(L_1)$. Thanks to classes of recognizable languages wider than the regular one, we get some new results. We pursue this method to tackle the problem of computing the set of descendants of a regular tree language by a rewrite system.

1 Introduction

This paper tackles decidability questions $\mathcal{R}el(L_1) \subseteq L_2$ where L_1 and L_2 are regular tree languages and $\mathcal{R}el$ is a relation based on a term rewriting system. For instance, we want to decide whether the set of direct successors of terms in L_1 by a term rewriting system is a subset of L_2. In this case $\mathcal{R}el$ is "one-step rewriting". For example, this enables us to decide whether $\mathcal{R}^*(L_1) = L_1$ for any term rewriting system $\mathcal{R}$. We essentially investigate cases where $\mathcal{R}el$ is one-pass rewriting with several strategies: one-step rewriting, parallel rewritings, IO or OI pass rewriting ([Eng78]) or leaf or root started rewriting ([FJSV98]). All these relations have a common principle: redexes are sufficiently spread out in terms to avoid creating overlaps between two rewritings. We say in this case that rewritings can be done concurrently. Basically, we exploit this property in our constructions.

Decidability problems $\mathcal{R}el(L_1) \subseteq L_2$ are difficult to solve because $\mathcal{R}el(L_1)$ is not regular. So we suffer from lack of usual useful tools to manipulate these sets, and to decide for instance, inclusion or emptiness. Nonetheless, related works in the literature often partially solve this problem with tree automata manipulations, *e.g.* by adding or transforming rules. Authors reduce these decidability questions about complex sets to decidability questions about regular tree languages. We propose in this paper an approach that reveals why this can be done and somehow unifies them in a standard understanding. Thanks to classes of recognizable languages larger than the regular one, we obtain new decidability results.

The main principle is to reduce the inclusion $\mathcal{R}el(L_1) \subseteq L_2$ to some inclusion between regular sets in such a way the latter holds if and only if the former also holds. To do that, we use transformations via inverse tree homomorphisms and intersections by regular tree languages[1]. Inverse homomorphisms fix redexes and in-

[1] In the sequel, we will write just "language" and "homomorphisms" instead of "tree language" and "tree homomorphism", respectively.

tersections control that the relation $\mathcal{R}el$ is well applied. For instance, let us consider that $\mathcal{R}el$ is one-step rewriting with a given rule $l \to r$. We denote by $\mathcal{R}el(L_1)$ the set $\{t \mid u \to t,\ u \in L_1\}$. We consider a new symbol $\bullet$ and two tree homomorphisms Ψ_r and Ψ_l defined by the mappings $\phi_l(\bullet) = l$ and $\phi_r(\bullet) = r$. In this case one can show that $\mathcal{R}el(L_1)$ is exactly the set $\Psi_r(\Psi_l^{-1}(L_1) \cap \mathcal{R}_c)$, where $\mathcal{R}_c$ is the regular language of terms containing exactly one symbol $\bullet$. So we have two immediate consequences, from the closure properties of regular languages. First, $\mathcal{R}el(L_1) \subseteq L_2$ is decidable because it is equivalent to decide whether $\Psi_l^{-1}(L_1) \cap \mathcal{R}_c) \subseteq \Psi_r^{-1}(L_2)$ and this last expression only involves regular languages. Second, if Ψ_r is a linear homomorphism, then $\mathcal{R}el(L_1)$ is regular and therefore we can decide whether $\mathcal{R}el(L_1) = L_2$. It is important to note that this method leads to effective proofs.

The question is now: how this method can be extended and what are its limits? We give some ideas for the answer. Rewritings can be "imitated" by homomorphisms only in some cases. Homomorphisms are not helpful when redexes are too close, namely when some instances of left-hand sides of rules overlap in two rewriting steps. When more than one rewriting step are applied, we need to follow some strategies in order to keep the opportunity of using homomorphisms. For instance, IO and OI ([Eng75,ES77,ES78]) strategies and one-pass leaf- and root-started strategies ([FJSV98]) have this non-overlapping property but some new restrictions on linearities on left or right-hand sides of rules are now required.

Another improvement is to try to compute the set $\mathcal{R}^*(L)$ of descendants (or the set of normal forms) of a regular language L by a rewrite system $\mathcal{R}$. This topic has been studied in many papers ([Sal88,DT90,CDGV94]) and all constructions rely on similar arguments. We are able to compute $\mathcal{R}^*(L)$ with our method under some restrictions. In fact, we construct a sequence of languages $(L_k)_{k \in N}$ with $L_0 = L$ such that:

$$\forall k \in N,\ \mathcal{R}(L_k) \subseteq L_{k+1} \subseteq \mathcal{R}^*(L_k).$$

These languages are regular when $\mathcal{R}$ is right-linear and under some other restrictions on $\mathcal{R}$, there exists an integer i such that $\forall k \geq i$, $L_k = L_i$ hence $\mathcal{R}^*(L) = L_i$.

The paper is organized as follows. In section 3, we introduce our method. We apply it in Section 4 on one-step rewriting, parallel rewriting, computation of normal and sentential forms according to one-pass IO, one-pass OI, one-pass root-started or leaves started. Finally we study the language $\mathcal{R}^*(L)$ in Section 5.

The table in Section 5.1 summarizes our results.

2 Preliminaries

2.1 Signature, Terms and Rewriting

Let $n \in N$. We denote by $[n]$ the set $\{1, 2, \ldots, n\}$ and N^* denotes the set of finite-length strings over N.

Let us consider a signature Σ and a countable set of variables $\mathcal{X}$. A *term* over $\Sigma \cup \mathcal{X}$ is a partial function $t : N^* \to \Sigma \cup \mathcal{X}$ whose domain $\mathcal{P}os(t)$ satisfies:

- $\mathcal{P}os(t)$ is nonempty and prefix-closed;
- If $t(p) \in \Sigma_n$, then $\{i \mid pi \in \mathcal{P}os(t)\} = \{0, 1, \ldots, n-1\}$;
- If $t(p) \in \mathcal{X}$, then $\{i \mid pi \in \mathcal{P}os(t)\} = \emptyset$.

Each element of $\mathcal{P}os(t)$ is called a *position*. Two positions p_1 and p_2 are *comparable* if there exists p such that $p_1 = pp_2$ or $p_2 = pp_1$, otherwise they are *incomparable*. A *branch* of t is a prefix-closed subset of $\mathcal{P}os(t)$ in which all positions are pairwise comparable and whose greater position p holds a constant or a variable, *i.e.* $t(p) \in \mathcal{X} \cup \Sigma_0$. We denote by $|p|$ the length of a position p. The *height* of a term t, denoted by $Height(t)$, is the maximal length over all positions in t. We denote by $\mathcal{V}ar(t)$ the set of variables which occur in t, and $t|_p$ is the subterm of t rooted at position p, and $t[u]_p$ is the term obtained by replacing in t the subterm at position p by u. If $|p| = n$, then $t|_p$ is called a subterm of t at depth n. Given a position p, subterms rooted at position pw, $w \in N$, are called *brother terms*.

The set of all terms (or *trees*) is denoted by $T(\Sigma, \mathcal{X})$. If $\mathcal{X} = \emptyset$ then $T(\Sigma, \mathcal{X})$ is denoted by $T(\Sigma)$. Terms of $T(\Sigma)$ are called *ground terms*. A term t in $T(\Sigma, \mathcal{X})$ is *linear* if each variable occurs at most once in t. When each variable either occurs at positions of length 1 or only once in t, we say that t is *semi-linear*. We fix some typographical conventions for the rest of the paper. In what follows, x, y and z, possibly with subscripts, are variables, and the letters t, u and v will denote terms.

We denote by $\mathcal{X}_n$ a set of n distinct variables, $\mathcal{X}_n = \{x_1, \ldots, x_n\}$. A linear term C of $T(\Sigma, \mathcal{X}_n)$ is called a *context* and the expression $C[t_1, \ldots, t_n]$ denotes the term obtained from C by replacing for each i, x_i by t_i. We denote by $\mathcal{C}^n(\Sigma)$ the set of contexts on Σ and $\mathcal{X}_n$, and $\mathcal{C}(\Sigma)$ the set of context with one variable.

A substitution $\sigma : \mathcal{X} \to T(\Sigma, \mathcal{X})$ is extended to a mapping $\sigma : T(\Sigma, \mathcal{X}) \to T(\Sigma, \mathcal{X})$ so that $\sigma(f(t_1, \ldots, t_n)) = f(\sigma(t_1), \ldots, \sigma(t_n))$. A term $t \in T(\Sigma)$ *encompasses* a term $u \in T(\Sigma, \mathcal{X})$ if there exists a substitution σ such that σu is a subterm of t.

Let Γ be a signature and φ a mapping which, with $f \in \Sigma_n$, associates a term $t_f \in T(\Gamma, \mathcal{X}_n)$. The *tree homomorphism* Φ from $T(\Sigma)$ into $T(\Gamma)$ determined by φ is defined as follows:

- $\Phi(a) = t_a \in T(\Gamma)$ for each $a \in \Sigma_0$;
- $\Phi(f(t_1, \ldots, t_n)) = t_f[\Phi(t_1), \ldots, \Phi(t_n)]$.

A tree homomorphism is linear when for each symbol $f \in \Sigma_n$, $\varphi(f) = t_f$ is a linear term of $T_\Gamma(\mathcal{X}_n)$ and semi-linear if t_f is semi-linear.

A *term rewriting system* (TRS) $\mathcal{R}$ over Σ is a finite set of *rewrite rules* $l \to r$ such that $l, r \in T(\Sigma, \mathcal{X})$, $\mathcal{V}ar(r) \subseteq \mathcal{V}ar(l)$ and $l \notin \mathcal{X}$. The rewrite relation $\to_\mathcal{R}$ on $T(\Sigma)$ induced by $\mathcal{R}$ is defined so that $t \to_\mathcal{R} u$ if there exist $p \in \mathcal{P}os(t)$, a substitution σ and $l \to r \in \mathcal{R}$ such that $t|_p = \sigma l$ and $u = t[\sigma r]_p$. The reflexive, transitive closure of $\to_\mathcal{R}$ is denoted by $\to^*_\mathcal{R}$. Hence $s \to^*_\mathcal{R} t$ iff there exists a *derivation* $t_0 \to_\mathcal{R} t_1 \to_\mathcal{R} \ldots \to_\mathcal{R} t_n$ in $\mathcal{R}$ such that $n \geq 0$, $t_0 = s$ and $t_n = t$.

Given a rule $l \to r \in \mathcal{R}$, the term l is the left-hand side and r is the right-hand side of the rule. When every left-hand side of rule is linear, we say that $\mathcal{R}$ is left-linear. Similarly, we define right-linear, left-semi-linear and right-semi-linear TRS.

A term $s \in T(\Sigma, \mathcal{X})$ is *irreducible* with respect to $\mathcal{R}$ if $s \to_\mathcal{R} u$ for no u. And s is a *normal form* of $t \in T(\Sigma, \mathcal{X})$ if $t \to^*_\mathcal{R} s$ and s is irreducible with respect to $\mathcal{R}$. Finally s is *normalizable* with respect to $\mathcal{R}$ if there exists a normal form of s.

2.2 Tree Automata

Most of the constructions in this paper rely on three classes of tree automata: finite tree automata, automata with tests between brothers and reduction automata. General tree automata generalized these three classes.

A general tree automaton $\mathcal{A} = (\Sigma, \mathcal{Q}, \mathcal{Q}_f, \Delta)$ is given by a signature Σ, a finite set of unary symbols called *states* $\mathcal{Q}$ such that $\Sigma \cap \mathcal{Q} = \emptyset$, a finite set of *final states* $\mathcal{Q}_f \subseteq \mathcal{Q}$, a finite set of *rewrite rules* Δ of the form $f(q_1(x_1), \ldots, q_n(x_n)) \xrightarrow{c} q(f(x_1, \ldots, x_n))$ where $f \in \Sigma_n$, $q, q_1, \ldots, q_n \in \mathcal{Q}$ and c is a *constraint*, *i.e.* a conjunction of positive constraint $p = p'$ and negative constraint $p \neq p'$ where p and p' are positions of length greater than or equal to 1.

Rules in Δ are constrained rules, *i.e.* $t \rightarrow_{\mathcal{A}} s$ at position p with $t, s \in T(\Sigma \cup \mathcal{Q})$ if there exists a rule $f(q_1(x_1), \ldots, q_n(x_n)) \xrightarrow{c} q(f(x_1, \ldots, x_n)) \in \Delta$ such that: $t \rightarrow s$ by the rule $f(q_1(x_1), \ldots, q_n(x_n)) \rightarrow q(f(x_1, \ldots, x_n))$ at position p, and $f(t|_{p11}, \ldots, t|_{pn1})$ satisfies c.

A term $t \in T(\Sigma)$ satisfies a positive constraint $p = p'$ (respectively negative constraint $p \neq p'$) if $t|_p = t|_{p'}$ (respectively $t|_p \neq t|_{p'}$).

A term $t \in T(\Sigma)$ is *accepted* by $\mathcal{A}$ if $t \rightarrow^*_{\mathcal{A}} q(t)$ where q is a final state. The language *recognized* by $\mathcal{A}$ is the set of terms accepted by $\mathcal{A}$ and is denoted by $L(\mathcal{A})$.

$\mathcal{A}$ is *deterministic* when for each couple $f(q_1(x_1), \ldots, q_n(x_n)) \xrightarrow{c} q(f(x_1, \ldots, x_n)$ and $f(q_1(x_1), \ldots, q_n(x_n)) \xrightarrow{c'} q'(f(x_1, \ldots, x_n)$ of rules of $\mathcal{A}$, the constraint $c \wedge c'$ is unsatisfiable if $q \neq q'$.

A tree automaton is a general tree automaton where rules are without constraint. The class of tree languages recognized by tree automata is the class of regular languages.

A tree automaton with tests between brothers is a general tree automaton where the length of positions occurring in each constraint is 1, *i.e.* equalities and inequalities are imposed between brother terms. $REC_{\neq}$ denotes the class of languages recognized by tree automata with tests between brothers.

A reduction automaton $\mathcal{A}$ is a general tree automaton such that there is an ordering on the states of $\mathcal{A}$ such that, for each rule $f(q_1(x_1), \ldots, q_n(x_n)) \xrightarrow{c} q(x_1, \ldots, x_n))$, q is smaller than each q_i. Moreover if a positive constraint occurs in c, q is strictly smaller than each q_i. RA denotes the class of languages recognized by the class of reduction automata.

Classes of tree languages recognized by tree automata, tree automata with tests between brothers and deterministic reduction automata are closed under union, intersection, complementation. Moreover, the inclusion problem and the emptiness problem are also decidable.

The class of regular languages is closed under inverse homomorphisms and linear homomorphisms, *i.e.* $\Phi^{-1}(L)$ and $\Psi(L)$ are regular for any homomorphism Φ, linear homomorphism Ψ and regular language L. The image of a regular language L by a semi-linear homomorphism is a language of $REC_{\neq}$.

Finally the set of terms encompassing a linear term is regular and the set of terms encompassing any term is recognizable by deterministic reduction automata. The recognizability problem is decidable for $REC_{\neq}$, *i.e.* it is decidable whether L is regular for any $L \in REC_{\neq}$ [BST99].

3 Decision Problems and General Method of Resolution

Let us consider the decision questions "$\mathcal{R}el(L_1) \subseteq L_2$?" and "$\mathcal{R}el(L_1) = L_2$?" where L_1 and L_2 are regular languages, and $\mathcal{R}el(L_1)$ is the image of terms of L_1 by a relation $\mathcal{R}el$.

Our method for deciding such questions is based on the following remark: under some assumptions, we can associate with $\mathcal{R}el$ two homomorphisms Ψ_l, Ψ_r and a regular "checking" language R_c, i.e. a tree bimorphism [AD82], such that:

$$\mathcal{R}el(L_1) = \Psi_r(\Psi_l^{-1}(L_1) \cap R_c) \quad (1)$$

The construction is very simple: let $\mathcal{R}$ be a TRS of n rules $l_i \to r_i$.

We associate with each rule of the system $\mathcal{R}$ a new symbol $\bullet_i$ whose arity equals to the number of distinct variables of $l_i \to r_i$. Let Λ be the signature of these new symbols and Γ the signature $\Sigma \cup \Lambda$. Ψ_l and Ψ_r will be the tree homomorphisms from $T(\Gamma)$ into $T(\Sigma)$ defined by:

- $\forall f \in \Sigma_p$, $\Psi_r(f) = \Psi_l(f) = f(x_1, \ldots, x_p)$, and
- $\forall i \in [n]$, $\Psi_l(\bullet_i) = l_i$ and $\Psi_r(\bullet_i) = r_i$.

The set $\Psi_l^{-1}(t)$, can be viewed as the set of terms of $T(\Gamma)$ obtained by putting symbols of Λ in place of some occurrences of left-hand sides of rules in t. Homomorphism Ψ_r will apply the rewrite rules at this redexes.

The checking language R_c will be defined w.r.t. $\mathcal{R}el$. For example, when $\mathcal{R}el$ is just the one-step rewriting, R_c just checks that there is only one $\bullet_i$.

Of course this method has its limitations. Roughly speaking, we can simulate by this way rewriting which can be done concurrently; it implies non-overlapping and some conditions on linearity.

We show now how this construction gives easy decision algorithms for our questions, under some assumptions.

Let us first notice that when Ψ_r is linear (i.e. $\mathcal{R}$ is right-linear) and R_c is recognizable, $\Psi_r(\Psi_l^{-1}(L_1) \cap R_c)$ is recognizable (and you can effectively construct an automaton for it). So we can decide whether it is included in (resp. equal to) a given regular language. In fact we can reduce the right-linearity condition to the right-semi-linearity condition, since the image of a regular language by a semi-linear homomorphism is in $REC_{\neq}$ and the inclusion problem is decidable for $REC_{\neq}$.

So in this case, we get an easy decision algorithm for deciding "$\mathcal{R}el(L_1) \subseteq L_2$?" and "$\mathcal{R}el(L_1) = L_2$?" where L_1 and L_2 are regular languages.

When Ψ_r is not linear, $\Psi_r(\Psi_l^{-1}(L_1) \cap R_c)$ is no more (in general) recognizable; however, we can decide "$\mathcal{R}el(L_1) \subseteq L_2$?".

Indeed, $\Psi_r(\Psi_l^{-1}(L_1) \cap R_c)$ is included in L_1 if and only if $\Psi_l^{-1}(L_1) \cap R_c$ is included in $\Psi_r^{-1}(L_1)$. When R_c is regular, both $\Psi_l^{-1}(L_1) \cap R_c$ and $\Psi_r^{-1}(L_1)$ are (effectively) regular. Hence we just have to decide as whether a regular language is included in another one.

In fact, we don't need the recognizability of R_c; using the properties of the class of languages recognized by reduction automata (decidability of emptiness, closure under intersection), we just have to suppose that R_c is recognizable by a reduction automaton. This allows us for example to deal with computation of normal forms according to some strategies like in [FJSV98]. This works fine because, for any finite set of terms E_s, the language $R_s^{FN} = \{t \in T(\Gamma) \mid t \text{ encompasses no term of } E_s\}$ is recognizable by a reduction automaton [DCC95].

To summarize, our method is based on the two following propositions:

Proposition 1. *If $\mathcal{R}el(L_1) = \Psi_r(\Psi_l^{-1}(L_1) \cap R_c)$ with Ψ_r right-semi-linear and R_c regular, then "$\mathcal{R}el(L_1) = L_2$?" with L_1 and L_2 regular, is decidable.*

Proposition 2. *If $\mathcal{R}el(L_1) = \Psi_r(\Psi_l^{-1}(L_1) \cap R_c)$ with R_c recognizable by a reduction automaton, then "$\mathcal{R}el(L_1) \subseteq L_2$?" with L_1 and L_2 regular, is decidable.*

4 Applications

We use the method developed in the previous section in order to study the decision questions "$\mathcal{R}el(L_1) \subseteq L_2$?" and "$\mathcal{R}el(L_1) = L_2$?" where L_1 and L_2 are regular for different relations $\mathcal{R}el$ depending on a TRS $\mathcal{R}$ composed of n rules. We give for each studied relation, the checking language R_c and the conditions for which $\mathcal{R}el(L_1) = \Psi_r(\Psi_l^{-1}(L_1) \cap R_c)$.

4.1 One-Step Rewriting and Parallel Rewriting

One-step rewriting is the application of one rule of $\mathcal{R}$. The image of L by one-step rewriting is $\mathcal{R}(L) = \{t \in T(\Sigma) \mid \exists u \in L, u \to_{\mathcal{R}} t\}$. *Parallel rewriting* is the application of a set of rules on incomparable positions. Let u and t be two terms of $T(\Sigma)$. If parallel rewriting is denoted by $\Vdash\!\!\!\!\to$, $t \Vdash\!\!\!\!\to u$ if and only if there exists $k \in N$, $C \in \mathcal{C}^k(\Sigma)$, $i_1, \ldots, i_k \in [n]$ and $\sigma_1, \ldots, \sigma_k$ substitutions on $T(\Sigma, \mathcal{X})$ such that $t = C[l_{i_1}\sigma_1, \ldots, l_{i_k}\sigma_k]$ and $u = C[r_{i_1}\sigma_1, \ldots, r_{i_k}\sigma_k]$. The image of L by application of parallel rewriting is $\mathcal{R}_{\|}(\mathcal{L}) = \{t \in T(\Sigma) \mid \exists u \in L, u \Vdash\!\!\!\!\to t\}$.

Example 1. Let us consider the signature $\Sigma = \{f/2, g/1, a/0, b/0\}$ and the TRS with two rules $f(x,x) \to g(x)$ and $b \to a$. If $L = \{f(g(b), g(b))\}$ then

$$\mathcal{R}(L) = \{g(g(b)),\ f(g(a), g(b)),\ f(g(b), g(a))\}$$
$$\mathcal{R}_{\|}(L) = \{f(g(b), g(b)),\ f(g(b), g(a)),\ f(g(a), g(b)),\ f(g(a), g(a)),\ g(g(b))\}$$

Let R_c^1 be the set of terms of $T(\Gamma)$ with only one symbol of Λ and $R_c^{\|}$ the set of terms of $T(\Gamma)$ with at most one symbol of Λ on each branch. These languages are regular and we can prove that:

$$\mathcal{R}(L_1) = \Psi_r(\Psi_l^{-1}(L_1) \cap \mathcal{R}_c^1) \qquad \mathcal{R}_{\|}(L_1) = \Psi_r(\Psi_l^{-1}(L_1) \cap \mathcal{R}_c^{\|})$$

Hence according to Propositions 2 and 1 we obtain:

Proposition 3. *"$\mathcal{R}(L_1) \subseteq L_2$?" and $\mathcal{R}_{\|}(L_1) \subseteq L_2$?" are decidable for any TRS $\mathcal{R}$, and "$\mathcal{R}(L_1) = L_2$?" and "$\mathcal{R}_{\|}(L_1) = L_2$?" are decidable for any right-semi-linear TRS $\mathcal{R}$.*

Moreover for any TRS $\mathcal{R}$ and for any regular language L, $\mathcal{R}(L) \subseteq L \Leftrightarrow \mathcal{R}^*(L) = L$. This statement can be used as a halting criterion for approximations of computations of the set $\mathcal{R}^*(L)$. For instance, see related works by J. Waldmann or T. Genet [Wal98,Gen98].

Corollary 1. *For any TRS $\mathcal{R}$ and regular language L, "$\mathcal{R}^*(L) = L$?" is decidable.*

4.2 One-pass IO and OI

If $\mathcal{R}$ is linear, a term t rewrites to a term u in *one-pass* if all rewritings in the derivation of t in u can be applied concurrently.

Example 2. Let $\Sigma = \{h/3, f/2, g/1, a/0\}$ and $\mathcal{R}$ be the TRS composed of the rules $h(x, g(y), z) \rightarrow f(y, x)$ and $g(x) \rightarrow a$. The following derivation can be done in one-pass:

$$h(f(g(a),a)\ ,g(g(a))\ ,a) \rightarrow f(g(a)\ ,f(g(a),a)\ ) \rightarrow f(g(a)\ ,f(a,a)\ ) \rightarrow f(a,f(a,a)\ )$$

In general, $\mathcal{R}$ is not linear, and hence we must indicate if subtrees are rewritten before or after generating or testing equalities defined by the rules. There are two usual strategies [Eng75]:

OI pass: Rewriting from the root (outermost) to the leaves (innermost).
IO pass: Rewriting from the leaves to the root.

If a term t rewrites to a term u in one-pass OI (respectively IO), then we denote $t \rightarrow^{oi} u$ (respectively $t \rightarrow^{io} u$).

Example 3. Let $\Sigma = \{f/2, g/2, a/0, b/0\}$ and consider the rewrites rules $f(x, x) \rightarrow g(x, x)$ and $a \rightarrow b$.

Hence $f(a, a) \rightarrow^{oi} g(a, b)$ but $f(a, a) \not\rightarrow^{io} g(a, b)$, and $f(a, b) \rightarrow^{io} g(b, b)$ but $f(a, b) \not\rightarrow^{oi} g(b, b)$.

Let us denote $\mathsf{SF}_{\mathsf{oi}}(L) = \{t \in T(\Sigma) \mid \exists s \in L, s \rightarrow^{oi} t\}$ and $\mathsf{SF}_{\mathsf{io}}(L) = \{t \in T(\Sigma) \mid \exists s \in L, s \rightarrow^{io} t\}$. When $\mathcal{R}el$ is one of these relations, there is no checking language and we can prove that $\Psi_r(\Psi_l^{-1}(L_1)) \subseteq \mathsf{SF}_{\mathsf{io}}(L_1)$ and $\Psi_r(\Psi_l^{-1}(L_1)) \subseteq \mathsf{SF}_{\mathsf{oi}}(L_1)$. Examples 4 and 5 show that these inclusions may be strict.

Example 4 (IO requires rules to be left-linear).
Let the signature $\Sigma = \{f/2, g/1, a/0, b/0\}$ and the rules $f(x, x) \rightarrow g(x)$ and $a \rightarrow b$.
We have $f(a, b) \rightarrow^{io} g(b)$ since $f(a, b) \rightarrow f(b, b) \rightarrow g(b)$. But there is no term u of $T(\Gamma)$ such that $\Psi_l(u) = f(a, b)$ and $\Psi_r(u) = g(b)$.

Example 5 (OI requires rules to be right-linear).
Let the signature $\Sigma = \{f/2, g/1, a/0, b/0\}$ and the rules $g(x) \rightarrow f(x, x)$, $g(x) \rightarrow a$ and $g(x) \rightarrow b$.
We have $g(g(a)) \rightarrow^{oi} f(a, b)$ since $g(g(a)) \rightarrow f(g(a), g(a)) \rightarrow f(a, g(a)) \rightarrow f(a, b)$. But there is no term u of $T(\Gamma)$ such that $\Psi_l(u) = g(g(a))$ and $\Psi_r(u) = f(a, b)$.

We can prove that for every left-linear (respectively right-linear) TRS $\mathcal{R}$, $\mathsf{SF}_{\mathsf{io}}(L_1) = \Psi_r(\Psi_l^{-1}(L_1))$ (respectively $\mathsf{SF}_{\mathsf{oi}}(L_1) = \Psi_r(\Psi_l^{-1}(L_1))$). Therefore,

Proposition 4. *For any left-linear TRS "$\mathsf{SF}_{\mathsf{io}}(L_1) \subseteq L_2$?" is decidable, and for any left-linear and right-semi-linear TRS "$\mathsf{SF}_{\mathsf{io}}(L_1) = L_2$?" is decidable.*
For any right-linear TRS, "$\mathsf{SF}_{\mathsf{oi}}(L_1) \subseteq L_2$?" and "$\mathsf{SF}_{\mathsf{oi}}(L_1) = L_2$?" are decidable.

4.3 One-Pass Root-Started Rewriting and One-Pass Leaf-Started Rewriting

Z. Fülöp *et al.* introduce variants of the one-pass IO or OI rewritings in [FJSV98]: *one-pass root-started and leaf-started rewritings*. The first (respectively second) is a OI pass (respectively OI) pass in which each rewriting concerns positions immediately adjacent to parts of the term rewritten in previous rewritings.

Because these strategies are very similar to IO and OI ones, we obtain here similar results.

One-pass root-started rewriting may be described as follows. Let t be the term of $T(\Sigma)$ to be rewritten. The portion of t first rewritten should include the root. Rewriting then proceeds towards the leaves so that each rewrite step applies to a root segment of a maximal unprocessed subtree but never involves any part of the tree produced by a previous step. For the formal definition a new TRS in which a new special symbol forces this kind of rewriting is associated with $\mathcal{R}$. The reader is reported to [FJSV98] for a formal definition of this rewriting strategy.

Z. Fülöp *et al.* obtain decidability results when rewrite systems are *left*-linear. Surprisingly, our method leads to decidability results in the case of *right*-linear rewrite systems. Moreover, our techniques are powerful enough to strengthen the result from inclusion problems to equality problems.

In the following, $\rightarrow^{rs}$ denotes the one-pass root-started rewriting relation.

Let us denote $\mathsf{SF}_{\downarrow}(L) = \{t \in T(\Sigma) \mid \exists s \in \mathcal{L}, s \rightarrow^{rs} t\}$ and $\mathsf{NF}_{\downarrow}(L)$ the set of all one-pass root-started normal forms of L. Z. Fülöp *et al.* [FJSV98] have shown that "$\mathsf{SF}_{\downarrow}(L_1) \subseteq L_2$?" and "$\mathsf{NF}_{\downarrow}(L_1) \subseteq L_2$?" are decidable for any left-linear TRS $\mathcal{R}$.

Let us denote, for any term of $T(\Gamma)$, by $l_\pi(t)$ the string defined by the symbols on the branch π of t.

Sentential forms In order to verify that a one-pass root-started rewriting applies to a term t of $T(\Gamma)$, each word $l_\pi(t)$ must belong to $\Lambda^*\Sigma^*$. Let $R_c^{\downarrow s}$ be the set of terms t of $T(\Gamma)$ such that for any branch π of t, $l_\pi(t) \in \Lambda^*\Sigma^*$. We can prove that $R_c^{\downarrow s}$ is regular and that for any right-linear TRS $\mathcal{R}$:

$$\mathsf{SF}_{\downarrow}(L_1) = \Psi_r(\Psi_l^{-1}(L_1) \cap R_c^{\downarrow s}).$$

Right linearity is necessary for the same reason than for one-pass OI.

Proposition 5. *For any right-linear TRS* "$\mathsf{SF}_{\downarrow}(L_1) \subseteq L_2$?" *and* "$\mathsf{SF}_{\downarrow}(L_1) = L_2$?" *are decidable.*

Normal forms The only difficulty here is to define the checking language. According to the strategy, when $u \in \Psi_l^{-1}(t)$ satisfies one of the two following properties a rewrite rule can be applied: there is an instance of a rule at the root of u or there is an instance of a rule just below a dotted symbol in Λ.

Let $E_{\downarrow}^{FN}$, $R_{\downarrow 1}^{FN}$ and $R_{\downarrow 2}^{FN}$ be the following sets of terms:

$$E_{\downarrow}^{FN} = \{\bullet_i[y_1, \ldots, y_{l-1}, l_k[x_1, \ldots, x_{p_k}], y_{l+1}, \ldots, y_{p_i}] \mid i, k \in [n], l \in [p_i] \text{ and } x_1, \ldots, x_{p_k}, y_1, \ldots, y_{l-1}, y_{k+1}, \ldots, y_{p_i} \in \mathcal{X}\}$$

$$R_{\downarrow 2}^{FN} = \{t \in T(\Gamma) \mid t \text{ encompasses no term of } E_{\downarrow}^{FN}\}$$
$$R_{\downarrow 2}^{FN} = \{t \in T(\Gamma) \mid \nexists i \in [n], \nexists \sigma \text{ substitution}, t = \sigma l_i\}$$

We have $\mathsf{NF}_{\downarrow}(L_1) = \Psi_r(\Psi_l^{-1}(L_1) \cap R_c^{\downarrow s} \cap R_{\downarrow 1}^{FN} \cap R_{\downarrow 2}^{FN})$. Moreover, $R_{\downarrow 2}^{FN}$ is recognizable by a deterministic reduction automaton. Hence,

Proposition 6. *For any right-linear TRS "$\mathsf{NF}_{\downarrow}(L_1) \subseteq L_2$?" is decidable.*
For any linear TRS, "$\mathsf{NF}_{\downarrow}(L_1) = L_2$?" is decidable.

One-pass leaf-started rewriting One-pass leaf-started rewriting starts from leaves (either from constants or just above constants) and proceeds upwards. Each rewrite step applies just at the border of unprocessed part of the tree.

Z. Fülöp *et al.* obtain decidability results when rewrite systems are *left*-linear. On the one hand, we need stronger restrictions, but on the other hand we are able to strengthen the result from inclusion problems to equality problems.

In the following, $\rightarrow^{ls}$ denotes the one-pass leaf-started rewriting relation, $\mathsf{SF}_{\uparrow}(L) = \{t \in T(\Sigma) \mid \exists s \in L, s \rightarrow^{ls} t\}$ and $\mathsf{NF}_{\uparrow}(L)$ is the set of all one-pass leaf-started normal forms of L. Z. Fülöp *et al.* [FJSV98] have shown that "$\mathsf{SF}_{\uparrow}(L_1) \subseteq L_2$?" and "$\mathsf{NF}_{\uparrow}(L_1) \subseteq L_2$?" are decidable for any left-linear TRS $\mathcal{R}$.

Sentential forms. In order to verify that a one-pass leaf-started rewriting applies on a term t of $T(\Gamma)$, each string $l_\pi(t)$ must belong to $\Sigma^* \Lambda^* \Sigma_0 \cup \Sigma^* \Lambda^*$. Let $R_c^{\uparrow s}$ be the set of terms t of $T(\Gamma)$ such that for any branch π of t, $l_\pi(t) \in \Sigma^* \Lambda^* \Sigma_0 \cup \Sigma^* \Lambda^*$. We can prove that $R_c^{\uparrow s}$ is regular and that for any left-linear TRS $\mathcal{R}$:

$$\mathsf{SF}_{\uparrow}(L_1) = \Psi_r(\Psi_l^{-1}(L_1) \cap R_c^{\uparrow s}).$$

Left linearity is necessary for the same reason than for one-pass IO.

Proposition 7. *For any left-linear and right-semi-linear TRS, "$\mathsf{SF}_{\uparrow}(L_1) = L_2$?" is decidable.*

Normal forms $\forall i \in [n]$, let p_i be the number of distinct variables of l_i. Let $E_{\uparrow}^{NF}$ and $\mathcal{R}_{\uparrow}^{NF}$ be the following sets of terms:

$$E_{\uparrow}^{NF} = \{l_i[u_1, \ldots, u_{p_i}] \mid i \in [n], \forall j \in [p_i], u_j \in \Sigma_0 \text{ or } \exists k \in [n], \exists y_1, \ldots, y_{p_k} \in \mathcal{X}, u_j = \bullet_k(y_1, \ldots, y_{p_k}) \text{ and variables of } \textstyle\bigcup_{j \in [p_i]} Var(u_j) \text{ are pairwise distinct}\}$$
$$\mathcal{R}_{\uparrow}^{NF} = \{t \in T(\Gamma) \mid t \text{ encompasses no term of } E_{\uparrow}^{NF}\}$$

We can prove that for any term $t \in R_c^{\uparrow s} \cap \mathcal{R}_{\uparrow}^{NF}$, one-pass leaf-started rewriting cannot be continued from t since no left-hand side of rule occur just above symbols of Λ and constants. We deduce that $\mathsf{NF}_{\uparrow}(L_1) = \Psi_r(\Psi_l^{-1}(L_1) \cap R_c^{\uparrow s} \cap \mathcal{R}_{\uparrow}^{NF})$. Moreover when $\mathcal{R}$ is left-linear, terms of $E_{\uparrow}^{NF}$ are linear. Hence,

Proposition 8. *For any left-linear and right-semi-linear TRS, "$\mathsf{NF}_\uparrow(L_1) = L_2$?" is decidable.*

5 Multiple Rewritings

In this section, we study the language $\mathcal{R}^*(L)$ where $\mathcal{R}$ is a TRS of n rules and L is a regular language: $\mathcal{R}^*(L) = \{t \mid \exists s \in L, s \xrightarrow{*}_{\mathcal{R}} t\}$.

Generally, this language is not regular even when the language $\mathcal{R}(L)$ is. However some authors proved that $\mathcal{R}^*(L)$ is regular for some classes of TRS: M. Dauchet et S. Tison [DT90] for ground TRS, *i.e.* systems whose left and right-hand sides of rules are ground terms; K. Salomaa [Sal88] for right-linear monadic TRS, a rule $l \to r$ being monadic if $Height(l) \geq 1$ and $Height(r) \leq 1$; J.-L. Coquidé *et al.* [CDGV94] for semi-monadic TRS, a rule $l \to r$ being semi-monadic if $height(l) \geq 1$ and r is a variable or a term whose variables occurs at depth one.

Jacquemard proves in [Jac96b,Jac96a] that the set of ground terms normalizable with respect to a growing TRS is regular. A rule $l \to r$ is growing if it is linear and if every variable x that occurs in both sides of the rule, x occurs at depth one in l. All these results are based on a similar construction that iteratively transforms a tree automaton.

Our result is more restrictive than results of K. Salomaa [Sal88] and J.-L. Coquidé *et al.* [CDGV94] since ground terms are not allowed at depth one of right-hand sides of rules. But our proof is directly issued from the previous section.

We are able to compute $\mathcal{R}(L)$ with our method. One can iterate the construction and compute $\mathcal{R}^*(L)$ under two restrictions: the construction preserves recognizability, and it terminates. Recognizability is preserved under right-linearity assumption and termination is obtained when the size of the associated automaton does not increase. Unfortunately, when the rewrite system is not left-linear, a determinization procedure is necessary at each step to compute inverse homomorphisms and hence, an exponential blow up step appears. Therefore, we restrict ourself to the case of linear rewrite systems.

Proposition 9. *Let $\mathcal{R}$ be a rewrite system. For each language L' on Σ, $L' \cup \mathcal{R}(L') \subseteq \Psi_r(\Psi_l^{-1}(L')) \subseteq \mathcal{R}^*(L')$.*

Let us consider now the sequence of languages $(L_k)_{k \in N}$ defined by $L_0 = L$ and for every $k > 0$, $L_k = \Psi_r(\Psi_l^{-1}(L_{k-1}))$. We deduce from Proposition 9 that:

$$\forall k \in N,\ L_k \cup \mathcal{R}(L_k) \subseteq L_{k+1} \subseteq \mathcal{R}^*(L_k).$$

Hence for each integer k, $L_k \subseteq L_{k+1} \subseteq \mathcal{R}^*(L_k)$, therefore $L_k \subseteq L_{k+1} \subseteq \mathcal{R}^*(L_0)$. Hence the sequence $(L_k)_{k \in N}$ is ordered by inclusion and dominated by $\mathcal{R}^*(L_0)$, i.e:

$$L_0 \subseteq L_1 \ldots \subseteq L_k \subseteq \ldots \subseteq \mathcal{R}^*(L_0).$$

We want this sequence to be consist of a finite number of regular languages in order to obtain for some natural number k, $L_k = \mathcal{R}^*(L_0)$. Since the construction is based on the computation of $\Psi_r(\Psi_l^{-1}(L_i))$, we first need Ψ_r to be linear. Moreover, because we want to build a finite sequence, we require the number of states in the automaton for L_{i+1} not to be greater than the number of states in the automaton for L_i. Hence, some supplementary restrictions are needed for Ψ_l and Ψ_r. The classical construction for $\Psi_l^{-1}(L_i)$ requires a deterministic automaton for L_i. But in the case of linear morphisms Ψ, this requirement can be dropped. Finally, if Ψ_r generates terms of height greater than one, additional states are necessary. Hence,

Definition 1. *A simple rewrite system is a linear TRS whose each right-hand side of rule is either a constant, or a variable, or a term of height one without ground subterms.*

Therefore, mappings Ψ_l and Ψ_r associated with *simple* rewrite systems are such that Ψ_l is linear and for each symbol f of Γ: either $\psi_r(f)$ is a constant or a variable, or $\psi_r(f)$ is a linear term of height one without any ground subterm, i.e. there exist $g \in \Sigma_p$, $p \neq 0$, and p distinct variables $x_1, \ldots, x_p$ such that $\psi_r(f) = g(x_1, \ldots, x_p)$.

Proposition 10. *Let $\mathcal{R}$ be a simple TRS and let Ψ_r and Ψ_l be the homomorphisms associated with $\mathcal{R}$. Let L' be a regular language and let $\mathcal{A}$ be a tree automaton recognizing L'. There exists a tree automaton $\mathcal{D}$ whose set of states is included in the set of states of $\mathcal{A}$ recognizing $\Psi_r(\Psi_l^{-1}(L'))$.*

Proof. The standard construction that proves regular languages are closed under linear homomorphisms and inverse morphisms applies here. The reader is referred for instance to [GS84,CDG+97].

According to Proposition 10, for each integer k, the language L_k is recognized by a tree automaton $\mathcal{A}_k$ such that the set of states of $\mathcal{A}_{k+1}$ is included in the set of states of $\mathcal{A}_k$. Because the sequence of languages is ordered by inclusion and because for a given number of states, the number of tree automata recognizing a distinct language is bounded, there exists an integer i such that $\forall k \geq i$, $L_k = L_i$.

But $\forall k \in N$, $\mathcal{R}^k(L_0) \subseteq L_k$ since $\mathcal{R}(L_k) \subseteq L_{k+1}$. Hence $\mathcal{R}^k(L_0) \subseteq L_i$ since the sequence is growing. We deduce that $\mathcal{R}^*(L_0) \subseteq L_i$. Finally $L_i = \mathcal{R}^*(L_0) = \mathcal{R}^*(L)$ since $L_i \subseteq \mathcal{R}^*(L_0)$. Hence $\mathcal{R}^*(L)$ is regular.

Proposition 11. *Let $\mathcal{R}$ be a simple TRS. If L is a regular language, then $\mathcal{R}^*(L)$ is regular.*

Last example shows that the TRS has to be left-linear.

Example 6. Let the signature $\Sigma = \{a/0, g/1, f/2\}$, the TRS $\mathcal{R}$ composed of the only rule $f(x,x) \to g(x)$ and the language:

$$L = \{t \in T(\Sigma) \mid t = f(f(\ldots f(a, t_1), \ldots, t_{n-1}), t_n) \text{ where } t_1, \ldots, t_n \in T(\{g, a\})\}.$$

We associate with the rule of $\mathcal{R}$ the symbol $\bullet$. The homomorphisms Ψ_l et Ψ_r are determined by:

$$\begin{array}{ll} \psi_l(a) = a & \psi_r(a) = a \\ \psi_l(g) = g(x_1) & \psi_r(g) = g(x_1) \\ \psi_l(f) = f(x_1, x_2) & \psi_r(f) = f(x_1, x_2) \\ \psi_l(\bullet) = f(x_1, x_1) & \psi_r(\bullet) = g(x_1) \end{array}$$

For each integer n, let us denote by t_n the term $f(f(\ldots f(a,a), \ldots, g^{n-1}(a)), g^n(a))$ of L. Let $n \in N$. The term t_n rewrites in $g^{n+1}(a)$ by $n+1$ applications of $\mathcal{R}$'s rule hence $t_n \in \mathcal{R}^*(L)$. Moreover t_n gives $g^{n+1}(a)$ by exactly $n+1$ applications of $\Psi_r(\Psi_l^{-1})$, i.e $t_n \in L_{k+1}$ and $t_n \notin L_k$. We deduce that the sequence $(L_k)_{k \in N}$ is infinite.

5.1 Summary

Our other results and those of Z. Fülöp *et al.* [FJSV98] are summarized in the following table.

	ordinary	right semi-lin.	right lin.	left lin.	left lin. right semi-lin.	linear
$\mathcal{R}$	$\subseteq$	$\subseteq, =$	$\subseteq, =$	$\subseteq$	$\subseteq, =$	$\subseteq, =$
$\mathcal{R}_{\parallel}$	$\subseteq$	$\subseteq, =$	$\subseteq, =$	$\subseteq$	$\subseteq, =$	$\subseteq, =$
SF_{io}				$\subseteq$	$\subseteq, =$	$\subseteq, =$
SF_{oi}			$\subseteq, =$			$\subseteq, =$
$SF_{\uparrow}$				$\subseteq$[FJSV98]	$\subseteq, =$	$\subseteq, =$
$NF_{\uparrow}$				$\subseteq$[FJSV98]	$\subseteq, =$	$\subseteq, =$
$SF_{\downarrow}$			$\subseteq, =$	$\subseteq$[FJSV98]	$\subseteq$	$\subseteq, =$
$NF_{\downarrow}$			$\subseteq$	$\subseteq$[FJSV98]	$\subseteq$	$\subseteq, =$

References

[AD82] A. Arnold and M. Dauchet. Morphismes et bimorphismes d'arbres. *Theorical Computer Science*, 20:33–93, 1982.

[BST99] B. Bogaert, F. Seynhaeve, and S. Tison. The recognizability problem for tree automata with comparisons between brothers. In W. Thomas, editor, *Proceedings, Foundations of Software Science and Computation Structures*, number 1578 in Lecture Notes in Computer Science, Amsterdam, 1999. Springer Verlag.

[CDG+97] H. Comon, M. Dauchet, R. Gilleron, , F. Jacquemard, D. Lugiez, S. Tison, and M. Tommasi. Tree automata techniques and applications. Available on: `http://www.grappa.univ-lille3.fr/tata`, 1997.

[CDGV94] J.L. Coquidé, M. Dauchet, R. Gilleron, and S. Vágvolgyi. Bottom-up tree pushdown automata : Classification and connection with rewrite systems. *Theorical Computer Science*, 127:69–98, 1994.

[DCC95] M. Dauchet, A.-C. Caron, and J.-L. Coquidé. Reduction properties and automata with constraints. *Journal of Symbolic Computation*, 20:215–233, 1995.

[DT90] M. Dauchet and S. Tison. The theory of ground rewrite systems is decidable. In *Proceedings, Fifth Annual IEEE Symposium on Logic in Computer Science*, pages 242–248. IEEE Computer Society Press, 1990.

[Eng75] J. Engelfriet. Bottom-up and top-down tree transformations. a comparision. *Mathematical System Theory*, 9:198–231, 1975.

[Eng78] J. Engelfriet. A hierarchy of tree transducers. In *Proceedings of the third Les Arbres en Algèbre et en Programmation*, pages 103–106, Lille, 1978.

[ES77] J. Engelfriet and E.M. Schmidt. IO and OI I. *Journal of Comput. and Syst. Sci.*, 15:328–353, 1977.

[ES78] J. Engelfriet and E.M. Schmidt. IO and OI II. *Journal of Comput. and Syst. Sci.*, 16:67–99, 1978.

[FJSV98] A. Fülöp, E. Jurvanen, M. Steinby, and S. Vágvölgy. On one-pass term rewriting. In L. Brim, J. Gruska, and J. Zlatusaksv, editors, *Proceedings of Mathematical Foundations of Computer Science*, volume 1450 of *Lecture Notes in Computer Science*, pages 248–256. Springer Verlag, 1998.

[Gen98] T. Genet. Decidable approximations of sets of descendants and sets of normal forms. In Nipkow [Nip98], pages 151–165.

[GS84] F. Gécseg and M. Steinby. *Tree Automata.* Akademiai Kiado, 1984.

[Jac96a] F. Jacquemard. *Automates d'arbres et réécriture de termes.* PhD thesis, Université de Paris XI, 1996.

[Jac96b] F. Jacquemard. Decidable approximations of term rewriting systems. In H. Ganzinger, editor, *Proceedings. Seventh International Conference on Rewriting Techniques and Applications*, volume 1103 of *Lecture Notes in Computer Science*, 1996.

[Nip98] T. Nipkow, editor. *Proceedings. Ninth International Conference on Rewriting Techniques and Applications*, volume 1379 of *Lecture Notes in Computer Science*, Tsukuba, 1998.

[Sal88] K. Salomaa. Deterministic tree pushdown automata and monadic tree rewriting systems. *Journal of Comput. and Syst. Sci.*, 37:367–394, 1988.

[Wal98] J. Waldmann. Normalization of s-terms is decidable. In Nipkow [Nip98], pages 138–150.

On Two-Sided Infinite Fixed Points of Morphisms

Jeffrey Shallit* and Ming-wei Wang

Department of Computer Science
University of Waterloo
Waterloo, Ontario, Canada N2L 3G1
shallit@graceland.uwaterloo.ca
m2wang@neumann.uwaterloo.ca

Abstract. Let Σ be a finite alphabet, and let $h : \Sigma^* \to \Sigma^*$ be a morphism. Finite and infinite fixed points of morphisms — i.e., those words w such that $h(w) = w$ — play an important role in formal language theory. Head characterized the finite fixed points of h, and later, Head and Lando characterized the one-sided infinite fixed points of h. Our paper has two main results. First, we complete the characterization of fixed points of morphisms by describing all two-sided infinite fixed points of h, for both the "pointed" and "unpointed" cases. Second, we completely characterize the solutions to the equation $h(xy) = yx$ in finite words.

1 Introduction and definitions

Let Σ be a finite alphabet, and let $h : \Sigma^* \to \Sigma^*$ be a morphism on the free monoid, i.e., a map satisfying $h(xy) = h(x)h(y)$ for all $x, y \in \Sigma^*$. If a word w (finite or infinite) satisfies the equation $h(w) = w$, then we call w a *fixed point* of h. Both finite and infinite fixed points of morphisms have long been studied in formal languages. For example, in one of the earliest works on formal languages, Axel Thue [12, 3] proved that the one-sided infinite word $\mathbf{t} = 0110100110010110\cdots$ is overlap-free, that is, contains no subword of the form $axaxa$, where $a \in \{0, 1\}$, and $x \in (0 + 1)^*$. Define a morphism μ by $\mu(0) = 01$ and $\mu(1) = 10$. The word $\mathbf{t}$, now called the Thue-Morse infinite word, is the unique one-sided infinite fixed point of μ which starts with 0. In fact, nearly every explicit construction of an infinite word avoiding certain patterns involves the fixed point of a morphism; for example, see [6]. One-sided infinite fixed points of uniform morphisms also play a crucial role in the theory of automatic sequences; see, for example, [1].

Because of their importance in formal languages, it is of great interest to characterize *all* the fixed points, both finite and infinite, of a morphism h. This problem was first studied by Head [7], who characterized the finite fixed points of h. Later, Head and Lando [8] characterized the one-sided infinite fixed points of

* Research supported in part by a grant from NSERC. A full version of this paper can be found at `http://math.uwaterloo.ca/~shallit/papers.html` .

h. In this paper we complete the description of all fixed points of morphisms by characterizing the *two-sided* infinite fixed points of h. Two-sided infinite words (sometimes called ***bi-infinite words*** or ***bi-infinite sequences***) play an important role in symbolic dynamics [10], and have also been studied in automata theory [11].

We first introduce some notation, some of which is standard and can be found in [9]. For single letters, that is, elements of Σ, we use the lower case letters a, b, c, d. For finite words, we use the lower case letters t, u, v, w, x, y, z. For infinite words, we use bold-face letters $\mathbf{t}, \mathbf{u}, \mathbf{v}, \mathbf{w}, \mathbf{x}, \mathbf{y}, \mathbf{z}$. We let ϵ denote the empty word. If $w \in \Sigma^*$, then by $|w|$ we mean the length of, or number of symbols in w. If S is a set, then by Card S we mean the number of elements of S. We say $x \in \Sigma^*$ is a *subword* of $y \in \Sigma^*$ if there exist words $w, z \in \Sigma^*$ such that $y = wxz$.

If h is a morphism, then we let h^j denote the j-fold composition of h with itself. If there exists an integer $j \geq 1$ such that $h^j(a) = \epsilon$, then the letter a is said to be *mortal*; otherwise a is *immortal*. The set of mortal letters associated with a morphism h is denoted by M_h. The *mortality exponent* of a morphism h is defined to be the least integer $t \geq 0$ such that $h^t(a) = \epsilon$ for all $a \in M_h$. We write the mortality exponent as $\exp(h) = t$. It is easy to prove that $\exp(h) \leq \text{Card } M_h$.

We let Σ^ω denote the set of all one-sided right-infinite words over the alphabet Σ. Most of the definitions above extend to Σ^ω in the obvious way. For example, if $\mathbf{w} = c_1 c_2 c_3 \cdots$, then $h(\mathbf{w}) = h(c_1)h(c_2)h(c_3)\cdots$. If $L \subseteq \Sigma^*$ is a language, then we define $L^\omega := \{w_1 w_2 w_3 \cdots \ : \ w_i \in L - \{\epsilon\} \text{ for all } i \geq 1\}$. Perhaps slightly less obviously, we can also define a limiting word $\overrightarrow{h^\omega}(a) := \lim_{n\to\infty} h^n(a)$ for a letter a, provided $h(a) = wax$ and $w \in M_h^*$. In this case, there exists $t \geq 0$ such that $h^t(w) = \epsilon$. Then we define $\overrightarrow{h^\omega}(a) := h^{t-1}(w) \cdots h(w)\, w\, a\, x\, h(x)\, h^2(x) \cdots$, which is infinite if and only if $x \notin M_h^*$. Note that the factorization of $h(a)$ as wax, with $w \in M_h^*$ and $x \notin M_h^*$, if it exists, is unique.

In a similar way, we let ${}^\omega\Sigma$ denote the set of all left-infinite words, which are of the form $\mathbf{w} = \cdots c_{-2}c_{-1}c_0$. We write $h(\mathbf{w}) = \cdots h(c_{-2})h(c_{-1})h(c_0)$. We define ${}^\omega L$ to be the set of left-infinite words formed by concatenating infinitely many words from L, that is, ${}^\omega L := \{\cdots w_{-2}w_{-1}w_0 \ : \ w_i \in L - \{\epsilon\} \text{ for all } i \leq 0\}$. If $h(a) = wax$, and $w \notin M_h^*$, $x \in M_h^*$, then we define the left-infinite word $\overleftarrow{h^\omega}(a) := \cdots h^2(w)\, h(w)\, w\, a\, x\, h(x) \ \cdots h^{t-1}(x)$, where $h^t(x) = \epsilon$. Again, if the factorization of $h(a)$ as wax exists, with $w \notin M_h^*$, $x \in M_h^*$, then it is unique.

We can convert left-infinite to right-infinite words (and vice versa) using the reverse operation, which is denoted $\mathbf{w}^R$. For example, if $w = c_0 c_1 c_2 \cdots$, then $\mathbf{w}^R = \cdots c_2 c_1 c_0$.

We now turn to the notation for two-sided infinite words. These have been much less studied in the literature than one-sided words, and the notation has not been standardized. Some authors consider 2 two-sided infinite words to be identical if they agree after applying a finite shift to one of the words. Other authors do not. (This distinction is sometimes called "unpointed" vs. "pointed" [2].) In this paper, we consider both the pointed and unpointed versions of the equation $h(\mathbf{w}) = \mathbf{w}$. As it turns out, the "pointed" version of this equation is quite easy to solve, based on known results, while the "unpointed" case is significantly more difficult. The latter is our first main result, which appears as Theorem 5.

We let $\Sigma^{\mathbb{Z}}$ denote the set of all two-sided infinite words over the alphabet Σ, which are of the form $\cdots c_{-2}c_{-1}c_0.c_1c_2\cdots$. In displaying an infinite word as a concatenation of words, we use a decimal point to the left of the character c_1, to indicate how the word is indexed. Of course, the decimal point is not part of the word itself. We define the *shift* $\sigma(\mathbf{w})$ to be the two-sided infinite word obtained by shifting $\mathbf{w}$ to the left one position, so that $\sigma(\cdots c_{-2}c_{-1}c_0.c_1c_2c_3\cdots) = \cdots c_{-1}c_0c_1.c_2c_3c_4\cdots$. Similarly, for $k \in \mathbb{Z}$ we define $\sigma^k(\cdots c_{-2}c_{-1}c_0.c_1c_2c_3\cdots) = \cdots c_{k-1}c_k.c_{k+1}c_{k+2}\cdots$. If $\mathbf{w}, \mathbf{x}$ are 2 two-sided infinite words, and there exists an integer k such that $\mathbf{x} = \sigma^k(\mathbf{w})$, then we call $\mathbf{w}$ and $\mathbf{x}$ *conjugates*, and we write $\mathbf{w} \sim \mathbf{x}$. It is easy to see that $\sim$ is an equivalence relation. We extend this notation to languages as follows: if L is a set of two-sided infinite words, then by $\mathbf{w} \sim L$ we mean there exists $\mathbf{x} \in L$ such that $\mathbf{w} \sim \mathbf{x}$.

If w is a nonempty finite word, then by $w^{\mathbb{Z}}$ we mean the two-sided infinite word $\cdots www.www\cdots$. Using concatenation, we can join a left-infinite word $\mathbf{w} = \cdots c_{-2}c_{-1}c_0$ with a right-infinite word $\mathbf{x} = d_0d_1d_2\cdots$ to form a new two-sided infinite word, as follows: $\mathbf{w}.\mathbf{x} := \cdots c_{-2}c_{-1}c_0.d_0d_1d_2\cdots$. If $L \subseteq \Sigma^*$ is a set of words, then we define $L^{\mathbb{Z}} := \{\cdots w_{-2}w_{-1}w_0.w_1w_2\cdots \ : \ w_i \in L - \{\epsilon\} \text{ for all } i \in \mathbb{Z}\}$. If $\mathbf{w} = \cdots c_{-2}c_{-1}c_0.c_1c_2\cdots$, and h is a morphism, then we define

$$h(\mathbf{w}) := \cdots h(c_{-2})h(c_{-1})h(c_0).h(c_1)h(c_2)\cdots \tag{1}$$

Finally, if $i = |wa|$, $h(a) = wax$, and $w, x \notin M_h^*$, then we define

$$\overleftrightarrow{h^{\omega;i}}(a) := \cdots h^2(w)\,h(w)\,w\,.a\,x\,h(x)\,h^2(x)\cdots,$$

a two-sided infinite word. Note that in this case the factorization of $h(a)$ as wax is *not* necessarily unique, and we use the superscript i to indicate which a is being chosen.

2 Finite and one-sided infinite fixed points

In this section we recall the results of Head [7] and Head and Lando [8]. We assume $h : \Sigma^* \to \Sigma^*$ is a morphism that is extended to the domains Σ^ω and ${}^\omega\Sigma$ in the manner discussed above. Define

$$A_h = \{a \in \Sigma \ : \ \exists\, x, y \in \Sigma^* \text{ such that } h(a) = xay \text{ and } xy \in M_h^*\}$$

and

$$F_h = \{h^t(a) \ : \ a \in A_h \text{ and } t = \exp(h)\}.$$

Note that there is at most one way to write $h(a)$ in the form xay with $xy \in M_h^*$.

Theorem 1 (Head). *A finite word $w \in \Sigma^*$ has the property that $w = h(w)$ if and only if $w \in F_h^*$.*

Theorem 2 (Head & Lando). *The right-infinite word* $\mathbf{w}$ *is a fixed point of* h *if and only if at least one of the following two conditions holds:*

(a) $\mathbf{w} \in F_h^\omega$*; or*

(b) $\mathbf{w} \in F_h^* \ \overrightarrow{h^\omega}(a)$ *for some* $a \in \Sigma$*, and there exist* $x \in M_h^*$ *and* $y \notin M_h^*$ *such that* $h(a) = xay$*.*

There is also an evident analogue of Theorem 2 for left-infinite words:

Theorem 3. *The left-infinite word* $\mathbf{w}$ *is a fixed point of* h *if and only if at least one of the following two conditions holds:*

(a) $\mathbf{w} \in {}^\omega F_h$*; or*

(b) $\mathbf{w} \in \overleftarrow{h^\omega}(a) F_h^*$ *for some* $a \in \Sigma$*, and there exist* $x \notin M_h^*$ *and* $y \in M_h^*$ *such that* $h(a) = xay$*.*

3 Two-sided infinite fixed points: the "pointed" case

In this section, we consider the equation $h(\mathbf{w}) = \mathbf{w}$ for two-sided infinite words. The next result follows immediately:

Proposition 4. *The equation* $h(\mathbf{w}) = \mathbf{w}$ *has a solution if and only if* $\mathbf{w} = \mathbf{x}.\mathbf{y}$ *for a left-infinite word* $\mathbf{x}$*, and a right-infinite word* $\mathbf{y}$*, where* $\mathbf{x}$ *is given by Theorem 3 and* $\mathbf{y}$ *is given by Theorem 2.*

Example. Let μ be the Thue-Morse morphism, which maps $0 \to 01$, and $1 \to 10$. Define $g = \mu^2$. Then $g(0) = 0110$, $g(1) = 1001$. Let $\mathbf{t} = 01101001\cdots$, the one-sided Thue-Morse infinite word. Then there are exactly 4 two-sided infinite fixed points of g, as follows:

$$\begin{aligned} \mathbf{t}^R.\mathbf{t} &= \cdots 10010110.01101001\cdots \\ \overline{\mathbf{t}}^R.\mathbf{t} &= \cdots 01101001.01101001\cdots \\ \overline{\mathbf{t}}^R.\overline{\mathbf{t}} &= \cdots 01101001.10010110\cdots \\ \mathbf{t}^R.\overline{\mathbf{t}} &= \cdots 10010110.10010110\cdots , \end{aligned}$$

where $\overline{0} = 1$, $\overline{1} = 0$. All of these fall under case (d) of Theorem 4. Incidentally, all four of these words are overlap-free.

4 Two-sided infinite fixed points: the "unpointed" case

We assume $h : \Sigma^* \to \Sigma^*$ is a morphism that is extended to the domain $\Sigma^{\mathbf{Z}}$ in the manner discussed above. In this section, we characterize the two-sided infinite fixed points of a morphism in the "unpointed" case. That is, our goal is to characterize the solutions to $h(\mathbf{w}) \sim \mathbf{w}$. The following theorem is one of our two main results.

Theorem 5. *Let h be a morphism. Then the two-sided infinite word $\mathbf{w}$ satisfies the relation $h(\mathbf{w}) \sim \mathbf{w}$ if and only if at least one of the following conditions holds:*

(a) $\mathbf{w} \sim F_h^{\mathbb{Z}}$*; or*

(b) $\mathbf{w} \sim \overleftarrow{h^\omega}(a) \,.\, F_h^\omega$ *for some $a \in \Sigma$, and there exist $x \notin M_h^*$ and $y \in M_h^*$ such that $h(a) = xay$; or*

(c) $\mathbf{w} \sim {}^\omega F_h \,.\, \overrightarrow{h^\omega}(a)$ *for some $a \in \Sigma$, and there exist $x \in M_h^*$ and $y \notin M_h^*$ such that $h(a) = xay$; or*

(d) $\mathbf{w} \sim \overleftarrow{h^\omega}(a) \,.\, F_h^* \, \overrightarrow{h^\omega}(b)$ *for some $a, b \in \Sigma$ and there exist $x, z \notin M_h^*$, $y, w \in M_h^*$, such that $h(a) = xay$ and $h(b) = wbz$; or*

(e) $\mathbf{w} \sim \overleftrightarrow{h^{\omega;i}}(a)$ *for some $a \in \Sigma$, and there exist $x, y \notin M_h^*$ such that $h(a) = xay$ with $|xa| = i$; or*

(f) $\mathbf{w} = (xy)^{\mathbb{Z}}$ *for some $x, y \in \Sigma^+$ such that $h(xy) = yx$.*

Before we prove Theorem 5, let us look at two examples.

Example 1. Consider the morphism f defined by a $\to$ bb, b $\to$ ϵ, c $\to$ aad, d $\to$ c. Let

$$\mathbf{w} = \cdots \mathtt{aadbbbbcaadbbbbc.aadbbbbcaadbbbbc} \cdots .$$

Then

$$f(\mathbf{w}) = \cdots \mathtt{bbbbcaadbbbbcaad.bbbbcaadbbbbcaad} \cdots .$$

This falls under case (f) of Theorem 5.

Example 2. Consider the morphism φ defined by $0 \to 201$, $1 \to 012$, and $2 \to 120$. Then if $\mathbf{w} = \overleftrightarrow{\varphi^{\omega;2}}(0) = \cdots c_{-2}c_{-1}.c_0c_1c_2 \cdots = \cdots 1202.01012 \cdots$, we have $\varphi(\mathbf{w}) \sim \mathbf{w}$. This falls under case (e) of Theorem 5. Incidentally, c_i equals the sum of the digits, modulo 3, in the balanced ternary representation of i.

We now prepare for the proof of Theorem 5 by stating three easy but useful lemmas without proof.

Lemma 6. *Suppose $\mathbf{w}$, $\mathbf{x}$ are 2 two-sided infinite words with $\mathbf{w} \sim \mathbf{x}$. Then $h(\mathbf{w}) \sim h(\mathbf{x})$.*

Our second lemma concerns periodicity of infinite words. We say a two-sided infinite word $\mathbf{w}$ is *periodic* if there exists an integer p (called a *period*) such that $\mathbf{w} = \sigma^p(\mathbf{w})$.

Lemma 7. *Suppose $\mathbf{w} = \cdots c_{-2}c_{-1}c_0.c_1c_2 \cdots$ is a two-sided infinite word such that there exists a one-sided right-infinite word $\mathbf{x}$ and infinitely many negative indices $0 > i_1 > i_2 > \cdots$ such that $\mathbf{x} = c_{i_j}c_{i_j+1}c_{i_j+2} \cdots$ for $j \geq 1$. Then $\mathbf{w}$ is periodic.*

Our third lemma concerns the growth functions of iterated morphisms.

Lemma 8. *Let $h : \Sigma^* \to \Sigma^*$ be a morphism. Then*

(a) there exist integers i, j with $0 \leq i < j$ and $|h^i(w)| \leq |h^j(w)|$ for all $w \in \Sigma^*$; and

(b) there exists an integer M depending only on $k = \text{Card } \Sigma$ such that for all $h : \Sigma^* \to \Sigma^*$, we have $j \leq M$.

We note that part (a) was asserted without proof by Cobham [4]. However, the proof easily follows from a result of Dickson [5] that $\mathbb{N}^k$ contains no infinite antichains under the usual partial ordering. For part (b), it is further known that we can take $M = 2^k$. See [13]. □

Now we can prove Theorem 5.

Proof. ($\Longleftarrow$): Suppose case (a) holds, and $\mathbf{w} \sim F_h^{\mathbb{Z}}$. Then there exists $\mathbf{x} \in F_h^{\mathbb{Z}}$ with $\mathbf{w} \sim \mathbf{x}$. Since $\mathbf{x} \in F_h^{\mathbb{Z}}$, we can write $\mathbf{x} = \cdots x_{-2}x_{-1}x_0.x_1x_2\cdots$, where $x_i \in F_h$ for all $i \in \mathbb{Z}$. Since $x_i \in F_h$, we have $h(x_i) = x_i$ for all $i \in \mathbb{Z}$. It follows that $h(\mathbf{x}) = \mathbf{x}$. Now, applying Lemma 6, we conclude that $h(\mathbf{w}) \sim h(\mathbf{x}) = \mathbf{x} \sim \mathbf{w}$.

Next, suppose case (b) holds, and $\mathbf{w} \sim \overleftarrow{h^{\omega}}(a).F_h^{\omega}$. Then $\mathbf{w} \sim \mathbf{x}$ for some $\mathbf{x}$ of the form $\mathbf{x} = \overleftarrow{h^{\omega}}(a).x_1x_2x_3\cdots$, where $x_i \in F_h$ for all $i \geq 1$, and $h(a) = xay$ with $x \notin M_h^*$ and $y \in M_h^*$. Then we have $h(\mathbf{x}) = \mathbf{x}$, and by Lemma 6, we conclude that $h(\mathbf{w}) \sim h(\mathbf{x}) = \mathbf{x} \sim \mathbf{w}$.

Cases (c), (d), and (e) are similar to case (b).

Finally, if case (f) holds, then $h(\mathbf{w}) = h(\cdots xyxy.xyxy\cdots) = \cdots yxyx.yxyx\cdots$, and so $h(\mathbf{w}) = \sigma^k(\mathbf{w})$ for $k = |x|$.

($\Longrightarrow$): Suppose $\mathbf{w} = \cdots c_{-2}c_{-1}c_0.c_1c_2\cdots$, and there exists k such that $h(\mathbf{w}) = \sigma^k(\mathbf{w})$. Let

$$s(i) := \begin{cases} |h(c_1c_2\cdots c_i)| + k, & \text{if } i \geq 0; \\ k - |h(c_{i+1}c_{i+2}\cdots c_0)|, & \text{if } i < 0. \end{cases} \tag{2}$$

Then it is not hard to see that

$$h(c_i) = c_{s(i-1)+1}\cdots c_{s(i)} \tag{3}$$

for $i \in \mathbb{Z}$; see Figure 1. Note that $s(0) = k$.

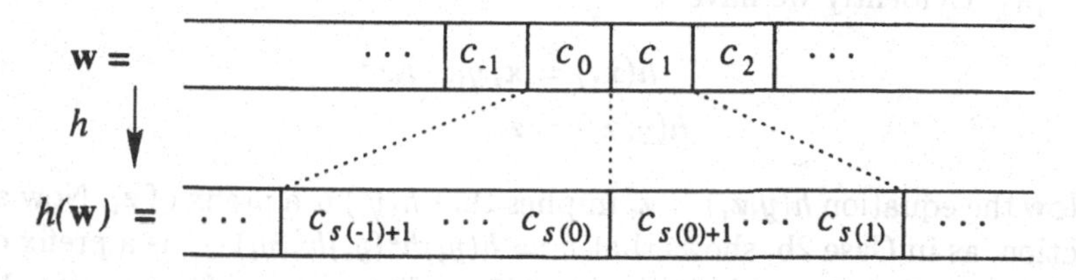

Fig. 1. Interpretation of the function s

We define the set C as follows: $C = \{i \in \mathbb{Z} \ : \ s(i) = i\}$. Our argument is divided into two major cases, depending on whether or not C is empty.

Case 1: $C \neq \emptyset$. In this there exists j such that $s(j) = j$. Now consider the pointed word $\mathbf{x} = \cdots c_{j-2}c_{j-1}c_j.c_{j+1}c_{j+2}\cdots$. We have $\mathbf{x} \sim \mathbf{w}$ and by Eq. (3) we have $h(\mathbf{x}) = \mathbf{x}$. Then, by Proposition 4, one of cases (a)–(d) must hold.

Case 2: $C = \emptyset$. There are several subcases to consider.

Case 2a: There exist integers i, j with $i < j$ such that

$$s(i) > i \text{ but } s(j) < j. \tag{4}$$

Then choose i, j satisfying (4) with $j - i$ minimal. Suppose there exists an integer k with $i < k < j$. If $s(k) < k$, then (i, k) is a pair with smaller difference, while if $s(k) > k$, then (k, j) is a pair with smaller difference, a contradiction. Hence $s(k) = k$. But this is impossible by our assumption. It follows that $j = i + 1$. Then $s(i) > i$, but $s(i+1) < i + 1$, a contradiction, since $s(i) \le s(i+1)$. Hence this case cannot occur.

Case 2b: There exists an integer r such that $s(i) < i$ for all $i < r$, and $s(i) > i$ for all $i \ge r$. Then $h(c_r) = c_{s(r-1)+1} \cdots c_{s(r)}$, which by the inequalities contains $c_{r-1}c_rc_{r+1}$ as a subword. Therefore, letting $a = c_r$, it follows that $\mathbf{w} \sim \mathbf{u}\, x\,.\,a\, y\, \mathbf{v}$, where $\mathbf{u} = \cdots c_{s(r-1)-1}c_{s(r-1)}$ is a left-infinite word, $x = c_{s(r-1)+1} \cdots c_{r-1}$ and $y = c_{r+1} \cdots c_{s(r)}$ are finite words, and $\mathbf{v} = c_{s(r)+1}c_{s(r)+2}\cdots$ is a right-infinite word. Furthermore, we have $h(\mathbf{u}x) = \mathbf{u}$, $h(a) = xay$, and $h(y\mathbf{v}) = \mathbf{v}$.

Now the equation $h(y\mathbf{v}) = \mathbf{v}$ implies that $h(y)$ is a prefix of $\mathbf{v}$, and by an easy induction we have $h(y)h^2(y)h^3(y)\cdots$ is a prefix of $\mathbf{v}$. Suppose this prefix is finite. Then $y \in M_h^*$, and so $h(y)h^2(y)h^3(y)\cdots = h(y)h^2(y)\cdots h^t(y)$, where $t = \exp(h)$. Define $z = h(y)h^2(y)\cdots h^t(y)$. Then $s(r+|y|+|z|) = r+|y|+|z|$, a contradiction, since we have assumed $C = \emptyset$. It follows that $\mathbf{z} := h(y)h^2(y)h^3(y)\cdots$ is right-infinite and hence $y \notin M_h^*$.

By exactly the same reasoning, we find that $\cdots h^3(x)\, h^2(x)\, h(x)$ is a left-infinite suffix of $\mathbf{u}$. We conclude that $\mathbf{w} \sim \overleftrightarrow{h^{\omega;i}}(a)$, and hence case (e) holds.

Case 2c: $s(i) > i$ for all $i \in \mathbb{Z}$. Let $\mathbf{w} = \cdots c_{-2}c_{-1}c_0.c_1c_2\cdots$.

Now consider the following factorization of certain conjugates of $\mathbf{w}$, as follows: for $i \le 0$, we have $\mathbf{w} \sim \mathbf{x}_i\, y_i\,.\,\mathbf{z}_i$, where $\mathbf{x}_i = \cdots c_{i-2}c_{i-1}$ (a left-infinite word), $y_i = c_i \cdots c_{s(i-1)}$ (a finite word), and $\mathbf{z}_i = c_{s(i-1)+1}c_{s(i-1)+2}\cdots$ (a right-infinite word). Note that $i - 1 < s(i-1)$ by assumption, so $i \le s(i-1)$; hence y_i is nonempty. Evidently we have

$$\begin{aligned} h(\mathbf{x}_i) &= \mathbf{x}_i\, y_i; \quad \text{and} \\ h(y_i\, \mathbf{z}_i) &= \mathbf{z}_i. \end{aligned} \tag{5}$$

Now the equation $h(y_i\mathbf{z}_i) = \mathbf{z}_i$ implies that $h(y_i)$ is a prefix of $\mathbf{z}_i$. Now an easy induction, as in Case 2b, shows that $v := h(y_i)h^2(y_i)h^3(y_i)\cdots$ is a prefix of $\mathbf{z}_i$. If v were finite, then we would have $y_i \in M_h^*$, and so $s(j) = j$ for $j = s(i-1) + |v|$, a contradiction, since $C = \emptyset$. Hence v is right-infinite, and so $y_i \notin M_h^*$. There are now two further subcases to consider: (i) $\sup_{i\leq 0}(s(i)-i) < +\infty$, and (ii) $\sup_{i\leq 0}(s(i)-i) = +\infty$.

Case 2ci: Suppose $\sup_{i\leq 0}(s(i)-i) = d < +\infty$. It then follows that $|y_i| \leq d$. Hence there is a finite word u such that $y_i = u$ for infinitely many indices $i \leq 0$. From the above argument we see that the right-infinite word $h(u)h^2(u)h^3(u)\cdots$ is a suffix of $\mathbf{w}$, beginning at position $s(i-1)+1$, for infinitely many indices $i \leq 0$. We now use Lemma 7 to conclude that $\mathbf{w}$ is periodic.

Thus we can write $\mathbf{w} = \cdots c_{-2}c_{-1}c_0.c_1c_2\cdots$, and $\mathbf{w} = \cdots vvv.vvv\cdots$, where $v = c_1c_2\cdots c_p$ for some integer $p \geq 1$. Without loss of generality, we may assume p is minimal.

We claim $|h(v)| = p$. For if not we must have $|h(v)| = q$, for $q \neq p$, and then since $h(\mathbf{w}) \sim \mathbf{w}$, we would have $\mathbf{w}$ is periodic with periods p and q, hence periodic of period $\gcd(p,q)$. But since p was minimal we must have $p\,|\,q$. Hence $q \geq 2p$. Now let $s(p) = l$; since $s(i) > i$ for all i we must have $l > 0$. Then $h(c_1c_2\cdots c_p) = c_{s(-1)+1}\cdots c_{s(p)} = c_{l-q+1}\cdots c_l$. It now follows that

$$s(ip) = l - q + iq \tag{6}$$

for all integers i. Now $p < q$, so $p \leq q-1$, and hence $p < q - 1 + q/l$. Hence, multiplying by $-l$, we get $-lp > l - ql - q$. Now take $i = -l$ in Eq. (6), and we have $s(-lp) = l - q - lq < -lp$, a contradiction, since $s(i) > i$ for all i. It follows that $|h(v)| = p$.

There exists k such that $h(c_1c_2\cdots c_p) = c_{k+1}c_{k+2}\cdots c_{k+p}$. Using the division theorem, write $k = jp + r$, where $0 \leq r < p$. Define

$$y = c_{k+1}\cdots c_{(j+1)p} = c_{r+1}\cdots c_p;$$
$$x = c_{(j+1)p+1}\cdots c_{k+p} = c_1\cdots c_r.$$

We have $h(xy) = yx$, and $v = xy$. Then $\mathbf{w} = v^{\mathbb{Z}} = (xy)^{\mathbb{Z}}$.

By above we know $|v| \geq 1$, so $xy \neq \epsilon$. Suppose $y = \epsilon$. Then $h(x) = x$, and so $x \in F_h^*$. It follows that $\mathbf{w} \in F_h^{\mathbb{Z}}$. A similar argument applies if $x = \epsilon$. However, if $\mathbf{w} \in F_h^{\mathbb{Z}}$, then $C \neq \emptyset$, a contradiction. Thus $x, y \neq \epsilon$, and case (f) holds.

Case 2cii: $\sup_{i\leq 0}(s(i)-i) = +\infty$. Recall that $s(i) > i$ for all $i \in \mathbb{Z}$ and $\mathbf{w} = \cdots c_{-2}c_{-1}c_0.c_1c_2\cdots$. Define

$$\mathbf{x} := \cdots c_{-2}c_{-1}c_0;$$
$$y := c_1c_2\cdots c_{s(0)};$$
$$\mathbf{z} := c_{s(0)+1}c_{s(0)+2}\cdots.$$

Then $\mathbf{w} = \mathbf{x}.y\mathbf{z}$ and $h(\mathbf{x}) = \mathbf{x}y$, $h(y\mathbf{z}) = \mathbf{z}$.

Define $B_j(k) = s^j(k) - s^{j-1}(k)$, where s^j denotes the j-fold composition of the function s with itself. First we state a technical lemma without proof.

Lemma 9. *For all integers $r \geq 1$ there exists an integer $n \leq 0$ such that $B_j(n) > r$ for $1 \leq j \leq t$.*

Now let M be the integer specified in Lemma 8, and define $r := \sup_{1 \le i \le M} B_i(0)$. By Lemma 9 there exists an integer $n \le 0$ such that $B_j(n) > r$ for $1 \le j \le M$. Define $w := c_{n+1} \cdots c_0$. We have $|h^j(w)| = s^j(0) - s^j(n)$; and $|h^{j-1}(w)| = s^{j-1}(0) - s^{j-1}(n)$. It follows that

$$\begin{aligned} |h^j(w)| &= (s^j(0) - s^{j-1}(0)) - (s^j(n) - s^{j-1}(n)) + |h^{j-1}(w)| \\ &= B_j(0) - B_j(n) + |h^{j-1}(w)| < B_j(0) - r + |h^{j-1}(w)| \\ &\le |h^{j-1}(w)| \end{aligned}$$

for $1 \le j \le M$. But this contradicts Lemma 8. This contradiction shows that this case cannot occur.

Case 2d: $s(i) < i$ for all $i \in \mathbb{Z}$. This case is the mirror image of Case 2c, and the proof is identical. The proof of Theorem 5 is complete. □

5 The equation $h(xy) = yx$ in finite words

It is not difficult to see that it is decidable whether any of conditions (a)–(e) of Theorem 5 hold for a given morphism h. However, this is somewhat less obvious for condition (f) of Theorem 5, which demands that the equation $h(xy) = yx$ possess a nontrivial[1] solution. We give a complete characterization of the solution set, which constitutes our second main result.

To do so it is useful to extend the notation $\sim$, previously used for two-sided infinite words, to finite words. We say $w \sim z$ for $w, z \in \Sigma^*$ if w is a cyclic shift of z, i.e., if there exist $x, y \in \Sigma^*$ such that $w = xy$ and $z = yx$. It is now easy to verify that $\sim$ is an equivalence relation. Furthermore, if $w \sim z$, and h is a morphism, then $h(w) \sim h(z)$. Thus condition (f) can be restated as $h(z) \sim z$.

It is easy to see that if $h(z) \sim z$, then there exist $i < j$ such that $h^i(z)$ is a fixed point of h^{j-i}. Since $h^i(z) \sim z$, we may restrict our attention to the set $S = T \cap (\bigcup_{i \ge 1} F^*_{h^i})$. Our set T then is the set of all cyclic permutations of words in S.

To describe S we introduce an auxiliary morphism $\tilde{h} : \tilde{\Sigma} \to \tilde{\Sigma}$, where $\tilde{\Sigma} \subseteq \Sigma$. A letter $a \in \tilde{\Sigma}$ if and only if the following three conditions hold:

(1) a is an immortal letter of h;
(2) $h^i(a)$ contains exactly one immortal letter for all $i \ge 1$; and
(3) $h^i(a)$ contains a for some $i \ge 1$.

We define the morphism $\tilde{h}$ by $\tilde{h}(a) = a'$ where a' is the unique immortal letter in $h(a)$.

The relation of $\tilde{h}$ to S is as follows. If $z \in S$, then $z \in F^*_{h^i}$ for some i. Hence there exists an integer p such that $z = z_1 \cdots z_p$ where $z_j = x_j a_j y_j \in F_{h^i}$, and a_j is an immortal letter for $1 \le j \le p$. It follows easily that $a_j \in \tilde{\Sigma}$. Hence h cyclically shifts z iff $\tilde{h}$ cyclically shifts $\tilde{z} = a_1 \cdots a_p$. (The words x_j and y_j are uniquely specified by i and a_j.)

[1] By nontrivial we mean $xy \ne \epsilon$.

Theorem 10. *We have* $\operatorname{Card} \bigcup_{i \geq 1} F_{h^i} < \infty$.

Proof. Suppose $a \in \tilde{\Sigma}$. Define a_j, x_j and y_j by $a_0 = a$ and $h(a_j) = x_j a_{j+1} y_j$ for $j \geq 0$, where $a_{j+1} \in \tilde{\Sigma}$. It is clear that there is a $t \leq \operatorname{Card} \tilde{\Sigma}$ such that if $j \equiv k \pmod{t}$ then $a_j = a_k$, $x_j = x_k$ and $y_j = y_k$. Define $e_i = \exp(h^i)$. By the definition of F_{h^i}, all words in F_{h^i} are of the form

$$h^{e_i-1}(x_{j_0}) h^{e_i-2}(x_{j_1}) \cdots h(x_{j_{e_i-2}}) x_{j_{e_i-1}} a_{e_i} y_{j_{e_i-1}} h(y_{j_{e_i-2}}) \cdots h^{e_i-2}(y_{j_1}) h^{e_i-1}(y_{j_0})$$

for some $a = a_0 \in \tilde{\Sigma}$. Since there are only finitely many a_j, x_j and y_j and $e_i \leq \operatorname{Card} \tilde{\Sigma}$ for all $i \geq 1$, the result follows. □

Therefore, we now concentrate on the set $\tilde{T}$ of words $\tilde{z}$ that are cyclically shifted by $\tilde{h}$.

Suppose $\tilde{\Sigma} = \{a_1, \ldots, a_s\}$. Since $\tilde{h}$ acts as a permutation P on $\tilde{\Sigma}$, there exists a unique factorization of P into disjoint cycles. Suppose $c = (d_0, \ldots, d_{t-1})$ is a cycle appearing in the factorization of P, and let $|c|$ denote the length t of the cycle c. Define the language $L(c)$ as follows:

$$L(c) = (d_0 d_1 d_2 \cdots d_{t-1})^* + (d_1 d_2 \cdots d_{t-1} d_0)^* + \cdots + (d_{t-1} d_0 d_1 \cdots d_{t-2})^*.$$

For example, if $c = (0, 1, 2)$ then $L(c) = (012)^* + (120)^* + (201)^*$. Note that the definition of $L(c)$ is independent of the particular representation chosen for the cycle.

Now define the finite collection $\mathcal{R}'$ of regular languages as follows:

$$\mathcal{R}' = \{L(c^v) \ : \ c \text{ is a cycle of } P \text{ and } 1 \leq v \leq |c| \text{ and } \gcd(v, |c|) = 1\}.$$

We now define a finite collection $\mathcal{R}$ of regular languages. Each language in $\mathcal{R}$ is the union of some languages of $\mathcal{R}'$. The union is defined as follows. Each language $L(c^v)$ in $\mathcal{R}'$ is associated with a pair (t, v) where $t = |c|$ and v is an integer relatively prime to t. Then the languages $L(c_1^{v_1}), \ldots, L(c_m^{v_m})$ in $\mathcal{R}'$ are each a subset of the same language of $\mathcal{R}$ if and only if the system of congruences

$$\begin{aligned} v_1 x &\equiv 1 \pmod{t_1} \\ v_2 x &\equiv 1 \pmod{t_2} \\ &\vdots \\ v_m x &\equiv 1 \pmod{t_m} \end{aligned} \tag{7}$$

possesses an integer solution x, where $t_j = |c_j|$ for $1 \leq j \leq m$. Note that a language in $\mathcal{R}'$ may be a subset of several languages of $\mathcal{R}$.

We say a word w is the *perfect shuffle* of words $w_1, \ldots, w_j$ if $|w_1| = \cdots = |w_j|$ and the first j symbols of w are the first symbols of $w_1, \ldots, w_j$ in that order, the second j symbols of w are the second symbols of $w_1, \ldots, w_j$ in that order, and so on. We write $w = ш(w_1, w_2, \ldots, w_j)$. The following theorem characterizes the set $\tilde{T}$, and is our second main result.

Theorem 11. *Let $\tilde{z} \in \tilde{\Sigma}^*$, and let $\tilde{h}$ permute $\tilde{\Sigma}$. Then $\tilde{h}(\tilde{z}) \sim \tilde{z}$ if and only if $\tilde{z}$ is the perfect shuffle of some finite number of words contained in some single language of $\mathcal{R}$.*

Proof. Let $\tilde{h}$ permute $\tilde{\Sigma}$, with induced permutation P. Let $\tilde{z} = b_0 b_1 \cdots b_{n-1}$.
($\Longleftarrow$): Suppose $\tilde{z}$ is the perfect shuffle of some finite number of words contained in a single language of $\mathcal{R}$. For simplicity of notation we consider the case where $\tilde{z}$ is the perfect shuffle of two such words; the general case is similar and is left to the reader.

Thus assume $\tilde{z} = \text{Ш}(w, \hat{w})$. Further, assume $w \in L(c^v)$ for some cycle c and integer v relatively prime to $t = |c|$, and $\hat{w} \in L(\hat{c}^{\hat{v}})$ for some cycle $\hat{c}$ and integer $\hat{v}$ relatively prime to $\hat{t} = |\hat{c}|$.

Then $w = (d_0 d_v d_{2v} \cdots d_{vt-1})^r$ for some cycle $(d_0, d_1, \ldots, d_{t-1})$ of P with $\tilde{h}(d_s) = d_{s+1}$ for $0 \le s < t$, where the indices are taken modulo t.) Similarly, $\hat{w} = (\hat{d}_0 \hat{d}_{\hat{v}} \hat{d}_{2\hat{v}} \cdots \hat{d}_{\hat{v}\hat{t}-1})^{\hat{r}}$ for some cycle $(\hat{d}_0, \hat{d}_1, \ldots, \hat{d}_{\hat{t}-1})$ of P with $\tilde{h}(\hat{d}_s) = \hat{d}_{s+1}$ for $0 \le s < \hat{t}$, where the indices are taken modulo $\hat{t}$.)

By hypothesis there exists an integer x such that $vx \equiv 1 \pmod{t}$, and $\hat{v}x \equiv 1 \pmod{\hat{t}}$. A simple calculation shows that we may assume $0 \le x < tr = \hat{t}\hat{r}$. Then $\tilde{z} = d_0 \hat{d}_0 \cdots$ and $\tilde{h}(\tilde{z}) = d_1 \hat{d}_1 \cdots = d_{vx} \hat{d}_{\hat{v}x} \cdots = b_{2x} b_{2x+1} \cdots$ (indices of a taken mod n), and so $\tilde{h}(\tilde{z}) \sim \tilde{z}$.

($\Longrightarrow$): Suppose $\tilde{h}(\tilde{z}) \sim \tilde{z}$. Then there exists an integer y such that $\tilde{h}(b_0 b_1 \cdots b_{n-1}) = b_y b_{y+1} \cdots b_{y-1}$, where the indices are taken modulo n. Define $g = \gcd(y, n)$ and $m = n/g$. Then, considering its action on $b_0 b_1 \cdots b_{n-1}$, the morphism $\tilde{h}$ induces a permutation of the indices $0, 1, \ldots, n-1$ sending $j \to j + y \pmod{n}$ which, by elementary group theory, factors into g disjoint cycles, each of length m.

Now, for $0 \le i < g$, define the words $w_i := b_i b_{g+i} b_{2g+i} \cdots b_{(m-1)g+i}$. It is clear that $\tilde{z} = \text{Ш}(w_0, w_1, \ldots, w_{g-1})$. Then

$$\begin{aligned} \tilde{h}(w_i) &= \tilde{h}(b_i\, b_{g+i}\, b_{2g+i} \cdots b_{(m-1)g+i}) \\ &= b_{i+y}\, b_{g+i+y}\, b_{2g+i+y} \cdots b_{(m-1)g+i+y} \\ &= b_{i+(\frac{y}{g})g}\, b_{i+(\frac{y}{g}+1)g}\, b_{i+(\frac{y}{g}+2)g} \cdots b_{i+(\frac{y}{g}+m-1)g}, \end{aligned}$$

and so it follows that $\tilde{h}$ cyclically shifts each w_i by y/g.

Now $\gcd(m, y/g) = 1$, so for each k there is a unique solution $t \pmod{m}$ of the congruence $t\frac{y}{g} \equiv k \pmod{m}$. Multiplying through by g, we find $ty \equiv kg \pmod{n}$ has a solution t, so $ty + i \equiv kg + i \pmod{n}$ has a solution t. But $\tilde{h}^t(b_i) = b_{ty+i}$, so each symbol b_{kg+i} of w_i is in the orbit of $\tilde{h}$ on z_i. It follows that each symbol of w_i is contained in the same cycle c_i of P. Suppose c_i has length t_i. Then $\tilde{h}^{t_i}(b_i) = b_i$, and furthermore t_i is the least positive integer with this property. However, we also have $\tilde{h}^m(b_i) = b_{i+ym} = b_{i+\frac{y}{g}n} = b_i$, and so $t_i \mid m$.

Since $\gcd(y/g, m) = 1$, there is a solution v to the congruence $v \cdot \frac{y}{g} \equiv 1 \pmod{m}$. Then $vy \equiv g \pmod{n}$. Using the division theorem, write $v = q_i t_i + v_i$, where $0 \le v_i < t_i$, for $0 \le i < g$. Since $\gcd(v, m) = 1$, and $t_i \mid m$, we must have $\gcd(v, t_i) = 1$. Thus $\gcd(v_i, t_i) = 1$.

Now

$$\tilde{h}^{v_i}(b_{kg+i}) = \tilde{h}^{v-q_i t_i}(b_{kg+i}) = \tilde{h}^{v}(b_{kg+i}) = b_{kg+i+vy} = b_{kg+i+g} = b_{(k+1)g+i}.$$

Then for $0 \leq i < g$ we have $w_i = (b_i \tilde{h}^{v_i}(b_i) \tilde{h}^{2v_i}(b_i) \cdots \tilde{h}^{(t_i-1)v_i}(b_i))^{m/t_i} \in L(c_i^{v_i})$. From $\tilde{h}(b_0 b_1 b_2 \cdots) = b_y b_{y+1} b_{y+2} \cdots$, it follows that $\tilde{h}^{\frac{y}{g}v_i}(b_i) = b_{y+i} = \tilde{h}(b_i)$, and so $\frac{y}{g}v_i \equiv 1 \pmod{t_i}$. Thus the system of equations (7) possesses a solution $x = y/g$. This completes the proof. □

References

1. J.-P. Allouche. Automates finis en théorie des nombres. *Exposition. Math.* **5** (1987), 239–266.
2. D. Beaquier. Ensembles reconnaissables de mots bi-infinis. In M. Nivat and D. Perrin, editors, *Automata on Infinite Words*, Vol. 192 of *Lecture Notes in Computer Science*, pp. 28–46. Springer-Verlag, 1985.
3. J. Berstel. *Axel Thue's Papers on Repetitions in Words: a Translation.* Number 20 in Publications du Laboratoire de Combinatoire et d'Informatique Mathématique. Université du Québec à Montréal, February 1995.
4. A. Cobham. On the Hartmanis-Stearns problem for a class of tag machines. In *IEEE Conference Record of 1968 Ninth Annual Symposium on Switching and Automata Theory*, pp. 51–60, 1968. Also appeared as IBM Research Technical Report RC-2178, August 23 1968.
5. L. E. Dickson. Finiteness of the odd perfect and primitive abundant numbers with distinct factors. *Amer. J. Math.* **35** (1913), 413–422.
6. D. Hawkins and W. E. Mientka. On sequences which contain no repetitions. *Math. Student* **24** (1956), 185–187.
7. T. Head. Fixed languages and the adult languages of $0L$ schemes. *Internat. J. Comput. Math.* **10** (1981), 103–107.
8. T. Head and B. Lando. Fixed and stationary ω-words and ω-languages. In G. Rozenberg and A. Salomaa, editors, *The Book of L*, pp. 147–156. Springer-Verlag, 1986.
9. J. E. Hopcroft and J. D. Ullman. *Introduction to Automata Theory, Languages, and Computation.* Addison-Wesley, 1979.
10. D. Lind and B. Marcus. *An Introduction to Symbolic Dynamics and Coding.* Cambridge University Press, 1995.
11. M. Nivat and D. Perrin. Ensembles reconnaissables de mots biinfinis. *Canad. J. Math.* **38** (1986), 513–537.
12. A. Thue. Über die gegenseitige Lage gleicher Teile gewisser Zeichenreihen. *Norske vid. Selsk. Skr. Mat. Nat. Kl.* **1** (1912), 1–67. Reprinted in *Selected Mathematical Papers of Axel Thue*, T. Nagell, editor, Universitetsforlaget, Oslo, 1977, pp. 413–478.
13. M.-w. Wang and J. Shallit. An inequality for non-negative matrices. *Linear Algebra and Its Applications* **290** (1999), 135–144.

Tiling Multi-dimensional Arrays

Jonathan P. Sharp*

Department of Computer Science
University of Warwick
Coventry CV4 7AL
United Kingdom
Jonathan.Sharp@dcs.warwick.ac.uk

Abstract. We continue the study of the tiling problems introduced in [KMP98]. The first problem we consider is: given a d-dimensional array of non-negative numbers and a tile limit p, partition the array into at most p rectangular, non-overlapping subarrays, referred to as *tiles*, in such a way as to minimise the *weight* of the heaviest tile, where the weight of a tile is the sum of the elements that fall within it. For one-dimensional arrays the problem can be solved optimally in polynomial time, whereas for two-dimensional arrays it is shown in [KMP98] that the problem is NP-hard and an approximation algorithm is given. This paper offers a new $(d^2+2d-1)/(2d-1)$ approximation algorithm for the d-dimensional problem ($d \geq 2$), which improves the $(d+3)/2$ approximation algorithm given in [SS99]. In particular, for two-dimensional arrays, our approximation ratio is $7/3$ improving on the ratio of $5/2$ in [KMP98] and [SS99]. We briefly consider the dual tiling problem where, rather than having a limit on the number of tiles allowed, we must ensure that all tiles produced have weight at most W and do so with a minimal number of tiles. The algorithm for the first problem can be modified to give a $2d$ approximation for this problem improving upon the $2d+1$ approximation given in [SS99]. These problems arise naturally in many applications including databases and load balancing.

1 Introduction

The partitioning of data is a problem that arises in many areas of computer science and other fields. We consider two particular partitioning problems which have applications to databases and load balancing. We call these problems RTILE (rectangular tiling) and DRTILE (dual rectangular tiling), following the terminology in [KMP98], where these problems were introduced and proven to be NP-hard. In this paper we consider the d-dimensional generalisations to these problems and give new approximation algorithms for them which improves those previously presented. We focus almost entirely on the RTILE problem and use our approximation algorithm for this problem to generate a new approximation

* This work was supported by the Engineering and Physical Sciences Research Council and in part by the ESPRIT LTR Project no. 20244 – ALCOM-IT.

algorithm for the dual DRTILE problem. In this section we define the problems, discuss our results and related work, and give some more details of motivation for the problems. Subsequent sections present the approximation algorithms for the problems.

1.1 Problem Definitions

The problems we consider must partition an array into subarrays, which we refer to as *tiles*. We assume that any partitioning into tiles completely cover the arrays without overlap. The sum of all the elements that fall within a given tile is called the *weight* of a tile. The arrays we consider are assumed to have a dimension of at least two, unless stated otherwise.

RTILE. Given a d-dimensional array A of size n in each dimension, containing non-negative integers, partition A into at most p rectangular tiles so that the maximum weight of any tile is minimised.

DRTILE. Given a d-dimensional array A of size n in each dimension, containing non-negative integers, partition A into rectangular tiles, with each tile having weight at most W, so that the number of tiles used is minimised.

1.2 Results and Related Work

Firstly, we note that the problems can be solved optimally by efficient algorithms when restricted to one-dimensional arrays, see [KMS97] for one such algorithm. In this paper we consider the problems for arrays with dimension at least two.

The main result of this paper is a new approximation algorithm for the RTILE problem. The RTILE problem, restricted to two dimensions, was introduced by Khanna et al in [KMP98], where it is shown to be NP-hard to approximate an optimal solution to within a factor of 5/4. They give a $O(n^2 + p \log n)$ time algorithm for the problem in two dimensions, that approximates an optimal solution to within a factor of 5/2. For the dual DRTILE problem, again in two dimensions, they give a series of approximation algorithms that trade quality of approximation for improved running time.

In [SS99] Smith and Suri extend the RTILE approximation algorithm given in [KMP98] to deal with the d-dimensional case. Their algorithm gives a $(d+3)/2$ approximation with a running time of $O(d!(n^d + pd \log n))$ and thus is only practical for small values of d. The same algorithm is then used to get a $(2d+1)$ approximation for the dual problem.

The main result of this paper is a new approximation algorithm for the RTILE problem giving a $(d^2 + 2d - 1)/(2d - 1)$ approximation which improves those described above. For comparison, in two dimensions this algorithm has an approximation ratio of $7/3$ compared with the previous best of $5/2$ for the approximation algorithms in [KMP98] and [SS99] and in three dimensions the new approximation ratio is 2.8 improving the ratio of 3 for the algorithm in [SS99]. The worst case running time for the algorithms is $O(n^d + 2^d pd(d + \log n))$.

Asymptotically this is better than the algorithm given in [SS99] but is still, never-the-less, exponential in d and thus only useful for small d. However, since the input to the problem is an array of size $O(n^d)$ any algorithm that examines the whole array will have running time exponential in d. It may be possible for some applications to exploit any known sparseness of the array in order to improve upon this. We briefly discuss this in relation to our algorithm in Sect. 2.4.

We use our RTILE approximation algorithm to generate an approximation algorithm for the dual, DRTILE, problem obtaining a $2d$ approximation with a running time of $O(n^d + 2^d pd \log n)$.

We note that all of the algorithms mentioned above produce a ***hierarchical*** tiling. That is, one in which there is a straight cut through one dimension of the array that partitions it into two disjoint regions, each of which is itself a hierarchical tiling. If we are restricted to this type of tiling, then we can solve the problem optimally by dynamic programming. The hierarchical tiling equivalent of the RTILE problem can be solved optimally in time $O(pdn^{2d+1})$ and the corresponding dual hierarchical problem can be solved optimally in $O(dn^{2d+1})$ time, both by the obvious dynamic programming algorithms. Due to the increased running time of the dynamic programming solutions to these problems the approximation algorithms discussed offer quicker alternatives, albeit at possible loss in optimality. However, the interesting question of how the optimal solutions to the RTILE and DRTILE problems compare with the optimal solutions to the hierarchical equivalents remains open.

The DRTILE problem and the hierarchical equivalent is considered in greater depth by Muthukrishnan et al in [MPS99]. In this paper they consider the hierarchical problem separately and also consider other methods of partitioning the array which have database applications as described in [Poo97]. They describe a series of approximation algorithms for the partitioning problems they consider and as in [KMP98] offer alternative algorithms which trade quality of approximation for improved running time.

1.3 Motivation

To motivate our work we shall briefly describe two of the application scenarios where the problems we consider arise.

Query Optimisation. In many relational database systems, users make requests for information retrieval in ***query languages*** such as SQL, which specify what information is desired, but not the specific details of how to retrieve it. With a complex query there may be many possible ***execution plans*** for the query which vary widely in the completion time. It is thus desirable to attempt to choose the execution plan which will execute the query the fastest. This work takes place in a component of the database system called the ***query optimiser***. Accurately picking the best execution plan is a difficult task without executing all the different plans and hence the query optimiser must make cost estimates for each execution plan. In order to achieve this, statistics on the data are stored. Disk access is of primary importance in executing a query since this will be slow and hence we would wish to choose the execution plan so as to reduce the results sizes of each intermediate operator in the query and thus reduce disk access. This requires knowledge of the data distributions of the ***attributes*** stored within each ***table*** or ***relation*** in the database. Since the distribution of data may not fit a particular probability distribution, ***histograms*** are stored to approximate the data distributions.

A majority of work in the databases community has concentrated on the construction of one-dimensional histograms. In this case a single attribute of a relation is considered in isolation. The frequency with which a particular value of the attribute occurs in the relation is stored in an array. There are many different strategies for partitioning this data into histograms which are discussed in [Poo97]. One such strategy is to partition the array so that attribute values are grouped into ranges such that the sum of frequencies within each range is 'nearly' equal over all ranges. These are known as ***equi-depth histograms.*** Once the partitioning has taken place the average frequency within each range is used to approximate the frequency of all values in the range.

For queries that involve more than one attribute at a time from the same relation, the results size depends on the joint frequency distribution. In this case, for each combination of attribute values the frequency with which that combination of values occurs is stored. This gives a multi-dimensional array which can be partitioned following the same strategies as for one-dimensional arrays. Multi-dimensional equi-depth histograms have been studied in [MD88] and [Poo97] among others.

The approximation algorithms for the RTILE and DRTILE problems will give approximate solutions to the problem of constructing equi-depth multi-dimensional histograms.

Data Partitioning. Many modern computer science applications are written to take advantage of parallelism in the architecture on which they run. To facilitate this, high performance computing languages, such as High Performance Fortran, allow the programmer to specify a partitioning of the data which is mapped onto different processors. The aim for the programmer is to achieve an even load balance on each processor. To give a specific example we describe one of the problems given in [Man93]. The problem is to compute the product Ax for a sparse $M \times N$ matrix A and vector x of length N on a SIMD computer. Matrix A is partitioned into $m \times n$ rectangular blocks which are mapped onto an $m \times n$ processor array ($m \ll M$ and $n \ll N$). Let $B = (a_{ij})$ be one such block, with $1 \leq s_1 \leq i \leq s_2 \leq M$ and $1 \leq t_1 \leq j \leq t_2 \leq N$, for some s_1, s_2, t_1 and t_2. If B is mapped on to processor p, then the elements x_j for $t_1 \leq j \leq t_2$ are also mapped to processor p. The processor will now compute the products $y_i = \sum_{j=t_1}^{t_2} a_{ij} x_j$ for $s_1 \leq i \leq s_2$. Each processor communicates the y values it has computed to the other processors, which sum the values to form the final answer for product Ax. The time spent by each processor in computing its y values will be proportional to the number of non-zero entries in the block of A it is given and thus we would like to make the number of non-zero entries in each block approximately the same. In [Man93] a scheme based upon a one-dimensional partitioning of the rows and columns of A individually is given. The RTILE problem in two dimensions would be another possible strategy to obtain this partitioning. It should be noted however, that the possibility of producing long thin tiles could increase the communication cost between processors and thus reduce the benefit of an even load balance. Experimental work similar to that in [Man93] is required to investigate the benefits of using the RTILE algorithm for this problem.

2 RTILE Approximation Algorithm

We assume that the input to the problem is a d-dimensional array A of size n in every dimension[1], containing non-negative integers and a tile limit p. The following algorithm gives a $(d^2 + 2d - 1)/(2d - 1)$ approximation to an optimal solution. For brevity we will let $\alpha = (d^2 + 2d - 1)/(2d - 1)$.

Before describing the algorithm we give some definitions and notation. Let $\mathrm{wt}(K)$ denote the weight of a subregion K of A, where the weight of K is the sum of all the elements within it. Let $w = \mathrm{wt}(A)/p$, which is the average tile weight assuming all possible tiles are used. If M is the value of the element(*s*) of maximal weight, it is clear that $M' = \max\{w, M\}$ gives a lower bound for the maximum tile weight in an optimal solution. We show that our algorithm produces tiles with weight no more than $\alpha M'$ and thus within α of an optimal solution.

In order to represent d-dimensional rectangular subregions of the array, we use the notation $[a_1, b_1] \times [a_2, b_2] \times \ldots \times [a_d, b_d]$, which will be the region from a_1 to b_1 in dimension one, from a_2 to b_2 in dimension two, etc. Individual elements will be represented by $(a_1, a_2, \ldots, a_d)$. Often we let $I_j = [a_j, b_j]$ and so regions are denoted by $I_1 \times I_2 \times \ldots \times I_d$.

The tilings produced by the algorithm are created in a top down fashion by *cutting* a region of the array along one of the dimensions. For example, region $[1, 9] \times [3, 8] \times [2, 3]$ can be cut in dimension two, into regions $[1, 9] \times [3, 5] \times [2, 3]$ and $[1, 9] \times [6, 8] \times [2, 3]$. We borrow some definitions from [SS99] to describe various types of region: a region is called *light* if it has weight no more than $\alpha M'$; a region is called *heavy* if it can be divided into two regions by cutting in any one of its dimensions with both subregions having weight at least M' - such a cut is called a *heavy cut*; in the remaining case a region is said to be *medium*.

The algorithm consists of three procedures, Heavy-Search, Heavy-Cut and Tile-Medium. It begins by passing the whole array to procedure Heavy-Search which starts dividing the array into regions with weight at least M'. The resulting regions are passed to Heavy-Cut which will recursively make heavy cuts until this is no longer possible. This procedure is identical to the Greedy-Tile procedure given in [SS99]. At this point the resulting medium regions are passed to Tile-Medium which will cut off as many light pieces of the region as it can, which will guarantee to make the weight of the remaining region no more than $\alpha M'$. Each of these procedures is described in detail below.

[1] This is purely to keep the analysis clean, the dimensions of A may be of different lengths.

2.1 Heavy-Search

The algorithm begins by passing the whole array A to this procedure together with the value $M' = \max\{\text{wt}(A)/p, M\}$. Array A is divided into subregions with weight at least M'. A subregion is created by making a cut in dimension one that slices off a minimal region with weight at least M'. The regions is then passed to procedure Heavy-Cut. The process is repeated with the remainder of the array until the whole array has been processed. The algorithm for the procedure is presented below.

Procedure 1 Heavy-Search

Input: Array A $= I_1 \times I_2 \times \ldots \times I_d$ where $I_j = [1, n]$ and M'.
1: $a_1 \leftarrow 1$, tiles $\leftarrow \phi$
2: **repeat**
3: $\quad m_1 \leftarrow$ minimal value in $[a_1, n]$ with $\text{wt}([a_1, m_1] \times I_2 \times \ldots \times I_d) \geq M'$
$\quad$ {Assume m_1 is set to n if m_1 is undefined.}
4: $\quad b_1 \leftarrow$ maximal value in $[m_1 + 1, n]$ with $\text{wt}([m_1 + 1, b_1] \times I_2 \times \ldots \times I_d) = 0$
$\quad$ {Assume b_1 is set to m_1 if b_1 is undefined.}
5: $\quad$ tiles $\leftarrow$ tiles $\cup$ Heavy-Cut$([a_1, b_1] \times I_2 \times \ldots \times I_d)$
6: $\quad a_1 \leftarrow b_1 + 1$
7: **until** $b_1 = n$
8: **return** tiles

Notice that when a region is found with weight at least M' the algorithm will expand this region to include any subsequent region with zero weight. This is necessary to prevent the algorithm from ever leaving a region of zero weight at the last iteration of the loop. It should be noted, however, that the final region produced may have weight less than M'.

In order to maintain the tile limit p we ensure that Heavy-Cut(K) will use at most $\lfloor \text{wt}(K)/w \rfloor$ tiles when $\text{wt}(K) \geq M'$ and one otherwise. Provided this is always the case, the final tiling will never use more than p tiles. To see this let us denote the regions $[a_1, b_1] \times I_2 \times \ldots \times I_d$ produced by Heavy-Search and passed to Heavy-Cut by K_i for $1 \leq i \leq l$. For all but K_l, Heavy-Cut will use at most $\lfloor \text{wt}(K_i)/w \rfloor$ tiles. In the case that $\text{wt}(K_l) \geq M'$ the total tile usage will be no more than,

$$\sum_{i=1}^{l} \lfloor \text{wt}(K_i)/w \rfloor \leq \sum_{i=1}^{l} \text{wt}(K_i)/w = p \, .$$

The second case to consider is when $\text{wt}(K_l) < M'$. Since this is the only region with weight less than M' and it does not have zero weight we will still meet our tile limit. In this case the total tile usage will be at most,

$$\sum_{i=1}^{l-1} \lfloor \text{wt}(K_i)/w \rfloor + 1 \leq \sum_{i=1}^{l-1} \text{wt}(K_i)/w + 1 = p - \gamma + 1 \; ,$$

where $\gamma = \text{wt}(K_l)/w$. Since $\gamma > 0$ and the tile usage is an integer it follows that the total tile usage is no more than p.

2.2 Heavy-Cut

The purpose of this procedure is to keep making heavy cuts on the region it is given until this is no longer possible. Should Heavy-Cut be given a light region, it is returned and will constitute an individual tile. Otherwise, each dimension of the region is examined, looking for a heavy cut in each one. Should a heavy cut be found, then it is made and Heavy-Cut recursively calls itself on the two subregions produced. If no heavy cut can be found then the region will be medium and is passed to Tile-Medium. We note that the procedure will meet the tile quota discussed above provided procedure Tile-Medium obeys this.

Procedure 2 Heavy-Cut

Input: Region K
1: **if** $\mathrm{wt}(K) \leq \alpha M'$ **then**
2: **return** $\{K\}$
3: **if** there is a heavy cut in some dimension j **then**
4: make the heavy cut giving regions K_1 and K_2 with $K = K_1 \cup K_2$.
5: **return** Heavy-Cut(K_1) $\cup$ Heavy-Cut(K_2)
6: **else** {there is no heavy cut, i.e. the tile is medium}
7: **return** Tile-Medium(K)

The search for heavy cuts and the properties of medium regions play an important role in the final part of the approximation algorithm and are examined further. We assume Heavy-Cut is given region $K = I_1 \times \ldots \times I_d$, where $I_j = [a_j, b_j]$. If the region K has weight more than $\alpha M'$ then the procedure must search for a heavy cut. For given dimension j, Heavy-Cut will search for a value $m_j \in [a_j, b_j]$ which induces the partitioning

$$
\begin{aligned}
L_j &= I_1 \times \ldots \times I_{j-1} \times [a_j, m_j - 1] \times I_{j+1} \times \ldots \times I_d \ , \\
C_j &= I_1 \times \ldots \times I_{j-1} \times [m_j, m_j] \times I_{j+1} \times \ldots \times I_d \ , \\
R_j &= I_1 \times \ldots \times I_{j-1} \times [m_j + 1, b_j] \times I_{j+1} \times \ldots \times I_d \ ,
\end{aligned}
$$

so that m_j is maximal with respect to the property that the weight of L_j is less than M'. Note that by construction of the region in Heavy-Search, no heavy cut will be found in dimension one and so there is no need to look there. Also value m_1 is computed in procedure Heavy-Search. Figure 1 shows an example of the process in three dimensions. Region L_1 is drawn in dotted lines, R_1 is drawn in dashed lines, L_2 and R_2 are the unshaded regions on the left and right of the diagram respectively, and finally C_2 is the shaded region in the centre.

We note that for dimension j, either L_j or R_j may be an empty region. Additionally if region R_1 is non-empty then it will have zero weight due to the construction of region K in Heavy-Search and it may be treated as if it were empty. Since m_j was chosen maximally the weight of $L_j \cup C_j$ is at least M'. If the weight of R_j is at least M' then there is a heavy cut which the procedure will make and then recurse on subregions $L_j \cup C_j$ and R_j. Otherwise there is no heavy cut in this dimension. Observe that the weight of $C_j \cup R_j$ is at least M'

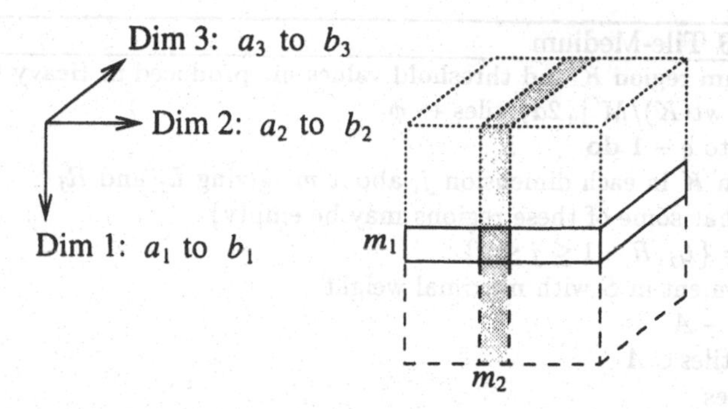

Fig. 1. Searching for a heavy cut

because the weight of K is at least $2M'$ ($\alpha M' > 2M'$ for $d \geq 2$). Therefore, the region L_j is the largest region on the left, with respect to dimension j from a_j to b_j, which has weight less than M' and R_j is the largest region on the right with respect to dimension j from a_j to b_j, which has weight less than M'. We call m_j a *threshold* value for dimension j.

If there is no heavy cut in any dimension then Heavy-Cut will have computed the threshold value m_j for each dimension j. In this case the region is medium and is passed to Tile-Medium along with the threshold values.

2.3 Tile-Medium

This procedure is given a medium region and the threshold values for each dimension, as computed in Heavy-Cut. We know that for a given dimension j, a threshold value divides the region into three subregions L_j, R_j and C_j. The procedure will choose the heaviest of the light regions, L_j or R_j, over all dimensions and slice it off. It will keep removing the heaviest of these regions until it has cut off $\lfloor \mathrm{wt}(K)/M' \rfloor - 1$ of them. The remaining region is guaranteed to have weight no more than $\alpha M'$ and will constitute one tile, making the total tile usage $\lfloor \mathrm{wt}(K)/M' \rfloor \leq \lfloor \mathrm{wt}(K)/w \rfloor$ as required. The proof of this follows and uses Lemma 1.

Lemma 1. *Let A be a finite set of objects with weight function $s : A \to \mathbb{N}$. If $\{B_1, \ldots, B_k\}$ is a collection of subsets of A satisfying $\bigcup_i B_i = A$, then $\sum_{x \in B_i} s(x) \geq \sum_{x \in A} s(x)/k$ for at least one set B_i.*

Proof. Straight forward and not given.

Theorem 1. *Procedure Tile-Medium will produce a tiling of a given medium region K using $\lfloor \mathrm{wt}(K)/M' \rfloor$ tiles with all but one tile having weight less than M' and the tile of maximal weight having weight no more than $(d^2+2d-1)/(2d-1)$ times the weight of the optimal solution.*

Procedure 3 Tile-Medium

Input: Medium region K and threshold values m_j produced in Heavy-Cut.

1: $q \leftarrow \min\{\lfloor \mathrm{wt}(K)/M' \rfloor, 2d\}$, tiles $\leftarrow \phi$
2: **for** $i \leftarrow 1$ to $q-1$ **do**
3: Partition K in each dimension j, about m_j giving L_j and R_j
{Note that some of these regions may be empty}
4: Let $S = \{L_j, R_j : 1 \le j \le d\}$
5: $A \leftarrow$ element in S with maximal weight
6: $K \leftarrow K - A$
7: tiles $\leftarrow$ tiles $\cup A$
8: **return** tiles

Proof. We shall let $q = \min\{\lfloor \mathrm{wt}(K)/M' \rfloor, 2d\}$ and we observe that the procedure produces at most q tiles and thus meets the tile quota. For reference sake let $K_0 = K$, and K_i be the remaining region K produced at the end of the ith iteration of the **for** loop (line 6 in Procedure 3). At each iteration a threshold value m_j creates a new partitioning of the region, giving new regions L_j and R_j. We let L_{j_i} and R_{j_i} be the regions L_j and R_j produced during the ith iteration of the loop. Note that $L_{j_i} \subseteq L_{j_1}$ and $R_{j_i} \subseteq R_{j_1}$ for all $1 \le i < q$ and $1 \le j \le d$ and hence the weight of L_{j_i} and the weight of R_{j_i} will be less than M' since K is a medium region.

At each iteration of the loop we pick the heaviest (non-empty) region L_{j_i} or R_{j_i} for $1 \le j \le d$ and cut it off. It will become an individual tile and we know the weight is less than M'. By the construction of K we know that R_1 is either empty or of zero weight and thus may be ignored. Therefore initially there are at most $2d-1$ non-empty regions in set S to be considered. Note that this implies $2d$ tiles is the maximum number required. Since we slice off one of these regions L_{j_i} or R_{j_i} during each iteration, at the start of the ith iteration there are at most $2d-i$ such (non-empty) regions remaining. The weight of the heaviest of these regions must be at least $(\mathrm{wt}(K_{i-1}) - \mathrm{wt}(C))/(2d-i)$, where $C = (m_1, m_2, \ldots, m_d)$ is the centre element. This follows from Lemma 1 since the union of all the regions L_{j_i} and R_{j_i} gives the region K_{i-1} less the centre element. Since we slice off the the heaviest of these regions at each iteration we obtain,

$$\mathrm{wt}(K_i) \le \mathrm{wt}(K_{i-1}) - \frac{\mathrm{wt}(K_{i-1}) - \mathrm{wt}(C)}{2d-i} .$$

Using the above inequality and induction gives,

$$\mathrm{wt}(K_i) \le \frac{(2d-i-1)\mathrm{wt}(K) + i\,\mathrm{wt}(C)}{2d-1} .$$

We check that the weight of the remaining region K after the $q-1$ iterations of the loop satisfies the theorem. We know that this weight will be no more than

$$\frac{(2d-q)\mathrm{wt}(K) + (q-1)\mathrm{wt}(C)}{2d-1},$$

and since we know that $\mathrm{wt}(K) < (q+1)M'$ and $\mathrm{wt}(C) \leq M'$ we find that the remaining region will have weight no more than,

$$\frac{(2d-q)(q+1)M' + (q-1)M'}{2d-1} .$$

This takes its maximum value when $q = d$ and on substitution we get the result. □

The theorem together with the preceding discussion proves that the algorithm does indeed achieve a $(d^2+2d-1)/(2d-1)$ approximation to the RTILE problem.

2.4 Running Time Analysis

We now examine the worst case running time for the RTILE approximation algorithm presented. A crucial component of the algorithm is the ability to calculate the weight of a given region of the array. As observed in [SS99] this can be achieved in $O(2^d)$ time for d dimensions given an appropriate data structure which is computed in a preprocessing stage in $O(n^d)$ time. Though exponential in d this is in fact linear in the size of the input. However, if we were given a sparse array we would hope to improve upon this. This is on-going work, but to allow our analysis to hold with any improved weight calculating function we shall assume that we can calculate the weight of any given rectangular region of the array in time $T_2(n,d)$ given a $T_1(n,d)$ time preprocessing stage.

We begin our analysis with the highest level loop in procedure Heavy-Search. Since we only look for divisions into subregions in dimension one, this part of the algorithm can be implemented in $O(nT_2(n,d))$ time. We note at this point, that if we do have to spend $O(n^d)$ time building a data structure for weight calculation, then we could at the same time build a $O(n)$ space data structure to reduce the time for this stage of the algorithm to $O(n)$.

We consider next the time for procedure Heavy-Cut. Searching for a heavy cut in a given dimension can be done by binary search in $O(T_2(n,d)\log n)$ time. For a given region there are at most d dimensions to be searched and since at most p tiles are produced there can be a maximum of p calls to this procedure giving a total of $O(pdT_2(n,d)\log n)$ time for this part of the algorithm.

Finally we consider the work done in procedure Tile-Medium. The running time will depend upon the weight of the region K given to the procedure. As in the algorithm we let $q = \min\{\lfloor \mathrm{wt}(K)/M' \rfloor, 2d\}$. On examination we see that the running time is bounded by,

$$\sum_{i=1}^{q-1}\sum_{j=1}^{2d-i}(T_2(n,d) + O(1)) ,$$

since there are at most $2d - i$ regions L_j and R_j whose weight needs to be computed in order to select the heaviest at the ith iteration. This solves to give a running time of $O(d^2T_2(n,d))$ and as for the previous procedure there can be at most p calls to this procedure giving a total running time of $O(pd^2T_2(n,d))$ for Tile-Medium.

Combining the above we get a total running time of $O(T_1(n,d) + (pd^2 + pd\log n + n)T_2(n,d))$. In the general case, assuming $T_1(n,d) = O(n^d)$ and $T_2(n,d) = O(2^d)$, this gives a running time of $O(n^d + pd2^d(d + \log n))$.

3 DRTILE Approximation Algorithm

In order to solve the dual DRTILE problem we use the same trick employed in [KMP98] and [SS99] and use a modified version of the RTILE algorithm. In the dual tiling problem we are given an array A and a weight limit W and must tile the array so that each tile produced has weight at most W using a minimal number of tiles. If p^* is the number of tiles used in an optimal solution to a given instance of the problem then $p^* \geq \lceil \mathrm{wt}(A)/W \rceil$, (we assume $W \geq M$).

To solve the problem we let $p = \lceil \mathrm{wt}(A)/W \rceil$ and use a modified version of the RTILE algorithm. If we run the RTILE algorithm with the array A and tile limit p, we will obtain a tiling with at most p tiles, with each tile having weight at most $\alpha M'$ where $M' = \mathrm{wt}(A)/p \leq W$. It is easy to modify the algorithm to use at most $2dp$ tiles and ensure each tile has weight at most M'. This then gives us a $2d$ approximation algorithm for the DRTILE problem. Procedure Heavy-Cut is modified to always pass a region with weight greater than M' to Tile-Medium (after computing the threshold values). Tile-Medium slices off all the non-empty regions L_j and R_j of which there are at most $2d - 1$ and we know each of these has weight less than M'. The remainder will be the centre element which will be the final tile making the total tile usage at most $2d$. Since there were at most p tiles produced in the original RTILE algorithm there are now at most $2dp$ tiles, which is less than $2dp^*$ and thus gives the $2d$ approximation for the DRTILE problem. This can be implemented in $O(n^d + pd2^d \log n)$ time.

4 Conclusions

The algorithms presented offer good approximations to optimal solutions for the problems considered. The approximation ratio of $\frac{d^2+2d-1}{2d-1} = \frac{d}{2} + \frac{5}{4} + \frac{1}{8d-4}$ for the new RTILE algorithm improves that of $\frac{d+3}{2}$ for the algorithm in [SS99]. Though only slight, this improvement is significant for small d. Additional benefits of the new algorithm include an improved running time and a far simpler and more direct analysis. It is also easy to construct instances of the RTILE problem that have optimal solutions with a maximum tile weight of $(d/2 + 1)M'$ and thus we can see that our algorithm has significantly reduced the gap between the approximation ratio of the RTILE algorithm in [SS99] and this value. In order for any algorithm to improve upon an approximation ratio of $d/2 + 1$ alternative techniques must be used which require an improvement in the the lower bound of M' used in our analysis.

Since the algorithms have running times which are exponential in the dimension d of the array, they will be practical for only small values of d. In order to make them practical for arbitrary d, further work must be done to take advantage of sparse data arrays. If the array is not sparse then any algorithm which examines an array of size $O(n^d)$ will only be practical for small values of d.

Another possibility for further work is to try to increase the lower bound of 5/4 for the best possible approximation for any RTILE algorithm. We would like to find a lower bound that is a function of d which would narrow the gap between the current lower bound of 5/4 and the $(d^2+2d-1)/(2d-1)$ approximation algorithm presented.

Another interesting problem is to examine what is lost in the RTILE and DRTILE algorithms, by restricting the type of tilings produced to hierarchical tilings. A relationship between the maximum tile weight produced by the RTILE approximation algorithm and the optimal maximum tile weight in the equivalent hierarchical tiling problem would be a useful result.

5 Acknowledgements

I would like to thank Mike Paterson for many discussions and suggestions about the problems covered in this paper and carefully reviewing my work.

References

[KMP98] Sanjeev Khanna, S. Muthukrishnan, and Mike Paterson. On approximating rectangular tiling and packing. In *Proc. 9th Annual Symposium on Discrete Algorithms (SODA)*, pages 384–393, 1998.

[KMS97] Sanjeev Khanna, S. Muthukrishnan, and S. Skiena. Efficient array partitioning. In *Proc. 24th International Colloquium on Automata, Languages and Programming (ICALP)*, pages 616–626, 1997.

[Man93] Fredrik Manne. *Load Balancing in Parallel Sparse Matrix Computations.* PhD thesis, Department of Informatics, University of Bergen, Norway, 1993.

[MD88] M. Muralikrishna and David J. Dewitt. Equi-depth histograms for estimating selectivity factors for multi-dimensional queriese. In *Proc. of the International Conference of Management of Data (SIGMOD)*, pages 28–36, 1988.

[MPS99] S. Muthukrishnan, Viswanath Poosala, and Torsten Suel. On rectangular partitions in two dimensions: Algorithms, complexity, and applications. In *7th International Conference on Database Theory (ICDT)*, pages 236–256, January 1999.

[Poo97] Viswanath Poosala. *Histogram-based estimation techniques in databases.* PhD thesis, Department of Computer Science, University of Wisconsin-Madison, US, 1997.

[SS99] Adam Smith and Subhash Suri. Rectangular tiling in multi-dimensional arrays. In *Proc. 10th Annual Symposium on Discrete Algorithms (SODA)*, 1999.

Modeling Interaction by Sheaves and Geometric Logic

Viorica Sofronie-Stokkermans[1] and Karel Stokkermans[2]

[1] Max-Planck Institut für Informatik, Saarbrücken, Germany
[2] Institut für Computerwissenschaften, Salzburg, Austria
email: sofronie@mpi-sb.mpg.de, stokkerm@cosy.sbg.ac.at

Abstract. In this paper we show that, given a family of interacting systems, many notions which are important for expressing properties of systems can be modeled as sheaves over a suitable topological space. In such contexts, geometric logic can be used to test whether "local" properties can be lifted to a global level. We develop a way to use this method in the study of interacting systems, illustrated by examples.

1 Introduction

Many properties of systems can be expressed as assertions about states, actions, transitions, behavior in time. In this paper we show that, given a family of interacting systems, under certain circumstances these notions can be modeled by sheaves over a suitable topological space (where the topology expresses how the interacting systems share the information). The main advantage is that this allows us to use geometric logic in order to study the links between the properties of the systems obtained by interconnecting families of interacting systems, and those of their components. This opens the way towards a possibility of verifying certain properties of complex systems in a modular way, thus increasing efficiency.

Among existing approaches to concurrency based on fiberings, sheaf and presheaf theory we mention [MP86], [Pfa91], [Gog92], [Lil93], [Mal94], [Win96], [CW96]. The starting point of our research is the work of Pfalzgraf [Pfa91] and the ideas of Goguen [Gog92] and Monteiro and Pereira [MP86]. The idea of modeling states, actions and transitions by sheaves with respect to a topological space, and of using geometric logic for studying the link between properties of the components and properties of the systems that arises from their interconnection is, to the best of our knowledge, new. Preliminary results (in a more theoretically involved framework) appear in [Sof96]. The results presented here simplify and considerably extend those in [Sof96].

The paper is structured as follows. In Section 2 we present some basic notions from sheaf theory and geometric logic. Section 3 discusses the systems and communication between subsystems we want to model. Section 4 describes our sheaf-theoretic model; then geometric logic is used to test whether local properties can be lifted to a global level.

2 Preliminaries

We present results from sheaf theory and geometric logic which we will use in our work. For definitions we refer to [Joh82] or [MLM92]. Notions from category theory and many-sorted logic are assumed known. Categories and sheaves will usually be denoted in sans-serif style, e.g. Set, $\mathsf{Sh}(I)$.

Sheaf theory. Let I be a topological space, and $\Omega(I)$ the topology on I. A *presheaf* on I is a functor $P : \Omega(I)^{\rm op} \to \mathsf{Set}$. Let $U \subseteq V$ be open sets in I, and $i^V_U : U \hookrightarrow V$ the corresponding morphism in $\Omega(I)$. The restriction to U, $P(i^V_U) : P(V) \to P(U)$ will also be denoted by ρ^V_U. A *sheaf* on I is a presheaf $F : \Omega(I)^{\rm op} \to \mathsf{Set}$ satisfying the following property:

for each open cover $(U_i)_{i \in I}$ of U and family of elements $s_i \in F(U_i)$ s.t. for all i, j we have $\rho^{U_i}_{U_i \cap U_j}(s_i) = \rho^{U_j}_{U_i \cap U_j}(s_j)$, there is a unique $s \in F(U)$ with $\rho^U_{U_i}(s) = s_i$ for all i.

The *stalk* of a sheaf F on I at a point $i \in I$ is the colimit $F_i = \varinjlim_{i \in U} F(U)$, where U ranges over all open neighborhoods of i. The morphisms of sheaves are natural transformations of functors. $\mathsf{Sh}(I)$ will denote the category of sheaves on I. The assignment $F \mapsto F_i$ defines the stalk functor at i, $\mathsf{Stalk}_i : \mathsf{Sh}(I) \to \mathsf{Set}$.

Interpreting many-sorted first order languages in $\mathsf{Sh}(I)$. Let $\mathcal{L}$ be a many-sorted first-order language consisting of a collection of sorts and collections of function and relation symbols. Terms and atomic formulae from $\mathcal{L}$ are defined in the standard way; compound formulae are constructed by using the connectives $\vee, \wedge, \Rightarrow, \neg$ and the quantifiers $\exists, \forall$, for every sort X. An *interpretation M of $\mathcal{L}$ in* $\mathsf{Sh}(I)$ is constructed by associating a sheaf X^M on I to every sort X, a subsheaf $R^M \subseteq X_1^M \times \cdots \times X_n^M$ to every relation symbol R of arity $X_1 \times \cdots \times X_n$ and an arrow $f^M : X_1^M \times \cdots \times X_n^M \to Y^M$ in $\mathsf{Sh}(I)$ to every function symbol f with arity $X_1 \times \cdots \times X_n \to Y$. Each term $t(x_1, \dots, x_n)$ of sort Y is (inductively) interpreted as an arrow $t^M : X_1^M \times \cdots \times X_n^M \to Y^M$; and every formula $\phi(x_1, \dots, x_n)$ with free variables $FV(\phi) \subseteq \{x_1, \dots, x_n\}$, where x_i is of sort X_i gives rise to a subsheaf $\{(x_1, \dots, x_n) \mid \phi(x_1, \dots, x_n)\}^M \subseteq X_1^M \times \cdots \times X_n^M$. For details we refer to [MLM92], Ch. X.

Geometric formulae and axioms. A *geometric formula* is a formula built up from atomic formulae using only the connectives $\vee$ and $\wedge$ and the quantifier $\exists$. A *geometric axiom* is a formula of the form $(\forall x_1, \dots, x_n)(\phi \Rightarrow \psi)$ where ϕ and ψ are geometric formulae. A geometric axiom $(\forall x_1, \dots, x_n)\ (\phi \Rightarrow \psi)$ is ***satisfied*** *in an interpretation M of $\mathcal{L}$ in* $\mathsf{Sh}(I)$ if $\{(x_1, \dots, x_n) \mid \phi\}^M$ is a subobject of $\{(x_1, \dots, x_n) \mid \psi\}^M$ in $\mathsf{Sh}(I)$.

Geometric morphisms; preservation properties. Let $f : I \to J$ be a continuous map between topological spaces. The *direct image functor* $f_* : \mathsf{Sh}(I) \to \mathsf{Sh}(J)$ corresponding to f associates with every $F \in \mathsf{Sh}(I)$ the sheaf $f_*(F) : \Omega(J)^{\rm op} \to \mathsf{Set}$ defined by $f_*(F)(U) = F(f^{-1}(U))$; the *inverse image functor* $f^* : \mathsf{Sh}(J) \to \mathsf{Sh}(I)$ associated to f is defined as follows: for every $G \in \mathsf{Sh}(J)$, let $p : \coprod_{x \in J} G_x \to J$ be the bundle associated to G; $f^*(G) \in \mathsf{Sh}(I)$ is the sheaf associated to the bundle $f^*(p) : f^*(\coprod_{x \in J} G_x) \to I$ as obtained when constructing the pullback of f and p. f^* preserves finite limits and arbitrary colimits

(f^* is left adjoint to f_*); hence it preserves the interpretation of any geometric formula. Moreover, f^* preserves the interpretation of all geometric axioms. Since f_* is right adjoint to f^*, it preserves limits, but not unions or images (in general). However, f_* preserves, to some extent, existential quantification, as follows. Let $\mathbb{T}$ be a theory in the language $\mathcal{L}$. A variable is called $\mathbb{T}$-*provably unique* if its value in every model of $\mathbb{T}$ is uniquely determined by the values of the remaining free variables. A *cartesian formula relative to* $\mathbb{T}$ is a formula built up from atomic formulae using only the connective $\wedge$ and the quantifier $\exists$ over $\mathbb{T}$-provably unique variables. A *cartesian axiom relative to* $\mathbb{T}$ is a formula of the form $(\forall x)(\phi(x) \Rightarrow \psi(x))$ where ϕ and ψ are cartesian formulae relative to $\mathbb{T}$. A *cartesian theory* is a theory whose axioms can be ordered such that each is cartesian relative to the preceding ones. Direct image functors preserve models of cartesian theories.

3 Systems

Our aim is modeling interconnected systems, whether hardware, software, or both. We assume a system S to be described by: a set X of control variables of the system, a set Γ of constraints on X, a set A of atomic actions, and a set C of constraints on A. This way of describing systems was influenced by the fact that, in many practical applications, the state of a system is determined by the values of certain control variables, among which dependencies may exist. An example and a detailed motivation can be found in [Sof96].

Definition 1. *Let $\Sigma = (\mathsf{Sort}, O, P)$ be a signature, consisting of a set* Sort *of sorts, a set O of operation symbols and a set P of predicate symbols, A Σ-*structure *is a structure $M = ((M_s)_{s \in \mathsf{Sort}}, \{f_M\}_{f \in O}, \{R_M\}_{R \in P})$ where if $f \in O$ has arity $s_1 \times \ldots \times s_n \rightarrow s$ then $f_M : M_{s_1} \times \ldots \times M_{s_n} \rightarrow M_s$ and if $R \in P$ has arity $s_1 \times \ldots \times s_n$ then $R_M \subseteq M_{s_1} \times \ldots \times M_{s_n}$. The class of Σ-structures is denoted* Str_Σ.

For a (many-sorted) set of variables $X = \{X_s\}_{s \in \mathsf{Sort}}$ let $\mathsf{Fma}_\Sigma(X)$ be the set of formulae over Σ. If $M \in \mathsf{Str}_\Sigma$, $s : X \rightarrow M$ is a sort-preserving assignment, and $\phi \in \mathsf{Fma}_\Sigma(X)$, $(M, s) \models \phi$ (abbreviated in what follows by $s \models \phi$) is defined in the usual way (cf. e.g. [CK90], Ch. 1).

Definition 2. *A* system *S is a tuple $(\Sigma, X, \Gamma, M, A, C)$, where*

- *$\Sigma = (\mathsf{Sort}, O, P)$ and $X = \{X_s\}_{s \in \mathsf{Sort}}$ are as above; together they form the* language *$\mathcal{L}_S$ of the system S;*
- *$\Gamma \subseteq \mathsf{Fma}_\Sigma(X)$ is a set of constraints on variables, which is closed with respect to the semantical consequence relation[1] $\models_M$;*
- *$M \in \mathsf{Str}_\Sigma$;*
- *A is a set of actions; for every $a \in A$ we have a set $X^a \subseteq X$ of variables on which a depends, and a transition relation $Tr^a \subseteq Sat^a \times Sat^a$, where $Sat^a = \{s_{|X^a} \mid s : X \rightarrow M, s \models \Gamma\}$;*

[1] The relation $\models_M$ is defined by $\Gamma \models_M \phi$ if and only if for every assignment of values in M to the variables in X, $s : X \rightarrow M$, if $s \models \gamma$ for every $\gamma \in \Gamma$, then $s \models \phi$.

– *C is a set of constraints on actions, expressed by boolean equations over $F_B(A)$ (the free boolean algebra generated by A) stating e.g. which actions can (or have to) be executed in parallel, and which cannot; C must contain all boolean equations that can be deduced from C.*

In what follows, we may refer to any of the components of a system S by adding S as a subscript, e.g. Σ_S for its signature. X^a_S will denote the minimal set of variables on which $a \in A_S$ depends, and Tr^a_S the transition relation associated with a.

Definition 3 (States; Parallel actions). *Let $S = (\Sigma, X, \Gamma, M, A, C)$ be a system. A* state *of S is an assignment $s : X \to M$ satisfying all formulae in Γ. The* set of states *of the system S is $St(S) = \{s : X \to M \mid s \models \Gamma\}$.*

The admissible parallel actions *of S are sets of actions, represented by maps $f : A \to \{0, 1\}$ that satisfy all constraints in C. The* set of admissible parallel actions *of S is the set is $Pa(S) = \{f : A \to \{0, 1\} \mid f$ satisfies $C\}$.*

Below we restrict our attention to *finite* systems, i.e. systems whose signatures, sets of control variables and sets of actions are finite; this suffices for practical applications and avoids modeling the undesirable case of infinitely many actions occurring in parallel.

Transitions. Let $S = (\Sigma, X, \Gamma, M, A, C)$ be a system. Let $Tr_S(a) = \{(s_1, s_2) \mid s_1, s_2 \in St(S), (s_{1|X^a}, s_{2|X^a}) \in Tr^a, s_1(x) = s_2(x)$ if $x \notin X^a\}$. In order to extend this notion of transition to parallel actions we present two non-equivalent properties of transitions that express compatibility of the actions in an admissible parallel action, **(Disj)** and **(Indep)**.

(Disj) Let $f \in Pa(S), s \in St(S)$ such that for every $a \in A$ with $f(a) = 1$ there is an $s^a \in St(S)_{|X^a}$ with $(s_{|X^a}, s^a) \in Tr^a$. Then for all $a, b \in A_S$ and $x \in X^a \cap X^b$, $s^a(x) = s^b(x)$, i.e. the new local states "agree on intersections". The transition induced by f is $Tr_S(f) = \{(s, t) \mid s, t \in St(S), (s_{|X^a}, t_{|X^a}) \in Tr^a$ for every a such that $f(a) = 1$, and $s(x) = t(x)$ if $x \notin \bigcup_{a, f(a)=1} X^a\}$.

(Indep) We assume that if $a = b \in C$ then $X^a = X^b$ and $Tr^a = Tr^b$, and a and b can be both identified with one action: the parallel execution of a, b. Let $f \in Pa(S), s \in St(S)$. We identify all elements $a, b \in A$ with $a = b \in C$ and $f(a) = f(b) = 1$. Let now $f^{-1}(1) = \{a_1, \dots, a_n\}$. Let $\{b_1, \dots, b_m\}$ be an arbitrary subset of $f^{-1}(1)$. We assume that:
(i) $g : A \to \{0, 1\}$, defined by $g(a) = 1$ if and only if $a \in \{b_1, \dots, b_m\}$, is in $Pa(S)$;
(ii) $s \overset{b_1}{\to} s_1 \overset{b_2}{\to} s_2 \cdots \overset{b_{m-1}}{\to} s_{m-1} \overset{b_m}{\to} t$ implies that for every permutation σ of $\{1, \dots, m\}$, there exist states $t^\sigma_1, \dots, t^\sigma_{m-1}$ such that $s \overset{b_{\sigma(1)}}{\to} t^\sigma_1 \overset{b_{\sigma(2)}}{\to} t^\sigma_2 \cdots \overset{b_{\sigma(m-1)}}{\to} t^\sigma_{m-1} \overset{b_{\sigma(m)}}{\to} t$ (the final state t is the same).
The transition induced by f is $Tr_S(f) = \{(s, t) \mid s, t \in St(S)$, and there exist $s = s_0, s_1, \dots, s_{n-1}, s_n = t \in St(S)$ such that $(s_{i-1}, s_i) \in Tr_S(a_i)$, for all $1 \leq i \leq n\}$. It is easy to see that if $(s, t) \in Tr_S(f)$ then for every $x \notin \bigcup_{a, f(a)=1} X^a$, $s(x) = t(x)$.

The property (**Disj**) applies when a parallel action $f : A \to \{0,1\}$ is admissible if and only if its components do not consume common resources. This happens for example if for all $a_1, a_2 \in A$ with $f(a_1) = f(a_2) = 1$, either $a_1 = a_2 \in C$ or X^{a_1} and X^{a_2} are disjoint. The property (**Indep**) reflects how transitions of parallel actions are interpreted when actions to be performed in parallel do consume common resources. It specifically applies if the state reached after executing an action is uniquely determined. In this case, the fact that all components of a parallel action $f : A \to \{0,1\}$ can be applied at a state s is a necessary condition for f to be applicable at state s, but in general not sufficient: in addition, one has to be sure that there are enough resources to perform all actions. Condition (**Indep**)(i) holds for instance if C is the set of all consequences of a set C_0 consisting only of formulae of the form $a_1 = a_2$ and $a_1 \wedge a_2 = 0$. Condition (**Indep**)(ii) states that the order in which the actions are executed is not relevant for determining the final state.

3.1 Communication between subsystems

Systems are usually related to other systems. We assume that, in order for two systems to be able to *communicate* they need a *"dictionary"*. Elements that are equal according to the dictionary are identified, so communicating systems are here supposed to share *common control variables* and *common actions*. We further assume that the values of the common control variables "sensed" simultaneously by two or more systems are the same. Essential to our model is that systems have common subsystems by which communication is handled (not the use of "dictionaries" or "translations"), and we focus on this aspect.

Definition 4. *Let S, T be two systems. We say that S is a* subsystem *of T (denoted $S \rightarrowtail T$) if $\Sigma_S \subseteq \Sigma_T$, $X_S \subseteq X_T$, $A_S \subseteq A_T$, the constraints in Γ_S (resp. C_S) are consequences of the constraints in Γ_T (resp. C_T), and $M_S = U^{\Sigma_T}_{\Sigma_S}(M_T)$ (where $U^{\Sigma_T}_{\Sigma_S} : \mathsf{Str}_{\Sigma_T} \to \mathsf{Str}_{\Sigma_S}$ is the forgetful functor).*

Let $S \rightarrowtail T$. If we regard a transition in T from the perspective of S, some variables in S may change their values with no apparent cause, namely if some action in A_T but not in A_S is performed, which depends on variables in X_S. If this cannot be the case, we call the subsystem $S \rightarrowtail T$ *transition-connected*. Formally, we have the following definition.

Definition 5. *S is a* transition-connected (t.c.) subsystem of T *(denoted $S \hookrightarrow T$) if $S \rightarrowtail T$ and the following two conditions hold:*

(T1) *for every $a \in A_T$, if $X^a_T \cap X_S \neq \emptyset$ then $a \in A_S$, and $X^a_S = X^a_T \cap X_S$;*
(T2) *for every $a \in A_S$ and every $s_1, s_2 \in St(T)$, if $(s_{1|X^a_T}, s_{2|X^a_T}) \in Tr^a_T$ then $(s_{1|X^a_S}, s_{2|X^a_S}) \in Tr^a_S$.*

It is easy to see that the relation $\hookrightarrow$ is a partial order on systems. We define a category TcSys with as objects systems and a morphism $S \hookrightarrow T$ between S and T whenever S is a t.c. subsystem of T.

Theorem 1. TcSys *has* pullbacks *(infimums with respect to this order of t.c. subsystems of a given system; we will denote this operation by $\wedge$) and* colimits *of diagrams of t.c. subsystems of a given system.*

In concrete applications, we tend to be interested in a subcategory of TcSys, containing only the systems relevant for the application. To this end, we assume a family InSys of interacting systems is specified, fulfilling:

- All $S \in$ InSys are t.c. subsystems of a system $\overline{S}$ with $A_{\overline{S}}$ finite.
- InSys is closed (in TcSys) under pullbacks of subsystems of $\overline{S}$.
- (InSys, $\wedge$) is a meet-semilattice.

The first condition enforces the compatibility of models on common sorts and the finiteness of A_S for every $S \in$ InSys; the second and third condition make sure that all systems by which communication is handled are taken into account.

A system obtained by interconnecting some elements of InSys can be seen either as the set of all elements of InSys by whose interaction it arises (a subset of InSys which is downwards-closed with respect to $\hookrightarrow$) or as the colimit of such a family of elements. We define Ω(InSys) as consisting of all families of elements of InSys which are closed under t.c. subsystems. It is a topology on InSys.

4 Modeling by sheaves

We show that the notions of states, (parallel) actions, behavior, and time can be represented as sheaves over the topological space (InSys, Ω(InSys)) previously defined. The fact that these notions can be expressed as sheaves with respect to an appropriate Grothendieck topology on a category of systems was already noticed in [Sof96]. We now show that the abstract framework presented there can be simplified. The main advantage of this simplification is that we can now express many properties of systems in the internal logic of the topos of sheaves over InSys. Geometric logic is then used to study how properties are preserved when interconnecting systems: interpretations corresponding to individual systems are obtained by using stalk functors, and interpretations corresponding to their interconnection are obtained by using the global section functors and colimits.

Definition 6 (States; Parallel actions).

(St) St : $\Omega(\mathsf{InSys})^{op} \rightarrow$ Set *is defined on objects by* $\mathsf{St}(U) = \{(s_i)_{S_i \in U} \mid s_i \in St(S_i)$, *and if* $S_i \hookrightarrow S_j$ *then* $s_i = s_{j|X_i}\}$, *and is such that for* $U_1 \overset{\iota}{\subseteq} U_2$, $\mathsf{St}(\iota) : \mathsf{St}(U_2) \rightarrow \mathsf{St}(U_1)$ *is defined by* $\mathsf{St}(\iota)((s_i)_{S_i \in U_2}) = (s_i)_{S_i \in U_1}$.

(Pa) Pa : $\Omega(\mathsf{InSys})^{op} \rightarrow$ Set *is defined on objects by* $\mathsf{Pa}(U) = \{(f_i)_{S_i \in U} \mid f_i \in Pa(S_i)$, *and if* $S_i \hookrightarrow S_j$ *then* $f_i = f_{j|A_i}\}$*; moreover, for* $U_1 \overset{\iota}{\subseteq} U_2$, $\mathsf{Pa}(\iota) : \mathsf{Pa}(U_2) \rightarrow \mathsf{Pa}(U_1)$ *is* $\mathsf{Pa}(\iota)((f_i)_{S_i \in U_2}) = (f_i)_{S_i \in U_1}$.

Theorem 2. *The functors* St *and* Pa *are sheaves on* InSys. *For every* $S_i \in$ InSys, *the stalk at* S_i *is in bijection with* $St(S_i)$ *resp.* $Pa(S_i)$. *Moreover, for every* $U \in \Omega$(InSys), St(U) *(resp.* Pa(U)*) is in bijection with* $St(S_U)$ *(resp.* $Pa(S_U)$*), where* S_U *is the colimit of the diagram defined by* U.

Proof: *(Sketch)* The fact that St and Pa are sheaves follows immediately from the definition of a sheaf. The fact that there exists a bijection between $\mathsf{St}(U)$ (resp. $\mathsf{Pa}(U)$) and $St(S_U)$ (resp. $Pa(S_U)$), where S_U is the colimit of the diagram defined by $U \in \Omega(\mathsf{InSys})$, follows from the definition of the colimit, taking into account that (i) if S_1 is a t.c. subsystem of S_2 and ϕ is a formula of S_1 then a state s of S_2 satisfies ϕ if and only if its restriction to S_1 satisfies ϕ; and (ii) for every $U \in \Omega(\mathsf{InSys})$, every family $\{s_i\}_{S_i \in U}$ of states which are compatible on the common variables can be "glued" to a (unique) state of the colimit S_U of the diagram defined by U. It can be shown that the stalk of St at S_i is $\mathsf{St}(\downarrow S_i)$. Since the colimit of the diagram defined by $\downarrow S_i = \{S_j \in \mathsf{InSys} \mid S_j \hookrightarrow S_i\}$ is S_i, $\mathsf{St}(\downarrow S_i)$ is in bijection with $St(S_i)$. The same results hold also for Pa. □

For every S_i in InSys and $f_i \in Pa(S_i)$, let $Trs_{S_i}(f_i)$ be the transition relation associated to f_i in S_i as explained in Section 3.

Definition 7 (Transition).

(Tr) $\mathsf{Tr} : \Omega(\mathsf{InSys})^{op} \to \mathsf{Set}$ *is defined on objects by* $\mathsf{Tr}(U) = \{(f, s, s') \mid f = (f_i)_{S_i \in U} \in \mathsf{Pa}(U), s = (s_i)_{S_i \in U}, s' = (s'_i)_{S_i \in U} \in \mathsf{St}(U), (s_i, s'_i) \in Trs_{S_i}(f_i),$ *for all* $S_i \in U\}$ *and is such that for* $U_1 \overset{\iota}{\subseteq} U_2$, $\mathsf{Tr}(\iota) : \mathsf{Tr}(U_2) \to \mathsf{Tr}(U_1)$ *is defined by* $\mathsf{Tr}(\iota)((f, s, s')) = (\mathsf{Pa}(\iota)(f), \mathsf{St}(\iota)(s), \mathsf{St}(\iota)(s'))$.

Theorem 3. *The functor* $\mathsf{Tr} : \Omega(\mathsf{InSys})^{op} \to \mathsf{Set}$ *is a subsheaf of* $\mathsf{Pa} \times \mathsf{St} \times \mathsf{St}$. *For every* $S_i \in \mathsf{InSys}$, *the stalk at* S_i *is in bijection with* $Tr(S_i) = \{(f, s, s') \mid (s, s') \in Trs_{S_i}(f)\}$. *If the transitions obey either* (**Disj**) *or* (**Indep**), *then, for every* $U \in \Omega(\mathsf{InSys})$, $\mathsf{Tr}(U)$ *is in bijection with* $Tr(S_U) = \{(f, s, s') \mid (s, s') \in Trs_{S_U}(f)\}$, *where* S_U *is the colimit of the diagram defined by* U.

Proof: *(Idea)* The fact that Tr is a subsheaf of $\mathsf{Pa} \times \mathsf{St} \times \mathsf{St}$ follows immediately. A careful analysis shows that if the transitions obey either (**Disj**) or (**Indep**), then (i) for every $U \in \Omega(\mathsf{InSys})$, every family of local (compatible) transitions in the systems $S_i \in U$ can be glued to a global transition of S_U, where S_U is the colimit of the diagram defined by U; and (ii) for every $S_i \in U$, the restrictions of a transition in S_U to S_i is a transition in S_i. The fact that the stalk at S_i is in bijection with $Tr(S_i)$ can be proved as in Theorem 2. □

We now define the behavior in time of a complex system. Our ideas are influenced by those in [Gog92], where objects are modeled by functors $F : \mathcal{T}^{op} \to \mathsf{Set}$, where $\mathcal{T}$ is a "base for observation", and behavior is described componentwise. Here, we propose a slightly different way of modeling behavior. In what follows, time is assumed to be discrete, and all actions take a constant, unit amount of time. We will assume that $\mathcal{T}$ is the basis for the topology on $\mathbb{N}$ consisting of $\mathbb{N}$ together with all sets $\{0, 1, \ldots, n\}, n \in \mathbb{N}$. The behavior in the interval $T \in \mathcal{T}$ of a complex system is modeled by all successions of pairs (state, action) of the component subsystems that can be observed during T. We show that behavior over an arbitrary but fixed time interval T, can be modeled by a sheaf. It may be interesting to combine sheaf conditions with respect to time and with respect to the structure of a system; this is planned for future work.

Definition 8 (Behavior). *Let $T \in \mathcal{T}$ be arbitrary but fixed. The behavior in the time interval T is modeled by $\mathsf{B}_T : \Omega(\mathsf{InSys})^{op} \to \mathsf{Set}$ defined for every $U \in \Omega(\mathsf{InSys})$ by $\mathsf{B}_T(U) = \{h : T \to \mathsf{St}(U) \times \mathsf{Pa}(U) \mid K(h,T)\}$ and for every $U_1 \overset{\iota}{\subseteq} U_2$ by $\mathsf{B}_T(\iota) : \mathsf{B}_T(U_2) \to \mathsf{B}_T(U_1)$, where for every $h \in \mathsf{B}_T(U_2)$, $\mathsf{B}_T(\iota)(h) = (\mathsf{St}(\iota) \times \mathsf{Pa}(\iota)) \circ h : T \overset{h}{\to} \mathsf{St}(U_2) \times \mathsf{Pa}(U_2) \xrightarrow{\mathsf{St}(\iota) \times \mathsf{Pa}(\iota)} \mathsf{St}(U_1) \times \mathsf{Pa}(U_1)$. Here $K(h,T)$ expresses the fact that for every n, if $n, n+1 \in T$ and $h(n) = (s,f)$, $h(n+1) = (s',f')$ then $(f,s,s') \in Tr(U)$.*

Let $B_T(S) = \{h : T \to St(S) \times Pa(S) \mid K_S(h,T)\}$, where $K_S(h,T)$ expresses the fact that for every n, if $n, n+1 \in T$ and $h(n) = (s,f)$, $h(n+1) = (s',f')$ then $(s,s') \in Tr_S(f)$.

Theorem 4. *For any $T \in \mathcal{T}$, $\mathsf{B}_T : \Omega(\mathsf{InSys})^{op} \to \mathsf{Set}$ is a sheaf. For every $S_i \in \mathsf{InSys}$, the stalk at S_i is in bijection with $B_T(S_i)$. If the transitions obey either* (**Disj**) *or* (**Indep**)*, then, for every $U \in \Omega(\mathsf{InSys})$, $\mathsf{B}_T(U)$ is in bijection with $B_T(S_U)$, where S_U is the colimit of the diagram defined by U.*

Proof: *(Idea)* The fact that B_T is a sheaf follows from the definition of B_T and the fact that St, Pa, and Tr are sheaves. The existence of a bijection between $\mathsf{B}_T(U)$ and $B_T(S_U)$ follows from results used in Theorems 2 and 3 when proving that $F(S_U)$ is in bijection with $\mathsf{F}(U)$ for $F \in \{St, Pa, Tr\}$. □

In order to reason about the evolution of systems in time, it may be useful to express time internally in the category Sh(InSys).

Definition 9 (Time). *Time is modeled by the sheafification of the constant presheaf $\mathcal{N} : \Omega(\mathsf{InSys})^{op} \to \mathsf{Set}$ (defined for every U by $\mathcal{N}(U) = \mathbb{N}$), which by abuse of notation we denote $\mathbb{N}$ as well.*[2]

We can also define functors $\mathsf{B^{St}}_T$ and $\mathsf{B^{Pa}}_T$ which only encode information about states (resp. actions). Various other sheaves and natural transformations can be defined by using standard categorical constructions in Sh(InSys). We can e.g. define a natural transformation $\mathsf{B}_{\mathbb{N}} \times \mathbb{N} \overset{\mathsf{a}}{\to} \mathsf{St} \times \mathsf{Pa}$ whose components $\mathsf{B}_{\mathbb{N}}(U) \times \mathbb{N}(U) \overset{\mathsf{a}_U}{\to} \mathsf{St}(U) \times \mathsf{Pa}(U)$ are defined by $\mathsf{a}_U(h, (n_i)_{S_i \in U}) = ((s^i_i)_{S_i \in U}, (f^i_i)_{S_i \in U})$, for every $U \in \Omega(\mathsf{InSys})$, where for every $S_i \in U$, $h(n_i) = ((s^i_j)_{S_j \in U}, (f^i_j)_{S_j \in U})$.

Theorem 5. *For every $S_i \in \mathsf{InSys}$, $\mathsf{Stalk}_{S_i}(\mathsf{a})$ is (up to isomorphism) the map $B_T(S_i) \times \mathbb{N} \overset{\mathsf{a}_{S_i}}{\to} St(S_i) \times Pa(S_i)$, defined by $\mathsf{a}_{S_i}(h,n) = h(n)$.*

Proof: *(Sketch)* Follows immediately from the way the stalk functors are defined on morphisms, and from the fact that, by Theorems 2 and 4, $F_{S_i} \simeq F(\downarrow S_i) \simeq F(S_i)$ for $F \in \{\mathsf{St}, \mathsf{Pa}, \mathsf{B}_T\}$, and from the fact that $\mathbb{N}_{S_i} \simeq \mathbb{N}(\downarrow S_i) \simeq \mathbb{N}$. □

An alternative way to describe behavior is by traces of execution. We obtained results which extend those given in [MP86], and which give a sheaf-theoretic formalization to results of Diekert [Die90]. Due to space limitations we cannot present these results here; for details cf. [SS97].

[2] It is denoted $(\mathcal{N}^+)^+$ in [MLM92], p. 130; it can be shown that for every $U \in \Omega(\mathsf{InSys})$, $(\mathcal{N}^+)^+(U) = \{i : U \to \mathbb{N} \mid i \text{ locally constant}\}$. $\mathbb{N} = (\mathcal{N}^+)^+$ is the natural number object in Sh(InSys); maps $1 \overset{0}{\to} \mathbb{N} \overset{s}{\to} \mathbb{N}$ and a subobject $\leq \subseteq \mathbb{N} \times \mathbb{N}$ can be defined.

4.1 Using geometric logic to express properties of systems

Let $\mathcal{L}$ be a fixed many-sorted language including at least sorts like st(ate), pa(rallel-action), b(ehavior), t(ime); constants like s_0 : st (initial state), 0 : t (initial moment of time); function symbols like appl : b×t → st×pa, p_1 : st×pa → st, p_2 : st × pa → pa; relation symbols like tr(ansition) ⊆ pa × st × st, $=_X \subseteq X \times X$ for every sort X, etc. Let M be an interpretation of $\mathcal{L}$ in Sh(InSys) such that $\mathsf{st}^M = \mathsf{St}$, $\mathsf{pa}^M = \mathsf{Pa}$, $\mathsf{b}^M = \mathsf{B}_{\mathbb{N}}$, $\mathsf{t} = \mathbb{N}$, $\mathsf{appl}^M = \mathsf{a}$, ${\mathsf{p}_1}^M = \pi_1, {\mathsf{p}_2}^M = \pi_2$ (the canonical projections), $\mathsf{tr}^M = \mathsf{Tr}$. For every sort X, we interpret $=_X: X \times X \to \Omega$ as usual.

Stalk functors. For every $S_i \in$ InSys let $f_i : \{*\} \to$ InSys be defined by $f_i(*) = S_i$. The inverse image functor corresponding to f_i, the stalk functor $\mathsf{Stalk}_{S_i} = f_i^*$: Sh(InSys) → Set, associates to every sheaf $F \in$ Sh(InSys) the stalk at S_i, F_{S_i}. For all $S_i \in$ InSys, f_i^* preserves the validity of geometric axioms. The stalk functors f_i^* : Sh(InSys) → Set are collectively faithful, so they reflect the validity of geometric axioms.

Global section functor. Consider the unique map g : InSys → $\{*\}$. The corresponding direct image functor, g_* : Sh(InSys) → Set, is the global section functor $g_*(F) = F(\mathsf{InSys})$ for every $F \in$ Sh(InSys). Thus, the global section functor preserves the interpretation of every cartesian axiom.

Theorem 6. Sh(InSys) *satisfies a geometric axiom in the interpretation M if and only if* Set *satisfies it in all interpretations $f_i^*(M)$. If* Sh(InSys) *satisfies a cartesian axiom, this is also true in* Set *in the interpretation $g_*(M)$ ($f_i^*(M)$ and $g_*(M)$ interpret a sort X as $f_i^*(X^M)$ resp. $g_*(X^M)$).*

From Theorems 2, 3, and 4 we know that for every $S_i \in$ InSys, $f_i^*(\mathsf{St}) = \mathsf{St}_{S_i} \simeq St(S_i)$ and $f_i^*(\mathsf{Pa}) = \mathsf{Pa}_{S_i} \simeq Pa(S_i)$; if S is the system obtained by interconnecting all elements in InSys, $g_*(\mathsf{St}) = \mathsf{St}(\mathsf{InSys}) \simeq St(S)$, and $g_*(\mathsf{Pa}) = \mathsf{Pa}(\mathsf{InSys}) \simeq Pa(S)$. The same holds for Tr and B_T. Moreover, $f_i^*(\mathbb{N}) = \mathbb{N}$, $g^*(\mathbb{N}) = \mathbb{N}(\mathsf{InSys})$, and, by Theorem 5, $f_i^*(\mathsf{appl}) = \mathsf{a}_{S_i} : B_{\mathbb{N}}(S_i) \times \mathbb{N} \to St(S_i) \times Pa(S_i)$. Hence, statements about states, actions and transitions in Sh(InSys) are translated by f_i^* (resp. g_*) to corresponding statements about states, actions and transitions in S_i (resp. S).

We illustrate the ideas above by several examples. We consider classes of properties of systems (adapted from [Krö87]) and express them in the language $\mathcal{L}$. For instance, if h is a possible behavior and j a moment of time, then $h(j)$ can be expressed in $\mathcal{L}$ by appl(h, j); the state of h at j can be expressed by s(h, j), where $\mathsf{s} = \mathsf{p}_1 \circ \mathsf{appl} : \mathsf{b} \times \mathsf{t} \overset{\mathsf{appl}}{\to} \mathsf{st} \times \mathsf{pa} \overset{\mathsf{p}_1}{\to} \mathsf{st}$.

(a) **Safety properties** are of the form $(\forall h : \mathsf{b})(\forall j : \mathsf{t})(P(\mathsf{s}(h,0)) \Rightarrow Q(\mathsf{s}(h,j)))$, where P and Q are formulae in $\mathcal{L}$. As examples we mention: *partial correctness:* $(\forall h : \mathsf{b})(\forall j : \mathsf{t})[(P(\mathsf{s}(h,0)) \wedge \mathsf{Final}(\mathsf{s}(h,j))) \Rightarrow Q(\mathsf{s}(h,j))]$; *global invariance of Q:* $(\forall h : \mathsf{b})(\forall j : \mathsf{t})[P(\mathsf{s}(h,0)) \Rightarrow Q(\mathsf{s}(h,j))]$.

(b) Liveness properties have the form $(\forall h : \mathsf{b})[P(\mathsf{s}(h,0)) \Rightarrow (\exists j : \mathsf{t})Q(\mathsf{s}(h,j))]$. With s_0 denoting the initial and s_f a final state, examples are: *total correctness and termination:* $(\forall h : \mathsf{b})[P(\mathsf{s}(h,0)) \Rightarrow (\exists j : \mathsf{t})(\mathsf{Final}(\mathsf{s}(h,j)) \wedge Q(\mathsf{s}(h,j)))]$; *accessibility:* $(\forall h : \mathsf{b})[(\mathsf{s}(h,0) = s_0) \Rightarrow (\exists j : \mathsf{t})(\mathsf{s}(h,j) = s_f)]$.

(c) Precedence properties are of the form $(\forall h : \mathsf{b})(\forall j : \mathsf{t})[(P(\mathsf{s}(h,0)) \wedge A(\mathsf{s}(h,j))) \Rightarrow Q(\mathsf{s}(h,j))]$.

Theorem 7. *Assume that the following conditions are fulfilled:*

(1) The final states form a subsheaf $\mathsf{St}_f \subseteq \mathsf{St}$ *interpreting a sort* $\mathsf{st_f}$ *of* $\mathcal{L}$. *(This happens e.g. if in the definition of a system final states are specified by additional constraints, and in defining colimits this information is also used.)*

(2) The properties P, Q, A *can be expressed in* $\mathcal{L}$ *(using the sorts, constants, function and relation symbols mentioned at the beginning of Section 4.1), and can be interpreted in both* $\mathsf{Sh}(\mathsf{InSys})$ *and* Set *(to express, for every* S_i *in* InSys, *the corresponding property of* S_i, *or* S*).*

Then, all formulae considered above (safety, liveness and precedence properties) are preserved under inverse image functors if in the definitions of the property P *(c.q.* Q, A*) only conjunction, disjunction and existential quantification occur. They are additionally preserved by direct image functors if only conjunction and unique existential quantification occur.*

Proof: *(Sketch)* Assume that (1) holds, and $\mathsf{st_f}$ is a subsort of st. Let $i : \mathsf{st_f} \to \mathsf{st}$ be the inclusion. Then $\mathsf{Final}(s)$ is expressed in $\mathcal{L}$ by $(\exists s' : \mathsf{st_f})(i(s') = s)$, and if s' exists, it is unique. If (1) and (2) hold, the formulae above can be expressed in the language $\mathcal{L}$. Therefore, the conclusion follows, since all given formulae are geometric if P, Q, A are (resp. cartesian if P, Q, A are, and only unique existential quantification occurs). □

Examples.

1. Let $\phi_1 = (\forall s, s', s'' : \mathsf{st})(\forall a : \mathsf{pa})[(\mathsf{tr}(a, s, s') \wedge \mathsf{tr}(a, s, s'')) \Rightarrow s' = s'']$ express *determinism*. Since ϕ_1 is a cartesian axiom, if all systems in InSys satisfy ϕ_1, then ϕ_1 is true (internally) in $\mathsf{Sh}(\mathsf{InSys})$. Moreover, it follows that ϕ_1 is true in the system obtained by interconnecting the systems in InSys.
2. Let $\phi_2 = (\forall h : \mathsf{b})(\forall a : \mathsf{pa})(\forall i : \mathsf{t})[(\exists s : \mathsf{st})(\mathsf{tr}(a, \mathsf{p}_1(\mathsf{appl}(h, i)), s)) \Rightarrow (\exists j : \mathsf{t})(j \geq i \wedge \mathsf{p}_2(\mathsf{appl}(h, j)) = a)]$ express *fairness of execution.* Since ϕ_2 is a geometric axiom, it is preserved and reflected by the stalk functors. Since an existential quantifier occurs, ϕ_2 may not be preserved by the global section functor. If ϕ_2 is part of a cartesian theory $\mathbb{T}$ (i.e. the existence of s and j is $\mathbb{T}$-provably unique), then its validity (as part of $\mathbb{T}$) is preserved by g_*. The validity of ϕ_2 is also preserved by g_* if all systems in InSys are independent, i.e. $S_i \wedge S_j = \emptyset$ if $S_i \neq S_j$ (If InSys is finite, $(\mathsf{InSys}\backslash\emptyset, \Omega(\mathsf{InSys}\backslash\emptyset))$ is then a Stone space, cf. also *Remark 1*).
3. Let $\phi_3 = (\forall s : \mathsf{st})(\exists a : \mathsf{pa})(\exists s' : \mathsf{st})(\mathsf{tr}(a, s, s') \wedge (s \neq s'))$ express *deadlock freedom.* Since ϕ_3 contains the negation sign, it is not geometric and may not be preserved by direct and inverse geometric morphisms; in particular neither by the global section nor by the stalk functors.

Remark 1. The empty system $\emptyset$ can be excluded from consideration, as follows. Let $\Omega_1(\mathsf{InSys})$ be the family of all subsets of $\mathsf{InSys}\backslash\emptyset$ closed under t.c. subsystems. (If no element in InSys is the colimit of other elements in InSys, then $\Omega_1(\mathsf{InSys})$ is the free frame freely generated by InSys together with the constraint that the empty family of systems covers the empty system.) All the considerations above remain valid when $\Omega(\mathsf{InSys})$ is replaced by $\Omega_1(\mathsf{InSys})$. The space $(\mathsf{InSys}\backslash\emptyset, \Omega_1(\mathsf{InSys}))$ is totally disconnected if for every $S_1, S_2 \in \mathsf{InSys}$, their largest common t.c. subsystem $S_1 \wedge S_2$ is empty; the space is compact if additionally InSys is finite. In this situation a larger class of axioms is preserved by the global section functor (uniqueness in existential quantification is not required, cf. e.g. [Joh82], Ch. V.1.12). Then, the definition of time as a sheaf $\mathbb{N}$ expresses the fact that independent systems may have independent clocks.

5 Conclusion

We showed that a family InSys of interacting systems closed under pullbacks can be endowed with a topology which models the way these systems interact. States, parallel actions, transitions, and behavior can be described as sheaves on this topological space. We then used geometric logic to determine which kind of properties of systems in InSys are preserved when interconnecting these systems. Our results are influenced by the results of Goguen in [Gog92], where a sheaf-theoretic framework for modeling concurrent interaction is presented. There, objects are taken to be sheaves, and then the behavior of systems (diagrams in the framework) corresponds to constructing a limit, while interconnecting systems amounts to taking colimits. At the end, Goguen suggests to look at the more elaborate framework of topos theory and see what kind of reasoning can be achieved using the internal logic of a topos of sheaves. This is the direction we have explored in this paper. The main advantage of our approach is that it opens the possibility to verify properties of complex systems in a modular way.

In recent papers on model checking, decomposition of systems was used to avoid the state explosion problem. We refer for instance to [CGL96], where systems are modeled by finite Kripke structures. In that context, it is shown that formulae in universal computation tree logics ($ACTL^*$) can be checked in a modular way; for this (i) certain fairness assumptions are made; and (ii) it may be necessary to make additional assumptions about the environment when verifying properties of individual components. Both in formulae in $ACTL^*$ and in geometric and cartesian formulae, as defined in this paper, restrictions are imposed in the use of existential quantification and negation. We would like to gain a better understanding about the possible links between the results presented in this paper and the methods from model checking mentioned above.

We plan to continue our research in several directions. First, we can consider categories with more special morphisms (e.g. conservative extensions, definitional extensions). Second, we can consider more general morphisms expressing "translations" between languages of different systems. For this, the theory of institutions may be a suitable theoretical framework. Third, since we showed that transitions define subsheaves Tr ⊆ Pa × St × St, we can associate a "generic transition system" to a given category of systems, where both states and actions are sheaves. The results of Adámek and Trnková [AT90] on defining automata in a category could then be applied to the concrete category Sh(InSys); this would allow us to carry over general constructions like minimal realization.

References

[AT90] J. Adámek and V. Trnková. *Automata and Algebras in Categories.* Kluwer Academic Publishers, 1990.

[CGL96] E.M. Clarke, O. Grumberg, and D.E. Long. Model checking. In *Nato ASI Series F*, volume 152, New York, Heidelberg, Berlin, 1996. Springer-Verlag.

[CK90] C.C. Chang and H.J. Keisler. *Model Theory.* North-Holland, Amsterdam, 3rd edition, 1990.

[CW96] G.L. Cattani and G. Winskel. Presheaf models for concurrency. In D. van Dalen and M. Bezem, editors, *Proceedings of Computer Science Logic '96*, LNCS 1258, pages 58–75. Springer Verlag, Berlin, 1996.

[Die90] V. Diekert. Combinatorics on Traces. In *LNCS 454*. Springer Verlag, 1990.

[Gog92] J.A. Goguen. Sheaf semantics for concurrent interacting objects. *Mathematical Structures in Computer Science*, 11:159–191, 1992.

[Joh82] P. Johnstone. *Stone Spaces.* Cambridge Studies in Advanced Mathematics 3. Cambridge University Press, 1982.

[Krö87] F. Kröger. *Temporal Logic of Programs*, volume 8 of *EATCS Monographs on Theoretical Computer Science.* Springer Verlag, 1987.

[Lil93] J. Lilius. A sheaf semantics for Petri nets. Technical Report A23, Dept. of Computer Science, Helsinki University of Technology, 1993.

[Mal94] G. Malcolm. Interconnections of object specifications. In R. Wieringa and R. Feenstra, editors, *Working Papers of the International Workshop on Information Systems - Correctness and Reusability*, 1994. Appeared as internal report IR-357 of the Vrije Universiteit Amsterdam.

[MLM92] S. Mac Lane and I. Moerdijk. *Sheaves in Geometry and Logic.* Universitext. Springer Verlag, 1992.

[MP86] L. Monteiro and F. Pereira. A sheaf theoretic model for concurrency. *Proc. Logic in Computer Science (LICS'86)*, 1986.

[Pfa91] J. Pfalzgraf. Logical fiberings and polycontextural systems. In P. Jorrand and J. Kelemen, editors, *Proc. Fundamentals of Artificial Intelligence Research*, volume 535 of *LNCS (subseries LNAI)*, pages 170–184. Springer Verlag, 1991.

[Sof96] V. Sofronie. Towards a sheaf theoretic approach to cooperating agents scenarios. In J. Calmet, J.A. Campbell, and J. Pfalzgraf, editors, *Proceedings of Artificial Intelligence and Symbolic Mathematical Computation, International Conference, AISMC-3, Steyr*, LNCS 1138, pages 289–304. Springer-Verlag, 1996.

[SS97] V. Sofronie-Stokkermans. *Fibered Structures and Applications to Automated Theorem Proving in Certain Classes of Finitely-Valued Logics and to Modeling Interacting Systems.* PhD thesis, RISC-Linz, J. Kepler University Linz, 1997.

[Win96] G. Winskel. A presheaf semantics of value-passing proceses. In Montanari and Sassone, editors, *Concurrency Theory: 7th International Conference, CONCUR '96 Proceedings*, LNCS 1119, pages 98–114, 1996.

The Operators minCh and maxCh on the Polynomial Hierarchy

Holger Spakowski[1] and Jörg Vogel[2]

[1] Ernst Moritz Arndt University, Dept. of Math. and Computer Science
D-17487 Greifswald, Germany
e-mail: spakow@rz.uni-greifswald.de
[2] Friedrich Schiller University, Computer Science Institute
D-07740 Jena, Germany
e-mail: vogel@minet.uni-jena.de

Abstract. In this paper we introduce a new acceptance concept for nondeterministic Turing machines with output device which allows a characterization of the complexity class $\Theta_2^p = \mathrm{P}^{\mathrm{NP[log]}}$ as a polynomial time bounded class. Thereby the internal structure of the output is essential: it looks at output with maximal number of mind changes instead of output with maximal value which was realized for the first time by Krentel [Kre88].
Motivated by this characterization we define in a general way two operators, the so called maxCh- and minCh- operator, respectively which are special types of optimization operators.
Following a paper by Hempel/Wechsung [HW96] we investigate the behaviour of these operators on the polynomial hierarchy. We prove a collection of relations regarding the interaction of operators maxCh, minCh, $, ∃, ∀, ⊕, Sig, C and U. So we get a tool to show that the maxCh- and minCh- classes are distinct under reasonable structural assumptions.
Finally, our proof techniques allow to solve one of the open questions of Hempel/Wechsung.

1 Introduction

Abstract operators play a central role in structural complexity. There are various attempts to relate complexity classes and function classes by defining operators which map complexity classes to function classes and vice versa. A starting point of this area is given by Toda [Tod91] with the #-operator which captures the essence of counting. E.g., this operator allows a characterization of the function class span-P introduced by Köbler/Schöning/Toran [KST89] as $\# \cdot \mathrm{NP}$.

Another central point of interest in complexity theory is the complexity of maximization (minimization) problems. Krentel investigated in [Kre88] optimization problems, e.g. TSP and maxClique. In terms of the function class $\mathrm{OptP} = \text{min-P} \cup \text{max-P}$ he described a classification of such problems: thereby a function f belongs to max-P, if there is a nondeterministic polynomial time bounded Turing machine with output device such that for any input x the maximum output of M for any accepting path β of $M(x)$ equals $f(x)$. Krentel has proved that, e.g. TSP (as function problem) where the length of a optimal tour equals the value of the function is complete in OptP by metric reduction.

Furthermore he has shown that any function of FP^{NP} can be described as an OptP-problem followed by a deterministic polynomial time computation.

As a consequence it stated a characterization of the complexity class $\Delta_2^p = P^{NP}$ by the so called MAX-acceptance concept:

Given a nondeterministic polynomial time bounded Turing machine with output device M and an input x then M accepts x in the sense of MAX iff any computation path with (quasilexicographically) maximum output accepts x.

Inspired by Krentels result and the abstract operator technique Hempel/Wechsung defined as optimization operators the max- and min- operator, respectively which allow an application of the MAX-acceptance concept to other complexity classes, e.g. P and coNP. Krentels result stated in terms of max/min-operators yield the following identities:

$\Delta_2^p = \oplus \cdot \min \cdot \mathrm{P}$ and $\Delta_2^p = \oplus \cdot \max \cdot \mathrm{P}$.

In [HW96] was proved a number of powerful relations regarding the interaction of the operators max and min with formerly used operators as U, $\oplus$, $\exists$ and $\forall$. In this way they proved an evidence for a strict hierarchy of the corresponding max and min function classes.

The complexity class $\Theta_2^p = P^{NP[\log]}$ was established by Wagner in [Wag90] as a constitutional part of the polynomial hierarchy, e.g. there is proven that $\Theta_2^p = L^{NP}$.

This paper is organized as follows:
We start with a further qualitative characterization of Θ_2^p by a special acceptance concept:

Let M be a nondeterministic polynomial bounded Turing machine with output device and let x be an input. M accepts x in the sense of MAX-CH iff at any computation path β of M on x with maximal number of mind chances of the output x is accepted. For $w \in \{0,1\}^*$ ch(w) denotes the number of mind-changes in w. It holds $\mathrm{ch}(0) = \mathrm{ch}(1) = \mathrm{ch}(00) = \mathrm{ch}(11) = 0, \mathrm{ch}(10) = \mathrm{ch}(01) = 1$ and e.g. $\mathrm{ch}(10010) = 3$, $\mathrm{ch}(10101) = 4$.

This concept means that the internal structure of the output is essential. It allows a characterization of Θ_2^p as a polynomial time bounded complexity class: $\Theta_2^p = \text{MAX-CH-P}$.

In section 3 we give the formal definitions of the operators maxCh and minCh (and \$) respectively which are caused by the MAX-CH acceptance type. We point out the identities $\Theta_2^p = \oplus \cdot \mathrm{maxCh} \cdot \mathrm{P}$ and $\Theta_2^p = \oplus \cdot \mathrm{minCh} \cdot \mathrm{P}$.

In section 4 we follow the way of Hempel/Wechsung to prove a number of relations regarding the interaction of our new operators maxCh and minCh and the known operators $\exists$, $\forall$, $\oplus$, Sig, U and C. Thus we are able to investigate the hierarchy of maxCh- and minCh- function classes and get structural evidence that the corresponding classes are distinct (or we have collapses in the polynomial hierarchy).

In the last section we refer an open question from Hempel/Wechsung. Our proof techniques established in section 4 allow an answer to this question.

We adopt the notations commonly used in structural complexity. For details we refer the reader to a standard book, e.g. [Pap94].

2 A Machine Based Characterization of Δ_2^p and Θ_2^p

Throughout this work our basic machine model is the nondeterministic polynomial-time bounded Turing machine with output device (NPTM). Every computation path writes an output over $\{0,1\}^*$ and accepts or rejects. The output of such a TM M on a path β on input x is denoted by $\mathrm{out}_M(x,\beta)$.

A machine M is said to be *normalized* if every paths β_1 and β_2 with $\mathrm{ch}(\mathrm{out}_M(x,\beta_1)) = \mathrm{ch}(\mathrm{out}_M(x,\beta_2))$ have the same acception behaviour.

Subsequently we consider only normalized machines.

As usual we have $\Delta_i^p = \mathrm{P}^{\Sigma_{i-1}^p}$ $(i \geq 1)$ – the class of sets decidable by a deterministic polynomial-time oracle-machine (DPOM) with an oracle from Σ_{i-1}^p and $\Theta_i^p = \mathrm{P}^{\Sigma_{i-1}^p[\log]}$ – where the number of queries is bounded by $\mathrm{O}(\log n)$.

In this section we present a characterization of Θ_2^p by the MAX-CH acceptance concept, i.e. we prove $\mathrm{P}^{\mathrm{NP}[\log]} = \text{MAX-CH-P}$, where MAX-CH-P is the class of all sets decidable in the sense of MAX-CH by polynomial time bounded machines.

As mentioned above Krentel has proved

Theorem 1. $\Delta_2^p = \mathit{MAXP}$.

Following the idea of Krentel we show

Theorem 2. $\Theta_2^p = \mathit{MAX\text{-}CH\text{-}P}$.

Proof. "$\subseteq$"
Let $A \in \mathrm{P}^{\mathrm{NP}[\log]}$, and let M a DPOM and $C \in \mathrm{NP}$, such that $A = \mathrm{L}\left(M^{(C)}\right)$ where M is asking w.l.o.g. exactly $z(n) = O(\log(n))$ queries to the oracle on inputs of length n.

Since $C \in \mathrm{NP}$ there exist $D \in \mathrm{P}$ and $p \in \mathit{Pol}$ such that for all $x \in \Sigma^*$

$$x \in C \Longleftrightarrow \bigvee_{y,|y| \leq p(|x|)} (\langle x,y \rangle \in D).$$

We construct a NPTM N as follows:

On input x N computes $z(|x|)$ and guesses nondeterministically $b_1 \dots b_{z(|x|)} \in \{0,1\}^{z(|x|)}$.

After that on each path β the following steps are carried out:

1. N constructs the oracle queries $q_1, \dots, q_{z(|x|)}$ simulating $M^{(C)}(x)$ substituting $b_1, \dots, b_{z(|x|)}$ for the answers to M's queries.
2. For each q_i such that $b_i = 1$ N guesses nondeterministically a y_i with $|y_i| \leq p(|q_i|)$.
   If not for all these y_i $\langle q_i, y_i \rangle \in D$ $(D \in \mathrm{P})$, then N outputs the word 0 and rejects on β.
   Otherwise N continues at **3.**

3. N outputs a word $w = 0101\ldots0(1)$ on β, where $\text{ch}(w)-1$ equals the natural number whose binary representation forms the string $b_1\ldots b_{z(|x|)}$. (Since $z(n) = O(\log n)$ this is possible in polynomial-time.)
4. N accepts if and only if the computation simulated in **1.** was accepting.

Let $\beta_{\max}$ be a path reaching **3.** where $b_1\ldots b_{z(|x|)}$ is lexicographically maximal among all paths reaching **3.** For the w output on this path $\text{ch}(w)$ is maximal too.

The $b_1,\ldots,b_{z(|x|)}$ of such a path $\beta_{\max}$ represent the correct oracle answers of $M^{(C)}$ on input x.

Hence $M^{(C)}(x)$ accepts if and only if $N(x)$ accepts in **4.** on $\beta_{\max}$.

"$\supseteq$"

Let $A \in$MAX-CH-P and M a NPTM such that $A = \text{L}_{\text{MAX-CH}}(M)$.
We define two auxiliary sets $H_1, H_2 \in \text{NP}$:

$H_1 =_{df} \{\langle x,y\rangle : \text{there exists a path } \beta \text{ with } \text{ch}(\text{out}_M(x,\beta)) \geq y\}$
$H_2 =_{df} \{\langle x,k\rangle : \text{there exists an accepting path } \beta \text{ with } \text{ch}(\text{out}_M(x,\beta)) = k\}$

For an input $x \in \Sigma^*$

$$k = \max\{\text{ch}(\text{out}_M(x,\beta)) : \beta \text{ is a path of } M(x)\}$$

is deterministically computable with $\text{O}(\log|x|)$ queries to H_1.

A single further question $\langle x,k\rangle \overset{?}{\in} H_2$ suffices to determine if $M(x)$ accepts on paths with the maximal change number k. □

3 Operators

Using the max/min - operators in [HW96] were shown $\Delta_2^p = \oplus\cdot\max\cdot\text{P}$ and $\Delta_2^p = \oplus\cdot\min\cdot\text{P}$.

One aim of this paper is to get a similar operator-based, machine-independent characterization for $\text{P}^{\text{NP[log]}}$. Therefore we define in the next subsections the operators used in this paper.

3.1 Operators Mapping Complexity Classes to Function Classes

We define for a complexity class $\mathcal{K}$ function classes $\max\cdot\mathcal{K}$, $\min\cdot\mathcal{K}$, $\#\cdot\mathcal{K}$, $\text{maxCh}\cdot\mathcal{K}$, $\text{minCh}\cdot\mathcal{K}$ and $\$\cdot\mathcal{K}$.

We recall from [Tod91] and [HW96]:

$$f \in \#\cdot\mathcal{K} \iff \bigvee_{A\in\mathcal{K}}\ \bigvee_{p\in Pol}\ \bigwedge_{x\in\Sigma^*} f(x) = \left\|\left\{y : 0 \leq y < 2^{p(|x|)} \wedge \langle x,y\rangle \in A\right\}\right\|$$

$$f \in \max\cdot\mathcal{K} \iff \bigvee_{A\in\mathcal{K}}\ \bigvee_{p\in Pol}\ \bigwedge_{x\in\Sigma^*} f(x) = \max\left\{y : 0 \leq y < 2^{p(|x|)} \wedge \langle x,y\rangle \in A\right\}$$

(and if this set is empty let $f(x) = 0$)

$$f \in \min\cdot\mathcal{K} \iff \bigvee_{A\in\mathcal{K}}\bigvee_{p\in Pol}\bigwedge_{x\in\Sigma^*} f(x) = \min\left\{y : 0 \le y < 2^{p(|x|)} \wedge \langle x,y\rangle \in A\right\}$$

(and if this set is empty let $f(x) = 2^{p(|x|)}$)

Our definitions motivated by the characterization of Θ_2^p are:

$$f \in \$\cdot\mathcal{K} \iff \bigvee_{A\in\mathcal{K}}\bigvee_{p\in Pol}\bigwedge_{x\in\Sigma^*} f(x) = \|\{\mathrm{ch}(y) : |y| \le p(|x|) \wedge \langle x,y\rangle \in A\}\|$$

$$f \in \mathrm{maxCh}\cdot\mathcal{K} \iff \bigvee_{A\in\mathcal{K}}\bigvee_{p\in Pol}\bigwedge_{x\in\Sigma^*} f(x) = \max\{\mathrm{ch}(y) : |y| \le p(|x|) \wedge \langle x,y\rangle \in A\}$$

(and if this set is empty let $f(x) = 0$)

$$f \in \mathrm{minCh}\cdot\mathcal{K} \iff \bigvee_{A\in\mathcal{K}}\bigvee_{p\in Pol}\bigwedge_{x\in\Sigma^*} f(x) = \min\{\mathrm{ch}(y) : |y| \le p(|x|) \wedge \langle x,y\rangle \in A\}$$

(and if this set is empty let $f(x) = p(|x|)$)

3.2 Operators Mapping Function Classes to Complexity Classes

For every function class $\mathcal{F}$ are $\mathrm{U}\cdot\mathcal{F}$, $\mathrm{Sig}\cdot\mathcal{F}$, $\mathrm{C}\cdot\mathcal{F}$ und $\oplus\cdot\mathcal{F}$ defined by

$$A \in \mathrm{U}\cdot\mathcal{F} \iff c_A \in \mathcal{F}$$

$$A \in \mathrm{Sig}\cdot\mathcal{F} \iff \bigvee_{f\in\mathcal{F}}\bigwedge_{x\in\Sigma^*} (x \in A \iff f(x) > 0)$$

$$A \in \mathrm{C}\cdot\mathcal{F} \iff \bigvee_{f\in\mathcal{F}}\bigvee_{g\in\mathrm{FP}}\bigwedge_{x\in\Sigma^*} (x \in A \iff f(x) \ge g(x))$$

$$A \in \oplus\cdot\mathcal{F} \iff \bigvee_{f\in\mathcal{F}}\bigwedge_{x\in\Sigma^*} (x \in A \iff f(x) \equiv 1 \pmod 2)$$

3.3 Operator Based Characterization of Θ_2^p

Using the identity MAX-CH-P $= \mathrm{P}^{\mathrm{NP}[\log]}$ we can prove

Theorem 3. $\Theta_2^p = \oplus\cdot\mathrm{maxCh}\cdot\mathrm{P}$ and $\Theta_2^p = \oplus\cdot\mathrm{minCh}\cdot\mathrm{P}$

Proof. "$\supseteq$"

The obvious construction using binary search is omitted due to space restrictions.

"$\subseteq$"

It suffices to show MAX-CH-P$\subseteq \oplus\cdot\mathrm{maxCh}\cdot\mathrm{P}$. Let $A \in$ MAX-CH-P.
Hence there is a NPTM M with $A \in \mathrm{L}_{\mathrm{MAX-CH}}(M)$. Let the computation time of M be bounded by the polynomial p.
We define $\beta_{\max}(x)$ to be a path such that

$$\mathrm{ch}\left(\mathrm{out}_M\left(x, \beta_{\max}(x)\right)\right) = \max\left\{\mathrm{ch}\left(\mathrm{out}_M(x,\beta')\right) : \beta' \text{ is a path of } M\right\}$$

Let $\mathrm{h} : \{0,1,\#\}^* \to \{0,1\}^*$ be the homomorphism defined by $\mathrm{h}(0) = 00$, $\mathrm{h}(1) = 11$, and $\mathrm{h}(\#) = 01$.

Hence holds $|h(\mathrm{out}_M(x,\beta)\#\beta)| \leq 4p(|x|) + 6$.

Let α and γ be sufficiently large such that $g(n) =_{df} 2\gamma n^\alpha \geq 4p(n) + 6$ for all natural numbers n.

It follows that

$$g(|x|) \geq |\mathrm{h}(\mathrm{out}_M(x,\beta)\#\beta)|$$

and hence

$$\begin{aligned} g(|x| + \mathrm{ch}(\mathrm{out}_M(x,\beta))) &\geq |\mathrm{h}(\mathrm{out}_M(x,\beta)\#\beta)| \\ &\geq \mathrm{ch}(\mathrm{h}(\mathrm{out}_M(x,\beta)\#\beta)) \end{aligned} \quad (1)$$

We define D to be

$D = \{\langle x, w\rangle : x, w \in \Sigma^*$ and

1. w has the form $h(\mathrm{out}_M(x,\beta)\#\beta\#)\underbrace{1010\ldots}_{\mu}$ and

2. The NPTM M has on input x on path β the output $\mathrm{out}_M(x,\beta)$ and

3. The μ in **1.** is chosen such that

$$\mathrm{ch}(w) = \begin{cases} g(|x| + \mathrm{ch}(\mathrm{out}_M(x,\beta))) & \text{if } M \text{ rejects on } \beta \\ g(|x| + \mathrm{ch}(\mathrm{out}_M(x,\beta))) + 1 & \text{if } M \text{ accepts on } \beta \end{cases}$$

is satisfied.$\}$

There is ever such a μ because of **(2)**.

From the definition we see that for a suitable chosen polynomial s holds

$$\bigwedge_{w \in \Sigma^*} (\langle x, w\rangle \in D \longrightarrow |w| \leq s(|x|)) .$$

Since $D \in \mathrm{P}$, the function f defined by

$$f(x) = \max\{\mathrm{ch}(w) : |w| \leq s(|x|) \wedge \langle x, w\rangle \in D\}$$

is in $\mathrm{maxCh} \cdot \mathrm{P}$.

Because of the monotonicity of g follows

$$f(x) \equiv 1 \pmod 2 \text{ iff } M \text{ accepts } x \text{ on } \beta_{\max}(x).$$

Hence we have shown $A \in \oplus \cdot \mathrm{maxCh} \cdot \mathrm{P}$.

$\Theta_2^p = \oplus \cdot \mathrm{minCh} \cdot \mathrm{P}$ can be shown analogously. □

4 The Operators minCh, maxCh and $ on PH

Now we want to investigate the hierarchy of function classes we obtain by applying the operators minCh, maxCh and $ to the polynomial hierarchy. In the following we will refer to it as maxChange-hierarchy. Under the assumption of an infinite polynomial hierarchy we get an inclusion structure given in figure 1.

The main difference between the maxChange-hierarchy and the hierarchy of min/max-classes investigated in [HW96] lies in the characterizability of the $-classes by the maxCh-classes.

4.1 The Operators Sig, C and U on Function Classes

To give evidence that certain inclusion relations between function classes are not valid we use monotone operators mapping function classes to complexity classes. This method was already used e.g. in [Vol94].

In this subsection are collected some auxiliary results regarding the operators Sig, C and U on minCh/maxCh - classes. The proofs are more or less straightforward and are omitted here.

Theorem 4. *For every complexity class $\mathcal{K}$ closed under $\leq_m^p$,*

$$\mathrm{Sig} \cdot \mathrm{maxCh} \cdot \mathcal{K} = \exists \cdot \mathcal{K}.$$

For every complexity class $\mathcal{K}$ closed under $\leq_{dtt}^p$,

$$\mathrm{Sig} \cdot \mathrm{minCh} \cdot \mathcal{K} = \mathrm{co}\mathcal{K}.$$

For every complexity class $\mathcal{K}$ closed under $\leq_m^p$,

1. $\mathrm{C} \cdot \mathrm{maxCh} \cdot \mathcal{K} = \exists \cdot \mathcal{K}$, 2. $\mathrm{C} \cdot \mathrm{minCh} \cdot \mathcal{K} = \forall \cdot \mathrm{co}\mathcal{K}$.

For every complexity class $\mathcal{K}$ closed under $\leq_{dtt}^p$,

1. $\mathrm{U} \cdot \mathrm{maxCh} \cdot \mathcal{K} = \mathcal{K}$, 2. $\mathrm{U} \cdot \mathrm{minCh} \cdot \mathcal{K} = \mathrm{co}\mathcal{K}$.

For every $i \geq 1$,

$$U \cdot (F\Theta_i^p)_{pol} = Sig \cdot (F\Theta_i^p)_{pol} = C \cdot (F\Theta_i^p)_{pol} = \Theta_i^p.$$

In [HW96] the same results can be found for the operators min and max except of that we need stronger assumptions for the closure properties.

4.2 General Relationships

In order to analyse the inclusion structure of the maxChange hierarchy we state in the current subsection some more general results regarding operators on complexity classes having resonable closure properties.

The $ - Classes.

It will turn out that $\$ \cdot \mathcal{K}$ can be expressed with the help of the operators maxCh and $\exists$ if $\mathcal{K}$ is closed under $\leq^p_{ctt}$. The essential part is done in the proof of the next theorem.

Theorem 5. *For every complexity class $\mathcal{K}$ closed under $\leq^p_m$,*

$$\$ \cdot \mathcal{K} = \$ \cdot \exists \cdot \mathcal{K}$$

Proof. It suffices to show $\$ \cdot \exists \cdot \mathcal{K} \subseteq \$ \cdot \mathcal{K}$.

Let $f \in \$ \cdot \exists \cdot \mathcal{K}$. Hence

$$\bigvee_{A \in \exists \cdot \mathcal{K}} \bigvee_{p \in Pol} \bigwedge_{x \in \Sigma^*} f(x) = \|\{\mathrm{ch}(y) : |y| \leq p(|x|) \wedge \langle x, y \rangle \in A\}\|.$$

Therefore exist $B \in \mathcal{K}$ and polynomials p and q such that

$$\bigwedge_{x \in \Sigma^*} f(x) = \left\| \left\{ \mathrm{ch}(y) : |y| \leq p(|x|) \wedge \bigvee_{z, |z| \leq q(|x|)} \langle x, \mathrm{h}(y\#z) \rangle \in B \right\} \right\|$$

where $\mathrm{h} : \{0,1,\#\}^* \rightarrow \{0,1\}^*$ is the homomorphism defined by $\mathrm{h}(0) = 00$, $\mathrm{h}(1) = 11$, $\mathrm{h}(\#) = 01$.

We define $r(n) =_{df} 2p(n) + 2q(n) + 2$.

Therefore $|h(y\#z)| = 2|y| + 2|z| + 2 \leq r(|x|)$.

Let α and γ be sufficiently large such that $g(n) =_{df} \gamma n^\alpha \geq r(n)$ is satisfied for all natural n.

Hence

$$g(|x|) \geq |h(y\#z)|$$

and therefore

$$g(|x| + \mathrm{ch}(y)) \geq |h(y\#z)|.$$

Since $\mathrm{ch}(x) < |x|$ follows

$$g(|x| + \mathrm{ch}(y)) \geq \mathrm{ch}(h(y\#z)). \tag{2}$$

Now we define C to be

$C = \{\langle x, w \rangle : x, w \in \Sigma^*$ and

1. w has the form $h(y\#z\#)\underbrace{1010\ldots}_{\mu}$ and

2. $\langle x, h(y\#z) \rangle \in B$ and

3. The μ in **1.** is chosen such that

$$\mathrm{ch}(h(y\#z\#)\underbrace{1010\ldots}_{\mu}) = g(|x| + \mathrm{ch}(y))$$

is satisfied. $\}$

There is ever such a μ in **3.** because of (2).

By the definition of C follows for a suitable chosen s:

$$\bigwedge_{w \in \Sigma^*} (\langle x, w \rangle \in C \longrightarrow |w| \leq s(|x|))$$

Since $\mathcal{K}$ is closed under $\leq_m^p$ holds $C \in \mathcal{K}$.

Because g is one-to-one we get for all x

$$f(x) = ||\{\mathrm{ch}(w) : |w| \leq s(|x|) \wedge \langle x, w\rangle \in C\}||$$

Hence $f \in \$ \cdot \mathcal{K}$. □

Now we can prove:

Theorem 6. *For all $\mathcal{K}$ closed under $\leq_{ctt}^p$,*

$$\mathrm{maxCh} \cdot \exists \cdot \mathcal{K} = \$ \cdot \mathcal{K}$$

Proof. "$\subseteq$"

We show $\mathrm{maxCh} \cdot \mathcal{K} \subseteq \$ \cdot \exists \cdot \mathcal{K}$ for all $\mathcal{K}$ closed under $\leq_m^p$.
$\mathrm{maxCh} \cdot \mathcal{K} \subseteq \$ \cdot \mathcal{K}$ and $\mathrm{maxCh} \cdot \exists \cdot \mathcal{K} \subseteq \$ \cdot \mathcal{K}$ follow then due to theorem 5.
Let $f \in \mathrm{maxCh} \cdot \mathcal{K}$, hence there exist $A \in \mathcal{K}$ and a polynomial p such that

$$\bigwedge_{x \in \Sigma^*} f(x) = \max\{\mathrm{ch}(y) : |y| \leq p(|x|) \wedge \langle x, y\rangle \in A\}$$

We define B to be

$B = \{\langle x, z\rangle : x, z \in \Sigma^*$ und

1. $z = \underbrace{0101\ldots0(1)}_{l}$ und

2. $\bigvee_{y,|y| \leq p(|x|)} (\langle x, y\rangle \in A \wedge \mathrm{ch}(y) \geq l)\}$

We conclude $B \in \exists \cdot \mathcal{K}$ since $\mathcal{K}$ is closed under $\leq_m^p$.
Since for all $x \in \Sigma^*$

$$f(x) = ||\{\mathrm{ch}(z) : |z| \leq p(|x|) \wedge \langle x, z\rangle \in B\}||,$$

we have proven $f \in \$ \cdot \exists \cdot \mathcal{K}$.

"$\supseteq$"

Let $f \in \$ \cdot \mathcal{K}$. Hence there are $A \in \mathcal{K}$ and a polynomial p such that

$$\bigwedge_{x \in \Sigma^*} f(x) = ||\{\mathrm{ch}(y) : |y| \leq p(|x|) \wedge \langle x, y\rangle \in A\}||$$

We define B to be $B = \{\langle x, z\rangle : x, z \in \Sigma^*$ and

1. $z = \underbrace{0101\ldots0(1)}_{l}$ $(l \leq p(|x|))$, and

2. There exist $y_1, \ldots, y_{l-1}$ satisfying $y_i \leq p(|x|)$ and $\mathrm{ch}(y_i) \neq \mathrm{ch}(y_j)$ for all $i \neq j$, and

3. $\langle x, y_1\rangle \in A \wedge \cdots \wedge \langle x, y_{l-1}\rangle \in A$ $\}$

We conclude $B \in \exists \cdot \mathcal{K}$ since $\mathcal{K}$ is closed under $\leq_{ctt}^p$.
We see that for all $x \in \Sigma^*$

$$f(x) = \max\{\mathrm{ch}(z) : |z| \leq p(|x|) \wedge \langle x, z\rangle \in B\}.$$

This proves $f \in \mathrm{maxCh} \cdot \exists \cdot \mathcal{K}$. □

Relationships among the minCh- and maxCh- classes.

Theorem 7. *For every complexity classes $\mathcal{K}$ and $\mathcal{C}$ closed under $\leq_{dtt}^{p}$,*

1. $\text{minCh} \cdot \mathcal{K} \subseteq \text{minCh} \cdot \mathcal{C} \iff \mathcal{K} \subseteq \mathcal{C}$
2. $\text{maxCh} \cdot \mathcal{K} \subseteq \text{maxCh} \cdot \mathcal{C} \iff \mathcal{K} \subseteq \mathcal{C}$
3. $\text{maxCh} \cdot \mathcal{K} \subseteq \text{minCh} \cdot \mathcal{C} \iff \exists \cdot \mathcal{K} \subseteq \text{co}\mathcal{C}$

Of special interest is the third statement. Unfortunately, for the inclusion in the vice versa direction we don't know a structural equivalence. We have

Theorem 8. *For every complexity classes $\mathcal{K}$ und $\mathcal{C}$ closed under $\leq_{dtt}^{p}$,*

1. $\text{minCh} \cdot \mathcal{K} \subseteq \text{maxCh} \cdot \mathcal{C} \Longleftarrow \exists \cdot \mathcal{K} \subseteq \text{co}\mathcal{C}$
2. $\text{minCh} \cdot \mathcal{K} \subseteq \text{maxCh} \cdot \mathcal{C} \Longrightarrow \mathcal{K} \subseteq \text{co}\mathcal{C} \wedge \exists \cdot \mathcal{K} \subseteq \forall \cdot \text{co}\mathcal{C}$

Nevertheless, the fact that we don't have a structural equivalence for $\text{minCh} \cdot \mathcal{K} \subseteq \text{maxCh} \cdot \mathcal{C}$ causes no problem for our main goal to analyse the maxChange hierarchy under the assumption of an infinite polynomial hierarchy. Differently it will be in the case of the investigation of certain collapse-events in the PH.

The $(F\Theta_i^p)_{pol}$-Classes.

We summarize here our results for the $(F\Theta_i^p)_{pol}$-classes

$\text{F}\Delta_i^p$ denotes the set of all functions computable in deterministic polynomial-time with the help of an oracle from Σ_{i-1}^p.

A function f from $\text{F}\Delta_i^p$ belongs to $\text{F}\Theta_i^p$ if $f(x)$ is computable with at most $O(\log |x|)$ queries to the oracle.

For a function class $\mathcal{F}$ we define $\mathcal{F}_{pol}$ to be the subset of polynomially bounded functions.

Theorem 9. *For every complexity classs $\mathcal{C}$ closed under $\leq_{dtt}^{p}$ and all $i \geq 1$,*

1. $(F\Theta_i^p)_{pol} \subseteq \text{maxCh} \cdot \mathcal{C} \iff \Theta_i^p \subseteq \mathcal{C}$
2. $(F\Theta_i^p)_{pol} \subseteq \text{minCh} \cdot \mathcal{C} \iff \Theta_i^p \subseteq \mathcal{C}$

Theorem 10. *For every complexity class $\mathcal{K}$ closed under $\leq_{m}^{p}$ and all $i \geq 1$,*

1. $\text{maxCh} \cdot \mathcal{K} \subseteq (F\Theta_i^p)_{pol} \Longrightarrow \exists \cdot \mathcal{K} \subseteq \Theta_i^p$
2. $\text{maxCh} \cdot \mathcal{K} \subseteq (F\Theta_{i+1}^p)_{pol} \Longleftarrow \exists \cdot \mathcal{K} \subseteq \Sigma_i^p$
3. $\text{minCh} \cdot \mathcal{K} \subseteq (F\Theta_i^p)_{pol} \Longrightarrow \exists \cdot \mathcal{K} \subseteq \Theta_i^p$
4. $\text{minCh} \cdot \mathcal{K} \subseteq (F\Theta_{i+1}^p)_{pol} \Longleftarrow \exists \cdot \mathcal{K} \subseteq \Sigma_i^p$

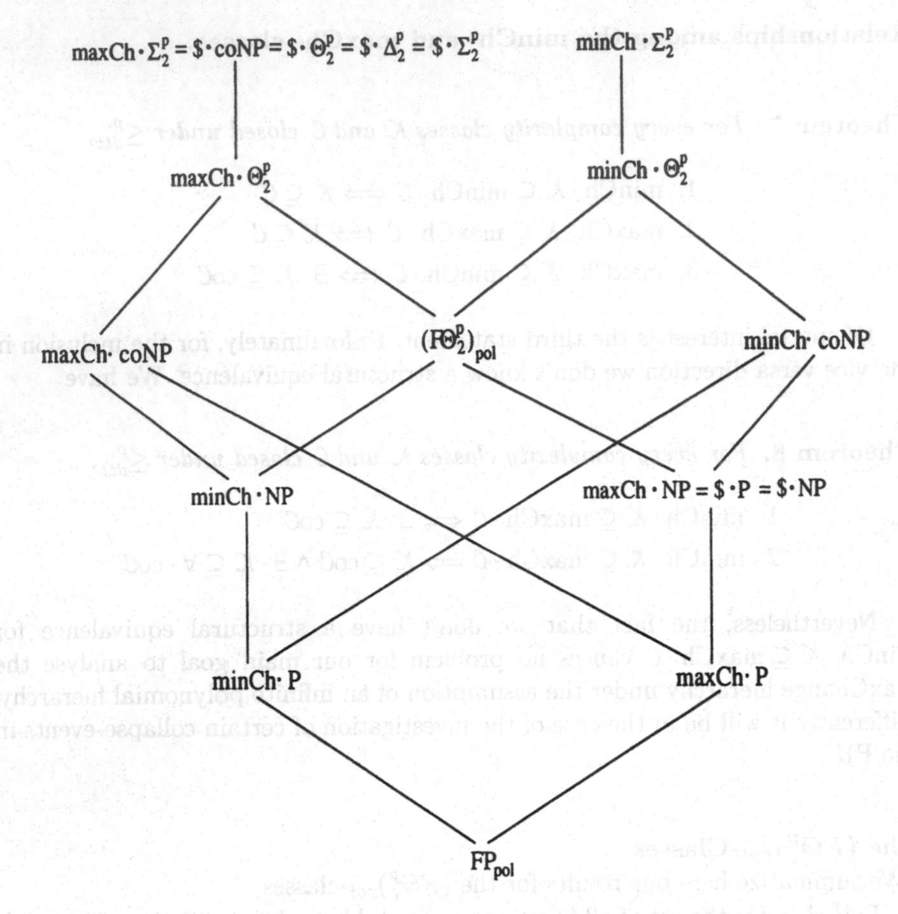

Fig. 1. The lowest levels of the maxChange hierarchy

4.3 Consequence for the maxChange-Hierarchy

The statements proved in 4.2 imply immediately

1. The inclusions shown in figure 1 are valid.
2. There are not any more inclusions unless the polynomial hierarchy collapses.

4.4 Investigation of Collapse Events in the Polynomial Hierarchy

Case P=NP.
P=NP clearly implies that the maxChange-hierarchy collapses to FP_{pol}.

Case P $\neq$ NP, NP = coNP.
Here remains as open question if minCh·P $\subseteq$ maxCh·P or minCh·P $\not\subseteq$ maxCh·P, because there is not known a structural equivalence for minCh $\cdot\, \mathcal{K} \subseteq$ maxCh $\cdot\, \mathcal{C}$ in 4.2.

Due to theorem 10.2 from NP=coNP follows $\text{maxCh} \cdot \mathcal{K} \subseteq (\text{F}\Theta_2^p)_{pol}$ for all $\mathcal{K} \in \text{PH}$. Furthermore from theorem 9 we conclude

$$\text{NP=coNP} \Longrightarrow \text{F}\Theta_2^p \subseteq \text{minCh} \cdot \text{NP} \cap \text{maxCh} \cdot \text{NP}.$$

5 An Open Question for max/min-Operators

In [HW96] was left open the problem to find structural equivalences for the inclusions

$$\# \cdot \text{P} \subseteq \max \cdot \text{coNP} \text{ and } \# \cdot \text{P} \subseteq \min \cdot \text{coNP}.$$

As a more general result we can state

Theorem 11. *For every $\mathcal{K}$ closed under $\leq_m^p$, intersection and complement and every $\mathcal{C}$ closed under $\leq_m^p$,*

$$\# \cdot \mathcal{K} \subseteq \max \cdot \mathcal{C} \Longleftrightarrow \mathcal{C}_= \cdot \mathcal{K} \subseteq \mathcal{C}$$
$$\# \cdot \mathcal{K} \subseteq \min \cdot \mathcal{C} \Longleftrightarrow \mathcal{C}_= \cdot \mathcal{K} \subseteq \mathcal{C}$$

As a consequence we get the relations

$$\# \cdot \text{P} \subseteq \max \cdot \text{coNP} \Longleftrightarrow \text{C}_=P \subseteq \text{coNP}$$
$$\# \cdot \text{P} \subseteq \min \cdot \text{coNP} \Longleftrightarrow \text{C}_=P \subseteq \text{coNP}$$

The proof of theorem 11 is available in the full version of the paper. (This is also true for all other proofs which are omitted here due to the lack of space.)

References

[HW96] H. Hempel, G. Wechsung. *The Operators and max on the Polynomial Time Hierarchy*. Proceedings of STACS 97, LNCS 1200, 93-104

[KST89] J. Köbler, U. Schöning, J. Torán. *On counting and approximation*. Acta Informatica, 26 (1989), 363-379

[Kre88] M. Krentel. *The complexity of optimization problems*. Journal of Computer and System Sciences, 36 (1988), 490-509

[Pap94] C. H. Papadimitriou. *Computational Complexity*. Addison-Wesley, 1994

[Tod91] S. Toda. *Computational Complexity of Counting Complexity Classes*. PhD thesis, Tokyo Institute of Technology, Department of Computer Science, Tokyo, Japan, 1991

[Vol94] H. Vollmer. *Komplexitätsklassen von Funktionen*. PhD thesis, Universität Würzburg, Institut für Informatik, Würzburg, Germany, 1994

[Wag90] K.W. Wagner. *Bounded query classes*. SIAM Journal on Computing, 19(1990), 833-846

The Kolmogorov Complexity of Real Numbers

Ludwig Staiger*

Institut für Informatik, Martin-Luther-Universität Halle-Wittenberg
Kurt-Mothes-Str. 1, D-06099 Halle, Germany
email: staiger@informatik.uni-halle.de

Abstract. We consider for a real number α the Kolmogorov complexities of its expansions with respect to different bases. In the paper it is shown that, for usual and self-delimiting Kolmogorov complexity, the complexity of the prefixes of their expansions with respect to different bases r and b are related in a way which depends only on the relative information of one base with respect to the other.
More precisely, we show that the complexity of the length $l \cdot \log_r b$ prefix of the base r expansion of α is the same (up to an additive constant) as the $\log_r b$-fold complexity of the length l prefix of the base b expansion of α.
Then we use this fact to derive complexity theoretic proofs for the base independence of the randomness of real numbers and for some properties of Liouville numbers.

Kolmogorov Complexity is mainly attributed to finite strings over a finite alphabet. As a function or, more coarsely, as a limit it measures the complexity of infinite strings.

Real numbers are described by their (infinite) r-ary expansions. Thus, choosing the base r, we may attribute Kolmogorov complexity also to real numbers, however, relative to the chosen base. Consequently, it might happen that the Kolmogorov complexity of a real number depends on the chosen base r.

Particular cases, where a property of a real number depends on the base r are disjunctiveness and Borel normality. An infinite r-ary expansion ξ of the real number $\nu_r(\xi) := 0.\xi$ is called disjunctive provided every finite r-ary string appears as an infix of ξ. Borel normality is defined in a similar way, taking into account also the relative frequencies of the infixes. For more detailed information see, e. g., [Ca94,He96]. It was already shown in [Cs59,Sc60] that Borel normality and disjunctiveness are not invariant under changes of the base r. On the other hand, it was shown in [CJ94], and in another context in [HW98], that the property of randomness of an infinite expansion of a real number is invariant under base change. Besides that it was claimed in [CH94] that the Kolmogorov complexity (as a limit) does not depend on the chosen base r.

In this note we investigate in more detail the Kolmogorov complexities, $K_r(\xi/n)$ of expansions ξ of a real number with respect to different bases r.

* The paper was prepared during my stay at the Centre for Discrete Mathematics and Theoretical Computer Science, University of Auckland, New Zealand, August 1998

We show that, if a real number is expanded in the scales of r and b, respectively, then complexity of the length $l \cdot \log_r b$ prefix of the base r expansion of is the same (up to an additive constant) as the $\log_r b$-fold complexity of the length l prefix of the base b expansion.

This result provides a third proof of the fact that randomness is base invariant for real numbers. Next we investigate the complexity of Liouville numbers, a kind of real numbers famous for an elegant and constructive proof of the existence of transcendental real numbers. Finally, utilizing our complexity theoretic arguments, we calculate the Hausdorff dimension of the set of Liouville numbers and investigate disjunctive Liouville numbers.

1 Notation and Preliminaries

By $\mathbb{N} = \{0, 1, 2, \ldots\}$ we denote the set of natural numbers. In order to treat the Kolmogorov complexities for arbitrary alphabets we let $X_r := \{0, \ldots, r-1\}$ be our alphabet of cardinality card $X_r = r$ for $r \in \mathbb{N}$ $r \geq 2$. By X_r^* we denote the set of finite strings (words) on X_r, including the *empty* word e. We consider also the space X_r^ω of infinite sequences (ω-words) over X_r. For $w \in X_r^*$ and $\eta \in X_r^* \cup X_r^\omega$ let $w \cdot \eta$ be their *concatenation*. This concatenation product extends in an obvious way to subsets $W \subseteq X_r^*$ and $F \subseteq X_r^* \cup X_r^\omega$.

By $w \sqsubseteq \eta$ we denote the prefix relation, that is, $w \sqsubseteq \eta$ if and only if there is an η' such that $w \cdot \eta' = \eta$.

For $\eta \in X_r^* \cup X_r^\omega$ we denote by $\nu_r(\eta) := 0.\eta$ the real number with (finite or infinite) base r expansion η.

We will consider the self-delimiting as well as the non self-delimiting complexity (cf. [Ca94,LV93]). To this end we fix for every $r \in \mathbb{N}$ a universal algorithm $U_r : X_r^* \to X_r^*$ and a universal self-delimiting algorithm $C_r : X_r^* \to X_r^*$, the domain of the latter is a prefix-free subset of X_r^*. Moreover we fix a recursive standard bijection between $\mathbb{N}$ and X_r^*, r-**string** : $\mathbb{N} \to X_r^*$. For the sake of convenience we agree that r-**string**(n) is the nth string in the quasilexicographical order of X_r^*. Then $|r\text{-}\mathbf{string}(n)| = \lfloor \log_r(n(r-1)+1) \rfloor \leq 1 + \log_r \max\{n, 1\}$.

The *Kolmogorov complexity* of a word $w \in X_r^*$ is defined as $K_r(w) := \inf\{|\pi| : \pi \in X_r^* \wedge U_r(\pi) = w\}$. Accordingly, the *self-delimiting Kolmogorov complexity* of a word $w \in X_r^*$ is $H_r(w) := \inf\{|\pi| : \pi \in X_r^* \wedge C_r(\pi) = w\}$.

In order to prove our results we need the following slight modifications of Theorem 5.1.b.ii in [Ca94] and Theorem 3.5 in [CC96]. We call a function $f : M \to M'$ of *bounded ambiguity* provided there is a $k \in \mathbb{N}$ such that for every $m \in M'$ the preimage $f^{-1}(m)$ has no more than k elements, and we call a function $h : \mathbb{N} \to \mathbb{N}$ *semi-computable from above* if the set $M_h := \{(n, j) : h(n) \leq j\}$ is recursively enumerable.

Theorem 1. *1. Let $f : \mathbb{N} \to X_r^*$ be a recursive function of bounded ambiguity. Then $\sum_{n \in \mathbb{N}} r^{-H_r(f(n))} < \infty$.*

2. If $g : \mathbb{N} \to X_r^*$ is recursive and $h : \mathbb{N} \to \mathbb{N}$ is semi-computable from above such that $\sum_{n \in \mathbb{N}} r^{-h(n)} < \infty$ then

$$\exists c (c \in \mathbb{N} \wedge \forall n (n \in \mathbb{N} \to H_r(g(n)) \leq h(n) + c)) .$$

Proof. 1. It is well-known that the self-delimiting complexity satisfies the inequality $\sum_{w\in X_r^*} r^{-H_r(w)} < \infty$ (see [Ca94,LV93]). Let $\mathrm{card} f^{-1}(w) \le k$ for every $w \in X_r^*$. Then

$$\sum_{n\in\mathbb{N}} r^{-H_r(f(n))} = \sum_{w\in X_r^*} \mathrm{card} f^{-1}(w)\cdot r^{-H(w)} \le k\cdot \sum_{w\in X_r^*} r^{-H_r(w)} < \infty\,.$$

2. If $\sum_{n\in\mathbb{N}} r^{-h(n)} < \infty$ then also $\sum_{n\in\mathbb{N}} \sum_{j\ge h(n)} r^{-j} = \frac{r}{r-1}\cdot \sum_{n\in\mathbb{N}} r^{-h(n)} < \infty$. Consequently, there is an $m\in\mathbb{N}$ such that $\sum_{n\in\mathbb{N}}\sum_{j\ge h(n)} r^{-h(n)-m} \le 1$.

Let $f_h : \mathbb{N} \to X_r^* \times \mathbb{N}$ be a recursive function enumerating the recursively enumerable set $M_h := \{(r\text{-}\mathbf{string}(n), j) : j \ge h(n)+m\}$. Above we derived the inequality $\sum_{(r\text{-}\mathbf{string}(n),j)\in M_h} r^{-j} \le 1$. Thus, according to the Kraft-Chaitin Theorem (Theorem 4.17 in [Ca94]) there is a mapping $C : X_r^* \to X_r^*$ with prefix-free domain such that $C(w_{n,j}) = r\text{-}\mathbf{string}(n)$ for some word $w_{n,j} \in X_r^*$ with $|w_{n,j}| = j$ whenever $(r\text{-}\mathbf{string}(n), j) \in M_h$.

Then $C' := g \circ r\text{-}\mathbf{string}^{-1} \circ C : X_r^* \to X_r^*$ is a partial recursive function with the same prefix-free domain as C and $C'(w_{n,j}) = g(n)$ for all $n, j \in \mathbb{N}$. Since $H_r(g(n)) \le H_{C'}(g(n)) + c$ where $H_{C'}(w) := \inf\{|\pi| : \pi \in X_r^* \wedge C'(\pi) = w\}$, we have $H_r(g(n)) \le h(n) + m + c$. □

The next theorem relates the complexities K_r and H_r to their counterparts for alphabets of different size $\mathrm{card}\, X_b = b$, K_b and H_b, respectively. To this end we denote by $(b,r)\text{-}\mathbf{trans} := b\text{-}\mathbf{string} \circ r\text{-}\mathbf{string}^{-1} : X_r^* \to X_b^*$ the standard bijection between r-ary and b-ary words.

Theorem 2. *Let $f : \mathbb{N} \to X_b^*$ be a recursive function of bounded ambiguity, and let $g : \mathbb{N} \to X_r^*$ be a recursive function. Then there is a constant $c > 0$ such that for all $n \in \mathbb{N}$ the following inequalities hold true*

$$K_r(g(n)) \le \log_r b \cdot K_b(f(n)) + c \text{ and}$$
$$H_r(g(n)) \le \log_r b \cdot H_b(f(n)) + c\,.$$

Proof. Let $\mathrm{card} f^{-1}(w) \le k$ for all $w \in X_b^*$. We define a function $\phi : X_r^* \to X_r^*$ in the following way:

If $|\pi| \le k$ let $\phi(\pi) := e$ (the empty word). Otherwise split the input $\pi \in X_r^*$ in two parts $\pi_1 \cdot \pi_2$ such that $|\pi_1| = k$.

Set $m := (|r\text{-}\mathbf{string}^{-1}(\pi_1)| \pmod k) \in \{1, \ldots, k\}$.

Then translate π_2 via the standard bijection $(b,r)\text{-}\mathbf{trans} : X_r^* \to X_b^*$ into a program $\sigma := (b,r)\text{-}\mathbf{trans}(\pi_2) \in X_b^*$. Compute $U_b(\sigma)$ for a universal computer w.r.t. X_b^*. If $U_b(\sigma)$ is defined then take from the set $\{i : f(i) = U_b(\sigma)\}$ the m-th element, n (say), and compute $g(n)$.

Thus, if $f(n) = U_b(\sigma)$ then $\mathrm{card} f^{-1}(f(n)) \le k$ and there is a prefix π_1 such that we have $\phi(\pi) = g(n)$ for $\pi := \pi_1 \cdot (r,b)\text{-}\mathbf{trans}(\sigma)$.[1] Finally observe that $K_\phi(g(n)) \le |\pi| \le k + \lceil |\sigma| \cdot \log_r b \rceil$.

In the case of self-delimiting complexity, the assertion follows from the previous theorem, because $\sum_{n\in\mathbb{N}} r^{-\log_r b\cdot H_b(f(n))} < \infty$. □

---

[1] Observe that $(r,b)\text{-}\mathbf{trans} = (b,r)\text{-}\mathbf{trans}^{-1}$.

2 Base Independence

In this section we consider expansions of real numbers with respect to different bases. It is well known that the mappings converting real numbers from scale r to scale b are not continuous functions mapping the r-ary expansion $\xi \in X_r^\omega$ of a real number $\alpha \in [0,1]$ to a b-ary expansion $\Phi(\xi) \in X_b^\omega$ of the same number. For instance, in the case $r = 3$ and $b = 2$ for $\alpha = \frac{1}{2}$, that is, $\xi = 111\ldots \in \{0,1,2\}^\omega$ we do not know the first bit of the ω-word $\Phi(\xi) \in \{0,1\}^\omega$ until we know the whole infinite ω-word ξ. For a more detailed account see [We92].

Despite this fact, we can show that the Kolmogorov complexities of the expansions of the same real number α do not differ too much. To this end we denote by $K_r(\xi/l)$ ($H_r(\xi/l)$) the (self-delimiting) Kolmogorov complexity of the prefix of length l of the ω-word $\xi \in X_r^\omega$, that is, $K_r(\xi/l) := K_r(w)$ ($H_r(\xi/l) := H_r(w)$) where $w \sqsubset \xi$ and $|w| = l$.

The aim of this section is to prove the following theorem.

Theorem 3. *Let $\alpha \in [0,1]$ be a real number, and let $\xi \in X_r^\omega$ and $\beta \in X_b^\omega$ be its base r and base b expansions, respectively.*

Then there is a constant c such that for every $l \in \mathbb{N}$ the following equations hold true:

$$|K_r(\xi/\lfloor l \cdot \log_r b \rfloor) - \log_r b \cdot K_b(\beta/l)| \le c \text{ , and}$$
$$|H_r(\xi/\lfloor l \cdot \log_r b \rfloor) - \log_r b \cdot H_b(\beta/l)| \le c \text{ .}$$

In order to prove Theorem 3, it suffices to show the inequalities

$$K_r(\xi/\lfloor l \cdot \log_r b \rfloor) \le \log_r b \cdot K_b(\beta/l) + c \text{ , and} \tag{1}$$
$$H_r(\xi/\lfloor l \cdot \log_r b \rfloor) \le \log_r b \cdot H_b(\beta/l) + c \text{ .} \tag{2}$$

To this end we derive the following facts establishing some connections between the prefixes of an r-ary expansion and a b-ary expansion of the same real number.

Fact 1. *Let $0 \le a_1 < a_2 \le 1$ for some real numbers $a_1, a_2 \in \mathbb{R}$ and let $r \in \mathbb{N}$, $r \ge 2$. Then there is at least one $a \in \mathbb{N}$ such that the interval $[a_1, a_2]$ is contained in the interval $\left[\frac{a-1}{r^m}, \frac{a+1}{r^m}\right]$ where $m := \lfloor \log_r \frac{1}{a_2 - a_1} \rfloor$.*

This fact is illustrated in the following picture.

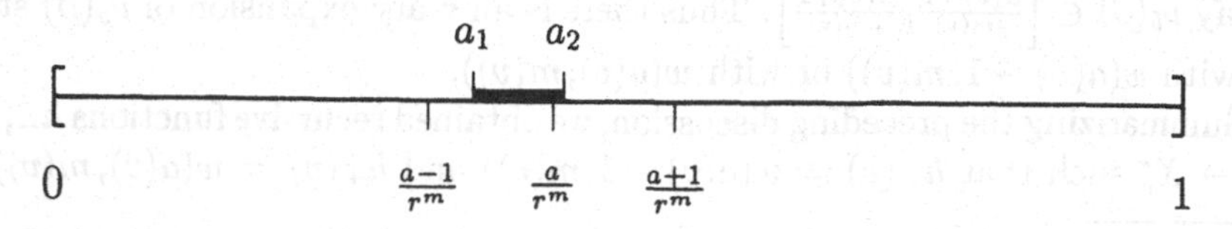

Remark. Observe that for $a_2 - a_1 \le r^{-m}$ it is not always possible to cover the interval $[a_1, a_2]$ by a single r-ary interval $\left[\frac{a}{r^m}, \frac{a+1}{r^m}\right]$. Fact 1 shows that, however, it is possible to cover $[a_1, a_2]$ by two adjacent r-ary intervals.

We note still that for $0 \leq a < r^m$ every real $\alpha \in [\frac{a}{r^m}, \frac{a+1}{r^m}]$ has an r-ary expansion which starts with the same prefix $w(a,m)$ of length m, that is, has an expansion between $w(a,m) \cdot 0^\omega$ and $w(a,m) \cdot (r-1)^\omega$. Here $w(a,m)$ is obtained by writing the integer $a \in \mathbb{N}$ in r-ary notation and filling with leading zeros up to the length m provided $a < r^m$.

The following fact summarizes our considerations about the containment of real intervals in r-ary intervals. To this end let $\mathrm{Cov}_r(a_1, a_2; a)$ denote the above illustrated fact that $[a_1, a_2] \subseteq [\frac{a-1}{r^m}, \frac{a+1}{r^m}]$ where $m := \lfloor \log_r \frac{1}{a_2 - a_1} \rfloor$.

Fact 2. *The relation*

$$R_r := \{(a_1, a_2, a) : a_1, a_2 \in \mathbb{Q} \cap [0,1] \wedge a \in \mathbb{N} \wedge \mathrm{Cov}_r(a_1, a_2; a)\}$$

is recursive and contains for every pair $a_1, a_2 \in \mathbb{Q} \cap [0,1]$ *such that* $a_1 < a_2$ *at least one triple* (a_1, a_2, a) *where* $a \in \mathbb{N}$.

As a consequence of Fact 2 we obtain that the functions h_r, $r \geq 2$ defined by

$$h_r\colon\ (\mathbb{Q} \cap [0,1])^2 \to \mathbb{N} \text{ where}$$
$$h_r(a_1, a_2) = \begin{cases} \mu a(a \in \mathbb{N} \wedge \mathrm{Cov}_r(a_1, a_2; a)), & \text{if } a_1 < a_2 \\ 0, & \text{otherwise} \end{cases} \tag{3}$$

are computable and satisfy the following properties.

Property 1. Let $0 \leq a_1 < a_2 \leq 1$ and $m := \lfloor \log_r \frac{1}{a_2 - a_1} \rfloor$. Then

$$h_r(a_1, a_2) \leq r^m \quad \text{and} \tag{4}$$
$$[a_1, a_2] \subseteq \left[\frac{h_r(a_1, a_2) - 1}{r^m}, \frac{h_r(a_1, a_2) + 1}{r^m}\right]. \tag{5}$$

Proof of Theorem 3. Let $v = \beta(1) \ldots \beta(l)$, that is, $|v| = l$. Then $0 \leq \nu_b(\beta) - \nu_b(v) \leq b^{-|v|}$.

According to Fact 1 and Property 1 the numbers

$$a(v) := h_r\left(\nu_b(v), \nu_b(v) + b^{-|v|}\right) \text{ and}$$
$$m(v) := \lfloor \log_r b^{|v|} \rfloor$$

satisfy $\nu_b(\beta) \in \left[\frac{a(v)-1}{r^{m(v)}}, \frac{a(v)+1}{r^{m(v)}}\right]$. Thus there is an r-ary expansion of $\nu_b(\beta)$ starting with $w(a(v)-1, m(v))$ or with $w(a(v), m(v))$.

Summarizing the preceding discussion, we obtained recursive functions $h_-, h_+ : X_b^* \to X_r^*$ such that $h_-(v) := w(a(v)-1, m(v))$ and $h_+(v) := w(a(v), m(v))$.[2]

[2] The choice between the two functions h_-, h_+ provides the missing information which prevented us, in the general case, from a continuous conversion between b-ary and r-ary expansions of real numbers. Observe, that the information we need to accomplish the choice between $w(a(v)-1, m(v))$ and $w(a(v), m(v))$ is only *one* bit.

The proof is now finished by applying Theorem 2 in the following way:

$$\left.\begin{array}{ll} w(a(v),m(v)) \stackrel{g}{\longleftarrow} & 2n \\ w(a(v)-1,m(v)) \stackrel{g}{\longleftarrow} & 2n+1 \end{array}\right\} \stackrel{f}{\longrightarrow} \quad v = b\text{-}\mathbf{string}(n)\ ,$$

that is, we associate with every word $v \in X_b^*$ two natural numbers $2n, 2n+1$ via

$$f(2n) := f(2n+1) := b\text{-}\mathbf{string}(n)\ ,$$

and, on the other hand, the function g maps the natural numbers $2n$ and $2n+1$ to the words $w(a(v)-1, m(v)) \in X_r^*$ and $w(a(v), m(v)) \in X_r^*$, respectively:

$$g(2n) := h_-(b\text{-}\mathbf{string}(n)) \text{ and}$$
$$g(2n+1) := h_+(b\text{-}\mathbf{string}(n))\ .$$

It is obvious that f is of bounded ambiguity, so Eqs. (1) and (2) follow from Theorem 2.
□

3 The Complexity of Real Numbers

In this section we consider the Kolmogorov complexity of real numbers with certain properties: the first class is the mentioned in the introduction class of random real numbers, and the second is the class of Liouville numbers, well-known as constructive examples of transcendental numbers.

To this end we introduce the lower and upper limit of the relative complexity of an ω-word $\xi \in X_r^*$.

$$\underline{\kappa}(\xi) := \liminf_{n\to\infty} \frac{K_r(\xi/n)}{n} \text{ and } \kappa(\xi) := \limsup_{n\to\infty} \frac{K_r(\xi/n)}{n} \tag{6}$$

Since $|H_r(\xi/n) - K_r(\xi/n)| \le o(n)$ it is of no importance whether we use the usual or self-delimiting complexity.

From Theorem 3 above we conclude that for a real number $\alpha \in [0,1]$ we can define its lower and upper limit of complexity in the same way as in Eq. (6):

$$\underline{\kappa}(\nu_r(\xi)) := \underline{\kappa}(\xi) \text{ and } \kappa(\nu_r(\xi)) := \kappa(\xi)\ .$$

3.1 Random reals

It was widely believed that the notion of randomness of a real number α is independent of the base of the expansion in which α is represented. Sound proofs of this fact were given only recently by different means [CJ94,HW98]. Here we give a third proof relying on the following definition of random sequences by self-delimiting Kolmogorov complexity (cf. [Ca94,Ch87,LV93]).

Definition 1. *An ω-word $\xi \in X_r^\omega$ is called random provided*

$$\lim_{l\to\infty} H_r(\xi/l) - l = \infty .$$

Now from Theorem 3 the proof of the independence result is immediate.

Lemma 1. *Let $\alpha \in [0,1]$ be a real number which is random in the scale of r. Then for every $b \in \mathbb{N}$, $b \geq 2$ the ω-word $\beta \in X_b^\omega$ with $\nu_b(\beta) = \alpha$ is random.*

3.2 The Kolmogorov complexity of Liouville numbers

The real numbers we deal with in this section are named after Liouville who invented them to demonstrate the existence of transcendental numbers. They are characterized by the fact that they have, although in a nonconstructive way, very tight rational approximations.

Definition 2. *A real number $\alpha \in \mathbb{R}$ is called a Liouville number provided*

1. *α is irrational.*
2. *$\forall n(n \in \mathbb{N} \to \exists p,q(p,q \in \mathbb{N} \wedge q > 1 \wedge |\alpha - \frac{p}{q}| < \frac{1}{q^n}))$.*

It should be noted that every Liouville number is transcendental (see [Ox71]).[3]

We obtain our first result.

Lemma 2. *If $\alpha \in [0,1]$ is a Liouville number then $\underline{\kappa}(\alpha) = 0$.*

Proof. We show that for the binary expansion $\eta \in \{0,1\}^\omega$ of α for every $n \in \mathbb{N}$ there is an $l \geq n$ such that

$$\frac{K_2(\eta/l)}{l} \leq \frac{\log_2 n}{n}$$

.

Let $|\alpha - \frac{p}{q}| < \frac{1}{q^n}$, where $0 \leq p \leq q$. We use the function h_2 defined in Eq. (3). Since $\alpha \in \left(\frac{p}{q} - \frac{1}{q^n}, \frac{p}{q} + \frac{1}{q^n}\right)$, we obtain for $a := h_2(\frac{p}{q} - \frac{1}{q^n}, \frac{p}{q} + \frac{1}{q^n})$ the restriction $a < r^m$ where $m := \log_2 \frac{q^n}{2} = n \cdot \log_2 q - 1$. As in the discussion following Fact 1 we define words $w(a-1,m)$ and $w(a,m)$ of length $m = \lfloor n \cdot \log_2 q \rfloor - 1$, one of them being a prefix of η.

Both words $w(a-1,m)$ and $w(a,m)$ can be specified by the numbers n,p,q. Utilizing a prefix-free binary encoding $code : \mathbb{N} \to \{0,1\}^*$ of the natural numbers, where $|code(n)| \leq 2 \cdot \log_2 n$ for $n \geq 4$, we obtain programs of the form

$$\pi_{n,p,q}(i) := i \cdot code(n) \cdot code(p) \cdot code(q),\ i \in \{0,1\} ,$$

and a computable function $\psi : \{0,1\}^* \to \{0,1\}^*$ such that

$$\psi(\pi_{n,p,q}(0)) = w(a-1,m) \text{ and } \psi(\pi_{n,p,q}(1)) = w(a,m) .$$

[3] Moreover, since $n \geq l \cdot (\log_2 k + 1)$ and $|\alpha - \frac{p}{q}| < \frac{1}{q^n}$ imply $|(\alpha + \frac{m}{k}) - (\frac{p}{q} + \frac{m}{k})| < \frac{1}{(q\cdot k)^l}$, the sum of a Liouville number and a rational number is again a Liouville number, whereas, as we shall see below, every real is the sum of at most two Liouville numbers.

Consequently, $K_\psi(w(a-1,m)), K_\psi(w(a,m)) \leq 1+2\log_2 n+2\log_2 p+2\log_2 q \leq 1+2\log_2 n+4\log_2 q$, and hence $K_2(w(a-1,m)), K_2(w(a,m)) \leq c+2\log_2 n+4\log_2 q$ for all triples (n,p,q) such that $|\alpha - \frac{p}{q}| < \frac{1}{q^n}$ and $n, q \geq 4$. Now, observe that in view of Definition 2 the values of the denominator q grow with the value of the exponent of precision n. Thus

$$\frac{K_2(\eta/\lfloor n \cdot \log_2 q - 1 \rfloor)}{\lfloor n \cdot \log_2 q - 1 \rfloor} \leq \frac{\log_2 n}{n}$$

if n (and hence q) is large enough. □

In connection with Definition 1 we obtain the following.

Corollary 1. *No Liouville number is a random real.*

Though Liouville numbers are not random, we show that the upper limit of complexity reaches its maximum value $\kappa(\alpha) = 1$ also for certain Liouville numbers α. We consider the following set constructed similar to the one in Example 3.18 of [St93].

Example 1. Define

$$F := X_r \cdot \prod_{i \in \mathbb{N}} X_r^{2i\cdot(2i)!} \cdot 0^{(2i+1)\cdot(2i+1)!} .$$

It is interesting to note that the set of finite prefixes of F, $\mathbf{A}(F) := \{w : w \in X_r^* \wedge \exists \xi(\xi \in F \wedge w \sqsubset \xi)\}$, is recursive.

If we consider ω-words $\beta = 0 \cdot \prod_{i \in \mathbb{N}} w_i \cdot 0^{(2i+1)\cdot(2i+1)!}$ where $|w_i| = 2i\cdot(2i)!$ and $K_r(w_i) \geq |w_i| - c$ for some $c \in \mathbb{N}$ then Daley's [Da74] diagonalization argument shows $\kappa(\beta) = 1$. Since the set $\{w : w \in X_r^* \wedge K_r(w) \geq |w| - 2\}$ contains at least two elements, F contains uncountably many ω-words β having $\kappa(\beta) = 1$.

The following consideration verifies that the set of numbers $\{\nu_r(\xi) : \xi \in F\} \setminus \mathbb{Q}$ consists entirely of Liouville numbers:

Let $\xi \in F$, $n \in \mathbb{N}$ and consider the prefix $w \sqsubset \xi$ of length $(2n+1)! = 1 + \sum_{i=0}^{2n} i \cdot i!$. Then $w \cdot 0^{(2n+1)\cdot(2n+1)!} \sqsubset \xi$, whence $\nu_r(\xi) \leq \nu_r(w) + \frac{1}{r^{(2n+2)!-1}}$, and $\nu_r(w) = p \cdot r^{-(2n+1)!}$ for some $p \in \mathbb{N}$. Consequently, $0 \leq \nu_r(\xi) - \nu_r(w) = \nu_r(\xi) - \frac{p}{r^{(2n+1)!}} < \frac{1}{(r^{(2n+1)!})^n}$. Thus, either $\nu_r(\xi)$ is rational or a Liouville number. □

Remark. In the same way one proves that $F' := \{0\} \cdot \prod_{i \in \mathbb{N}} 0^{2i\cdot(2i)!} \cdot X_r^{(2i+1)\cdot(2i+1)!}$ contains only rational or Liouville numbers. It is readily seen that every number $\alpha = \nu_r(\zeta) \in [0,1]$ can be represented as the sum $\nu_r(\zeta) = \nu_r(\xi) + \nu_r(\xi')$ where $\xi \in F$ and $\xi' \in F'$ are the letter-by-letter projections of ζ onto F or F', respectively. The numbers $\nu_r(\xi)$ and $\nu_r(\xi')$ in the above sum are rational or Liouville numbers, thus according to Footnote 3 $\nu_r(\zeta)$ is a Liouville number or the sum of two Liouville numbers. □

In the subsequent parts, we use the results obtained so far to give a complexity-theoretic proof of Theorem 2.4 in [Ox71] and to prove the existence of disjunctive Liouville numbers.

3.3 The Hausdorff dimension of Liouville numbers

First we consider the Hausdorff dimension of the set of Liouville numbers, $L \subseteq [0,1]$. It was mentioned in [MS94] that the Hausdorff dimension of a subset $M \subseteq [0,1]$ coincides with the one of $\{\xi : \xi \in X_r^\omega \wedge \nu_r(\xi) \in M\}$. The latter can be defined as follows.

Definition 3. The *Hausdorff dimension* of a set $F \subseteq X_r^\omega$, $\dim F$, is the smallest real number $\alpha \geq 0$ such that for all $\gamma > \alpha$ it holds

$$\forall \varepsilon \left(\varepsilon > 0 \to \exists W \left(W \subseteq X_r^* \wedge F \subseteq W \cdot X_r^\omega \wedge \sum_{w \in W} \left(r^{-\gamma}\right)^{|w|} < \varepsilon\right)\right).$$

From the definition it is evident that Hausdorff Dimension is monotone with respect to set inclusion and that $\dim\{\xi\} = 0$. We mention still that Hausdorff Dimension is also countably stable.

$$\dim \bigcup_{i \in \mathbb{N}} F_i = \sup_{i \in \mathbb{N}} \dim F_i \tag{7}$$

Thus every countable subset $F \subseteq X_r^\omega$ has $\dim F = 0$. For further properties of the Hausdorff dimension see, e.g., [Fa90].

We are going to give a complexity-theoretic proof of the fact (cf. [Ox71, Theorem 2.4]) that the set of Liouville numbers L is an uncountable set of Hausdorff dimension $\dim L = 0$.

In the papers [Ry86,St93,St98] close connections between Hausdorff dimension and Kolmogorov complexity are derived. We need here the following one (see [Ry86, Theorem 2] or [St93, Corollary 3.14]).

Lemma 3. *For every $F \subseteq X_r^\omega$ the following bound is true.*

$$\dim F \leq \sup\{\underline{\kappa}(\xi) : \xi \in F\}$$

Now Lemmas 2 and 3 yield the announced result.

Corollary 2. *The set of Liouville numbers $L \subseteq [0,1]$ has Hausdorff dimension* $\dim L = 0$.

3.4 Disjunctive Liouville numbers

In this last part we turn to disjunctive ω-words. As it was mentioned above, an ω-word $\xi \in X_r^\omega$ is called *disjunctive* provided every word $w \in X_r^*$ appears as an infix of ξ, that is, $\forall w(w \in X_r^* \to \exists v(v \sqsubset \xi \wedge v \cdot w \sqsubset \xi)$.

Proposition 8 of [JT88] proves that for every $r \geq 2$ there are uncountably many disjunctive $\xi \in X_r^\omega$ such that $\nu_r(\xi)$ is a Liouville number. The paper [He96] presents examples of Liouville numbers whose expansions are disjunctive with respect to one base, but not to with respect to all bases (e.g. $\sum_{i=1}^{\infty} r^{-i!-i}$ which is not disjunctive in the scale of r). We prove the existence of Liouville numbers disjunctive with respect to all bases.

Lemma 4. *There are uncountably many Liouville numbers α such that for every $r \in \mathbb{N}$, $r \geq 2$ the ω-word $\xi \in X_r^\omega$ with $\nu_r(\xi) = \alpha$ is disjunctive.*

Proof. Eq. (5.3) of [St93] shows that an ω-word $\xi \in X_r^\omega$ with $\kappa(\xi) = 1$ is disjunctive. In fact, if $\xi \in X_r^\omega$ does not contain a word $w \in X_r^*$ of length $|w| = l$ as infix then $\kappa(\xi) \leq l^{-1} \cdot \log_r(r^l - 1) < 1$. Now Example 1 yields the existence of Liouville numbers which are disjunctive in every scale r. □

References

[CH94] J.-Y. Cai and J. Hartmanis, On Hausdorff and topological dimensions of the Kolmogorov complexity of the real line. *J. Comput. System Sci.* **49** (1994) 3, 605 - 619.

[Ca94] C. Calude, *Information and Randomness. An Algorithmic Perspective.* Springer-Verlag, Berlin, 1994.

[CC96] C. Calude and C. Câmpeanu, Are Binary Codings Universal?, *Complexity* **1** (1996) 15, 47 - 50.

[CJ94] C. Calude and H. Jürgensen, Randomness as an invariant for number representations. In: *Results and Trends in Theoretical Computer Science* (H. Maurer, J. Karhumäki, G. Rozenberg, Eds.), Lecture Notes in Comput. Sci., Vol. **812**, Springer-Verlag, Berlin, 1994, 44 - 66.

[Cs59] J.W.S. Cassels, On a problem of Steinhaus about normal numbers. *Colloquium Math.* **7** (1959), 95 - 101.

[Ch87] G. J. Chaitin, *Information, Randomness, & Incompleteness. Papers on Algorithmic Information Theory.* World Scientific, Singapore, 1987.

[Da74] R.P. Daley, The extent and density of sequences within the minimal-program complexity hierarchies. *J. Comput. System Sci.* **9** (1974), 151 - 163.

[Fa90] K.J. Falconer, *Fractal Geometry*. Wiley, Chichester, 1990.

[He96] P. Hertling, Disjunctive ω-words and Real Numbers, *Journal of Universal Computer Science* **2** (1996) 7, 549 - 568.

[HW98] P. Hertling and K. Weihrauch, Randomness Spaces. in: Automata, Languages and Programming, Proc. 25th Int. Colloq. ICALP'98, (K. G. Larsen, S. Skyum and G. Winskel, Eds.), Lecture Notes in Comput. Sci., Vol. **1443**, Springer-Verlag, Berlin, 1998, 796 - 807.

[JT88] H. Jürgensen and G. Thierrin, Some structural properties of ω-languages. in: Sbornik 13th Nat. School "*Applications of Mathematics in Technology*", Sofia, 1988, 56 - 63.

[LV93] M. Li and P.M.B. Vitányi, *An Introduction to Kolmogorov Complexity and its Applications.* Springer–Verlag, New York, 1993.

[MS94] W. Merzenich and L. Staiger, Fractals, dimension, and formal languages. *RAIRO–Inform. Théor.* **28** (1994) 3–4, 361 - 386.

[Ox71] J.C. Oxtoby, *Measure and Category.* Springer-Verlag, Berlin 1971.

[Ry86] B.Ya. Ryabko, Noiseless coding of combinatorial sources, Hausdorff dimension and Kolmogorov complexity. *Problems of Information Transmission* **22** (1986) 3, 170 - 179.

[Sc60] W.M. Schmidt, On normal numbers. *Pac. J. Math.* **10** (1960), 661 - 672.

[St93] L. Staiger, Kolmogorov complexity and Hausdorff dimension. *Inform. and Comput.* **103** (1993) 2, 159 - 194. *Preliminary version in*: "Fundamentals of Computation Theory" (J. Csirik, J. Demetrovics and F. Gécseg Eds.), Lecture Notes in Comput. Sci., No. **380**, Springer-Verlag, Berlin 1989, 334 - 343.

[St98] L. Staiger, A tight upper bound on Kolmogorov complexity and uniformly optimal prediction. *Theory of Computing Systems* **31** (1998), 215 – 229.

[We92] K. Weihrauch, The Degrees of Discontinuity of some Translators Between Representations of the Real Numbers. Informatik-Bericht **129**, FernUniversität Hagen 1992.

A Partial Order Method for the Verification of Time Petri Nets *

I. Virbitskaite and E. Pokozy

Institute of Informatics Systems
Siberian Division of the Russian Academy of Sciences
6, Acad. Lavrentiev av., 630090, Novosibirsk, Russia
{virb,pokozy}@iis.nsk.su

Abstract. The intention of the paper is to develop an efficient model checker for real time systems represented by safe time Petri nets and the real time branching time temporal logic TCTL. Our method is based on the idea of, (1), using the known region graph technique [1] to construct a finite representation of the state-space of a time Petri net and, (2), further reducing the size of this representation by exploiting the net concurrency. To show the correctness of the reduction we introduce a notion of a timed stuttering equivalence. Some experimental results which demonstrate the efficiency of the method are also given.

1 Introduction

One of the most successful techniques for automatic verification of finite-state systems has been model-checking: a property is given as a formula of a propositional temporal logic and automatically compared with a state-graph representing the system behaviour. One of the advantages of this method is its efficiency: model checking is linear in the product of the size of the state-graph and the size of the formula, when the logic is the branching time temporal logic CTL (Computation Tree Logic) [7]. Unfortunately, the verification of large concurrent systems suffers from the so-called state explosion problem. An approach to confine this problem is to use partial orders and thus to avoid the construction of equivalent states reachable by different interleaving of atomic events. Several methods (see [11, 18] among others) based on this approach have been proposed for analysis of reachability and various other properties of concurrent systems. So far, these methods have been developed and implemented for linear time temporal logics model checking. A first step to adapt the partial order approach for checking properties expressed in branching time temporal logics is presented in [10].

More recently, a few attempts (see [3, 16]) have been made to extend the success of partial order reductions to the setting of real time systems represented

* This work is supported in part by the Russian Ministry of High Education, Grant for Basic Research in Mathematics.

by timed automata [2]. However, concurrency can not be modelled directly by such timed state-graphs. On the other hand, the paper [14] proposed time Petri nets as an adaquate model of timed concurrent systems, generalizing other models in a natural way. The paper [13] proposes to exploit McMillan's reduction techniques based on unfolding in state-space search of time Petri nets. A partial order verification algorithm for time Petri nets and a linear time temporal logic was proposed in [19].

The present paper shows how a partial order approach can be applied to TCTL [1], a real time extension of the branching time temporal logic CTL, that models the behaviour of a system as a continuous computation tree. A given real time system is represented by a safe time Petri net. Automatic verification is achieved by generating a reduced state space of the net, which is big enough to evaluate a given formula, and by traversing the reduced state space with the formula.

The rest of the paper is organized as follows. The basic definitions concerning time Petri nets are given in the next section. Section 3 recalls the syntax and semantics of TCTL. Both the basic algorithm model checking and its partial order improvement are developed in the following two sections. A notion of a timed stuttering equivalence is also introduced to show the correctness of the reduction method. Some remarks about experimental results are finally given. The proofs are relegated to an Appendix when they disturb the exposition.

2 Time Petri Nets

In this section we define some terminology concerning time Petri nets [14]. 'Time Petri net' is a Petri net whose transitions are labelled by two temporal constraints that indicate their earliest and latest firing times. Let $\mathbf{N}$ be the set of natural numbers, $\mathbf{R}_0^+$ the set of nonnegative real numbers, and $\mathbf{R}^+$ the set of positive real numbers.

Definition 1. A *time Petri net* is a tuple $\mathcal{N} = (P, T, F, Eft, Lft, m_0)$, where

- $P = \{p_1, p_2, \ldots, p_m\}$ is a finite set of places;
- $T = \{t_1, t_2, \ldots, t_n\}$ is a finite set of transitions ($P \cap T = \emptyset$);
- $F \subseteq (P \times T) \cup (T \times P)$ is the flow relation;
- $Eft, Lft : T \to \mathbf{N}$ are functions for the *earliest* and *latest firing times* of transitions, satisfying $Eft(t) \leq Lft(t)$ for all $t \in T$;
- $m_0 \subseteq P$ is the initial marking.

Fig. 1 shows an example of a time Petri net, where a pair of numbers near by a transition corresponds to its earliest and latest firing times.

For $t \in T$, ${}^\bullet t = \{p \in P \mid (p,t) \in F\}$ and $t^\bullet = \{p \in P \mid (t,p) \in F\}$ denote the *preset* and *postset* of t, respectively. To simplify the presentation, we assume that ${}^\bullet t \cap t^\bullet = \emptyset$ for every transition t. For the sake of convenience, we fix a time Petri net $\mathcal{N} = (P, T, F, Eft, Lft, m_0)$ and work with it throughout what follows.

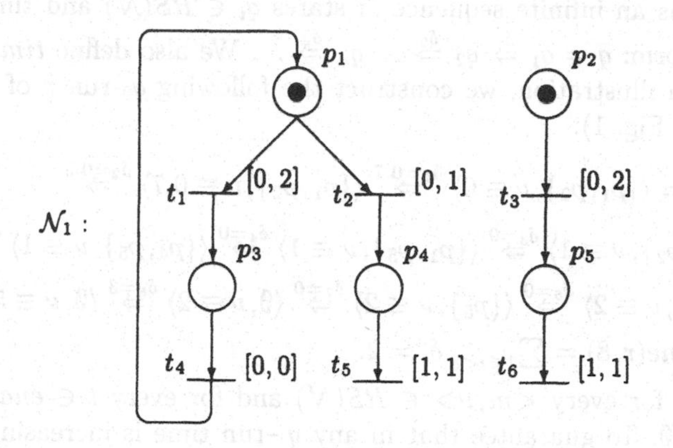

Fig. 1. An example of a time Petri net: $\mathcal{N}_1$

A *marking* m of $\mathcal{N}$ is any subset of P. A transition t is *enabled* in a marking m if ${}^\bullet t \subseteq m$ (all its input places have tokens in m), otherwise it is *disabled.* Let *enable*(m) be the set of transitions, enabled in m.

Let $\Gamma(\mathcal{N}) = [T \to \mathbf{R}_0^+]$ be the set of *time assignments* for transitions from T. Given $\nu \in \Gamma(\mathcal{N})$ and $\delta \in \mathbf{R}_0^+$, we let $\nu + \delta$ denote the time assignment of the value $\nu(t) + \delta$ to each t from T.

A *state* q of $\mathcal{N}$ is a pair $<m, \nu>$, where m is a marking and $\nu \in \Gamma(\mathcal{N})$. The *initial state* of $\mathcal{N}$ is a pair $q_0 = <m_0, \nu_0>$, where $\nu_0(t) = 0$ for all $t \in T$.

The states of $\mathcal{N}$ change, if time passes or if a transition fires. In a state $q = <m, \nu>$, a time $\delta \in \mathbf{R}_0^+$ *can pass* if for all $t \in enable(m)$ holds: $\nu(t) + \delta \leq Lft(t)$. In this case, the state $q' = <m', \nu'>$ is *obtained by passing* δ from q (written $q \overset{\delta}{\Rightarrow} q'$), if

- $m' = m$, and
- $\nu'(t) = \nu(t) + \delta$ for all $t \in T$.

In a state $q = <m, \nu>$, a transition $t \in T$ is *fireable* if $t \in enable(m)$ and $\nu(t) \geq Eft(t)$. In this case, the state $q' = <m', \nu'>$ is *obtained by firing* t from q (written $q \overset{0}{\Rightarrow} q'$), if

- $m' = (m \setminus {}^\bullet t) \cup t^\bullet$, and
- $\forall t' \in T \,.\, \nu'(t') = \begin{cases} 0, & \text{if } t' \in enable(m') \setminus enable(m), \\ \nu(t'), & \text{otherwise.} \end{cases}$

In the case when it is essential that q' is obtained from q by firing a concrete transition t we shall also write $q \overset{t}{\Rightarrow} q'$.

A state q is *reachable* if $q = q_0$ or there exists a reachable state q' such that

$q' \overset{\delta}{\Rightarrow} q$ for some $\delta \in \mathbf{R}_0^+$. Let $RS(\mathcal{N})$ denote the set of all reachable states of $\mathcal{N}$. A *q-run* r in $\mathcal{N}$ is an infinite sequence of states $q_i \in RS(\mathcal{N})$ and time values $\delta_i \in \mathbf{R}_0^+$ of the form: $q = q_1 \overset{\delta_1}{\Rightarrow} q_2 \overset{\delta_2}{\Rightarrow} \dots q_n \overset{\delta_n}{\Rightarrow} \dots$ We also define $time(r,n) = \sum_{1 \le i < n} \delta_i$. As an illustration, we construct the following q_0-run r of the time Petri net $\mathcal{N}_1$ (see Fig. 1):

$$r = \langle\{p_1,p_2\}, \nu \equiv 0\rangle \overset{\delta_1 = 0.7}{\Rightarrow} \langle\{p_1,p_2\}, \nu \equiv 0.7\rangle \overset{\delta_2 = 0.3}{\Rightarrow}$$

$$\overset{\delta_2 = 0.3}{\Rightarrow} \langle\{p_1,p_2\}, \nu \equiv 1\rangle \overset{\delta_3 = 0}{\Rightarrow} \langle\{p_1,p_5\}, \nu \equiv 1\rangle \overset{\delta_4 = 0}{\Rightarrow} \langle\{p_4,p_5\}, \nu \equiv 1\rangle \overset{\delta_5 = 1}{\Rightarrow}$$

$$\overset{\delta_5 = 1}{\Rightarrow} \langle\{p_4,p_5\}, \nu \equiv 2\rangle \overset{\delta_6 = 0}{\Rightarrow} \langle\{p_5\}, \nu \equiv 2\rangle \overset{\delta_7 = 0}{\Rightarrow} \langle\emptyset, \nu \equiv 2\rangle \overset{\delta_8 = 3}{\Rightarrow} \langle\emptyset, \nu \equiv 5\rangle \dots$$

We then have $time(r,8) = \sum_{1 \le i < 8} \delta_i = 2$.

$\mathcal{N}$ is *one-safe*, if for every $<m,\nu> \in RS(\mathcal{N})$ and for every $t \in enable(m)$ it holds: $t^\bullet \cap m = \emptyset$. To guarantee that in any q−run time is increasing beyond any bound, we need the following *progress condition*: for every set of transitions $\{t_1, t_2, \dots, t_n\}$ such that $\forall 1 \le i < n \,.\, t_i^\bullet \cap {}^\bullet t_{i+1} \ne \emptyset$ and $t_n^\bullet \cap {}^\bullet t_1 \ne \emptyset$ it holds $\sum_{1 \le i \le n} Eft(t_i) > 0$. In the sequel, $\mathcal{N}$ will always denote a one-safe time Petri net satisfying the progress condition.

3 TCTL: Syntax and Semantics

Timed Computation Tree Logic (TCTL) was introduced by R. Alur, C. Courcoubetis, D. Dill [1] as a specification language for real time systems. We now review the syntax and semantics of TCTL. Let AP be a set of atomic propositions. For our purpose, it is convenient to take $AP = P$.

Definition 2. A *formula* ϕ of TCTL is inductively defined as follows:

$$\phi := p \mid \neg\phi_1 \mid \phi_1 \rightarrow \phi_2 \mid \forall\phi_1\mathcal{U}_{\sim c}\phi_2 \mid \exists\phi_1\mathcal{U}_{\sim c}\phi_2,$$

where $p \in AP$, $c \in \mathbf{N}$, ϕ_1 and ϕ_2 are formulas of TCTL, $\sim$ stands for one of the binary relations $\{<, \le, =, \ge, >\}$. For a TCTL-formula ϕ, let c_ϕ denote the maximal constant appearing in ϕ. We shall use $I(\sim c)$ to denote the interval of real numbers, corresponding to $'\sim c'$.

Informally, $\exists\phi_1\mathcal{U}_{<c}\phi_2$ means that for some computation run there exists an initial prefix of time duration less than c such that ϕ_2 holds in the last state of the prefix, and ϕ_1 holds in all its intermediate states.

The other logical connectives can be defined as usual: $\forall\Diamond_{\sim c}\phi \equiv \forall true\ \mathcal{U}_{\sim c}\phi$, $\exists\Diamond_{\sim c}\phi \equiv \exists true\ \mathcal{U}_{\sim c}\phi$, $\forall\Box_{\sim c}\phi \equiv \neg\exists\Diamond_{\sim c}\neg\phi$, $\exists\Box_{\sim c}\phi \equiv \neg\forall\Diamond_{\sim c}\neg\phi$.

Definition 3. Given a TCTL-formula ϕ and $q = <m,\nu> \in RS(\mathcal{N})$, we define the *satisfaction relation* $q \models \phi$ inductively as follows:

$$\begin{array}{lcl} q \models p & \Longleftrightarrow & m(p) > 0; \\ q \models \neg\phi_1 & \Longleftrightarrow & q \not\models \phi_1; \\ q \models \phi_1 \rightarrow \phi_2 & \Longleftrightarrow & q \not\models \phi_1 \text{ or } q \models \phi_2; \\ q \models \exists\phi_1\mathcal{U}_{\sim c}\phi_2 & \Longleftrightarrow & \text{for some } q\text{-run } r \text{ of } \mathcal{N}, r \models \phi_1\mathcal{U}_{\sim c}\phi_2; \\ q \models \forall\phi_1\mathcal{U}_{\sim c}\phi_2 & \Longleftrightarrow & \text{for every } q\text{-run } r \text{ of } \mathcal{N}, r \models \phi_1\mathcal{U}_{\sim c}\phi_2. \end{array}$$

For a q-run $r = (q = q_1 \overset{\delta_1}{\Rightarrow} q_2 \overset{\delta_2}{\Rightarrow} \ldots)$, the relation $r \models \phi_1 \mathcal{U}_{\sim c} \phi_2$ holds iff there exists k and $\delta \leq \delta_k$ such that: (1) $(\delta + time(r, k)) \sim c$; (2) $<m_k, \nu_k + \delta> \models \phi_2$; (3) $\forall 1 \leq i < k . (<m_i, \nu_i> \models \phi_1 \wedge \forall 0 < \delta' < \delta_i . <m_i, \nu_i + \delta'> \models \phi_1)$; (4) $\forall\, 0 \leq \delta' < \delta . <m_k, \nu_k + \delta'> \models \phi_1$. $\mathcal{N}$ satisfies a TCTL-formula ϕ (written $\mathcal{N} \models \phi$) iff $q_0 \models \phi$. A TCTL-formula ϕ is *satisfiable* iff there is a time Petri net $\mathcal{N}$ such that $\mathcal{N} \models \phi$. Notice, the TCTL satisfiability problem is undecidable [1].

As an example consider the time Petri net $\mathcal{N}_1$ (see Fig. 1) and the TCTL-formula $\phi_1 = \exists \Diamond_{\geq 2} (p_4 \wedge p_5)$. $\mathcal{N}_1$ satisfies ϕ, since there exists a q_0-run r in $\mathcal{N}_1$ (see above) along which the places p_4 and p_5 contain tokens at time moment equal to 2. So, ϕ_1 is satisfiable.

4 Model Checking

In this section we present an algorithm for deciding whether a time Petri net meets its specification given as a TCTL-formula. Since a time Petri net constitutes a dense time model, the number of its states is infinite. In order to get a finite representation of the state-space of the net, we use the concepts of regions (equivalence classes of states) and region graphs [1].

For any $\delta \in \mathbf{R}_0^+$ $\{\delta\}$ denotes the fractional part of δ, and $\lfloor \delta \rfloor$ denotes the integral part of δ. Let $\nu, \nu' \in \Gamma(\mathcal{N})$. Then $\nu \simeq \nu'$ iff the following conditions are met:

- for each $t \in T$ either $\lfloor \nu(t) \rfloor = \lfloor \nu'(t) \rfloor$ or both $\nu(t)$ and $\nu'(t)$ are greater than $Lft(t)$;
- for each $t, t' \in T$ such that $\nu(t) \leq Lft(t)$ and $\nu(t') \leq Lft(t')$:
  - $\{\nu(t)\} \leq \{\nu(t')\}$ iff $\{\nu'(t)\} \leq \{\nu'(t')\}$;
  - $\{\nu(t)\} = 0$ iff $\{\nu'(t)\} = 0$.

In order to simplify the checking of temporal constraints of a given TCTL-formula ϕ, we introduce an additional transition $t^* \notin T$ which is disabled in any marking of $\mathcal{N}$, and its time assignment therefore keeps time elapsed since some fixed initial moment. Let $T^* = T \cup \{t^*\}$ and $\Gamma^*(\mathcal{N}) = [T^* \to \mathbf{R}_0^+]$. We define an equivalence relation $\simeq^*$ of $\Gamma^*(\mathcal{N})$ similar to the relation $\simeq$ of $\Gamma(\mathcal{N})$, supposing that $Lft(t^*)$ is equal to the maximal temporal constant, appearing in the formula ϕ. We use $[\nu]_\phi^*$ to denote the class of equivalence of $\Gamma^*(\mathcal{N})$ to which ν belongs. For a time assignment ν and $x \in \mathbf{R}_0^+$, let $[[x]\nu]$ denote the time assignment from $\Gamma(\mathcal{N})^*$ that assigns $\nu(t^*) = x$ and agrees with ν on the values of the remaining assignments.

A region of $\mathcal{N}$ w.r.t. a TCTL-formula ϕ is a pair $v = <m, [\nu]_\phi^*>$, where $<m, \nu> \in RS(\mathcal{N})$. We consider $v_0 = <m_0, [[0]\nu_0]_\phi^*>$ as the *initial region of* $\mathcal{N}$ *w.r.t.* ϕ. We use $GS(\mathcal{N}, \phi)$ to denote the set of regions of $\mathcal{N}$ w.r.t. ϕ. When the context of a TCTL-formula ϕ is obvious, we shall write $v = <m, [\nu]^*>$, for simplicity. Let $<m, [\nu]^*>, <m', [\nu']^*> \in GS(\mathcal{N}, \phi)$. Then $<m', [\nu']^*>$ is said to be a δ-*successor* for $<m, [\nu]^*>$ (written $succ_\delta(<m, [\nu]^*>) = <m', [\nu']^*>$)

if $m = m'$ and there exists $\delta \in \mathbf{R}^+$ such that $\nu + \delta \in [\nu']^*$ and $\{\nu + \delta' \mid 0 \leq \delta' \leq \delta\} \subseteq [\nu]^* \cup [\nu']^*$. Further, $<m', [\nu']^*>$ is called a *t-successor* for $<m, [\nu]^*>$ (written $succ_t(<m, [\nu]^*>) = <m', [\nu']^*>$) if there is $\delta \in \mathbf{R}_0^+$ such that $\{ \nu + \delta' | 0 \leq \delta' \leq \delta\} \subseteq [\nu]^*$ and $<m, \nu + \delta> \stackrel{t}{\Rightarrow} <m', \nu'>$. We let $fireable(<m, [\nu]^*>) = \{t \in T \mid succ_t(<m, [\nu]^*>) \in GS(\mathcal{N}, \phi)\} \cup \{\delta \mid succ_\delta(<m, [\nu]^*>) \in GS(\mathcal{N}, \phi)\}$.

For a given $\mathcal{N}$ and TCTL-formula ϕ, we construct the *region graph* $G(\mathcal{N}, \phi) = (V, E, l)$ as follows:

```
Init:  St := ∅, H := ∅;
       V := {v0}; E := ∅; l := ∅;
       push v0 into St;
   while St ≠ ∅ do {
       pop v from St;
       if v is NOT already in H then {
           push v into H;
           TS:= fireable(v);
           ∀t ∈ TS do {
               v' := { succ_t(v),  if t ∈ T,
                     { succ_δ(v),  otherwise
               V := V ∪ {v};
               E := E ∪ {(v, v')};
               l((v, v')) = t;
               push v' into St}}}.
```

St indicates a stack recording the region states that remain to be explored. H is an array recording the regions already visited.

A decision procedure for model checking is the following: given $\mathcal{N}$ and a TCTL-formula ϕ, first construct the region graph $G(\mathcal{N}, \phi) = (V, E, l)$. Then label all its vertices with the subformulas of ϕ using the labelling algorithm from [1]. $\mathcal{N}$ satisfies ϕ iff $<m_0, [\nu_0]^*>$ is labelled with ϕ.

Using the ideas above, one can implement an algorithm for model checking which run in time linear in the qualitative part and exponential in the timing part of the input.

Theorem 1. Given a TCTL-formula ϕ, there is a decision procedure for checking whether or not $\mathcal{N} \models \phi$ bounded by:

$$O[|\phi| \cdot c_\phi \cdot \mid T \mid ! \cdot 2^{|P|+2|T|} \cdot \prod_{t \in T} (Lft(t) + 1)].$$

Proof. According to [1], the number of equivalence classes of $\Gamma^*(\mathcal{N})$ induced by $\simeq^*$ is bounded by $|T^*|! \cdot 2^{2|T^*|} \cdot \prod_{t \in T^*}(Lft(t) + 1)$. The maximal number of markings of $\mathcal{N}$ is equal to $2^{|P|}$. Thus the number of vertices of $G(\mathcal{N}, \phi)$ is bounded by $O[c_\phi \cdot \mid T \mid ! \cdot 2^{|P|+2|T|} \cdot \prod_{t \in T}(Lft(t)+1)]$. For each vertex v in $G(\mathcal{N}, \phi)$, there exist at most $\mid T \mid$ output edges, representing transition firings, and one edge, representing the passage of time. Hence, $|E| = O[c_\phi \cdot |T|! \cdot 2^{|P|+2|T|} \cdot \prod_{t\in T}(Lft(t)+1)]$. So, $G(\mathcal{N},\phi)$ can be constructed in time $O[|V|+|E|]$. Thus the labelling algorithm takes time $O[|\phi| \cdot (|V|+|E|)] = O[|\phi| \cdot c_\phi \cdot |T|! \cdot 2^{|P|+2|T|} \cdot \prod_{t\in T}(Lft(t)+1)]$. □

5 A Partial Order Reduction

In this section we show how to reduce the size of the region graph $G(\mathcal{N},\phi)$ without affecting the correctness of the model checking procedure. The idea of the reduction is based on exploring for each visited vertex of $G(\mathcal{N},\phi)$ only a subset of its successors and, hence, a logical formula can be verified in less space and time. In order to show the correctness of the reduction, we introduce a notion of a timed stuttering equivalence. Before doing so, we need to define some auxiliary notations. For a TCTL-formula ϕ, let $P(\phi)$ be the set of atomic propositions and $C(\phi)$ the set of temporal constraints, appearing in ϕ.

Definition 4. Let ϕ be a TCTL-formula and $G' = G(\mathcal{N},\phi') = (V',E',l')$, $G'' = G(\mathcal{N},\phi'') = (V'',E'',l'')$ the region graphs such that $P(\phi) = P(\phi') = P(\phi'')$ and $C(\phi) = C(\phi') = C(\phi'')$. Let v_0' and v_0'' are the initial vertices of G' and G'', respectively. A relation $\approx_\phi \subseteq V' \times V''$ is *a timed stuttering equivalence w.r.t.* ϕ (w.r.t. atomic propositions and temporal constraints, appearing in ϕ) between G' and G'', if $(v_0', v_0'') \in \approx_\phi$ and for all $(v' = <m',[\nu']^*>, v'' = <m'',[\nu'']^*>) \in \approx_\phi$ it holds:

1. $m' \cap P(\phi) = m'' \cap P(\phi)$, and $\nu'(t^*) \sim c \Leftrightarrow \nu''(t^*) \sim c$ for all $' \sim c'$ in $C(\phi)$;
2. (a) if $(v', v_1') \in E'$ then there exists $v^0, \cdots, v^n \in V''(n \geq 0)$ such that $v'' = v^0$, $((v^i, v^{i+1}) \in E'' \wedge (v', v^i \in\approx_\phi)$ for all $0 \leq i < n$, and $(v_1', v^n) \in \approx_\phi$;
   (b) similar to item (a) but the sets V' and E' are exchanged by V'' and E'', respectively.

G' and G'' are *timed stuttering equivalent w.r.t.* ϕ (written $G' \approx_\phi G''$) if there exists a relation $\approx_\phi$ between them.

Theorem 2. Let ϕ be a TCTL-formula and G', G'' region graphs such that $G' \approx_\phi G''$. Let v' and v'' be vertices of G' and G'', respectively. If $(v',v'') \in \approx_\phi$ then v' is labelled with ϕ iff v'' is labelled with ϕ.

Proof. It follows similar lines as other standard proofs of that a stuttering equivalence is a necessary and sufficient condition for ensuring that the two stuttering equivalent structures satisfy the same logical formulas (see e.g., [5]). □

We now try to find the proper constraints on the subset that is chosen to be explored at each visited vertex of $G(\mathcal{N},\phi)$. Let t be a transition in T and $v = <m,[\nu]^*>$ a vertex in $G(\mathcal{N},\phi)$. Then

- a transition t' is in *conflict with* t if $^\bullet t \cap {}^\bullet t' \neq \emptyset$ and $t \neq t'$. Let *conflict*(t) denote the set of transitions conflicting with t;

- a transition t' is a ***predecessor*** of t in v if $t'F^*t$ and ${}^\bullet t' \subseteq m$, where F^* is the transitive reflexive closure of F. Let $pred(t, v)$ denote the set of predecessors of t in v;
- a transition t' is a ***+-predecessor*** of t in v if t' is a predecessor of a transition t'' conflicting with a predecessor t''' of t and $dist(t', t'') \leq Lft(t'')$, where $dist(\tilde{t}, \hat{t})$ is the minimal value of sums of the earliest firing times of transitions in the paths from $\hat{t}$ to $\tilde{t}$, not including $Eft(\hat{t})$. Let $pred^+(t, v)$ denote the set of +-predecessors of t in v;
- a transition t' is a ***visible*** w.r.t. a TCTL-formula ϕ if $({}^\bullet t \cup t^\bullet) \cap P(\phi) \neq \emptyset$; Let $visible(\phi)$ denote the set of transitions visible w.r.t. ϕ;
- $visible^+(v, \phi) = \cup_{t \in visible(\phi)} (pred(t, v) \cup pred^+(t, v))$;
- a time δ is ***visible*** w.r.t. to a TCTL-formula ϕ in v, if the value of t^*'s time assignment is in the interval $I'(\sim c)$, where
$$I'(\sim c) = \begin{cases} [0, c), & \text{if } \sim = <, \\ [0, c] \cup I(\sim c), & \text{otherwise,} \end{cases}$$
- $dependent(v, \phi) = visible^+(v, \phi) \cup visible^+_\delta(v, \phi)$.

For the time Petri net $\mathcal{N}_1$ (see Fig. 1) and the TCTL-formula $\phi_2 = \exists\Diamond_{\geq 1}(p_4)$ we have: $conflict(t_2) = \{t_1\}$, $pred(t_2, v_0) = \{t_2\}$, $pred^+(t_2, v_0) = \{t_1\}$, $visible(\phi_2) = \{t_2, t_5\}$, $visible^+(v_0, \phi_2) = \{t_1, t_2\}$, $visible^+_\delta(v_0, \phi_2) = \{\delta\}$ and $dependent(v_0, \phi_2) = \{\delta, t_1, t_2\}$. Whereas, for $\mathcal{N}_1$ and the TCTL-formula $\phi_3 = \exists\Diamond_{\geq 1} p_5$ we get: $conflict(t_6) = \emptyset$, $pred(t_6, v_0) = \{t_3\}$, $pred^+(t_6, v_0) = \emptyset$, $visible(\phi_3) = \{t_3, t_6\}$, $visible^+(v_0, \phi_3) = \{t_3\}$, $visible^+_\delta(v_0, \phi_3) = \{\delta\}$ and $dependent(v_0, \phi_3) = \{\delta, t_3\}$.

We define a set $ready(v, \phi)$ as a minimal subset of t- and δ-successors from $fireable(v)$ such that $fireable(v) \cap dependent(v, \phi) \subseteq ready(v, \phi)$.

For the time Petri net $\mathcal{N}_1$ and the TCTL-formula ϕ_3 we get $fireable(v_0) = \{\delta, t_1, t_2, t_3\}$ and $ready(v_0, \phi_3) = \{\delta, t_3\}$.

The complexity of the construction of $ready(v, \phi)$ is $O(| P | \cdot | T |^2)$. The algorithm of the construction of a reduced region graph $G_R(\mathcal{N}, \phi)$ is that of the construction of the region graph $G(\mathcal{N}, \phi)$ except for the set TS which is defined as $ready(v, \phi)$.

Theorem 3. Let the region graph $G' = G(\mathcal{N}, \phi)$ and a reduced region graph $G'' = G(\mathcal{N}, \phi)$ be given. Then $G' \approx_\phi G''$.

Proof. See Appendix.

Thus from Theorems 2 and 3 it follows that the labelling algorithm for $G(\mathcal{N}, \phi)$ can be reduced to the labelling algorithm for $G_R(\mathcal{N}, \phi)$.

Currently we have found in the literature only two papers developing partial order techniques for time Petri nets: the paper [13] by Lilius and the paper [19] by Yoneda et al. Lilius proposes to exploit McMillan's reduction techniques based on unfolding in the state-space search of time Petri nets. In the approach by Yoneda et al., the notion of independence between transitions is structural like ours, because ready sets are calculated using the structure of the net. However, our method allows us to reduce not only the number of states obtained by transition firings but also those obtained by the passage of time. The price is that we have to control the values of time assignments for transitions but it does not have any noticeable influence on the complexity of the method. Further, the proof of the correctness of our algorithm is novel in that instead of using traces, i.e. equivalence classes of sequences, we show a timed stuttering equivalence between the full region graph and a reduced one.

We have implemented both the basic model checking algorithm and its partial order improvement on a Pentium 166 MHz with 128 MBytes of memory in C++ as a part of the system PEP (Programming Environment based on Petri Nets) [4]. The performance of both the algorithms can be demonstrated with an example of concurrent n-buffer from [9] extended with timing constraints. The corresponding time Petri net has $2n$ places, $n+1$ transitions, and the first and $(n+1)$th transitions have the time interval $[1,1]$, whilst the others are associated with the time interval $[1,2]$. As an example property, we verify, if the nth slot of n-buffer is not empty at least once during the time interval $(n, 2n-1)$ along any computation run. The table below shows the impact of reduction on the effectiveness of the model checking. The 'n' column refers to the capacity of buffer. In the 'G' and 'G_R' columns, we list the numbers of vertices ('$\mid V \mid$'), edges ('$\mid E \mid$') and times ('τ') in seconds measured for checking the fixed TCTL-formula in G and G_R, respectively.

	G			G_R		
n	$\mid V \mid$	$\mid E \mid$	τ	$\mid V \mid$	$\mid E \mid$	τ
2	36	43	0.02	33	39	0.01
3	130	168	0.10	117	153	0.08
4	1368	1813	0.39	923	1187	0.30
5	10928	14632	29.3	7341	10015	21.1
6	117745	151250	5020	85136	99814	3127
7	—	—	—	506138	679254	18812

As a future work, we suppose to approach the possibility of increasing an effectiveness of behaviour analysis algorithms by means of using different methods of reduction: symmetry [8], 'unfolding' [15], symbolic states [6], and also to implement experimental researches with real-life communication protocols.

References

1. ALUR R., COURCOUBETIS C., DILL D. Model-checking for real-time systems. Proc. *5th IEEE LICS* (1990) 414–425.
2. ALUR, R., DILL, D.: The theory of timed automata. Theoretical Computer Science **126** (1994) 183–235.

3. Bengtsson, J., Jonsson, B., Lilius, J., Yi, W. Partial order reductions for timed systems. Proc. *CONCUR'98* (1998) 485–496.
4. Best, E., Grahlmann, B.: PEP — more than a Petri net tool. *Lecture Notes in Computer Science* **1055** (1996) 397–401.
5. Browne, M.C., Clarke, E.M., Grumberg, O.: Characterizing finite Kripke structures in propositional temporal logic. *Theoretical Computer Science* **59** (1988) 115–131.
6. Burch, J.R., Clarke, E.M., McMillan, L., Dill, D., Hwang, J.: Symbolic model checking: 10^{20} and beyond. Proc. *5th IEEE LICS* (1990) 428–439.
7. Clarke, E.M., Emerson, E.A., Sistla, A.P.: Automatic verification of finite-state systems using temporal logic specifications. *ACM TOPLAS* **8(2)** (1986) 244–263.
8. Emersen, E., Sistla, A.: Symmetry and model checking. *Lecture Notes in Computer Science* **697** (1993) 463–478.
9. J. Esparza. Model Checking Using Net Unfoldings. *Science of Computer Programming*, 23:151–195, 1994.
10. Gerth, R., Kuiper, R., Peled, D., Penczek, W.: A partial order approach to branching time logic model-checking. Proc. *3rd Israel Symposium on Theory of Computing and Systems* (1994).
11. Godefroid, P.: Partial-order methods for the verification of concurrent systems. An approach for state-explosion problem. *Lecture Notes in Computer Science* **1032** (1996) 143 p.
12. Henzinger, T.A., Manna, Z., Pnueli, A.: Timed transition systems. *Lecture Notes in Computer Science* **600** (1991) 226–251.
13. Lilius, J. *Efficient state space search for time Petri nets.* Proc. MFCS'98 Workshop on Concurrency, August 1998, Brno (Czech Republic), FIMU Report Series, FIMU RS-98-06 (1998) 123–130.
14. Merlin, P., Faber, D.J.: Recoverability of communication protocols. *IEEE Trans. of Communication* **24(9)** (1976) 1036–1043.
15. McMilan, K.: Using unfolding to avoid the state explosion problem in the verification of asynchronous circuits. *Lecture Notes in Computer Science* **663** (1992) 164–177.
16. Pagani F.: Partial orders and verification of real-time systems *Lecture Notes in Computer Science* **1135** (1996) 327–346.
17. Roux, J-L., Berthomieu, B.: Verification of local area network protocol with Tine, a software package for time Petri nets. Proc. *7th European Workshop on Application and Theory of Petri Nets* (1986) 183–205.
18. Valmari, A.: A stubborn attack on state explosion. *Lecture Notes in Computer Science* **535** (1990).
19. Yoneda, T., Shibayama, A., Schlingloff, B.H., Clarke, E.M.: Efficient verification of parallel real-time systems. *Lecture Notes in Computer Science* **697** (1993) 321–333.

Appendix

Proof Sketch of Theorem 3.

We construct the relation $\approx^*_\phi \subseteq V' \times V''$ as follows: $\approx^*_\phi = \{(v' = <m', [\nu']^*>, v'' = <m'', [\nu'']^*>) \mid dependent(v', \phi) = dependent(v'', \phi)$, $m' \cap P(\phi) = m'' \cap P(\phi)$, and $\nu'(t^*) \sim c \Leftrightarrow \nu''(t^*) \sim c$ for all $' \sim c'$ in $C(\phi)\}$. Let us show that $\approx^*_\phi$ is a timed stuttering equivalence w.r.t. ϕ between G' and G''. Before doing so, we need to consider the following

Claim A.
Let $(v', v'') \in \approx_\phi^*$, $(v', v_1') \in E'$ and $l'((v', v_1')) = t'$. If $t' \notin \textit{dependent}(v', \phi)$ then $(v_1', v'') \in \approx_\phi^*$.

Claim B.
Let $(v', v'') \in \approx_\phi^*$, $(v'', v_1'') \in E''$ and $l''((v'', v_1'')) = t''$. If $t'' \notin \textit{dependent}(v'', \phi)$ then $(v', v_1'') \in \approx_\phi^*$.

Claim C.
Let $(v', v'') \in \approx_\phi^*$. Let $(v', v_1') \in E'$, $l'((v', v_1')) = t'$ and $(v'', v_1'') \in E''$, $l''((v'', v_1''))$ $= t''$ such that $t' \in \textit{dependent}(v', \phi)$ and $t'' \in \textit{dependent}(v'', \phi)$. Then

(i) $t' \notin \textit{fireable}(v'') \wedge t'' \notin \textit{fireable}(v') \wedge t' \neq t'' \Rightarrow (v_1', v_1'') \in \approx_\phi^*$.
(ii) $t' \in \textit{fireable}(v'') \wedge t'' \in \textit{fireable}(v') \wedge t' = t'' \Rightarrow (v_1', v_1'') \in \approx_\phi^*$.

The proofs of the claims are too technical to appear in the paper.

We proceed the proof of that $\approx_\phi^*$ is a timed stuttering equivalence w.r.t. ϕ between G' and G''. Suppose v_0' and v_0'' to be the initial vertices of G' and G'', respectively. According to the construction of G' and G'', we have $(v_0', v_0'') \in \approx_\phi^*$, because $v_0' = v_0''$. Assume $(v', v'') \in \approx_\phi^*$. The validity of point 1 of Definition 4 follows from the construction of $\approx_\phi^*$. Consider point 2(a) of Definition 4 (point 2(b) is symmetrical). Suppose $(v', v_1') \in E'$. We have to show that there exist $v^0, \cdots, v^n \in V''(n \geq 0)$ such that $v'' = v^0$, $((v^i, v^{i+1}) \in E'' \;\wedge\; (v', v^i) \in \approx_\phi^*)$ for all $0 \leq i < n$, and $(v_1', v^n) \in \approx_\phi^*$. Two cases are admissible.

$n = 0$. If $t' \notin \textit{dependent}(v', \phi)$ then the result immediatly follows from Claim A. Consider the case $t' \in \textit{dependent}(v', \phi)$. Let us show that $\textit{fireable}(v'') = \emptyset$. Suppose a contrary, i.e. $\textit{fireable}(v'') \neq \emptyset$. W.l.o.g. assume $|\, \textit{fireable}(v'') \,| = 1$. Let $l''((v'', v^1)) = t''$. If $t'' \notin \textit{dependent}(v'', \phi)$ then using Claim B we get $(v', v^1) \in \approx_\phi^*$, contradicting our assumption. Consider the case $t'' \in \textit{dependent}(v'', \phi)$. We distinguish two cases.

1. $t' \in \textit{visible}^+(v', \phi) \vee t'' \in \textit{visible}^+(v'', \phi)$. Then $t' \notin \textit{fireable}(v'')$ and $t'' \notin \textit{fireable}(v')$. Due to Claim C(i), we have $(v_1', v^1) \in \approx_\phi^*$, contradicting our assumption.
2. $t' \in \textit{visible}_\delta^+(v', \phi) \wedge t'' \in \textit{visible}_\delta^+(v'', \phi)$. Then $t' \in \textit{fireable}(v'')$ and $t'' \in \textit{fireable}(v')$. Due to Claim C(ii), we get $(v_1', v^1) \in \approx_\phi^*$, contradicting our assumption.

So, $\textit{fireable}(v'') = \emptyset$. We now show that $t' \notin \textit{visible}^+(v', \phi)$. Assume a contrary, i.e. $t \in \textit{visible}^+(v', \phi)$. Then $t \in \textit{visible}^+(v'', \phi)$, by the construction of $\approx_\phi^*$. Hence it holds: $\exists \tilde{t} \in \textit{visible}(\phi)\,.\,[t\, F^*\, \tilde{t} \wedge {}^\bullet t \subseteq m''] \vee [\exists \hat{t}\; (\hat{t}\, F^*\, \tilde{t} \wedge {}^\bullet \hat{t} \subseteq m''), \exists \bar{t}\; ({}^\bullet \hat{t} \cap {}^\bullet \bar{t} \neq \emptyset \;\wedge\; \hat{t} \neq \bar{t})\,.\,(t\, F^*\bar{t} \;\wedge\; {}^\bullet t \subseteq m'' \;\wedge\; dist(t, \bar{t}) \leq Lft(\hat{t}))]$. Further, since $t \notin \textit{fireable}(v'')$, it also holds: $\forall \delta \in \mathbf{R}^+(\{\nu'' + \delta' \mid 0 \leq \delta' \leq \delta\} \subseteq [\nu'']^*)\,.\,({}^\bullet t \not\subseteq m'' \;\vee\; \nu''(t) + \delta < Eft(t))$. Hence, $\nu''(t) + \delta < Eft(t)$ for all $\delta \in \mathbf{R}^+$ such that $\{\nu'' + \delta' \mid 0 \leq \delta' \leq \delta\} \subseteq [\nu'']^*$. We get a contradiction, because $\nu_0''(t') > Lft(t')$ for all $t' \in T^*$.

So, we have $t \in \textit{visible}_\delta^+(v', \phi)$. Since $v_1' = \textit{succ}_\delta(v')$, then $m' = m_1'$ and there exists $\delta \in \mathbf{R}^+$ such that $\nu' + \delta \in [\nu_1']^*$ and $\{\nu' + \delta' \mid 0 \leq \delta' \leq \delta\} \subseteq [\nu']^* \cup [\nu_1']^*$. Hence $m' \cap P(\phi) = m_1' \cap P(\phi)$.

Let us next show that $\nu'(t^*) \sim c \Leftrightarrow \nu_1'(t^*) \sim c$ for all $' \sim c'$ in $C(\phi)$. Supposing a contrary, we get two cases: (a) $\nu'(t^*) \sim c \wedge \nu_1'(t^*) \not\sim c$ for some $' \sim c'$ in ϕ, and (b) $\nu'(t^*) \not\sim c \wedge \nu_1'(t^*) \sim c$ for some $' \sim c'$ in ϕ. The case (a) contradicts either $\nu_1'(t^*) > \nu'(t^*)$, if $\sim \in \{>, \geq\}$, or $\nu'(t^*) \sim c \Leftrightarrow \nu''(t^*) \sim c$, because $\nu''(t^*) > c_\phi$, if $\sim \in \{<, \leq, =\}$. The case (b) obtain similary.

We finally show $\delta \in \textit{visible}_\delta^+(v_1', \phi)$. Suppose a contrary, i.e., $\nu'(t^*) \in I'(\sim c)$ for some $' \sim c'$ in ϕ and $\nu_1'(t^*) \notin I'(\sim c)$ for all $' \sim c'$ in ϕ. This contradicts either the definition of $I'(\sim c)$, because $\nu_1'(t^*) > \nu'(t^*)$, if $\sim \in \{>, \geq\}$, or $\nu'(t^*) \sim c \Leftrightarrow \nu_0''(t^*) \sim c$, because $\nu_0''(t^*) > c_\phi$, if $\sim \in \{<, \leq, =\}$.

Thus, $\textit{dependent}(v', \phi) = \textit{dependent}(v_1', \phi)$, $m' \cap P(\phi) = m_1' \cap P(\phi)$, and $\nu'(t^*) \sim c \Leftrightarrow \nu_1'(t^*) \sim c$ for all $' \sim c'$ in $C(\phi)$. Since $(\nu', \nu_0'') \in \approx_\phi^*$, then $(\nu_1', \nu_0'') \in \approx_\phi^*$, due to the construction $\approx_\phi^*$.

$n > 0$. W.l.o.g. assume $n = 1$. Let $l''((v'', v^1)) = t''$. If $(t' \in \textit{dependent}(v', \phi) \wedge t'' \notin \textit{dependent}(v'', \phi))$ or $(t' \notin \textit{dependent}(v', \phi) \wedge t'' \in \textit{dependent}(v'', \phi))$, then using Clams A and B we get a contradiction to our assumption. It remains to consider two cases.

1. $t' \notin \textit{dependent}(v', \phi) \wedge t'' \notin \textit{dependent}(v'', \phi)$. The result follows from Claims A nad B, because $(v', v'') \in \approx^*$.
2. $t' \in \textit{dependent}(v', \phi) \wedge t'' \in \textit{dependent}(v'', \phi)$. The result immediately follows from Claim C(i, ii).

So, there exist $v^0, \cdots, v^n \in V''(n \geq 0)$ such that $v'' = v^0$, $((v^i, v^{i+1}) \in E'' \wedge (v', v^i) \in \approx_\phi^*)$ for all $0 \leq i < n$, and $(v_1', v^n) \in \approx_\phi^*$. Thus, $\approx_\phi^*$ is a timed stuttering equivalence w.r.t. ϕ between G' and G''. $\square$

Deriving Formulas for Domination Numbers of Fasciagraphs and Rotagraphs *

Janez Žerovnik

Faculty of Mechanical Engineering
University of Maribor
Smetanova 17
SI-2000 Maribor, Slovenia
and
Department of Theoretical Computer Science
Institute of Mathematics, Physics and Mechanics
Jadranska 19
SI-1111 Ljubljana, Slovenia
Email: janez.zerovnik@imfm.uni-lj.si.

Abstract. Recently, an algebraic approach which can be used to compute distance-based graph invariants on fasciagraphs and rotagraphs was given in [Mohar, Juvan, Žerovnik, Discrete Appl. Math. 80 (1997) 57-71]. Here we give an analogous method which can be employed for deriving formulas for the domination number of fasciagraphs and rotagraphs. In other words, it computes the domination numbers of these graphs in constant time, i.e. in time which depends only on the size and structure of a monograph and is independent of the number of monographs. Some further generalizations of the method are discussed, in particular the computation of the independent number and the k-coloring decision problem. Examples of fasciagraphs and rotagraphs include complete grid graphs. Grid graphs are one of the most frequently used model of processor interconnections in multiprocessor VLSI systems.

1 Introduction

The notion of a polygraph was introduced in chemical graph theory as a generalization of the chemical notion of polymers [3]. Polygraphs are not only of interest in chemistry. For example, grid graphs are one of the most frequently used model of processor interconnections in multiprocessor VLSI systems [9]. An important class of polygraphs form fasciagraphs and rotagraphs. For example, complete grid graphs are fasciagraphs and Cartesian products of cycles are rotagraphs.

In general, problems related to domination in graphs are widely studied [17]. The problem of computing the domination number of grid graphs is NP–complete while the complexity is open for complete grid graphs (with both k and n variable), cf. [17]. Hence it is worthwhile to look for algorithms that compute the

* This work was partially supported by the Ministry of Science and Technology of Slovenia under the grant J2-1015.

domination numbers of these graphs. Among many applications of domination numbers of grid graphs let us only mention the link between the existence of perfect Lee codes and minimum dominating sets of Cartesian products of paths and cycles [13].

An algebraic approach yielding $O(\log n)$ algorithms for various problems on fasciagraphs and rotagraphs was proposed in [23]. The problems treated include the domination number, the independent number and the k-coloring problem. Analogous method was successfully applied to computing graph invariants of interest in chemistry, including computation of the Wiener index [33], the Szeged index [15], the determinant and the permanent (see [20, 11, 12], respectively). Existence of constant time algorithms on fasciagraphs and rotagraphs for distance-based graph invariants was proved in [21]. Recently, a computer program in Mathematica for computing the formulas for Szeged index of infinite families of fasciagraphs was developed [26]. In this paper, we prove that it is possible to use analogous approach for deriving formulas for domination numbers on fasciagraphs and rotagraphs. Furthermore, we sketch how the same approach can be used to compute the independence number of fasciagraphs and rotagraphs and how we can decide k-colorability of such graphs by a constant time algorithm.

The rest of the paper is organized as follows. In the next section a concept of a polygraph is introduced and two special subclasses of graphs, the fasciagraphs and the rotagraphs are defined. In Section 3, the concept of a path algebra is introduced and an algorithm is recalled (from [23]) which can be used to solve various problems on fasciagraphs and rotagraphs. In Section 4 we give an instance of the algorithm which computes the domination number of a fasciagraph and a rotagraph. We then prove that the powers of the matrices which correspond to the solution have a special structure, which implies existence of a constant time algorithm for computing any power. Together with the results of Section 4 this implies the main result, the Theorem 4. In the last section we discuss two more examples, the independence number and the k-colorability decision problem.

We finally observe that the approach can also be extended to a wider class of graphs including polygraphs, but in this case the algorithms become linear. However, since these graphs have bounded tree-width, linear algorithms are already known.

2 Polygraphs

We consider finite undirected and directed graphs. A graph will always mean an undirected graph, a digraph will stand for a directed graph. P_n and C_n will denote the path on n vertices and the cycle on n vertices, respectively. An edge $\{u, v\}$ of a graph will be denoted uv (hence uv and vu mean exactly the same edge). An arc from u to v in a digraph will be denoted (u, v).

Let G_1, G_2, ..., G_n be arbitrary, mutually disjoint graphs, and let X_1, X_2, ..., X_n be a sequence of sets of edges such that an edge of X_i joins a vertex of $V(G_i)$ with a vertex of $V(G_{i+1})$. For convenience we also set $G_0 = G_n$,

$G_{n+1} = G_1$ and $X_0 = X_n$. This in particular means that edges in X_n join vertices of G_n with vertices of G_1. A *polygraph*

$$\Omega_n = \Omega_n(G_1, G_2, \dots, G_n; X_1, X_2, \dots, X_n)$$

over *monographs* $G_1, G_2, \dots, G_n$ is defined in the following way:

$$V(\Omega_n) = V(G_1) \cup V(G_2) \cup \dots \cup V(G_n),$$
$$E(\Omega_n) = E(G_1) \cup X_1 \cup E(G_2) \cup X_2 \cup \dots \cup E(G_n) \cup X_n.$$

For a polygraph Ω_n and for $i = 1, 2, \dots n$ we also define

$$D_i = \{u \in V(G_i) \mid \exists v \in G_{i+1} : uv \in X_i\},$$
$$R_i = \{u \in V(G_{i+1}) \mid \exists v \in G_i : uv \in X_i\}.$$

In general $R_i \cap D_{i+1}$ need not be empty.

Assume that for $1 \le i \le n$, G_i is isomorphic to a fixed graph G and that we have identified each G_i with G. Let in addition the sets X_i, $1 \le i \le n$, be equal to a fixed edge set $X \subseteq V(G) \times V(G)$. Then we call the polygraph a *rotagraph* and denote it $\omega_n(G; X)$. A *fasciagraph* $\psi_n(G; X)$ is a rotagraph $\omega_n(G; X)$ without edges between the first and the last copy of a monograph. Formally, in $\psi_n(G; X)$ we have $X_1 = X_2 = \dots = X_{n-1}$ and $X_n = \emptyset$. Since in a rotagraph all the sets D_i and the sets R_i are equal, we will denote them by D and R, respectively. The same notation will be used for fasciagraphs as well, keeping in mind that R_n and D_0 are empty.

3 Path algebras and the algorithm

In this section a general framework for solving different problems on the class of fasciagraphs and rotagraphs is given [23]. The essence of the method is a computation of powers of matrices over certain semirings. We wish to remark that similar ideas are implicitly used in [14, 28]. Before giving the algorithm, a concept of path algebras is introduced. We follow the approach given in [5], see also [31, 32].

A *semiring* $\mathcal{P} = (P, \oplus, \circ, 0, 1)$ is a set P on which two binary operations, $\oplus$ and $\circ$, are defined such that

(i) $(P, \oplus)$ forms an commutative monoid with 0 as unit,
(ii) $(P, \circ)$ forms a monoid with 1 as unit,
(iii) operation $\circ$ is left- and right-distributive over operation $\oplus$,
(iv) for all $x \in P$, $x \circ 0 = 0 = 0 \circ x$.

An idempotent semiring (for all $x \in P$, $x \oplus x = x$) is called a *path algebra.* It is easy to see that a semiring is a path algebra if and only if $1 \oplus 1 = 1$ holds. Examples of path algebras include (for more examples we refer to [5]):

$\mathcal{P}_1$: $(\mathbb{N}_0 \cup \{\infty\}, \min, +, \infty, 0)$,
$\mathcal{P}_2$: $(\mathbb{N}_0 \cup \{-\infty\}, \max, +, -\infty, 0)$,
$\mathcal{P}_3$: $(\{0, 1\}, \max, \min, 0, 1)$.

Let $\mathcal{P} = (P, \oplus, \circ, 0, 1)$ be a path algebra and let $\mathcal{M}_n(\mathcal{P})$ be the set of all $n \times n$ matrices over P. Let $A, B \in \mathcal{M}_n(\mathcal{P})$ and define operations $A \oplus B$ and $A \circ B$ in the usual way:

$$(A \oplus B)_{ij} = A_{ij} \oplus B_{ij},$$
$$(A \circ B)_{ij} = \sum_{k=1}^{n} A_{ik} \circ B_{kj}.$$

$\mathcal{M}_n(\mathcal{P})$ equipped with the above operations is a path algebra itself with the zero and the unit matrix as units of the semiring.

Let $\mathcal{P}$ be a path algebra and let G be a *labeled* digraph, i.e., a digraph together with a labeling function ℓ which assigns to every arc of G an element of P. Let $V(G) = \{v_1, v_2, \ldots, v_n\}$. The labeling ℓ of G is extended to paths as follows. For a path $Q = (x_{i_0}, x_{i_1})(x_{i_1}, x_{i_2}) \cdots (x_{i_{k-1}}, x_{i_k})$ of G let

$$\ell(Q) = \ell(x_{i_0}, x_{i_1}) \circ \ell(x_{i_1}, x_{i_2}) \circ \cdots \circ \ell(x_{i_{k-1}}, x_{i_k}).$$

Let S_{ij}^k be the set of all paths of order k from x_i to x_j in G and let $A(G)$ be the matrix defined by $A(G)_{ij} = \ell(x_i, x_j)$ if (x_i, x_j) is an arc of G and $A(G)_{ij} = 0$ otherwise. Now we can state the following well-known result (see, for instance, [5, p. 99]):

Theorem 1. $(A(G)^k)_{ij} = \sum_{Q \in S_{ij}^k} \ell(Q)$.

Let $\psi_n(G; X)$ and $\omega_n(G; X)$ be a fasciagraph and a rotagraph, respectively. Set $W = D_i \cup R_i = D \cup R$ and let $N = 2^{|W|}$. Define a labeled digraph $\mathcal{G} = \mathcal{G}(G; X)$ as follows. The vertex set of $\mathcal{G}$ is formed by the subsets of W, which will be denoted by C_i; in particular we will use C_0 for the empty subset. An arc joins a subset C_i with a subset C_j if C_i is not in a "conflict" with C_j. Here a "conflict" of C_i with C_j means that using C_i and C_j as a part of a solution in consecutive copies of G would violate a problem assumption. For instance, if we search for a largest independent set, such a conflict would be an edge between a vertex of C_i and a vertex of C_j. Let finally $\ell : E(\mathcal{G}) \to P$ be a labeling of $\mathcal{G}$ where $\mathcal{P}$ is a path algebra on the set P. The general scheme for our algorithm is the following:

Algorithm 2

1. Select an appropriate path algebra $\mathcal{P} = (P, \oplus, \circ, 0, 1)$.
2. Determine an appropriate labeling ℓ of $\mathcal{G}(G; X)$.
3. In $\mathcal{M}_N(\mathcal{P})$ calculate $A(\mathcal{G})^n$.
4. Among admissible coefficients of $A(\mathcal{G})^n$ select one which optimizes the corresponding goal function.

It is well known that, in general, Step 3 of the algorithm can be done in $O(\log n)$ steps. In Section 4 we will show that in some cases it is possible to compute the powers of the matrix $A(\mathcal{G})^n$ in constant time, i.e. with an algorithm of time complexity which is independent of n. Hence if we assume that the size of G is a given constant (and n is a variable), then the algorithm will run in constant time. However, the algorithm is useful for practical purposes only if the number of vertices of the monograph G is relatively small, since the time complexity is in general exponential in the number of vertices of the monograph G.

4 Domination numbers of fasciagraphs and rotagraphs

A set S of vertices of a graph G is a *dominating set* if every vertex from $V(G)\backslash S$ is adjacent to at least one vertex in S. The *domination number*, $\gamma(G)$, is the smallest number of vertices in a dominating set of G.

Let $\psi_n(G;X)$ and $\omega_n(G;X)$ be a fasciagraph and a rotagraph, respectively. Let $C_i, C_j \in V(\mathcal{G}(G;X))$, i.e., $C_i, C_j \subseteq D \cup R$, and consider for a moment $\psi_3(G;X)$. Let $C_i \subset D_1 \cup R_1$ and $C_j \subset D_2 \cup R_2$, where $D_1 = D_2 = D$ and $R_1 = R_2 = R$.

Let $\gamma_{ij}(G;X)$ be the size of a smallest dominating set $S \subseteq G_2\backslash((C_i \cap R_1) \cup (D_2 \cap C_j))$, such that G_2 is dominated by $C_i \cup S \cup C_j$. Then set

$$\ell(C_i, C_j) = |C_i \cap R| + \gamma_{ij}(G;X) + |D \cap C_j| - |C_i \cap C_j| \tag{1}$$

The labeling implies that (C_i, C_j) is an arc of $\mathcal{G}(G;X)$ if $C_i \cap R \cap D \cap C_j = \emptyset$.

Algorithm 3

1. For a path algebra select $\mathcal{P}_1 = (\mathbb{N}_0 \cup \{\infty\}, \min, +, \infty, 0)$.
2. Label $\mathcal{G}(G;X)$ as defined in (1).
3. In $\mathcal{M}_N(\mathcal{P}_1)$ calculate $A(\mathcal{G})^n$.
4. Let $\gamma(\psi_n(G;X)) = (A(\mathcal{G})^n)_{00}$ and $\gamma(\omega_n(G;X)) = \min\limits_i (A(\mathcal{G})^n)_{ii}$.

We first discuss the correctness of the algorithm.

Lemma 1. *Algorithm 3 correctly computes* $\gamma(\psi_n(G;X))$.

Lemma 1 was proved in [23], while the rotagraphs part was left to the reader. For completeness, we give here a proof of the next Lemma.

Lemma 2. *Algorithm 3 correctly computes* $\gamma(\omega_n(G;X))$.

Proof. By Theorem 1, for any C_j,

$$\begin{aligned}(A(\mathcal{G})^n)_{jj} &= \min_{Q \in S^n_{jj}} \ell(Q) \\ &= \min_{i_1, i_2, \ldots, i_{n-1}} (\ell(C_j, C_{i_1}) + \ell(C_{i_1}, C_{i_2}) + \cdots + \ell(C_{i_{n-1}}, C_j)).\end{aligned}$$

Assume now that the minimum is attained on indices $i_1, i_2, \ldots, i_{n-1}$. Then $(A(\mathcal{G})^n)_{jj}$ is equal to

$(|C_j \cap R_0| + \gamma_{j,i_1} + |D_1 \cap C_{i_1}|) + (|C_{i_1} \cap R_1| + \gamma_{i_1,i_2} + |D_2 \cap C_{i_2}|) + \cdots +$
$(|C_{i_{n-1}} \cap R_{n-1}| + \gamma_{i_{n-1},j} + |D_n \cap C_j|)$.

By the definition of γ_{ij}, the above expression is the size of a dominating set of $\omega_n(G;X)$. On the other hand, a smallest dominating set of $\omega_n(G;X)$ gives rise to such an expression for some C_j, thus

$$\min_j (A(\mathcal{G})^n)_{jj}$$

is the size of a smallest dominating set of $\omega_n(G;X)$. □

We now prove a Lemma which will imply the existence of a constant time algorithm for computing the powers of $A(\mathcal{G})$ in step 3 of the Algorithm 3.

Let us denote $A_l = A(\mathcal{G})^l$. Note that the meaning of the value of $(A_l)_{ij}$ is the size of the dominating set of a subgraph of $\psi_l(G;X)$. More precisely, provided that the sets C_i and C_j are not conflicting, $(A_l)_{ij}$ is the domination number of the subgraph induced on the vertices

$$(V(G_1) \setminus C_i) \cup V(G_2) \cup \ldots \cup V(G_{l-1}) \cup (V(G_l) \setminus C_j).$$

It can be shown that for large enough indices l, the matrices A_l have a special structure that enables us to compute them efficiently. The following proposition is a variant of the "cyclicity" theorem for the "tropical" semiring $(\mathbb{N}_0 \cup \infty, \min, +, \infty, 0)$, see, e.g., [4, Theorem 3.112]. By a constant matrix we mean a matrix with all entries equal.

Lemma 3. *Let* $k = |V(\mathcal{G}(G;X))|$, *and* $K = |V(G)|$. *Then there is an index* $q \leq (2K+2)^{k^2}$ *such that* $D_q = D_p + C$ *for some index* $p < q$ *and some constant matrix* C. *Let* $P = q - p$. *Then for every* $r \geq p$ *and every* $s \geq 0$ *we have*

$$A_{r+sP} = A_r + sC\,.$$

Proof. First we prove the claim: *For any* $l \geq 1$, *the difference between any pair of entries of* A_l, *both different from* ∞, *is bounded by* $2K$.

Assume $(A_l)_{ij} \neq \infty$. Then clearly,

$$(A_l)_{ij} \leq \gamma(\psi_l(G;X))$$

and, since $|C_i \cap R| + |D \cap C_j| \leq 2|V(G)|$,

$$(A_l)_{ij} \geq \gamma(\psi_l(G;X)) - 2|V(G)|$$

Hence the claim follows.

For $l \geq 1$, define $K_l = \min\{(A_l)_{ij}\}$ and let $A'_l = A_l - (K_l)J$, where J is the matrix with all entries equal to 1. Since the difference between any two elements of A_l, different from ∞, cannot be greater than $2K$. (Note that $\infty - x = \infty$ for any x.) The entries of A'_l can therefore have only values $0, 1, \ldots, 2K, \infty$ and hence there are indices $p < q \leq (2K+2)^{k^2}$ such that $A'_p = A'_q$. This proves the first part of the proposition.

The equality $A_{r+sP} = A_r + sC$ follows from the fact that for arbitrary matrices D, E and a constant matrix C we have

$$(D \oplus C) \circ E = D \circ E \oplus C.$$

This can easily be seen by computing the values of ij-th entries of both sides of the equality: $(D \oplus C) \circ E)_{ij} = \min_k \{((D)_{ik} + (C)_{ik}) + (E)_{kj}\} = \min_k \{(D)_{ik} + (E)_{kj}\} + (C)_{ij}$ and $(D \circ E \oplus C)_{ij} = \min_k \{(D)_{ik} + (E)_{kj}\} + (C)_{ij}$. □

Note that since $k \leq |V(\mathcal{G}(G; X))| = 2^{|W|} = N \leq 2|V(G)|$, and $K = |V(G)|$, the constants p, q and P in Lemma 3 depend only on the size of the monograph and are thus independent of n.

Lemma 4. *Algorithm 3 can be implemented to run in constant time.*

Proof. First note that for Step 2 of Algorithm 3 any procedure for computing the domination number can be used since the time complexity is clearly constant in n. With the same argument Step 4 can be computed in constant time.

Finally, the time complexity of Step 3 is constant in n because of Lemma 3. For any n, only constant number, q, of powers of the matrix has to be computed. Then the formula $A_n = A_{n-jP} + jC$ can be used, where $j = (n-p)$ div P. □

Combining Lemmas 1, 2 and 4 we have

Theorem 4. *Algorithm 3 correctly computes $\gamma(\psi_n(G; X))$ and $\gamma(\omega_n(G; X))$ and can be implemented to run in $O(C)$ time.*

Example. The *Cartesian product* $G = H \,\square\, K$ of graphs H and K is the graph with vertex set $V(G) = V(H) \times V(K)$. Vertices (x_1, x_2) and (y_1, y_2) are adjacent in $H \,\square\, K$ if either $x_1 y_1 \in E(H)$ and $x_2 = y_2$ or $x_2 y_2 \in E(K)$ and $x_1 = y_1$. Note that $P_k \,\square\, P_n = \psi_n(P_k; X)$, $P_k \,\square\, C_n = \omega_n(P_k; X)$, $C_k \,\square\, C_n = \omega_n(C_k; X)$ and $C_k \,\square\, P_n = \psi_n(C_k; X)$, where X is the matching defined by the identity isomorphism between two copies of P_k and C_k, respectively.

One of the motivations for this work was a widely studied problem of determining the domination number of complete grid graphs and Cartesian products of cycles [7, 9, 16, 19, 22]. For complete grid graphs, i.e. graphs $P_k \,\square\, P_n$, algorithms were given in [16] which for a fixed k compute $\gamma(P_k \,\square\, P_n)$ in $O(n)$ time. An $O(\log n)$ algorithm was proposed in [23]. Of course, the Theorem 4 implies that the domination number problem for $k \times n$ grids, where k is fixed, has a constant time solution, which was claimed already in [25].

It may be interesting to note that in [10] formulas are given for families $\{P_k \,\square\, P_n \mid n \in \mathbb{N}\}$ for k up to 19. For $k \geq 20$, it is conjectured [6] that

$$\gamma(P_k \,\square\, P_n) = \left\lceil \frac{(k+2)(n+2)}{5} \right\rceil - 4$$

and the problem is still open. □

5 Two more applications

The size of a largest independent set of vertices of a graph G is called the *independence number* of G, $\alpha(G)$. Select $\mathcal{P}_2 = (\mathbb{N}_0 \cup \{-\infty\}, \max, +, -\infty, 0)$ as a path algebra and define a labeling of $\mathcal{G}(G;X)$ similarly as in (1). The difference is that two vertices are in conflict (and hence the corresponding arc must be labeled $-\infty$) if $(C_i \cap R) \cup (C_j \cap D)$ is not an independent set in G.

We omit the proof of a lemma saying that the difference in size of independent sets of subgraphs induced on

$$(V(G_1) \setminus C_i) \cup V(G_2) \cup \ldots \cup V(G_{l-1} \cup (V(G_l) \setminus C_j).$$

for fixed l and variable C_i and C_j can only differ at most for $2|V(G)|$.

Consequently we have

Theorem 5. *One can compute $\alpha(\psi_n(G;X))$ and $\alpha(\omega_n(G;X))$ in constant time.*

As a second application we consider the k-coloring problem. To solve it on fasciagraphs and rotagraphs we first select $\mathcal{P}_3 = (\{0,1\}, \max, \min, 0, 1)$ as a path algebra. We next define a labeled digraph $\mathcal{G}(G;X)$ slightly different as we did by now. The vertex set of $\mathcal{G}$ is formed by the k-colorings of $W = D \cup R$ or, equivalently, by the k-partitions of W with parts being independent sets. An arc joins a k-coloring C_i with a k-coloring C_j if and only if the corresponding partitions coincide on their (possible) intersection in G_2 and can be extended to a k-coloring of G_2. The labelling of $\mathcal{G}(G;X)$ is then defined just by the adjacency relation. Finally, in $\mathcal{M}_N(\mathcal{P}_3)$ we calculate $A(\mathcal{G})^n$ and conclude that that $\psi_n(G;X)$ or $\omega_n(G;X)$ is k-colorable if and only if $(A(\mathcal{G})^n)_{00} = 1$ or $\max_i (A(\mathcal{G})^n)_{ii} = 1$, respectively.

By observing that the maximal differences between pairs of entries of matrix $A(\mathcal{G})$ is bounded, we have:

Theorem 6. *The k-coloring problem of the graphs $\psi_n(G;X)$ and $\omega_n(G;X)$ is solvable in constant time.*

Remark. There are many further examples on which the method can be used. A good starting list may be the domination-type problems studied in [29]. An interesting question is to find a general properties which a problem must fulfill.

Remark. The algebraic approach used here can also be generalized to more general graphs, for example to graphs which are obtained from trees by expanding nodes to arbitrary (small) graphs. The matrix must then be replaced by a tensor of dimension equal to the degree of the vertex (monograph) in the original tree. Computing the matrix products, starting from leaves to the center of the tree would yield an algorithm, linear in n. Again, these are graphs of bounded tree-width, for which existence of linear algorithms for many problems is already known.

Even more easily, our algebraic approach can be extended to polygraphs as well. Instead of computing a single graph $\mathcal{G}(G;X)$ and calculating the n-th power of $A(\mathcal{G})$, we must determine n graphs and calculate the matrix product of the corresponding matrices over an appropriate path algebra. This yields to $O(n)$ algorithms for polygraphs. However, the tree-width of a polygraphs can be bounded by a constant depending on the size of a monograph. (For definitions of a tree-width see, for example, [24, 27, 30], cf. also [1].) Arnborg and Proskurowski, [2], see also [1], obtained linear time algorithms for different problems of graphs with bounded tree-width, including dominating set, independent set and k-colorability problem. Their algorithms are linear in the size of the problem instance, but are exponential in the tree-width of the involved graphs – the case analogous to the present approach.

References

1. S. Arnborg, Efficient algorithms for combinatorial problems on graphs with bounded decomposability - a survey, BIT 25 (1985) 2–23.
2. S. Arnborg and A. Proskurowski, Linear time algorithms for NP-hard problems on graphs embedded in k-trees, TRITA-NA-8404, Dept. of Num. Anal. and Comp. Sci, Royal Institute of Technology, Stockholm, Sweden (1984).
3. D. Babić, A. Graovac, B. Mohar and T. Pisanski, The matching polynomial of a polygraph, Discrete Appl. Math. 15 (1986) 11-24.
4. F. Baccelli, G. Cohen, G. J. Olsder and J. P. Quadrat, Synchronization and Linearity (Wiley, 1992).
5. B. Carré, Graphs and Networks (Clarendon Press, Oxford, 1979).
6. T.Y.Chang, Domination Numbers of Grid GRaphs, Ph.D. Thesis, Dept. of Mathematics, Univ. of South Florida, 1992.
7. T.Y. Chang and W.E. Clark, The domination numbers of the $5 \times n$ and $6 \times n$ grid graphs, J. Graph Theory 17 (1993) 81-107.
8. B.N. Clark, C.J. Colbourn and D.S. Johnson, Unit disc graphs, Discrete Math. 86 (1990) 165-177.
9. E.J. Cockayne, E.O. Hare, S.T. Hedetniemi and T.V. Wimer, Bounds for the domination number of grid graphs, Congr. Numer. 47 (1985) 217-228.
10. D.C. Fisher, The domination number of complete grid graphs, J. Graph Theory, submitted.
11. A. Graovac, M. Juvan, B. Mohar, S. Vesel and J. Žerovnik, The Szeged index of polygraphs, in preparation.
12. A. Graovac, M. Juvan, B. Mohar and J. Žerovnik, Computing the determinant and the algebraic structure count on polygraphs, submitted.
13. S. Gravier and M. Mollard, On domination numbers of Cartesian product of paths, Discrete Appl. Math. 80 (1997) 247-250.
14. I. Gutman, N. Kolaković, A. Graovac and D. Babić, A method for calculation of the Hosoya index of polymers, in: A. Graovac, ed., Studies in Physical and Theoretical Chemistry, vol. 63 (Elsevier, Amsterdam, 1989) 141–154.
15. I. Gutman, Formula for the Wiener Number of trees and its extension to graphs containing cycles, Graph Theory Notes New York **27** (1994) 9-15.
16. E.O. Hare, S.T. Hedetniemi and W.R. Hare, Algorithms for computing the domination number of $k \times n$ complete grid graphs, Congr. Numer. 55 (1986) 81-92.
17. S.T. Hedetniemi and R.C. Laskar, Introduction, Discrete Math. 86 (1990) 3-9.
18. M.S. Jacobson and L.F. Kinch, On the domination number of products of graphs: I, Ars Combin. 18 (1983) 33–44.

19. M.S. Jacobson and L.F. Kinch, On the domination of the products of graphs II: trees, J. Graph Theory 10 (1986) 97–106.
20. M. Juvan, B. Mohar, A. Graovac, S. Klavžar and J. Žerovnik, Fast computation of the Wiener index of fasciagraphs and rotagraphs, J. Chem. Inf. Comput. Sci. 35 (1995) 834–840.
21. M. Juvan, B. Mohar and J. Žerovnik, Distance-related invariants on polygraphs, Discrete Appl. Math. 80 (1997) 57-71.
22. S. Klavžar and N. Seifter, Dominating Cartesian products of cycles, Discrete Appl. Math. 59 (1995) 129–136.
23. S. Klavžar and J. Žerovnik, Algebraic approach to fasciagraphs and rotagraphs, Discrete Appl. Math. 68 (1996) 93–100.
24. J. van Leeuwen, Graph algorithms, in: J. van Leeuwen, ed., Handbook of Theoretical Computer Science, Volume A, Algorithms and Complexity (Elsevier, Amsterdam, 1990) 525–631.
25. Livingston and Stout, 25th Int. Conf. on Combinatorics, Graph Theory and Computing, Boca Racon, FL, 1994.
26. M. Petkovšek, A. Vesel and J. Žerovnik, Computing the Szeged index of fasciagraphs and rotagraphs using *Mathematica*, manuscript in preparation.
27. N. Robertson and P.D. Seymour, Graph minors II. Algorithmic aspects of tree-width, J. Algorithms, 7 (1986) 309–322.
28. W.A. Seitz, D.J. Klein and A. Graovac, Transfer matrix methods for regular polymer graphs, in: R.C. Lacher, ed., Studies in Physical and Theoretical Chemistry, vol. 54 (Elsevier, Amsterdam, 1988) 157–171.
29. J.A. Telle, Complexity of domination-type problems in graphs, Nordic Journal of Computing 1 (1994) 157-171.
30. J.A. Telle and A.Proskurowski, Practical algorithms on partial k-trees with an application to domination-like problems, Lecture Notes in Computer Science 709 (1993) 610-621.
31. A. Wongseelashote, Semirings and path spaces, Discrete Math. 26 (1979) 55–78.
32. U. Zimmermann, Linear and Combinatorial Optimization in Ordered Algebraic Structures (Ann. Discrete Math. 10, North Holland, Amsterdam, 1981).
33. H. Wiener, Structural determination of paraffin boiling points. J. Am. Chem. Soc. 69 (1947) 17–20.

Author Index

Lecture Notes in Computer Science

For information about Vols. 1–1584 please contact your bookseller or Springer-Verlag

Vol. 1587: J. Pieprzyk, R. Safavi-Naini, J. Seberry (Eds.), Information Security and Privacy. Proceedings, 1999. XI, 327 pages. 1999.

Vol. 1589: J.L. Fiadeiro (Ed.), Recent Trends in Algebraic Development Techniques. Proceedings, 1998. X, 341 pages. 1999.

Vol. 1590: P. Atzeni, A. Mendelzon, G. Mecca (Eds.), The World Wide Web and Databases. Proceedings, 1998. VIII, 213 pages. 1999.

Vol. 1592: J. Stern (Ed.), Advances in Cryptology – EUROCRYPT '99. Proceedings, 1999. XII, 475 pages. 1999.

Vol. 1593: P. Sloot, M. Bubak, A. Hoekstra, B. Hertzberger (Eds.), High-Performance Computing and Networking. Proceedings, 1999. XXIII, 1318 pages. 1999.

Vol. 1594: P. Ciancarini, A.L. Wolf (Eds.), Coordination Languages and Models. Proceedings, 1999. IX, 420 pages. 1999.

Vol. 1595: K. Hammond, T. Davie, C. Clack (Eds.), Implementation of Functional Languages. Proceedings, 1998. X, 247 pages. 1999.

Vol. 1596: R. Poli, H.-M. Voigt, S. Cagnoni, D. Corne, G.D. Smith, T.C. Fogarty (Eds.), Evolutionary Image Analysis, Signal Processing and Telecommunications. Proceedings, 1999. X, 225 pages. 1999.

Vol. 1597: H. Zuidweg, M. Campolargo, J. Delgado, A. Mullery (Eds.), Intelligence in Services and Networks. Proceedings, 1999. XII, 552 pages. 1999.

Vol. 1598: R. Poli, P. Nordin, W.B. Langdon, T.C. Fogarty (Eds.), Genetic Programming. Proceedings, 1999. X, 283 pages. 1999.

Vol. 1599: T. Ishida (Ed.), Multiagent Platforms. Proceedings, 1998. VIII, 187 pages. 1999. (Subseries LNAI).

Vol. 1600: M. Wooldridge, M. Veloso (Eds.), Artificial Intelligence Today. VIII, 489 pages. 1999. (Subseries LNAI).

Vol. 1601: J.-P. Katoen (Ed.), Formal Methods for Real-Time and Probabilistic Systems. Proceedings, 1999. X, 355 pages. 1999.

Vol. 1602: A. Sivasubramaniam, M. Lauria (Eds.), Network-Based Parallel Computing. Proceedings, 1999. VIII, 225 pages. 1999.

Vol. 1603: J. Vitek, C.D. Jensen (Eds.), Secure Internet Programming. X, 501 pages. 1999.

Vol. 1604: M. Asada, H. Kitano (Eds.), RoboCup-98: Robot Soccer World Cup II. XI, 509 pages. 1999. (Subseries LNAI).

Vol. 1605: J. Billington, M. Diaz, G. Rozenberg (Eds.), Application of Petri Nets to Communication Networks. IX, 303 pages. 1999.

Vol. 1606: J. Mira, J.V. Sánchez-Andrés (Eds.), Foundations and Tools for Neural Modeling. Proceedings, Vol. I, 1999. XXIII, 865 pages. 1999.

Vol. 1607: J. Mira, J.V. Sánchez-Andrés (Eds.), Engineering Applications of Bio-Inspired Artificial Neural Networks. Proceedings, Vol. II, 1999. XXIII, 907 pages. 1999.

Vol. 1608: S. Doaitse Swierstra, P.R. Henriques, J.N. Oliveira (Eds.), Advanced Functional Programming. Proceedings, 1998. XII, 289 pages. 1999.

Vol. 1609: Z. W. Raś, A. Skowron (Eds.), Foundations of Intelligent Systems. Proceedings, 1999. XII, 676 pages. 1999. (Subseries LNAI).

Vol. 1610: G. Cornuéjols, R.E. Burkard, G.J. Woeginger (Eds.), Integer Programming and Combinatorial Optimization. Proceedings, 1999. IX, 453 pages. 1999.

Vol. 1611: I. Imam, Y. Kodratoff, A. El-Dessouki, M. Ali (Eds.), Multiple Approaches to Intelligent Systems. Proceedings, 1999. XIX, 899 pages. 1999. (Subseries LNAI).

Vol. 1612: R. Bergmann, S. Breen, M. Göker, M. Manago, S. Wess, Developing Industrial Case-Based Reasoning Applications. XX, 188 pages. 1999. (Subseries LNAI).

Vol. 1613: A. Kuba, M. Šámal, A. Todd-Pokropek (Eds.), Information Processing in Medical Imaging. Proceedings, 1999. XVII, 508 pages. 1999.

Vol. 1614: D.P. Huijsmans, A.W.M. Smeulders (Eds.), Visual Information and Information Systems. Proceedings, 1999. XVII, 827 pages. 1999.

Vol. 1615: C. Polychronopoulos, K. Joe, A. Fukuda, S. Tomita (Eds.), High Performance Computing. Proceedings, 1999. XIV, 408 pages. 1999.

Vol. 1616: P. Cointe (Ed.), Meta-Level Architectures and Reflection. Proceedings, 1999. XI, 273 pages. 1999.

Vol. 1617: N.V. Murray (Ed.), Automated Reasoning with Analytic Tableaux and Related Methods. Proceedings, 1999. X, 325 pages. 1999. (Subseries LNAI).

Vol. 1618: J. Bézivin, P.-A. Muller (Eds.), The Unified Modeling Language. Proceedings, 1998. IX, 443 pages. 1999.

Vol. 1619: M.T. Goodrich, C.C. McGeoch (Eds.), Algorithm Engineering and Experimentation. Proceedings, 1999. VIII, 349 pages. 1999.

Vol. 1620: W. Horn, Y. Shahar, G. Lindberg, S. Andreassen, J. Wyatt (Eds.), Artificial Intelligence in Medicine. Proceedings, 1999. XIII, 454 pages. 1999. (Subseries LNAI).

Vol. 1621: D. Fensel, R. Studer (Eds.), Knowledge Acquisition Modeling and Management. Proceedings, 1999. XI, 404 pages. 1999. (Subseries LNAI).

Vol. 1622: M. González Harbour, J.A. de la Puente (Eds.), Reliable Software Technologies – Ada-Europe'99. Proceedings, 1999. XIII, 451 pages. 1999.

Vol. 1623: T. Reinartz, Focusing Solutions for Data Mining. XV, 309 pages. 1999. (Subseries LNAI).

Vol. 1625: B. Reusch (Ed.), Computational Intelligence. Proceedings, 1999. XIV, 710 pages. 1999.

Vol. 1626: M. Jarke, A. Oberweis (Eds.), Advanced Information Systems Engineering. Proceedings, 1999. XIV, 478 pages. 1999.

Vol. 1627: T. Asano, H. Imai, D.T. Lee, S.-i. Nakano, T. Tokuyama (Eds.), Computing and Combinatorics. Proceedings, 1999. XIV, 494 pages. 1999.

Col. 1628: R. Guerraoui (Ed.), ECOOP'99 - Object-Oriented Programming. Proceedings, 1999. XIII, 529 pages. 1999.

Vol. 1629: H. Leopold, N. García (Eds.), Multimedia Applications, Services and Techniques - ECMAST'99. Proceedings, 1999. XV, 574 pages. 1999.

Vol. 1631: P. Narendran, M. Rusinowitch (Eds.), Rewriting Techniques and Applications. Proceedings, 1999. XI, 397 pages. 1999.

Vol. 1632: H. Ganzinger (Ed.), Automated Deduction – Cade-16. Proceedings, 1999. XIV, 429 pages. 1999. (Subseries LNAI).

Vol. 1633: N. Halbwachs, D. Peled (Eds.), Computer Aided Verification. Proceedings, 1999. XII, 506 pages. 1999.

Vol. 1634: S. Džeroski, P. Flach (Eds.), Inductive Logic Programming. Proceedings, 1999. VIII, 303 pages. 1999. (Subseries LNAI).

Vol. 1636: L. Knudsen (Ed.), Fast Software Encryption. Proceedings, 1999. VIII, 317 pages. 1999.

Vol. 1637: J.P. Walser, Integer Optimization by Local Search. XIX, 137 pages. 1999. (Subseries LNAI).

Vol. 1638: A. Hunter, S. Parsons (Eds.), Symbolic and Quantitative Approaches to Reasoning and Uncertainty. Proceedings, 1999. IX, 397 pages. 1999. (Subseries LNAI).

Vol. 1639: S. Donatelli, J. Kleijn (Eds.), Application and Theory of Petri Nets 1999. Proceedings, 1999. VIII, 425 pages. 1999.

Vol. 1640: W. Tepfenhart, W. Cyre (Eds.), Conceptual Structures: Standards and Practices. Proceedings, 1999. XII, 515 pages. 1999. (Subseries LNAI).

Vol. 1642: D.J. Hand, J.N. Kok, M.R. Berthold (Eds.), Advances in Intelligent Data Analysis. Proceedings, 1999. XII, 538 pages. 1999.

Vol. 1643: J. Nešetřil (Ed.), Algorithms – ESA '99. Proceedings, 1999. XII, 552 pages. 1999.

Vol. 1644: J. Wiedermann, P. van Emde Boas, M. Nielsen (Eds.), Automata, Languages, and Programming. Proceedings, 1999. XIV, 720 pages. 1999.

Vol. 1645: M. Crochemore, M. Paterson (Eds.), Combinatorial Pattern Matching. Proceedings, 1999. VIII, 295 pages. 1999.

Vol. 1647: F.J. Garijo, M. Boman (Eds.), Multi-Agent System Engineering. Proceedings, 1999. X, 233 pages. 1999. (Subseries LNAI).

Vol. 1648: M. Franklin (Ed.), Financial Cryptography. Proceedings, 1999. VIII, 269 pages. 1999.

Vol. 1649: R.Y. Pinter, S. Tsur (Eds.), Next Generation Information Technologies and Systems. Proceedings, 1999. IX, 327 pages. 1999.

Vol. 1650: K.-D. Althoff, R. Bergmann, L.K. Branting (Eds.), Case-Based Reasoning Research and Development. Proceedings, 1999. XII, 598 pages. 1999. (Subseries LNAI).

Vol. 1651: R.H. Güting, D. Papadias, F. Lochovsky (Eds.), Advances in Spatial Databases. Proceedings, 1999. XI, 371 pages. 1999.

Vol. 1652: M. Klusch, O.M. Shehory, G. Weiss (Eds.), Cooperative Information Agents III. Proceedings, 1999. XI, 404 pages. 1999. (Subseries LNAI).

Vol. 1653: S. Covaci (Ed.), Active Networks. Proceedings, 1999. XIII, 346 pages. 1999.

Vol. 1654: E.R. Hancock, M. Pelillo (Eds.), Energy Minimization Methods in Computer Vision and Pattern Recognition. Proceedings, 1999. IX, 331 pages. 1999.

Vol. 1656: S. Chatterjee, J.F. Prins, L. Carter, J. Ferrante, Z. Li, D. Sehr, P.-C. Yew (Eds.), Languages and Compilers for Parallel Computing. Proceedings, 1998. XI, 384 pages. 1999.

Vol. 1661: C. Freksa, D.M. Mark (Eds.), Spatial Information Theory. Proceedings, 1999. XIII, 477 pages. 1999.

Vol. 1662: V. Malyshkin (Ed.), Parallel Computing Technologies. Proceedings, 1999. XIX, 510 pages. 1999.

Vol. 1663: F. Dehne, A. Gupta. J.-R. Sack, R. Tamassia (Eds.), Algorithms and Data Structures. Proceedings, 1999. IX, 366 pages. 1999.

Vol. 1664: J.C.M. Baeten, S. Mauw (Eds.), CONCUR'99. Concurrency Theory. Proceedings, 1999. XI, 573 pages. 1999.

Vol. 1666: M. Wiener (Ed.), Advances in Cryptology – CRYPTO '99. Proceedings, 1999. XII, 639 pages. 1999.

Vol. 1668: J.S. Vitter, C.D. Zaroliagis (Eds.), Algorithm Engineering. Proceedings, 1999. VIII, 361 pages. 1999.

Vol. 1671: D. Hochbaum, K. Jansen, J.D.P. Rolim, A. Sinclair (Eds.), Randomization, Approximation, and Combinatorial Optimization. Proceedings, 1999. IX, 289 pages. 1999.

Vol. 1678: M.H. Böhlen, C.S. Jensen, M.O. Scholl (Eds.), Spatio-Temporal Database Management. Proceedings, 1999. X, 243 pages. 1999.

Vol. 1684: G. Ciobanu, G. Păun (Eds.), Fundamentals of Computation Theory. Proceedings, 1999. XI, 570 pages. 1999.

Vol. 1685: P. Amestoy, P. Berger, M. Daydé, I. Duff, V. Frayssé, L. Giraud, D. Ruiz (Eds.), Euro-Par'99. Parallel Processing. Proceedings, 1999. XXXII, 1503 pages. 1999.

Vol. 1688: P. Bouquet, L. Serafini, P. Brézillon, M. Benerecetti, F. Castellani (Eds.), Modeling and Using Context. Proceedings, 1999. XII, 528 pages. 1999. (Subseries LNAI).

Vol. 1689: F. Solina, A. Leonardis (Eds.), Computer Analysis of Images and Patterns. Proceedings, 1999. XIV, 650 pages. 1999.